KAPLAN

ASVAB TOTAL PREP 2024-2025

7 PRACTICE TESTS + PROVEN STRATEGIES + VIDEO + FLASHCARDS

ACKNOWLEDGMENTS

Editor, 2024–2025 edition

Paula L. Fleming, MA, MBA

Special thanks to Matt Belinkie, Brian Carlidge, Michael Collins, Lola Disparte, Joanna Graham, Allison Harm, Craig Harman, Jack Hayes, Elizabeth Horowitz, Rebecca Houck, Melissa McLaughlin, Camellia Mukherjee, Rachel Pearsall, Tammi Rice, Anne Marie Salloum, Amit Schlesinger, Gordon Spector, Sascha Strelka, Bonnie Wang, Ethan Weber, Michael Wolff, Jacob Zampier, and many others.

Published by Kaplan North America, LLC
1515 W. Cypress Creek Road
Fort Lauderdale, FL 33309

10 9 8 7 6 5 4 3

ISBN: 978-1-5062-9079-9

CONTENTS

FOREWORD

Congratulations on taking an important step toward enlisting in one of the United States Armed Services! The Air Force, Army, Coast Guard, Marines, Navy, and National Guard all emphasize the importance of mission preparedness. By obtaining this Kaplan ASVAB book, you are on your way toward being well prepared to take the ASVAB test that is a prerequisite for enlisting in one of the services. The United States Military Entrance Processing Command administers the ASVAB for all the services. That organization calls the enlistment process "Freedom's Front Door."

The Armed Services give our marines, sailors, soldiers, guardsmen and -women, and airmen and -women the opportunity to grow personally and professionally, to assume major responsibilities, and to work in integrated teams to achieve their goals. The military needs bright, talented young individuals to serve our country, and the ASVAB is one of the factors used to enlist the most qualified applicants. Being selected for the military is very competitive; not only do you have to meet certain threshold standards, including ASVAB scores, but you are also competing with thousands of other applicants for these coveted openings.

My efforts as a member of the team that prepares new editions of the Kaplan ASVAB book have brought back many memories of my four and a half years as a proud member of the U.S. Navy. When I was in the engineering department on a destroyer, an unusual combination of circumstances resulted in our ship being about 40 percent below the proper staffing level just three months before we were scheduled for a major deployment. Our crew had to be augmented quickly, but, as the saying goes, it takes three years to get three years' experience. Consequently, sailors just graduating from advanced training ("A School" in Navy parlance) were slotted into positions on our ship normally staffed by third-class petty officers with two or so years under their belts and sometimes even second-class petty officers who might have been in their second term of enlistment.

Fortunately, almost all of our new crew members were well suited for their shipboard duties thanks, in part, to the version of the ASVAB that was being used at that time. Consequently, they learned quickly and assumed their responsibilities well. During their first full year on board, we were deployed 290 days and we never missed a beat. In fact, many of us were awarded the Navy Commendation Medal for restaffing so successfully that we even covered a couple of operational commitments for other ships.

Needless to say, my experience convinced me that the process of selecting the right people for the right jobs was very well executed. I made many lifetime friendships in the Navy, and I have stayed in touch with some of my shipmates since I left. Some went on to very successful careers in the Navy. Several others used the G.I. Bill to help further their education and went on to civilian careers.

Enough of my reminiscing! I mention my experiences to encourage you to get the most that you can from this Kaplan ASVAB book so that you score up to your maximum potential and are selected for the service and occupational specialty that you want and that will enable you to excel.

Kaplan is the premier test preparation company in the world, and we have applied our expertise and experience to this new edition. Because practice is so important to your preparation, we have included three full-length practice ASVAB tests in this book and scores of practice problems and explanations. Your purchase of this book will also give you access to Kaplan's online resources, including four computer-based ASVAB practice tests, 1,000 more practice questions, and videos covering key concepts. When you complete an online test, your Performance Summary will give you a detailed performance breakdown by skill or concept.

You have chosen the right tool to help you score as high as possible on the ASVAB; the rest is up to you. Merely reading the book might help some, but the real payoff will come from studying the material, taking the practice assessments, and focusing on the areas where you can realize the most improvement. Remember, your hard work will pay off with rewards on Test Day and beyond! I wish you the best as you begin your journey and hope that you find your time in the military as rewarding as I did.

Jack Hayes
Former LT., USN (and a former recruiter)

WELCOME TO YOUR ASVAB STUDIES!

Thank you for choosing Kaplan to help you study for the ASVAB! We are honored to be part of your preparation for a military career. Your Kaplan book and online resources contain all the tools you need to succeed on Test Day and earn your place in a branch of the armed services.

YOUR KAPLAN RESOURCES

Your Kaplan book contains:

- a full-length diagnostic test
- tools to help you plan your studies
- chapters that cover each of the ASVAB subject tests, including:
    - Kaplan Methods for every question type
    - worked examples that display the expert approach
    - practice questions with explanations
- two more full-length practice tests
- 500 flashcards to help you remember important concepts for the technical subtests

Your Kaplan online resources offer:

- 40 instructional videos with key test-taking strategies
- four additional full-length practice tests, to prepare you for Test Day
- a Qbank with 1,000 items, to help you hone your skills
- Performance Summaries, to help you keep track of your strengths and areas of opportunity
- Kaplan's flashcard app, so you can drill with your cards on any device

Between the Kaplan book and online resources, you have all the resources you'll need to achieve your goals on the ASVAB.

THREE LEVELS OF STUDY

Perhaps you're already pretty good at the ASVAB, but you need to brush up on specific skills or to review all of the skills lightly before Test Day. Or maybe you feel moderately confident about the ASVAB, but you need significant review on some or all of the topics. Or perhaps, like many people, you've forgotten much of what you learned in high school and need a comprehensive review.

Kaplan's book and online resources will help you prepare for the ASVAB regardless of which of those groups you fall into. The chart below outlines three approaches you might take depending on your needs. Of course, your individual situation may be best served by a combination of these approaches.

If you need a quick brush-up or light overall review . . .	If you need a moderate review of some or all of the topics . . .	If you need to (re)learn many of the skills from scratch . . .
Start by devoting one week to each of the subject tests. During each of those weeks, read the appropriate chapter and do the practice items in the book and online.	Use the results from your diagnostic to identify your areas of greatest need. Start by devoting two weeks to each of those subject tests. During each of those weeks, read the appropriate chapter and do the practice items in the book and online.	Give yourself plenty of time to work through this book, chapter by chapter. Periodically review the earlier chapters so that those skills stay fresh. Use the practice questions in the book and online to reinforce what you've learned and review the explanations carefully.
After you've completed your comprehensive review, take a full-length test in the book or online. If you haven't seen significant improvement, continue to work on the areas you find most challenging. Take another full-length test every week or two, depending on your test date.	After you've done so, take a full-length test in the book or online to gauge your progress. Continue to study and take a full-length test every week or two, depending on your Test Day. In the last few weeks before Test Day, give a week or several days to each of the subtests on which you were already strong.	Don't take many full-length tests until you've reviewed most of the subject tests. In the last few weeks before Test Day, take a full-length practice test (either in the book or online) once a week or once every two weeks, depending on your test date.

GETTING STARTED

HOW TO GET STARTED

1. Register your Kaplan online resources.
2. Take the diagnostic test in your Kaplan book.
3. Review your performance and read through the explanations.
4. Contact a recruiter and do some research about your desired military career.
5. Learn, practice, review.
6. Take full-length practice tests in your Kaplan book and in your online resources.

1. Register your Kaplan online resources.

To obtain access to your Kaplan online resources, visit **kaptest.com/booksonline**. Create your account by choosing "ASVAB," selecting your Kaplan ASVAB book, and answering the question or questions that appear.

Once you have created your username and password, log in to your resources at **kaptest.com/login**. Enter your username and password. Click on the title of your book to see your resources.

2. Take the diagnostic test in your Kaplan book.

Chapter 2 of this book offers a full-length diagnostic test. As a first step in your studies, set aside about three uninterrupted hours when you can take the diagnostic test. Work through the subject tests in the order they're presented, with one break at most.

3. Review your performance and read through the explanations.

After you've taken the complete diagnostic test, check your answers using the explanations that appear immediately after the tests. When you check your answers, don't just check whether you got the question right or not. Rather, read the explanations for all the questions. That's because you can learn a great deal from reviewing questions you've already done, even if you got those questions right. As you review your performance on the diagnostic, really think about why the right answers are right, why the incorrect answers are incorrect, and what drew you to the answer you chose.

4. Contact a recruiter and do some research.

You're going to want to set some goals regarding your ASVAB score, but different enlistees will have different priorities in studying for the ASVAB, depending on their career aspirations. For example, a test taker who wants to go into a technical career (such as equipment or computer technician) will likely need high scores in Word Knowledge, Paragraph Comprehension, Mathematics Knowledge, Arithmetic Reasoning, General Science, and Mechanical Comprehension. However, a test taker who wants to go into a career repairing structures or vehicles should likely emphasize Auto and Shop Information, Electronic Information, and Mechanical Comprehension over General Science.

Your recruiter is the best source of information about careers in the military and the scores you will need to earn to be competitive for those careers. Use information from your recruiter as well as information available on the various branches' websites to inform your ASVAB score goals.

5. Learn, practice, review.

Once you have taken your diagnostic and established some goals regarding your ASVAB score, you're ready to get down to serious studying. Effective studying has three phases: learn, practice, review.

Learn: Each subject test on the ASVAB has its own chapter in this book. Read these chapters carefully, paying particular attention to these features:

- Concepts are introduced with **learning objectives**. Use these learning objectives as a checklist of skills, to keep track of which you have mastered and which you need to work on more.
- **Key terms** you'll need to understand are introduced in bold type.
- Important strategies, takeaways, and shortcuts are highlighted in **gray boxes** sprinkled throughout the book.
- **Worked examples** appear throughout the book: each one shows, step-by-step, how an expert test taker would approach ASVAB questions.

Practice: Practice items are sprinkled throughout each chapter. Do each one as you come to it and then carefully review the explanation that follows. Moreover, each chapter ends with a practice set. Don't time yourself on these practice questions. It's more important to really understand how the questions work; you can work on timing later.

When you do these practice sets, don't just tally up how many you got right. Rather . . .

Review: Always review the explanation for every practice problem. Think about why the right answer is right and why the wrong answer is wrong. If you answered the question incorrectly, think carefully about where your thinking went astray and use that to inform your next steps. (For example, in an algebra problem, perhaps you understood the underlying concept but you made a simple addition error. In that case, the remedy would not be to study more algebra; rather, the remedy would be to practice adding and subtracting quickly while avoiding errors.) If you answered the question correctly, ask yourself whether you got it right for the right reasons and whether you could have arrived at the same answer more efficiently.

Repeat this process every time you take a full-length test. Review all the answers and explanations for every item in the test and think about how you performed overall.

6. Take full-length practice tests.

Once you have learned, practiced, and reviewed all of the test concepts, use the full-length tests available in your book and online resources. Taking a full-length test helps you in four ways:

- It reinforces the skills and strategies you've learned.
- It helps you work on timing.
- It gives you a sense of how you're doing on the various topics and what you need to work on more.
- It helps you learn to cope with test fatigue.

Now, you'll notice that one item not on that list is "It helps you learn concepts." You don't learn how to approach questions correctly by taking a full-length test: that learning should be done in an untimed fashion and accompanied by deep thinking, as described above. That's why we don't recommend that you start your studies by taking a bunch of full-length tests. However, the last few weeks before Test Day are a perfect time to take full-length tests.

In addition to the diagnostic test in chapter 2, your Kaplan book has two full-length practice tests in the back. These follow the ASVAB's paper-and-pencil format. You also have four practice tests in your online resources. These are computer-based tests that will help you get used to testing using a computer under the time constraints used by the CAT-ASVAB. While the computer-based tests in your online account are not adaptive, their structure does allow you to prepare for the CAT-ASVAB's format and content.

MANAGING STRESS

You have a lot riding on the ASVAB. However, you're also doing the work you need to do to reach your goals. Unfortunately, though, simply knowing that you're working hard won't make your test anxiety go away. Thus, here are some stress management tips from our long experience of helping students prepare for standardized tests.

Clock in and out: Once you've set up a study schedule for yourself, treat it like a job. That is, imagine clocking yourself in and out of ASVAB studies according to that schedule. Do your best to stick to your schedule, and when you're not "clocked in," don't let yourself think about the ASVAB. That will help you release your stress about the test in between study sessions.

Don't punish yourself: If you get tired or overwhelmed or discouraged when studying, don't respond by pushing yourself harder. Rather, step away and engage in a relaxing activity like going for a walk, watching a movie, or playing with your cat or dog. Then, when you're ready, return to your studies with fresh eyes.

Breathe: Remember to breathe into your stomach. That forces some of the muscles that tense up when you're stressed to relax.

Set small, manageable goals: Each week, set manageable goals related to your ASVAB progress and reward yourself when you've achieved them. Examples of small goals might be:

- This week, memorize and practice the Kaplan Method for Assembling Objects questions until I no longer have to think about what the steps of the method are.
- This week, try 40 math questions and practice choosing a strategy for solving each (such as Backsolving, Picking Numbers, estimating, straight-up math).
- This week, review all Paragraph Comprehension question types until I can identify each question's type and the appropriate strategy.

Keep yourself healthy: Poor health, fatigue, and isolation make it harder to cope with stress and anxiety. Get on a regular sleep schedule as much as possible during your studies, eat well, continue to exercise, and spend time with those you care about. Also, don't fuel your studies with caffeine and sugar. Those substances may make you feel alert, but they can also damage focus.

Keep the right mindset: Most importantly, keep telling yourself that you can do this. Don't fall into the trap of thinking that you're not "allowed" to feel confident yet. That's a self-punishing attitude that will only hurt you. Rather, remember that confidence breeds success. So let yourself be confident about your abilities. You're obviously ambitious and intelligent, so walk into the ASVAB knowing that about yourself.

If you get discouraged, make a list: If you ever start to wonder if you'll ever reach your ASVAB goals, stop what you're doing and make a list of everything you're good at. List *every specific skill* that you are bringing to the ASVAB. Examples of specific skills might include:

- researching a detail in a paragraph
- using suffixes to tell whether a vocabulary word is a noun, verb, or adjective
- identifying what a math question is asking for
- factoring a quadratic equation
- identifying whether a circuit is built in series or in parallel
- identifying types of hammers

Post that list of things you're good at somewhere you'll see it every day, then add to it as you continue to study. We at Kaplan recommend this because many people focus too heavily on their weaknesses while preparing for a standardized test. But if you only focus on your weaknesses, you aren't seeing an objective picture. There *are* ASVAB skills you're good at. Keep that in mind and focus on building on those strengths.

FINAL PREPARATIONS FOR TEST DAY

In the last few weeks before Test Day, do a comprehensive review of all ASVAB topics. Pay particular attention to the subject tests that are most important to your preferred career(s) and to the four subject tests that are most important for enlistment. (Those are Mathematics Knowledge, Arithmetic Reasoning, Paragraph Comprehension, and Word Knowledge; see chapter 1 for more information.) Take a full-length test once or twice a week and use the results of those practice tests to inform your review.

Be sure to contact your recruiter to find out more about what to expect from Test Day—what you should bring with you, when you can expect your scores (if you're taking the paper version of the ASVAB), and other specifics.

In the Week Before Test Day

Your activities in the last week before Test Day should include:

Rest: Make sure you're on a regular sleep schedule.

Rehearse: Find out where your testing center is located and consider doing a "dry run." That is, drive or commute to the testing center around the same time of day as your testing appointment. You don't want to be surprised by traffic or road construction on Test Day.

Review: Do a very brief, high-level review. In other words, flip through the lessons and rework a few practice problems here and there to reinforce all of the good habits you've developed in your preparation. (Redoing practice problems you've already done is fine: you can actually learn a lot that way about how to approach those types of questions more efficiently in the future.)

Stop: Two days before the test, stop studying. No studying at all: you're not likely to learn anything new in those two days, and you'll get a lot more out of walking into the test feeling rested.

Relax: The evening before the test, do something fun but not crazy or tiring. Maybe you could have a nice dinner (without alcohol), watch a movie, or do something else relaxing.

Go to bed at your usual time the night before the exam.

On Test Day Itself

On the day of the test, be sure to follow the guidance below:

Warm up: Before you take the test, do an ASVAB warm-up. This can help your brain get ready to function at its best. You probably can't take any practice materials into the testing center, but you can do a few easy practice problems at home or in the car before you go into the testing center.

Don't let nerves derail you: If you feel nervous while taking the test, remember to breathe deeply into your stomach. Take a few deep breaths and focus your eyes on something other than the computer screen or test booklet for a moment.

Keep moving: Don't let yourself get bogged down on any one question. If you're taking the paper version, you can come back to questions that you weren't sure about, so skip questions whenever they threaten to slow you down or to steal time from the other questions. If you're taking the CAT-ASVAB, you can't return to previous questions, so you will have to decide to make a guess whenever a question is threatening to take too much time.

Don't assess yourself: This is very important. As you're testing, don't let yourself stop and think about how you *feel* you're doing. Taking a standardized test hardly ever *feels* good. Your own impressions of how it's going are totally unreliable. So, instead of focusing on that, remind yourself that you're prepared and that you are going to succeed, even if you feel discouraged as the test is underway.

After the test, celebrate!

You've prepared, practiced, and performed like a champion. Now that the test is over, it's time to congratulate yourself on a job well done. Celebrate responsibly with friends and family, and enjoy the rest of your day, knowing that you just took an important step toward reaching your goals.

GOOD LUCK!

Everyone at Kaplan wishes you the very best in your studies, on the ASVAB, and in your military career. We're rooting for you!

CHAPTER 1

ABOUT THE ASVAB

WHAT IS THE ASVAB?

The ASVAB (Armed Services Vocational Aptitude Battery) is the most widely used multiple-aptitude test battery in the world. It measures a test taker's suitability to enlist in the United States Armed Forces and assesses his or her abilities to be trained in specific civilian or military jobs.

When you take the ASVAB officially, you will be given either a paper-and-pencil version of the test or a computer version (also referred to as a Computer Adaptive Test ASVAB, or CAT-ASVAB). About 70 percent of prospective recruits take the CAT-ASVAB, so that's likely to be the version you'll encounter.

In this book, you'll learn how the ASVAB is structured and what it tests. You'll also learn strategies and methods that will help you improve your score significantly in each section.

What Does the ASVAB Test?

In addition to evaluating math and reading skills, the ASVAB assesses performance in categories such as science, electronics, auto repair, and the ability to assemble objects. While the section topics and question types are the same on both the paper-and-pencil and CAT-ASVAB versions of the test, the amount of time and number of questions will differ slightly on each version. In the table below, you'll see the order in which the ASVAB subtests are arranged, the material tested in each section, and the differences between the paper-and-pencil and CAT-ASVAB versions. Additionally, note the sections that are highlighted in gray. These subtests make up the AFQT, which is discussed in more depth in the following section. In this table, the subtests are listed in the order in which you'll take them on Test Day.

Subtest	Questions / Time Limit in Minutes (CAT-ASVAB)	Questions / Time Limit in Minutes (Paper & Pencil)	What's Tested
General Science (GS)	15 questions / 10 minutes	25 questions / 11 minutes	Knowledge of general concepts from life, Earth and space, and physical sciences
Arithmetic Reasoning (AR)	15 questions / 42 minutes	30 questions / 36 minutes	Ability to answer word problems that involve basic arithmetic calculations
Word Knowledge (WK)	15 questions / 9 minutes	35 questions / 11 minutes	Ability to recognize synonyms of words
Paragraph Comprehension (PC)	10 questions / 27 minutes	15 questions / 13 minutes	Ability to answer questions based on short passages (of 30–120 words)
Mathematics Knowledge (MK)	15 questions / 23 minutes	25 questions / 24 minutes	Knowledge of math concepts, including arithmetic, algebra, and geometry
Electronics Information (EI)	15 questions / 10 minutes	20 questions / 9 minutes	Knowledge of electricity principles and terminology and of basic electronic circuitry
Auto and Shop Information (AS)*	10 questions / 7 minutes and 10 questions / 6 minutes	25 questions / 11 minutes	Knowledge of automobiles, and of tool and shop practices and terminology
Mechanical Comprehension (MC)	15 questions / 22 minutes	25 questions / 19 minutes	Knowledge of basic mechanical and physical principles
Assembling Objects (AO)	15 questions / 17 minutes	25 questions / 15 minutes	Ability to determine how a disassembled object will look when it is put back together
Totals	135 questions / 173 minutes	225 questions / 149 minutes	

*On the CAT-ASVAB, Auto and Shop Information is split into two parts (Auto Information [AI] and Shop Information [SI]), but one score is reported.

WHAT IS THE AFQT?

While the ASVAB is designed to determine a potential recruit's occupational fit in the military, there is no actual "overall" ASVAB score. When people talk about getting a score of, say, a 75 or 80 on the ASVAB, they are really talking about something called the AFQT (Armed Forces Qualification Test) score. A candidate's score on the AFQT determines that candidate's eligibility for all branches of the Armed Services. The AFQT score is derived from your performance on just the verbal and math subtests of the ASVAB, as explained below.

Your AFQT Score

The military determines your AFQT score by first adding your Word Knowledge and Paragraph Comprehension scores together to get your Verbal Expression or VE score. The formula to derive the AFQT raw score is 2VE + AR (Arithmetic Reasoning) + MK (Mathematics Knowledge) = AFQT score.

2 × [Word Knowledge + Paragraph Comprehension]

+

[Arithmetic Reasoning + Mathematics Knowledge]

= AFQT RAW SCORE

The AFQT raw score you receive is then translated into a percentile score that tells you how well you did on the AFQT compared to a base group of approximately 6,000 other test takers ages 18–23. For instance, if your percentile score is 68, you scored as well as or better than 68 percent of the base group. This AFQT percentile score is used to determine your eligibility for the armed forces.

AFQT QUALIFYING SCORES FOR DIFFERENT BRANCHES OF THE MILITARY

The requirements listed below are **minimum** standards for high school graduates and are subject to change. The scores necessary for many occupations and for enlistment bonuses are usually significantly higher. Also, these are the requirements for high school graduates; the requirements are more stringent for those who have earned their high school equivalency diplomas by taking the GED® test, the TASC test, or the HiSET® exam. An applicant without a high school degree or equivalent can be accepted into military service only in special circumstances.

Applicants who possess special skills or experience (for example, fluency in a certain language or computer programming experience) *may* be eligible for waivers of AFQT minimum scores. Also, if you score well on any subtest(s) that relate to particular occupations that are recruiting targets, you *may* be able to get a waiver.

You should refer to the services' websites or publications for more specific information or contact a military recruiter.

Air Force—minimum AFQT score of 31

Army—minimum AFQT score of 31

Marine Corps—minimum AFQT score of 31

Navy—minimum AFQT score of 35

Coast Guard—minimum AFQT score of 40

ASVAB LOGISTICS

Registering for and Taking the ASVAB

Unless you are taking the ASVAB at your high school as part of the Department of Defense Career Exploration Program, your first step toward registering to take the ASVAB is to contact the local recruiter for the service branch that you wish to join. The recruiter will help you complete your enlistment application; you will need to provide necessary documentation. Once you have met the basic qualifications for enlistment, the recruiter will schedule either a proctored ASVAB test or an unproctored form of the ASVAB called the PiCAT.

The Pending Internet Computerized Adaptive Test (PiCAT) is an unproctored, full-length ASVAB that you take on your own time. Your local recruiter will get you registered and give you an access code. Your recruiter can also tell you your score after you've finished the test. Then, if you choose to enlist, you'll take a proctored verification test that is much shorter than the ASVAB, only 25–30 minutes, to confirm your PiCAT score.

The Armed Forces have been piloting another potential military entrance assessment, the Tailored Adaptive Personality Assessment System (TAPAS). Based on the studies conducted so far, this computer-adaptive personality test is thought to provide information about recruits' motivation and other noncognitive characteristics that indicate whether they will be successful after enlistment. The TAPAS consists of 120 questions and, as of this writing, continues to be offered to certain candidates who also take the ASVAB as a way of continuing to evaluate the test's usefulness. The military's goal is to use the additional information provided by the TAPAS to identify recruits who will perform well, despite not quite meeting threshold AFQT scores.

ASVAB results are valid for two years. After taking an initial ASVAB test, you may retake it after one calendar month. After the first retest, you may take another retest after another calendar month has passed. If you have had two retests, you must wait at least six months before taking the test again.

How to Read Your ASVAB Scores

Your official ASVAB scores will come in a variety of styles and will be fully explained to you by your guidance counselor or recruiter. All of the scores matter, though some may matter more than others. Here's a quick breakdown of what you will find.

AFQT Score—You will receive a single numerical score for the AFQT.

Standard Scores—You will receive a Standard Score for each of the subtests. These scores are calculated using a comparison of your raw scores to the raw scores of a standard national sample. According to the Department of Defense, roughly half the population achieves a Standard Score of 50 or above but only about 16 percent scores higher than 59.

Service Composite Scores—These score combinations, sometimes called Line Scores, are used to determine whether a test taker has the necessary vocational aptitude to be trained for different job assignments in all the military branches. For example, a Navy Engineering Aid composite score (abbreviated EA) is the sum of twice the Mathematics Knowledge score plus the Arithmetic Reasoning and General Science scores. An enlistee must achieve a minimum composite score for the vocation of interest to be able to qualify to be trained for the job. To cite another example, to qualify for electronics training and occupations in the Army, you must attain a certain score that combines your results on the General Science, Arithmetic Reasoning, Mathematics Knowledge, and Electronics Information tests. For more detailed information, contact your local recruiter or visit the website for your service branch of interest.

Career Exploration Scores—In addition to the Standard Scores, students who take the ASVAB in their high schools as part of the Career Exploration Program receive three Career Exploration Scores in the composite areas of Verbal Skills, Math Skills, and Science and Technical Skills, which are all reported as standard scores and as percentiles relative to grade in school and gender.

Different Versions of the ASVAB

Depending upon your reasons for taking the ASVAB and your stage in the career decision-making process, you will take on one of the following three versions of the test battery:

Enlistment Testing Program ASVAB

This version of the ASVAB, sometimes referred to as the *Production Version*, is used for enlistment purposes only and is administered to potential enlistees in all branches of the military. A potential recruit's performance on the ASVAB subtests is used to determine whether the candidate has the necessary aptitudes to enlist in a desired branch of service and for which military jobs the candidate is best suited.

Career Exploration Program ASVAB

This version of the ASVAB is administered, along with an interest inventory, to high school and postsecondary students as part of the Department of Defense's Career Exploration Program. The content and format of this ASVAB are the same as the enlistment version, with one exception: in the paper-and-pencil version, the Assembling Objects subtest is not given.

Armed Forces Classification Test (AFCT)

Also known as the "In-Service" ASVAB, the AFCT is administered to those already in the military who are looking to switch jobs within the military. It is identical to the paper-and-pencil version of the ASVAB given prior to enlistment.

The Paper-and-Pencil Test vs. the CAT

Computer-adaptive testing, or CAT for short, is just a fancy way of saying that the difficulty of the questions you get on the test adapts based on your performance up to that point.

The diagram below models how a computer-adaptive test works. In the diagram, each dot represents a question. Each question is labeled with whether the test taker got the question correct (*C*) or incorrect (*I*). The graph also shows the difficulty of each question.

The first question you will see will be of medium difficulty—that is, it will be aligned to the average ability of test takers. Notice that each time you answer a question correctly, you "earn" a more difficult question, which is aligned to a higher scoring level. Each time you give an incorrect answer, your next question will be easier and thus aligned to a lower scoring level. The CAT continues to adapt until you are getting roughly half of the questions you see correct. Once your performance has stabilized in this way, the CAT determines your score based on the difficulty level around which your answers are hovering.

How a CAT Works

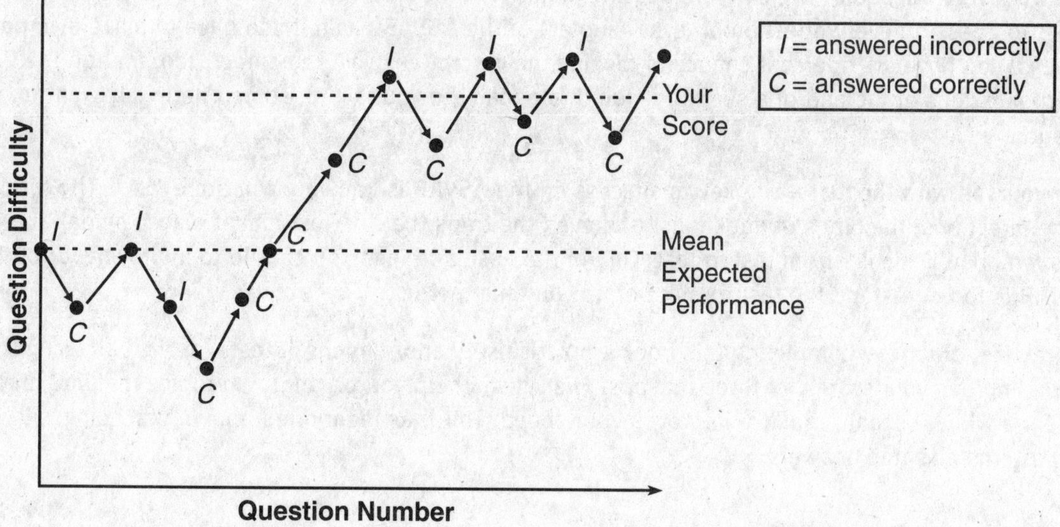

Some test takers think they can get a high score by working slowly, making sure to get questions right and driving up their score, even if that means they run out of time before answering all the questions. However, the makers of the ASVAB have thought of this, and the scoring includes a penalty for every question left unanswered, eliminating any advantage that might be gained. Therefore, it's important to stay on pace so you finish each section with enough time to do your best on the last question, even if that means guessing on some questions along the way.

Studies have shown that, overall, people perform the same on the ASVAB whether they take the paper or CAT version of the test. There are, however, some individuals who will tend to do better on one version of the test than the other. Now, you may not have a choice regarding which version of the ASVAB you take. If you do have a choice, here are some of the advantages and disadvantages of taking the CAT versus the paper test:

Advantages of Taking the CAT-ASVAB

- You need to answer far fewer questions than on the paper-and-pencil ASVAB, and you get more time per question to answer.
- The test can be scored immediately. You will know how well you did as soon as you finish the test.
- The test administration is very flexible, so you don't have to wait for the next scheduled test date to take the test.
- There's no chance of losing points by filling out your answer sheet incorrectly.
- The CAT format gives you the chance to work methodically on one question at a time with no other questions there to distract you.

Disadvantages of Taking the CAT-ASVAB

- You cannot skip around on this test; you must answer the questions one at a time in the order the computer gives them to you.
- If you realize later that you answered a question incorrectly, you cannot go back and change your answer.
- You can't cross off an answer choice, so you'll have to use your scrap paper to keep track of the answers you've eliminated.

KAPLAN'S ASVAB STRATEGIES

Always answer every question—There is no guessing penalty on the ASVAB. This means that it is absolutely in your interest to guess on every question on every subtest of the ASVAB! Even if you have to make a completely random guess, you have a 25 percent chance of picking the correct answer. Remember, too, that on the CAT-ASVAB, unanswered questions at the end of a section result in a penalty to your score, so you definitely want to get to every question.

Familiarize yourself with the test—One key to success on the ASVAB is knowing what to expect. The format—which includes the directions, the types of questions, and even the traps that the test maker places among the answer choices—is remarkably similar from test to test. One of the easiest things you can do to improve your performance on the ASVAB is to understand the test format before you take the test.

Practice, practice, practice—Completing this book's practice sets and full-length tests, as well as using the Qbank, will help you improve your scores for three reasons. First, it will help you brush up on topics that you may not have seen for a while. Second, practice improves your speed. Third, as mentioned above, practicing will help you recognize patterns and trap answers.

Take advantage of the multiple-choice format—You start with a 25 percent chance of getting the correct answer by random guessing, but eliminating any of the answer choices improves your odds. If you know that one of the four choices cannot be right and eliminate it accordingly, you now have a 33 percent chance of getting the correct answer. Remove one more wrong answer choice, and your chance of getting the question right is now 50–50. On certain ASVAB questions, you will find that eliminating wrong choices is as effective as spotting the correct answer the straightforward way. Often, even if you are completely confused by a question, you can still make a solid guess by eliminating answer choices that run counter to the other three choices.

CAT-ASVAB Strategies

If you are taking the CAT-ASVAB, applying certain CAT-specific strategies will have a direct, positive impact on your score:

- Work at the same steady pace throughout each section. Questions earlier in the section tend to affect your score more than later questions. However, if you spend extra time getting these questions right, you'll begin to get much harder questions *and* you'll have less time to do them. This will result in missing a string of questions and maybe even leaving some questions unanswered, undoing all of your early gains.
- As you progress through the middle part of a section, try to avoid getting several questions in a row wrong, as this can sink your score. If you know that you answered the previous question with a random guess, spend a little extra time trying to get the next one right.
- The CAT does not allow you to skip questions. So if you are given a question you cannot answer, you'll have to guess. Guess strategically by eliminating any choices that you know are wrong and choosing from those remaining. Once you know you'll need to guess, do so quickly and move on; if you don't know the answer, staring at the question longer is unlikely to produce enlightenment.
- Don't get rattled if you see difficult questions. Because the CAT increases in difficulty when you get correct answers, earning difficult questions just means that you are doing well. Moreover, keep in mind that the CAT will continue to adapt until you are answering roughly half of the questions you see correctly. Thus, toward the end of the test, you likely will feel that you are getting about half the questions you see wrong. That means the test is working the way it should. So don't get discouraged—just keep doing your best and keep your confidence up!

Getting Ready for the ASVAB

This book is divided into chapters that offer specific test-taking strategies for dealing with each of the subtests, with emphasis on those subtests that constitute the AFQT. Your online resources are organized in the same way as the book. Kaplan recommends that you take the first practice test, in chapter 2, before you work through this book and the online videos and Qbank. By doing so, you can pay special attention to the areas that were difficult for you as you progress through the book. You've made the right move in deciding to prepare for the ASVAB. It is a highly coachable test, and we will give you the tools you need to score high on the ASVAB and qualify for the military career you desire.

CHAPTER 2

ASVAB DIAGNOSTIC TEST

ABOUT THE DIAGNOSTIC TEST

This diagnostic test is intended to help guide you in your ASVAB preparation. You'll learn several things at once by taking this diagnostic test:

What's on the ASVAB: This diagnostic test will help you become familiar with the topics and question types that will appear on the ASVAB. After taking the diagnostic test, review your answers carefully, and make mental notes about the question types or topics that seem least familiar.

What your strengths and areas of opportunity are: The diagnostic mimics the paper-and-pencil ASVAB, rather than the CAT-ASVAB, which means it has more questions than you'll see on the CAT-ASVAB. There's a good reason for that: the greater number of questions on this diagnostic test will allow you to see more clearly which topics you need more work on.

How you react to test fatigue: Find a quiet space where you won't be interrupted for about three hours. Take this test under timed conditions, one subject after another in the order they appear here. You can take a five-minute bathroom break if needed, but don't let yourself take a long break. This experience will help you assess how ready you are to focus on test material for an extended time period. If you find that fatigue interferes with your performance, be sure to take a full-length test (either in the book or online) each week in the last few weeks before your test date.

Hint: If you run out of time before you finish any given section, be sure to fill in all the blanks on your answer sheet, as there's no wrong answer penalty and some of your guesses could turn out to be correct! Good luck!

ANSWER SHEET

Part 1: General Science (GS)

1 A B C D	6 A B C D	11 A B C D	16 A B C D	21 A B C D
2 A B C D	7 A B C D	12 A B C D	17 A B C D	22 A B C D
3 A B C D	8 A B C D	13 A B C D	18 A B C D	23 A B C D
4 A B C D	9 A B C D	14 A B C D	19 A B C D	24 A B C D
5 A B C D	10 A B C D	15 A B C D	20 A B C D	25 A B C D

Part 2: Arithmetic Reasoning (AR)

1 A B C D	7 A B C D	13 A B C D	19 A B C D	25 A B C D
2 A B C D	8 A B C D	14 A B C D	20 A B C D	26 A B C D
3 A B C D	9 A B C D	15 A B C D	21 A B C D	27 A B C D
4 A B C D	10 A B C D	16 A B C D	22 A B C D	28 A B C D
5 A B C D	11 A B C D	17 A B C D	23 A B C D	29 A B C D
6 A B C D	12 A B C D	18 A B C D	24 A B C D	30 A B C D

Part 3: Word Knowledge (WK)

1 A B C D	8 A B C D	15 A B C D	22 A B C D	29 A B C D
2 A B C D	9 A B C D	16 A B C D	23 A B C D	30 A B C D
3 A B C D	10 A B C D	17 A B C D	24 A B C D	31 A B C D
4 A B C D	11 A B C D	18 A B C D	25 A B C D	32 A B C D
5 A B C D	12 A B C D	19 A B C D	26 A B C D	33 A B C D
6 A B C D	13 A B C D	20 A B C D	27 A B C D	34 A B C D
7 A B C D	14 A B C D	21 A B C D	28 A B C D	35 A B C D

Part 4: Paragraph Comprehension (PC)

1 A B C D	4 A B C D	7 A B C D	10 A B C D	13 A B C D
2 A B C D	5 A B C D	8 A B C D	11 A B C D	14 A B C D
3 A B C D	6 A B C D	9 A B C D	12 A B C D	15 A B C D

Part 5: Mathematics Knowledge (MK)

1 A B C D	6 A B C D	11 A B C D	16 A B C D	21 A B C D
2 A B C D	7 A B C D	12 A B C D	17 A B C D	22 A B C D
3 A B C D	8 A B C D	13 A B C D	18 A B C D	23 A B C D
4 A B C D	9 A B C D	14 A B C D	19 A B C D	24 A B C D
5 A B C D	10 A B C D	15 A B C D	20 A B C D	25 A B C D

ANSWER SHEET

Part 6: Electronics Information (EI)

1	Ⓐ Ⓑ Ⓒ Ⓓ	5	Ⓐ Ⓑ Ⓒ Ⓓ	9	Ⓐ Ⓑ Ⓒ Ⓓ	13	Ⓐ Ⓑ Ⓒ Ⓓ	17	Ⓐ Ⓑ Ⓒ Ⓓ
2	Ⓐ Ⓑ Ⓒ Ⓓ	6	Ⓐ Ⓑ Ⓒ Ⓓ	10	Ⓐ Ⓑ Ⓒ Ⓓ	14	Ⓐ Ⓑ Ⓒ Ⓓ	18	Ⓐ Ⓑ Ⓒ Ⓓ
3	Ⓐ Ⓑ Ⓒ Ⓓ	7	Ⓐ Ⓑ Ⓒ Ⓓ	11	Ⓐ Ⓑ Ⓒ Ⓓ	15	Ⓐ Ⓑ Ⓒ Ⓓ	19	Ⓐ Ⓑ Ⓒ Ⓓ
4	Ⓐ Ⓑ Ⓒ Ⓓ	8	Ⓐ Ⓑ Ⓒ Ⓓ	12	Ⓐ Ⓑ Ⓒ Ⓓ	16	Ⓐ Ⓑ Ⓒ Ⓓ	20	Ⓐ Ⓑ Ⓒ Ⓓ

Part 7: Auto and Shop Information (AS)

1	Ⓐ Ⓑ Ⓒ Ⓓ	6	Ⓐ Ⓑ Ⓒ Ⓓ	11	Ⓐ Ⓑ Ⓒ Ⓓ	16	Ⓐ Ⓑ Ⓒ Ⓓ	21	Ⓐ Ⓑ Ⓒ Ⓓ
2	Ⓐ Ⓑ Ⓒ Ⓓ	7	Ⓐ Ⓑ Ⓒ Ⓓ	12	Ⓐ Ⓑ Ⓒ Ⓓ	17	Ⓐ Ⓑ Ⓒ Ⓓ	22	Ⓐ Ⓑ Ⓒ Ⓓ
3	Ⓐ Ⓑ Ⓒ Ⓓ	8	Ⓐ Ⓑ Ⓒ Ⓓ	13	Ⓐ Ⓑ Ⓒ Ⓓ	18	Ⓐ Ⓑ Ⓒ Ⓓ	23	Ⓐ Ⓑ Ⓒ Ⓓ
4	Ⓐ Ⓑ Ⓒ Ⓓ	9	Ⓐ Ⓑ Ⓒ Ⓓ	14	Ⓐ Ⓑ Ⓒ Ⓓ	19	Ⓐ Ⓑ Ⓒ Ⓓ	24	Ⓐ Ⓑ Ⓒ Ⓓ
5	Ⓐ Ⓑ Ⓒ Ⓓ	10	Ⓐ Ⓑ Ⓒ Ⓓ	15	Ⓐ Ⓑ Ⓒ Ⓓ	20	Ⓐ Ⓑ Ⓒ Ⓓ	25	Ⓐ Ⓑ Ⓒ Ⓓ

Part 8: Mechanical Comprehension (MC)

1	Ⓐ Ⓑ Ⓒ Ⓓ	6	Ⓐ Ⓑ Ⓒ Ⓓ	11	Ⓐ Ⓑ Ⓒ Ⓓ	16	Ⓐ Ⓑ Ⓒ Ⓓ	21	Ⓐ Ⓑ Ⓒ Ⓓ
2	Ⓐ Ⓑ Ⓒ Ⓓ	7	Ⓐ Ⓑ Ⓒ Ⓓ	12	Ⓐ Ⓑ Ⓒ Ⓓ	17	Ⓐ Ⓑ Ⓒ Ⓓ	22	Ⓐ Ⓑ Ⓒ Ⓓ
3	Ⓐ Ⓑ Ⓒ Ⓓ	8	Ⓐ Ⓑ Ⓒ Ⓓ	13	Ⓐ Ⓑ Ⓒ Ⓓ	18	Ⓐ Ⓑ Ⓒ Ⓓ	23	Ⓐ Ⓑ Ⓒ Ⓓ
4	Ⓐ Ⓑ Ⓒ Ⓓ	9	Ⓐ Ⓑ Ⓒ Ⓓ	14	Ⓐ Ⓑ Ⓒ Ⓓ	19	Ⓐ Ⓑ Ⓒ Ⓓ	24	Ⓐ Ⓑ Ⓒ Ⓓ
5	Ⓐ Ⓑ Ⓒ Ⓓ	10	Ⓐ Ⓑ Ⓒ Ⓓ	15	Ⓐ Ⓑ Ⓒ Ⓓ	20	Ⓐ Ⓑ Ⓒ Ⓓ	25	Ⓐ Ⓑ Ⓒ Ⓓ

Part 9: Assembling Objects (AO)

1	Ⓐ Ⓑ Ⓒ Ⓓ	6	Ⓐ Ⓑ Ⓒ Ⓓ	11	Ⓐ Ⓑ Ⓒ Ⓓ	16	Ⓐ Ⓑ Ⓒ Ⓓ	21	Ⓐ Ⓑ Ⓒ Ⓓ
2	Ⓐ Ⓑ Ⓒ Ⓓ	7	Ⓐ Ⓑ Ⓒ Ⓓ	12	Ⓐ Ⓑ Ⓒ Ⓓ	17	Ⓐ Ⓑ Ⓒ Ⓓ	22	Ⓐ Ⓑ Ⓒ Ⓓ
3	Ⓐ Ⓑ Ⓒ Ⓓ	8	Ⓐ Ⓑ Ⓒ Ⓓ	13	Ⓐ Ⓑ Ⓒ Ⓓ	18	Ⓐ Ⓑ Ⓒ Ⓓ	23	Ⓐ Ⓑ Ⓒ Ⓓ
4	Ⓐ Ⓑ Ⓒ Ⓓ	9	Ⓐ Ⓑ Ⓒ Ⓓ	14	Ⓐ Ⓑ Ⓒ Ⓓ	19	Ⓐ Ⓑ Ⓒ Ⓓ	24	Ⓐ Ⓑ Ⓒ Ⓓ
5	Ⓐ Ⓑ Ⓒ Ⓓ	10	Ⓐ Ⓑ Ⓒ Ⓓ	15	Ⓐ Ⓑ Ⓒ Ⓓ	20	Ⓐ Ⓑ Ⓒ Ⓓ	25	Ⓐ Ⓑ Ⓒ Ⓓ

ASVAB DIAGNOSTIC TEST

Part 1: General Science (GS)

Time: 11 minutes; 25 questions

Directions: In this section, you will be tested on your knowledge of concepts in science generally reviewed in high school. For each question, select the best answer and mark the corresponding oval on your answer sheet.

1. ______ are necessary for the body's maintenance, growth, and repair.
   (A) Proteins
   (B) Carbohydrates
   (C) Fats
   (D) Vitamins

2. On a Fahrenheit thermometer, the boiling point of water at sea level is
   (A) 100°
   (B) 180°
   (C) 212°
   (D) 373°

3. The process by which the body's cells use oxygen and glucose to produce energy, while releasing carbon dioxide and water vapor as waste products, is known as
   (A) decomposition
   (B) photosynthesis
   (C) oxidation
   (D) respiration

4. ______ carry blood back to the heart from the capillaries.
   (A) Arteries
   (B) Veins
   (C) Ventricles
   (D) Red blood cells

5. Which of the following is an example of a chemical process?
   (A) Helium mixes with neon.
   (B) Iron forms rust.
   (C) Sugar dissolves in water.
   (D) Ice melts.

6. A universal donor is a person with which of the following blood types?
   (A) O negative
   (B) A positive
   (C) AB positive
   (D) B negative

7. Blood enters the right atrium of the heart from the
   (A) aorta
   (B) left ventricle
   (C) pulmonary vein
   (D) vena cava

8. Which of the following substances has the highest pH?
   (A) ammonia
   (B) battery acid
   (C) isopropyl alcohol
   (D) water

9. Which of the following organs does the most work to break down food using enzymes?
   (A) pancreas
   (B) stomach
   (C) small intestine
   (D) large intestine

10. Which of the following is NOT part of the female reproductive system?
   (A) oviduct
   (B) uterus
   (C) ovary
   (D) testes

11. The most basic unit of inheritance is known as a

(A) phenotype
(B) genotype
(C) chromosome
(D) gene

12. Which of the following is a sedimentary rock?

(A) granite
(B) marble
(C) shale
(D) slate

13. A producer is also known as a(n)

(A) heterotroph
(B) saprotroph
(C) autotroph
(D) scavenger

14. A vulture would be considered a

(A) producer
(B) decomposer
(C) scavenger
(D) parasite

15. What type of rock is obsidian?

(A) sedimentary
(B) igneous
(C) metamorphic
(D) sandstone

16. During a lunar eclipse

(A) the Moon lies between the Earth and Sun
(B) the Sun lies between the Moon and Earth
(C) the Earth lies between the Moon and Sun
(D) the Sun lies outside the Moon's umbra

17. A centimeter is

(A) one hundredth of a meter
(B) one tenth of a meter
(C) ten meters
(D) one hundred meters

18. Momentum is

(A) the push or pull that forces an object to change its speed or direction
(B) the rate of change of velocity
(C) the rate at which an object changes position
(D) the tendency of an object to continue moving in the same direction

19. Which color's light waves have the highest frequency?

(A) yellow
(B) green
(C) red
(D) violet

20. The major portion of an atom's mass consists of

(A) neutrons and protons
(B) electrons and protons
(C) electrons and neutrons
(D) neutrons and positrons

21. Which of the following is NOT an example of an arthropod?

(A) crab
(B) centipede
(C) sea urchin
(D) spider

22. Table salt is considered a(n)

(A) ionic compound
(B) semi-ionic compound
(C) covalent compound
(D) element

23. Over the course of 24 hours

(A) the Earth rotates 360° around the Sun
(B) the Moon rotates 360° around the Earth
(C) the Earth rotates 360° about its axis
(D) the Moon rotates 360° about its axis

24. Which of the following kingdoms is composed of prokaryotic life forms such as bacteria, and as such is considered the most primitive?
    (A) Fungi
    (B) Protista
    (C) Monera
    (D) Plantae

25. Human beings belong to the phylum
    (A) Animalia
    (B) Chordata
    (C) Mammalia
    (D) Primata

STOP. IF YOU FINISH BEFORE THE TIME IS UP, YOU MAY CHECK OVER YOUR WORK ON THIS PART ONLY.

Part 2: Arithmetic Reasoning (AR)

Time: 36 minutes; 30 questions

Directions: In this section, you are tested on your ability to use arithmetic. For each question, select the best answer and mark the corresponding oval on your answer sheet.

1. John bought a camera on sale that normally costs $160. If the price was reduced 20% during the sale, what was the sale price of the camera?

(A) $120
(B) $124
(C) $128
(D) $140

2. A subway car passes 3 stations every 10 minutes. At this rate, how many stations will it pass in one hour?

(A) 15
(B) 18
(C) 20
(D) 30

3. On a certain map, $\frac{3}{4}$ inch represents one mile. What distance, in miles, is represented by $1\frac{3}{4}$ inches?

(A) $1\frac{1}{2}$
(B) $2\frac{1}{3}$
(C) $2\frac{1}{2}$
(D) $5\frac{1}{4}$

4. A certain box contains baseballs and golf balls. If the ratio of baseballs to golf balls is 2:3 and there are 30 baseballs in the box, how many golf balls are in the box?

(A) 18
(B) 20
(C) 36
(D) 45

5. Four people shared a taxi to the airport. The fare was $36.00, and they gave the driver a tip equal to 25% of the fare. If they equally shared the cost of the fare and tip, how much did each person pay?

(A) $9.75
(B) $10.25
(C) $10.75
(D) $11.25

6. If a car travels $\frac{1}{100}$ of a kilometer each second, how many kilometers does it travel in an hour?

(A) 36
(B) 60
(C) 72
(D) 100

7. $20 - (-5) =$

(A) −25
(B) 25
(C) 15
(D) −15

8. Ms. Smith drove a total of 700 miles on a business trip. If her car averaged 35 miles per gallon of gasoline and gasoline cost $1.25 per gallon, what was the cost in dollars of the gasoline for the trip?

(A) $20.00
(B) $24.00
(C) $25.00
(D) $40.00

9. After eating 25% of the jelly beans, Brett had 72 left. How many jelly beans did Brett have originally?

(A) 90
(B) 94
(C) 95
(D) 96

10. A student finishes the first half of an exam in $\frac{2}{3}$ the time it takes him to finish the second half. If the entire exam takes him an hour, how many minutes does he spend on the first half of the exam?

(A) 20
(B) 24
(C) 27
(D) 36

11. A 25-ounce solution is 20% alcohol. If 50 ounces of water are added to it, what percent of the new solution is alcohol?

(A) $6\frac{2}{3}\%$
(B) $7\frac{1}{2}\%$
(C) 10%
(D) $13\frac{1}{3}\%$

12. Marty has exactly 5 blue pens, 6 black pens, and 4 red pens in his backpack. If he pulls out one pen at random from his backpack, what is the probability that the pen is either red or black?

(A) $\frac{2}{3}$
(B) $\frac{3}{5}$
(C) $\frac{2}{5}$
(D) $\frac{1}{3}$

13. From 1980 through 1990, the population of Country X increased by 100%. From 1990 to 2000, the population increased by 50%. What was the combined increase for the period 1980–2000?

(A) 150%
(B) $166\frac{2}{3}\%$
(C) 175%
(D) 200%

14. If a worker earns $200 for the first 40 hours of work in a week and then is paid one-and-one-half times her regular rate for any additional hours, how many hours must she work to make $230 in a week?

(A) 43
(B) 44
(C) 45
(D) 46

15. If 50% of x is 150, what is 75% of x?

(A) 225
(B) 250
(C) 275
(D) 300

16. The total fare for two adults and three children on an excursion boat is $14. If each child's fare is one half of each adult's fare, what is the adult fare?

(A) $2.00
(B) $3.00
(C) $3.50
(D) $4.00

17. What is the prime factorization of 140?

(A) 2×70
(B) $2 \times 3 \times 5 \times 7$
(C) $2 \times 2 \times 5 \times 7$
(D) $2 \times 2 \times 2 \times 5 \times 7$

18. A painter charges $12 an hour while his son charges $6 an hour. If the father and son worked the same amount of time together on a job, how many hours did each of them work if their combined charge for their labor was $108?

(A) 6
(B) 9
(C) 12
(D) 18

19. $4! =$

(A) 4
(B) 16
(C) 24
(D) 256

20. At garage A, it costs $8.75 to park a car for the first hour and $1.25 for each additional hour. At garage B, it costs $5.50 to park a car for the first hour and $2.50 for each additional hour. What is the difference between the cost of parking a car for 5 hours at garage A and parking it for the same length of time at garage B?

(A) $2.25
(B) $1.75
(C) $1.50
(D) $1.25

21. Jan types at an average rate of 12 pages per hour. At that rate, how long will it take Jan to type 100 pages?

(A) 8 hours and 10 minutes
(B) 8 hours and 15 minutes
(C) 8 hours and 20 minutes
(D) 8 hours and 30 minutes

22. Two large sodas contain the same amount as three medium sodas. Two medium sodas contain the same amount as three small sodas. How many small sodas contain the same amount as eight large sodas?

(A) 24
(B) 18
(C) 16
(D) 12

23. If each digit 5 in the number 258,546 is replaced with the digit 7, by how much will the number be increased?

(A) 2,020
(B) 2,200
(C) 20,020
(D) 20,200

24. Michael bought $2\frac{1}{4}$ pounds of lumber at $4.00 per pound. If a 7% sales tax was added, how much did Michael pay?

(A) $9.63
(B) $9.98
(C) $10.70
(D) $11.77

25. The ratio of $3\frac{1}{4}$ to $5\frac{1}{4}$ is equivalent to the ratio of

(A) 3 to 5
(B) 4 to 7
(C) 8 to 13
(D) 13 to 21

26. A cat is fed $\frac{3}{8}$ of a pound of cat food every day. For how many days will 72 pounds of this cat food feed the cat?

(A) 160
(B) 172
(C) 180
(D) 192

27. After spending $\frac{5}{12}$ of her salary, Eva has $420 left. What is her salary?

(A) $175
(B) $245
(C) $720
(D) $1,008

28. A stock decreases in value by 20%. By what percent must the stock price increase to reach its former value?

(A) 15%
(B) 20%
(C) 25%
(D) 40%

29. Joan can shovel a certain driveway in 50 minutes. If Mary can shovel the same driveway in 20 minutes, how long will it take them, to the nearest minute, to shovel the driveway if they work together?

(A) 12
(B) 13
(C) 14
(D) 15

30. June's weekly salary is $70 less than Kelly's, which is $50 more than Eileen's. If Eileen earns $280 per week, how much does June earn per week?

(A) $160
(B) $260
(C) $280
(D) $300

STOP. IF YOU FINISH BEFORE THE TIME IS UP, YOU MAY CHECK OVER YOUR WORK ON THIS PART ONLY.

Part 3: Word Knowledge (WK)

Time: 11 minutes; 35 questions

Directions: In this section, you are tested on the meaning of words. Each of the following questions has an underlined word. Select the answer that most nearly means the same as the underlined word and mark the corresponding oval on your answer sheet.

1. Noble most nearly means

(A) comely
(B) loose
(C) aristocratic
(D) lackadaisical

2. John initially disagreed with Megan, but then he had to concede her point about the budget.

(A) argue
(B) understand
(C) counter
(D) admit

3. Goad most nearly means

(A) listen
(B) provoke
(C) pacify
(D) ignore

4. Teenagers at the mall love to roam in a herd.

(A) jacket
(B) line
(C) pack
(D) ratio

5. Panicking in an emergency is not a viable response for an EMT.

(A) total
(B) collective
(C) lucid
(D) workable

6. Judicious most nearly means

(A) accessible
(B) cold
(C) wise
(D) talkative

7. Hesitating in a time of crisis can often lead to failure.

(A) broadening
(B) creating
(C) leaving
(D) pausing

8. Applicants who insist on falsifying information ruin the process for everyone.

(A) fabricating
(B) listing
(C) furthering
(D) taking on

9. Hollow most nearly means

(A) dangerous
(B) potent
(C) empty
(D) superb

10. Coax most nearly means

(A) advise
(B) trade
(C) plead
(D) grace

11. The monotonous speech left them all feeling sleepy.

(A) telling
(B) boring
(C) caustic
(D) hilarious

12. As Mrs. Higgins attempted to teach the difficult material, she sensed growing consternation among her students.

(A) desires
(B) inability
(C) frustration
(D) behavior

13. With the holidays approaching, Dave looked forward to a savory feast or two.

(A) tasty
(B) guilty
(C) heroic
(D) skimpy

14. A man of some renown, the mayor walked with his chest puffed out.

(A) size
(B) fame
(C) confusion
(D) toil

15. Raconteur most nearly means

(A) believer
(B) storyteller
(C) standout
(D) pedant

16. She felt that nothing could really quench her curiosity.

(A) justify
(B) break
(C) illuminate
(D) satisfy

17. The committee was polarized on the issue.

(A) split
(B) disgusted
(C) grateful
(D) cold

18. The need to be precise was clear to everyone.

(A) after the fact
(B) cautious
(C) exact
(D) barren

19. Drucker surveyed the terrain before him for water.

(A) oversight
(B) landscape
(C) river
(D) goal

20. Terminal most nearly means

(A) easy
(B) glittering
(C) busy
(D) final

21. Augment most nearly means

(A) craft
(B) end
(C) throw away
(D) enhance

22. Her involuntary spasm knocked over a lamp.

(A) unintentional
(B) painful
(C) listless
(D) binary

23. The other owners accused the brothers of collusion.

(A) coalition
(B) secret agreement
(C) hoping
(D) pretending

24. The mongoose shows great tenacity in the face of danger.

(A) fear
(B) candor
(C) determination
(D) speed

25. Tactile most nearly means

(A) ghastly
(B) easy
(C) patient
(D) tangible

26. The dark clouds seemed to portend a gloomy weekend ahead.

(A) fake
(B) lose
(C) predict
(D) edit

27. Germinate most nearly means
   (A) sprout
   (B) oppress
   (C) adulate
   (D) foster

28. Restore most nearly means
   (A) trip up
   (B) invigorate
   (C) care for
   (D) toughen

29. Filament most nearly means
   (A) horse
   (B) triage
   (C) nightmare
   (D) thread

30. Whatever the original intent, the focus has clearly mutated at this point.
   (A) disappeared
   (B) reiterated
   (C) altered
   (D) intensified

31. To be elected president, one must be a calm yet still dynamic figure.
   (A) reassuring
   (B) exciting
   (C) manic
   (D) terrible

32. Congeal most nearly means
   (A) fade
   (B) swirl
   (C) harden
   (D) undulate

33. Reconnoiter most nearly means
   (A) advance
   (B) posit
   (C) grade
   (D) scout

34. Whether one can accrue enough money to live on is always the question.
   (A) accumulate
   (B) acquiesce
   (C) trap
   (D) magnify

35. His innate ability to make the correct turn was amazing.
   (A) unsure
   (B) discussed
   (C) creative
   (D) natural

STOP. IF YOU FINISH BEFORE THE TIME IS UP, YOU MAY CHECK OVER YOUR WORK ON THIS PART ONLY.

Part 4: Paragraph Comprehension (PC)

Time: 13 minutes; 15 questions

Directions: This section contains paragraphs followed by incomplete statements or questions. For each question, read the paragraph and select the answer that best completes the statements or answers the question that follows. Mark the corresponding oval on your answer sheet.

The first detective stories, written by Edgar Allan Poe and Arthur Conan Doyle, emerged in the mid-nineteenth century, at a time when there was enormous public interest in science. The newspapers of the day continually publicized the latest scientific discoveries, and scientists were acclaimed as the heroes of the age. Poe and Conan Doyle shared this fascination with the methodical, logical approach used by scientists in their experiments, and instilled their detective heroes with outstanding powers of scientific reasoning.

1. The main idea of this passage is
   - (A) science fiction was not popular among nineteenth-century readers
   - (B) scientific progress made its way into the fiction of the time
   - (C) newspapers detailed detective work each day
   - (D) the first detective stories were written by scientists

Children have an amazing talent for learning vocabulary. Between the ages of one and seventeen, the average child learns the meaning of about 80,000 words—about 14 per day. Dictionaries and traditional classroom vocabulary lessons only account for part of this knowledge growth. More important are individuals' reading habits and their dialogues with people whose vocabularies are larger than their own. Reading shows students how words are used in sentences. Conversation offers students the chance to ask questions about the language.

2. According to the passage, reading is valuable to students because
   - (A) children learn differently than adults
   - (B) words used in stories are generally harder
   - (C) reading provides vocabulary clues within sentences
   - (D) vocabulary is learned mostly through conversation

The first truly American art movement was formed by a group of landscape painters that emerged in the early nineteenth century called the Hudson River School. The first works in this style were created by Thomas Cole, Thomas Doughty, and Asher Durand, a trio of painters who worked during the 1820s in the Hudson River Valley and surrounding locations. Heavily influenced by European artists, these painters set out to convey the remoteness and splendor of the American wilderness. The strongly patriotic tone of their paintings caught the spirit of the times, and within a generation the movement had grown to include landscape painters from all over the United States.

3. The passage is primarily concerned with which of the following?
   - (A) the history of the Hudson River School of painters
   - (B) American art movements of the nineteenth century
   - (C) how American landscape painters were influenced by European painters
   - (D) the artistic origins of nationalism in the United States

Different people have different approaches to choosing a personal computer. Some people pick a new computer at random, falling victim to the latest trend or advertisement. These people often regret their decisions in the long run. On the other hand, people who do thorough research before purchasing a computer are much happier with their decisions in the long run. When you are shopping for a new computer, conducting research is an important step.

4. Which of the following statements best expresses the main idea of the passage?
   - (A) People should base computer purchase decisions on advertisements.
   - (B) People should not base computer purchase decisions on advertisements.
   - (C) People should conduct research to be happy in life.
   - (D) People should conduct research before purchasing a computer.

Questions 5 and 6 refer to the following passage.

The painter Georgia O'Keeffe was born in Wisconsin in 1887 and grew up on her family's farm. At seventeen she decided she wanted to be an artist and left the farm for schools in Chicago and New York, but she never lost her bond with the land. Like most painters, O'Keeffe painted the things that were most important to her, and nearly all her works are portrayals of nature. O'Keeffe became famous when her paintings were discovered in New York by the photographer Alfred Stieglitz, whom she married in 1924. During a visit to New Mexico in 1929, O'Keeffe was so moved by the bleak landscape and broad skies of the Western desert that she began to paint its images.

5. In this context, the word *bleak* most nearly means
   (A) empty
   (B) moody
   (C) cold
   (D) vivid

6. Georgia O'Keeffe's work generally shows
   (A) an ability to paint something complex accurately
   (B) her love for Alfred Stieglitz
   (C) her desire for fame
   (D) her love of the land

The four brightest moons of Jupiter were the first objects in the solar system discovered with the use of the telescope. This proof played a central role in Galileo's famous argument in support of the Copernican model of the solar system, in which the planets are described as revolving around the Sun. For several hundred years, scientific understanding of these moons was slow to develop. But spectacular close-up photographs sent back by the 1979 Voyager missions forever changed our perception of these moons.

7. Which best describes the Copernican model of the solar system?
   (A) Planets move counterclockwise as they rotate.
   (B) The Sun and other planets revolve around the Earth.
   (C) The planets move in orbit around the Sun.
   (D) The four brightest moons of Jupiter used to be planets.

As the sky opened up and sun at last rushed into the room, Toby smiled knowing that the game would proceed as planned. It had to, if only because his father would be there and it might be the last opportunity he would have to see Toby play. Now the birds began to appear here and there. Toby got out his baseball glove and ball and waited for his dad to arrive to take him to the game.

8. The mood of the character in the passage is
   (A) sad
   (B) careless
   (C) uneasy
   (D) eager

A human body can survive without water for several days and without food for as many as several weeks. If breathing stops for as little as three to six minutes, however, death is likely. All animals require a constant supply of oxygen to the body tissues, and especially to the heart or brain. In the human body, the respiratory and circulatory systems perform this function by delivering oxygen to the blood, which then transports it to tissues throughout the body. Respiration in large animals involves more than just breathing in oxygen. It is a complex process that delivers oxygen while eliminating carbon dioxide produced by cells.

9. Which bodily function, according to the passage, is least essential to the immediate survival of a human being?
   (A) eating
   (B) drinking
   (C) breathing
   (D) excretion

The media are really out of control. When the press gets a story, it seems that within minutes it has produced flashy moving graphics and sound effects to entice viewers and garner ratings. Real facts and unbiased coverage of an issue are totally abandoned in exchange for an overly sentimental or one-sided story that too often distorts the truth. Viewers need to learn to distinguish real reporting from the junk on nearly every television channel these days.

10. The author would be most likely to agree with which of the following?
    (A) Newspapers should have more editorials.
    (B) Flashy graphics add substance to television news reporting.
    (C) Objective news reporting is a dying art.
    (D) Television news anchors are valuable sources of information.

The poems of the earliest Greeks, like those of other ancient societies, consisted of magical charms, mysterious predictions, prayers, and traditional songs of work and war. These poems were intended to be sung or recited, not written down, since they were created before the Greeks began to use writing for literary purposes. The different forms of early Greek poetry all had something in common: they described the way of life of the Greek people. Poetry expressed ideas and feelings that were shared by everyone in a community—their folktales, their memories of historical events, and their religious speculation.

11. Early Greek poetry was which of the following?

(A) mainly an oral form
(B) a departure from poetic traditions in other societies
(C) widely thought to be an act of the gods
(D) usually about lost love and sadness

In computer design, the effectiveness of a program generally depends on the ability of the programmer. Still, remarkable progress has been made in the development of artificial intelligence. This progress has scientists wondering whether it will eventually be possible to develop a computer capable of intelligent thought. When a computer defeated Garry Kasparov, considered by many the greatest chess player of all time, it was taken to be a vindication of the claims of the strongest supporters of artificial intelligence. Despite this accomplishment, others argue that while computers may imitate the human mind, they will never possess the capacity for true intelligence.

12. The main idea of this passage is

(A) computers can never learn to think
(B) chess is a game in which computers are superior
(C) great strides have been made in artificial intelligence
(D) artificial intelligence is a scientific miracle

Questions 13 and 14 refer to the following passage.

Coral reefs are created over the course of hundreds or even thousands of years. The main architect in coral reef formation is the stony coral, a relative of the sea anemone that lives in tropical climates and secretes a skeleton of almost pure calcium carbonate. Its partner is the green alga, a tiny unicellular plant that lives within the tissues of the coral. The two organisms form a mutually beneficial relationship, with the algae consuming carbon dioxide given off by the corals, and the corals thriving on the abundant oxygen produced photosynthetically by the algae. When the coral dies, its skeleton is left, and other organisms grow on top of it. Over the years, the mass of coral skeletons together with those of associated organisms combine to form the petrified underwater forest that divers find so fascinating.

13. Which of the following best describes what this passage is about?

(A) the varieties of animal life that live in coral reefs
(B) the formation of coral reefs
(C) the life and death cycles of coral reefs
(D) the physical beauty of coral reefs

14. The relationship between the coral and the algae is best described as

(A) parasitic
(B) competitive
(C) predatory
(D) cooperative

For do-it-yourself types, the cost of getting regular oil changes seems unnecessary. After all, the steps are fairly easy as long as you are safe. First, make sure that the car is stationary and on a level surface. Always use the emergency brake to ensure that the car does not roll on top of you. Next, locate the drain plug for the oil under the engine. Remember to place the oil drain pan under the plug before you start. When it is drained fully, wipe off the drain plug and the plug opening and then replace the drain plug. Next, simply place your funnel in the engine and pour in new oil. Be sure to return the oil cap when you're done. Finally, run the engine for a minute, and then check the dipstick to see if you need more oil in your engine.

15. After draining the old oil from the engine, you should

(A) replace the oil cap
(B) run the engine for a moment and check the dipstick
(C) wipe off and replace the drain plug
(D) engage the emergency brake

STOP. IF YOU FINISH BEFORE THE TIME IS UP, YOU MAY CHECK OVER YOUR WORK ON THIS PART ONLY.

Part 5: Mathematics Knowledge (MK)

Time: 24 minutes; 25 questions

Directions: In this section, you will be tested on your knowledge of basic mathematics. For each question, select the best answer and mark the corresponding oval on your answer sheet.

1. If 48 is divided by 0.08, the result is

(A) 0.06
(B) 0.6
(C) 60
(D) 600

2. If the number 9,899,399 is increased by 2,082, the result will be

(A) 9,901,471
(B) 9,901,481
(C) 9,902,471
(D) 9,902,481

3. The cube of 9 is

(A) 27
(B) 81
(C) 243
(D) 729

4. What is the value of $(-ab)(a)$ when $a = -2$ and $b = 3$?

(A) -12
(B) -6
(C) 6
(D) 12

5. $(x - 4)(x - 4) =$

(A) $x^2 + 8x - 16$
(B) $x^2 - 8x - 16$
(C) $x^2 - 8x + 16$
(D) $x^2 - 16x + 8$

6. $0.123 \times 10^4 =$

(A) 0.0000123
(B) 0.00123
(C) 1.23
(D) 1,230

7. $\sqrt{100} - \sqrt{64} =$

(A) 2
(B) 4
(C) 6
(D) 8

8. A circle has a diameter of 6, an area of b square units, and a circumference of c units. What is the value of $b + c$?

(A) 9π
(B) 15π
(C) 18π
(D) 42π

9. A bag contains 8 white, 4 red, 7 green, and 5 blue marbles. Eight marbles are withdrawn at random. How many of the withdrawn marbles are white if the chance of drawing a white marble is now 1 in 4?

(A) 3
(B) 4
(C) 5
(D) 6

10. Liza has 40 fewer than 3 times the number of books that Janice has. If B is equal to the number of books that Janice has, which of the following expressions shows the number of books that Liza and Janice have together?

(A) $3B - 40$
(B) $3B + 40$
(C) $4B - 40$
(D) $4B + 40$

11. If the perimeter of a square is 32 meters, then what is the area of the square, in square meters?

(A) 16
(B) 32
(C) 48
(D) 64

12. If $x \neq 0$, then $\frac{6x^6}{2x^2} =$

(A) $4x^4$
(B) $4x^3$
(C) $3x^4$
(D) $3x^3$

13. A number is considered "blue" if the sum of its digits is equal to the product of its digits. Which of the following is "blue"?

(A) 111
(B) 220
(C) 321
(D) 422

14. If $x = \frac{1}{8}$, what is the value of y when $\frac{2}{x} = \frac{y}{4}$?

(A) $\frac{1}{4}$
(B) 4
(C) 16
(D) 64

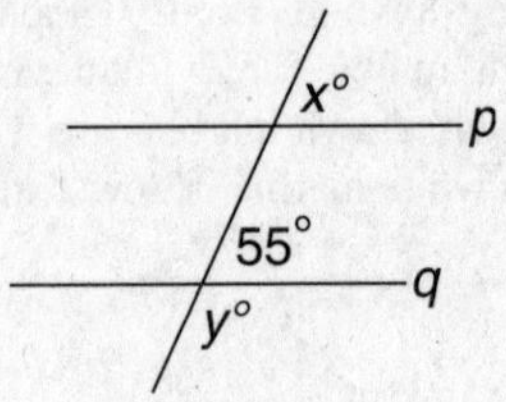

15. If line p above is parallel to line q, what is the value of $x + y$?

(A) 90
(B) 110
(C) 125
(D) 180

16. If $3ab = 6$, what is the value of a in terms of b?

(A) $\frac{2}{b}$
(B) $\frac{2}{b^2}$
(C) $2b$
(D) $2b^2$

17. For what value of y is $4(y - 1) = 2(y + 2)$?

(A) 0
(B) 2
(C) 4
(D) 6

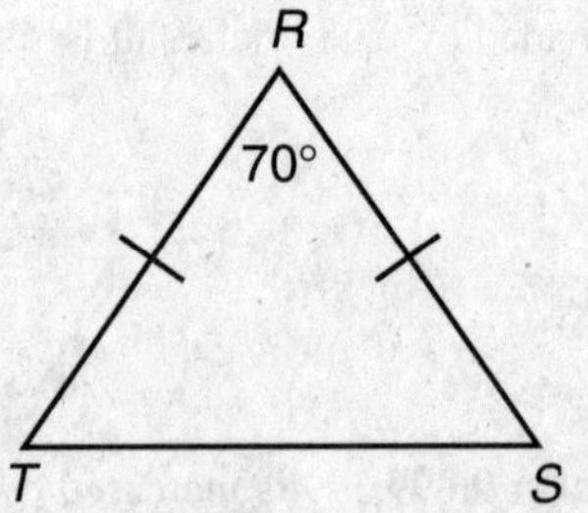

18. In triangle RST above, if $RS = RT$, what is the degree measure of angle S?

(A) 40
(B) 55
(C) 70
(D) cannot be determined from the information given

19. When D is divided by 15, the result is 6 with a remainder of 2. What is the remainder when D is divided by 6?

(A) 0
(B) 2
(C) 3
(D) 4

20. If the average of 7 consecutive even numbers is 24, then the largest number is

(A) 26
(B) 28
(C) 30
(D) 34

21. A box that has dimensions of 2 inches by 3 inches by 4 inches has a total surface area of

(A) 24 square inches
(B) 26 square inches
(C) 48 square inches
(D) 52 square inches

22. If $100 \div x = 10n$, then which of the following is equal to nx?

(A) 10
(B) $10x$
(C) 100
(D) $10xn$

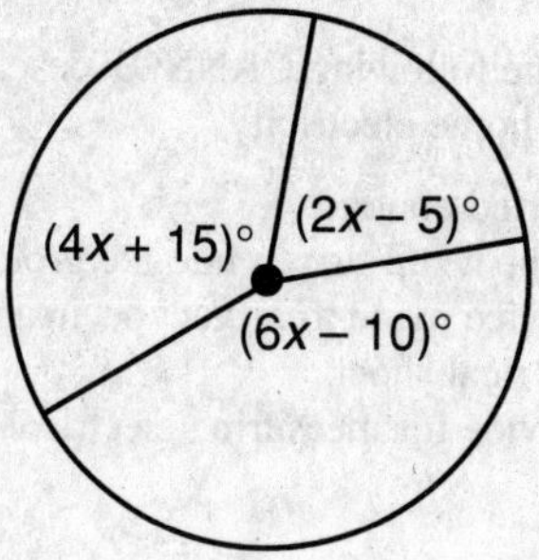

23. In the figure above, what is the value of x?

(A) 15
(B) 30
(C) 55
(D) 70

24. If $7! = 7 \times 6 \times 5 \times 4 \times 3 \times 2 \times 1$, then $5! =$

(A) 15
(B) 75
(C) 120
(D) 125

25. Melissa took $5n$ photographs on a certain trip. If she gives n photographs to each of her three friends, how many photographs will she have left?

(A) $2n$
(B) $3n$
(C) $4n - 3$
(D) $4n + 3$

STOP. IF YOU FINISH BEFORE THE TIME IS UP, YOU MAY CHECK OVER YOUR WORK ON THIS PART ONLY.

Part 6: Electronics Information (EI)

Time: 9 minutes; 20 questions

Directions: In this section, you will be tested on your knowledge of electronics basics. For each question, select the best answer and mark the corresponding oval on your answer sheet.

1. A load
   (A) has very low resistance and conducts current throughout the circuit
   (B) is a device that converts electrical energy into heat, light, or motion
   (C) is a voltage source
   (D) switches electrical current off and on

2. Which of the following symbols represents a photosensitive diode?

(A) 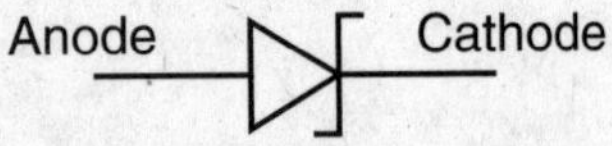

(B) 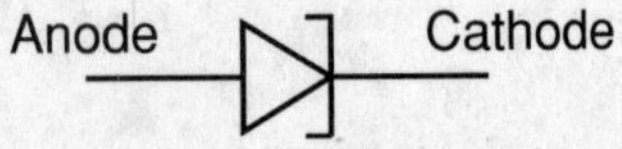

(C) 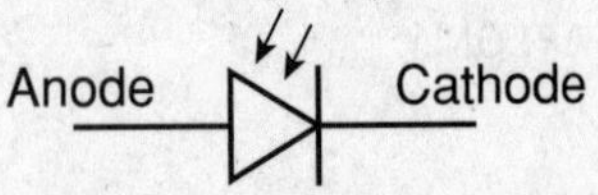

(D) 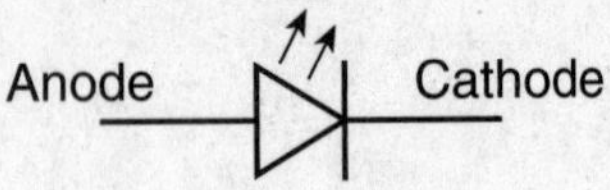

3. One hertz is equivalent to
   (A) one cycle per second of any continuous process
   (B) an acceleration of 1 m/s^2
   (C) a change in frequency of one cycle per second per second
   (D) the negative of the period

4. A(n) ____________ is an element that freely conducts electricity.
   (A) insulator
   (B) conductor
   (C) semiconductor
   (D) molecule

5. Which of the following CANNOT describe an "earth ground" in home electricity?
   (A) a buried conduit
   (B) a copper rod driven into the ground
   (C) a device made to protect occupants from electrical shock
   (D) a device for measuring electrical resistance

6. Electron flow theory states that
   (A) electrons flow best through liquids
   (B) electrons flow from areas of excess negative charge to areas of less negative charge
   (C) electrons flow from areas of excess positive charge to areas of less positive charge
   (D) electrons can only flow from one area to another if there is no resistance

7. The "electrical pressure" that causes electrons to flow in one direction through a conducting path is a result of
   (A) a voltage
   (B) a difference in resistance
   (C) parallel paths
   (D) a wire moving downhill

8. This is the symbol for which type of meter?

   (A) voltmeter
   (B) ammeter
   (C) ohmmeter
   (D) galvanometer

9. What type of circuit does this symbol represent?

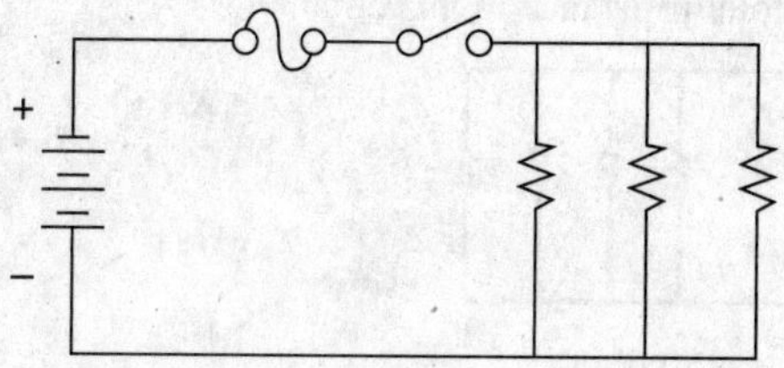

(A) parallel circuit
(B) series-parallel circuit
(C) series circuit
(D) short circuit

10. Under a constant voltage, increasing resistance results in current flow

(A) dropping
(B) rising
(C) staying the same
(D) changing direction

11. Increasing the voltage in a circuit and keeping resistance the same will result in

(A) increased current flow
(B) decreased current flow
(C) current flow staying the same
(D) zero current flow

12. Several loads in series have different resistances. Given that the same current flows through each of them, what relationship does Ohm's law predict between resistance and voltage drop?

(A) Larger voltage drops occur across loads with greater resistances.
(B) Smaller voltage drops occur across loads with greater resistances.
(C) Larger voltage drops occur across loads with lesser resistances.
(D) An equal voltage drop occurs across each load in series, independent of resistance.

13. Capacitive reactance decreases as electrical frequency

(A) decreases
(B) increases
(C) varies
(D) gets closer to DC

14. Which is the emitter in the transistor symbol below?

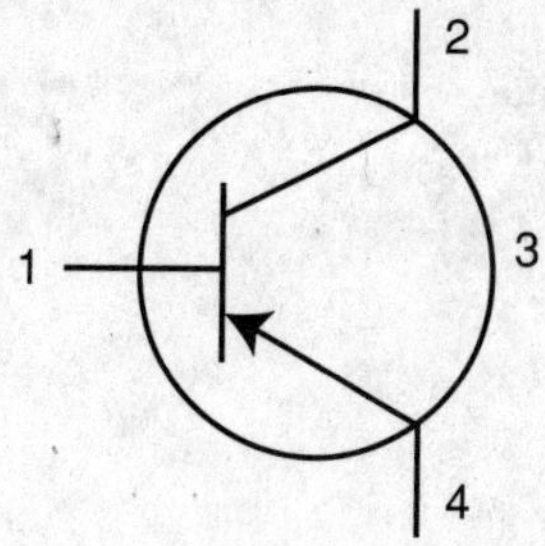

(A) 1
(B) 2
(C) 3
(D) 4

15. Which of the following CANNOT be used to make the magnetic field in a coil of wire stronger?

(A) increasing the number of turns of wire in the coil
(B) increasing the current flowing through the coil
(C) inserting an iron core into the middle of the coil
(D) inserting a dielectric between each coil

16. Whenever current passes through a resistance, ______ is most often generated.

(A) voltage
(B) capacitance
(C) heat
(D) light

17. The formula for Ohm's law is

(A) $V = I \times R$
(B) $V = I - R$
(C) $V = I + R$
(D) $V = I \div R$

18. If current is able to pass through a diode, the diode must be

(A) reverse-biased
(B) forward-biased
(C) open
(D) grounded

19. Transistors are turned on and off by voltages applied to their

(A) collector
(B) emitter
(C) base
(D) cathode

20. If each resistor in this circuit equals 1,000 ohms, what is the total resistance in this circuit?

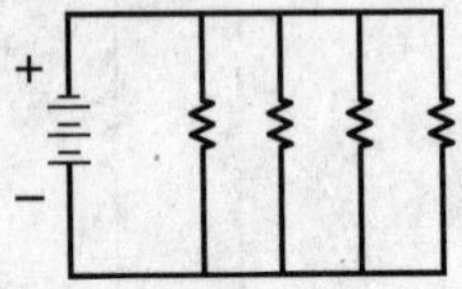

(A) 250 ohms
(B) 500 ohms
(C) 1,000 ohms
(D) 4,000 ohms

STOP. IF YOU FINISH BEFORE THE TIME IS UP, YOU MAY CHECK OVER YOUR WORK ON THIS PART ONLY.

Part 7: Auto and Shop Information (AS)

Time: 11 minutes; 25 questions

Directions: In this section, you will be tested on your knowledge of automotive and shop basics. For each question, select the best answer and mark the corresponding oval on your answer sheet.

1. High-intensity ultraviolet light is generated during ____________.

   (A) welding
   (B) sawing
   (C) soldering
   (D) drilling

2. Most drill bits are made to cut when

   (A) rotating to the right when viewed from the top
   (B) rotating to the left when viewed from the top
   (C) rotating to the right when viewed from the bottom
   (D) oscillating

3. A motor oil with a ____ prefix for its quality rating would be suited for a diesel engine.

   (A) CD
   (B) SA
   (C) SJ
   (D) DM

4. Coil-on-plug ignition systems eliminate the need for

   (A) secondary coil winding
   (B) spark plug wires
   (C) spark plugs
   (D) primary coil winding

5. A DOHC V-8 engine would have a total of __________ camshafts.

   (A) two
   (B) three
   (C) four
   (D) eight

6. What type of file is depicted below?

   (A) smoothing file
   (B) bastard file
   (C) round rasp
   (D) flat rasp

7. Cutting torches use a mixture of ____________ to produce a high-temperature flame.

   (A) nitrogen and acetylene
   (B) nitrogen and oxygen
   (C) acetylene and helium
   (D) acetylene and oxygen

8. All cars are currently being built with ____________ electrical systems.

   (A) direct current
   (B) alternating current
   (C) negative ground
   (D) both direct and alternating current

9. As lead-acid batteries discharge, their electrolyte gradually turns to

   (A) sulphuric acid
   (B) water
   (C) lead peroxide
   (D) none of the above

10. The camshaft turns at __________ the speed of the engine's crankshaft.

   (A) one-fourth
   (B) one-half
   (C) twice
   (D) triple

11. An outside micrometer can be used to measure all of the following EXCEPT

(A) the distance between two wooden posts
(B) the thickness of flat objects
(C) the outside diameter of small cylindrically shaped objects
(D) the outside diameter of small spherical objects

12. Which of the following is the most common type of pliers?

(A) 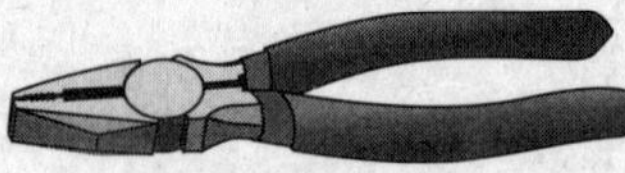

(B) 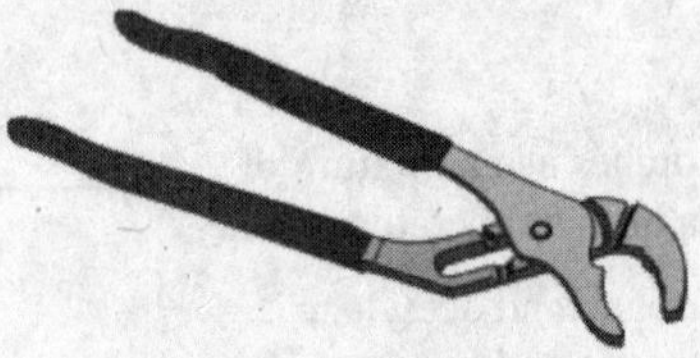

(C) 

(D) 

13. What type of saw is depicted below?

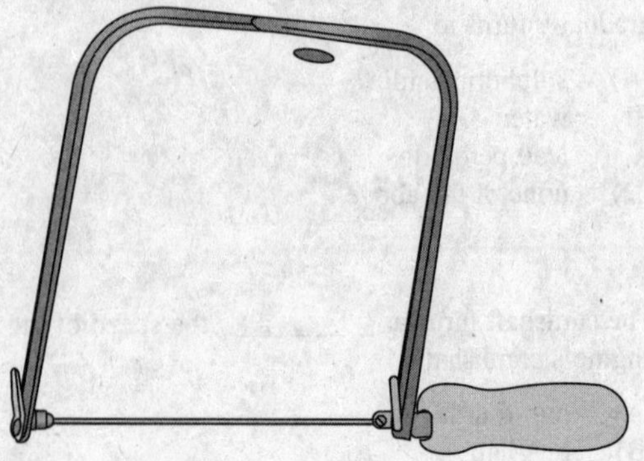

(A) rip saw
(B) crosscut saw
(C) coping saw
(D) back saw

14. Engine temperature is controlled by the

(A) electrical system
(B) water pump
(C) radiator
(D) thermostat

15. This image depicts what stroke in the four-stroke cycle?

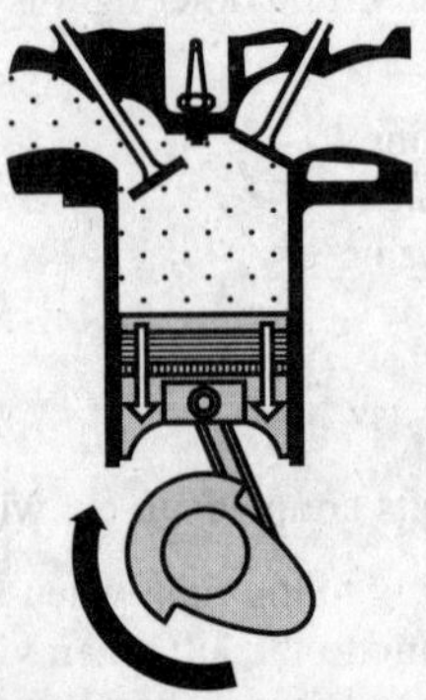

(A) intake stroke
(B) compression stroke
(C) power stroke
(D) exhaust stroke

16. The three elements needed to initiate combustion are

(A) air, light, and fuel
(B) air, fuel, and an ignition source
(C) air, compression, and an ignition source
(D) air, heat, and compression

17. A four-cylinder engine's firing order always starts with cylinder number

(A) 1
(B) 2
(C) 4
(D) 8

18. _______ cylinders will fire in one revolution of a six-cylinder engine.

(A) Two
(B) Three
(C) Four
(D) Eight

19. Which of the following is the LEAST likely to be found in an auto mechanic's toolbox?

(A) 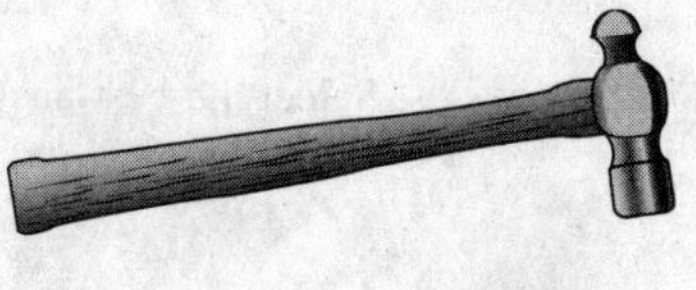

(B) 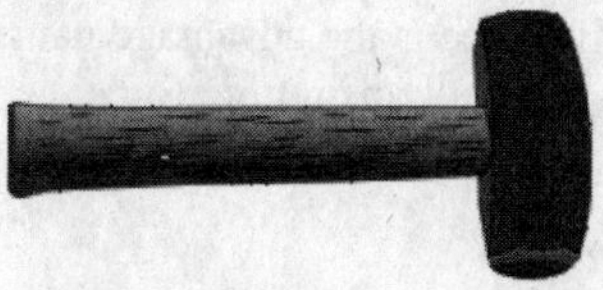

(C) 

(D) 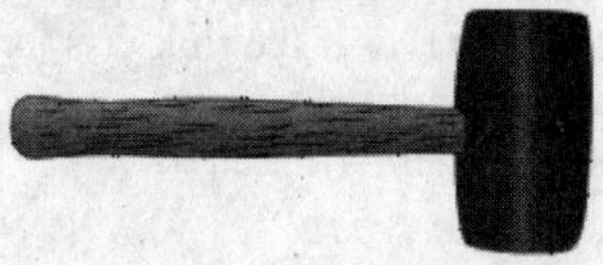

20. Most solders are an alloy of __________.

(A) tin and copper
(B) tin and lead
(C) copper and lead
(D) brass and copper

21. A lead-acid battery has lead plates immersed in electrolyte composed of __________ and water.

(A) citric acid
(B) hydrochloric acid
(C) carbolic acid
(D) sulphuric acid

22. The starter motor's drive gear engages with the engine's

(A) flywheel ring gear
(B) crankshaft
(C) vibration damper
(D) timing chain

23. Sockets come in both __________ point designs

(A) 6 and 12
(B) 7 and 13
(C) 5 and 10
(D) 1 and 2

24. The stoichiometric, or ideal, air-fuel ratio is

(A) 10:1
(B) 17:1
(C) 14.7:1
(D) 17:4.2

25. With disc brakes, the __________ rotates with the vehicle's wheels.

(A) brake rotor
(B) brake caliper
(C) brake drum
(D) wheel cylinder

STOP. IF YOU FINISH BEFORE THE TIME IS UP, YOU MAY CHECK OVER YOUR WORK ON THIS PART ONLY.

Part 8: Mechanical Comprehension (MC)

Time: 19 minutes; 25 questions

Directions: In this section, you will be tested on your knowledge of mechanics and basic physics. Select the best answer for each question and mark the corresponding oval on your answer sheet.

1. Torque is
   (A) the degree to which a force causes an object to rotate
   (B) the same as horsepower
   (C) a push or pull
   (D) a force that travels in a circle

2. Speed is different from velocity because
   (A) speed is measured in metric units while velocity is measured in English units
   (B) velocity involves both speed and direction
   (C) speed involves both velocity and direction
   (D) velocity is a scalar quantity

3. Compared to a smaller mass, a larger mass requires ________ force to achieve the same acceleration rate.
   (A) less
   (B) more
   (C) the same
   (D) varying

4. Which of the following statements about force is NOT true?
   (A) Force is a scalar quantity.
   (B) Force is a push or pull.
   (C) Greater force results in greater acceleration.
   (D) Smaller masses require less force to achieve the same acceleration as larger masses.

5. Mechanical advantage is the advantage gained by the use of __________ in transmitting force.
   (A) power
   (B) a transformer
   (C) a mechanism
   (D) an engine

6. A hockey puck sliding on the ice
   (A) has no net force acting on it
   (B) would slide forever if the rink was long enough
   (C) has speed, but not velocity
   (D) experiences kinetic friction

7. If gear A turns in a counterclockwise direction, how does gear B turn?

   (A) in a counterclockwise direction
   (B) in a clockwise direction
   (C) remains stationary
   (D) turns more slowly than gear A

8. Which of the following statements about weight is NOT true?
   (A) Weight increases closer to Earth's surface.
   (B) Weight is totally dependent on mass.
   (C) Weight is greater on planets with greater mass.
   (D) Weight varies from location to location.

9. While attempting to push a heavy box across the floor,
   (A) the amount of force required to start the box sliding is less than that required to keep it sliding
   (B) the amount of work being done is not dependent on how far the box moves
   (C) the coefficient of static friction is dependent on the nature of the surface the box is resting on
   (D) the mass of the box does not affect the amount of force required

10. Which of the following represents the mechanical advantage of a wheel and axle system where the driven wheel has a radius of 10 inches and the drive wheel has a diameter of 5 inches?
   (A) 4:1
   (B) 2:1
   (C) 1:2
   (D) 1:4

11. If it takes a force of 20 pounds to stretch the spring one inch, how much force must be applied to stretch the spring three inches?

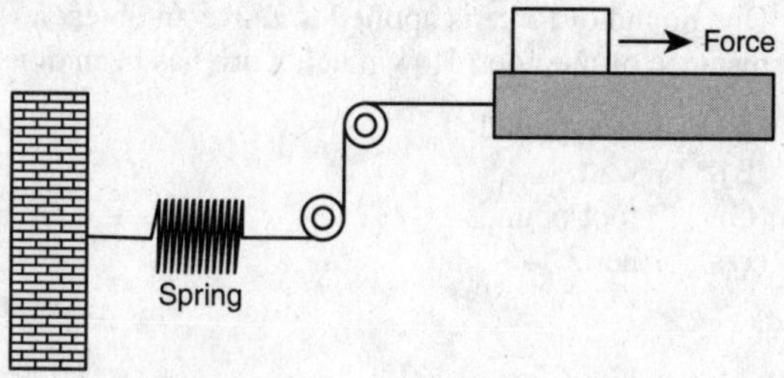

   (A) 180 pounds
   (B) 60 pounds
   (C) 20 pounds
   (D) 6.67 pounds

12. In order to apply more torque to a bolt, a mechanic could
   (A) use a longer wrench
   (B) apply less force to the wrench
   (C) use a shorter wrench
   (D) move the wrench more quickly

13. A vehicle travels at a constant speed on the highway. It can be said that
   (A) its acceleration rate is less than zero
   (B) the net force acting on the vehicle is zero
   (C) the force applied by the vehicle's drive wheels is greater than the forces that act to slow the vehicle
   (D) it is accelerating at a constant rate

14. The two water towers below are of equal size and height above the ground. If both towers are completely full and their valves are opened, which of the following statements is true?

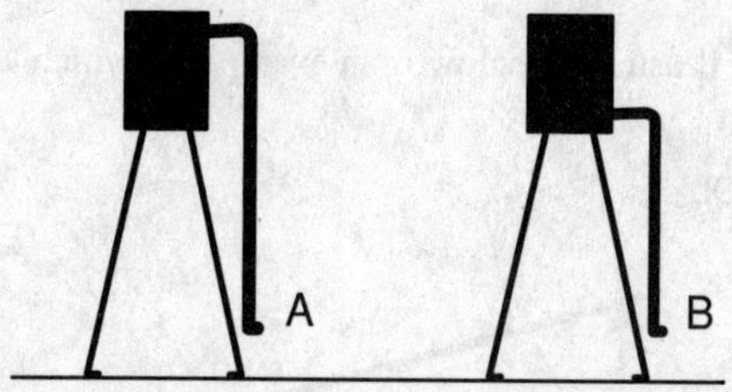

   (A) The two towers will release the same amount of water.
   (B) Tower B will release more water than tower A.
   (C) Tower A will release water three times as fast as tower B.
   (D) Tower A will release more water than tower B.

15. Two signs of equal weight are attached to a beam, using cords of equal length. Which of the following statements is true?

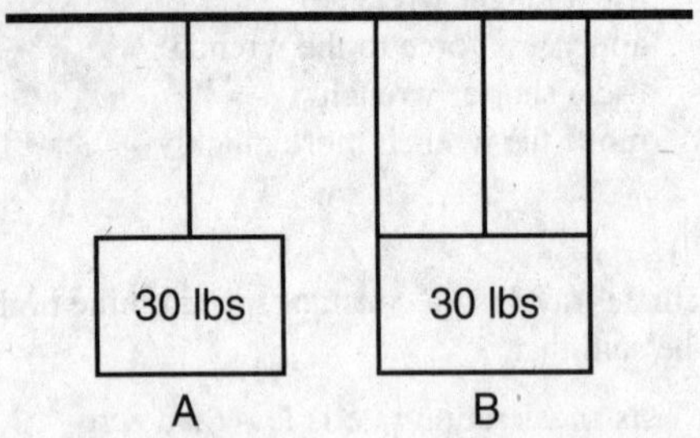

(A) All cords have the same amount of tension because the signs are of equal weight.
(B) The cord holding sign A is under twice the tension of the cords holding sign B.
(C) The cords holding sign B are under $\frac{1}{3}$ the tension of the cord holding sign A.
(D) The cords holding sign B are under $\frac{1}{9}$ the tension of the cord holding sign A.

16. The illustration below is an example of which kind of lever?

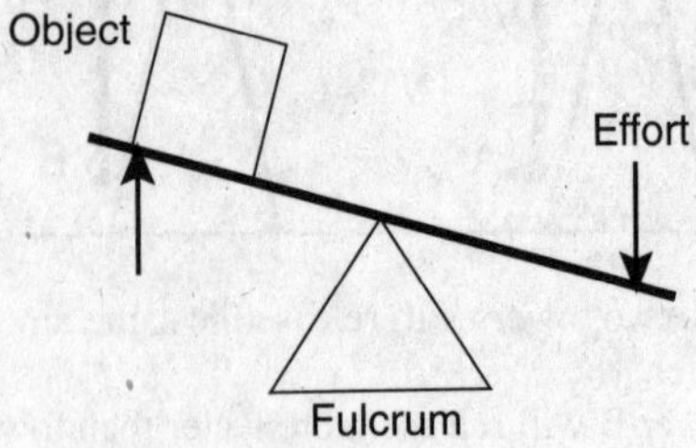

(A) first-class lever
(B) second-class lever
(C) third-class lever
(D) this is not a lever

17. If a vehicle accelerates from a standstill at a rate of $1\frac{m}{s^2}$, its velocity after 10 seconds will be

(A) $0.10\frac{m}{s}$

(B) $0.10\frac{m}{s^2}$

(C) $10\frac{m}{s}$

(D) $10\frac{m}{s^2}$

18. A wheelbarrow, as pictured below, is an example of

(A) a second-class lever
(B) a pulley system
(C) a third-class lever
(D) a crank

19. The gears in the illustration all have the same number of teeth. If gear X moves clockwise, which statement about gear Y is true?

(A) Gear Y moves counterclockwise.
(B) Gear Y moves twice as fast as gear X.
(C) Gear Y moves $\frac{1}{3}$ more slowly than gear X.
(D) Gear Y moves clockwise.

20. One pound of force is applied to move an object a distance of one foot. How much work has been done?

(A) 1 foot-pound
(B) 1 watt
(C) 2 foot-pounds
(D) 1 hertz

21. Which of the following statements about energy is NOT true?

(A) Energy cannot be created.
(B) The amount of energy in the universe is slowly diminishing.
(C) Energy cannot be destroyed.
(D) Energy can be converted from one form into another.

22. How many feet must the right end of the rope be pulled to raise weight W by 2 feet?

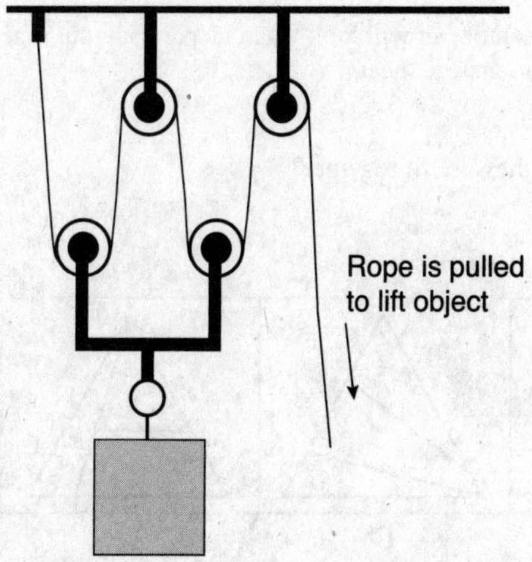

(A) 1
(B) 2
(C) 4
(D) 8

23. Which of the following is NOT a true statement about the principles that underlie hydraulics?

(A) Liquids conform to the shape of their container.
(B) A liquid can be dramatically compressed in order to increase the amount of force the liquid can transfer.
(C) A liquid is effectively incompressible.
(D) When pressure is applied to a completely enclosed fluid, this pressure is transmitted to all parts of the fluid and the enclosing walls.

24. Efficiency of a machine is determined by

(A) how much horsepower it can produce
(B) how much energy it consumes
(C) how much of the source energy is converted into usable energy
(D) how long the machine can operate at full output

25. In order to hit a baseball so that the ball has greater velocity, the player must

(A) hit the ball with less force
(B) make contact with the ball for a longer period of time
(C) apply more torque to the ball
(D) hit the ball so that it travels at roughly a 45° angle relative to the ground

STOP. IF YOU FINISH BEFORE THE TIME IS UP, YOU MAY CHECK OVER YOUR WORK ON THIS PART ONLY.

Part 9: Assembling Objects (AO)

Time: 15 minutes; 25 questions

<u>Directions</u>: In this section, you will be tested on your ability to determine how an object will look when its parts are put together. For each question, select the best answer and mark the corresponding oval on your answer sheet.

For questions 1–12:
Which figure best shows how the objects in the box on the left will appear if they are fit together?

1.

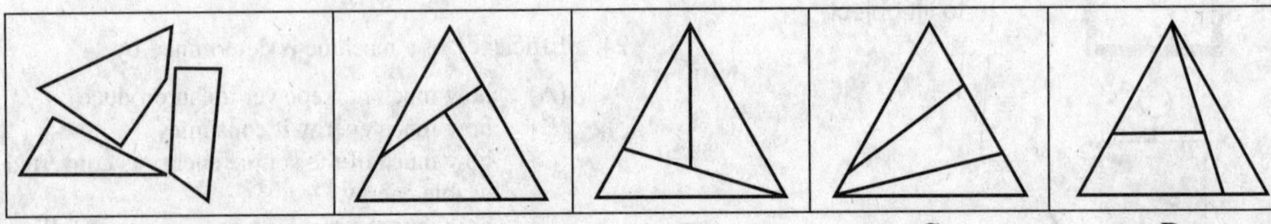

A B C D

2.

A B C D

3.

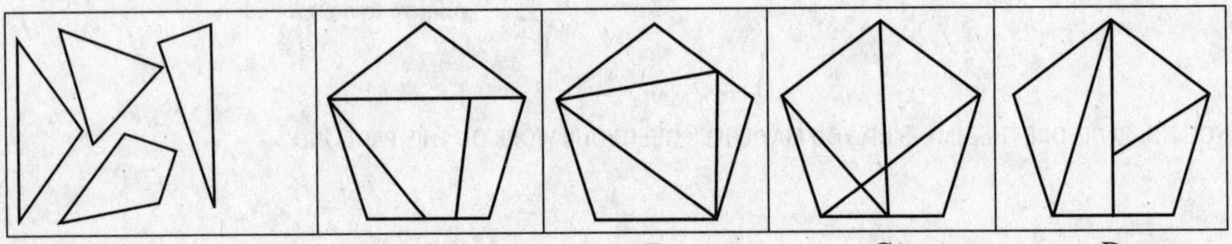

A B C D

4.

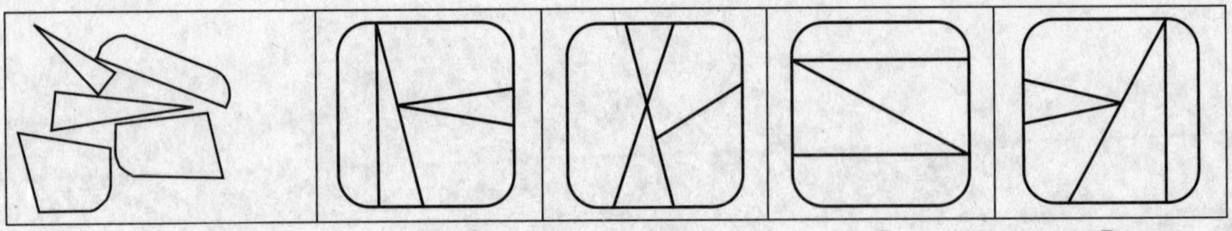

A B C D

5.

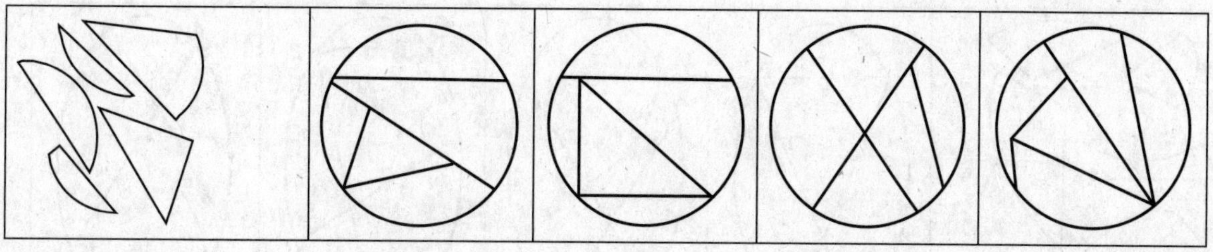

A B C D

6.

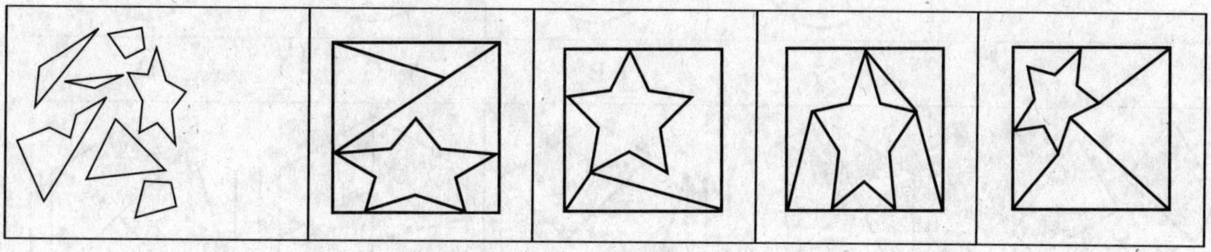

A B C D

7.

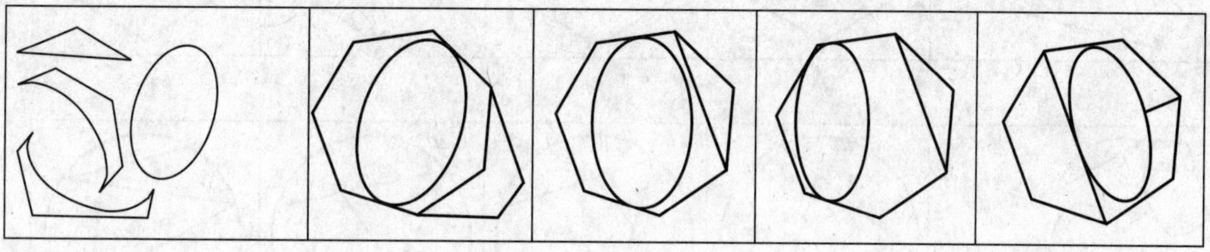

A B C D

8.

A B C D

9.

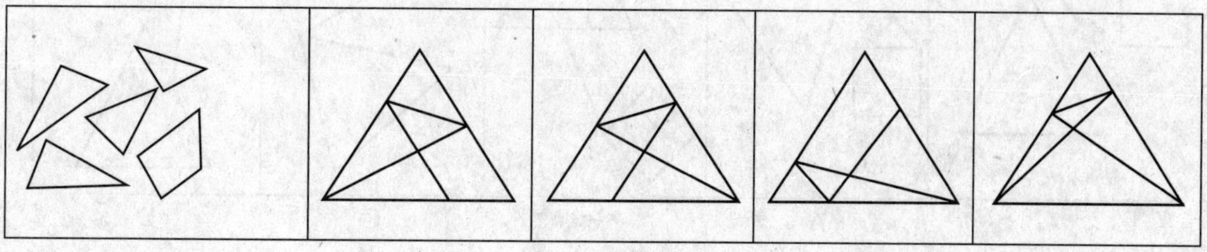

A B C D

10.

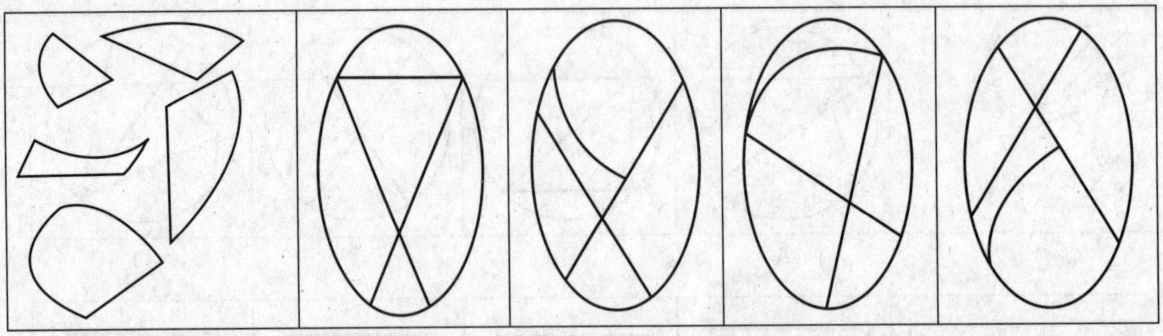

A B C D

11.

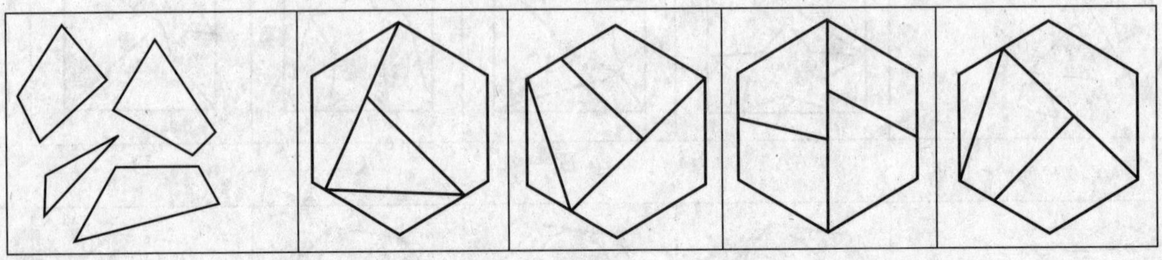

A B C D

12.

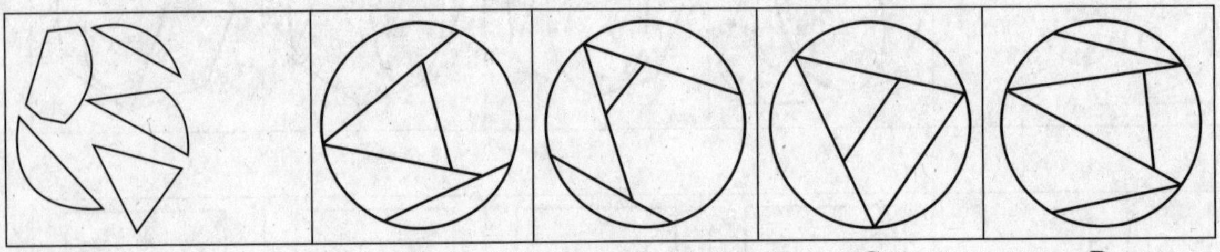

A B C D

For questions 13–25:
Which figure best shows how the objects in the box on the left will touch if the letters for each object are matched?

13.

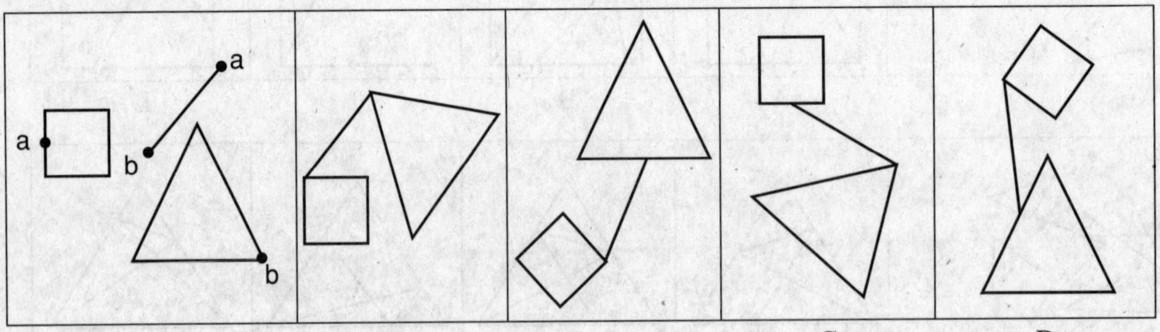

A B C D

14.

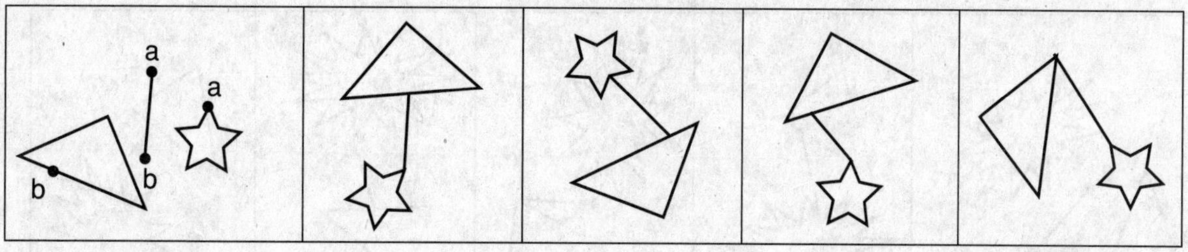

A B C D

15.

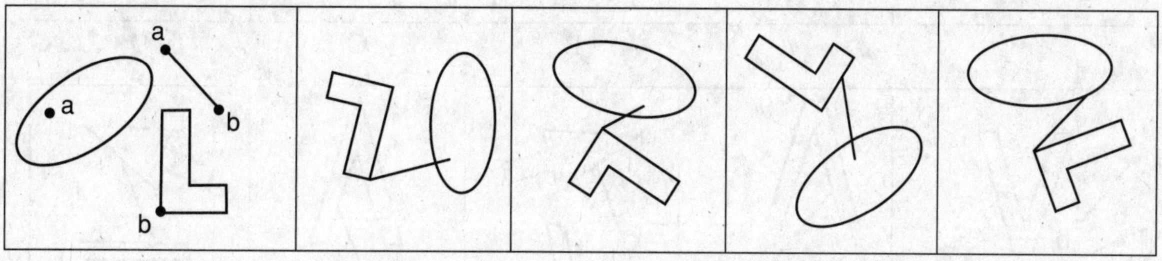

A B C D

16.

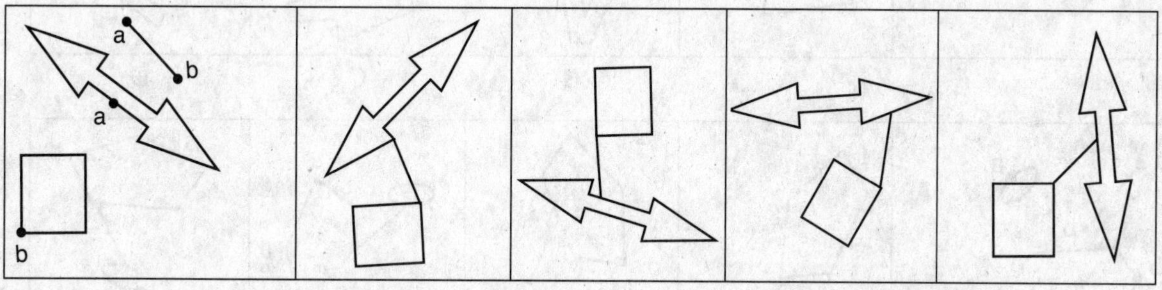

A B C D

17.

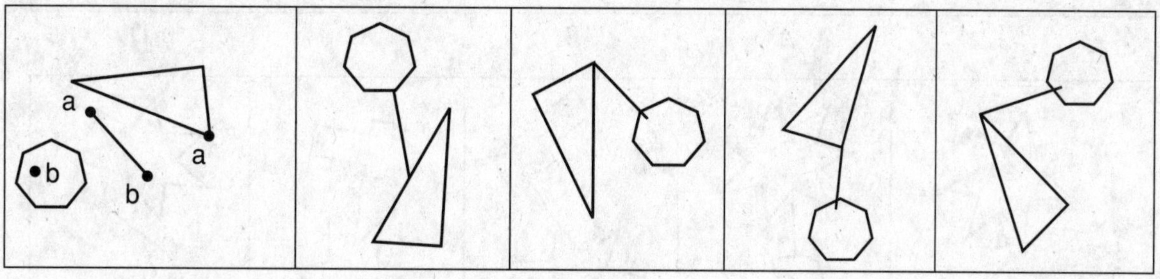

A B C D

18.

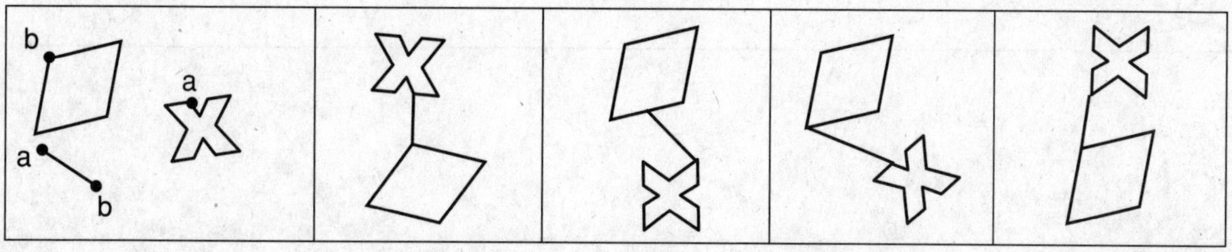

A B C D

19.

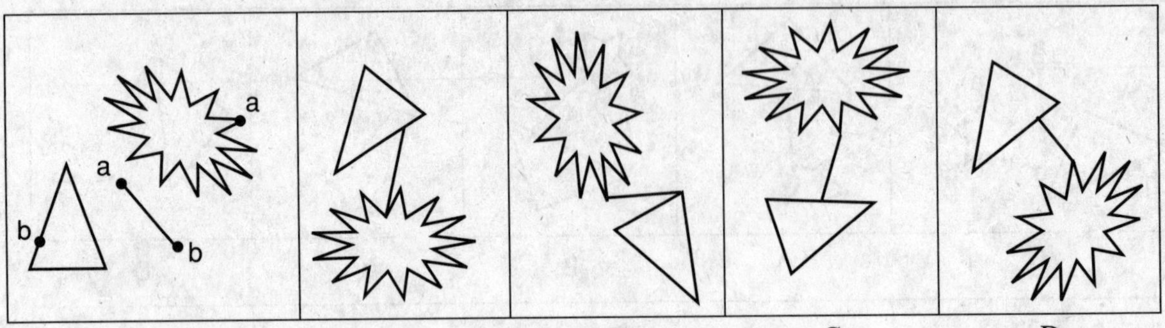

A B C D

20.

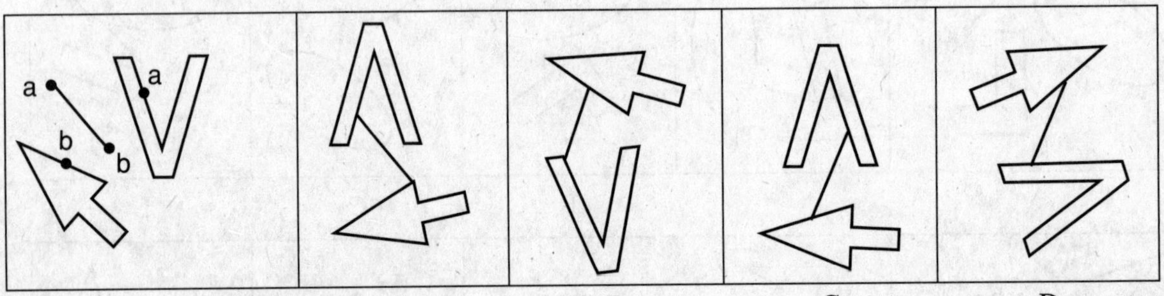

A B C D

21.

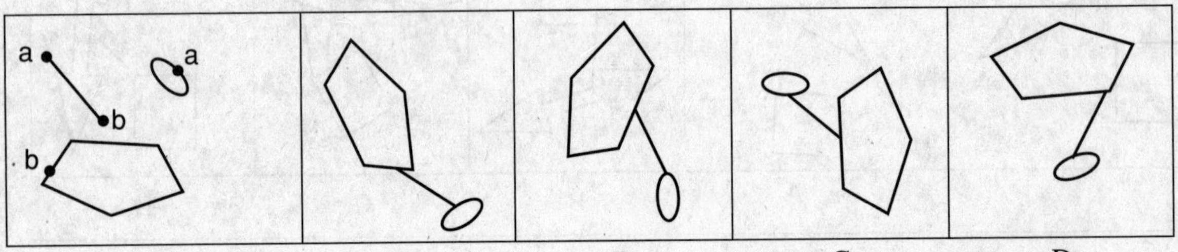

A B C D

22.

A B C D

23.

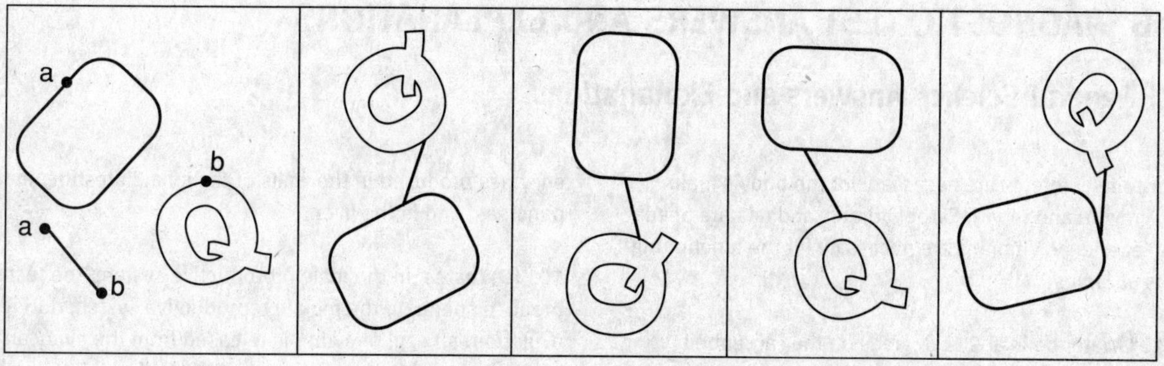

A B C D

24.

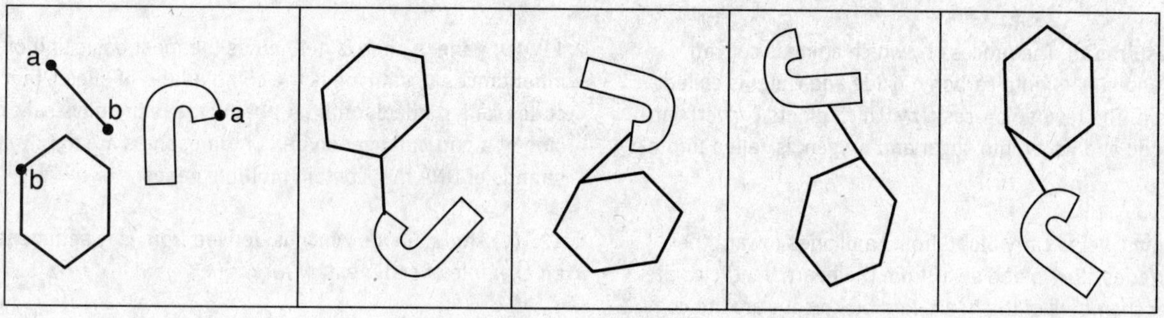

A B C D

25.

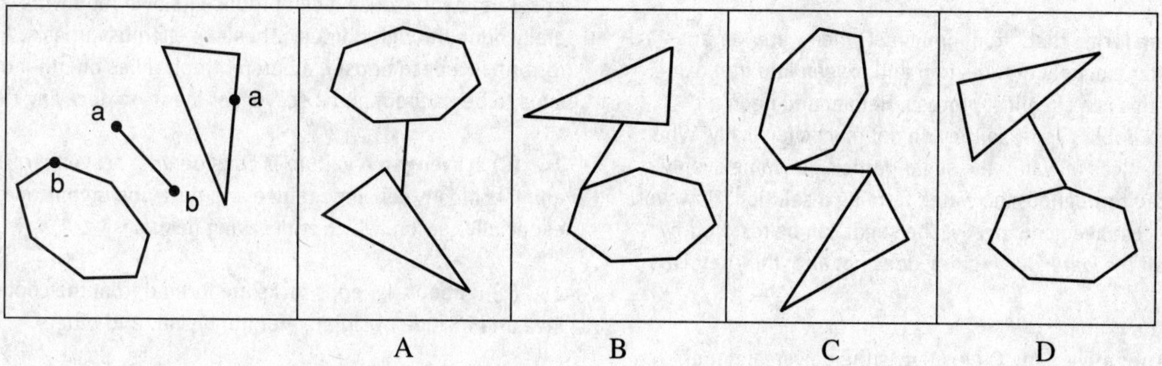

A B C D

STOP. IF YOU FINISH BEFORE THE TIME IS UP, YOU MAY CHECK OVER YOUR WORK ON THIS PART ONLY.

CONGRATULATIONS! YOU HAVE COMPLETED THE DIAGNOSTIC TEST.

ASVAB DIAGNOSTIC TEST ANSWERS AND EXPLANATIONS

Part 1: General Science Answers and Explanations

1. **(A) Proteins** Proteins are necessary for the body's maintenance, growth, and repair. Carbohydrates and fats are used primarily for energy. Vitamins are necessary for the functioning of various enzymes.

2. **(C) 212°** Water boils at 212 degrees on the Fahrenheit scale at sea level. On the Celsius scale, it boils at 100 degrees and on the Kelvin scale at 373 K.

3. **(D) respiration** The process by which animals convert oxygen (and sugars) into carbon dioxide and water is called respiration. The reverse process, by which plants convert carbon dioxide and water into sugar and oxygen, is called photosynthesis.

4. **(B) Veins** Veins carry blood from capillaries toward the heart. Arteries carry blood away from the heart. Ventricles are the lower chambers of the heart. Red blood cells are the component of blood that carries oxygen.

5. **(B) Iron forms rust.** Iron forms rust when water (or an even better electrolyte) turns iron and oxygen into iron oxide (Fe_2O_3); this is a chemical process. Helium and neon are both inert noble gases, so they do not react chemically. When sugar dissolves in water, the sugar particles become evenly distributed throughout the water, forming a solution. However, neither is changed chemically. The sugar can be restored by boiling off the water. Ice melting does not alter the chemistry of H_2O.

6. **(A) O negative** Type O negative is the universal donor, which means that type O negative blood can be given to anybody without an adverse reaction.

7. **(D) vena cava** Blood enters the right atrium of the heart from the vena cava.

8. **(A) ammonia** The more basic (that is, alkaline) a substance is, the higher the pH. A substance with a pH of 7 is neutral, like pure water. A substance with a pH of less than 7 is acidic, and a substance with a pH of greater than 7 is basic/alkaline. Look in the answer choices for a base. Of the substances listed, ammonia is the most basic or alkaline.

9. **(C) small intestine** Of the various digestive organs, the small intestine does the most work in breaking the food down into usable nutrients. Food is broken down completely by enzymes produced in the walls of the small intestine, in the pancreas, and in the liver.

10. **(D) testes** In the male reproductive system, the testes produce sperm. In the female reproductive system during ovulation, an egg, or ovum, is released from the ovary and begins to travel through the oviduct (fallopian tube) and into the uterus.

11. **(D) gene** A gene is defined as the most basic unit of inheritance. A genotype is the combination of alleles that codes for a particular trait. A phenotype is the physical expression of a particular genotype. Chromosomes are tightly coiled strands of DNA that contain multiple genes.

12. **(C) shale** Shale, which is derived from clay sediments, is an example of sedimentary rock.

13. **(C) autotroph** An autotroph is known as a producer or self-feeder because it can produce its own food. This is generally done through photosynthesis or chemosynthesis. A saprotroph is a decomposer, a hetereotroph relies on other organisms to be its food, and a scavenger feeds on decaying flesh.

14. **(C) scavenger** A vulture is considered a scavenger. These animals mostly consume refuse and decaying organic matter, especially carrion, which is decaying flesh.

15. **(B) igneous** Igneous rocks are formed from the cooling of lava and include obsidian, granite, basalt, and others.

16. **(C) the Earth lies between the Moon and Sun** A lunar eclipse happens when the Moon passes through the Earth's shadow. During a lunar eclipse, the Earth lies between the Moon and Sun.

17. **(A) one hundredth of a meter** The prefix *centi–* indicates *hundredth.* One thousandth of a meter is called a millimeter. 10 meters is a decameter and 100 meters is a hectometer.

18. **(D) the tendency of an object to continue moving in the same direction** Momentum is the tendency of an object to continue moving in the same direction. Velocity is the rate at which an object changes position. Acceleration is the rate of change of velocity. Force is the push or pull that forces an object to change its speed or direction.

19. **(D) violet** Visible light is composed of different colors each with a different frequency. Red has the lowest frequency, and violet has the highest frequency.

20. **(A) neutrons and protons** The major portion of an atom's mass consists of neutrons and protons. Electrons, positrons, neutrinos, and other subatomic particles have negligible masses.

21. **(C) sea urchin** Arthropods represent a large phylum of the animal kingdom characterized by chitinous exoskeletons, segmented bodies, and jointed legs. Examples include crabs, centipedes, and spiders, but not the sea urchin, which belongs to the phylum Echinodermata.

22. **(A) ionic compound** Table salt (NaCl) is an ionic compound because each chlorine atom borrows an electron from each sodium atom. This results in very strong ionic bonds that enable salt to form a tightly bound crystalline structure when in the solid form. When salt is placed in water, however, the crystalline structure breaks down, and the individual sodium and chlorine ions are both able to dissociate into the water.

23. **(C) the Earth rotates 360° about its axis** Over the course of 24 hours, the Earth rotates 360 degrees, or one complete rotation, about its axis.

24. **(C) Monera** The Monera kingdom is considered the most primitive kingdom because its organisms are prokaryotic—that is, their cells lack nuclei.

25. **(B) Chordata** The phylum Chordata contains animals with vertebrae. (In human beings, the vertebrae are the bones that make up the spine.) Animalia is the kingdom that includes humans. Mammalia is the class that includes humans. Primata is an order that includes primates (such as humans).

Part 2: Arithmetic Reasoning Answers and Explanations

1. **(C) $128** This question asks you to determine the sale price of a camera that normally sells at $160 and is discounted 20%. To solve, determine what 20% of $160 equals. Rewrite 20% as a decimal.
20% = 0.20. So 20% of $160 = 0.20 × $160 = $32
The sale price of the camera would be $160 – $32 = $128, choice (C).

2. **(B) 18** First, set up the rate as a proportion, where *x* is the number of stations.

$$\frac{3 \text{ stations}}{10 \text{ minutes}} = \frac{x \text{ stations}}{1 \text{ hour}}$$

Then, convert the units.

$$\frac{3 \text{ stations}}{10 \text{ minutes}} = \frac{x \text{ stations}}{60 \text{ minutes}}$$

Cross-multiply and solve for *x*.

$$180 = 10x$$

$$18 = x$$

3. **(B) $2\frac{1}{3}$** In this question, the ratio is implied: for every $\frac{3}{4}$ inch of map there is 1 real mile, so the ratio of inches to the miles they represent is always $\frac{3}{4}$ to 1. Therefore, you can set up the proportion:

$$\frac{\text{Number of inches}}{\text{Number of miles}} = \frac{\frac{3}{4}}{1} = \frac{3}{4}$$

Now $1\frac{3}{4}$ inches $= \frac{7}{4}$ inches.

Set up a proportion:

$\frac{7}{4}$ inches

$$\frac{\frac{7}{4}\text{ inches}}{\text{Number of miles}} = \frac{3}{4}$$

Cross-multiply:

$$\frac{7}{4}(4) = 3(\text{Number of miles})$$

$$7 = 3(\text{Number of miles})$$

$$\frac{7}{3} = \text{Number of miles}$$

$$\text{or } 2\frac{1}{3} = \text{Number of miles}$$

4. **(D) 45** You can express the ratio of baseballs to golf balls as $\frac{2}{3}$. Since you know the number of baseballs, you can set up a proportion: $\frac{2}{3} = \frac{30}{x}$ where *x* is the number of golf balls. To solve, cross-multiply to get $2x = 90$, or $x = 45$.

5. **(D) $11.25** The total cost of the taxi ride equals $36 + (25% of $36), or $36 + (.25 × $36) = $36 + $9 = $45. If four people split the cost equally, then each person paid $\frac{\$45}{4}$, or $11.25 each.

6. **(A) 36** Find the number of seconds in an hour and then multiply this by the distance the car is traveling each second. There are 60 seconds in a minute and 60 minutes in one hour; therefore, there are 60 × 60, or 3,600, seconds in an hour. In one second the car travels $\frac{1}{100}$ kilometers; in one hour the car will travel $3{,}600 \times \frac{1}{100}$ or 36 kilometers.

7. **(B) 25** Subtracting a negative number is the same as addition, so $20 - (-5)$ is really $20 + 5 = 25$.

8. **(C) $25.00** If Ms. Smith's car averages 35 miles per gallon, she can go 35 miles on 1 gallon. To go 700 miles she will need $\frac{700}{35}$, or 20 gallons of gasoline. The price of gasoline was $1.25 per gallon, so she spent 20 × $1.25, or $25, for her trip.

9. **(D) 96** Be careful with a question like this one. You're given the percent decrease (25%) and the *new* number (72), and you're asked to reconstruct the original number. Don't just take 25% of 72 and add it on. That 25% is based not on the new number, 72, but on the original number—the number you're looking for. The best way to do a problem like this is to set up an equation:

(Original number) – (25% of Original number) = New number

$$x - 0.25x = 72$$

$$0.75x = 72$$

$$x = 96$$

Alternatively, you can use the answer choices to determine the correct answer. The original number of jelly beans has to be reducible by 25%, or $\frac{1}{4}$. That means the original number of jelly beans has to be a multiple of 4 (or else you'd be reducing by pieces of jelly beans). Only the correct answer, 96, is a multiple of 4.

10. **(B) 24** The time it takes to complete the entire exam is the sum of the time spent on the first half of the exam and the time spent on the second half. The time spent on the first half is $\frac{2}{3}$ of the time spent on the second half. If S represents the time spent on the second half, then the total time spent is $\frac{2}{3}S + S$ or $\frac{5}{3}S$. You know this total time is one hour, or 60 minutes. Set up a simple equation and solve for S.

$$\frac{5}{3}S = 60$$
$$\frac{3}{5} \times \frac{5}{3}S = \frac{3}{5} \times 60$$
$$S = 36$$

So the second half takes 36 minutes. The first half takes $\frac{2}{3}$ of this, or 24 minutes. You could also find the first half by subtracting 36 minutes from the total time, 60 minutes.

11. **(A) $6\frac{2}{3}$%** You're asked what percent of the new solution is alcohol. The *part* is the number of ounces of alcohol; the *whole* is the total number of ounces of the new solution. There were 25 ounces originally. Then 50 ounces were added, so there are 75 ounces of new solution. How many ounces are alcohol? 20% of the original 25-ounce solution was alcohol. 20% is $\frac{1}{5}$, so $\frac{1}{5}$ of 25, or 5 ounces are alcohol. Now you can find the percent of alcohol in the new solution:

$$\%\text{ alcohol} = \frac{\text{alcohol}}{\text{total solution}} \times 100\%$$
$$= \frac{5}{75} \times 100\%$$
$$= \frac{20}{3}\% = 6\frac{2}{3}\%$$

12. **(A) $\frac{2}{3}$** To find probability, determine the number of desired outcomes and divide that by the number of possible outcomes. The probability formula looks like this:

$$\text{Probability} = \frac{\text{\# of desired outcomes}}{\text{\# of possible outcomes}}$$

In this case, Marty is pulling one pen at random from his knapsack, and you want to determine the probability that the pen is either red or black. There are 5 blue pens, 6 black pens, and 4 red pens in the knapsack. Let's return to the probability formula:

$$\text{Probability} = \frac{\text{\# of desired outcomes}}{\text{\# of possible outcomes}}$$
$$= \frac{\text{Number of red + black pens}}{\text{Number of red + black + blue pens}}$$
$$= \frac{4+6}{4+6+5} = \frac{10}{15} = \frac{2}{3}$$

13. **(D) 200%** Be careful with combined percent increase. You cannot just add the two percents, because they're percents of different bases. In this instance, the 100% increase is based on the 1980 population, but the 50% increase is based on the larger 1990 population. If you just added 100% and 50% to get 150%, you would have chosen a wrong answer.

The best way to do a problem like this one is to pick a number for the original whole and just see what happens. The best number to pick here is 100. (That may be a small number for the population of a country, but reality is not important—all that matters is the math.)

If the 1980 population was 100, then a 100% increase would put the 1990 population at 200. And a 50% increase over 200 would be 200 + 100 = 300.

Since the population went from 100 to 300, that's a percent increase of 200%.

$$\frac{300 - 100}{100} \times 100\% = \frac{200}{100} \times 100\% = 200\%$$

14. **(B) 44** To learn the worker's overtime rate of pay, first figure out her regular rate of pay. Divide the amount of money made, \$200, by the time it took to make it, 40 hours. \$200 ÷ 40 hours = \$5 per hour. That is the normal rate. The worker is paid $1\frac{1}{2}$ times her regular rate during overtime, so when working more than 40 hours, she makes $\frac{3}{2}$ × \$5 per hour = \$7.50 per hour. Now figure out how long it takes the worker to make \$230. It takes her 40 hours to make the first \$200. The last \$30 are made at the overtime rate. Since it takes one hour to make \$7.50 at this rate, you can figure out the number of extra hours by dividing \$30 by \$7.50 per hour: \$30 ÷ \$7.50 per hour = 4 hours. The total time needed is 40 hours plus 4 hours, or 44 hours.

15. **(A) 225** The calculations aren't too bad on this one. The most important thing to keep in mind is that you're solving for 75% of x and not for *x itself*. First, you are told that 50% of x is 150. That means that half of x is 150, and that x is 300. So 75% of $x = 0.75 \times 300 = 225$.

16. **(D) \$4.00** This is a question where Backsolving (plugging in an answer choice to see if it's correct) can save you a lot of time. Let's start with choice (B) and see if it works. If (B) is correct, an adult's ticket would cost \$3.00, and a child's ticket would cost \$1.50. The total fare you're asked for is for two adults and three children. If an adult's fare was \$3.00, that total fare would be 2(\$3.00) + 3(\$1.50) = \$6.00 + \$4.50 = \$10.50. That's too low, since the question states that the total fare is \$14.00.

Now see what happens if an adult fare was more expensive. If (D) was correct, an adult's ticket would cost \$4.00 and a child's ticket would cost \$2.00. The total fare would equal 2(\$4.00) + 3(\$2.00) = \$8.00 + \$6.00 = \$14.00. That's the total fare you're looking for, so (D) is correct.

17. **(C) $2 \times 2 \times 5 \times 7$** To find the prime factorization of a number, find one prime that will go into the number (here 2 is a good place to start). Express the number as that prime multiplied by some other number.

$140 = 2 \times 70$

Then keep breaking down the larger factor until you are left with only prime numbers.

$140 = 2 \times 2 \times 35$

$140 = 2 \times 2 \times 5 \times 7$

18. **(A) 6** When the painter and his son work together, they charge the sum of their hourly rates, \$12 + \$6, or \$18 per hour. Their bill equals the product of this combined rate and the number of hours they worked. Therefore \$108 must equal \$18 per hour times the number of hours they worked. Divide \$108 by \$18 per hour to find the number of hours. \$108 ÷ \$18 = 6.

19. **(C) 24** The exclamation mark indicates a factorial. A factorial is an integer multiplied by every smaller integer, down to the number 1, like this: $4! = 4 \times 3 \times 2 \times 1 = 24$.

20. **(B) \$1.75** Compute the cost of parking a car for 5 hours at each garage. Since the two garages have a split-rate system of charging, the cost for the first hour is different from the cost of each remaining hour.

The first hour at garage A costs \$8.75.

The next 4 hours cost 4 × \$1.25 = \$5.00.

The total cost for parking at garage A = \$8.75 + \$5.00 = \$13.75.

The first hour at garage B costs \$5.50.

The next 4 hours cost 4 × \$2.50 = \$10.00.

The total cost for parking at garage B = \$5.50 + \$10.00 = \$15.50.

So the difference in cost = \$15.50 − \$13.75 = \$1.75, (B).

21. **(C) 8 hours and 20 minutes** Set up a proportion:

$$\frac{12 \text{ pages}}{1 \text{ hour}} = \frac{100 \text{ pages}}{x \text{ hours}}$$

$$12x = 100$$

$$x = \frac{100}{12} = 8\frac{1}{3}$$

An hour is 60 minutes; one third of that is 20 minutes. So $8\frac{1}{3}$ hours is 8 hours and 20 minutes.

22. **(B) 18** This problem sets up relationships among large, medium, and small sodas—2 large sodas are equal to 3 medium sodas, and 2 medium sodas are equal to 3 small sodas. How many small sodas equal 8 large sodas? Well, 2 larges equal 3 mediums, so 12 mediums must equal 4 × 2 or 8 large sodas. You now can find how many small sodas represent 12 mediums. Since 2 mediums are the same as 3 small sodas, 12 mediums must equal 6 × 3 or 18 small sodas.

23. **(D) 20,200** If you change each digit 5 into a 7 in the number 258,546, the new number would be 278,746. The difference between these two numbers would be 278,746 − 258,546 = 20,200.

24. **(A) \$9.63** Since 1 pound of lumber costs \$4.00, $2\frac{1}{4}$ pounds of lumber cost 2.25 × \$4.00 = \$9.00. Then add 7% sales tax to \$9.00. Find 7% of \$9.00 by multiplying 0.07 × \$9.00 = \$0.63. Add \$0.63 to \$9.00 to get \$9.63, choice (A).

25. **(D) 13 to 21** The question asks which of five ratios is equivalent to the ratio of $3\frac{1}{4}$ to $5\frac{1}{4}$. Since the ratios in the answer choices are expressed in whole numbers, turn this ratio into whole numbers. Start by turning the ratio into improper fractions:

$$3\frac{1}{4} : 5\frac{1}{4}$$

$$= \frac{13}{4} : \frac{21}{4}$$

Multiply both sides of the ratio by 4.

$$= 13 : 21$$

26. **(D) 192** Set up the proportion.

$$\frac{\frac{3}{8}\text{lb}}{1 \text{ day}} = \frac{72 \text{ lbs}}{x \text{ days}}$$

Cross-multiply.

$$\frac{3}{8}x = 72$$

$$x = 72 \times \frac{8}{3}$$

$$x = 192$$

27. **(C) \$720** You can save valuable time by estimating on this one. Pay special attention to how much you have left and how much you've already spent. If Eva spent $\frac{5}{12}$ of her salary and was left with \$420, that means that she had $\frac{7}{12}$ left, and if Eva's salary is x dollars, then $\frac{7}{12}x = \$420$. That means that \$420 is a little more than half of her salary. So her salary would be a little less than 2(\$420) = \$840. Choice (C), \$720, is a little less than \$840. So (C) works perfectly, and it's the correct answer here.

28. **(C) 25%** The key to this question is that while the value of the stock decreases and increases by the same *amount*, it doesn't decrease and increase by the same *percent*. When the stock first decreases, that amount of change is part of a larger whole. If the stock were to increase to its former value, that same amount of change would be a larger percent of a smaller whole.

Pick a number for the original value of the stock, such as $100. (Since it's easy to take percents of 100, it's usually best to choose 100.) The 20% decrease represents $20, so the stock decreases to a value of $80. Now in order for the stock to reach the value of $100 again, there must be a $20 increase. What percent of $80 is $20? It's $\frac{\$20}{\$80} \times 100\%$, or $\frac{1}{4} \times 100\%$, or 25%.

29. **(C) 14** This is a combined work problem. Joan can shovel the whole driveway in 50 minutes, so each minute she does $\frac{1}{50}$ of the driveway. Mary can shovel the whole driveway in 20 minutes; in each minute she does $\frac{1}{20}$ of the driveway. In one minute they do:

$$\frac{1}{50} + \frac{1}{20} = \frac{2}{100} + \frac{5}{100} = \frac{7}{100}$$

If they do $\frac{7}{100}$ of the driveway in one minute, they do the entire driveway in $\frac{100}{7}$ minutes. (If you do $\frac{1}{2}$ of a job in 1 minute, you do the whole job in the reciprocal of $\frac{1}{2}$, or 2 minutes.) So all that remains is to round $\frac{100}{7}$ off to the nearest integer. Since $\frac{100}{7} = 14\frac{2}{7}$, $\frac{100}{7}$ is approximately 14. It takes about 14 minutes for both of them to shovel the driveway.

30. **(B) $260** You're told that Eileen earns $280 per week. Kelly earns $50 more than Eileen, so Kelly earns $280 + $50 = $330 per week. June's salary is $70 less than Kelly's, so June earns $330 − $70 = $260 per week, and (B) is correct.

Part 3: Word Knowledge Answers and Explanations

1. **(C) aristocratic** *Noble* means "related to high rank or social class." The *aristocracy* is composed of the highest classes in society; therefore, choice (C) is correct.

2. **(D) admit** The sentence draws a contrast between John initially disagreeing with Megan and then conceding her point. That context suggests that *concede* likely means the opposite of *disagree*. *Admit* is the best match: to admit a point is to agree with it.

3. **(B) provoke** The verb form of *goad* means to "prod into action or coerce." Of the answer choices, only *provoke* approximates this meaning.

4. **(C) pack** The word *herd* refers to a group of people or animals. Similarly, a *pack* is a group of people or animals, like a wolf pack or a scout pack.

5. **(D) workable** If you know that an EMT is a rescue technician, then you may grasp the idea that panic is not allowed. *Viable* then can mean "capable of survival or success." Only (D), *workable*, which means "able to produce the desired result," suggests that meaning.

6. **(C) wise** *Judicious* means "possessing or displaying good judgment," or in other words, being "wise."

7. **(D) pausing** *Hesitating* can mean "pausing before doing something," or sometimes "reluctant or indecisive." Among the answer choices available, the only possible correct answer is (D), *pausing*.

8. **(A) fabricating** You probably know the root word *false* as meaning "wrong," or in reference to a lie. The textbook definition is to *misrepresent*. Looking at the possible answer choices, choice (A), *fabricating*, which means "making up," is the best possible answer.

9. **(C) empty** The word *hollow* means "lacking a center or empty."

10. **(C) plead** The verb *to coax* means "to try to persuade." Of the answer choices, only (C), *plead*, approaches being correct.

11. **(B) boring** The root of the word *monotonous* is *mono* meaning *one*. Something *monotone* is in one flat tone and is completely lacking in variety. *Boring* would be another way to say this.

12. **(C) frustration** The word *consternation* means "anxiety or dismay." Of the answer choices, *frustration* is the closest match. The context of the sentence hints at this meaning, because the material the students are learning is difficult.

13. **(A) tasty** *Savory* in reference to food means "appetizing to taste." Choice (A), *tasty*, is the best answer.

14. **(B) fame** *Renown* means "fame or reputation."

15. **(B) storyteller** A *raconteur* is someone who tells stories. The word even sounds a little like "recount."

16. **(D) satisfy** To *quench* something is to "sate or satisfy" it.

17. **(A) split** The word *polarized* has the same root as *polar*. Think of "polar opposites" and you will be on the right track. In this case, choice (A), *split*, is closest in meaning to *polarized*.

18. **(C) exact** *Precise* means "exact or specific."

19. **(B) landscape** *Terrain* comes from the root *terra* or *earth*. So of the answer choices, only *landscape* would be appropriate. *Landscape* is the ground in view.

20. **(D) final** *Terminal* means "relating to an end, limit, or boundary." Choice (D), *final*, is most similar in meaning.

21. **(D) enhance** *Augment* means to "enhance something already developed."

22. **(A) unintentional** Something *involuntary* is something done without plan or accidentally–in other words, *unintentionally*.

23. **(B) secret agreement** The noun *collusion* refers to an agreement of an illicit or secret nature. The prefix *con–* (which here appears as *col–*) is a useful hint here, since it means *together*: people who *collude* agree to act together.

24. **(C) determination** *Tenacity* means showing "persistence and determination." The word root *ten* relates to holding, so you could guess that *tenacity* might mean something like "holding on."

25. **(D) tangible** *Tactile* means "relating to the sense of touch." Of the answer choices, (D), *tangible*, is the most similar in meaning to the original word. In fact, *tactile* and *tangible* come from the same root: in Latin, *tangere* means "to touch."

26. **(C) predict** To *portend* is to "foreshadow" or "foretell," as an omen or advance warning sign. Here, *predict* is the closest match.

27. **(A) sprout** To *germinate* means to "sprout or bud, as a plant."

28. **(B) invigorate** *Restore* has the prefix *re–*, meaning "again." Thus, *restore* means "bring back something that had been lost." Of the choices given, both (B) and (D) refer to affecting change on something. But of the two, only (B), *invigorate*, means to "renew."

29. **(D) thread** *A filament* is a "thread or string that is very thin."

30. **(C) altered** *Mutate* is a word you've probably seen before. Science-fiction movies and comic books often have mutated characters (like the Teenage Mutant Ninja Turtles or Godzilla). The word means "something that has changed from its original state." Of the choices given, (C), *altered*, is the best choice.

31. **(B) exciting** The adjective *dynamic* means "constantly changing or exciting." The word *yet* in the sentence is a clue that you are looking for a word that contrasts with *calm*, making *exciting* a good choice.

32. **(C) harden** *Congeal* means to "solidify, coagulate, or harden."

33. **(D) scout** To *reconnoiter* means to "make a preliminary inspection," or "see before others." Of the answer choices, only (D), *scout*, means the same thing as the given word.

34. **(A) accumulate** The verb *accrue* means to "accumulate over time as a result of growth."

35. **(D) natural** An *innate* trait is one that is *inherent* or *natural*.

Part 4: Paragraph Comprehension Answers and Explanations

1. **(B) scientific progress made its way into the fiction of the time** The main idea of a passage is the most important idea conveyed by the author. Choice (B) pinpoints the author's main point. Choice (C) distorts a detail from the passage. Choice (A) is not mentioned in the passage. The author states that science was extremely popular in the nineteenth century, which implies that science fiction was as well, but no direct information is given about the popularity of science fiction. Choice (D) is incorrect because there is no mention of Poe and Doyle being scientists.

2. **(C) reading provides vocabulary clues within sentences** The answer to a Detail question such as this will be a paraphrase of what you find in the passage. You are told that reading shows students how words are used in sentences, so (C) is correct. (A) is incorrect because adults are never mentioned. (B) also was not stated in the passage. (D) does not answer the question and is not necessarily true based on the information given.

3. **(A) the history of the Hudson River School of painters** Wrong answer choices on Global questions (that is, questions that ask about the passage's topic or main idea) are often too broad or too specific. (B) is too broad because the passage is not about all American art movements, only the Hudson River School movement. (C) is too specific because the influence of European painters is just one detail mentioned, while (A), *the history of the Hudson River School of painters*, is being described. (D) is incorrect because nationalism is not the focus of the passage.

4. **(D) People should conduct research before purchasing a computer.** This is the central idea of the passage. (If a passage begins or ends with a recommendation, that recommendation will often be the main idea.) The author does imply that people should not base their decisions on advertisements, choice (B), but that is a detail rather than the main idea. Choice (A) is the opposite of the detail stated in the passage. Choice (C) is too broad. The passage is primarily concerned with computer purchase decisions, not life in general.

5. **(A) empty** It is important on Vocabulary-in-Context questions not to rely on your vocabulary knowledge alone for the word in question, but instead to choose an answer that best represents how the word is being used in the paragraph. *Bleak* can be used to mean *cold*, choice (C), or *depressing*, which might make choice (B) seem appealing. But only choice (A) does not alter the meaning of the sentence when plugged in: "the empty landscape and broad skies."

6. **(D) her love of the land** Since the answer to a Detail question can be found in the text, the clause "nearly all her works are portrayals of nature" would direct you to choice (D). Choices (A), (B), and (C) all state things that were never mentioned in the passage, a characteristic typical of wrong answer choices to Detail questions.

7. **(C) The planets move in orbit around the Sun.** The text states that the Copernican model of the solar system describes that planets are revolving around the Sun, choice (C). This is contradicted by choice (B). (A) and (D) are not mentioned.

8. **(D) eager** Toby is smiling while he waits for his father, already holding his baseball glove and ball. He is *eager*, choice (D), to get to the game. Choice (A) is distracting since it is sad that it might be the last game that Toby's father will be able to attend, but the descriptive language of the passage supports eagerness and not sadness: the sun rushes into the room, now the birds are appearing. Toby is smiling, so there is no evidence of him being *uneasy*, choice (C). He is anxiously waiting for his father, which contradicts his being *careless*, choice (B).

9. **(A) eating** Since the body can survive without food longer than it can survive without (B) *drinking* or (C) *breathing*, *eating* is the least essential to survival. Choice (A) is correct. (D) is incorrect because *excretion* is not mentioned in the passage.

10. **(C) Objective news reporting is a dying art.** For Inference questions you should read the entire passage, and it is often best to attack each answer choice and eliminate those that do not follow from the passage. (B) clearly contradicts the author's words. You have no information about how the author feels about television anchors, so (D) is unsupported. (A) goes beyond the scope of the passage, and may or may not be true based on what you have read. The correct answer will be something that must be true based on what's given, like choice (C).

11. **(A) mainly an oral form** This is a Detail question, which means that you can look up the answer in the text. The poems "were intended to be sung or recited, not written down," so they were *mainly an oral form*, choice (A). (B), which states that Greek poetry was unlike those of other societies, contradicts the passage. (C) and (D) are not mentioned.

12. **(C) great strides have been made in artificial intelligence** The correct choice for a Global question will express what the author believes. The passage states that remarkable progress has been made in artificial intelligence, so choice (C) is correct. While the author discusses the difference of opinion between those who believe that there will eventually be a computer capable of intelligent thought and those who do not, she does not assert the truth of either statement, so choice (A) is incorrect. Be wary of answer choices such as (D) that use extreme language. Chess is not the main focus of the passage, so choice (B) is incorrect.

13. **(B) the formation of coral reefs** This is a Global question, so either the correct answer will make so much sense you will want to pick it, or you can eliminate wrong answer choices because they are too broad, too specific, or otherwise don't properly describe the passage. Here the correct answer choice does make a lot of sense. The passage describes how coral reefs are created, so choice (B), *the formation of coral reefs*, describes the passage well. Choice (A) is out because "varieties" of animal life are nowhere described. (C) is wrong because "death cycles" of coral reefs are never touched upon. And (D) is far too narrow; there's only the barest reference in the passage to "the physical beauty" of coral reefs.

14. **(D) cooperative** For this Detail question, you just want to pick the answer choice that best paraphrases the relationship between the coral and algae as described in the passage. The passage states that the "two organisms form a mutually beneficial relationship"; in other words, the relationship is *cooperative*, choice (D).

15. **(C) wipe off and replace the drain plug** For correct sequence questions, look up the answer in the text. According to the passage, when the engine is fully drained, you should *wipe off and replace the drain plug*, choice (C).

Part 5: Mathematics Knowledge Answers and Explanations

1. **(D) 600** A question like this one tests your ability to work with decimals. $\frac{48}{0.08}$ is the same as $\frac{4,800}{8} = 600$. (D) is correct. When dividing by a decimal, be sure to move the decimal place the same number of spaces for both numbers.

2. **(B) 9,901,481** Be careful with your number crunching here. $9,899,399 + 2,082 = 9,901,481$, choice (B).

3. **(D) 729** The cube of a number is that number multiplied by itself three times. So the cube of 9 would be $9 \times 9 \times 9 = 729$, choice (D).

4. **(A) −12** Plug in the values for a and b and remember your order of operations when working through your calculations. When $a = -2$ and $b = 3$,

$$(-ab)(a) = [-(-2) \times 3](-2)$$
$$= (-[-6])(-2)$$
$$= (6) \times (-2) = -12, \text{ choice (A).}$$

5. **(C) $x^2 - 8x + 16$** This is a classic product of two binomials. Remember to FOIL, and you're good to go.

$$(x - 4)(x - 4)$$
$$= (x)(x) + (x)(-4) + (-4)(x) + (-4)(-4)$$
$$= x^2 - 4x - 4x + 16$$
$$= x^2 - 8x + 16$$

Choice (C) is correct. If you recognized this as one of the classic quadratics, a binomial squared, you could have answered this correctly without using FOIL.

6. **(D) 1,230** $10^4 = 10,000$, indicating that you should move the decimal four places to the right.
So $0.123 \times 10^4 = 0.123 \times 10,000 = 1,230$, choice (D).

7. **(A) 2** Know your perfect squares. $\sqrt{100} = 10$ and $\sqrt{64} = 8$. So $\sqrt{100} - \sqrt{64} = 10 - 8$, choice (A).

8. **(B) 15π** Given a diameter of 6, the radius must equal $\frac{1}{2}$ of 6, or 3. Next, the circumference $(c) = 2\pi r = 2\pi(3) = 6\pi$. The area $(b) = \pi r^2 = \pi(3^2) = 9\pi$. Add those two: $9\pi + 6\pi = 15\pi$.

9. **(B) 4** To find the probability of something occurring, divide the number of desired outcomes by the number of total outcomes. In the example of the bag of marbles, you begin with $8 + 4 + 7 + 5 = 24$ marbles, and draw out 8, leaving you with 16 marbles. Out of those 16 marbles, 4 must be white since the chance of drawing a white marble is now $\frac{4}{16}$ or $\frac{1}{4}$. If you are left with 4 white marbles, you must have already withdrawn 4 white marbles. (B) is correct.

10. **(C) 4*B* − 40** If B is equal to the number of books that Janice has, and you know that Liza has 40 fewer than 3 times the number of books that Janice has, Liza has $3B - 40$ books, and Janice has B books. Together they have $4B - 40$ books, choice (C).

11. **(D) 64** The perimeter of a square is $4s$ where s is the length of a side. If a square has a perimeter of 32, then it has a side length of 8. The area of the square is $s^2 = 8^2 = 64$, choice (D).

12. **(C) $3x^4$** You can simplify this expression as follows:

$$\frac{6x^2}{2x^2} = \frac{6}{2}\left(\frac{x^6}{x^2}\right) = 3x^4$$

Remember to subtract exponents when dividing.

13. **(C) 321** Go through the answer choices one at a time, and select the choice whose digits have a sum and product that are equal.

(A)	111. Product = (1)(1)(1) = 1	Sum = 1 + 1 + 1 = 3 Eliminate.
(B)	220. Product = (2)(2)(0) = 0	Sum = 2 + 2 + 0 = 4 Eliminate.
(C)	321. Product = (3)(2)(1) = 6	Sum = 3 + 2 + 1 = 6

Choice (C) is correct.

14. **(D) 64** If $x = \frac{1}{8}$, and we are asked to solve for y when $\frac{2}{x} = \frac{y}{4}$, begin by plugging in $\frac{1}{8}$ for x. $\frac{2}{\frac{1}{8}} = \frac{y}{4}$. Cross-multiply and solve. $\frac{1}{8}y = 8$, so $y = 64$, (D).

15. **(D) 180** When parallel lines are crossed by a transversal, all acute angles formed are equal, and all acute angles are supplementary to all obtuse angles. So in this diagram, obtuse angle y is supplementary to the acute angle of 55°. Angle x is an acute angle, so it is equal to 55°. Therefore, angle x is supplementary to angle y, and the two must sum to 180°.

16. **(A) $\frac{2}{b}$** If you're looking for a in terms of b, isolate the a on one side of the equation.

$$3ab = 6$$
$$ab = 2$$
$$a = \frac{2}{b}$$

17. **(C) 4** Distribute the numbers outside the parentheses and solve for y.

$$4(y-1) = 2(y+2)$$
$$4y - 4 = 2y + 4$$
$$2y = 8$$
$$y = 4$$

18. **(B) 55** Since RS and RT are equal, the angles opposite them must be equal. Therefore, angle T = angle S. Since the degree measures of the three interior angles of a triangle sum to 180, 70 + angle measure S + angle measure T = 180, and angle measure S + angle measure T = 110. Since the two angles, S and T, are equal, each must have angle measures half of 110, or 55.

19. **(B) 2** When D is divided by 15, the result is 6 with a remainder of 2. That means that $D = 6(15) + 2$ or 92. When 92 is divided by 6, the remainder is 2. (B) is correct.

20. **(C) 30** The average of 7 consecutive even numbers is 24. In a set of evenly spaced numbers, the median is the same as the average. That means that 24 must be the middle number the set of numbers. So the set must be {18, 20, 22, 24, 26, 28, 30}. The largest number is 30, choice (C).

21. **(D) 52 square inches** The surface area of a rectangular solid is $2lw + 2lh + 2wh$. In this case that would be $(2 \times 2 \times 3) + (2 \times 2 \times 4) + (2 \times 3 \times 4) = 12 + 16 + 24 = 52$ square inches, choice (D).

22. **(A) 10** If $100 \div x = 10n$, that can be rewritten as $\frac{100}{x} = 10n$. Cross-multiply and you get $100 = 10nx$. Divide both sides by 10 to solve for nx.

$$nx = \frac{100}{10} = 10.$$ (A) is correct.

23. **(B) 30** A circle contains 360°, so:

$$(4x + 15) + (2x - 5) + (6x - 10) = 360$$
$$4x + 2x + 6x + 15 - 5 - 10 = 360$$
$$12x = 360$$
$$x = 30$$

24. **(C) 120** $5! = 5 \times 4 \times 3 \times 2 \times 1 = 120$

25. **(A) 2*n*** If Melissa has $5n$ photographs, and she gives n photographs to each of three friends, she would have given away $3n$ photographs. $5n - 3n = 2n$. (A) is correct.

Part 6: Electronics Information Answers and Explanations

1. **(B) is a device that converts electrical energy into heat, light, or motion** Loads convert electrical energy into some other form of energy. Examples of loads include heating elements (heat), light bulbs (light), and solenoids (motion).

2. **(C)**

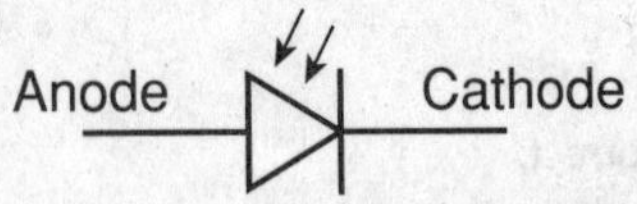

The choices represent four different diode symbols: (A) a zener diode, (B) a tunnel diode, (C) a photosensitive diode, and (D) a light-emitting diode (LED).

3. **(A) one cycle per second of any continuous process** A hertz is a unit of measurement used to express the frequency (cycles per second) of alternating current. One hertz (Hz) is the same as one cycle per second. Choice (C) is incorrect because it indicates the rate of change in frequency rather than the frequency itself.

4. **(B) conductor** A conductor is an element that freely conducts electricity, whereas an insulator does not conduct electricity at all. A semiconductor is neither a good conductor nor insulator, but has some remarkable properties that make it very useful for making electronic components.

5. **(D) a device for measuring electrical resistance** An earth ground is found outside a building, and normally utilizes conductors such as conduit or pipe that is already in the ground. All of the ground connectors in a residential wiring system will be attached to an earth ground, which is used to "funnel" away stray electricity in appliances and prevent it from causing electrical shock. A ground does not measure resistance.

6. **(B) electrons flow from areas of excess negative charge to areas of less negative charge** In electron flow theory, electrons flow away from areas of excess negative charge to those with a deficiency of negative charge. When a conductor is connected across the terminals of a battery, the electrons in the conductor will be forced away from the negative terminal of the battery and toward the positive terminal. Therefore (C) is clearly wrong. Electrons can flow through solids and gases as well as liquids, so (A) is incorrect. While resistance can reduce the flow of electrons, it does not necessarily stop the flow completely, so (D) is incorrect.

7. **(A) a voltage** Electrical pressure is known as voltage, and it is measured in volts (symbolized by the letter V).

8. **(C) ohmmeter** This is the circuit symbol for an ohmmeter, (C), which is used to measure resistance.

9. **(B) series-parallel circuit** This symbol represents a series-parallel circuit, which would have some components, such as an on/off switch, wired in series with a number of loads that are connected in parallel.

10. **(A) dropping** Ohm's law tells us that $V = I \times R$. If voltage stays constant, and resistance R rises, current I would have to drop.

11. **(A) increased current flow** Ohm's law tells us that $R = V \div I$. If resistance R stays constant, and voltage V rises, current I will also have to increase.

12. **(A) Larger voltage drops occur across loads with greater resistances.** Since $V = I \times R$, and the current is the same across all the resistors, an increase in resistance must result in an increase in the voltage drop.

13. **(B) increases** A capacitor is made to block DC (direct current), but allow AC (alternating current) to flow. Capacitive reactance is a capacitor's "opposition" to the flow of current, and this tends to diminish as the frequency of alternating current increases.

14. **(D) 4** The symbol for the transistor has an arrow that identifies the emitter. The direction of the arrow tells us what type of transistor it is; this is a PNP transistor.

15. **(D) inserting a dielectric between each coil** An electromagnet's magnetic field becomes stronger when more turns of wire are added to it, more current is passed through the coil, or an iron core is placed in the middle of the coil.

16. **(C) heat** When current passes through a resistance, a voltage drop will take place. This represents an energy loss, and this energy is normally dissipated in the form of heat.

17. **(A) $V = I \times R$** Ohm's law states that voltage in volts is equal to the current in amperes multiplied by the resistance in ohms, or $V = I \times R$. V represents voltage, and I represents current. Current is the rate of flow of electrons, or the *intensity* of the flow. (Specifically, I is the rate of charge flow.) Finally, R represents *resistance*.

18. **(B) forward-biased** When current flows freely through a diode, this is known as "forward bias." If the orientation of the diode is such that it blocks current flow, that would be reverse bias. *Open* and *grounded* are irrelevant to the question that was asked.

19. **(C) base** A transistor has three connections: the base, the emitter, and the collector. The transistor is switched off and on by voltages applied to its base. When a voltage appears across the base-emitter junction, the transistor switches on and allows current to flow between the collector and emitter.

20. **(A) 250 ohms** For a simple parallel circuit with four resistors of equal value, divide the resistance of a single component by the total number of components. For this parallel circuit, you have four 1,000 ohm resistors, so 1,000 ohms $\div$ 4 = 250 ohms.

Part 7: Auto and Shop Information Answers and Explanations

1. **(A) welding** Soldering does not produce the electric arc that welding does. Welders must cover all exposed skin with protective clothing and wear face shields with light filters for protection.

2. **(A) rotating to the right when viewed from the top** The vast majority of drill bits are made to cut while rotating in a clockwise direction (as viewed from above). These are known as *right-hand* drill bits. (Notice that choices (B) and (C) are logically equivalent and therefore must be incorrect, as each question has only one correct answer.)

3. **(A) CD** A motor oil with a "C" prefix for its quality rating would be suited for diesel engine use (i.e., a CD rating). Motor oil that had both an "S" and a "C" rating would be suited for either gasoline or diesel engine use. An example of this would be motor oil with a rating of SJ/CD.

4. **(B) spark plug wires** Coil-on-plug ignition systems eliminate the need for spark plug wires because the ignition coil is mounted directly over the spark plugs. *Secondary coil winding,* (A), *spark plugs,* (C), and *primary coil winding,* (D), are still necessary components in a coil-on-plug ignition system.

5. **(C) four** A double overhead cam arrangement puts two camshafts into each cylinder head, and makes it so one cam operates the exhaust valves in that head, and the other operates all the intake valves. Since the configuration of a V-8 engine has two heads, there would be a total of four camshafts.

6. **(C) round rasp** This is a round rasp. Carpenters would use a round rasp for cleaning out holes in wood. Round rasps are useful for cleaning up holes, whereas a flat rasp would be used to smooth flat surfaces.

7. **(D) acetylene and oxygen** Using an oxyacetylene cutting torch involves the burning of oxygen and acetylene to produce a flame that is hot enough to melt steel.

8. **(A) direct current** All automotive electrical systems are designed to utilize direct current (DC). Some vehicles may use a negative ground as part of their electrical systems, but not all cars do, so (C) is incorrect.

9. **(B) water** As the lead-acid battery discharges, the sulphuric acid in the electrolyte is reduced to water. The lead plates then become lead sulphate. Charging the battery restores the chemical composition of the lead plates and the electrolyte.

10. **(B) one-half** The camshaft is responsible for the opening and closing of the engine's intake and exhaust valves. The camshaft turns at one-half the speed of the engine's crankshaft.

11. **(A) the distance between two wooden posts** An outside micrometer is used to measure the outside dimensions of small things, such as cylinders, spheres, or relatively flat, thin objects.

12. **(D)**

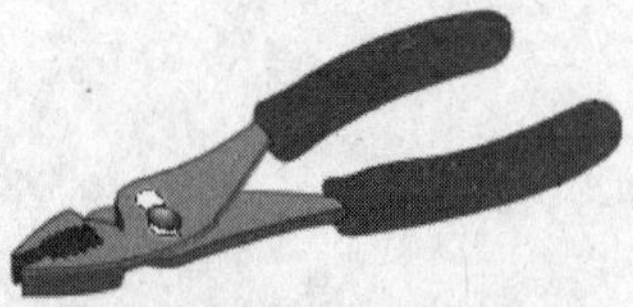

The most common type of pliers is the combination slip-joint. These are adjustable at the joint of the two handles of the pliers. With two different positions to choose from, these pliers can grip objects in a wide range of sizes.

13. **(C) coping saw** A coping saw is used to make fine, curving cuts. This saw uses a thin, flexible blade that is held tight on a wide frame.

14. **(D) thermostat** The thermostat controls engine temperature by allowing coolant to flow into the radiator when the coolant temperature rises above a certain level.

15. **(A) intake stroke** Since the piston is moving downward and the intake valve is open, this image depicts the intake stroke, which is the first stroke in the four-stroke cycle.

16. **(B) air, fuel, and an ignition source** A specific mixture of air and fuel plus an ignition source to get the whole thing going is required for combustion.

17. **(A) 1** The firing order for four-cylinder engines is 1-3-4-2. Therefore, (A) is the answer.

18. **(B) Three** It takes two full revolutions of the crankshaft to complete one cycle of events in a four-stroke cycle engine. This means that all of the cylinders in the engine must complete a power stroke in two revolutions of the crankshaft. So in *one* revolution, only three cylinders will fire. Thus, the correct answer is (B).

19. **(C)**

Claw hammers are a more specialized tool often preferred by carpenters. Their purpose is twofold. The hammer head has two ends: One drives nails and the other removes them. This is the tool least likely to be found in a mechanic's tool box.

20. **(B) tin and lead** Most solders are an alloy of lead and tin. The percentages of each metal in the solder will vary depending on the desired properties of the solder, i.e., melting point.

21. **(D) sulphuric acid** An automobile battery, or lead-acid battery, is made up of lead plates immersed in an electrolyte made up of sulphuric acid and water.

22. **(A) flywheel ring gear** Moving the ignition switch to the "start" position sends an electrical current to the starter solenoid. This engages the starter drive gear onto the engine's ring gear, which is located on the flywheel.

23. **(A) 6 and 12** Sockets come in both 6- and 12-point designs. Six-point is a stronger design, and is usually the mechanic's first choice in the smaller socket drive sizes. However, 12-point is definitely the most popular in large drive sizes.

24. **(C) 14.7:1** The stoichiometric, or ideal, air-fuel ratio is 14.7:1. This means that 14.7 pounds of air is combined with 1 pound of fuel to create an ideal air-fuel mix.

25. **(A) brake rotor** It is the brake rotor attached to the wheel that rotates. Then the brake caliper clamps to slow the car wheels down. While a brake drum does rotate with the wheel, it is not a component of a disc brake system.

Part 8: Mechanical Comprehension Answers and Explanations

1. **(A) the degree to which a force causes an object to rotate** Torque results in a twisting motion in an object. This is very different from (B) *horsepower* (the rate that work is done), or force (either (C) *a push or pull*, or (D) *a force that travels in a circle*).

2. **(B) velocity involves both speed and direction** Speed is different from velocity, in that velocity (which is a vector quantity) implies both speed (a scalar quantity) and direction.

3. **(B) more** The relationship between force, mass, and acceleration is described using the formula $F = ma$. If mass increases, more force is required to achieve the same acceleration rate.

4. **(A) Force is a scalar quality.** Force is a vector quantity. This means that it expresses both magnitude and direction.

5. **(C) a mechanism** The definition of *mechanical advantage* is the advantage gained by the use of a mechanism in transmitting force. For example, a lever and fulcrum can be used to multiply the force applied to an object.

6. **(D) experiences kinetic friction** A hockey puck experiences kinetic friction (however small) that causes it to lose velocity.

7. **(B) in a clockwise direction** This is an example of meshed gears. Meshed gears always revolve in opposite directions.

8. **(B) Weight is totally dependent on mass.** This is not a true statement because weight is dependent on both mass and acceleration due to gravity ($W = mg$).

9. **(C) the coefficient of static friction is dependent on the nature of the surface the box is resting on** The coefficient of static friction is always greater than the coefficient of kinetic friction. If a force is applied, but the box does not move, no work has been done ($W = Fd$). The nature of the surface the box rests on will define the coefficient of friction between the box and that surface.

10. **(A) 4:1** The mechanical advantage of a wheel and axle system is determined by the ratio of the radius of the wheel where the force is applied to the radius of the wheel where the force is transferred. In this case, the ratio of the radii is 10:2.5 or 4:1.

11. **(B) 60 pounds** Three inches of movement multiplied by 20 pounds per inch is 60 pounds of force.

12. **(A) use a longer wrench** Torque (twisting force) can be increased by increasing the length of the wrench, or by increasing the force applied to the wrench.

13. **(B) the net force acting on the vehicle is zero** In accordance with Newton's first law of motion, a vehicle traveling at a constant speed has no net force acting on it.

14. **(B) Tower B will release more water than tower A.** The outlet pipe on tower B is nearer to the bottom of the cistern than the outlet pipe for tower A. Therefore, it will release more water than tower A. While tower B will release water faster due to increased pressure lower in the cistern, there is no way of knowing how much faster from the information that is provided.

15. **(C) The cords holding sign B are under $\frac{1}{3}$ the tension of the cord holding sign A.** Because there are three cords holding sign B, they are under $\frac{1}{3}$ the tension of the cord holding sign A.

16. **(A) first-class lever** The illustration shows a first-class lever because the fulcrum is between the load and the effort (force).

17. **(C) $10\frac{m}{s}$** Accelerating at the rate of $1\frac{m}{s^2}$ will result in a velocity of $10\frac{m}{s}$ at the end of ten seconds.

$$1\frac{m}{s^2} \times 10s = 10\frac{m}{s}$$

18. **(A) a second-class lever** A wheelbarrow is an example of a second-class lever because the fulcrum is at one end, the effort is at the other, and the load is in between.

19. **(A) Gear Y moves counterclockwise.** Gears that are an odd number away from the indicated gear move in the opposite direction, while gears that are an even number away move in the same direction as the indicated gear. In this case, gear Y is three gears away from gear X, which means it will move in the opposite direction as gear X. Gear X is moving clockwise, so gear Y must move counterclockwise.

20. **(A) 1 foot-pound** Using the formula $W = Fd$, it can be seen that 1 pound of force applied through a distance of 1 foot will result in 1 foot-pound of work being done.

21. **(B) The amount of energy in the universe is slowly diminishing.** The principle of conservation of energy tells us that the amount of energy in the universe is constant.

22. **(D) 8** Since there are 4 pulleys in the block and tackle shown, the mechanical advantage is 4:1. Therefore, the rope must be pulled 4 times the distance that the weight is raised, which is $4 \times 2 = 8$.

23. **(B) A liquid can be dramatically compressed in order to increase the amount of force the liquid can transfer.** Choice (B) is not a true statement; the use of hydraulic force relies on the fact that liquids are effectively incompressible.

24. **(C) how much of the source energy is converted into usable energy** A machine's efficiency is expressed as the percentage of the source energy that it converts into usable energy.

25. **(B) make contact with the ball for a longer period of time** The velocity at which the ball travels will depend upon the impulse that was applied to it. Impulse is determined by multiplying the force by the amount of time that the force was applied. To increase the impulse, the player should hit the ball with greater force and make contact with the ball for a longer period of time.

Part 9: Assembling Objects Answers and Explanations

For answers 1–12:
In the answers below, we have numbered the pieces, in order to help you see how the correct answer relates to the instructions box.

1. **A**

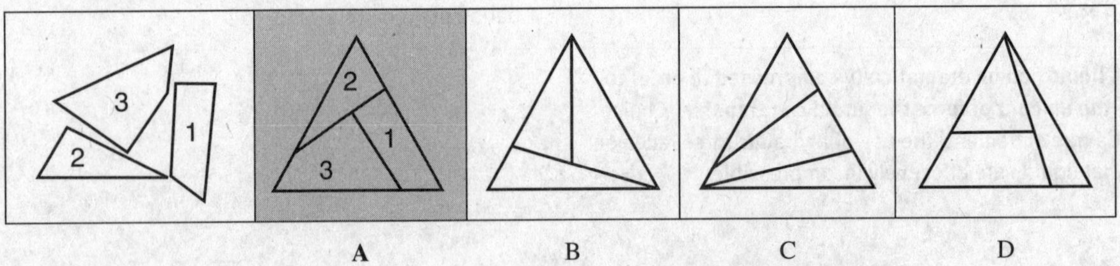

2. **C**

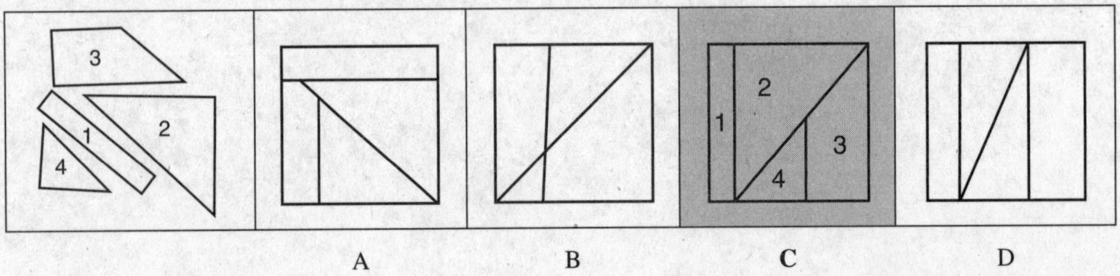

3. **D**

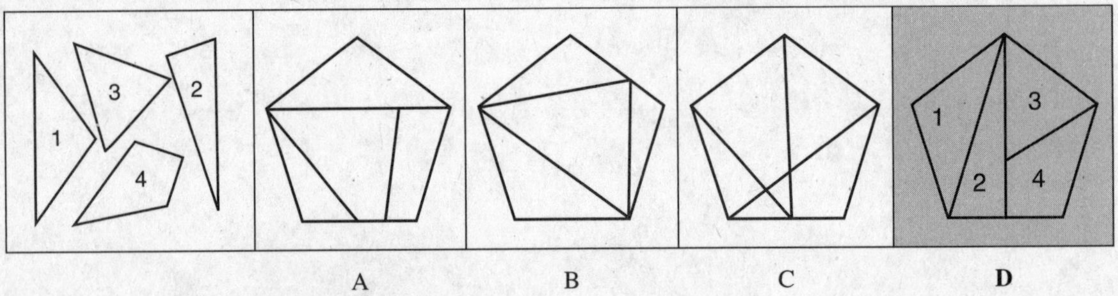

4. **A**

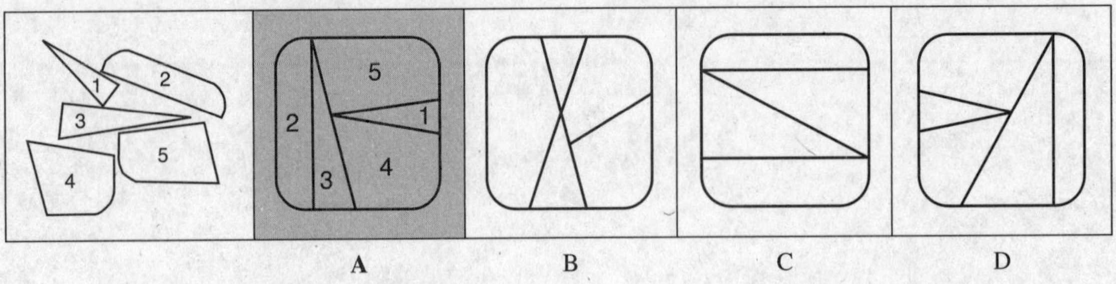

5. **B**

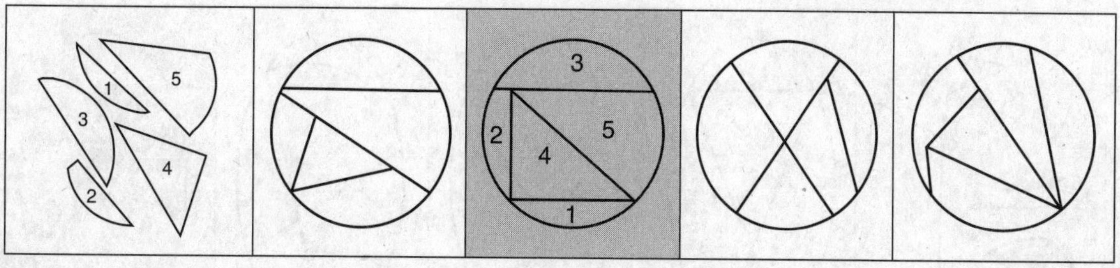

A **B** C D

6. **A**

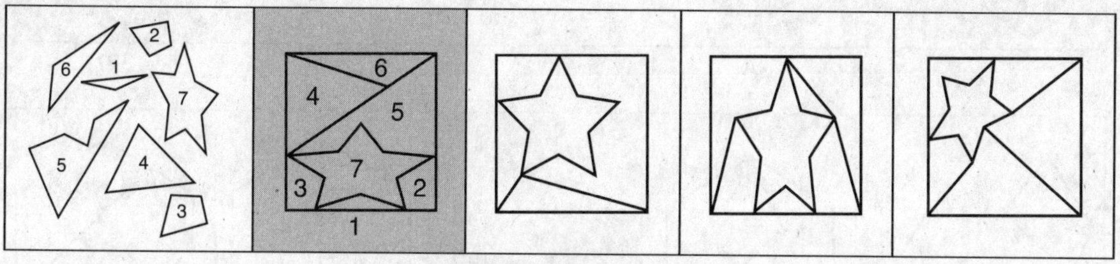

A B C D

7. **B**

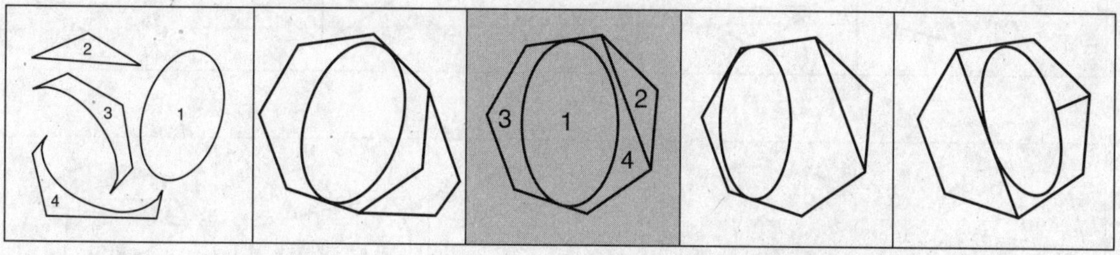

A **B** C D

8. **C**

A B **C** D

9. **A**

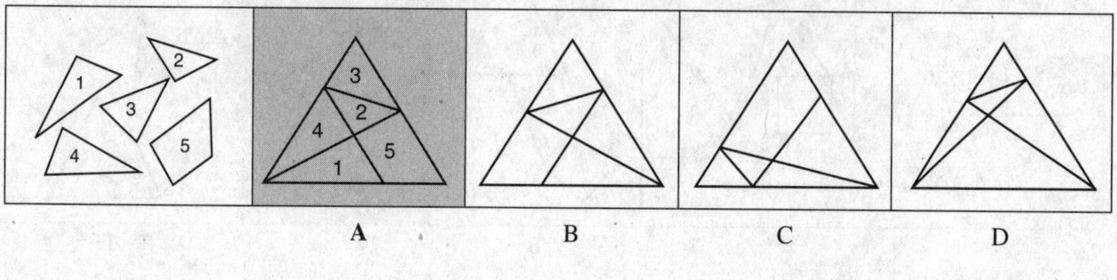

A B C D

10. **D**

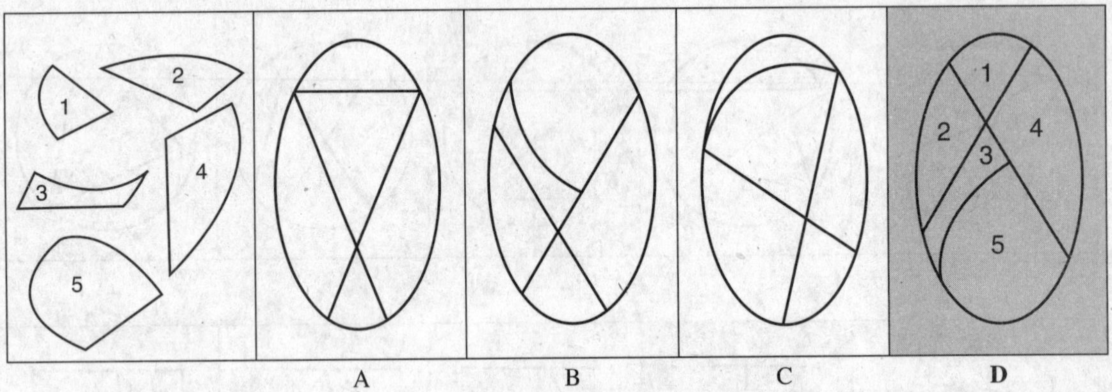

11. **B**

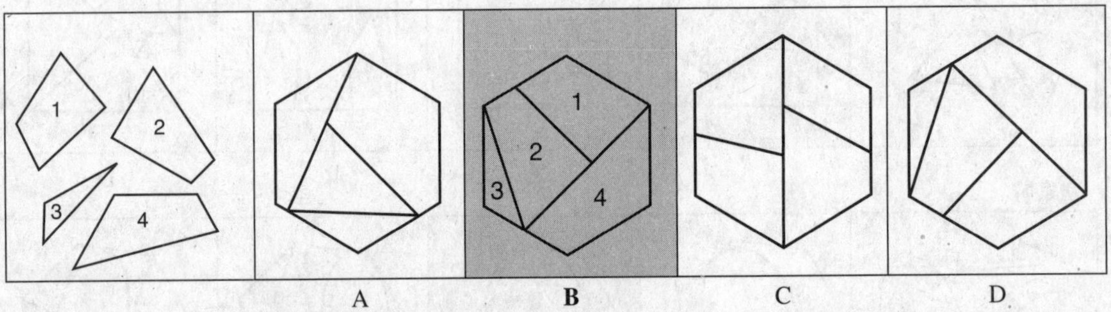

12. **A**

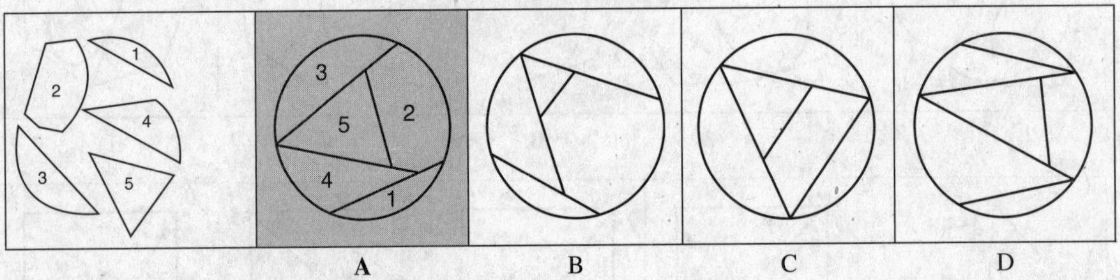

For questions 13–25:
In the answers below, we have placed dots in the answer choices, in order to help you see how the correct answer relates to the instructions box.

13. **C**

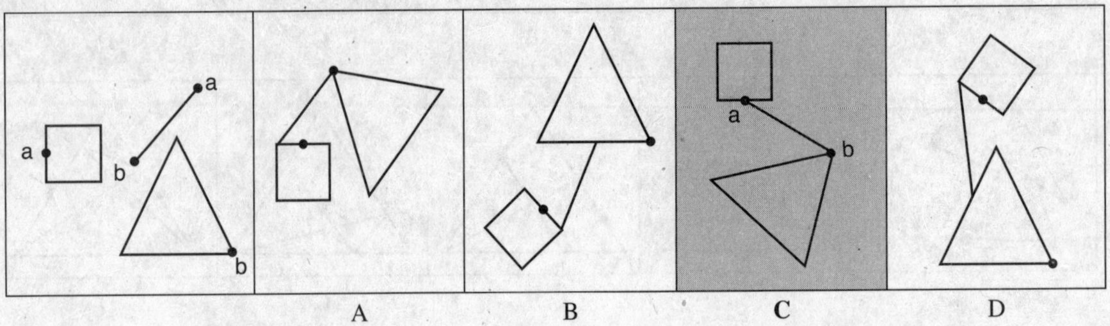

14. **C**

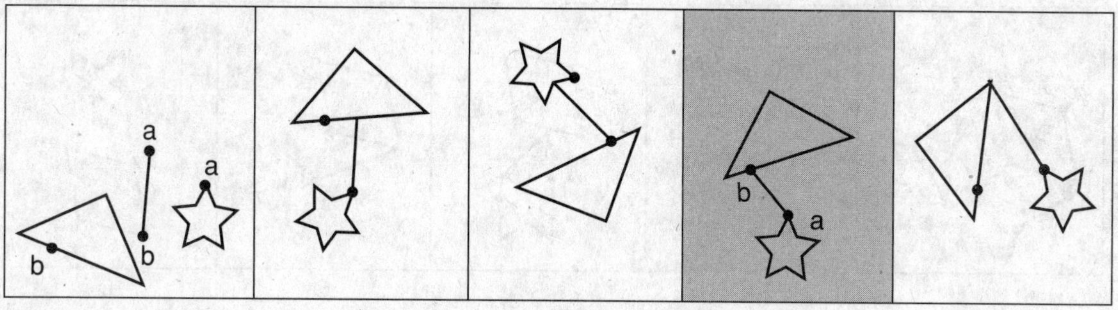

A B **C** D

15. **B**

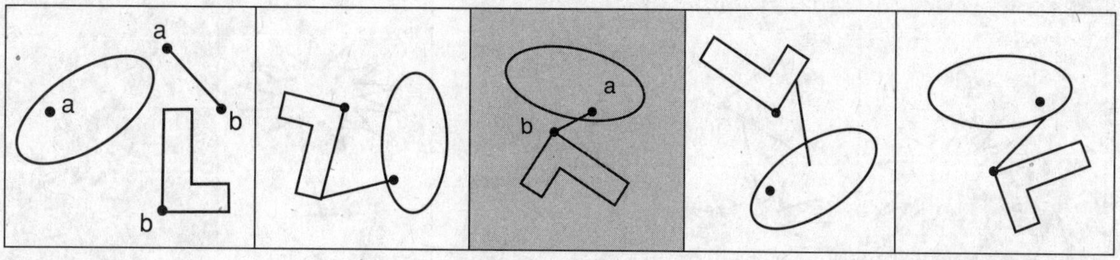

A **B** C D

16. **D**

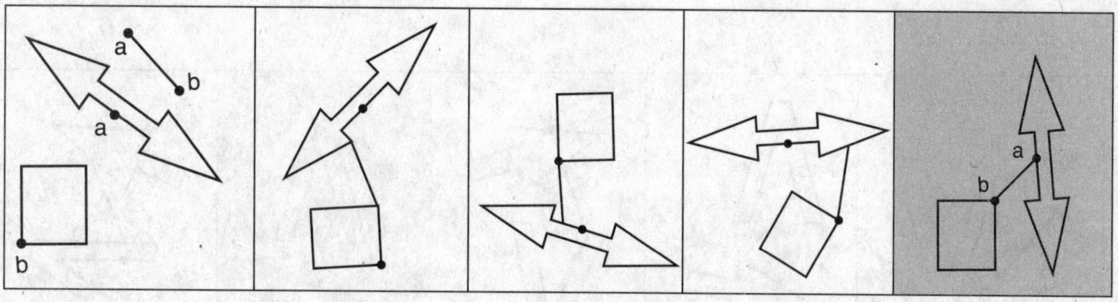

A B C **D**

17. **B**

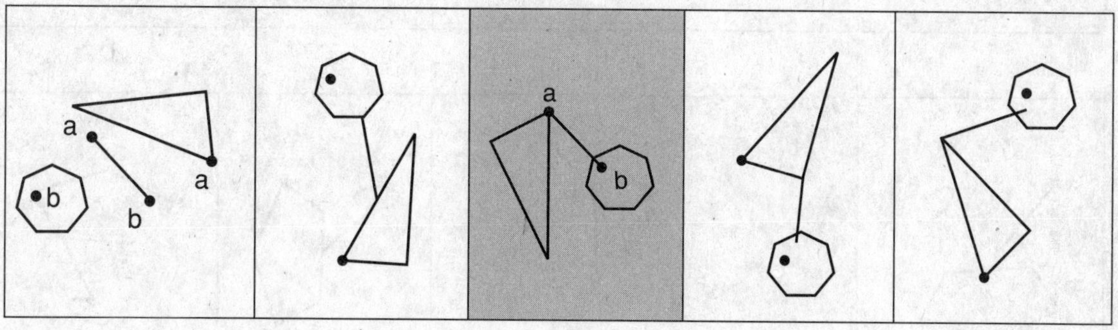

A **B** C D

18. **A**

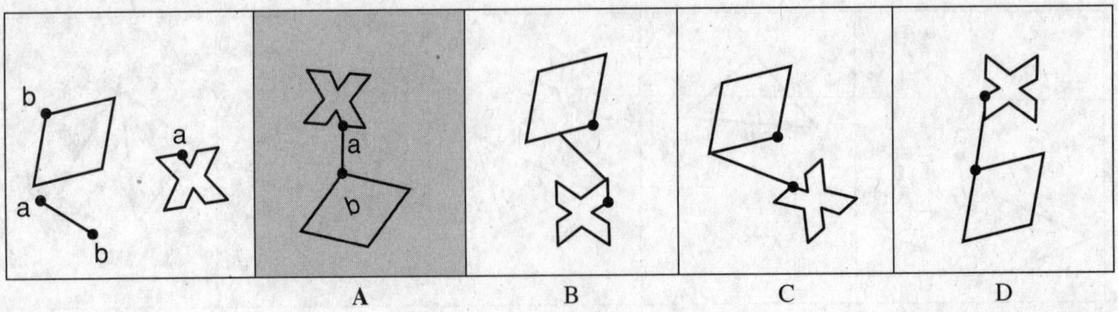

19. **D**

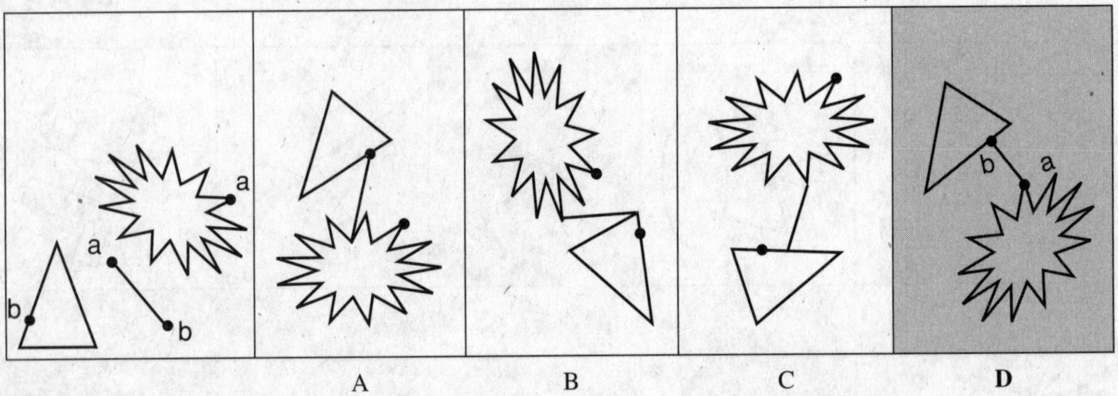

20. **C**

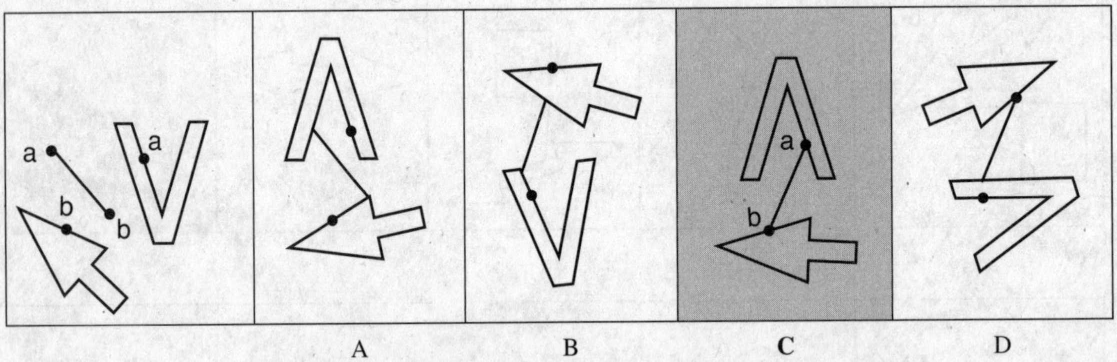

21. **A**

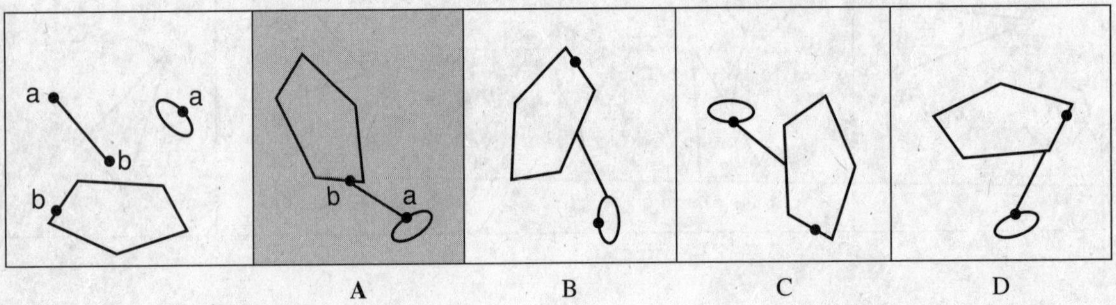

22. **B**

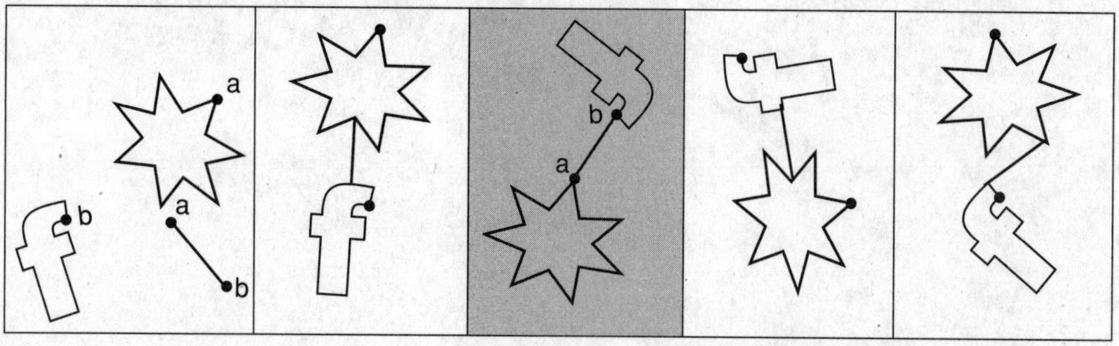

A B C D

23. **C**

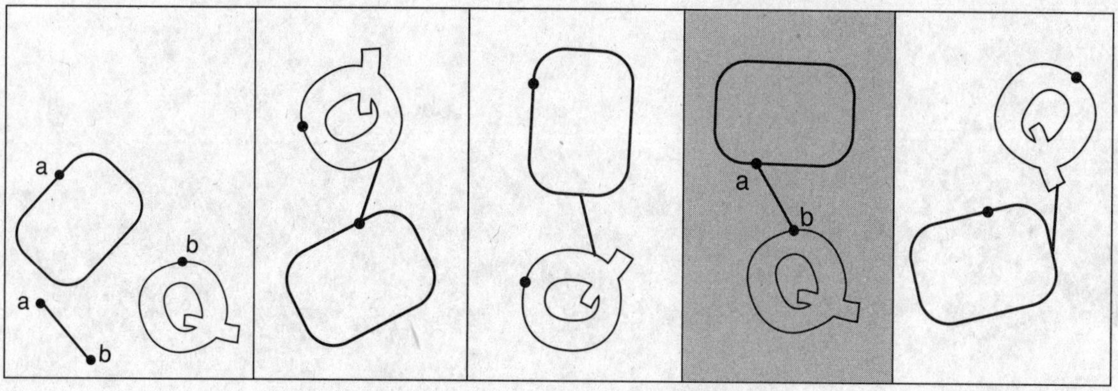

A B C D

24. **A**

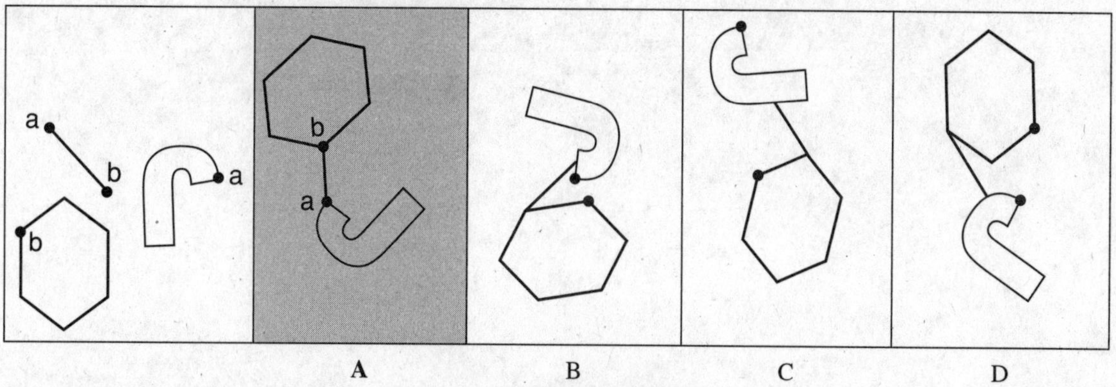

A B C D

25. **B**

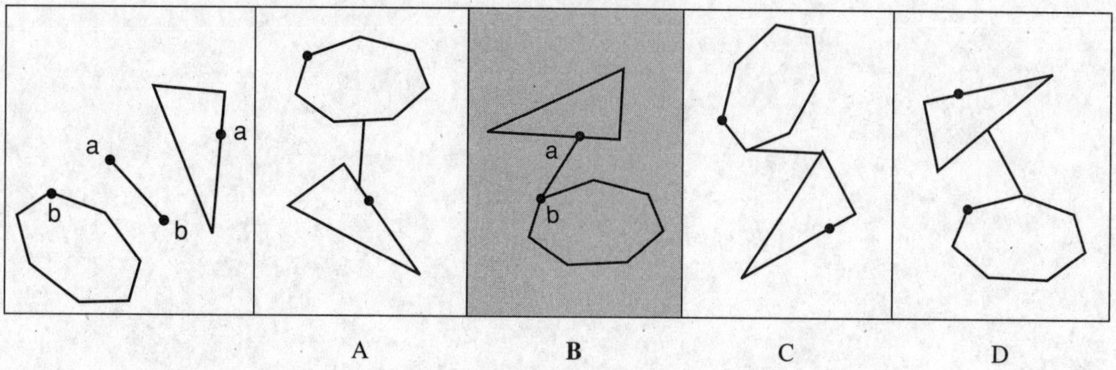

A B C D

CHAPTER 3

WORD KNOWLEDGE

KNOW WHAT TO EXPECT

One of the subsections of the ASVAB evaluates your ability to identify the meanings of words. This section, called the **Word Knowledge** (WK) section, is essentially a test of your vocabulary. However, that's not the entire story. While knowing the definitions of many words will certainly help you in this section, there are other ways to identify the correct answer and eliminate the wrong answers in WK questions. This chapter will not only focus on helping you improve your vocabulary, it will also help you develop strategies to guess effectively when you may not know the meanings of the words in the question.

On the WK section of the CAT-ASVAB, you will have 9 minutes to answer 15 vocabulary questions. If you are taking the paper-and-pencil ASVAB, you will be given 11 minutes to complete a total of 35 questions. Whichever version of the test you're taking, your goal is the same: find the synonym for a given word in a fraction of a minute.

A little more than half of the questions in the WK section will ask you to define words with no context. We'll call these **No-Context** WK questions. Here's an example of this type of question.

Gregarious most nearly means

(A) conspicuous
(B) twisting
(C) outgoing
(D) dark

If you happen to know the meaning of the underlined word, fantastic. If not, don't get discouraged. Although questions of this type do not provide you with context clues, there are **decoding strategies,** or ways of guessing a word's meaning, that can help you answer these questions correctly. Decoding strategies that involve understanding words' prefixes, suffixes, and roots will be a major emphasis of this chapter.

The other type of question you're likely to see in this section will ask you to define a word that appears in the context of a sentence. We'll call these **In-Context** WK questions. Take a look at an example of this type of question:

Nomadic tribes often move their villages when the seasons change.

(A) warlike
(B) wandering
(C) exclusive
(D) hasty

Notice that in this type of WK question, you can use clues in the sentence to help you find the word's meaning. Later in this chapter, we'll discuss ways to get the most out of each sentence's context to help you more effectively predict the underlined word's meaning. Oh, and don't worry—we'll be sure to revisit the two questions above later in this chapter.

This chapter includes:

- the Kaplan Method for Word Knowledge Questions
- paraphrasing a word's meaning
- decoding strategies for guessing a word's meaning, including:
  - prefixes, suffixes, and word roots
  - loose meaning associations
  - positive and negative charge
  - remembered context
- using context to guess a word's meaning
- eliminating answer choices based on logic
- strategies for building your vocabulary
- practice questions

Regardless of the type of WK question you're facing, having a consistent and repeatable method will help you tremendously in this section. Use Kaplan's 3-Step Method for WK to attack every question you see in your practice; by Test Day, it will be second nature.

THE KAPLAN METHOD FOR WORD KNOWLEDGE QUESTIONS

LEARNING OBJECTIVES

In this section, you'll learn to:

- apply the Kaplan Method for Word Knowledge questions
- paraphrase the meaning of a word you are already familiar with

The sooner you learn and apply this method to your studies, the sooner it can help you conquer the WK subtest.

THE KAPLAN METHOD FOR WORD KNOWLEDGE QUESTIONS

Step 1: Identify the word's meaning, or apply decoding strategies to guess its meaning.

Step 2: Make a prediction.

Step 3: Look for your prediction among the answer choices, or strategically eliminate incorrect answers.

Let's see how you would apply this to a question in which you know the meaning of the underlined word.

Question	Analysis
The unusually cold weather was an <u>anomaly</u>.	**Step 1:** *Anomaly* means "something unusual," or "something that doesn't seem to fit expectations."
	Step 2: Make a prediction before looking at the choices: the answer choice will mean "out of the ordinary" or "unexpected."
(A) oddity	**Step 3:** Correct. *Oddity* closely matches that prediction.
(B) ally	An *ally* is a friend. Incorrect.
(C) anonymity	*Anonymity* is the state of not being known. Incorrect.
(D) heat wave	A *heat wave* is not, by definition, out of the ordinary. Also, in the context of the sentence, a heat wave does not make sense. Incorrect.

Try one now on your own. Follow the method and try to make a prediction if you can.

Bob shows excellent <u>judgment</u> in his choice of friends.

(A) dispensation
(B) wisdom
(C) jurist
(D) ability

Explanation

Your prediction may have differed but should have resembled the following:

Step 1: *Judgment*, the way it's used in this sentence, means "the ability to make good decisions."

Step 2: The answer choice will mean something like "discernment," or "good decision making."

Step 3: Only choice **(B)** matches that meaning: *wisdom* also means "the ability to make good decisions."

You'll notice that Step 2 requires coming up with your own paraphrase of what the answer choice will say. That can take some practice. So here's a chance to try it on your own. Use the following list of common words to practice paraphrasing their meaning. If you don't know what any of these words mean, first look them up in a dictionary and then write down a paraphrase in your own words. You can use casual language to describe them if you like; you don't have to sound like a dictionary when you're paraphrasing.

Word	Your Paraphrase
decision	______________
anxiety	______________
platform	______________
invention	______________
disgusting	______________

How easy did you find it to paraphrase the words' meanings? It can take a while to get the hang of paraphrasing, even when you know what a word means, but it's well worth it. Knowing what you're looking for in the answer choices will help you zero in on the answer much more efficiently and accurately.

Your paraphrases may have varied; the following are sample answers using casual, everyday language:

Word	Paraphrase
decision	*choice*
anxiety	*nervousness, fear*
platform	*basis, thing that something is built on*
invention	*something new, or a lie*
disgusting	*nauseating, stomach-turning*

Of course, there will be times when you will not know the underlined word's meaning. In those cases you'll have to use decoding strategies or context to make a guess about it. Let's begin with decoding strategies.

DECODING STRATEGIES FOR GUESSING A WORD'S MEANING

LEARNING OBJECTIVES

In this section, you will learn to:

- break apart words into prefix, suffix, and root
- learn common prefixes, suffixes, and Greek and Latin word roots
- guess at a word's meaning using the word's positive or negative charge

You have many tools at your disposal for guessing a word's meaning, and this section will cover them each in turn.

Breaking Words Apart

One of the most effective decoding strategies is to break a word into parts. We can do this because many long words in English are composed of a **prefix** (a first part that affects the word's meaning) and/or a **suffix** (a final part that can also affect the word's meaning), plus a **word root** (a middle part that often derives from a Greek or Latin word).

For example, consider the word *convolution*. It breaks into parts like this:

con	+	*volu*	+	*tion*
prefix	+	word root	+	suffix

Suppose you didn't know what the word *convolution* means, but you did know that the prefix *con–* means "together" and that the word root *volu* means "turning" or "rolling." The suffix *–tion* indicates that the word is a noun. So you could guess that a *convolution* is something rolled up together. And that's a pretty good guess: In fact, *convolution* means a "coil or twist" or "the act of coiling." It can also refer to a complicated thought process that might be described as "twisty."

As a first step, get some practice breaking words apart into prefix, suffix, and root. Be aware that not all the words below will have all of those parts. The first one has been done for you.

Word	Prefix	Root	Suffix
dejection	de	ject	ion
paraphrase	________	________	________
multilingual	________	________	________
regenerate	________	________	________
ossify	________	________	________

Check your work against the answers below.

Word	Prefix	Root	Suffix
dejection	*de*	*ject*	*ion*
paraphrase	*para*	*phrase*	*(none)*
multilingual	*multi*	*lingu*	*al*
regenerate	*re*	*gener*	*ate*
ossify	*(none)*	*oss*	*ify*

So far so good, but what do all those word parts *mean*? Let's start with the meanings of common prefixes.

Prefixes

There are many prefixes you might see on Test Day. **We offer a comprehensive list of prefixes in the Appendix of this book.** If you are struggling to raise your Word Knowledge score, we would encourage you to make flashcards out of the word parts in the Appendix.

For our purposes here, we're going to present a few prefixes to enable you to get some practice working with them. Here are some extremely common prefixes with examples:

Prefix	Meaning	Examples
a, an	not, without	amoral: not related to morality; lacking regard for morality atypical: not typical
co, col, com, con	together, completely	collaborate: to work with another; cooperate compatible: able to exist together with someone or something else; capable of harmonious coexistence
de	away, off, down, reversal	defame: to slander; to publicly speak ill of descend: to move from a higher to a lower place

Prefix	Meaning	Examples
e, ex	out of, from, former	efface: to rub or wipe out; surpass; eclipse exclude: to shut out; to leave out
in, im	not, without	immoral: not moral; evil impartial: unbiased; fair
pre	before, in front	premonition: a feeling that an event may happen in the future presentiment: foreboding
pro	in front, before, much, for	proceed: to go forward propound: to set forth for consideration
re	back, again	recline: to lean back; to lie down regain: to gain again; to take back
sub, sup	below, under	subliminal: existing beneath consciousness substandard: inadequate; below expectations or requirements

Try working with prefixes. Use the list above to predict the meaning of the following words:

Word	Your Prediction
preset	________________
subset	________________
reset	________________

The word root *gno* relates to knowing or discerning. Use that information and the list of prefixes above to predict the meanings of the following words.

Word	Your Prediction
agnostic	________________
prognosis	________________
ignoramus	________________

The word root *ject* relates to throwing or throwing downward. Use that information and the list of prefixes above to predict the meanings of the following words.

Word	Your Prediction
eject	________________
dejected	________________
conjecture	________________

How did you do? Your answers may vary somewhat, but they should resemble the definitions below.

Word	Your Prediction
preset	set before: that is, arranged beforehand
subset	something set under: that is, a small group within a larger group

Word	Your Prediction
reset	set again
agnostic	a without-knower: that is, a person that believes that the nature of God or the universe is unknowable
prognosis	the act of knowing before: that is, a prediction about how a problem will progress
ignoramus	a not-knower: that is, a very ignorant person, an idiot (formed by combining *in* + *gno* and then dropping the first *n*)
eject	throw out
dejected	thrown off or down: that is, sad, depressed
conjecture	something thrown together: that is, the formation of an opinion or guess without enough information

Great work! The next step in decoding a long word involves looking at the end of the word. That brings us to suffixes.

Suffixes

Unlike prefixes, suffixes usually do two jobs at once. They affect the meaning of a word, and they *also* typically tell you what part of speech the word is. Here's a reminder about parts of speech:

PARTS OF SPEECH INDICATED BY SUFFIXES

Noun: a word that names a person, place, thing, or idea. For example, the underlined words below are all nouns:

Hegel, a famous philosopher who lived in Germany, believed in an idea called *dialectic*.

Verb: a word that expresses an action or state of being. Examples:

Julius constructed his own bookshelves.

I am normally a cheerful person.

Adjective: a word that modifies a noun. For example, the underlined words below are all adjectives:

The quick brown fox jumps over the lazy dog.

Adverb: a word that modifies a verb, adjective, or another adverb. For example, the underlined words below are all adverbs:

The second time he played poker, his skills were greatly improved and he played his cards very carefully.

The other parts of speech are **pronouns** (*he, she, it*), **interjections** (*wow, ouch*), **prepositions** (*of, to, in, on*), and **conjunctions** (*but, nor, for*). However, long words that end in suffixes don't usually belong to those parts of speech.

There are many suffixes you might see on Test Day. **We offer a comprehensive list of suffixes in the Appendix of this book.** If you are struggling to raise your Word Knowledge score, we would encourage you to make flashcards out of the word parts in the Appendix.

For our purposes here, we're going to present a few suffixes to enable you to get some practice working with them. Below is a list of some extremely common suffixes with examples.

Suffix	Part of Speech It Usually Indicates	Meaning	Examples
able, ible	adjective	capable of, worthy of	changeable: able to be changed combustible: capable of catching fire easily
ic	adjective	having the quality of, related to	robotic: like a robot or having to do with robots melodic: having to do with melody; sweetly musical
ion, tion, sion	noun	act of doing, act of being, result of action	notation: act or process of making notes sensation: act of feeling or perceiving
ist	noun	one who does an action or has a trait	pianist: one who plays the piano anesthetist: a medical professional who administers anesthesia (that is, drugs that lessen pain or cause unconsciousness)
ity	noun	state or quality of	novelty: state of being new; a new thing solemnity: state of being solemn or serious
logy	noun	study of	geology: study of Earth's structure zoology: study of animals
oid	adjective or noun	resembling or almost	ovoid: resembling an egg, having an oval or egg-like shape asteroid: a small body orbiting the Sun (literally, an almost-star)
ous	adjective	having the quality of, having to do with	bulbous: having a bulb-like shape venous: related to the veins

The word root *am* or *amo* relates to love or liking. Use that information and your knowledge of suffixes to predict the parts of speech and the meanings of the following words:

Word	Your Prediction: Part of Speech	Your Prediction: Meaning
amiable	____________	____________
amorous	____________	____________
amity	____________	____________

The word root *anthro* or *andro* means *man* or *human*. Use that information and your knowledge of suffixes to predict the parts of speech and the meanings of the following words:

Word	Your Prediction: Part of Speech	Your Prediction: Meaning
anthropology	______	______
anthropologist	______	______
anthropomorphize	______	______
android	______	______
misanthropic	______	______

How did you do? Your answers may have varied, but they should resemble the definitions below.

Word	Your Prediction: Part of Speech	Your Prediction: Meaning
amiable	adjective	capable of being liked or loved; likable
amorous	adjective	displaying love or attraction
amity	noun	liking, friendship
anthropology	noun	study of human beings
anthropologist	noun	one who studies human beings
anthropomorphize	verb	to make into something that resembles humans
android	noun	something (such as a robot) that resembles humans
misanthropic	adjective	having a hatred of people (*mis* means bad, wrong, or hateful)

Of course, prefixes and suffixes are only part of the story. You'll also need to be able to decode what's in the middle of a word.

Greek and Latin Word Roots

Many, though not all, of the long words in English derive from Greek or Latin, which means that learning some Greek and Latin word roots can help you decode English words. In your practice with prefixes and suffixes, above, you have already seen that a single word root can produce many words in English.

This section will examine a selection of word roots and how they form the basis for English words. This discussion will be followed by in-format exercises designed to help you put your decoding strategies to work.

There are many word roots you might see on Test Day. **We offer a comprehensive list of word roots in the Appendix of this book.** If you are struggling to raise your Word Knowledge score, we would encourage you to make flashcards out of the word parts in the Appendix.

For our purposes here, we're going to present a few word roots to enable you to get some practice working with them. Below is a list of some extremely common roots with examples.

Root	Meaning	Examples
ced, ceed, cess	go, hold back, yield	cessation: stoppage incessant: continuous; without stopping
centr	center	central: in the center; important concentrate: to bring to the center; to direct to one point
pat, path	feeling, suffering	empathy: identification with the feelings of another impassive: showing no feeling or emotion
vi, viv	life, living	vital: concerned with or necessary to life

Practice with Word Roots

Use your knowledge of prefixes and suffixes, along with the word root list above, to predict the part of speech and meaning of the following words:

Word	Your Prediction: Part of Speech	Your Prediction: Meaning
concentric	______________	______________________
viable	______________	______________________
conceded	______________	______________________
procession	______________	______________________
apathetic	______________	______________________
pathology	______________	______________________
vitality	______________	______________________
eccentricity	______________	______________________

How did you do? Your answers may have varied, but they should resemble the definitions below.

Word	Your Prediction: Part of Speech	Your Prediction: Meaning
concentric	adjective	together with the center: that is, having a common center
viable	adjective	able to live: possible or workable
conceded	verb	reluctantly accepted or gave up something
procession	noun	act of moving forward
apathetic	adjective	without feeling or emotion
pathology	noun	study of suffering; study of diseases
vitality	noun	capacity for life or survival; state of being strong or full of life
eccentricity	noun	state of being out of the center: that is, oddity; state of being odd or unusual

<u>Full lists of word parts are in the back of this book, in the Appendix.</u> Be sure to get more practice with word parts before Test Day. The best way to do this might be to read a variety of sources and, when you come across a long, unfamiliar word, see if you can break it into parts and analyze the parts.

Using Positive or Negative Charge

One of the secrets to success in the WK section of the ASVAB is knowing that you don't actually need precise definitions of the words that are being tested. As long as you can confidently eliminate three answer choices as being incorrect, you'll find your way to the correct answer. That means that many times, just knowing if a word is positive, negative, or neutral will be enough to answer the question correctly. For example, let's return to a question you've seen previously:

<u>Gregarious</u> most nearly means

(A) conspicuous
(B) twisting
(C) outgoing
(D) dark

Do you know what the word *gregarious* means? If so, then great, this is an easy question for you. But let's say you don't know what it means. What do you do then? Well, you can start by breaking the word down into parts, as you've seen demonstrated earlier in the chapter.

But let's imagine you aren't familiar with the root *greg*. While you may be tempted to throw in the towel and give up, try something else: do you think *gregarious* has a positive, negative, or neutral charge? If you've ever heard the word *gregarious* used in conversation, then you probably heard it used to describe a person. And was the person using the word *gregarious* delivering a compliment or an insult?

Now, you may be thinking to yourself, "Hmm, okay, I don't really know what *gregarious* means, but I'm pretty sure I've heard it used to compliment someone, so this is a positive word." Now look through the answer choices. Which of those, if applied to a person, would be considered positive? Three answer choices are either negative or neutral. Only (C), *outgoing*, is positive when applied to a person. Select it and move on.

Of course, using a word's positive or negative charge should never be your first strategy when tackling a WK question. Use this approach only when you have a general but not specific understanding of the word.

(For the record, the root *greg* means "to flock to," and the suffix *-ious* means "to be full of." Therefore, *gregarious* means "living with others, like a flock of animals," and the closest choice is (C) *outgoing*.)

Try this method out on a few words below. For each word you see, rate it as positive, negative, or neutral.

Word	Positive, negative, or neutral?
diabolic	________________
harmony	________________
fanfare	________________
destitute	________________
burdensome	________________
zestful	________________
decision	________________
swampy	________________
currently	________________

Think about how you made the predictions you listed above. How did you come to that determination? Was it because you recognized word parts or remembered situations in which you've heard the word? Or was it more that the word simply "sounded" positive or negative? Either one can work, depending on how reliable your memory is. Now, take a look below. We've included each word's meaning, simply to confirm why it has a positive or negative charge.

Word	Positive, negative, or neutral?	Meaning
diabolic	negative	devilish or evil
harmony	positive	agreement, or a pleasing combination of tones in music
fanfare	positive	song played in celebration, or advertisement
destitute	negative	poverty-stricken
burdensome	negative	very heavy; tiresome; troublesome
zestful	positive	full of enjoyment
decision	neutral	choice
swampy	negative	unpleasantly humid; resembling a swamp
currently	neutral	at the present time; now

Using Real-World Context to Guess at a Word's Meaning

One of the most common—and effective—ways of guessing a word's meaning is to place the word in context. Even if you may not know a textbook definition of the word, perhaps you remember it being used in a specific phrase. Ask yourself, "where have I heard this word before?" Sometimes, you will realize that you did know the meaning of the word.

Take this question, for example:

3. Broach most nearly means

    (A) poke
    (B) introduce
    (C) sermonize
    (D) cook

Unfortunately, *broach* is one of those words that comes from Middle English, so knowing Latin and Greek roots won't be very helpful. And if you can't rattle off a specific definition of the word right away, you may think that this is a lost point. But ask yourself if you have heard this word used in a common phrase before. Does "broaching the subject" ring a bell? Perhaps you've heard a parent, teacher, or television character use this phrase. Something along the lines of: "Well, I didn't want to talk about it, but Donny broached the subject anyway." Using the context of how the word is used in a phrase you've heard before is a remarkable way to understand the "fuzzy definition" of a word. In this example, *broach* would mean something like "bring up" or "start to discuss." The only word that is a close fit is answer choice (B), *introduce*.

Try this approach yourself. For each of the words below, try to place the word in a common phrase that you may have heard before. How does the context of the phrase help you determine the meaning of the word?

Word	Think of a phrase you've heard the word used in, and then use that to guess at the meaning of the word:
vicarious	______________________
jiffy	______________________
guzzle	______________________
hybrid	______________________

Did those words sound familiar? When you tried to place them in phrases or a real-world context, what did you come up with? Below, take a look at how a person might have used common usage to help them determine the right definition. Your answers, of course, may vary.

Word	Think of a phrase you've heard the word used in, and then use that to guess at the meaning of the word (sample answers):
vicarious	"I've heard the phrase 'So and so lived vicariously through someone else.' I guess that means vicariously is like experiencing something through someone else's eyes."
jiffy	"My dad used to say 'I'll be back in a jiffy.' I guess a jiffy is really fast."
guzzle	"I have a friend who is always talking about his truck being a real gas guzzler. He's always filling it up, so I guess being a guzzler means you consume a lot."
hybrid	"There are a bunch of new cars out there that are considered to be hybrids because they use a little bit of gas and a little bit of electric. Maybe hybrid means doing two things at the same time."

Notice that the predictions are not specific definitions or predictions that point to the word's meaning. Instead, the phrases allow you to form a fuzzy definition of the word in your mind; that fuzzy definition will often be more than enough to select the right answer.

MAKING THE MOST OF CONTEXT

LEARNING OBJECTIVES

In this section, you will learn to:

- use the meaning of a sentence to predict an answer on In-Context WK questions
- use the meaning of a sentence to eliminate answer choices on In-Context WK questions

In addition to the decoding strategies discussed above, you can also use the sentence itself to help you answer In-Context WK questions. **Context** simply means the words or phrases surrounding an underlined word, and using context involves looking for clues in those other words in order to help you guess at the underlined word's meaning.

There are several ways context can be helpful. Sometimes the sentence signals that the underlined word *resembles* another idea in the sentence. Consider the following:

Question	Analysis
The children thronged around the table like moths around a flame.	The word *like* signals that, whatever the children were doing, they resembled moths swarming around a flame. So *thronged* must convey that same sense.

Sure enough, *thronged* means *swarmed* or *gathered in a crowd.*

In each of the examples below, identify the word that tells you that the underlined word is similar to some other idea in the sentence, and then predict the meaning of the underlined word.

Sentence	What word or words signal similarity?	Predict the underlined word's meaning
The crowd greeted the famous chef with an eagerness amounting to fervor.	____________	____________
The budget meeting was a difficult conversation; indeed, it was almost a quagmire.	____________	____________
Today's creative child may be tomorrow's virtuoso.	____________	____________

Explanation

Your paraphrases may differ somewhat but should resemble the following:

Sentence	What word or words signal similarity?	Predict the underlined word's meaning
The crowd greeted the famous chef with an eagerness amounting to fervor.	*amounting to*	*tremendous enthusiasm*
The budget meeting was a difficult conversation; indeed, it was almost a quagmire.	*indeed*	*extremely difficult or awkward situation*
Today's creative child may be tomorrow's virtuoso.	parallel drawn between *today* and *tomorrow*	*someone very skilled at a creative activity like art or music*

In other cases, sentences may contain words that signal that the underlined word *contrasts with* or *forms an unexpected combination with* another idea in the sentence. Consider the following example:

Question	Analysis
The best coaches are supportive but exacting.	The word *but* tells you that *exacting* either contrasts with *supportive* or else that you would not expect to see *supportive* and *exacting* linked together. It is unlikely that *exacting* is the direct opposite of *supportive*, since the writer of this sentence thinks good coaches display both traits. So it must be the case that *supportive* and *exacting* are somehow a surprising pair.

In fact, *exacting* means "making large demands." Thus, the prediction above is correct: while *exacting* does not mean the opposite of *supportive,* it is perhaps a surprising idea to find linked with *supportive.*

In each of the examples below, identify the word that tells you that the underlined word contrasts with or forms a surprising combination with some other idea in the sentence, and then predict the meaning of the underlined word.

Sentence	What word signals contrast or an unexpected combination of ideas?	Predict the underlined word's meaning
Some human behavior is learned, while other human behavior is instinctive.	________________	________________
Trisha found the task to be enjoyable rather than onerous.	________________	________________
I prefer action films but had to sit through a cloying love story last Friday night.	________________	________________

Explanation

Your paraphrases may differ somewhat but should resemble the following:

Sentence	What word signals contrast or an unexpected combination of ideas?	Predict the underlined word's meaning
Some human behavior is learned, while other human behavior is instinctive.	*while*	contrasts with *learned*, so *innate* or *inborn*
Trisha found the task to be enjoyable rather than onerous.	*rather*	contrasts with *enjoyable*, so *tiresome* or *burdensome*
I prefer action films but had to sit through a cloying love story last Friday night.	*but*	contrasts with *action films* and has a negative charge, so *too sentimental* or *sugary*

Sometimes context merely tells you whether an underlined word has a negative or positive connotation. In many cases, this deduction may be enough to answer the WK question. Consider the following example:

Question	Analysis
Jalisa was looking forward to a well-earned respite.	**Step 1:** A *respite* must be a good thing, because Jalisa is *looking forward* to it. Also, the adjective *well-earned* has a positive connotation.
	Step 2: *Respite* must be something enjoyable or beneficial.
(A) deadline (B) incarceration (C) rest (D) prediction	**Step 3:** Of the answer choices, only (C) *rest* is likely to be enjoyable or beneficial. Choose **(C).**

In fact, a *respite* is a period of rest or relief from work or from something unpleasant.

Try it on your own. For each of the sentences below, identify the word or words that signal whether the underlined word has a positive or negative connotation. Then try, if possible, to predict the underlined word's meaning.

Sentence	What word or words signal that the underlined word has a positive or negative connotation?	Predict the underlined word's meaning
I struggled to stay awake as the presentation droned into evening.	________	________
Your injudicious proposal would have serious consequences for our town's future.	________	________
Wasting more money on the machine would only exacerbate our difficulties.	________	________

Explanation

Your paraphrases may have differed but should have resembled the following:

Sentence	What word or words signal that the underlined word has a positive or negative connotation?	Predict the underlined word's meaning
I struggled to stay awake as the presentation droned into evening.	*struggled*	*continued in a boring or monotonous way*
Your injudicious proposal would have serious consequences for our town's future.	*serious consequences*	*unwise*
Wasting more money on the machine would only exacerbate our difficulties.	*wasting*	*worsen*

Finally, sometimes the context surrounding an underlined word's meaning simply defines that word or hints at its meaning. Remember this example from the beginning of the chapter?

Question	Analysis
Nomadic tribes often move their villages when the seasons change.	**Step 1:** Whatever *nomadic* tribes may be, they move around a good deal.
	Step 2: Check the answer choices for a word that means *moving around*.
(A) warlike (B) wandering (C) exclusive (D) hasty	**Step 3: (B)** *wandering* matches the prediction best.

In fact, the sentence in the question above simply defines *nomadic*, which means "moving from place to place; having no fixed dwelling-place."

Try the following examples:

Sentence	Predict the meaning of the underlined word
As Lydia struggled to understand the difficult material, her <u>bewilderment</u> became obvious to the teacher.	____________________
The thieves took pains to cover their tracks on the way to their <u>clandestine</u> meeting.	____________________
The incoming freshman found the Advanced Econometrics course material to be <u>opaque</u>.	____________________

Explanations

Your paraphrases may differ somewhat but should resemble the following.

Sentence	Predict the meaning of the underlined word
As Lydia struggled to understand the difficult material, her <u>bewilderment</u> became obvious to the teacher.	She *struggled to understand*, so *bewilderment* likely means *confusion* or *lack of understanding*.
The thieves took pains to cover their tracks on the way to their <u>clandestine</u> meeting.	The thieves were clearly trying to keep their meeting a secret, so *clandestine* likely means *secret*.
The incoming freshman found the Advanced Econometrics course material to be <u>opaque</u>.	Whatever *Advanced Econometrics* is, it sounds too hard for a freshman. So the freshman likely found it to be *very difficult to understand.*

Of course, sometimes you will struggle to glean clues from the context of a sentence. You may, in these cases, be able to eliminate some answer choices by mentally replacing the underlined word with each of the answer choices to try to "hear" which choices sound wrong. Study the following example to learn how a test taker might go about this:

Question	Analysis
Everyone on the committee agreed that acting with <u>circumspection</u> was important.	**Steps 1 and 2.** The sentence doesn't give many hints aside from suggesting that *circumspection* is probably a good thing.
(A) wisdom (B) voting (C) caution (D) individuality	**Step 3.** More than one of the answer choices has a positive connotation. Try each in the original sentence: (A) *Everyone on the committee agreed that acting with wisdom was important.* That seems like an odd thing to say. Isn't wisdom always important? Try the others. (B) *Everyone on the committee agreed that acting with voting was important.* That doesn't seem to make much sense. (C) *Everyone on the committee agreed that acting with caution was important.* That makes a great deal of sense. (D) *Everyone on the committee agreed that acting with individuality was important.* That seems to go against the idea of a committee, especially one where everyone agrees. Choice **(C)** is the best fit.

Not surprisingly, *circumspection* means "caution or the act of carefully considering."

Try the example below. If you are having trouble predicting what the underlined word means, mentally reread the sentence, substituting each answer choice, to find the one that sounds most sensible.

Joanna feigned happiness at the wedding.

(A) created
(B) passed out
(C) pretended
(D) concealed

Explanation

Steps 1 and 2. Perhaps you had trouble applying decoding strategies or using context here. If so, go directly to the choices.

Step 3. Try the answer choices one by one in the sentence:

(A) *Joanna created happiness at the wedding.* That doesn't make a lot of sense. Weddings might make people happy, but people don't typically say that one individual *creates* happiness.

(B) *Joanna passed out happiness at the wedding.* Unless happiness is a new kind of party favor, this makes no sense. (This answer choice was probably intended to tempt test takers who think *feign* looks like *faint.*)

(C) *Joanna pretended happiness at the wedding.* This makes sense: if Joanna had some reason to be unhappy that day, she may have had to pretend to be happy.

(D) *Joanna concealed happiness at the wedding.* Although it's possible that someone might have a reason to conceal, or hide, their happiness at a wedding, that seems far less likely than choice (C).

ELIMINATING ANSWER CHOICES BASED ON LOGIC

If you really can't come up with a prediction, or if context is proving to be no help, then you may in some cases be able to eliminate answer choices by thinking logically about them. For example, if two answer choices mean nearly the same thing, they must be wrong, since the ASVAB has only one right answer for each question. Consider the following example:

Question	Analysis
Legerdemain most nearly means	**Steps 1 and 2.** Hard to see recognizable prefixes, suffixes, or word parts in this word. Also, there's no context. Difficult to predict. Look at the choices.
(A) monster (B) massive amount (C) magic trick (D) ogre	**Step 3.** Now, *monster* and *ogre* are not exactly synonyms, but they both mean creatures that are frightening and dangerous. Therefore it's unlikely that either of them are correct. Eliminate (A) and (D). At this point, either make a guess between (B) and (C) or try to remember situations in which you've heard the word *legerdemain.*

In fact, *legerdemain* means *sleight of hand*, so *magic trick* is the best fit.

If two answer choices have opposite meanings, it may be likely that one of them is correct. That's because test makers often include an answer choice that means the opposite of the correct answer. Study the following example:

Question	Analysis
Hauteur most nearly means	**Steps 1 and 2.** Hard to see recognizable prefixes, suffixes, or word parts in this word. Also, there's no context. Difficult to predict. Look at the choices.
(A) author (B) arrogance (C) moisture (D) humility	**Step 3.** Choices (B) and (D) have opposite meanings, so it may be the case that one of them is correct.

Once you have eliminated (A) and (C), you may have to simply guess between the other two. Alternatively, if you happen to have a sense that *hauteur* has a negative charge, you could use that to guess that it probably means *arrogance*. You could also think about the fact that *hauteur* sounds like *haughty*, which means arrogant. So **(B)** is indeed correct.

In conclusion, eliminating choices based on logic usually helps you to narrow choices down to two. You may need to use another strategy or simply guess in order to select from the remaining two choices.

Try to eliminate wrong answer choices using logic in the following questions.

1. Demur most nearly means
   (A) trust
   (B) investigate
   (C) believe
   (D) disagree

2. Logy most nearly means
   (A) oblong
   (B) lively
   (C) sluggish
   (D) chubby

Explanations

1. Steps 1 and 2. Perhaps you don't know the meaning of *demur* and are having trouble applying decoding strategies to the word.

Step 3. Scan the choices. (A) *trust* and (C) *believe* are very similar in meaning, and therefore it's unlikely that one of them is correct. Eliminate them.

Either guess between (B) and (D) or apply another strategy. It may be helpful to note that the prefix *de* can mean *away from*, and the prefix *dis* can mean *not* or *opposed to*. That may signal that these words are similar. Choose **(D)** *disagree*. (In fact, to *demur* means "to raise objections, disagree, or hesitate.")

2. Steps 1 and 2. Perhaps you don't know the meaning of *logy* and are having trouble applying decoding strategies to the word.

Step 3. Scan the choices. (B) *lively* and (C) *sluggish* are opposites, so it may be that one of them is correct. Either hazard a guess or try to remember if you've heard *logy* anywhere. You may remember that it has a slightly negative charge, which makes **(C)** a better fit. (In fact, *logy* means "sluggish.")

STRATEGIES FOR BUILDING YOUR VOCABULARY

You now have a variety of strategies for conquering the WK section of the ASVAB. But those tips aren't helpful if your vocabulary remains poor. To learn more words between now and your ASVAB Test Day, make this commitment: **learn seven new words a day**. If that sounds like too few, rest assured, it's plenty—in fact, that's nearly 50 words a week.

The best and most efficient way to improve your vocabulary is to consume as many books, magazines, and newspapers that you can get your hands on. The reason for this is pretty straightforward: learning words in context, as you read, helps you retain those words longer and with greater precision than learning those words by themselves, with no context. If you have a smartphone, download a dictionary app and look up any words that you don't know. If you're at your computer, simply use a search engine. Always look up unfamiliar words that you encounter in your reading of books and magazines, as well as words you see on the Internet, hear in movies, or encounter in daily conversation.

Another tried and true method for improving your vocabulary is to **make and use vocabulary flashcards**. While the "making" part of flashcards is often overlooked, its value is significant—the act of writing a word and its definition on an index card will help you remember that word's definition both as a visual image and as an action. So, here's your task: as you encounter and learn new words, write them down on index cards and, as soon as you're able, write the definitions of those words on the backs of the cards. Then, when you have some spare time, browse through your pile. Once you feel like you've really "got" a word, and that you won't lose its definition, remove it from the pile. If you follow through with this method, you'll find that you're constantly adding and removing words from your pile of vocabulary cards.

Some people find that **creating a vocabulary notebook** is more convenient than flashcards. The idea is similar: jot down words you don't know as you find them in your readings and daily interactions, and look up their definitions. List words in the left-hand column and their meanings in the right-hand column. Cover up or fold over the page to test yourself. See how many words you can define from memory.

Finally, there are a number of **apps and websites** you might use to improve your vocabulary. To find out about Kaplan vocabulary apps, call 1-800-KAPTEST or visit your virtual app store.

After you learn new words, try to use them in conversation. This is an excellent way to confirm that you truly understand a word's meaning and the proper context in which to use it. If you have an ASVAB study partner, try to have a conversation in which each person must use an ASVAB vocabulary word in a sentence. You'll be surprised by how well the words' definitions "stick" once you've correctly used them in conversation.

ASVAB TIPS

Improve your vocabulary by:

- reading books, magazines, newspapers, and articles on the Internet
- looking up words you don't know and jotting down their definitions either:
  - on flashcards or
  - in a vocabulary notebook
- making use of vocabulary-building websites or apps
- committing to learning seven words every day
- using recently learned words in daily conversation

WORD KNOWLEDGE PRACTICE SET 1

For each question, select the answer choice that most closely matches the meaning of the underlined word. This question set has 15 practice questions, which is the number of Word Knowledge questions you will see if you take the CAT-ASVAB.

1. Ghastly most nearly means
   (A) fun
   (B) lazy
   (C) torrid
   (D) awful

2. The brothers ran away in cowardice.
   (A) pain
   (B) fear
   (C) hopelessness
   (D) temperance

3. Resignation most nearly means
   (A) losing
   (B) waste
   (C) acceptance
   (D) pride

4. She promised to cooperate with the authorities.
   (A) fight
   (B) talk
   (C) work with
   (D) placate

5. Everyone says he lost the election due to lack of initiative.
   (A) satisfaction
   (B) irritation
   (C) money
   (D) ambition

6. The city council sought reparations for the oil spill.
   (A) compensation
   (B) sadness
   (C) thanks
   (D) antipathy

7. Many times, the older sibling holds dominion over her younger siblings.
   (A) authority
   (B) safety
   (C) ability
   (D) guilt

8. The professor did his best to appear universally erudite, even when asked questions about subjects he knew little about.
   (A) civil
   (B) progressive
   (C) scholarly
   (D) amoral

9. Tangible most nearly means
   (A) real
   (B) open
   (C) graphic
   (D) costly

10. Alleviate most nearly means
    (A) elevate
    (B) improve
    (C) encompass
    (D) make foreign

11. Trite most nearly means
    (A) snug
    (B) correct
    (C) modern
    (D) stale

12. Imitating a marching band, the children used pots and pans to make a loud <u>clamor</u>.

    (A) noise
    (B) protest
    (C) music
    (D) harmony

13. <u>Candid</u> most nearly means

    (A) able
    (B) whimsical
    (C) valid
    (D) honest

14. After careful study, he was able to <u>discern</u> which model would be the best car for his needs.

    (A) avoid
    (B) spoil
    (C) perceive
    (D) savor

15. <u>Dupe</u> most nearly means

    (A) double
    (B) delay
    (C) deceive
    (D) delight

WORD KNOWLEDGE PRACTICE SET 2

For each question, select the answer choice that most closely matches the meaning of the underlined word. This question set has 15 practice questions, which is the number of Word Knowledge questions you will see if you take the CAT-ASVAB.

1. The athlete trained for the competition with vigor.
   (A) stiffness
   (B) intensity
   (C) swelling
   (D) concern

2. Enigma most nearly means
   (A) puzzle
   (B) motor
   (C) energy
   (D) clarity

3. She displayed fortitude in finishing the race despite the pain in her foot.
   (A) speed
   (B) cheerfulness
   (C) courage
   (D) vulnerability

4. The frugal man was always reluctant to spend money.
   (A) fearful
   (B) thrifty
   (C) imprudent
   (D) lavish

5. Finding just the right part for her old car at the swap meet was fortuitous.
   (A) expected
   (B) interesting
   (C) lucky
   (D) exciting

6. By his third fall, Jack's enthusiasm for roller skating had abated.
   (A) waned
   (B) solidified
   (C) renewed
   (D) flourished

7. Congeniality most nearly means
   (A) agreeability
   (B) coldness
   (C) competence
   (D) beauty

8. His brusque responses discouraged further questions from the students.
   (A) amusing
   (B) curt
   (C) empathetic
   (D) articulate

9. Eminence most nearly means
   (A) mischief
   (B) devotion
   (C) greatness
   (D) dishonor

10. The unwitting clerk was absolved when her manager confessed to embezzling the money.
    (A) impressed
    (B) accused
    (C) incarcerated
    (D) exonerated

11. Disdain most nearly means
    (A) sympathy
    (B) dishonesty
    (C) contempt
    (D) flattery

12. Abnegate most nearly means
    (A) approve
    (B) hope
    (C) deny
    (D) recognize

13. Although she was not exactly friendly, she treated us with civility.

(A) honesty
(B) courtesy
(C) compliance
(D) rudeness

14. Deranged most nearly means

(A) unapologetic
(B) tamed
(C) relocated
(D) crazy

15. Although he never accused her outright, he insinuated that she had lied about the incident.

(A) implied
(B) insisted
(C) concealed
(D) disavowed

WORD KNOWLEDGE PRACTICE SET 1

Answers and Explanations

1. **D** Even if you didn't know that *ghastly* means "awful," if you had a sense that *ghastly* has a negative charge, you could have gotten the correct answer that way. *Torrid*, by the way, means hot or passionate, as in a "torrid love affair."

2. **B** Context is useful here: the brothers *ran away*, so ask yourself what might make someone run away. While *pain* and *hopelessness* are both bad things, *fear* is far more likely to produce the response of running away.

3. **C** The suffix *–ation* tells you that this is a noun made out of a verb. The verb here is *resign*. A good strategy at this point might be to remember the contexts in which you've heard the word *resign* in life: people *resign* from their jobs, but none of the answer choices convey that sense. You may also have heard people say that someone was *resigned* to his fate. That means the person had accepted his fate. So the noun form, *resignation*, means *acceptance*.

4. **C** This is a good candidate for using word roots: *co–* means *with* and *operate* means *work*.

5. **D** A good strategy here might be to mentally reread the sentence, substituting each of the choices in turn. Choices (A) and (B) make no sense in the context of this sentence. So now you're left with (C) and (D). Think about what words *initiative* sounds like: *initial* and *initiate* might come to mind. They both have to do with starting things. That doesn't seem to have much to do with money, so eliminate (C). In fact, *initiative* means "resourcefulness" or "drive," so *ambition* is the best match. (You could say that someone who has *initiative* or *ambition* is a self-starter; hence the link with *initial*.)

6. **A** Try decoding this one using word parts: *Re–* means "back" or "again." The root *par* means "equal", and the suffix *–ation* tells you that the word is a noun that describes the act of doing something. So *reparation* means something like "the act of making something equal again" or perhaps *the act of making something right*. Thus, choice (A) *compensation* is the best match. In fact, *reparations* means "repayment for loss suffered."

7. **A** An excellent strategy here would be to predict a meaning based on the context of the sentence. What might an older sibling do to younger siblings? Perhaps she would influence them or even boss them around. Indeed, *dominion* means "control," "sway," or "authority." (You could also have used word parts here: *dom* relates to ruling or authority.)

8. **C** The professor wants to appear erudite even when he doesn't know much about a subject. You can predict from the context that *erudite* means "learned." Choice (C) *scholarly* means "learned" and is the correct answer. *Civil, progressive,* and *amoral* don't relate to knowing a lot about a certain subject.

9. **A** You might be able to remember other circumstances in which you've heard the word *tangible*. And from those, you might have a sense that something *tangible* can be felt or seen. From that, you should be able to pick (A) *real* as the closest match. You could also use word roots here: the root *tang* or *tac* relates to touch; something *tangible* is *touchable* or *real*.

10. **B** You may have heard the word *alleviate* in the context of "alleviating pain," meaning to make the pain less and help someone feel better. *Alleviate* is a positive word that means "to ease or allay a bad situation," so the correct answer here is (B) *improve*. *Elevate* means "to raise" (think *elevator*), and *encompass* means "to surround."

11. **D** If something is *trite*, it has been used or done so many times that it is no longer fresh or original. For example, you might say that your friend's story about being stuck in an elevator was funny the first hundred times she told it but is now *trite*. The correct answer is (D) *stale*.

12. **A** Use context to predict an answer: what are children likely to make with pots and pans? Probably a loud, unpleasant noise. (B) *protest* is not supported by the context of the sentence. And choices (C) and (D) both have far too positive a connotation. It is unlikely the children produced pleasant *music* or *harmony*.

13. **D** This question is a good candidate for loose meaning associations. *Candid* comes from the noun *candor*, which means "honesty." *Candid* means "sincere" or (D) *honest*. Although *candid* might sound a little like "can do," it does not mean (A) *able*. Choice (C), *valid*, is another positive word that relates to truth, so it might have been tempting here, but *valid* relates to facts or logic whereas *candid* and *honest* refer to people. Choice (B), *whimsical*, means "playful" or "not serious"; if you buy a new phone on a *whim*, you purchase it impulsively.

14. **C** Use the context to make a prediction: *study* gives a person more information, and more information allows that person to make better choices. So after study, he was able to see or decide which car to choose. This matches choice (C), *perceive*.

You could also mentally reread the sentence, substituting each answer choice. He is unlikely to want to (A) *avoid* or (B) *spoil* the best car. He might (D) *savor*, or "enjoy," his car after picking it out, but this sentence does not concern what happens after he finds a good vehicle.

15. **C** You may have a sense that *dupe* has a negative charge. If so, you can eliminate (A) *double*, which is neither negative nor positive, and (D) *delight*, which has a strong positive charge. In fact, to *dupe* someone is to cheat or lie to that person, and (C) *deceive* matches that meaning. Related words are *betray*, *delude*, and *dissimulate*. *Dupe* can also be a noun, meaning "someone who is easily taken advantage of." You don't want to be someone's *dupe*!

WORD KNOWLEDGE PRACTICE SET 2

Answers and Explanations

1. **B** Use context to predict the meaning of this word. An athlete will work hard to prepare for a competition, and *vigor* means "strength," "force," or (B) *intensity*. The athlete may feel concerned about the competition, but the underlined word here deals with the athlete's training, not his feelings. If you thought the underlined word was *rigor*, you might have picked (A) *stiffness*: be sure to read carefully.

2. **A** An *enigma* is a mystery or (A) *puzzle*. You may have heard of the Enigma machine, which was a famous encryption device used by the Germans in World War II.

3. **C** Because she finished the race even though she was in pain, you know she was pretty tough. This aligns with her showing (C) *courage*. If anything, a sore foot implies that she ran more slowly than usual, so *speed* doesn't fit. You don't know whether she showed (B) *cheerfulness* or was gritting her teeth with every step. Choice (D) *vulnerability*, or "weakness," is the opposite of what you need here.

4. **B** On this question, the sentence actually defines the word for you. Someone who is frugal avoids spending money; in other words, he's (B) *thrifty*. Choice (A) *fearful* is too strong; you don't know that he is actually afraid to spend money, just that he's not eager to do so. Choice (C) *imprudent* means "not careful," and is the opposite of the correct answer. Choice (D) *lavish* means "very generous" or "abundant" and is also the opposite of *frugal*.

5. **C** The word *fortuitous* has the same root as *fortune* and *fortunate*, and it means "based on luck." The clue that the find was (C) *lucky* is the fact that "just the right part" was found at a swap meet, where one may encounter a random assortment of items and not necessarily those one desires. Choice (A) *expected* is the opposite of the author's meaning. While the find may have been (B) *interesting* and (D) *exciting*, these choices do not fit the root *fortu*, which conveys luck.

6. **A** After Jack had fallen several times, his enthusiasm was bound to be less, not more. The words *abated* and (A) *waned* both mean "lessened." If you were not sure of the word *waned*, you might have noticed that two of the other choices—(C) *renewed* and (D) *flourished*—suggest more enthusiasm, not less. Choice (B), *solidified*, means "made solid or firm." All of these are the opposite of the meaning needed here.

7. **A** The prefix *con*– means "together," so the word *congeniality* is likely to indicate people or things coming together. This alone might lead you to eliminate (B) *coldness*, meaning "lack of warmth" or "unfriendliness." Choice (C) *competence* and (D) *beauty* are both positive characteristics but do not relate to coming together. However, (A) *agreeability* suggests two people getting along. Thus, the prefix alone could help you to strategically eliminate three choices and arrive at (A) as the correct answer. In fact, the root word *genial* means "warm and friendly."

8. **B** The clue in the sentence is that the students' questions were discouraged by his remarks. This tells you that the responses were not helpful and perhaps even rude. This eliminates (A) *amusing* (funny) and (D) *articulate* (spoken clearly) as answers. Choice (C) *empathetic* means "sharing feelings"; the root *path*– connotes feelings, as in *sympathy* (feeling for) and *apathy* (without feeling). Empathetic replies would not discourage students. So you are left with (B) *curt* as the correct answer, meaning "short and unresponsive."

9. **C** The word *eminence* is often used as a title of respect for leaders, especially religious leaders, and is used to honor a person of very high rank or great distinction. As such, it can best be understood as meaning (C) *greatness*. Choice (A) *mischief* and choice (D) *dishonor* contradict the positive connotation of the word. Choice (B) *devotion* might be tempting if you are familiar with the word's frequent association with religious leaders; however, the word *eminence* describes the high rank of the individual, not the strength of belief.

10. **D** The clue word *unwitting* tells you that the clerk was unaware of the embezzlement. Also, it makes sense that when the manager confessed to the crime, the clerk was found to be innocent. This is the meaning of *absolved* and its synonym (D) *exonerated*. If you were unsure of the meaning of *exonerated*, you could eliminate (B) *accused* because it is the opposite of what is needed here. Likewise, (C) *incarcerated*—meaning "jailed"—would give the sentence the opposite meaning. Choice (A) *impressed* simply does not fit the context of the sentence.

11. **C** The prefix *dis*– gives this word a clear negative connotation, allowing you to eliminate (A) *sympathy*. And although (D) *flattery* is often used in a negative context, it is not an inherently negative word, so it can be eliminated. While it is true that (B) *dishonesty* is negative, the word *disdain* has no association with a lack of honesty. *Disdain* suggests simply a negative judgment, and this is the meaning of (C) *contempt*.

12. **C** The root word *negate* within *abnegate* means "to deny the existence of something." The prefix *ab–* means "away from," so to *abnegate* means "to turn one's back on or reject." This makes choice (C), *deny*, the correct answer.

13. **B** The contrast clue *although* tells you that the word *civility* has a meaning opposite to "not exactly friendly." So look for an answer choice that says she was not unfriendly. Choice (B) *courtesy* means "politeness," which is consistent with an absence of unfriendliness. Choice (A) *honesty* is not related to the clue about friendliness. Choice (C) *compliance* means "obedience" and is also not related to friendliness. Choice (D) *rudeness* is the opposite of the needed meaning.

14. **D** The prefix *de–*, meaning *not*, gives the word a negative connotation and allows you to rule out the neutral words (C) *relocated* and (B) *tamed* with some confidence. Another form of the word *deranged* is *disarranged*, meaning "disordered." *Deranged* can be used to describe a person with a disordered mind, making (D) *crazy* an appropriate synonym.

15. **A** The contrast word *although* tells you that even though he did not accuse her outright, or directly, he did accuse her. So the word *insinuated* must mean that he accused her indirectly, and (A) *implied* means "stated indirectly." Choice (B) *insisted* would mean that he made his accusation forcefully, which is the opposite of what he did. Choice (C) *concealed* means "hid." This is incorrect because he did not hide or cover up her lie; instead, he accused her of lying. Choice (D) *disavowed* means "denied or rejected," which again contradicts the fact that he did accuse her.

Review and Reflect

As you review your work on these practice questions, ask yourself some questions:

- Did you remember to apply strategies, such as using word parts or using real-life remembered context?
- How often were you able to sense whether a word had a positive or negative charge?
- How well did you use context on the In-Context questions?
- Use your thoughts about these questions to guide your review of this chapter.

Want more instruction and practice with Word Knowledge? Check out the companion video for this section. Also, try more questions in the Qbank online.

COMPANION VIDEO

kaptest.com/login

CHAPTER 4

PARAGRAPH COMPREHENSION

KNOW WHAT TO EXPECT

The ASVAB includes a reading section among its several subtests. This section, called the **Paragraph Comprehension** (PC) section, presents short passages for you to read and then asks questions based on those passages. While the subject matter of each passage will vary, no outside knowledge is required to answer the questions correctly. For each question in this section, everything you need to answer the question will be contained in the passage above it.

On the paper-and-pencil version of the ASVAB, the PC section will ask 15 questions about 13 or 14 short passages. Timing is extremely important in this version of the test, as you are only given 13 minutes to answer these 15 questions. On the CAT-ASVAB, you are given 27 minutes to answer 10 questions. The extra time allows for the fact that question difficulty can increase as you accumulate correct answers.

THE KAPLAN METHOD FOR PARAGRAPH COMPREHENSION QUESTIONS

LEARNING OBJECTIVE

In this section, you'll learn to:

- apply the Kaplan Method for Paragraph Comprehension Questions

To be successful in both the paper and pencil and CAT versions of the PC subtest, you will need to read the passages in an effective and efficient manner. Keep in mind that your goal in this section is not to learn the subject matter of the passages but simply to answer the questions correctly. Therefore, your focus should be on understanding each question's task rather than absorbing every detail in the passage. In this chapter, you'll learn how to use Kaplan's approach to PC questions to help you answer each type of question you'll see on the PC subtest.

Having a consistent and repeatable method will help you quickly and confidently answer questions in this section. Use Kaplan's 4-Step Method for Paragraph Comprehension Questions to attack every question you see in your practice and on Test Day.

THE KAPLAN METHOD FOR PARAGRAPH COMPREHENSION QUESTIONS

Step 1: Read the question stem to identify your task.

Step 2: Read the passage strategically.

Step 3: Make a prediction.

Step 4: Find the correct answer.

Step 1: Read the question stem to identify your task.

How you read and analyze a passage will vary depending upon the task defined by the question stem, so it's important to always read the question before you look at the passage. After you read the question stem and identify the specific type of question you're being asked, it's time to read the passage strategically. In this chapter, we will cover all the different types of questions you might see on Test Day.

Step 2: Read the passage strategically.

Once you know your task, read as much of the passage as necessary to answer the question. In a question that asks you to identify the main idea of a passage, separate supporting details from opinions and recommendations to determine the author's overall point. Other PC questions may ask you to identify specific details, or to determine the meaning of a word. The key in this step is to read the passage in the way that is appropriate to answer the specific question that is being asked.

Step 3: Make a prediction.

As mentioned earlier, every question in this section can be answered by the information in the passage that precedes it; no outside information is necessary. For each question, take a few seconds to predict what the correct answer will look like. Sometimes you'll know the answer without even looking at the answer choices. Other times, you'll have only a general idea of the phrasing of the correct answer. Either way, taking the time to predict will make it easier to find the correct answer.

Step 4: Find the correct answer.

Once you come up with a prediction of the correct answer, simply find the answer choice that is the closest match. If you are unable to make a prediction, you'll still have an opportunity to find the correct answer by eliminating answer choices that you know are incorrect. Many will be incorrect because they stray beyond the scope of the subject matter in the passage. Other answer choices are incorrect because they are too extreme, or they distort details in the passage. After you've eliminated the wrong answer choices, select the one that remains and move on to the next question.

CONQUERING PARAGRAPH COMPREHENSION QUESTIONS

LEARNING OBJECTIVES

In this section, you'll learn to:

- identify PC question types
- read PC passages strategically
- make accurate predictions
- match your prediction to the correct answer

Global Questions

A common question type in the PC section is one that asks you to identify a passage's main idea or theme, the author's purpose, or the tone of the passage. Here's what a Global question stem might look like:

Which of the following is the main idea of the passage?

The author's tone in the passage above can be characterized as

The purpose of the passage above is to

The passage above is primarily concerned with

In order to find the main idea in a passage, you must be able to distinguish the values an author assigns to different statements. In each passage that asks you to determine a main idea, the author will use supporting details to establish a main idea. Things like simple facts, other people's opinions, and background information can all operate as supporting details in a passage. Those details are then used to support an author's claim, which often comes in the form of a strong opinion, a recommendation, a prediction, or a rebuttal to another person's position.

As an example, imagine a passage with the following main idea: "Purchasing gold is a wise investment." This claim is not a fact but rather a recommendation. In turn, the author needs to support this claim by offering reasons why gold is such a good investment. Perhaps it's because "the amount of gold available worldwide is set to plateau," or maybe it's because "gold is the worldwide monetary standard." Those two statements, alone or together, help to explain why gold would make a wise investment. But the recommendation that gold is a good investment does not explain why gold is the worldwide monetary standard.

So, one method for separating the main idea from the rest of the passage is to realize that the supporting details and the main idea answer two different types of questions, as you can see:

This:	***Answers the question:***
Main idea	What does the author believe?
Supporting details	Why does the author believe what he or she believes?

Sometimes, like when the author expresses a strong opinion, the main idea will be clear. Words and phrases like *thus, therefore, I suggest,* and *I believe* all indicate that the author is providing a strong conclusion. Other times, a passage's main idea will be more difficult to identify. In those passages, be prepared to work a little harder to separate supporting ideas from the main idea. Often, one of the tricks to zeroing in on the main idea of a passage is to pay attention to structural clues that indicate a contrast. Words such as *but, though, however, although,* and *yet* provide subtle clues into an author's point of view: the statement after the contrast word is frequently a reflection of the author's opinion, especially in passages where the author disagrees with someone else's opinion.

Finally, on Global questions it is often easy to eliminate a few answer choices immediately. Wrong answer choices for Global questions usually do one of the following:

- They're too specific, dealing with just one small detail of the passage.
- They're too general, going beyond the scope of the passage.
- They're contradictory to the information presented in the passage.
- They're too extreme; that is, they distort the author's opinion by overstating it.

Below, take a look at how a Kaplan-trained test taker approaches a typical Paragraph Comprehension Global question. Remember to start with the question stem, labeled with "Step 1" below.

Question	Analysis
Many countries around the world have instituted laws that mandate phasing out the use of incandescent light bulbs and replacing them with alternative light sources that consume less energy. However, many incandescent bulbs have been replaced with fluorescent lights containing the extremely toxic element, mercury. Similarly, some governments have required gasoline for automobiles to contain ethanol, a fuel that generates less pollutants than gasoline when burned. However, ethanol is produced from crops such as corn. Cultivating the crops and converting them to ethanol require large amounts of energy, and the former also increases the use of fertilizers, which can leach into streams and rivers.	**Step 2:** The passage starts with background information that describes what countries are doing about light bulbs. Then a possible conflict is introduced: the new type of bulb contains mercury, which is toxic. More background information is provided: some governments are pushing to add ethanol to gasoline. The fourth and fifth sentences introduce a possible drawback to the proposed change: using corn for fuel uses a lot of energy.
The main theme of the passage is that	**Step 1:** This is a Global question. Read the passage to determine the author's main point.
	Step 3: The author discusses two things governments are trying to do to reduce energy and pollutants. After each example, she uses the contrast keyword "however" to highlight the fact that *there are drawbacks to these plans*. That's a solid prediction.
(A) environmental regulations can have negative consequences	**Step 4:** Correct. This matches the prediction.
(B) light bulbs and ethanol regulations are misguided	Just because the regulations have some drawbacks does not mean they are necessarily misguided. Incorrect.
(C) environmental concerns should take precedence over economic issues	This is an irrelevant comparison between two considerations that are not explicitly discussed in the passage.
(D) fluorescent lights contain more mercury than incandescent lights	While this is a valid inference that can be made from the passage, this detail is not the main theme of the passage.

Notice that the well-trained test taker follows a specific method. She is not in a rush and takes her time in Step 3. She knows that a well-phrased prediction will help her answer the question efficiently and confidently.

Now it's your turn. Follow the Kaplan Method for PC by working through each step deliberately. Focus on reading the passage strategically and making a strong prediction. When you're finished, check your work against the explanation below.

> Alchemy is the name given to the attempt to change lead, copper, and other metals into silver or gold. Today alchemy is regarded as a pseudoscience. It is associated with astrology and the occult in the modern mind, and the alchemist is viewed in retrospect as a charlatan obsessed with dreams of impossible wealth. But for many centuries, alchemy was a respected art. In the search for the elusive secret to making gold, alchemists helped to develop many of the apparatuses and procedures used in laboratories today, and the results of their experiments laid the basic conceptual basis for the modern science of chemistry.
>
> The central point of the passage is that
>
> (A) alchemy is a pseudoscience
> (B) alchemists tried, but failed, to make gold from other metals
> (C) many alchemists dreamt of becoming rich
> (D) modern chemistry evolved out of alchemy

Explanation

Step 1: This Global question asks for the passage's central point, or main idea.

Step 2: The first sentence defines the topic of the passage: alchemy. The second and third sentences state that alchemy is no longer recognized as a real science; an analogy is drawn to astrology and the occult. The author then uses the keyword "but" to draw a contrast: for a long time, alchemy was a respected art. The passage then finishes by describing all of the things that alchemists did that helped pave the way for modern chemists.

Step 3: There is no clear opinion from the author, but she does indicate contrast in the middle of the passage. Focus on what the author discusses after the contrast: alchemy used to be a respected art, and the work of alchemists laid the foundation for modern chemistry.

Step 4: Answer choice **(D)** matches the prediction that "the work of alchemists laid the foundation for modern chemistry." The other answer choices don't match the prediction, so eliminate them.

ASVAB STRATEGY

To correctly answer Global questions:

- Identify any opinions, recommendations, suggestions, or predictions made by the author.
- Look to contrast words to help separate supporting details from the author's view.
- Always make a prediction before looking at the answer choices.

Detail Questions

Detail questions ask you to find specific information that is explicitly stated within the passage. Here are some examples of Detail question stems:

According to the author, which of the following is a type of rug found in French castles?

The second step in constructing a picket fence is to

Which of the following is cited in the passage as an advantage of a retirement account?

Notice how these questions are very different from Global questions. In these types of questions, you do not have to read for the author's main idea. In fact, it is crucial in Detail questions not to overanalyze or read too much into the question. The correct answer to these questions will almost always be a paraphrase of something found directly in the passage.

Beware of the difference between simply recognizing text from the passage and recognizing text from the passage that answers the question. Wrong answer choices will sometimes be pulled from irrelevant areas of the given text. For Detail questions, always research the text, make a prediction, and then find a match for that prediction among the answer choices.

Below, take a look at how a Kaplan-trained test taker approaches a typical Paragraph Comprehension Detail question. Remember to start with the question stem, labeled with "Step 1" below.

Question	Analysis
The dancer and choreographer Martha Graham is regarded as one of the outstanding innovators in the history of dance. In a career that lasted over 50 years, Graham created more than 170 works ranging from solos to large-scale pieces, and danced in most of them herself. Trained in a variety of different international styles of dance, she began in the early 1920s to break away from the rigid traditions of classical ballet. She wanted to create a new dance form that would reflect the transformed atmosphere of the postwar period.	**Step 2:** Read the passage, looking for *why* Graham introduced new techniques. The passage starts with background information regarding Martha Graham and her accomplishments. Then, her dancing background and training are discussed. The final sentence discusses *why* she created a new dance style.
Martha Graham's motivation for introducing new dance techniques was to	**Step 1:** This is a Detail question. You are looking for a reason why Martha Graham introduced new dance techniques.
	Step 3: Rephrase the last sentence of the passage: Graham introduced new dance techniques because she wanted to come up with a new, original dance form that expressed the way the world felt after the war.
(A) break away from the traditions of classical ballet	**Step 4:** This choice describes *what* Graham did, but is incorrect because it does not answer the given question of *why* she did it.
(B) attract attention to her dance troupe	This is a distortion of what the passage said. While attracting attention may have been a nice side effect, it's not the reason *why* Graham introduced new dance techniques. Incorrect.

Question	Analysis
(C) express the changed mood of her time	Correct. This matches the prediction.
(D) emphasize the rigidity of conventional dance movement	This is incorrect because it distorts Graham's intention. She was not commenting on conventional dance movement. She was updating dance to reflect the current times.

Now, you take a turn! Use the example above to help you through the following practice item. Make sure to follow the Kaplan Method to secure great habits for Test Day.

Golden retrievers, one of the most popular dog breeds in the United States, have earned their wide acceptance for a variety of reasons. As hunting dogs, they will retrieve game carefully and willingly bring it back to the hunter without damage. As family pets, goldens are gentle and patient additions to the household, willingly abiding unintentional abuse by toddlers. Goldens are intelligent and can be readily trained to respond to a wide variety of commands. For this reason, they are among the favorite breeds chosen to be trained as assistance dogs for disabled persons. Also, their friendly nature has made them perfect visitors to communities for the elderly. A golden retriever will do its utmost to please its human master.

The author states that golden retrievers are excellent assistance dogs because they are

(A) friendly
(B) intelligent
(C) gentle
(D) available

Explanation

Step 1: The question asks for what the passage states is the reason why golden retrievers are excellent assistance dogs. This is a Detail question.

Step 2: The passage discusses various reasons for why golden retrievers are so popular. Sentences four and five describe the attributes that make golden retrievers excellent assistance dogs.

Step 3: Combine sentences four and five to formulate your prediction: *goldens are excellent assistance dogs because they are intelligent and easily trained.*

Step 4: Answer choice **(B)** matches the prediction that goldens make great assistance dogs because they are smart and easily trained. Because the other answer choices don't match the prediction, we can eliminate them.

ASVAB STRATEGY

To correctly answer Detail questions:

- Identify specific information in the passage that answers the question.
- Don't fall for answer choices that rephrase statements in the passage but that don't answer the question.

Inference Questions

Sometimes the ASVAB will ask you questions about the passage to which the answers are not directly stated in the text but are instead implied by the passage. Here are just a few of the ways an Inference question might be worded:

Which of the following is implied by the passage?

The author apparently feels that

It can be inferred from the passage that

As you can see, the wording in Inference questions rarely points you to a specific part of the passage, as Detail questions do. And Inference questions won't necessarily ask you about the ideas that are most important to the author, as Global questions do. Instead, Inference questions require you to consider multiple statements in a passage and, from them, determine what else the author must believe. Because there are many inferences that can be made from a given piece of text, it's not necessary to predict before considering the answer choices. Rather, paraphrase the passage and move through the answer choices, selecting the one that is supported by the given text.

Because the correct answer to an Inference question is simply the one answer choice that is fully supported by what is stated in the passage, you can also eliminate wrong answer choices because they commit one of the following errors:

- They contradict information in the passage.
- They bring in outside information that is not discussed in the passage.
- They distort the information presented in the passage.
- They make an extreme claim that is not completely supported by the passage.

Let's see how a Kaplan-trained test taker handles a Paragraph Comprehension Inference question. Remember to start with the question stem, labeled with "Step 1" below.

Question	Analysis
Dan and Sonya are married and have exactly three children—Betty, George, and Tara—and exactly three grandchildren. George usually babysits Betty's twin daughters, while Tara usually babysits George's child.	**Step 2:** The passage starts with background on Dan and Sonya, who are married and have three children: Betty, George, and Tara. They also have three grandchildren. The passage states that Betty has twin daughters, and that George has a child.
If these statements are true, which of the following must be true?	**Step 1:** This question asks what must be true based on what the passage stated, so it's an Inference question. Because there are no clues in the question stem pointing to a single area of the text, read and paraphrase the entire passage.
	Step 3: Mentally or on scratch paper, create a "family tree" to visualize the relationships: Dan & Sonya Betty — George — Tara Betty: 2 Daughters; George: 1 Child This is a summary of the given information. Since there are only three grandchildren, this means that Tara must not have any children.
(A) All of Dan's grandchildren are female.	**Step 4:** Though Betty has two daughters, the sex of George's child is not known. Therefore, it's not necessarily true that all of the grandchildren are female. Eliminate.
(B) Tara has no children.	Correct. This must be true according to the statements in the passage.
(C) Tara sometimes babysits for Betty's children.	Though the passage states that Tara usually babysits George's child, it does not mention whether she ever babysits for Betty or not.
(D) Sonya has at least one grandson.	From the statements given in the passage, all that is known for sure is that Sonya has two granddaughters and another grandchild, whose sex is unknown.

Now, you try it. Follow the steps used in the example above to help you master your technique for tackling Inference questions on Test Day.

> In each of the last three years, a court in this country has awarded a settlement in excess of $300 million. This is a travesty of justice, and it unfairly burdens the court system. To alleviate the strain on the nation's court system, _______________.
>
> Which of the following best completes the above passage?
>
> (A) there should be fewer lawsuits
> (B) lawsuits should name more than one defendant
> (C) courts should cease awarding excessive settlements
> (D) settlements should be awarded based solely on need

Explanation

Step 1: The question asks what would complete the passage. Since the correct answer will be based on what was said in the rest of the passage, this is an Inference question.

Step 2: The passage states that, in the past three years, courts have been awarding extremely high settlements. The author feels that these awards are outrageous, saying that they are a "travesty of justice" and they "unfairly burden the court system."

Step 3: The gist of the argument is that awards are too big, so it would seem logical that if courts stopped going overboard in awarding settlements, it would help lessen the burden on the courts.

Step 4: The correct answer here is choice **(C)**, that courts should stop awarding excessive settlements. Answer choices (A) and (B) are not issues discussed in this passage. Choice (D), likewise, is not discussed in the passage, and the use of "solely" makes it too extreme as well. There is nothing in the passage to support that settlements should be awarded only based on need.

ASVAB STRATEGY

To correctly answer Inference questions:

- Remember that the correct answer is supported by statements in the passage.
- Don't read between the lines too much.
- Eliminate choices that introduce details or opinions the author doesn't mention.

Vocabulary-in-Context Questions

One final type of question found in the PC section will ask you for the meaning of a word used in the paragraph. These questions, which we call Vocabulary-in-Context questions, are pretty straightforward: the correct answer will be a word that can replace the word in question without altering the meaning of the sentence.

To handle this question type, focus on the word in the question stem while you are reading the passage. Then, predict an answer by defining the word as it is used in context. Finally, attack the answer choices by looking for a word that matches your prediction. Once you have selected an answer choice, reread the initial sentence with the answer choice in place of the vocabulary word to be sure the meaning of the sentence is the same.

If you feel as though your vocabulary could use some improvement, be sure to utilize the vocabulary-building resources available to you in chapter 3 (Word Knowledge) and in the Appendix.

Below, take a look at how an expert uses the Kaplan Method to efficiently and effectively attack a Vocabulary-in-Context passage. Remember to start with the question stem, labeled with "Step 1" below.

Question	Analysis
When voters choose a candidate in an election, they cast one vote for one candidate. However, there are other ways that votes can be cast in different situations. One of the more common alternatives is rank voting. In this scheme, each voter ranks the choices from first to last. Thus, if there were ten positions to fill, each voter would assign ten points to her favorite choice, nine points to the second favorite, and so on. This system is used by the Associated Press to rank college sports teams and a modified version is used in Australian elections.	**Step 2:** The word *scheme* is used in the fourth sentence of the passage. Leading up to the word, the passage is talking about the process of choosing a candidate in an election.
A word that could be properly substituted for *scheme* in the passage is	**Step 1:** The question asks you to determine the meaning of a word used in the passage. This is a Vocabulary-in-Context question.
	Step 3: There are different approaches to elections listed. This tells us the word is being used to describe a *procedure* or a *process*. One of those words, or a synonym, would be a good prediction for the meaning of the word *scheme* in this passage.
(A) plot	**Step 4:** While a common usage of the word *scheme* is to describe a plot, no one is plotting anything in the passage.
(B) collection	*Collection* does not make sense in the context of the passage. Eliminate.
(C) vision	*Vision* does not make sense in the context of the passage. Eliminate.
(D) method	Correct. *This scheme* refers to rank voting, which is a *method* of voting.

Now it's your turn. Follow the steps used in the example above to correctly and confidently answer the Vocabulary-in-Context question below.

Who was the first man to discover the North Pole? Soon after the turn of the twentieth century, two men each claimed the honor individually. While Frederick Cook claimed he was the first to discover the North Pole in 1908, Robert Peary disputed Cook's account and set out on his own expedition in 1909. Today, it is unclear which man, if either, actually reached the North Pole. After poring through the written accounts of Cook and Peary's expeditions, researchers believe that while both men came close to the North Pole, neither man actually found it.

As used in the passage, "poring through" most nearly means

(A) reading carefully
(B) filtering out
(C) searching for
(D) making notes in

Explanation

Step 1: How is the phrase "poring through" being used in the passage in this Vocabulary-in-Context question?

Step 2: The researchers had access to Cook and Peary's written accounts, so they must have evaluated them by reading through them.

Step 3: Anything that expresses the idea of "reading" the accounts will be correct.

Step 4: Only answer choice **(A)**, *reading carefully*, matches. Answer choice (B), *filtering out*, changes the meaning of the sentence: the researchers weren't removing some of the written accounts. Similarly, the researchers weren't *searching for* the accounts, (C), since it's inferred that they already had the documents. Same with answer choice (D): the researchers were not *making notes in* the explorers' accounts.

ASVAB STRATEGY

To correctly answer Vocabulary-in-Context questions:

- Many words and phrases have different meanings depending on their context. Think about the term's specific meaning in the passage.
- Always make a prediction before you look through the answer choices.
- The correct choice will not change the meaning of the sentence.

PARAGRAPH COMPREHENSION PRACTICE SET 1

For each question, read the paragraph and select the answer that best completes the statement or answers the question that follows. This question set has 10 questions, which is the number of Paragraph Comprehension questions you will see if you take the CAT-ASVAB.

Four years ago, the governor came into office seeking to change the way politics were run in this state. Now, it appears he has been the victim of his own ambitious political philosophy. Trying to do too much has given him a reputation as being pushy, and the backlash in the state has let him accomplish little. He may very well lose in his reelection bid.

1. The governor's approach to politics was
   (A) business as usual
   (B) overly idealistic
   (C) careless and sloppy
   (D) influenced by her critics

Since its first official documentation by Sir George Everest in 1865, Mount Everest in Nepal has been the "Holy Grail" of mountaineers. Sir Edmund Hillary of New Zealand and Tenzing Norgay of Nepal were the first men to successfully complete the ascent to the peak in May 1953. This feat won them international acclaim, not to mention knighthood for Hillary. But much less celebrated is the first successful ascent of Everest by a woman, which did not take place until May 16, 1975. Junko Tabei of Japan was the first woman to reach the summit of the world's most famous single peak. The first American woman to scale its heights successfully was Stacy Allison of Portland in 1988.

2. A member of the first successful Mount Everest expedition was
   (A) Sir George Everest
   (B) Junko Tabei
   (C) Sir Edmund Hillary
   (D) Stacy Allison

It is without question a travesty that our children are no longer given healthy, nutritious food options for lunch in our public schools. Hamburgers, pizza, and chocolate are not only giving our kids bigger waistlines, but these junk foods are also helping teach them poor eating habits. It is imperative that we change the mindset that any food is good food and start offering students better meals at the same prices. Otherwise, a new generation of obese Americans is a given.

3. According to the passage, over the past few years, school lunches have gotten
   (A) more expensive
   (B) more exotic
   (C) healthier
   (D) less nutritious

Many times families choose to replace old furniture when just a minor amount of maintenance is all that is needed. To fix a wooden chair that is wobbly, follow these easy steps. First, check the joints of the chair to see if the chair is structurally sound. There should be small dowel rods and a corner block to keep the chair together. Next, use a ripping chisel to remove the corner block. Once the block is free from the chair, you should be able to glue the joints back together. Finally, once the glue looks dry, place the corner block back on the chair and gently mallet the block onto the dowels. In no time, you will have saved not only the chair, but also your hard-earned money!

4. After removing the corner block, you should
   (A) mallet the dowels into place
   (B) check the joints for damage
   (C) glue the joints together
   (D) replace the corner block

James felt the pulse of the crowd. There was a low murmur just under the house music. Backstage, his bandmates were tuning or drumming lightly on tabletops. In a few moments, the whole country would watch the band play. What a change from those dingy bars and clubs a few years ago. Maybe all the hard work had finally paid off. Looking down at his callused hands, he wondered if maybe this would be the break they had been working so hard for.

5. The tone of this passage is one of
   (A) sadness
   (B) anticipation
   (C) anger
   (D) ambivalence

Packaging on many popular foods is deceiving to consumers. Too often, the print is small and hard to read. And if you can read it, it's often confusing or intentionally vague. This is especially true on the nutrition label. The government really ought to do something about the nutrition labels on food because the existing laws just don't go far enough.

6. The author would probably support which of the following?
   (A) magazine advertisements for cigarettes
   (B) allergy information prominently listed on food labels
   (C) fine print on a contract
   (D) food ads in the Sunday paper

In an age where we have pills for depression, dysfunction, and aggression, not to mention headaches, it is important that we not forget that many drugs can have serious side effects. These may range from internal bleeding, vomiting, or soreness in the limbs to, in more extreme cases, loss of consciousness or even coma. If you experience unwanted side effects, it is important to get to a hospital immediately and seek treatment. Drinking alcohol or smoking cigarettes may also contribute to violent side effects.

7. According to the passage, one of the possible side effects of drugs is

(A) drinking alcohol
(B) dysfunction
(C) internal bleeding
(D) aggression

First created at the height of atomic postwar paranoia, the Incredible Hulk stories offer a fascinating look at the dual nature of human beings. On the one hand, he is a mild-mannered, bespectacled scientist. On the other, he is a raging, rampaging beast. More than a statement about nuclear dangers, the Hulk is a reflection of the two sides in each of us—the calm, logical human and the raging animal.

8. According to the author, the comic book character of the Hulk is

(A) a reflection of humanity
(B) really an animal
(C) mild mannered
(D) a protest about atomic power

Once considered the best high school player in the country, the onetime prodigy now spends his days working as a bricklayer for a local construction company. Asked if he is bitter about the way his life turned out, he replies, "Not at all." In fact, he says, his only regret is that he didn't study hard enough and go to college. He still gets recognized on occasion, but an extra 80 pounds and bad knees keep him from reliving his former glory on the court.

9. The word *prodigy* in the passage most nearly means

(A) depressed loner
(B) shy scientist
(C) gifted youngster
(D) bitter malcontent

Celebrated as one of the greatest film directors of the twentieth century, Alfred Hitchcock made his name creating some of the most critically acclaimed suspense films of all time. But while other directors of the genre seemed content telling stories of domestic intrigue, Hitchcock was not afraid to make films centering on unconventional subjects. In fact, one of his most celebrated films, *The Birds*, has no typical antagonist: the suspense comes not from a person who is out to do wrong, but rather from nature itself.

10. The main idea of the passage above is that

(A) Alfred Hitchcock is one of the greatest film directors ever
(B) most suspense films tell stories of domestic intrigue
(C) Alfred Hitchcock was not afraid to tell unconventional stories
(D) *The Birds* is scarier than traditional suspense films

PARAGRAPH COMPREHENSION PRACTICE SET 2

For each question, read the paragraph and select the answer that best completes the statement or answers the question that follows. This question set has 10 questions, which is the number of Paragraph Comprehension questions you will see if you take the CAT-ASVAB.

Bassoons are double-reeded wood instruments that produce a deep baritone sound. A bassoon's pitch can be altered by adjusting a curved tube called a bocal. The deeper the bocal is inserted into the bassoon, the higher the pitch. Conversely, pulling the bocal out of the instrument lowers the pitch.

1. It can be inferred from the passage that the bassoon

   (A) is a larger instrument than the oboe
   (B) is the deepest-sounding wood instrument
   (C) is the only instrument that utilizes a bocal
   (D) is capable of a range of pitches

Investors who believe that a company's stock price is overvalued can perform an action called a "short sale." To short a stock, an investor sells shares that she does not currently own; typically, shares for short selling are borrowed from a broker. If the stock goes down in price, the investor is able to buy the shares at a lower price. Shorting stocks is risky, though; if the stock suddenly rises in price, the broker can force the investor to purchase the shares at the higher price, resulting in a significant loss.

2. According to the passage, an investor who shorts stocks first

   (A) buys shares of a company's stock, then sells those shares for a loss
   (B) sells shares of a company's stock, then buys those shares at a later date
   (C) buys shares of a company's stock, then sells those shares whenever he wishes
   (D) sells shares of a company's stock, then uses that money to buy stock in another company

As rays of sunlight peeked through the tops of the trees, Mark felt at peace. It had been years since he had walked these trails. Everything was quiet and calm. It seemed to him that he finally saw the woods as they truly were: a refuge, a place to be renewed, a respite from the clamor and chaos of the city.

3. In this passage, *respite* most nearly means

   (A) relief
   (B) opposite
   (C) valley
   (D) hope

The bus system in the city is in need of a drastic overhaul. Last week, the Main Street bus line was reduced from one bus every 15 minutes to one bus an hour. The week before that, the Uptown bus was canceled completely. Each change occurred with no notice! If the city expects people to continue to use public transportation, it needs to reduce the bus system issues.

4. Which of the following is mentioned in the passage?

   (A) The city needs to notify customers of bus line interruptions.
   (B) The city should allocate money from other programs to revamp the bus system.
   (C) The Main Street bus line has experienced reduced service in the past.
   (D) The Main Street bus line has been shut down in the past.

The European hedgehog is found in a wide range of habitats in Western Europe. It is quite popular because of its charming appearance and because of its appetite for many of the pests that plague European gardens. Unlike warmer-climate species of hedgehog, the European hedgehog is known to hibernate in the winter. It is a solitary animal, although occasionally a male and female will share a hibernation nest. While currently stable across much of continental Europe, the population of European hedgehogs is thought to be declining severely in Great Britain.

5. Which of the following best describes the topic of this passage?

   (A) the decline of the European hedgehog in Great Britain
   (B) why the European hedgehog is popular among European gardeners
   (C) the characteristics and behavior of one species of hedgehog in Europe
   (D) the hibernation habits of European hedgehogs

Questions 6 and 7 refer to the following paragraph.

In 1918, as the world prepared to celebrate the end of World War I, a stealthier form of death appeared: the so-called "Spanish influenza." It is said that this strain of influenza killed more in a single year than the bubonic plague killed in a century. The outbreak gave modern scientists their first close look at a worldwide pandemic and paved the way for great advances in medicine. Furthermore, the unprecedented number of patients led to a boom in the medical field. One lasting result was an increase in pay for doctors, encouraging many to enter the profession. It could be said that the Spanish flu introduced the idea of "medicine for profit" to the world.

6. It can be inferred from the passage that

   (A) advances in medicine in the early twentieth century would not have happened without the outbreak of the Spanish flu
   (B) scientists know little about the cause of the bubonic plague
   (C) a doctor today might not have chosen to enter the medical profession before World War I
   (D) the huge numbers of Spanish flu patients overwhelmed the available medical care and resulted in drafting more doctors to deal with the epidemic

7. In the context of the paragraph, the word *unprecedented* most nearly means

   (A) extraordinary
   (B) tragic
   (C) perplexing
   (D) advantageous

Before the Great Chicago Fire of 1871 was extinguished, rumors circulated that it had been caused by the now infamous "Mrs. O'Leary's cow." Yet, after nine days of questioning fifty people, investigating authorities issued an inconclusive report about the fire's cause. "Whether it originated from a spark blown from a chimney on that windy night," the report read, "or was set on fire by human agency, we are unable to determine." Besides Catherine O'Leary and her cow, suspects at the time included several colorful residents of Chicago's immigrant community, including "Peg Leg" Sullivan, who had first alerted the O'Leary family to the fire. In the end, it is likely that the true cause or culprit will never be known.

8. Which of the following statements best expresses the main idea of the passage?

   (A) Although the investigation was inconclusive, it is now considered likely that Catherine O'Leary's cow was the true cause of the Chicago fire.
   (B) Investigative authorities believed that the Chicago fire was set by an unknown suspect rather than being an accidental event.
   (C) Investigators considered "Peg Leg" Sullivan to be a more likely suspect than Catherine O'Leary in causing the Chicago fire.
   (D) The cause of the 1871 Chicago fire will probably never be determined.

Despite his famous theories about how children are sexual beings who may develop into adults with unconscious psychological conflicts, Freud had only one patient during his lifetime who was actually a child. But this fact alone is not sufficient reason to dismiss his entire theory. Many of his ideas were based on information gained from reliable case histories of his adult patients. While some of his conclusions have been questioned in light of recent findings in neuropsychology, his most noteworthy contributions to psychology stand up against the criticisms of those who accuse him of fabricating his evidence to confirm his preconceived theories.

9. The author would most likely agree with which of the following?

(A) Neuropsychology has recently proven that Freud used unreliable evidence to support his theories.
(B) Although recent science has weakened some of Freud's theories, he should still be seen as having made significant contributions to psychology.
(C) Freud's most famous theories are called into question by the fact that he based them all on one child patient.
(D) Freud's theories of child sexuality have been confirmed by numerous case histories.

At the beginning of the universe, temperatures were incredibly high. During this period of high energy, vast amounts of hydrogen, helium, and lithium were created. Although hydrogen and helium are still abundant, scientists believe that the amount of lithium currently measured comprises only about a third of what we should expect to see. There are a wide variety of explanations for why this might be, including some involving hypothetical elementary particles known as axions. Others believe that lithium is trapped in the core of stars, making it undetectable by our current instruments. Of the many theories proposed, there is no clear front-runner to explain the absence of lithium in the universe.

10. According to the passage, axions have been hypothesized in order to

(A) show how lithium was created at the beginning of the universe
(B) explain why scientists detect less lithium in the universe than expected
(C) explain why hydrogen and helium make up most of the mass of the universe
(D) support the belief that lithium has become trapped in the core of stars

PARAGRAPH COMPREHENSION PRACTICE SET 1

Answers and Explanations

1. **B** From the information given, you should have seen that the governor was very determined to get things done his way. So his approach could hardly be described as *business as usual*, choice (A). Nowhere in the passage is his approach described as *careless and sloppy*, choice (C). And since the mayor insisted on doing things his way, his approach was not *influenced by his critics*, choice (D). Only (B) is addressed in the paragraph. Seeking to change the way politics are run indicates an *idealistic* approach to politics.

2. **C** This Detail question asks you for a member of the first expedition to conquer Mount Everest. Researching the passage, you must be careful not to assume that the mountain is named for its conqueror. While *Sir George Everest*, (A), first documented and recorded the height of Everest in 1865, it was (C) *Sir Edmund Hillary* who, along with Tenzing Norgay, completed the first ascent to the peak in 1953. The question does not ask for the first woman, or the first American woman, to reach the summit, so (B) *Junko Tabei* and (D) *Stacy Allison* are out. The correct answer is (C).

3. **D** The author feels strongly that meals in school cafeterias have become more and more similar to junk food. Choice (A) is not applicable, regardless of its validity, because it is not the central point of the passage. Choice (B) is nowhere indicated in the passage, and choice (C) is the opposite of the correct answer. Of the answer choices given, only choice (D), *less nutritious*, correctly answers the question.

4. **C** This Detail question asks you to identify a specific step in a process, so your first task is to locate the step that discusses removing the corner block. Removing the corner block is mentioned in the fifth sentence. After using a ripping chisel to remove the block from the chair, the worker is free to glue the joints back together to tighten them, choice (C).

5. **B** To correctly gauge the tone of a passage, you should pay attention not only to the details, but also to the language and description. In this passage, James is clearly waiting to go onstage. Words like *pulse*, *murmur*, and *maybe* invite the reader to feel excitement and nervousness as James does. Of the answer choices given, (D), *ambivalence*, is clearly wrong, as James definitely cares about what is going to happen. He seems wistful, but never *sad*, (A), or *angry*, (C). The only answer that successfully captures James's mood is choice (B), *anticipation*.

6. **B** Judging from the critical tone of the author and the subject matter at hand, one can safely assume that the author is interested in public safety. Clearly small print is not going to be favored by this author, so choice (C) is out. Choice (D) doesn't make sense and choice (A) is not something thought of as good for public safety. Choice (B), *allergy information prominently listed on food labels*, fits the author's passion for consumer labels.

7. **C** According to the details of the passage, one possible side effect from prescription drug usage is *internal bleeding*, answer choice (C). *Drinking alcohol*, (A), can exacerbate side effects, but it is not a side effect itself. *Dysfunction*, (B), and *aggression*, (D), are discussed as conditions treated by drugs, not as side effects of drug use.

8. **A** From the details of the passage, it is clear that the author sees the character of the Hulk as a symbol. The passage states that the Hulk is a reflection of the two sides of each of us, which matches choice (A) *a reflection of humanity*. The passage states that the Hulk varies between *an animal mentality*, (B), and *a mild-mannered person*, (C), but it does not claim that he is one over the other. While the passage mentions "atomic postwar paranoia," it does not suggest that the Hulk is used as a protest of any type, as in choice (D).

9. **C** According to the passage, the onetime star athlete is now a local bricklayer. There is nothing in the passage to indicate that the person in question was ever a *depressed loner* (A), *shy scientist* (B), or *bitter malcontent* (D), but it only makes sense that he was once a *gifted youngster* (C). The term "prodigy" refers to a person with natural ability, often at a young age.

10. **C** Remember to look to contrast keywords to help determine an author's main point. Here, the passage makes a distinction between the work of typical directors and that of Hitchcock: Hitchcock, unlike the others, dared to make films that were unconventional, which is answer choice (C). While answer choices (A) and (B) are supported by the passage, they are too narrow to encompass the main point. Answer choice (D) is out of scope and is not stated or implied in the passage.

PARAGRAPH COMPREHENSION PRACTICE SET 2

Answers and Explanations

1. **D** While it may be true that the bassoon is larger than the oboe, that claim is not supported by information in the passage, so (A) is out. The passage does state that the bassoon produces a deep sound, but claiming that it is the deepest-sounding wood instrument is too extreme, so (B) is out. Nothing in the passage supports the claim that the bassoon is the only instrument that utilizes a bocal, which means (C) is incorrect. The passage does state that a bassoon's pitch can be altered by adjusting the bocal, so (D) is correct.

2. **B** In this Detail question that asks about a process, refer back to the passage to find the correct answer. You will find the information in the second, third, and fourth sentences. The second sentence states that an investor sells shares she does not currently own, while the third and fourth sentences indicate that she then purchases those shares at a later date. That sequence is described in answer choice (B). Both (A) and (C) are incorrect because a short seller does not first buy shares of stock. (D) is out because although a short seller does first sell shares of stock, nothing in the passage indicates that that money is then used to purchase stock in another company.

3. **A** Always make a prediction in Vocabulary-in-Context questions. The tone of this passage is one of peacefulness and calm. Therefore, you could infer that Mark's quiet, relaxing walk in the woods is a break, or a period of *relief*, from the clamor and chaos of the city. While (B) *opposite* might sound appealing, since the passage contrasts the forest with the city, *opposite* doesn't make sense if inserted into the sentence. Mark might be in a (C) *valley*, but that also doesn't make sense if inserted into the sentence. Finally, (D) *hope* doesn't match the idea that Mark has simply found a quiet place away from the city.

4. **C** This statement is a specific detail taken from the second sentence of the passage, "the Main Street bus line was reduced from one bus every 15 minutes to one bus an hour." There is no evidence that the Main Street bus line has ever been completely shut down, (D). The author does mention not being given any notification, but he does not state choice (A), that the city needs to notify customers. Choice (B) might be an idea the author would agree with, but it is not mentioned in the passage.

5. **C** The author introduces the European hedgehog as "a species" of hedgehog in the first sentence, comparing it to other species. The passage then provides information about the European hedgehog's eating and hibernation habits, as well as the males' aggressive behavior. Choice (C) captures the paragraph's focus and is the correct answer to this Global question. Choices (A), (B), and (D) all mention facts given in the passage but miss the big picture by narrowly focusing on only one idea.

6. **C** The question asks for an inference that can be drawn from the passage but provides no clues about the nature of the inference. Each answer choice must be compared with the information given to identify the one choice that must be true based on the passage. The correct answer is supported by the fifth sentence, "One lasting result was an increase in pay for doctors, encouraging many to enter the profession." If many people were encouraged to become doctors because of higher pay, we can infer that lower pay before World War I was a deterrent to entering the profession. The word "lasting" implies that higher pay is still motivating more people to become doctors today. Choice (A) is extreme; even though the flu brought about advances, you cannot infer that advances would not have happened anyway. Choice (B) is unsupported by any facts presented. Choice (D) distorts the facts given. Doctors were encouraged by higher pay, not drafted. Also, there is no evidence that medical care was overwhelmed by the unprecedented numbers of patients. In fact, the patients caused a "boom" in the field.

7. **A** This Vocabulary-in-Context question asks about the author's use of the word *unprecedented*. The author uses the word to describe the increase in the number of patients, which had a lasting impact on the medical profession. To have such a significant effect, the number of patients must have been unusual, or *extraordinary*. Choice (B), *tragic*, might be tempting because of the number of people who died. But this meaning does not fit the point of the sentence. Choice (C) can be eliminated because the text does not suggest that the number of patients was *perplexing*, or puzzling. Choice (D), *advantageous*, would suggest that the author thinks the number of patients was beneficial, which is not the author's point at all. The change in the medical profession may or may not have been *advantageous*; the author gives no opinion.

8. **D** This Global question asks for the author's main idea. The word "Yet" in the second sentence signals the author's key point: authorities were unable to establish the cause of the fire. The final sentence reinforces this point. The rest of the text discusses the investigation of the fire and speculation about its causes. Thus, choice (D) is correct. Choice (A) is never stated or implied in the passage. Choice (B) contradicts the quote from the report, which states that whether the cause was accidental or intentional is unknown. Choice (C) is a distortion; "Peg Leg" Sullivan was considered as a suspect, but no likelihood of his guilt is proposed.

9. **B** The correct answer to this Inference question will reflect an opinion given by the author in the text. The first sentence points out a perceived problem with the data supporting one of Freud's theories. The author then states an opinion in the second sentence that this weakness in the data is not enough to dismiss Freud's ideas. The rest of the paragraph supports this claim, ending with the opinion that Freud's contributions to the field of psychology still stand up. Choice (B) sums up the author's opinion. Choice (A) brings up an idea presented, but the contrast word "While" in the passage indicates that this is not the author's opinion. Choice (C) brings up the problem with Freud's evidence mentioned in the first sentence. This, however, is the point that the author dismisses in the second sentence. Choice (D) distorts the author's defense of Freud; the word "confirmed" is extreme and does not match the author's concession that some of Freud's conclusions have been questioned.

10. **B** This Detail question asks about something specifically mentioned in the passage. Axions are mentioned in the fourth sentence, which begins "There are a wide variety of explanations for why this might be . . ." to introduce one proposed explanation for the unexpected scarcity of lithium in the universe. Choices (A) and (C) both refer to the creation of various elements, and the author does not mention axions in relation to how the elements came into being. Choice (D) distorts the idea of axions by joining it with another explanation, set apart in the next sentence by the words "Others believe."

Review and Reflect

So, how did you do? What was challenging about these problems? Did you notice that you were performing better on some question types than others?

Going back over these problems may be helpful. As you review your performance, ask yourself a few questions:

- Were you able to identify different Paragraph Comprehension question types?
- Do you have a strategy for each of the different question types?
- Are you reading Paragraph Comprehension passages strategically?
- Can you separate background information, supporting details, and main points?
- Do you make accurate predictions?
- Are you able to match your prediction to the correct answer?
- Can you identify and eliminate common wrong answer types?

In future Paragraph Comprehension questions, apply what you've learned in this chapter. Approach each question with Kaplan's 4-step method, and you'll see your performance in this section continue to improve.

Want more instruction and practice with Paragraph Comprehension? Check out the companion video for this section. Also, try more questions in the Qbank online.

COMPANION VIDEO

kaptest.com/login

CHAPTER 5

MATH STRATEGIES FOR THE ASVAB

KNOW WHAT TO EXPECT

Often the most efficient way to get the correct answer to a math problem is to use a strategy rather than just to "do the math." That's especially true on the ASVAB, since you cannot use a calculator on the exam. On some math tests, if you get the wrong answer but show that you set up at least some of the math correctly, you'll still get partial credit. On the ASVAB, you only get credit for a right answer. One advantage of this is that you will get credit for the correct answer regardless of how you get to it.

The two math sections on the ASVAB are called "Arithmetic Reasoning" and "Mathematics Knowledge." Together they form the quantitative half of the Armed Forces Qualifying Test (AFQT), so you'll want to do well on these sections no matter what your ultimate vocational aim in the military.

The **Arithmetic Reasoning** (AR) section tests your ability to handle arithmetic word problems. The CAT version of this section gives you 55 minutes to answer 15 questions, while the paper-and-pencil version gives you 36 minutes for 30 questions. This section is designed to measure your ability to apply reasoning to solve problems involving common math concepts. Many of the questions will be in the format of word problems. Typical AR topics involve number properties, rates, percentages, ratios, proportions, averages, and unit conversions.

The **Mathematics Knowledge** (MK) section tests your understanding of a wide range of concepts in applied arithmetic, algebra, and geometry. The CAT version gives you 23 minutes to answer 15 questions; the paper-and-pencil version gives you 24 minutes to answer 25 questions. This section is designed to measure general mathematical knowledge. You may see the occasional word problem on the MK section of the ASVAB, but in general the questions are more direct than the word problems found on the AR section. For this reason, you are given less time per question on this section than on the AR section.

THE KAPLAN METHOD FOR ASVAB MATH QUESTIONS

LEARNING OBJECTIVES

In this section, you will learn to:

- identify your task on math problems
- determine the most efficient strategy for getting the correct answer
- apply the Kaplan Method for ASVAB Math Questions

Working quickly and efficiently is essential to maximizing your score on these sections. To accomplish this, use the Kaplan Method for ASVAB Math Questions.

THE KAPLAN METHOD FOR ASVAB MATH QUESTIONS

Step 1: Analyze the information given.

Step 2: Identify what you are being asked for.

Step 3: Solve strategically.

Step 4: Confirm your answer.

Step 1: Analyze the information given.

Read the entire question carefully *before* you start solving the problem. If you don't read the question carefully, you may make a careless mistake or overlook the simplest approach to answering the question.

Step 2: Identify what you are being asked for.

Before you choose your approach, make sure you know what you're solving for. In other words, what does the correct answer choice represent? This is an important step to keep you from falling for tempting wrong answer choices. For example, if you are given an equation with two variables, *x* and *y*, identify whether you are solving for *x*, for *y*, or for something else. This step is important because the ASVAB may give you wrong answer choices that represent the "right answer to the wrong question." That is, if you are asked to solve for *x*, one wrong answer choice might represent the value of *y*.

Step 3: Solve strategically.

Once you understand what the question is asking for, it's time to look for the most strategic approach. Use your analysis from Steps 1 and 2 to find the most efficient route to the correct answer. This step might involve performing calculations (that is, "doing the math"), or it might be the case that applying a strategy would get you to the correct answer more quickly. This chapter will discuss these strategic approaches.

Step 4: Confirm your answer.

Reread the question after you select your answer. Make sure you've answered the question asked. If you notice that you missed something earlier, rework the problem and change your answer if necessary.

Here's an example of the Kaplan Method for ASVAB Math in action.

Question	Analysis
One bag contains 6 pieces of candy, another bag contains 8 pieces, and a third bag contains 16 pieces. What is the average number of pieces of candy in a single bag?	**Step 1:** The question tells you the number of pieces of candy in three bags.
	Step 2: Your task is to calculate the average number of candies in one bag.
(A) 6 (B) 8 (C) 10 (D) 30	**Step 3:** Use the average formula: $\frac{\text{Sum of terms}}{\text{Number of terms}} = \text{Average}$ $\frac{6+8+16}{3} =$ $\frac{30}{3} = 10$ The average is 10, answer choice **(C)**.
	Step 4: Don't rework the math from scratch; rather, ask yourself whether the answer makes sense given what you were asked for. Does 10 seem like a likely average given the numbers 6, 8, and 16? Yes: Since the average of a set of numbers has a value that's between the smallest and largest numbers in the set, 10 makes sense as an answer.

STRATEGIES FOR SOLVING MATH PROBLEMS

ASVAB MATH STRATEGIES

- Backsolving
- Picking Numbers
- Strategic Guessing Using Logic
- Combination of approaches

Several methods are extremely useful when you don't know—or don't have time to use—the textbook approach to solving the question. In addition, performing all the calculations called for in the question can often be more time-consuming than using a strategic approach and can increase the potential for mistakes.

Two problem-solving strategies that may be new to you are **Backsolving** and **Picking Numbers**. These strategies are a great way to make confusing problems more concrete. If you know how to apply these strategies, you'll nail the correct answer every time you use them.

Strategic Guessing Using Logic and using a **combination of approaches** are other useful shortcuts to getting more correct answers more quickly. Remember, you get points for correct answers, not for how you got those answers, so efficiency is key to maximizing your score. This section will discuss each of these strategies in turn.

Backsolving

Sometimes it's easiest to work backward from the answer choices. Since many Arithmetic Reasoning questions are word problems with numbers in the answer choices, you can often use this to your advantage by using **Backsolving**. After all, the test gives you the correct answer—it's just mixed in with the wrong answer choices. If you try an answer choice in the question and it fits with the information given, then you've got the right answer.

Here's how it works. When the answer choices are numbers, you can expect them to be arranged from small to large (or occasionally from large to small). Start by trying either choice (B) or (C). If that number works when you plug it into the problem, you've found the correct answer. If it doesn't work, you can usually figure out whether to try a larger or smaller answer choice next. Even better, if you deduce that you need a smaller (or larger) number, and only one such smaller (or larger) number appears among the answer choices, that choice must be correct. You do not have to try that answer choice: simply select it and move on to the next question.

By backsolving strategically this way, you won't have to try out more than two answer choices before you zero in on the correct answer. To see an example of Backsolving, check out the following problem and explanation.

Question	Analysis
An appliance store reduced the price of a refrigerator by 20% and then raised the price by 10% from the lower price. What was the original price of the refrigerator, if the final price was $70.40? Answer choices:	**Step 1:** The price of a refrigerator is reduced 20% and then that reduced price is raised 10%. The final price is $70.40.
	Step 2: The correct answer represents the original price, before the changes.

Question	Analysis
(A) \$50 (B) \$70 (C) \$80 (D) \$100	**Step 3:** To answer this question using algebra would be complex and time-consuming and would afford many opportunities for errors. Instead, since all the answer choices are numbers, backsolve. Start by trying out (B) \$70. \$70 reduced by 20%: \$70 – \$14 = \$56 \$56 raised by 10%; \$56 + \$5.60 = \$61.60. That's lower than the final price of \$70.40, so choice (B) is too low. Eliminate both answer choices (B) and (A). Now try either (C) or (D). (D) is easier for a percent problem. Reduce \$100 by 20%: \$100 – \$20 = \$80 Raise that \$80 by 10%: \$80 + \$8 = \$88. That final price is far too high, so **(C)** must be the correct answer.
	Step 4: The answer of \$80 is the only one that is neither too large nor too small to yield the final price of \$70.40 specified in the question. Done.

In Backsolving, when you start with (B) or (C) and that answer doesn't work, you'll usually know which direction to go. For example, if the answer choices are listed smallest to largest and (B) is too large when you plug it in, you will know that (A) is the correct answer. If, on the other hand, (B) had been too small, you would know that the answer was (C) or (D).

Now, try this one on your own. Use Backsolving.

(Hint: The hypotenuse (or longest side) of a right triangle is equal to the square root of the sum of the squares of the other two sides. In other words, if the other two sides are a and b and the hypotenuse is c, then $a^2 + b^2 = c^2$.)

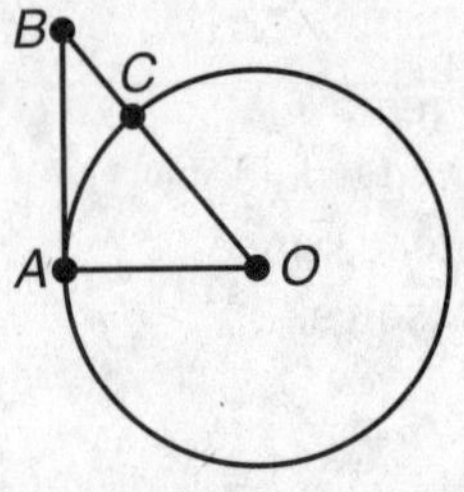

In the figure above, the circle with center O has a radius of 6 (in other words, $AO = 6$). If $AB = 10$ and $\angle OAB$ is a right angle, what is the length of BC?

(A) $2(\sqrt{17} - 3)$

(B) 4

(C) $2(\sqrt{34} - 3)$

(D) 6

Explanation

Step 1: You are told that the length of the circle's radius is 6, and you are given the length of one leg of a right triangle that is partly embedded in the circle. Because the other leg of the triangle, *OA*, is a radius of the circle, you know it has a length of 6. Also, a portion of the hypotenuse of the triangle, *OC*, is another a radius of the circle and, likewise, has a length of 6.

Step 2: Your task is to determine the part of the hypotenuse outside the circle, or the entire hypotenuse minus 6.

Step 3: You could use the Pythagorean theorem here: $a^2 + b^2 = c^2$, where a and b are the lengths of the legs of a right triangle and c is the length of the hypotenuse. However, a look at the answer choices reveals that you might have to do some really burdensome calculations.

Instead, backsolve. Choosing between answer choices (B) or (C), start with (B) because 4 is much easier to work with than $2(\sqrt{34} - 3)$. If $BC = 4$, the hypotenuse would equal $4 + 6 = 10$. Since one of the legs has that same length, and the hypotenuse is always the longest side of a right triangle, this is too short. Eliminate (A) and (B).

Next, try (D) 6. If $BC = 6$, the hypotenuse would equal $6 + 6 = 12$. Using the Pythagorean theorem, test to see if $6^2 + 10^2$ equals 12^2. In fact, $36 + 100$ does not equal 144, so the correct answer must be **(C)**.

Step 4: Because the other answer choices cannot be correct, choice (C) must represent the length of BC.

Picking Numbers

Another strategy that comes in handy on many Mathematics Knowledge questions and also on some Arithmetic Reasoning questions is **Picking Numbers**. Just because the question contains numbers in the answer choices, that doesn't mean that you can always backsolve. There may be numbers in the answer choices, but sometimes you won't have enough information in the question to easily match up an answer choice to a specific value in the question stem. For example, a problem might present an equation with many variables, or it might give you information about percentages of some unknown quantity and ask you for another percent. If the test maker hasn't provided you with a quantity that would be really helpful to have in order to solve the problem, you may be able to simply pick a value to assign to that unknown. The other case in which you can pick numbers is when there are variables in the answer choices.

When you are picking numbers, be sure that the numbers you select are permissible (follow the rules of the problem) and manageable (easy to work with). In general, it's a good idea to avoid picking −1, 0, or 1 because they have unique number properties that can skew your results.

Here's a great example showing how Picking Numbers can make an abstract problem concrete.

Question	Analysis
When n is divided by 14, the remainder is 9. What is the remainder when n is divided by 7?	**Step 1:** An unknown number, n, is 9 larger than a multiple of 14.
	Step 2: The correct answer represents the remainder when n is divided by 7.
(A) 1 (B) 2 (C) 3 (D) 4	**Step 3:** To make this abstract question concrete, pick a number for n that leaves a remainder of 9 when divided by 14. The most manageable number to pick that is also permissible in the problem is $n = 23$ (because $14 + 9 = 23$). Now try out your number: $23 \div 7 = 3$ with a remainder of 2. **(B)** is the correct answer.
	Step 4: Briefly look back over the math to check that you are solving for the correct value. Done.

Now, try this one on your own using the Picking Numbers strategy.

If a bicyclist in motion increases his speed by 30 percent and then increases this speed by 10 percent, what percent of the original speed is the total increase in speed?

(A) 10%
(B) 40%
(C) 43%
(D) 140%

Explanation

Step 1: The question gives information about two increases to the speed of a bicyclist.

Step 2: The question asks for the total percent increase.

Step 3: You are not told how fast the bicyclist is going when he starts his trip, so this is a great opportunity for Picking Numbers. The easiest number to use in percent problems is 100, so start the bicyclist at a speed of 100 miles per hour. Remember, the numbers you pick should be permissible and manageable, but they do not have to be realistic.

The first increase in speed is 30%. Because 30% of 100 is 30, the new speed is $100 + 30 = 130$. The second increase in speed is 10%, but not 10% of the original speed—10% of the new speed. Because 10% of 130 is 13, the final speed is $130 + 13 = 143$. Subtract the original speed from the final speed, and the increase in speed is $143 - 100 = 43$. Since 100 was the initial number, 43 is simply 43% of the original, answer choice **(C)**.

Step 4: When confirming the answer for questions that ask about multiple percent changes, make sure that you calculated the percent changes appropriately and didn't simply add the percentages given in the question, as (B) does. Also, check to make sure that you solved for the correct value, here the *increase* in speed as a percentage of the *original speed*. Answer choice (D) makes the same mistake as (B) but may look tempting to some test takers, because it is close to the final speed of 143.

When there are variables in the problem and in the answer choices, you can pick numbers for those variables. Evaluate the expression in the question stem using your chosen numbers and then evaluate each answer choice using the same numbers. Your goal is to find the answer that yields the same numerical result as the one you calculated using your chosen numbers. When you use this method, you must evaluate *all* of the answer choices. If more than one yields the same numerical result, choose a different set of numbers to evaluate only the remaining choices that gave matching solutions with the first set of numbers that you chose.

To solve problems containing variables in the question stem and answer choices using Picking Numbers, start by picking permissible and manageable numbers for the variables. Answer the question using the numbers you've picked. This answer is your target number. Then, substitute the numbers you picked for the variables into the answer choices. You are looking for the answer choice that gives you the target number.

Take a look at this example.

Question	Analysis
Camilla spent d dollars on groceries each week for w weeks, and p percent of the amount she spent on groceries was spent on fresh vegetables. How much money did she spend over the whole time period on groceries other than fresh vegetables?	**Step 1:** The question contains no numbers; only variables: $d =$ the amount Camilla spent on groceries each week $w =$ the number of weeks she shopped $p =$ the percent of that money that went to fresh vegetables.
	Step 2: The correct answer represents the amount of grocery money not spent on fresh vegetables.

Question	Analysis
(A) $dw\left(1-\dfrac{p}{100}\right)$ (B) $\dfrac{dwp}{100}$ (C) $100p(1-dw)$ (D) $\dfrac{100p}{dw}$	**Step 3:** Make this problem concrete by replacing all the variables with numbers. Pick manageable numbers: $d = 20$ $w = 5$ $p = 10$ With those numbers, Camilla spent \$20 a week on groceries for 5 weeks, for a total of \$100. 10% of that, or \$10, was on fresh veggies. This means that \$100 − \$10 = \$90 was on other groceries. \$90 is the target number. See what number results from plugging the same values into choice (A): $dw\left(1-\dfrac{p}{100}\right) = (20 \times 5)\left(1-\dfrac{10}{100}\right) = 100 \times (1-0.1)$ $= 100 \times (0.9) = 90.$ That matches! However, because you chose numbers for the variables, more than one answer might match the target number. Keep testing the other answer choices. (B) $\dfrac{dwp}{100} = \dfrac{20 \times 5 \times 10}{100} = \dfrac{1{,}000}{100} = 10.$ Incorrect. (C) $100p(1-dw) = 100 \times 10(1-20 \times 5) = 10.$ This will be negative, which is incorrect. (D) $\dfrac{100p}{dw} = \dfrac{100 \times 10}{20 \times 5} = 10.$ Incorrect. Only one answer choice, **(A)**, matches the target number, so it is correct.
	Step 4: Briefly look back over the math you did when checking choice (A). If you plugged in the numbers correctly, you're done.

Try this one on your own. Use the Picking Numbers strategy.

For all r, s, t, and u, what does $r(t+u) - s(t+u)$ equal?

(A) $(r+s)(t+u)$
(B) $(r-s)(t-u)$
(C) $(r+s)(t-u)$
(D) $(r-s)(t+u)$

Explanation

Step 1: You are given an algebraic expression and asked to find an equivalent expression.

Step 2: The correct answer simplifies the expression $r(t + u) - s(t + u)$.

Step 3: Since you are given no values for any of the four variables, you can pick numbers for each of them. Some good numbers to pick here are $r = 5$, $s = 4$, $t = 3$, and $u = 2$. You can, however, use any permissible and manageable numbers you wish.

Replacing the variables in the expressions with the numbers picked, you get:

$5(3 + 2) - 4(3 + 2) = 25 - 20 = 5.$

Then replace the variables in each answer choice to see which choice gives the target number of 5:

(A) $(r + s)\ (t + u)$	$(5 + 4)(3 + 2) = (9)(5) = 45$	Incorrect.
(B) $(r - s)\ (t - u)$	$(5 - 4)(3 - 2) = (1)(1) = 1$	Incorrect.
(C) $(r + s)\ (t - u)$	$(5 + 4)(3 - 2) = (9)(1) = 9$	Incorrect.
(D) $(r - s)\ (t + u)$	$(5 - 4)(3 + 2) = (1)(5) = 5$	Correct!

Step 4: In this example, only choice **(D)** works. If more than one choice had worked, you would need to pick another set of numbers and try only those answer choices again.

Not only does Picking Numbers make some problems easier to understand, but also you can be sure you got the right answer because you've already proven the answer works with real numbers.

Strategic Guessing Using Logic

Sometimes, you can determine the characteristics of a correct answer without doing a lot of calculations. Study the example below.

Question	Analysis
After eating 25 percent of the pretzels, Sonya had 42 left. How many pretzels did Sonya have originally?	**Step 1:** Original # of pretzels − 25% of pretzels = 42.
	Step 2: The correct answer represents the original number of pretzels, before Sonya ate any.
(A) 50 (B) 54 (C) 56 (D) 58	**Step 3:** Because 25% is the same as one-quarter, the correct answer must be divisible by 4 with no remainder. If the number of pretzels Sonia started with was not divisible by 4, and she then ate $\frac{1}{4}$ of the pretzels, she'd be left with fractions of pretzels left over. Of the answer choices, only **(C)** 56 is evenly divisible by 4, so it has to be the correct answer.
	Step 4: One-quarter of 56 is 14, and $56 - 14 = 42$, the number of pretzels Sonya had left.

Now, try this one on your own. Try to avoid doing extensive calculations; rather, see if you can eliminate some answer choices based on logic.

Mark can paint a room in 3 hours, and Kevin can paint an identical room in 4 hours. How many hours would it take Mark and Kevin to paint the room if they work together at their respective rates?

(A) $\frac{3}{2}$

(B) $\frac{12}{7}$

(C) $\frac{10}{3}$

(D) 4

Explanation

Step 1: You are told the rates at which two people can paint a room separately.

Step 2: The correct answer represents how long it would take the two people to paint the room together, each working at the same rate at which he works alone.

Step 3: If Mark, the faster painter, works completely on his own, he will complete the task in 3 hours. With help, he will take even less time to complete the job, so you can eliminate answer choices (C) and (D), which are greater than 3. Since Kevin is slower than Mark, working together will take longer than half of 3 hours. Since (A) is half of 3, you can eliminate it. That leaves only answer choice **(B)**, which must be correct.

Step 4: Check to make sure that the question asked for how long it took to paint just one room. It did, so (B) is correct.

Combination of Approaches

There is no rule that says you have to use just one approach to get the correct answer. Study the example below.

Question	Analysis
Youssef can either walk from his home to his workplace or ride his bicycle. He walks at a pace of 1 block per minute, but he can travel 1 block in 20 seconds on his bicycle. If it takes Youssef 10 minutes longer to walk to work than to ride his bike, how many blocks away from work does he live?	**Step 1:** The question involves two different units of time. In order to be able to compare the rates, first convert Youssef's biking rate into minutes. Since a minute is 60 seconds, he bikes to work at a rate of $\frac{20}{60}$, or $\frac{1}{3}$, minute per block. He walks to work at 1 minute per block. Time to walk − time to bike = 10 minutes.
	Step 2: The correct answer represents the number of blocks Youssef lives from his workplace.

Question	Analysis
(A) 5 (B) 10 (C) 15 (D) 20	**Step 3:** This problem could be solved with algebra, but using some logic mixed with Backsolving can get the correct answer more quickly. If Youssef walks 1 block per minute and it takes him 10 minutes longer to walk than to ride his bike, the correct answer must be greater than 10 blocks. Otherwise his bike time would be zero or less. That leaves only two choices, 15 or 20, so pick one and see if it works. If Youssef lived 15 blocks from work, walking to work would take him 15 minutes. Riding his bike at the rate of 20 seconds per block, he can travel 3 blocks every minute, so he could ride to work in $15 \div 3 = 5$ minutes. Then 15 minutes − 5 minutes = 10 minutes longer to walk, so 15 is the correct answer. (Had you chosen to start with 20 instead, you could quickly have determined that 20 is incorrect, leaving 15 as the only possible correct answer.)
	Step 4: You tested the correct answer based on the information given in the question, and it worked. **(C)** is correct.

Now, try this one on your own. A combination of strategies may be helpful on this problem.

Shoshanna bought a new cell phone, cell phone case, and wall charger. The cell phone cost $149.99, the case cost $19.99, and the wall charger cost $29.99. If tax on each of these items was 9.5%, which of the following is closest to the total amount Shoshanna spent?

(A) $180
(B) $200
(C) $205
(D) $220

Explanation

Step 1: You are told the prices of three items that Shoshanna purchased, plus a sales tax rate.

Step 2: You are asked to find the answer that is closest to the total amount Shoshanna paid for her purchases.

Step 3: Because you are asked for the *closest* amount, you do not have to use an exact calculation to find the correct answer. Estimation will work well. Round the prices for each item before tax to the nearest dollar: Shoshanna spent $150 + $20 + $30 = $200 before tax. You can eliminate answer choices (A) and (B) because they are too small. Now round the tax rate to 10%. Because 10% of $200 is $20, Shoshanna spent about $200 + $20 = $220 in total. Answer choice **(D)** is correct.

Step 4: Check that you did the arithmetic correctly and that the answer makes sense. All the other answer choices are too low to be close to what she spent.

Sometimes, you'll find that you have to make a guess, but don't guess at random. Narrow down the answer choices to increase your odds of guessing the correct one. First, eliminate answer choices you know are wrong. Next, avoid answer choices that don't make logical sense. Finally, choose one of the remaining answer choices.

MATH STRATEGIES FOR THE ASVAB PRACTICE SET

On the ASVAB, either the question itself or the answer choices will signal which strategy will be most helpful. Look at the partial questions or answer choices below (note that complete problems are not shown). Then determine which strategy (Picking Numbers, Backsolving, Strategic Guessing Using Logic, or a combination of approaches) is likely to be useful.

1. An airplane uses 79% of a tank of fuel to fly 1,496 miles. If a full tank holds 996 gallons of fuel, how many gallons would the plane use to fly 3,016 miles?

   Which strategy is likely to be useful for this question, and why?

   ____

   ____

2. (A) 80
   (B) 100
   (C) 120
   (D) 150

   Which strategy is likely to be useful for this question, and why?

   ____

   ____

3. (A) $\frac{p(b + d)}{c}$
   (B) $pb + dc$
   (C) $\frac{d + p}{c + b}$
   (D) $\frac{c + b}{d + p}$

   Which strategy is likely to be useful for this question, and why?

   ____

   ____

4. (A) $\frac{1}{625}$
   (B) $\frac{26}{125}$
   (C) $\frac{64}{125}$
   (D) $\frac{523}{625}$

   Which strategy is likely to be useful for this question, and why?

   ____

   ____

5. If $|4x + 6| = 18$, what is one possible value for x?

   Which strategy is likely to be useful for this question, and why?

   ____

   ____

6. Jim spent 30% of his paycheck on rent, and he spent 40% of the remainder on car repairs. What percent of his paycheck was left after his rent and car repairs?

   Which strategy is likely to be useful for this question, and why?

   ____

   ____

MATH STRATEGIES FOR THE ASVAB PRACTICE SET

Answers and Explanations

1. **Combination of Approaches** This problem includes some unwieldy numbers, but rounding them for estimation should get you an approximate answer. Find 80% of 1,000 gallons to determine how many gallons are used for the first trip ($80\% \times 1{,}000 = 800$ gallons). Then, round 1,496 to 1,500 and round 3,016 to 3,000. Notice that the question really asks, "How many gallons of fuel would the airplane use to fly about twice as far as the first trip?" Just multiply $800 \times 2 = 1{,}600$ gallons and find the answer choice closest to 1,600. In other words, combining a little calculation with some estimation works well here.

2. **Backsolving** Even without a question stem, you can see that all the answer choices are numbers. That makes this an excellent candidate for Backsolving. Start by trying (B) or (C).

3. **Picking Numbers** There are variables in the answer choices. Picking Numbers will make the problem more concrete, and plugging the numbers you picked into the answer choices will allow you to find your target number.

4. **Strategic Guessing Using Logic** Although these fractions might look intimidating at first, closer examination will show that they are quite spread out. (A) is very tiny, (B) is about $\frac{1}{5}$, (C) is about $\frac{1}{2}$, and (D) is close to $\frac{5}{6}$. Depending on the question asked, you can probably get a rough idea of which answer choice is the right answer without doing a lot of complex calculating.

5. **Backsolving** Quickly plug answer choices into the equation to eliminate choices that are too large or too small.

6. **Picking Numbers** A problem with percents and no amount given for Jim's paycheck makes this a perfect opportunity for Picking Numbers. Use $100 for the amount of Jim's paycheck and then apply the percentages in the problem. He spends 30% of $100, or $30, on rent. This leaves him $70, of which he spends 40% on his car, or $40\% \times \$70 = \$28$. Subtract $30 for rent and $28 for the car repairs from his $100 paycheck, and he has $42 left. Since $100 represents the full paycheck, the dollar amount left is the same as the percent left, 42%.

Review and Reflect

How did you do? If you weren't sure which strategy to apply to some of the partial questions in the practice set, you might want to reread this chapter. Also, return to it from time to time throughout your ASVAB math practice to refresh your memory of the various strategies available to you.

The rest of the math section of this book will deal with math content review. Even if you feel comfortable with a particular subject, make sure to do the practice sets. There's no harm in practicing extra problems. Throughout the math section, you will be given many opportunities to "do the math." However, every time you begin a question, first check if you can use a strategy to minimize your time and effort toward getting the correct answer. This will allow you to move through the section more efficiently.

Want more instruction and practice with Math Strategies? Check out the companion video for this section. Also, try more questions in the Qbank online.

COMPANION VIDEO

kaptest.com/login

CHAPTER 6

ARITHMETIC REASONING

KNOW WHAT TO EXPECT

For the Arithmetic Reasoning (AR) section of the paper-and-pencil ASVAB, you will have 36 minutes to complete 30 arithmetic word problems. For the CAT-ASVAB, you will have 42 minutes to complete 15 problems. This chapter will review the concepts you need to understand in order to succeed on the Arithmetic Reasoning test. As you work through this chapter, don't forget that many problems are best solved by using the strategies you read about in chapter 5.

The principles covered in this chapter will also serve you well as the building blocks for Mathematics Knowledge (MK), which will be the focus of chapter 7. Both the AR and MK subtests are particularly important because they are part of the AFQT score. Additionally, high AR and MK scores are required for you to be eligible to enter many of the specialized occupations in the military.

This chapter will take you through a review of arithmetic concepts including:

- key arithmetic definitions
- number properties
- absolute value
- primes, factors, multiples, and remainders
- divisibility rules
- fractions, decimals, and scientific notation
- exponents and radicals
- factorials

The chapter will then review applied arithmetic concepts, such as:

- order of operations and calculation rules
- ratios, proportions, and rates
- combined work
- percentages
- averages, means, medians, and modes
- sequences and probability

The chapter will also discuss Arithmetic word problems including:

- translation
- word problems with formulas

The chapter concludes with a practice set.

ARITHMETIC REVIEW

LEARNING OBJECTIVES

In this section, you will learn to:

- use the same definitions of key concepts as the ASVAB uses
- apply the rules that govern number properties
- identify and utilize factors and prime factorization
- perform arithmetic operations with decimals, fractions, exponents, radicals, and factorials
- convert numbers to and from the scientific notation format

Arithmetic Definitions

It is important that you use certain critical arithmetic terms in exactly the same manner as the ASVAB. Familiarize yourself with these key terms and concepts.

Concept	Definition	Examples
Integers	All whole numbers, including zero, and their negative counterparts.	−900, −3, 0, 1, 54
Fractions	A fraction is a number that is written in the form $\frac{A}{B}$, where A is the numerator and B is the denominator.	$-\frac{5}{6}, -\frac{3}{17}, \frac{1}{2}, \frac{899}{901}$
	An improper fraction has a numerator with a greater absolute value than that of the denominator.	$-\frac{65}{64}, \frac{9}{8}, \frac{57}{10}$
	A mixed number consists of a whole number and a fraction.	$-1\frac{1}{64}, 1\frac{1}{8}, 5\frac{7}{10}$
	An improper fraction can be converted to a mixed number and vice versa.	$2\frac{3}{5} = \frac{13}{5}$

Concept	Definition	Examples
Positive/Negative	Numbers greater than 0 are positive numbers; numbers less than 0 are negative numbers. 0 is neither positive nor negative.	Positive: $\frac{7}{8}$, 1, 5.6, 900 Negative: −64, −40, −1.11, $-\frac{6}{13}$
Even/Odd	An even number is an integer that is a multiple of 2. An odd number is an integer that is not a multiple of 2. Fractions and mixed numbers are neither even nor odd.	Even numbers: −8, −2, 0, 12, 188 Odd numbers: −17, −1, 3, 9, 457
Factor	A factor is a positive integer that divides evenly into a given number.	The complete list of factors of 12: 1, 2, 3, 4, 6, 12
Prime number	An integer greater than 1 that has no factors other than 1 and itself. 2 is the only even prime number.	2, 3, 5, 7, 11, 59, 83
Consecutive numbers	Numbers that follow one after another, in order, without skipping any. In a series of consecutive numbers, the differences between any consecutive numbers are equal.	Consecutive integers: 3, 4, 5, 6 Consecutive even integers: 2, 4, 6, 8, 10 Consecutive multiples of −9: −9, −18, −27, −36
Multiple	A multiple of a number is the product of that number and an integer.	Some multiples of 12: 0, 12, 24, 60

Number Properties

Certain number properties follow rules that will never vary. Often, you can use these rules as an effective way to help you solve problems on the ASVAB.

There are only a few things to remember about **positive** and **negative** numbers.

Adding a negative number is the same as subtracting a positive number with the same absolute value.

$6 + (-4)$ is equal to $6 - 4$ which equals 2.

$4 + (-6)$ is equal to $4 - 6$ which equals −2.

Subtracting a negative number is the same as adding a positive number with the same absolute value.

$6 - (-4)$ is equal to $6 + 4$ which equals 10.

$-6 - (-4)$ is equal to $-6 + 4$ which equals −2.

Multiplying and dividing positive and negative numbers is the same as all other multiplication and division, with one catch: you need to figure out whether the solution is positive or negative. To do this, count the number of negative numbers. If you had an odd number of negatives, the answer will be negative. If you started with an even number of negative numbers, the answer will be positive.

$(-6) \times (-4) = 24$ (2 negatives → positive product)

$(-1) \times (-6) \times (-4) = -24$ (3 negatives → negative product)

Similarly,

$-24 \div 6 = -4$ (1 negative → negative quotient)

$-24 \div (-4) = 6$ (2 negatives → positive quotient)

The rules for working with integers are very brief. If you add, subtract, or multiply two integers, the result will always be an integer. However, if you divide two integers, the result *may or may not be* an integer.

$8 \div 4 = 2$ which is an integer, but $8 \div 3 = \frac{8}{3} = 2\frac{2}{3}$ which is *not* an integer.

Working with odd and even numbers can be described with a few rules as well. When you add or subtract two odd numbers or two even numbers, the result will always be an even number.

$6 + 4 = 10$ (even + even = even)

$7 - 3 = 4$ (odd − odd = even)

When you add or subtract an even with an odd, the result will be odd.

$2 + 7 = 9$ (even + odd = odd)

$5 - 4 = 1$ (odd − even = odd)

When multiplying two numbers, the rule for determining whether the result is even or odd is that any integer times an even integer will result in an even number. The only way the product of two integers will be an odd number is if you multiply two odd numbers.

$6 \times 4 = 24$ (even × even = even)

$3 \times 4 = 12$ (odd × even = even)

$3 \times 5 = 15$ (odd × odd = odd)

There are no rules predicting whether the outcome of division will be even or odd because division does not always result in an integer. Sometimes dividing produces a fraction, and fractions are neither odd nor even.

Study how a well-prepared test taker would apply these rules to an AR question.

Question	Analysis
If n is an integer, what is $2(n) + 1$?	**Step 1:** The question tells you that n is an integer.
	Step 2: The question asks for the value of a formula with a variable, n.
(A) an even integer (B) an odd integer (C) an even non-integer (D) possibly either even or odd	**Step 3:** Even though you don't know whether n is odd or even, it is an integer. When any integer is multiplied by an even number the result is also an even number, so the term $2(n)$ must be even. If the odd number 1 is added to an even number, then the result is an odd number. Answer choice **(B)** is correct.
	Step 4: Check that you used number properties correctly throughout the problem. Note that (C) is impossible no matter what the result of the equation because only integers can be odd or even.

Absolute Value

When you see a number or expression bracketed by two vertical lines like this $|-3|$, you are seeing the symbol for **absolute value**. The absolute value of what is between the vertical lines is the *positive magnitude* of the number, regardless of whether it is positive or negative.

If $|x| = 5$, then $x = ?$

Clearly x can equal 5, but x could also equal -5 since absolute value is the positive magnitude of what is between the vertical lines. Therefore $x = +5$ or -5.

If $|x - 5| = 5$, then $x = ?$

The definition of absolute value says that the expression between the lines can be either 5 or -5. Therefore, in order to find the possible values of x you must write two equations:

$x - 5 = 5$ and $x - 5 = -5$

If $x - 5 = 5$, then $x = 10$, but if $x - 5 = -5$, then $x = 0$. This makes the two possible values of x: 10 and 0.

When you encounter an **absolute value** problem on the ASVAB you must solve for **both** the positive and negative values of whatever is between the vertical lines. Test your understanding of the absolute value concept on the following problem:

Question	Analysis
If $\|x - 2\| + 3 = 7$ which of the following is a possible value of x?	**Step 1:** The question gives you a formula with variable x.
	Step 2: The question asks for one possible value of x.

Question	Analysis
(A) 6 (B) −6 (C) 2 (D) 4	**Step 3:** First, subtract 3 from both sides. $\lvert x - 2\rvert + 3 = 7$ becomes $\lvert x - 2\rvert = 4$. Thus, $x - 2$ must equal either $+4$ or -4. If $x - 2 = 4$, then $x = 6$ If $x - 2 = -4$, then $x = -2$. Since -2 is not among the answer choices, **(A)**, 6, is the only correct choice.
	Step 4: Check your math and check to make sure you solved for the proper variable. In this case, the only variable was *x*, but other problems might contain more than one variable and require you to choose correctly.

Factors, Multiples, and Prime Numbers

In order to determine the **factors** of a number, find the pairs of positive integers which can be multiplied together to produce the number that is being factored. Start with 1 and the number itself, which are *always* factors of any integer. Then examine 2, 3, 4 and so on, as shown in this example of finding the factors of 36:

1×36

2×18

3×12

4×9

6×6

Examining the two columns shows that the factors of 36 are 1, 2, 3, 4, 6, 9, 12, 18, and 36. When creating the list of factors, as soon as you reach a factor on the left which is equal to or greater than a factor already listed on the right, you can stop, since you're about to start repeating factor pairs. In this table, you need go no higher than 6 because any number greater than 6 that is a factor of 36 is already in the right column.

Multiples are, in a way, the opposite of factors. For instance, in the example above, 1, 2, 3, 4, 6, 9, 12, and 18 are factors of 36. Conversely, 36 is a multiple of 1, 2, 3, 4, 6, 9, 12, 18, and 36.

If you become confused between the terms *factor* and *multiple*, you can remember that *multiples* are larger because you can produce them by *multiplying*. There are a finite number of factors of any number, but there are an infinite number of multiples.

Prime numbers, by their very definition, cannot be multiples of any integers other than 1 and the prime number itself. However, prime numbers can be factors of other numbers.

$3 \times 13 = 39$ (3 and 13 are prime numbers, but 39 is not because it is a multiple of 3 and a multiple of 13.)

A number's **prime factors** are factors that are prime. The **prime factorization** of a number is the number expressed using multiplication containing only primes, even if some of those primes repeat.

To determine the prime factorization of a number, keep dividing by prime numbers until you can no longer divide because you are left with only prime numbers.

For example, to find the prime factorization of 168:

Since 168 is an even number, the prime number 2 is a factor of 168.

$168 = 84 \times 2$, so **2** is a prime factor of 168.

84 is still an even number equal to 42×2, so **2** is a prime factor for a second time.

42 is still an even number equal to 21×2, so **2** is a prime factor for the third time.

21 is the product of **3** and **7**.

Since these are both prime numbers, the process of finding the prime factors is complete.

An efficient way to determine the prime factors of any number is to use a "tree" as shown below.

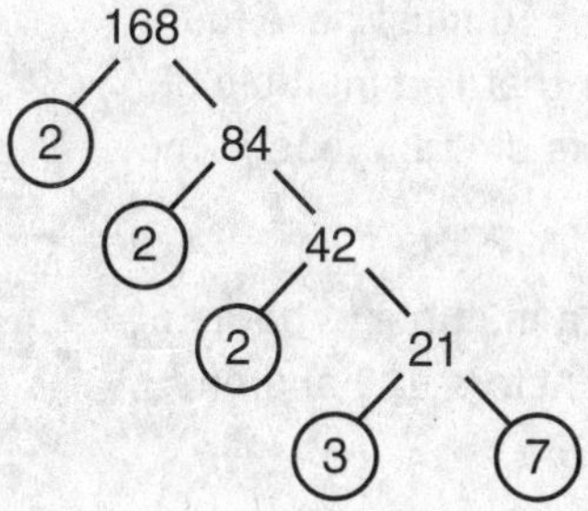

Based on the above operations and tree, the prime factorization of 168 is $2 \times 2 \times 2 \times 3 \times 7$. All the prime factors, including repeated numbers, must be listed because the original number is the product of all of these numbers.

Prime factorization can be a very useful tool to help answer some questions on the ASVAB. For instance, if you need to find the **greatest common factor (GCF)** of two integers, break down both integers into their prime factorizations and multiply all prime factors they have in common, as shown in the example below:

Question	Analysis
Find the greatest common factor of 40 and 140.	**Step 1:** You are given two numbers, 40 and 140.
	Step 2: The question asks for the greatest common factor of the two numbers.
	Step 3: The prime factors of 40 are 2, 2, 2, and 5. The prime factors of 140 are 2, 2, 5, and 7. Both numbers share two 2s and a 5 as prime factors, so the GCF is $2 \times 2 \times 5 =$ **20**.
	Step 4: Double-checking the result, $40 = 20 \times 2$ and $140 = 20 \times 7$.

This can be helpful when you need to reduce a fraction, because you can find the GCF and then divide the top and bottom of the fraction by that number.

If you need to find a **common multiple** of two integers, you can just multiply them. However, in other cases you may need to find the **least common multiple (LCM)**. You can use prime factors to efficiently identify LCMs. Take a look at this example to see how that works.

Question	Analysis
Find the least common multiple of 20 and 16.	**Step 1:** You are given two numbers, 20 and 16.
	Step 2: The question asks for the least common multiple of the two numbers.
	Step 3: First, identify the prime factors of each number: $16 = 2 \times 2 \times 2 \times 2$ $20 = 2 \times 2 \times 5$ Any integer that is a multiple of 16 must have four prime factors of 2. Any integer that is a multiple of 20 must have two prime factors of 2 and one prime factor of 5. Therefore, any number that is a multiple of both 16 and 20 must have four prime factors of 2 and one prime factor of 5. The least common multiple equals $2 \times 2 \times 2 \times 2 \times 5 =$ **80.**
	Step 4: Double-check that all of the prime factors for each number were included in the least common multiple.

Practice the technique on this question:

Marcus needs to order some parts for his machine shop. His supplier sometimes fills orders with boxes that contain 15 of these parts and sometimes with boxes of 18, depending upon what he happens to have in stock. The supplier will only ship complete boxes, and will ship only one size of box in a given order. What is the minimum quantity of parts that Marcus can order and be assured that he will receive exactly that many parts regardless of which boxes his supplier uses?

(A) 15
(B) 18
(C) 90
(D) 270

Explanation

Step 1: The question gives you a situation in which Marcus is ordering parts. He wants to receive the exact number of parts he needs. You are told that the company ships parts in boxes of 15 and 18.

Step 2: You are asked to find the smallest number of parts that can be filled with full boxes of 15 parts or full boxes of 18 parts. Thus, you need to find the least common multiple of 15 and 18.

Step 3: The prime factors of 18 are $2 \times 3 \times 3$ and the prime factors of 15 are 5×3. Therefore, the least common multiple must have two factors of 3, one factor of 2, and one factor of 5. This is $2 \times 3 \times 3 \times 5 = 90$. Answer choice **(C)** is correct.

Notice that (D), 270, is a common multiple of 15 and 18, but it is not the least common multiple.

Step 4: Confirm that 15 and 18 both divide evenly into 90.

Divisibility

Whether evaluating integers to determine if they are prime numbers, finding factors, or finding prime factors, you can check whether a number is a factor of another number more quickly and effectively if you use the following tips to check for divisibility.

Divisibility Rules

Number	Divisibility Rule
2	All even numbers are divisible by 2.
3	Add up the individual digits of the number. If the total is divisible by 3, then the number itself is divisible by 3; for example, 243 is divisible by 3 because the sum of its digits is $2 + 4 + 3 = 9$, but 367 is not because the sum of its digits is $3 + 6 + 7 = 16$ and 16 is not a multiple of 3.
4	Take the last two digits and divide them by 2. If the result is even, the number is divisible by 4. If the result is odd, then the number is not divisible by 4.
5	All numbers ending in 5 or 0 are divisible by 5.
6	All even numbers that meet the test for divisibility by 3 are divisible by 6.
8	Divide the number by 2 twice; if the result is even, then the number is divisible by 8.
9	Add up the digits of the number; if the total is divisible by 9, then the number is divisible by 9.

Practice divisibility rules on this question:

Question	Analysis
What are all the prime numbers between 60 and 70 inclusive?	**Step 1:** You are given a range of numbers.
	Step 2: The question asks for the number of primes within a given range.

Question	Analysis
(A) 61, 67 (B) 61, 63, 67 (C) 61, 63, 65, 67 (D) 62, 64, 66, 68	**Step 3:** Use divisibility tests to eliminate numbers that are not prime. Eliminate 60, 62, 64, 66, 68, and 70 because they are all divisible by 2. Eliminate 65 because it is divisible by 5. Checking for divisibility by 3 eliminates 63 (sum of the digits is 9) and 69 (sum of the digits is 15). The only remaining numbers are 61 and 67. Answer choice **(A)** is correct.
	Step 4: All of the other answer choices contain numbers that were eliminated by using divisibility rules.

Fractions and Decimals

Generally, it's a good idea to **reduce fractions** when solving math questions. To do this, use your knowledge of factors to cancel all factors that the numerator and denominator have in common.

$$\frac{28}{36}=\frac{\cancel{4}\times 7}{\cancel{4}\times 9}=\frac{7}{9}$$

When you perform operations on fractions, always remember to perform the same operation on both the numerator and the denominator. For instance, in the example above, both the numerator and denominator were divided by 4.

To **add or subtract fractions**, you must convert all the fractions so that they have a common denominator. Think of the process of converting to a common denominator as finding the least common multiple of all the denominators. Don't forget to multiply the individual numerators by the same number that was used to convert to a common denominator. Once you have converted the individual fractions so that they have common denominators, you merely add or subtract the numerators to obtain your result.

$$\frac{1}{4}+\frac{1}{3}=\frac{1\times 3}{4\times 3}+\frac{1\times 4}{3\times 4}=\frac{3}{12}+\frac{4}{12}=\frac{7}{12}$$

$$\frac{1}{4}-\frac{1}{3}=\frac{3}{12}-\frac{4}{12}=\frac{3-4}{12}=-\frac{1}{12}$$

To **multiply fractions**, multiply the numerators together and then multiply the denominators together.

$$\frac{1}{4}\times\frac{1}{3}=\frac{1\times 1}{4\times 3}=\frac{1}{12}$$

To **divide** one fraction by another you invert (flip) the second fraction and multiply. In other words, multiply the first fraction by the reciprocal of the second fraction.

$$\frac{1}{4} \div \frac{1}{3} = \frac{1}{4} \times \frac{3}{1} = \frac{1 \times 3}{4 \times 1} = \frac{3}{4}$$

Occasionally a question may require that you **compare fractions** to determine which has the greater value. The most straightforward way to do this is to convert the fractions being compared to a common denominator and compare their numerators.

Question	Analysis
Compare $\frac{2}{3}$ and $\frac{5}{8}$.	The lowest common denominator is $3 \times 8 = 24$.
	Multiple the first fraction by $\frac{8}{8}$ and the second fraction by $\frac{3}{3}$.
	$\frac{2}{3} = \frac{2 \times 8}{3 \times 8} = \frac{16}{24}$ $\frac{5}{8} = \frac{5 \times 3}{8 \times 3} = \frac{15}{24}$
	16 is greater than 15, so $\frac{2}{3}$ is greater than $\frac{5}{8}$.

Practice the operations for fractions to answer the following question:

$$\frac{\frac{1}{4} + \frac{3}{5}}{\frac{2}{3}} = ?$$

(A) $\frac{2}{15}$

(B) $\frac{17}{30}$

(C) $1\frac{11}{40}$

(D) $2\frac{1}{40}$

Explanation

Step 1: You are given an equation.

Step 2: You need to solve the equation using fraction rules.

Step 3: Dividing by a fraction is the same as multiplying by its reciprocal, so you can get rid of the cumbersome double-stack fraction by rewriting: $\left(\frac{1}{4} + \frac{3}{5}\right) \times \left(\frac{3}{2}\right) = ?$

You need to add the two fractions in the left parentheses before multiplying. In order to add the two fractions, convert them to have a common denominator of 20. $\left(\frac{5}{20}+\frac{12}{20}\right)\times\left(\frac{3}{2}\right)=?$

Now add the first two fractions. $\left(\frac{17}{20}\right)\times\left(\frac{3}{2}\right)=?$

Multiplying the numerators and denominators results in an answer of $\frac{51}{40}$. But this answer is not in the answer choices as written. For this problem, you must convert an improper fraction to a mixed number. $\frac{40}{40}+\frac{11}{40}=1\frac{11}{40}$. Select answer choice **(C)**.

Step 4: Check to be sure you used the proper order of operations and check your math.

To **convert a fraction to a decimal**, divide the denominator into the numerator.

To convert $\frac{8}{25}$ to a decimal, divide 25 into 8.

$$\frac{8}{25}=0.32$$

To **convert a decimal to a fraction**, first set the decimal as the numerator of a fraction with a denominator of 1. Then, move the decimal over as many places as it takes until it is immediately to the right of the units digit. Count the number of places that you moved the decimal. Then add that many 0s to the 1 in the denominator.

$$0.3=\frac{0.3}{1}=\frac{3.0}{10}=\frac{3}{10}$$

Try this one on your own:

$0.0025 = ?$

(A) $\frac{1}{4{,}000}$

(B) $\frac{1}{400}$

(C) $\frac{1}{40}$

(D) $\frac{1}{25}$

Explanation

Step 1: The question gives you a number in decimal format.

Step 2: The answer choices show you that you will need to convert to a fraction.

Step 3: First, convert to a fraction with a denominator of 1, $\frac{.0025}{1}$. In order to get rid of the decimal in the numerator, you will need to move it four places to the right and simultaneously add four 0s to the denominator. This results in the fraction $\frac{25}{10{,}000}$.

To reduce this fraction to its simplest terms, divide both the numerator and denominator by 5 to get $\frac{5}{2{,}000}$. Since both the numerator and denominator are still divisible by 5, repeat the process to get $\frac{1}{400}$. Select answer choice **(B)**.

Step 4: Since many of the answer choices vary by the number of zeros in the denominator, double-check that you added the appropriate number of zeros to the denominator before you started simplifying.

Decimal Division

One common type of arithmetic problem on the ASVAB involves dividing numbers that contain decimals. Keeping track of the decimal places can complicate a division problem, so get rid of the decimal places before you divide. You can accomplish this by multiplying both numbers by 10 as many times as necessary. When you multiply a number with decimals by 10, simply move the decimal point one place to the right. When multiplying numbers without decimals by ten, add a zero at the end.

As an example,

$1.22 \times 10 = 12.2$
$12.2 \times 10 = 122$
$122 \times 10 = 1{,}220$

When presented with a decimal division question you should write the division problem as a fraction:

$$7.2 \div 0.004 = \frac{7.2}{0.004}$$

Then you will move the decimal points in the denominator and the numerator the same number of places to the right, adding zeros as necessary, until you have completely gotten rid of any decimal points.

$$\frac{7.2}{0.004} = \frac{7{,}200}{4} = 1{,}800$$

The example below shows how to tackle an ASVAB question about decimal operations.

Question	Analysis
David is drawing a blueprint for a model home. Due to the scale he used, the width of the bathroom floor is 0.18 inches on his paper. He needs to draw 30 tiles across that width. What is the width of each individual tile in his sketch in inches?	**Step 1:** You are told that there are 0.18 inches to draw 30 tiles.
	Step 2: The question asks for the measurement of width in inches of each tile as drawn on the paper.
(A) 0.006 (B) 0.06 (C) 0.6 (D) 6	**Step 3:** You need to divide 30 into 0.18. Write the problem as a fraction: $\frac{0.18}{30}$. Eliminate the decimal point by moving it two places to the right and adding 2 zeros to the denominator: $\frac{18}{3{,}000}$. A quick glance at the answer choices reveals that they are in decimal format rather than fractions, so you need to convert to a decimal. Divide both the numerator and denominator by the common factor of 3 so that the numbers are easier to work with: $\frac{6}{1{,}000}$. You have to get rid of the 3 zeros in the denominator so that the denominator becomes 1. Move the decimal point in the numerator 3 places to the left, inserting zeros after the decimal as needed, resulting in $\frac{.006}{1} = 0.006$. Answer choice **(A)** is correct.
	Step 4: Check your math carefully. Double-check that you inserted the correct number of zeros.

Scientific notation is a method of writing very large and very small numbers that also involves moving decimal points. The first part of a number in scientific notation will be equal to or greater than 1 and less than 10. The second part of the number will be a power of 10. For powers of 10, the exponent is the number of zeroes the number has when written out. For example, $10^4 = 10{,}000$. For numbers written in scientific notation, where the first number always has exactly one digit to the left of the decimal point, the long version of the number can be written out by moving the decimal to the right by the same number of places as the value in the exponent of the power of 10.

$1.23 \times 10^4 = 12{,}300$

Scientific notation also uses negative exponents to indicate the proper placement of the decimal point in a very small number. A negative exponent in scientific notation means that you move the decimal point to the left. Once the decimal point has been moved as far left as possible, start adding zeros to the right of the decimal.

$4.321 \times 10^{-2} = 0.04321$

Use this problem to practice your understanding of scientific notation:

Question	Analysis
Sam writes for a scientific journal. In her research, she finds a number written as 6,483,000. How can she express that in scientific notation?	**Step 1:** You are given a number.
	Step 2: You are asked to rewrite the number in scientific notation.
(A) 6.483×10^3 (B) 6.483×10^5 (C) 6.483×10^6 (D) 6.483×10^7	**Step 3:** Move the decimal point 6 places to the left so 6,483,000 becomes 6.483000. If you move the decimal 6 places to the left, you must write 10^6. Thus, written in scientific notation the number becomes **(C)** 6.483×10^6.
	Step 4: Check to be sure you moved the decimal over the correct number of places.

Exponents and Radicals

Exponents are the small raised numbers written above and to the right of a variable or number (the **base**). They indicate the number of times that the variable or number is multiplied by itself. The rules for performing different operations on terms involving a base and an exponent will require you to pay close attention to the details. There are two important things to remember about exponents. First, any number or variable with an exponent of 1 is equal to the base itself. Second, any number or variable with an exponent of 0 equals 1.

$x^1 = x \quad 5^1 = 5 \quad x^0 = 1 \quad 5^0 = 1$

To **add** or **subtract** terms involving variables and exponents, *both* the variables and the exponents must be the same.

$2x^2 + x^2 = 3x^2$

$3x^4 - 2x^4 = x^4$

$x^2 + x^3$ cannot be combined.

$a^2 + b^2$ cannot be combined.

To **multiply terms** with the *same base*, merely add the exponents. This can be done because exponents represent how many times the base is multiplied by itself.

$2^3 \times 2^2 = (2 \times 2 \times 2) \times (2 \times 2) = 2^{3+2} = 2^5$

Similarly, to **divide terms** with the *same base*, subtract the exponent that is in the denominator from the exponent in the numerator.

$\frac{3^4}{3^2} = 3^{4-2} = 3^2$

To **raise a term involving an exponent to another exponent**, multiply the exponents.

$(x^2)^4 = x^{2\times4} = x^8$

A **coefficient** is a number multiplied by a variable in a term. For instance, in the term $2x^2$, the coefficient is 2. To multiply terms consisting of coefficients and exponents that have the same variable in the base, multiply the coefficients and add the exponents.

$6x^7 \times 2x^5 = (6 \times 2)\ (x^{7+5}) = 12x^{12}$

To **divide** terms consisting of coefficients and exponents that have the same variable in the base, divide the coefficients and subtract the exponents.

$6x^7 \div 2x^5 = (6 \div 2)\ (x^{7-5}) = 3x^2$

You might see a **negative exponent** in a problem on the ASVAB. Don't panic! Negative exponents are just the reciprocal of their positive counterparts.

$2^{-3} = \frac{1}{2^3} = \frac{1}{8}$

Practice working with exponents on this problem:

Question	Analysis
$\left(\frac{(n^5)\times(n^3)}{n^4}\right)^2 + \frac{2}{n^{-8}} = ?$	**Step 1:** You are given an equation.
	Step 2: You need to simplify the equation.
(A) n^8 (B) $2n^8$ (C) $3n^8$ (D) n^{20}	**Step 3:** Since there are variables in the question stem and the answer, you can use Picking Numbers here. Pick a number for *n* and solve for the equation in the question stem. The result is your target number. Then plug your initial number for *n* into all of the answer choices and the one that yields the target number would be correct. To solve with straightforward math, apply the exponent rules one at a time.

Question	Analysis
	Using the rule for multiplying two terms with the same base (in this case, *n*) the numerator of the first expression can be simplified to read: $\left(\frac{n^8}{n^4}\right)^2 + \frac{2}{n^{-8}}$. Next, apply the rule for division of terms with the same base so that the expression on the left becomes: $\left(n^4\right)^2 + \frac{2}{n^{-8}}$. Use the rule for raising an exponent to an exponent and you have $n^8 + \frac{2}{n^{-8}}$. A number to a negative exponent is the same as the reciprocal of that number to a positive exponent. Thus, the expression on the right can be rewritten as: $n^8 + 2n^8$. Since both terms have the same base and the same exponent, you can add them together to get the correct answer, $3n^8$, choice **(C)**.
	Step 4: Briefly confirm that you performed the steps correctly.

Try this problem both by Picking Numbers and by using straightforward math so that you get a sense of which method you will be more comfortable with for this type of problem on Test Day.

Perfect squares are integers that are the result of **squaring** (multiplying by itself) another integer. For instance, 25 is a perfect square because it is 5 × 5. You can save time and trouble on the ASVAB if you memorize the perfect squares up through 12 × 12 = 144.

Hone your skills with the following practice problem:

Question	Analysis
Which of the following numbers is NOT a perfect square?	**Step 1:** The question stem doesn't give much information, but you can infer that three of the answer choices ARE perfect squares.
	Step 2: You are looking for a number that is not equal to an integer times itself.
(A) 1 (B) 16 (C) 27 (D) 36	**Step 3:** $1 = 1 \times 1$, so it is a perfect square. $16 = 4 \times 4$, which is also a perfect square. $27 = 3 \times 3 \times 3$, which is not a perfect square because it is a number multiplied by itself three times rather than twice. Answer choice **(C)** is correct.
	Step 4: Check to see if the remaining answer choice is a perfect square. $36 = 6 \times 6$, so it is a perfect square.

A **square root** is a number that, when multiplied by itself, produces the given quantity. The **radical** sign $\sqrt{\ }$ is used to represent the positive square root of a number, so $\sqrt{25} = 5$, since $5 \times 5 = 25$. (Even though $(-5) \times (-5)$ is also 25, when you are asked for the square root, the answer will be a positive number.) To **add** or **subtract** radicals, make sure the numbers under the radical sign are the same. If they are, you can add or subtract the coefficients outside the radical signs.

$2\sqrt{2} + 3\sqrt{2} = 5\sqrt{2}$

$\sqrt{2} + \sqrt{3}$ cannot be combined.

To **simplify** radicals, factor out the perfect squares under the radical, calculate the square roots of the perfect squares, and put the results in front of the radical sign.

$\sqrt{32} = \sqrt{16 \times 2} = 4\sqrt{2}$

Note that when you have simplified to the point where the number under the radical does not contain any perfect squares, your result will be sufficient for the ASVAB.

To **multiply** or **divide** radicals, multiply or divide the coefficients outside and inside the radical separately.

$\sqrt{x} \times \sqrt{y} = \sqrt{xy}$

$3\sqrt{2} \times 4\sqrt{5} = 12\sqrt{10}$

$$\frac{\sqrt{x}}{\sqrt{y}} = \sqrt{\frac{x}{y}}$$

$$\frac{12\sqrt{10}}{3\sqrt{2}} = 4\sqrt{5}$$

To **take the square root of a fraction**, you can break the fraction into two separate roots and take the square root of the numerator and the denominator separately.

$$\sqrt{\frac{16}{25}} = \frac{\sqrt{16}}{\sqrt{25}} = \frac{4}{5}$$

Practice computations with radicals on this problem:

Question	Analysis
Simplify $\sqrt{\frac{(3\times27)}{(2\times32)}}$.	**Step 1:** The question gives an expression in the form of a fraction.
	Step 2: You are asked to simplify the fraction as much as possible.
(A) $\frac{2}{3}$ (B) $\frac{8}{9}$ (C) $1\frac{1}{8}$ (D) $1\frac{1}{3}$	**Step 3:** First, check to see if there is anything you can divide out of the top and bottom to make the fraction easier to work with. In this case, there is not. Since the example contains a radical sign, check to see if any of the numbers (as written) are perfect squares. None of the numbers are perfect squares, so complete the multiplication on the top and on the bottom. $\sqrt{\frac{81}{64}}$ Before performing the division, check to see if any of the numbers are perfect squares. $81 = 9 \times 9$ $64 = 8 \times 8$ Since both numbers are perfect squares, take the square root of the top and bottom independently. $\sqrt{\frac{81}{64}} = \frac{\sqrt{81}}{\sqrt{64}} = \frac{9}{8}$
	Scanning the answer choices, $\frac{9}{8}$ is not present as written. However, since the numerator is greater than the denominator, this is an improper fraction that can be converted to a mixed number. $\frac{9}{8} = \frac{8}{8} + \frac{1}{8} = 1 + \frac{1}{8} = 1\frac{1}{8}$ Answer choice **(C)** is correct.
	Step 4: Check your math and check to be sure you solved for the correct question.

Factorials

On the ASVAB math sections, you may see an occasional **factorial** question. You'll know you're dealing with a factorial when you see an integer followed by an exclamation point (!). A factorial is the product of the integer before the factorial sign and all the positive integers below it. For instance, when you see the following:

$7! = ?$

The question is actually asking you what "7 factorial" is, and the answer to that question is:

$7! = 7 \times 6 \times 5 \times 4 \times 3 \times 2 \times 1 = 5{,}040$

If you're ever given a fraction with factorials, there's often a lot of canceling that you can do before you try to multiply out the factorials.

Look at the following example.

Question	Analysis
Solve for $\frac{6!}{4!}$.	**Step 1:** You are given a fraction containing factorials.
	Step 2: You need to perform the calculations.
	Step 3: Write out the factorials $\frac{6!}{4!} = \frac{6 \times 5 \times 4 \times 3 \times 2 \times 1}{4 \times 3 \times 2 \times 1}$ Cancel out the numbers that appear on the top and bottom. $\frac{6 \times 5 \times \cancel{4 \times 3 \times 2 \times 1}}{\cancel{4 \times 3 \times 2 \times 1}} = \frac{6 \times 5}{1} = 6 \times 5 = 30$ The correct answer is **30**.
	Step 4: Double-check your calculations and your elimination.

If you see a problem with factorials in a fraction on Test Day, simplify before attempting to solve.

Check out the companion video for this section.

COMPANION VIDEO

kaptest.com/login

APPLIED ARITHMETIC REVIEW

LEARNING OBJECTIVES

In this section, you will learn to:

- calculate values for complex arithmetic expressions using the proper rules and order of operations
- utilize ratios, percentages, and proportions to solve problems involving comparisons
- solve rate and work problems
- apply the formulas for statistical terms to different situations
- compute numerical probabilities

While the arithmetic concepts you've looked at so far could appear on the Arithmetic Reasoning section of the ASVAB, it's worth noting that, due to the very nature of word problems, most of the questions in this section will involve some form of applied arithmetic. On the ASVAB you will use arithmetic principles to solve practical problems. Applied arithmetic includes subjects like ratios, percent problems, rates, averages, and probability.

The Order of Operations

You must perform arithmetic operations in the proper order to get the correct answer to complex arithmetic problems. The acronym **PEMDAS** can help you remember that order.

P = Parentheses

E = Exponents

M D = Multiplication and Division (from left to right)

A S = Addition and Subtraction (from left to right)

For example:

$3^3 - 8(4 - 2) + 60 \div 4$

$= 3^3 - 8(2) + 60 \div 4$ (P)

$= 27 - 8(2) + 60 \div 4$ (E)

$= 27 - 16 + 15$ (MD)

$= 26$ (AS)

One way to remember PEMDAS is to think of the phrase "**P**lease **E**xcuse **M**y **D**ear **A**unt **S**ally."

Try out PEMDAS on this problem:

Question	Analysis
Solve $3 \times 5 + (6 \times 2) - 3^3$.	**Step 1:** You are given an equation.
	Step 2: You are asked to solve that equation.
	Step 3: Use PEMDAS: $3 \times 5 + (6 \times 2) - 3^3$ (P) $3 \times 5 + 12 - 3^3$ (E) $3 \times 5 + 12 - 27$ (MD) $15 + 12 - 27$ (AS) 0
	Step 4: Double-check that you performed the operations in the correct order and that your math calculations were performed correctly.

Distributive and Commutative Properties

In addition to being certain to follow the correct order of operations when solving complex arithmetic problems, you may find that you need to utilize the distributive and commutative properties of mathematics to facilitate your computations. It's not important that you remember the specific terms "distributive" and "commutative," but it is important that you know how to *use* these properties.

The **distributive property** says that the result of multiplying one number by the sum or difference of two numbers can be obtained by multiplying each number individually and then totaling the results.

$5 \times (20 + 2) = 5 \times 20 + 5 \times 2 = 100 + 10 = 110$

$5 \times (20 - 2) = 5 \times 20 - 5 \times 2 = 100 - 10 = 90$

Similarly, a fraction with multiple terms in the numerator and only one term in the denominator can be subdivided.

$$\frac{x^3 + 4x^2}{x^2} = \frac{x^3}{x^2} + \frac{4x^2}{x^2} = x + 4$$

Do not attempt to use this property if the denominator has more than one term.

$\frac{x^2}{x^3 + 4x^2}$ cannot be split up.

Not only is the distributive property useful in solving problems on the ASVAB, it can also help you quickly and accurately multiply in the absence of a calculator.

Question	Analysis
Multiply 6×38 without a calculator.	You can convert that to $6 \times (30 + 8)$.
	Then, $6 \times 30 + 6 \times 8 = 180 + 48 =$ **228.**

Try this sample problem:

$$17 \times 8 + \frac{48x - 8}{4} = ?$$

(A) $10x + 34$
(B) $10x + 136$
(C) $12x + 134$
(D) $12x + 136$

Explanation

Step 1: You are given an equation.

Step 2: You are asked to solve that equation.

Step 3: You can simplify multiplying 17×8 by treating that as $(10 + 7) \times 8 = 80 + 56 = 136$.

The fraction is equivalent to $\frac{48x}{4} - \frac{8}{4} = 12x - 2$.

Now the simplified expression is $136 + 12x - 2 = 12x + 134$. Answer choice **(C)** is correct.

Step 4: Check your math and make sure you used the correct order of operations.

The **commutative property** says, in a nutshell, that order does not matter when adding or multiplying numbers. For example:

$3 \times 4 = 4 \times 3$

$3 + 4 = 4 + 3$

This property does not change PEMDAS; it merely gives you some flexibility when you get to the multiply or add part of the order of operations. The commutative property does *not* apply to subtraction or division.

$4 - 3$ does not equal $3 - 4$.

$\frac{4}{3}$ does not equal $\frac{3}{4}$.

This flexibility when dealing with addition and, in particular, multiplication can be a handy addition to your arithmetic toolbox as shown in the example below:

Question	Analysis
Solve for $2 \times 19 \times 5$ without a calculator.	**Step 1:** You are given an equation.
	Step 2: You need to solve this equation without a calculator.
	Step 3: You could answer this by first multiplying $2 \times 19 = 38$, then multiplying $38 \times 5 = 190$. However, before jumping straight into the math, use strategic logic. Note that $2 \times 5 = 10$. This is helpful because 10 is an easy number to multiply. If you rearrange the equation as $2 \times 5 \times 19$ before solving, the math becomes more manageable. $2 \times 5 \times 19 = 10 \times 19 =$ **190**
	Step 4: Double-check your math.

Ratios, Proportions, and Rates

Ratios represent the relationship of one quantity to another. Ratios can be presented in three different ways. They can be verbal, such as "the ratio of cats to dogs is 3 to 4." Another way of describing a ratio is to use a colon, as in "3:4." In order to use ratios in calculations, however, you will use the fractional form, $\frac{3}{4}$. Ratios do not necessarily represent the actual number of two different things. In this example, you don't know if there are actually 3 dogs and 4 cats, or 6 cats and 8 dogs, or 600 cats and 800 dogs.

Ratios are either **part to part** or **part to whole** relationships, depending upon what quantities are being compared:

A class contains 12 male students and 21 female students.

The part to part ratio of male students to female students is

$$\frac{12}{21} = \frac{4}{7}$$

The part to whole ratio of male students to all the students in the class is

$$\frac{12}{12+21} = \frac{12}{33} = \frac{4}{11}$$

Always simplify ratios to their lowest terms, as was done in the above example.

Here's how an expert test taker would apply these basic rules for ratios to an AR question:

Question	Analysis
Alicia is an avid video game player. She has a total of 21 video games. Her ratio of sports games to action games to mystery games is 4:2:1. How many action games does she have?	**Step 1:** The question gives the total number of video games and the ratio of different types of games in the collection.
	Step 2: The question asks for the number of action games in the collection.
(A) 2 (B) 6 (C) 8 (D) 12	**Step 3:** The part to whole ratio of action games to total video games can be represented as $\frac{2}{4+2+1}=\frac{2}{7}$. But Alicia has a total of 21 games, not 7. In order to determine how many action games Alicia has, multiply the part to whole ratio by the total number of games: $21\times\left(\frac{2}{7}\right)=6$. Answer choice **(B)** is correct.
	Step 4: You can double-check that $\frac{6}{21}$ does, in fact, equal the part to whole ratio of $\frac{2}{7}$ by dividing the numerator and denominator by their lowest common multiple, which is 3.

A **proportion** is an equation of two ratios that shows the comparative relationship between parts, things, or elements with respect to size, amount, or degree.

When working with proportions, you can use a helpful technique called **cross-multiplying** to help solve the equation. To cross-multiply an equation that consists of two fractions, you multiply the numerator of the first fraction by the denominator of the second and vice versa. Here is an example of cross-multiplying:

$$\frac{2x}{5}=\frac{3}{4}$$

$$2x(4)=3(5)$$

Here's an example of a proportion question in which you can use cross-multiplying:

Question	Analysis
A picture that is 4 inches wide and 6 inches long is enlarged so that it is $7\frac{1}{2}$ inches long. What is the width of the enlarged picture?	**Step 1:** The question gives you the dimensions of a picture and says that the picture was then enlarged.
	Step 2: The question asks for the new width of the picture.
	Step 3: Set the given information up as a proportion. The ratio of width to length in the original picture is 4:6 and the ratio of width to length of the enlargement will be the same. The length of the enlargement will be $7\frac{1}{2}$ inches, but the width is some unknown value (w). Since the width/length ratio of the two pictures must be the same, you can set the proportions equal to each other: $\frac{4}{6} = \frac{w}{7.5}$ $4 \times 7.5 = 6w$ $30 = 6w$ $5 = w$ So, the enlarged picture is **5 inches wide.**
	Step 4: Double-check your math. Also, be sure you put the correct measurements in the correct places in the ratio.

Practice working with ratios on this sample problem:

Micah can ride his bicycle 42 miles in 3 hours. Assuming that he travels at the same speed, how long would it take Micah to ride 70 miles?

(A) 5.0 hours
(B) 5.5 hours
(C) 7.0 hours
(D) 7.5 hours

Explanation

Step 1: The problem states that Micah rides at the same speed in both instances. You are told how long it takes to ride 42 miles.

Step 2: You are asked how long it will take him to ride 70 miles.

Step 3: Micah's distance to time relationship can be set up using a proportion.

$$\frac{3}{42}=\frac{t}{70}$$

Using the cross-multiplication technique, $3 \times 70 = 42t$.

Before multiplying large numbers together and then dividing by 42, check to see if you can reduce the equation. Since 42 is a multiple of 3, you can divide both sides of the equation by 3.

$70 = 14t$

$5 = t$

Answer choice **(A)** is correct.

Step 4: You can double-check your math by inserting your answer into the proportion and cross-multiplying.

$$\frac{3}{42}=\frac{5}{70}$$

$3(70) = 5(42)$

$210 = 210$

A **rate** is simply a ratio that compares two different but related quantities, such as distance divided by time (speed), or amount divided by time, or cost per unit.

In other words: $\text{Rate} = \frac{\text{Distance}}{\text{Time}}$ or $\text{Rate} = \frac{\text{Amount}}{\text{Time}}$ or $\text{Rate} = \frac{\text{Cost}}{\text{Units}}$

Another way to look at rates is to think of them as changes in the numerator per changes in the denominator. For instance, if your rate of pay is $15/hour, you know that if you work one more hour you will earn an additional $15.

The key to solving rate problems is to set them up as proportions. Convert the units if necessary and solve for the unknown value.

Take a look at the following rate question:

Question	Analysis
A waitress serves 3 diners every 5 minutes. At this rate, how many customers will she serve in one hour?	**Step 1:** The question gives the rate at which a waitress can serve people. The rate is given in minutes.
	Step 2: The question asks for the number of customers that will be served in one hour.

Question	Analysis
	Step 3: Set the rate up as a proportion. $\frac{3 \text{ diners}}{5 \text{ minutes}} = \frac{x \text{ diners}}{1 \text{ hour}}$ Convert units. $\frac{3 \text{ diners}}{5 \text{ minutes}} = \frac{x \text{ diners}}{60 \text{ minutes}}$ Cross-multiply and solve. $5x = 180$ $x = \mathbf{36}$
	Step 4: Double-check your math and that the units were converted correctly.

Pay close attention to units in ratios and proportions. Had hours not been converted to minutes in the example, the answer would have been wrong.

Try this sample problem to practice rates and units:

> The volume of a certain substance is directly proportional to its weight. If 48 cubic inches of the substance weigh 112 ounces, what is the volume, in cubic inches, of 63 ounces of the substance?
>
> (A) 27
> (B) 36
> (C) 42
> (D) 64

Explanation

Step 1: The problem gives the ratio of volume to weight of a certain substance.

Step 2: The problem asks for the volume of a different amount of the substance.

Step 3: Start by setting up the proportion $\frac{48}{112} = \frac{v}{63}$.

You might notice that cross-multiplying this proportion looks difficult without a calculator and wonder if there is a more efficient way to answer the question. There is, by using logic.

In the given ratio, 48 is less than half of 112, so the correct answer will be less than half of 63. Looking at the answer choices, only 27 is less than half of 63.

Answer choice **(A)** is correct.

Step 4: You can double-check your work by reducing the proportions to their lowest form and checking that they equal each other.

$\frac{48}{112} \div \frac{2}{2} = \frac{24}{56}$ $\frac{27}{63} \div \frac{3}{3} = \frac{9}{21}$

$\frac{24}{56} \div \frac{2}{2} = \frac{12}{28}$ $\frac{9}{21} \div \frac{3}{3} = \frac{3}{7}$

$\frac{12}{28} \div \frac{4}{4} = \frac{3}{7}$

Both proportions reduce to $\frac{3}{7}$, which means that the two proportions are equal and the answer choice is confirmed.

Combined Work

You may see a problem on the ASVAB that asks you to determine how long it takes the combined effort of two people or machines to complete a task, or something similar such as how long it takes for two pipes to fill a tank. These are just a different form of rate problem.

For instance, if it takes Joe 3 hours to paint a room, then his rate of painting is $\frac{1 \text{ room}}{3 \text{ hours}}$. The tasks to be completed that are on the ASVAB are additive; the **combined work** of the people (or machines or pipes) can be added together to obtain a combined rate. This approach also works for more than two entities working on the same task.

Question	Analysis
Jose can paint a room in 3 hours. Jill can paint the same room in 2 hours. How long will it take them to paint the room working together?	**Step 1:** The question gives the amount of time it takes two different people to paint a room.
	Step 2: The question asks for the length of time it will take them to paint the room together.
	Step 3: Express the individual times as rates. $\frac{1}{3} + \frac{1}{2} = ?$ Convert to a common denominator and add the rates. $\frac{2}{6} + \frac{3}{6} = \frac{5 \text{ rooms}}{6 \text{ hours}}$ Invert the fraction. $\frac{5 \text{ rooms}}{6 \text{ hours}} = \frac{6 \text{ hours}}{5 \text{ rooms}}$ If it takes 6 hours to paint 5 rooms, it would take $\frac{6}{5}$ hours to paint 1 room. $\frac{6}{5} = \frac{5}{5} + \frac{1}{5} = 1\frac{1}{5} =$ **1.2 hours**

Question	Analysis
	Step 4: Double-check your math. Also double-check that you are solving for the correct number of rooms, in this case, 1 room.

Try this problem (which does not have answer choices) on your own:

> There are three pipes that can be used to fill a tank. One of them can fill the tank in 2 hours, another can fill it in 3 hours, and the third can fill the tank in 6 hours. If all three pipes are used to fill the tank simultaneously, how long will it take?

Explanation

Step 1: The three times needed for each pipe to fill the tank individually are given.

Step 2: You are asked to calculate the rate at which they would be able to fill the tank together.

Step 3: The rates can be converted to $\frac{\text{tanks}}{\text{hour}}$, specifically $\frac{1}{2}$, $\frac{1}{3}$, and $\frac{1}{6}$. Convert their rates to a common denominator and add: $\frac{3}{6}+\frac{2}{6}+\frac{1}{6}=\frac{6}{6}=1\frac{\text{tank}}{\text{hour}}$. Therefore to fill the one tank, the three pipes working together can accomplish the task in **1 hour.**

Percent Problems

Percents are ratios of an amount to 100, so the techniques used to work with ratios and proportions are valuable tools when you work with percents.

Question	Analysis
A bag contains 8 marbles and 6 of them are green. What percent of the marbles in the bag are green?	**Step 1:** The question gives the total amount of marbles and the amount of green marbles.
	Step 2: The question asks for the percent of total marbles that are green.
	Step 3: To solve this, you can set up a proportion equating the numbers of marbles to the unknown percentage: $\frac{6}{8}=\frac{g}{100}$ $600 = 8g$ $75 = g$ So **75%** of the marbles are green.
	Step 4: Double-check to confirm your math is correct.

To convert the English into math on a percent problem, use the following conversion table.

English	Math Translation
% or percent	÷ 100 (or use decimal or fractional equivalent)
of	× (times)
what	x (or n, or any variable you like)
is	= (equals)

Using these translations, you can see, for instance, that 40% of 25 = 0.40 × 25 = 10.

Some ASVAB questions require you to increase or decrease a number by a given percent. To calculate such increases or decrease, take that percent of the original number and add it to or subtract it from the original number.

To increase 25 by 60%, first find 60% of 25.

$25 \times 0.6 = 15$

Then add the result to the original number.

$25 + 15 = 40$

To decrease 25 by the same percent, subtract the 15.

$25 - 15 = 10$

Other ASVAB questions, rather than giving you the percent increase or decrease, will ask you to calculate the percentage change between an original value and a new value. In order to calculate that change, always use this formula:

$$\text{Percent Change} = 100\% \times \frac{\text{New Value} - \text{Original Value}}{\text{Original Value}}$$

If the percent change is an increase, the result will be positive. If the change is a decrease, the result will be negative. ASVAB percentage change problems ask for the change relative to the original value; using the new value in the denominator will produce an incorrect answer.

Question	Analysis
A camera that originally cost $125 was sold on sale for $100. The sale price was what percent less than the original price?	**Step 1:** The question gives the original price and the sale price of a camera.
	Step 2: The question asks for the percentage decrease between the sale price and the original price.
	Step 3: Percent Change = $100\% \times \frac{100 - 125}{125} = 100\% \times \frac{-25}{125} = 100\% \times \frac{-1}{5} = -20\%$ Since the result is negative, you know that there was a price decrease. The answer is **–20%**.
	Step 4: Briefly confirm that the answer makes sense. 20% of 125 is 25, so your answer is correct.

Practice using the formula on this sample problem:

Ryan weighed 40 pounds when he finished kindergarten and 44 pounds when he started first grade. What was the percentage change in his weight between kindergarten and first grade?

(A) 10% decrease
(B) 4% increase
(C) 10% increase
(D) 40% increase

Explanation

Step 1: The problem gives Ryan's weight at various ages.

Step 2: The question asks for his percent change in weight.

Step 3:

$$\text{Percent Change} = 100\% \times \frac{44-40}{40} = 100\% \times \frac{4}{40} = 100\% \times \frac{1}{10} = 10\%$$

Since the result is positive, the answer is a 10% increase, choice **(C)**.

Step 4: Check your calculations and check that you solved for the percent asked for.

Now practice this skill on another problem.

Rachel made a certain investment in the stock market but the value of her investment declined by 20%. However, the value of that investment then increased by 50%. What was the overall percentage change in the value of the investment from the original purchase price?

(A) 30% decrease
(B) No change
(C) 20% increase
(D) 30% increase

Explanation

Step 1: The question gives information about the decrease and increase in the value of a stock.

Step 2: The question asks for the overall change.

Step 3: Notice that the question does not mention any specific prices. That makes this problem an excellent candidate for the Picking Numbers technique. Since the problem deals with percentages, start with an original value equal to 100.

If the investment lost 20% of its value, that would have been a loss of 20, so the value after the decrease would have been 80. Fortunately for Rachel, the value then increased 50%. Since the 50% increase was based upon a value of 80, that would be a gain of 40. This results in a final value of $80 + 40 = 120$, which is a 20% increase from the original assigned value of 100, answer choice **(C)**.

Step 4: Confirm that your math is correct and that you are solving for the correct piece of information.

Statistical Terms

Averages, like ratios, proportions, and rates, use fractions to find the answer to a problem. Here is the average formula:

$$\text{Average} = \frac{\text{Sum of the Terms}}{\text{Number of Terms}}$$

What is the average of 3, 4, and 8?

$$\text{Average} = \frac{\text{Sum of the Terms}}{\text{Number of Terms}} = \frac{3+4+8}{3} = \frac{15}{3} = 5$$

Sometimes the ASVAB may not present the information in such simple terms. The formula for averages, like other three-part formulas, can be rearranged as needed.

Question	Analysis
Six people went fishing and caught an average of 7 fish each. How many total fish did they catch?	**Step 1:** You are given the average amount of fish caught by 7 people.
	Step 2: You are asked for the total number of fish.
	Step 3: Plug the values that you are given into the formula and solve for the value needed. $7 = \frac{\text{Total}}{6}$ Multiply both sides of the equation by 6. $6 \times 7 = \text{Total} = \textbf{42}$ **fish**
	Step 4: Check to be sure you used the numbers given in the problem and that you performed the calculations correctly.

Here is a practice problem for you:

Lola was pleased to learn that her final grade in her history course, which was based upon her average for 4 tests, was 90. She knew that she had scored 93, 80, and 90 on her first three tests, but she could not remember her score on the last test. What was Lola's score on the fourth test?

(A) 87
(B) 90
(C) 93
(D) 97

Explanation

Step 1: You are given the score on 3 tests and the average score of 4 tests.

Step 2: The question asks for the score of the fourth test that would give the average listed.

Step 3: Use the average formula to write the equation $90 = \frac{93+80+90+t}{4}$ with t representing the unknown score. The denominator is 4 because the final grade of 90 was based on 4 tests even though Lola only knew the results of 3 tests.

Solving the average equation for t, add all the numbers in the numerator. $90 = \frac{263+t}{4}$

Multiply both sides of the equation by 4 to get $360 = 263 + t$. Then t is $360 - 263$ which equals 97, choice **(D)**.

Step 4: Check your math. Make sure you plugged the correct values into the average formula in the beginning.

The ASVAB will not require you to make complex statistical calculations, but there are a few basic statistical terms with which you should be familiar. **Mean** is the same as average and is calculated the same way. The **range** is how "wide" the group of numbers is and can be calculated by subtracting the smallest number from the largest. The **mode** is the number that appears most frequently. Finally, **median** is the middle value of a group of numbers.

Practice by identifying these terms using the number set: 1, 2, 7, −3, 0, 5, 2.

In order to apply the statistical terms to a group of numbers, the very first thing you must do is arrange the numbers in ascending order: −3, 0, 1, 2, 2, 5, 7.

The mean of these numbers is $\frac{-3+0+1+2+2+5+7}{7} = 2$.

The mode of the group is 2 since it is the only number that appears twice.

The range is $7 - (-3) = 10$.

The median is 2, since three numbers are smaller than that and three are larger.

Should you encounter a situation whereby there are two numbers in the middle, the average of those two middle numbers is the median.

For instance, the median of −3, 0, 1, 2, 2, 5 is $\frac{1+2}{2} = 1.5$.

Sequences

A **sequence** is merely a group of numbers placed in order, as in the example above. You are most likely to encounter a special type of sequence on the ASVAB called an **arithmetic sequence**. The property that creates an arithmetic sequence is that each number is equal to the number before it plus a constant number. The sequence 1, 3, 5, 7 is an arithmetic sequence because each number is 2 more than the previous number in the sequence. One property that makes arithmetic sequences special is that the mean equals the median.

Here's an example of the type of sequence problem that may be on the ASVAB:

Question	Analysis
What is the mean of a sequence of multiples of 3 that begins with −3 and ends with 15?	**Step 1:** The question gives you the rules of a sequence as well as the beginning and ending terms of that sequence.
	Step 2: The question asks for the mean.

Question	Analysis
	Step 3: Since the numbers are multiples of 3, the numbers will all be 3 apart. This is an arithmetic sequence. You can write down the sequence: −3, 0, 3, 6, 9, 12, 15. Because this is an arithmetic sequence, the median is equal to the mean, **6**.
	Step 4: Check that you defined the sequence properly and solved for the correct statistical term.

Probability

Probability is the numerical likelihood that a particular outcome will occur. To find the probability that something is going to happen, use this formula:

$$\text{Probability} = \frac{\text{Number of Outcomes of Interest}}{\text{Number of Possible Outcomes}}$$

Probability can be expressed as a fraction, a decimal, or a percent.

Question	Analysis
If there are 12 books on a shelf and 9 of them are mysteries, what is the probability of picking a mystery at random?	**Step 1:** The question gives the total number of books and the number of mystery books.
	Step 2: The question asks for the probability of picking a mystery book. In other words, it asks you for the proportion of mystery books to total books.
	Step 3: Probability $= \frac{9}{12} = \mathbf{\frac{3}{4}}$ This probability can also be expressed as 0.75 or 75%.
	Step 4: Check that you set up and simplified the fraction appropriately.

Probabilities are always between 0 and 1 (or between 0% and 100%). A probability of 0 means that there are no outcomes of interest and a probability of 1 means that all the possible outcomes result in the outcome of interest occurring.

To find the probability that both of two events will occur, find the probability that the first event occurs and multiply this by the probability that the second event occurs. The probability of two independent events both occurring will be less than the probability of either occurring by itself. When you see a probability question that deals with one event *and* another event both occurring, you will multiply the probabilities.

Study the following example.

Question	Analysis
If there are 12 books on a shelf and 9 of them are mysteries, what is the probability of picking a mystery first AND a non-mystery book second if exactly two books are selected and neither of them is replaced on the shelf?	**Step 1:** You are given the same information about the number of mystery books compared to the number of books overall.
	Step 2: This time, the question asks for the probability that two different events will both occur.
	Step 3: The probability of picking a mystery book first: $\frac{9}{12} = \frac{3}{4}$ Probability of picking a non-mystery book second (if a mystery is picked first): $\frac{3}{11}$ (Originally there were 9 mysteries and 3 non-mysteries. After the mystery is selected first, there are 8 mysteries and 3 non-mysteries, i.e., 11 books remaining.) Probability of picking a mystery book first and then a non-mystery book second: $\frac{3}{4} \times \frac{3}{11} = \mathbf{\frac{9}{44}}$
	Step 4: Check to be sure you set up the second proportion to incorporate the missing book that was already picked. Double-check your math and be sure you solved for the probability that was asked for.

A question might ask you to determine the probability of one **or** another event occurring. In this case, you **add** the individual probabilities. The probability of one **or** another event occurring will be greater than the probability of either event occurring alone.

Question	Analysis
What is the probability of rolling a 1 OR a 2 on one roll of a fair six-sided die?	**Step 1:** The question gives you background information that you are dealing with a die that has 6 faces and isn't weighted.
	Step 2: The question asks for the probability of either of two scenarios happening.
	Step 3: Probability of rolling a 1 $= \frac{1}{6}$ Probability of rolling a 2 $= \frac{1}{6}$ Probability of rolling a 1 or a 2 $= \frac{1}{6} + \frac{1}{6} = \frac{2}{6} = \mathbf{\frac{1}{3}}$
	Step 4: Briefly check your math before moving on to the next question.

Practice the principles of probability on the following problem:

If a fair coin is flipped 3 times, what is the probability that none of the occurrences will result in a head?

(A) 0

(B) $\frac{1}{8}$

(C) $\frac{1}{6}$

(D) $\frac{1}{3}$

Explanation

Step 1: The question sets up a scenario of flipping coins.

Step 2: The question asks for the probability of landing on tails 3 times in a row.

Step 3: In order to flip a coin 3 times with no heads, the first flip must be a tail *and* the second flip must also be a tail *and* the third flip must be a tail, too.

Therefore, the equation to determine the probability of no heads on 3 flips is $\left(\frac{1}{2}\right)\times\left(\frac{1}{2}\right)\times\left(\frac{1}{2}\right)=\frac{1}{8}$. Answer choice **(B)** is correct.

Step 4: Double-check that you solved for the correct probability and you performed the calculations correctly.

Check out the companion video for this section.

COMPANION VIDEO

kaptest.com/login

ARITHMETIC WORD PROBLEMS

LEARNING OBJECTIVES

In this section, you will learn to:

- convert word problems to equations
- use common predefined formulas to answer word problems

As you may have noticed, we already started sneaking word problems into the arithmetic review section you just completed. Word problems are simply math problems with another step added. Generally, all you have to do is translate the text into math and solve. In some questions, the translation will be embedded within a "story." Don't be put off by the details of the scenario—it's the numbers that matter. Focus on the math and translate.

Translation

Often, word problems seem tricky because it's hard to figure out precisely what they're asking. The following table lists some common words and phrases that turn up in word problems, along with their mathematical translations.

When you see:	Think:
sum, plus, more than, added to, combined, total	$+$
minus, less than, difference between, decreased by	$-$
is, was, equals, is equivalent to, is the same as, adds up to	$=$
times, product, multiplied by, of	$\times$
divided by, over, quotient, per, out of, into	$\div$
what, how much, how many, a number	x, n, etc.

Word Problems with Formulas

Some of the more difficult word problems may involve translations with mathematical formulas. The most common ones that you should know are $\text{Rate} = \frac{\text{Distance}}{\text{Time}}$, $\text{Average} = \frac{\text{Sum of the Terms}}{\text{Number of Terms}}$, and $\text{Probability} = \frac{\text{Number of Outcomes of Interest}}{\text{Number of Possible Outcomes}}$, all of which were discussed earlier in this chapter.

Question	Analysis
If a truck travels at 50 miles per hour for $6\frac{1}{2}$ hours, how far will the truck travel?	**Step 1:** The question gives you the rate of the truck and how long that truck will be traveling.
	Step 2: The question asks how far the truck will go in that time.
	Step 3: To answer this question, you need to use the Time-Rate-Distance Formula: $\text{Rate} = \frac{\text{Distance}}{\text{Time}}$. Multiplying both sides of the equation by time gives you: Distance $=$ Rate $\times$ Time Now, plug in the numbers. $D = 50 \times 6.5$ $D =$ **325 miles**
	Step 4: Check that you performed the math correctly.

Try this rate problem:

> Elisa rode her motorcycle for an hour and traveled 60.0 miles. Unfortunately she then encountered heavy traffic, and it took her 2 hours to travel an additional 45.0 miles. What was her overall speed for the entire trip?
>
> (A) 35.0 miles/hour
> (B) 45.0 miles/hour
> (C) 52.5 miles/hour
> (D) 60.0 miles/hour

Explanation

Step 1: The question gives you the time and distance of one portion of a trip and then the time and distance of another portion of the same trip.

Step 2: You are asked to come up with the average speed for both portions of the trip together.

Step 3: The total distance traveled was $60 + 45 = 105$ miles.

The total time was $1 + 2 = 3$ hours.

$\text{Rate} = \frac{\text{Distance}}{\text{Time}} = \frac{105}{3} = 35$ miles per hour. Answer choice **(A)** is correct.

Step 4: It is important to remember that you cannot simply take the average of the two speeds. This is because Elisa travels at each speed for a different length of time. Make sure you use the rate formula correctly.

Check out the companion video for this section.

COMPANION VIDEO

kaptest.com/login

ARITHMETIC REASONING PRACTICE SET 1

Select the best answer for each question. This section has 15 questions, which is the number of Arithmetic Reasoning questions you will see if you take the CAT-ASVAB.

1. If $|y - 3| = 3$, what is the greatest possible value of y?

   (A) 2
   (B) 4
   (C) 6
   (D) 8

2. Two teams are having a contest in which the prize is a box of candy that the members of the winning team will divide evenly. If team A wins, each player will get exactly 3 pieces of candy, and if team B wins, each player will get exactly 5 pieces of candy. Which of the following could be the number of pieces of candy in the box?

   (A) 325
   (B) 353
   (C) 425
   (D) 555

3. What is the least common multiple of 7, 9, and 21?

   (A) 21
   (B) 63
   (C) 147
   (D) 189

4. What is the value of $\sqrt{\dfrac{5 \times 35 \times 7 \times 16}{50 \times 8}} = ?$

   (A) 5
   (B) 7
   (C) 14
   (D) 35

5. A copier can make 150 copies per minute. At this rate, how many minutes would it take to make 4,500 copies?

   (A) 20
   (B) 25
   (C) 30
   (D) 35

6. The average (arithmetic mean) age of the members in a five-person choir is 34. If the ages of four of the members are 47, 31, 27, and 36, what is the fifth member's age?

   (A) 29
   (B) 32
   (C) 34
   (D) 37

7. In the past year, XYZ Corporation reported sales of $\$1.17 \times 10^6$. The year before, the company's sales were $\$9.00 \times 10^5$. What was the percentage change in sales?

   (A) 30% decrease
   (B) 27% decrease
   (C) 27% increase
   (D) 30% increase

8. What is the value of $30 - 5 \times 4 + (7 - 3)^2 \div 8$?

   (A) 4
   (B) 12
   (C) 36
   (D) 102

9. A group of five workers want to have a fair way of deciding who has to perform a difficult task, so they place 4 green balls and 1 red ball into an opaque box from which each worker will draw a ball at random. The person who draws the red ball will have to perform the task. What is the probability that the last (5th) person will get the red ball?

   (A) 0.10
   (B) 0.15
   (C) 0.20
   (D) 0.25

10. John's grade for a class is determined by the results of 3 tests and 1 final exam. If the exam counts twice as much as each of the tests, what fraction of the final grade is determined by the final exam?

(A) $\frac{1}{4}$
(B) $\frac{1}{3}$
(C) $\frac{2}{5}$
(D) $\frac{3}{5}$

11. If $5\sqrt{2x} = 40$, what is the value of $5x$?

(A) 160
(B) 200
(C) 400
(D) 500

12. A full 10-liter container of diluted orange juice contains 20% water and 80% pure juice. After half of the diluted orange juice is consumed, 1 liter of pure juice is added to sweeten the taste. What is the ratio of liters of water to liters of pure juice after the liter of pure juice is added?

(A) 1:5
(B) 1:4
(C) 4:5
(D) 1:1

13. For all integers x, which of the following must be true about $4x + 4$?

(A) Must be even
(B) Must be odd
(C) Must be a multiple of 3
(D) Must be a multiple of 5

14. A car is moving at 60 miles per hour. At this rate, how many hours will it take for the car to travel 420 miles?

(A) 5
(B) 6
(C) 7
(D) 8

15. A building has $\frac{2}{5}$ of its floors above ground. What is the ratio of the number of floors below ground to the number of floors above ground?

(A) 5:2
(B) 3:2
(C) 2:3
(D) 2:5

ARITHMETIC REASONING PRACTICE SET 2

Select the best answer for each question. This section has 15 questions, which is the number of Arithmetic Reasoning questions you will see if you take the CAT-ASVAB.

1. What is 28% of 25?
   (A) 4
   (B) 5
   (C) 6
   (D) 7

2. What is the average of $\frac{1}{2}$, $\frac{1}{4}$, and $\frac{1}{5}$?
   (A) $\frac{1}{40}$
   (B) $\frac{3}{13}$
   (C) $\frac{19}{60}$
   (D) 1

3. How many ways are there to choose an outfit from 3 shirts and 4 pairs of pants, assuming an outfit consists of one shirt and one pair of pants?
   (A) 4
   (B) 7
   (C) 12
   (D) 20

4. What is the value of $50 - 2(2 + 1)^2 \div 6$?
   (A) 0
   (B) 15
   (C) 47
   (D) 72

5. If $|x - 5| = 10$, what is the smallest possible value of x?
   (A) −15
   (B) −5
   (C) 15
   (D) 25

6. The average of x and y is 30. If $z = 15$, what is the average of the three terms x, y, and z?
   (A) 3
   (B) 15
   (C) 25
   (D) 30

7. What is the value of $\dfrac{\frac{3}{4}}{\frac{1}{2} + \frac{4}{5} - \frac{1}{10}}$?
   (A) $\frac{3}{5}$
   (B) $\frac{5}{8}$
   (C) 1
   (D) $\frac{27}{10}$

8. What is the greatest common factor of 50, 60, and 75?
   (A) 2
   (B) 5
   (C) 10
   (D) 25

9. A $200 watch is on sale for $160. What is the percent change in the price of the watch?
   (A) 25% decrease
   (B) 20% decrease
   (C) 20% increase
   (D) 25% increase

10. Michael can wash a car in two hours, and Jim can wash the same car in four hours. How long will it take the two of them working together to wash two of the same cars?
   (A) 1 hour, 20 minutes
   (B) 2 hours, 40 minutes
   (C) 3 hours
   (D) 6 hours

11. What is the prime factorization of 26^3?

(A) $2 \times 3 \times 13$
(B) $2 \times 2 \times 13 \times 13$
(C) $2 \times 2 \times 3 \times 13 \times 13$
(D) $2 \times 2 \times 2 \times 13 \times 13 \times 13$

12. In the set of numbers {1, 4, 7, 10, 13, 16, 19, 22, 25}, what is the arithmetic mean?

(A) 13
(B) 14.5
(C) 24.5
(D) 25

13. If $|4x - 6| = 18$, which of the following is a possible value of x?

(A) −4
(B) −3
(C) 4
(D) 24

14. For all positive values of $x < 50$, how many values are divisible by both 3 and 5?

(A) 0
(B) 1
(C) 2
(D) 3

15. Three teams are having a contest in which the winning team will split a cash prize equally among all the members of the team. If Team 1 wins, each member will receive $10. If Team 2 wins, each member will receive $15. If Team 3 wins, each member will receive $25. Which of the following could be the number of dollars in the cash prize?

(A) $120
(B) $375
(C) $750
(D) $1,250

ARITHMETIC REASONING PRACTICE SET 1

Answers and Explanations

1. **C** To simplify an absolute value problem by eliminating the absolute value function, write two equations, one of which sets whatever is inside the vertical lines to its positive value, and the other of which sets that equal to its negative value. So, either $y - 3 = 3$ or $y - 3 = -3$. Clearly, y has a greater value in the first equation. By adding 3 to both sides of the equation, $y = 6$.

2. **D** This problem tells you that the number of pieces of candy in the box can be divided evenly by 5 and 3. So the correct answer is the choice that has a 0 or 5 as its last digit, and the sum of whose digits is divisible by 3. Eliminate (B) because it doesn't end in either 0 or 5. Of the remaining choices, only 555, (D), is also divisible by 3, since $5 + 5 + 5 = 15$.

3. **B** To find the least common multiple, you should identify the prime factors of each number. The prime factor of 7 is 7, since it is a prime number. The prime factors of 9 are 3 and 3. The prime factors of 21 are 7 and 3. Therefore the LCM must have two factors of 3 and one factor of 7; $3 \times 3 \times 7 = 63$. To check your work, you can confirm that 7, 9 and 21 all divide evenly into 63.

4. **B** While this problem may look formidable at first glance, it can be solved by reducing the fraction to simpler terms before attempting to identify the square root. When you see a complex calculation that makes you wonder how you are going to do the math without a calculator, that is often a hint to use prime factorization. If you restate the problem as $\sqrt{\frac{(5 \times 5 \times 7 \times 7 \times 2 \times 2 \times 2 \times 2)}{5 \times 5 \times 2 \times 2 \times 2 \times 2}}$ you can see that many terms in the numerator and denominator of the radical cancel out: $\sqrt{\frac{(\cancel{5 \times 5} \times 7 \times 7 \times \cancel{2 \times 2 \times 2 \times 2})}{\cancel{5 \times 5} \times \cancel{2 \times 2 \times 2 \times 2}}}$, so that you are left with $\sqrt{7 \times 7} = 7$.

5. **C** Set up a proportion: $\frac{150 \text{ copies}}{1 \text{ minute}} = \frac{4{,}500}{x \text{ minutes}}$ Now cross-multiply and solve: $150x = 4{,}500$. So $x = \frac{4{,}500}{150} = 30$.

6. **A** Average $= \frac{\text{Sum of the Terms}}{\text{Number of Terms}}$, so here the average is $\frac{47 + 31 + 27 + 36 + n}{5} = 34$. Thus $34 = \frac{141 + n}{5}$; multiply both sides by 5 to get $170 = 141 + n$. Subtract 141 from both sides and $n = 29$.

7. **D** This is a percent change problem, so you will need to identify the original and new values that are being compared and use the formula for calculating percent change: $\frac{(1.17 \times 10^6) - (9.00 \times 10^5)}{9.00 \times 10^5}$. In order to subtract powers, they must have the same base and exponent, so you'll have to convert one of the terms $\frac{(11.7 \times 10^5) - (9.00 \times 10^5)}{9.00 \times 10^5} = \frac{2.7 \times 10^5}{9.00 \times 10^5}$. This can be simplified to $\frac{2.7}{9} = \frac{27}{90} = \frac{3}{10} = 30\%$.

8. **B** Follow PEMDAS to get the answer. Since there is a set of **P**arentheses, start there and change the expression to $30 - 5 \times 4 + 4^2 \div 8$. There is an **E**xponent, so calculate the value of the exponent term next: $30 - 5 \times 4 + 16 \div 8$. Now you can calculate the **M**ultiplication and **D**ivision parts of the expression from left to right to get $30 - 20 + 2$. Finish up with the **A**ddition and **S**ubtraction from left to right to get the answer, 12.

9. **C** This is a probability question with multiple events. In order for the last person to get stuck with the red ball, the first person must draw a green ball **and** so must the 2nd, 3rd, and 4th persons. The number of green balls will decrease with each successive pick, so the probability is $\frac{4}{5} \times \frac{3}{4} \times \frac{2}{3} \times \frac{1}{2} = \frac{\cancel{4 \times 3 \times 2} \times 1}{5 \times \cancel{4 \times 3 \times 2}} = \frac{1}{5} = 0.2$

10. **C** Each of the tests counts once and the final exam counts twice, which means there are $3 + 2 = 5$ total parts. The final exam is 2 parts, so it counts as $\frac{2}{5}$ of the final grade.

11. **A** First, solve $5\sqrt{2x} = 40$ for x. Divide by 5: $\sqrt{2x} = 8$. Square both sides, so $2x = 64$, then divide both sides by 2 so $x = 32$. Lastly, multiply by 5: $5 \times 32 = 160$.

12. **A** Since half of the 10 liters of diluted juice are consumed, there must be only 5 liters remaining. 20% is water and 80% is pure juice, so there is now 1 liter of water and 4 liters of pure juice. Another liter of pure juice is added, so there are 5 liters of pure juice. The ratio is 1:5.

13. **A** Since x is an integer, multiplying any integer by 4 will result in an even integer. So $4x$ must be even. Adding 4 to an even value must result in an even number. Therefore, $4x + 4$ must be even.

14. **C** The formula to use to find time when given the speed and the distance is $\text{Time} = \frac{\text{Distance}}{\text{Speed}}$. Since the distance given here is 420, and the speed is 60, $\frac{420}{60} = 7$.

15. **B** The question provides a part to whole ratio and asks for a part to part ratio. If 2 of the 5 floors are above ground, then $5 - 2 = 3$ floors must be below ground. So the ratio of floors below ground to floors above ground is 3:2.

ARITHMETIC REASONING PRACTICE SET 2

Answers and Explanations

1. **D** $28\% = \frac{28}{100}$, so multiply that by 25. However, there is a faster way to solve this. To find the value, you would calculate $\frac{28}{100} \times 25$, but since multiplication can be done in any order, this is the same as $\frac{25}{100} \times 28$. $\frac{25}{100}$ can be simplified to $\frac{1}{4}$, and multiplying that by 28 equals 7.

2. **C** The average of a set of numbers is found by adding the numbers, and then dividing by how many terms there are. So, add the three fractions, and then divide by 3, which is the number of terms. To add the three fractions, start by finding the common denominator, which is 20. Then convert each fraction to have a denominator of 20. $\frac{1}{2} = \frac{10}{20}$, $\frac{1}{4} = \frac{5}{20}$, and $\frac{1}{5} = \frac{4}{20}$. So the sum is $\frac{19}{20}$. Then divide by 3 or multiply by the reciprocal, $\frac{1}{3}$. The result will be $\frac{19}{20} \times \frac{1}{3} = \frac{19}{60}$.

3. **C** To find the number of possible combinations consisting of two different elements, multiply how much of each element you have. Since each outfit consists of one shirt and one pair of pants, take the number of shirts (3) and the number of pairs of pants (4), and multiply them together. $3 \times 4 = 12$.

4. **C** Follow the order of operations (PEMDAS) to simplify this expression correctly. First, simplify what is in the parentheses: in this case, $2 + 1$, which equals 3. Then apply the exponent to what's in the parentheses. $3^2 = 9$. So now the expression is $50 - 2(9) \div 6$. Next, perform the operations of multiplication and division, from left to right. $2 \times 9 = 18$ and $18 \div 6 = 3$. The expression is now $50 - 3$, which equals 47.

5. **B** For an equation with an expression inside the absolute value signs, create two equations: one with a positive value (in this case, 10) and one with a negative value (in this case, -10). Since the question asks for the smallest possible value of x, start with the equation with the negative value. Since $x - 5 = -10$, add 5 to both sides, and the resulting equation is $x = -5$.

6. **C** To find the average of the three terms, first find the sum of the terms and then divide by 3. There's no way to know the exact values of x and y, but since the average of the two terms is 30, the sum must be $2(30) = 60$. Since z equals 15, then the sum of the three terms is $60 + 15 = 75$. Lastly, divide 75 by 3, which equals 25.

7. **B** First, combine the fractions in the denominator. The common denominator of these is 10, so $\frac{1}{2} = \frac{5}{10}$ and $\frac{4}{5} = \frac{8}{10}$. Once all the denominators are the same, add or subtract the numerators: $5 + 8 - 1 = 12$. So the fraction in the denominator is $\frac{12}{10}$. Then, when dividing by a fraction (in this case, $\frac{12}{10}$), take the reciprocal and multiply that by the numerator. $\frac{3}{4} \times \frac{10}{12} = \frac{30}{48}$. Both the numerator and denominator have a common factor of 6, so divide both the numerator and denominator by 6, and the fraction simplifies to $\frac{5}{8}$.

8. **B** To find common factors of a number, first break down each number into its prime factors. Then see which factors they have in common, and find the product of those common factors. The prime factors of 50 are 2, 5, and 5. The prime factors of 60 are 2, 2, 3, and 5. The prime factors of 75 are 3, 5, and 5. Since one factor of 5 is in each of those sets of prime factors, but the prime factors 2 and 3 are not found in all of the numbers, the greatest common factor is 5.

9. **B** Since the price of the watch has gone down, the percent change must represent a decrease. Using the formula for percent change, divide the actual change by the original amount. Since the watch was originally \$200 and during the sale it is \$160, the watch decreased in price by \$40. Divide \$40 by the original amount of \$200 to get 0.2, or 20%.

10. **B** Since the problem states how long it takes each person to wash the car, the units of which are hours/car, take the reciprocal of the times to get the rates in terms of cars/hour and add them together to get the combined rate: $\frac{1}{2} + \frac{1}{4} = \frac{2}{4} + \frac{1}{4} = \frac{3}{4}$ cars/hour working together. Invert that combined rate to determine that they can wash 1 car in $\frac{4}{3}$ hour. However, the question asks how long it will take to wash two cars, so that would be $\frac{8}{3} = 2\frac{2}{3}$ hours.

11. **D** Don't calculate or even estimate the value of 26^3. The prime factorization of 26 is 2×13. So, since there are three factors of 26, there are three factors of 2 and three factors of 13. $26^3 = (2 \times 13)^3$, or $2 \times 2 \times 2 \times 13 \times 13 \times 13$.

12. **A** Avoid the temptation to use the formula for finding the arithmetic mean, or average. Instead, notice that each number in the set has the same difference from the previous number so that they form an arithmetic sequence. Whenever that is true of a set of numbers, the arithmetic mean is equal to the median. Since there are 9 numbers in this set, the 5th number (when all the numbers are in order) is the median. In this case, the 5th number is 13.

13. **B** For an equation with an absolute value sign, consider both the positive and the negative values for what is inside the signs. $4x - 6$ can equal either 18 or -18. If $4x - 6 = 18$, first add 6 to both sides: $4x = 24$. Then divide both sides by 4, and $x = 6$. However, 6 is not one of the answer choices. So try the other equation: $4x - 6 = -18$. Add -6 to both sides, so $4x = -12$. Divide both sides by 4, and $x = -3$.

Alternatively, the answer could be determined by backsolving. For (A), $|4(-4) -6| = |-16 - 6| = |-22| = 22$. For (B), $|4(-3) - 6| = |-12 - 6| = |-18| = 18$.

14. **D** Any number divisible by both 3 and 5 must be divisible by 3×5, since 3 and 5 are both prime numbers. That's 15, which is the smallest number. Add another 15 to 15 to find the next number, 30. Then add another 15, which equals 45. The next number would be 60, but that's above the range of values indicated in the question. So there are 3 possible values: 15, 30, and 45.

15. **C** Since the number of dollars in the cash prize must be equally divided among all members of each team and there is no remainder, the number of dollars must be a multiple of each. First, find the prime factors of each number: $10 = 2 \times 5$, $15 = 3 \times 5$, and $25 = 5 \times 5$. The correct answer will be a common multiple of the different cash prize amounts. The least common multiple is found by taking each distinct prime factor (in this case, 2, 3, and 5), and raising each to the highest power of each factor. Because 2 shows up once, 3 shows up once, and 5 shows up twice, the lowest common multiple is $2 \times 3 \times 5 \times 5$. The correct answer must have at least one 2, one 3, and two 5s to be a multiple. The factorization of 120 is $2 \times 2 \times 2 \times 3 \times 5$, so it's missing one factor of 5 to be a multiple. The factorization of 375 is $3 \times 5 \times 5 \times 5$, so it's missing a 2. The factorization of 750 is $2 \times 3 \times 5 \times 5 \times 5$, so this is a multiple. The factorization of 1,250 is $2 \times 5 \times 5 \times 5 \times 5$, so it's missing a factor of 3. Note that because $2 \times 3 \times 5 \times 5 = 150$, another way to find the correct answer would be to look for a choice that is a multiple of 150. $750 = 150 \times 5$, so 750 is correct.

Review and Reflect

After completing this chapter, think about the areas that you fully understand. Are there any areas still giving you trouble? Pay close attention to the themes that help you on multiple types of questions, such as finding the least common multiple and determining factors. Also, be sure that you practice simplifying fractions and performing basic calculations without a calculator.

Want more instruction and practice with Arithmetic? Check out the companion videos for this section. Also, try more questions in the Qbank online.

COMPANION VIDEOS

kaptest.com/login

CHAPTER 7

MATHEMATICS KNOWLEDGE

KNOW WHAT TO EXPECT

The Mathematics Knowledge subtest gives you 23 minutes to answer 15 questions, if you take the computer-adaptive test. If you take the paper-and-pencil version, then you have 24 minutes to answer 25 questions. Thus, the pace on the Mathematics Knowledge section is quicker than on the Arithmetic Reasoning section, giving you about a minute per problem.

You may see applied arithmetic problems on the MK subtest. Review these topics in chapter 6: Arithmetic Reasoning. Additionally, expect to see algebra and geometry on the MK subtest.

Much of what you learned in the previous chapter about arithmetic applies to algebra as well. The main difference between arithmetic and algebra is that algebra involves variables as well as numbers. In this chapter, you will learn how to work with

- algebraic expressions
- monomials and binomials
- quadratic equations
- solving for variables
- inequalities

You will also learn in this chapter about geometry on the ASVAB. You will learn the rules that apply to lines and angles and to two- and three-dimensional shapes, including

- triangles
- quadrilaterals
- circles
- solid geometry
- coordinate geometry

ALGEBRA REVIEW

The difference between algebra and arithmetic is that algebra uses symbols called **variables**, such as ***x*** and ***y***, to generalize arithmetic relationships.

Algebra problems on the ASVAB can appear in two forms: as straightforward math problems or as word problems. You will find that algebra problems on the Mathematics Knowledge section are often good candidates for a strategic approach, such as Picking Numbers or Backsolving. Nonetheless, if you are comfortable with algebra, using straightforward math may be faster than using a strategy on some problems. You want to have all these tools—Picking Numbers, Backsolving, Strategic Guessing Using Logic, a combination of approaches, and straightforward math—in your toolkit on Test Day.

Terms and Expressions

LEARNING OBJECTIVES

In this section, you will learn how to:

- recognize whether terms can be combined
- add, subtract, multiply, and divide terms
- multiply binomials and factor quadratic expressions

A **term** is a number, a variable, or one or more numbers multiplied by one or more variables. Here are some examples:

4 $\quad$ $3x$ $\quad$ $7ab^2$ $\quad$ $\frac{6z}{4}$

An algebraic **expression** on the ASVAB test could look like this:

$(11 + 3x) - (5 - 2y)$

If an expression consists of one term, it is a **monomial**. If it consists of two terms, it is a **binomial**. If it consists of more than two terms, like the one above, then it is a **polynomial**.

You can only add or subtract "like terms." **Like terms** have the same variables and the same exponents. The following pairs of terms are like terms:

$3a + 7a$ $\quad$ $10z^2 - 5z^2$

To combine like terms, add or subtract the coefficients (the numbers that come before the variables) of terms that have the same variable raised to the same exponent as explained in chapter 6. Do not change the variables. Here are some examples:

$6a + 5a = 11a$

$8b^2 - 2b^2 = 6b^2$

$3a + 2b - 8a = 3a - 8a + 2b = -5a + 2b$ or $2b - 5a$

Remember, you cannot combine unlike terms such as these:

$6a + 5a^2$

$3a + 2b$

Study this example:

Question	Analysis
Simplify: $(11 + 3x) - (5 - 2x)$	
$= 11 + 3x - 5 + 2x$	Distribute the minus sign across the parentheses.
$= 3x + 2x + 11 - 5$	Group like terms together.
$= 5x + 6$	Combine like terms.

Multiplying and dividing monomials is different from adding and subtracting them. In addition and subtraction, you can combine only like terms. With multiplication and division, however, you can multiply and divide terms that are different. When you multiply monomials, multiply the coefficients of each term, add the exponents of like variables, and multiply different variables together. Study this example:

Question	Analysis
Simplify: $(6a)(4b)$	
$= (6 \times 4)(a \times b)$ $= 24ab$	Multiply the coefficients by each other and the variables by each other.

Now try one on your own:

Simplify: $(6a)(4ab) + (3a^2)(8b) = ?$

Explanation

Step 1: The question presents an expression.

Step 2: You need to simplify the expression.

Step 3: Multiply the coefficients together, and multiply the variables together. Remember not to add or subtract unless you're dealing with like terms.

$(6a)(4ab) + (3a^2)(8b) =$

$(6 \times 4)(a \times a \times b) + (3 \times 8)\ (a^2 \times b) =$

$(6 \times 4)(a^{1+1} \times b) + (3 \times 8)\ (a^2 \times b) =$

$24a^2b + 24a^2b =$

$48a^2b$

The answer is $\mathbf{48a^2b}$.

Step 4: Review your work to ensure you multiplied correctly.

To multiply binomials, use the FOIL method. FOIL stands for **F**irst, **O**uter, **I**nner, **L**ast and refers to the position of each term in the parentheses. Here's an example:

Question	Analysis
Simplify: $(y + 1)(y + 2)$	
First terms: $(y \times y)$ Outer terms: $(y \times 2)$ Inner terms: $(1 \times y)$ Last terms: (1×2) $= y^2 + 2y + y + 2$	Use FOIL to multiply each term in the binomials.
$= y^2 + 3y + 2$	Combine like terms.

The final expression above, $y^2 + 3y + 2$, is known as a **quadratic expression**. A quadratic expression contains variables raised to the exponent of 2, in other words, variables that are squared. An expression without an exponent is a linear expression.

You may occasionally be called upon to **factor** a quadratic expression or equation on the ASVAB. Do this with "reverse-FOIL." Take a look:

Question		Analysis
Factor: $x^2 - 4x - 21$		
$(x \quad)(x \quad)$		To begin, build parentheses with the First terms that multiply to x^2.
1 3 7 21	−21 −7 −3 −1	What Last terms will multiply to produce −21?
3	−7	Which of these pairs will add to produce −4?
$(x + 3)(x - 7)$		Plug these into the parentheses.

Classic Quadratics

If you learn to recognize the following three common quadratics, you can multiply them and factor them without needing to FOIL or reverse-FOIL:

Classic Quadratic	Example
$(x+y)(x-y)=x^2-y^2$	$(x+3)(x-3)=x^2-9$
$(x+y)^2=x^2+2xy+y^2$	$(a+4)^2=a^2+8a+16$
$(x-y)^2=x^2-2xy+y^2$	$(y-6)^2=y^2-12y+36$

If you see a perfect square, such as x^2 or 16, on each end of a quadratic, it is probably one of these classic quadratics.

Study this example.

Question	Analysis
Which of the following is equivalent to the expression a^2-b^2?	**Step 1:** The question gives a quadratic expression. **Step 2:** You are asked to find the answer that is the same expression, expressed in a different form. **Step 3:** Use math knowledge to recognize that this is a classic quadratic. $a^2-b^2=(a+b)(a-b)$
(A) $(a-b)^2$ (B) $(a+b)^2$ (C) $(a+b)(a-b)$ (D) $a^2-2ab+b^2$	The answer is **(C).** **Step 4:** Briefly check your calculations.

Equations with One Variable

LEARNING OBJECTIVE

In this section, you will learn how to:

- find the value of a variable in an equation

An equation is two expressions separated by an equal sign. The key to **solving equations** is to do the same thing to both sides of the equation until you have your variable isolated on one side of the equal sign and everything else on the other side.

Question	Analysis
Solve for a: $12a + 8 = 23 - 3a$	**Step 1:** The question presents an equation with one variable, a. **Step 2:** The task is to find the value of a. **Step 3:** (B) is a fraction, which makes backsolving potentially awkward. Use straightforward math to isolate the variable. $12a + 8 = 23 - 3a$ $12a = 15 - 3a$ $15a = 15$ $\frac{15a}{15} = \frac{15}{15}$ $a = 1$
(A) 1 (B) $\frac{5}{3}$ (C) 2 (D) 11	**Step 4:** Plug 1 in for a in the equation: $(12)(1) + 8 = 23 - (3)(1)$ $12 + 8 = 23 - 3$ $20 = 20$ Answer choice **(A)**, $a = 1$, works and is correct.

Try it on your own.

Solve for x: $2x + 7 = \frac{x}{4}$.

Explanation

Step 1: This is a linear equation with one variable.

Step 2: You need to find the value of that variable.

Step 3: Multiply each side by 4 to eliminate the fraction:

$4 \times (2x + 7) = \frac{x}{4} \times 4$

$8x + 28 = x$

Subtract x from both sides:
$7x + 28 = 0$

Subtract 28 from both sides:
$7x = -28$

Divide both sides by 7:
$x = -4$

Step 4: To confirm, plug −4 into the original equation for *x*:

$$2(-4)+7=\frac{-4}{4}$$

$$-8+7=-1$$

$$-1=-1$$

The answer −**4** works and is correct.

Equations with Two Variables

LEARNING OBJECTIVES

In this section, you will learn how to:

- find the value of an expression when given the values of the variables
- solve for one variable in terms of another
- find the values of two different variables when given two equations

Sometimes a problem will give you the numerical value of the variables and ask you for the value of an expression with those variables in it. To solve, just plug the values into the equation. Be sure to follow the rules of PEMDAS (see chapter 6 on arithmetic) and be careful with your calculations.

Question	Analysis
If $x = 15$ and $y = 10$, what is the value of $4x(x - y)$?	
$4(15)(15 - 10)$	Plug 15 in for *x* and 10 in for *y*.
$= 4(15)(5)$ $= (60)(5)$ $= 300$	Use PEMDAS and solve.

Sometimes you are given an equation with two variables and asked to **solve for one variable in terms of the other.** This means that you must isolate the variable for which you are solving on one side of the equation and put everything else on the other side. In other words, when you're done, you'll have *x* (or whatever the variable is) on one side of the equation and an expression on the other side.

Question	Analysis
Solve $7x + 2y = 3x + 10y - 16$ for *x* in terms of *y*.	
$7x + 2y = 3x + 10y - 16$ $7x = 3x + 8y - 16$	Subtract 2*y* from both sides to get all the *y*'s on one side of the equation.
$4x = 8y - 16$ $\frac{4x}{4} = \frac{8y - 16}{4}$	Subtract 3*x* from both sides to get all the *x*'s on the other side of the equation.
$x = 2y - 4$	Divide both sides by 4 to isolate *x*. The solution for *x* in terms of *y* is $2y - 4$.

Try the next one on your own.

Solve $16t + 6u(3) + 8 = -4t - 2u - 6$ for t in terms of u.

Explanation

Step 1: You are given an equation with two variables.

Step 2: You need to solve for the variable t, but you will not end up with one numerical value. The variable t will be equal to an expression that contains u.

Step 3: First, use PEMDAS to simplify the left side of the equation.

$16t + 18u + 8 = -4t - 2u - 6$

Then, add $4t$ to both sides to get all of the t's on one side of the equation.

$20t + 18u + 8 = -2u - 6$

Subtract $18u$ from both sides.

$20t + 8 = -20u - 6$

Subtract 8 from both sides.

$20t = -20u - 14$

Divide both sides by 20.

$t = -u - \frac{14}{20} = -u - \frac{7}{10}$

Step 4: Check that you performed all of your calculations correctly.

Systems of Equations

On the ASVAB, you may be presented with two equations with two variables and be asked to solve for the value of one of the variables. There are two ways to do this: substitution and combination.

Substitution

Using the substitution technique, you will use one equation to express one variable in terms of the other. Then you'll substitute that expression for that variable into the other equation.

Question	Analysis
Find the value of x: $6x + y = 15$ $2x - 3y = -5$	**Step 1:** The question gives you two linear equations with two variables. **Step 2:** You need to solve for one variable, x. **Step 3:** It is easy to isolate y in the first equation, so use substitution to do so: $y = -6x + 15$. Plug in $-6x + 15$ for y in the second equation: $2x - 3y = -5$ $2x - 3(15 - 6x) = -5$ Solve for x: $2x - 45 + 18x = -5$ $20x = 40$ $x = 2$

Question	Analysis
(A) $-2\frac{1}{2}$ (B) $\frac{5}{4}$ (C) 2 (D) 3	Choice **(C)** is correct. **Step 4:** Briefly confirm that you followed PEMDAS and performed the calculations correctly.

Alternatively, you could use Backsolving on that problem. Here's how that might work: answer choice (C) is probably the easiest one to start with, since it's a whole number. Plug 2 in for x in $6x + y = 15$ and solve for y:

$$(6)(2) + y = 15$$
$$12 + y = 15$$
$$y = 3$$

Then check $x = 2$ and $y = 3$ in the other equation:

$$2x - 3y = -5$$
$$(2)(2) - (3)(3) = -5$$
$$4 - 9 = -5$$
$$-5 = -5$$

The answer, $x = 2$, works in both equations and is correct.

Combination

The other way to solve for the value of a variable when you have two variables and two equations is to combine the two equations by adding or subtracting them in such a way that you eliminate one of the variables. Then you can solve for the remaining variable.

Question	Analysis
Find the value of a: $4a + 2b = 44$ $6a - 2b = 46$	**Step 1:** The question gives you two linear equations with two variables. **Step 2:** You need to solve for one variable, a. **Step 3:** If you add the two equations together, you will eliminate b. This can be done by using combination. $4a + 2b = 44$ $+ [6a - 2b = 46]$ $10a = 90$ $a = 9$

Question	Analysis
(A) 4 (B) 9 (C) 10 (D) 12	Choice **(B)** is correct. **Step 4:** Briefly confirm that you performed the calculations correctly.

Alternatively, you could use Backsolving on that problem. Here's how that might work: if you decided to start with choice (B), you would plug 9 in for a in $4a + 2b = 44$ and solve for b:

$$(4)(9) + 2b = 44$$
$$36 + 2b = 44$$
$$2b = 8$$
$$b = 4$$

Then check $a = 9$ and $b = 4$ in the other equation:

$$6a - 2b = 46$$
$$(6)(9) - (2)(4) = 46$$
$$54 - 8 = 46$$
$$46 = 46$$

The answer, $a = 9$, works in both equations and is correct.

Try it on your own.

Solve for x:

$5y + 17x = 90$

$-2x - 5y = -30$

Explanation

Step 1: You are given two linear equations with two variables.

Step 2: You need to find the value of one of the variables, x.

Step 3: The term $5y$ appears in both equations, so combination will allow you to get rid of the y term efficiently. Rearrange the terms of the first equation to make them easier to add together.

$$\begin{array}{r} 17x + 5y = 90 \\ +[\,-2x - 5y = 30] \\ \hline 15x = 60 \\ x = 4 \end{array}$$

The correct answer is $x =$ **4**.

Step 4: To confirm your answer, check that you performed all of the calculations correctly.

Quadratic Equations with Variables

LEARNING OBJECTIVE

In this section, you will learn how to:

- solve for the value or values of a variable in a quadratic equation

When an equation has a quadratic expression on one side and zero on the other side, you can use factoring to solve for the value or values of the variable.

Question	Analysis
Solve for y: $y^2 + 4y + 3 = 0$	
$(y + 3)(y + 1) = 0$	Factor the left side of the equation.
If $y + 3 = 0$, then $y = -3$. If $y + 1 = 0$, then $y = -1$. Therefore, $y = -3$ or -1.	Whenever multiplication results in zero, at least one of the factors must equal zero. Use this fact to solve for *y*.

Note that the solution to a quadratic equation often has two values. The only time you will get a single value for the variable is when the factors are the same, as in this example that uses one of the classic quadratics you learned about above:

Question	Analysis
Solve for x: $x^2 - 6x + 9 = 0$.	
$(x - 3)(x - 3) = 0$	Either factor the left side of the equation using reverse-FOIL or use your knowledge of classic quadratics to factor.
If $x - 3 = 0$, then $x = 3$.	Whenever multiplication results in zero, at least one of the factors must equal zero. Since both of the factors are the same, you will only get one result for *x*.

If you see a quadratic expression set equal to a quantity other than zero, simply manipulate the equation so that the expression is set equal to zero. Similarly, if you see a quadratic expression presented in a different order, simply move the terms around in order to put it in the form $ax^2 + bx + c = 0$. Study this example:

Question	Analysis
Solve for x: $-4 + 2x = 31 - x^2$	This equation appears to have all the parts of a quadratic equation, but it will require some manipulation before you can use reverse-FOIL.

$-4 + 2x - 31 = -x^2$ $-4 + 2x - 31 + x^2 = 0$	First, subtract 31 from both sides, and add x^2 to both sides, so that the expression is set equal to zero.
$x^2 + 2x - 4 - 31 = 0$ $x^2 + 2x - 35 = 0$	Now rearrange and combine the terms, so that the equation has the form $ax^2 + bx + c = 0$.
	Now you're ready to perform reverse-FOIL. What are the factors of 35? 1 35 5 7 One of those factors will be negative, since the third term is negative. If so, which two numbers listed above can have the product of −35 and a difference of +2? −5 +7
$x^2 + 2x - 35 = 0$ $(x - 5)(x + 7) = 0$ $x = 5$ or -7	Therefore, $x =$ **5** or **−7**.

Inequalities

LEARNING OBJECTIVE

In this section, you will learn how to:

- solve for the range of values of a variable in an inequality

PLOTTING INEQUALITIES ON THE NUMBER LINE

An **inequality** is represented by one of the following four signs:

$x > 2$ x is greater than 2.

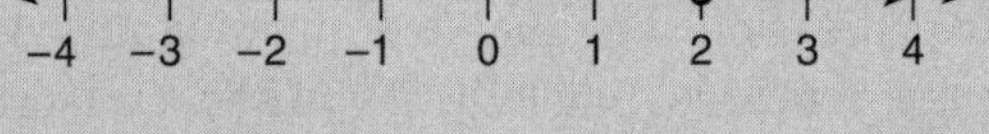

$x < -1$ x is less than −1.

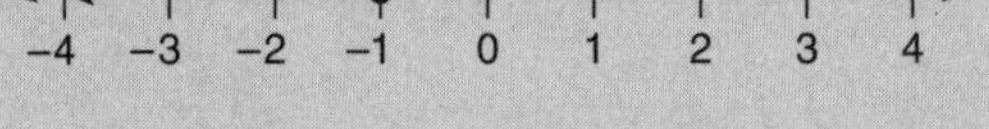

$x \geq 2$ x is greater than or equal to 2.

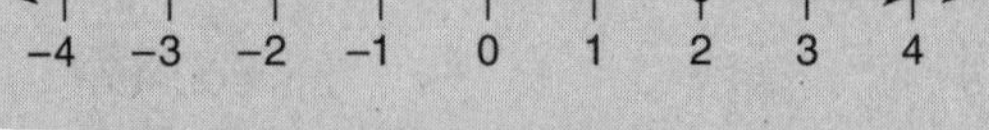

$x \leq -1$ x is less than or equal to −1.

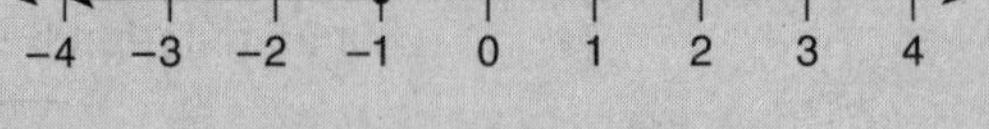

Solve inequalities in the same way you have solved equations. Isolate the variable that you are solving for on one side of the equation and put everything else on the other side of the equation. However, when you have solved an inequality, your solution will typically be a range of values rather than a single value. Study the example below.

Question	Analysis
Solve for a: $4a + 6 > 2a + 10$	
$4a - 2a + 6 > 10$	Subtract $2a$ from both sides.
$4a - 2a > 10 - 6$	Subtract 6 from both sides.
$2a > 4$	Combine like terms.
$a > 2$	Divide both sides by 2.

In this example, the range of possible values of a are all numbers larger than 2. That is, a can be any value greater than 2 but cannot equal 2. Most of the math for inequalities is the same as if you were solving an equation. There is, however, one crucial difference between solving equations and inequalities. When you multiply or divide an inequality by a *negative* number, you must change the direction of the inequality sign. So for example, when you divide both sides of the inequality $-5a > 10$ by -5 to isolate a, you change the direction of the sign: $a < -2$. Because of this rule, you *cannot* multiply or divide both sides of an inequality by a variable, since you do not know whether the variable has a positive or negative value.

To understand why this is so, consider this example, which uses integers:

$1 < 2$

If you multiply both sides by -1:

$-1 \; ? \; -2$

Which is greater? The number -1 is greater than -2, so you need to use a greater than sign:

$-1 > -2$

The inequality sign has flipped. You can test the previous example, in which a is less than -2, by plugging in a couple of numbers. Plug in -3 for a, and you'll find that $(-5)(-3) > 10$ becomes $15 > 10$, a true statement. Plug in a number greater than -2, say 0, and you'll find that $(-5)(0) > 10$ becomes $0 > 10$, which is not a true statement. Only by flipping the inequality sign do you get the correct answer.

Try an inequality question on your own. This question does not have answer choices. Express your answer as an inequality.

Solve for z:

$$-\frac{4z}{2} - 3 > 4z$$

Explanation

Step 1: You are given an inequality.

Step 2: You are asked to solve for the variable, z.

Step 3: First, simplify the fraction:

$$-\frac{4z}{2} - 3 > 4z$$

$$-2z - 3 > 4z$$

Isolate z on one side of the inequality by adding $2z$ to both sides:

$$-3 > 6z$$

Divide both sides by 6:

$$\frac{-3}{6} > \frac{6}{6}z$$

$$-\frac{1}{2} > z$$

If this were Test Day, you might see the correct answer stated that way, or it might be reversed, like this:

$$z < -\frac{1}{2}$$

Step 4: Double-check your math. Be sure to check that you used all of the appropriate signs.

Check out the companion video for this section.

COMPANION VIDEO

kaptest.com/login

GEOMETRY REVIEW

Throughout the geometry problems on the ASVAB, you will be presented with shapes and diagrams. Unless otherwise noted, you cannot assume the drawings are drawn to scale. Thus, work using the information given in a diagram; don't rely on how it looks.

Lines and Angles

LEARNING OBJECTIVES

In this section, you will learn how to:

- find the length of a line segment
- use the properties of straight lines, intersecting lines, and parallel lines to solve for the value of angles

Line Segments

A **line segment** is a piece of a line, and it has an exact measurable length. Questions may give you a segment divided into several pieces, give you the measurements of some of these pieces, and ask you for the length of another piece.

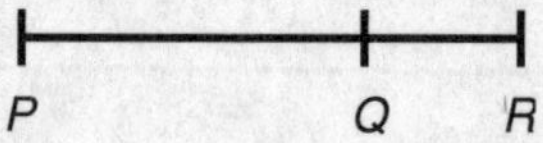

Question	Analysis
If $PR = 12$ and $QR = 4$, $PQ = ?$	
$PQ = PR - QR$	Subtract *QR* from the whole segment to find the remaining piece, *PQ*.
$PQ = 12 - 4$ $PQ = 8$	Plug in the lengths given and solve.

The point exactly in the middle of a line segment, halfway between the endpoints, is called the **midpoint** of the line segment. To **bisect** means to cut in half, so the midpoint of a line segment bisects that line segment.

M is the midpoint of AB, so $AM = MB$.

Angles

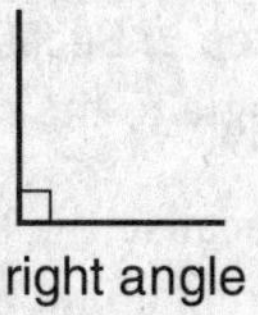

A **right angle** measures 90 degrees and is usually indicated in a diagram by a little box. The figure above is a right angle. Lines that intersect to form right angles are said to be **perpendicular.**

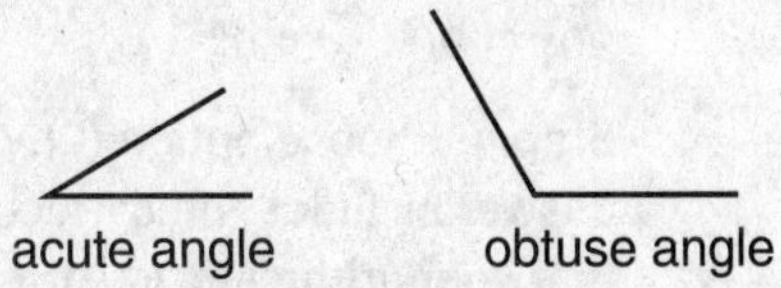

An **acute angle** measures less than 90 degrees. An **obtuse angle** measures more than 90 degrees.

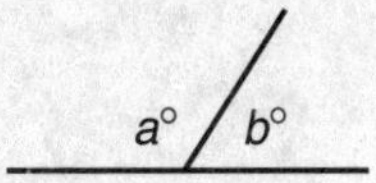

Angles that form a straight line add up to 180 degrees. In the figure above, $a + b = 180$.

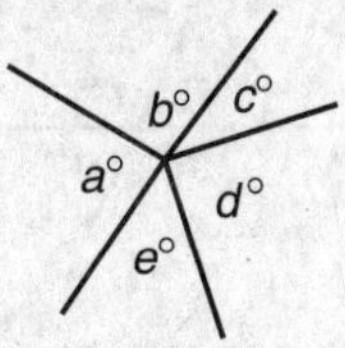

Angles around a point add up to 360 degrees. In the figure above, $a + b + c + d + e = 360$.

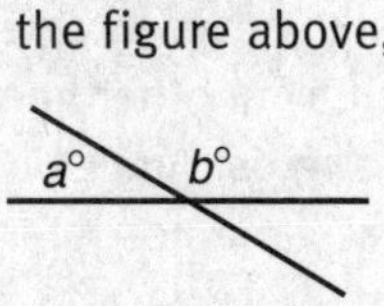

When two lines intersect, adjacent angles are **supplementary**, meaning they add up to 180 degrees. In the figure above, $a + b = 180$.

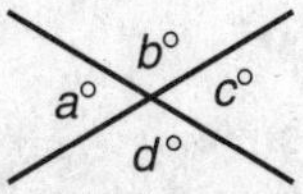

When lines intersect, angles across the vertex from each other are called **vertical angles** and are equal to each other. Above, $a = c$ and $b = d$.

Question	Analysis
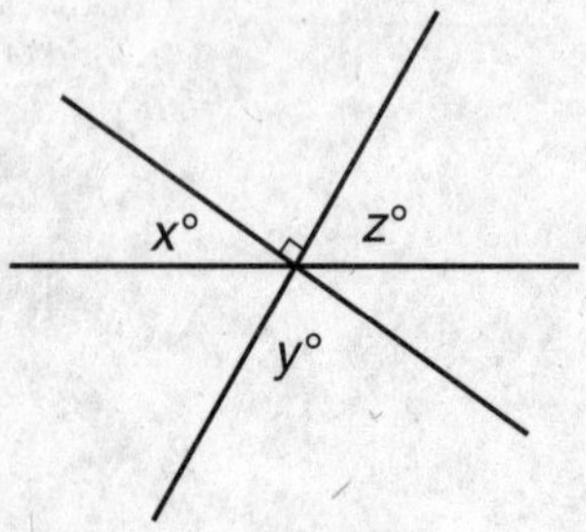 What is the value of $x + y + z$?	**Step 1:** The figure shows three intersecting lines that form six angles, which all together equal 360°. An unlabeled angle is 90°. Three labeled angles are *x*, *y*, and *z*. **Step 2:** The question asks for the sum of the three labeled angles. **Step 3:** Because the angle labeled *y* is a vertical angle to the right angle, *y* must equal 90°. Because the angles labeled *x* and *z* form a straight line with the right angle, together they equal $180° - 90° = 90°$. $y + (x + z) =$ $90° + 90° = 180°$
(A) 90 (B) 120 (C) 180 (D) 270	**Step 4:** Choose option **(C)**. Confirm that the answer includes the correct three angles and that your arithmetic is correct.

Parallel Lines

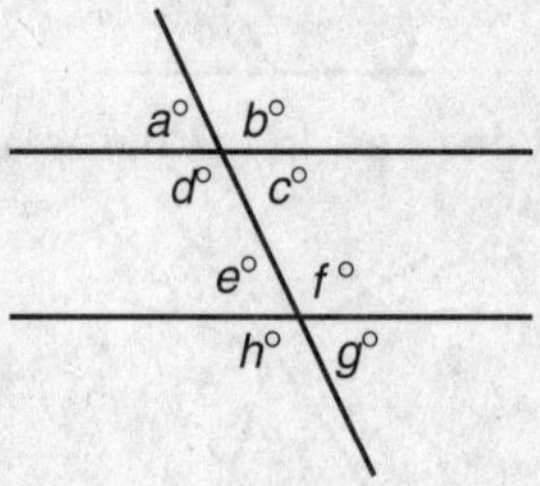

Parallel lines are lines that have the same slope as one another. The **slope** represents the steepness of the line and the direction it goes. Therefore, lines that are parallel never intersect each other. You can imagine them as being like railroad tracks. When parallel lines are crossed by another line, this line is called a **transversal**, and angles are formed. All of these angles are said to be **interior angles**. Angles with the same relationship to the transversal are called **corresponding angles**. In the gray box below, variables refer to the diagram above.

PROPERTIES OF ANGLES MADE BY A TRANSVERSAL CROSSING PARALLEL LINES

- Corresponding angles are equal (for example, $\boldsymbol{a = e}$).
- Alternate interior angles are equal ($\boldsymbol{d = f}$).
- Same-side interior angles are supplementary ($\boldsymbol{c + f = 180}$).
- All four acute angles are equal, as are all four obtuse angles (for example, $\boldsymbol{c = e}$ and $\boldsymbol{b = h}$).

Question	Analysis
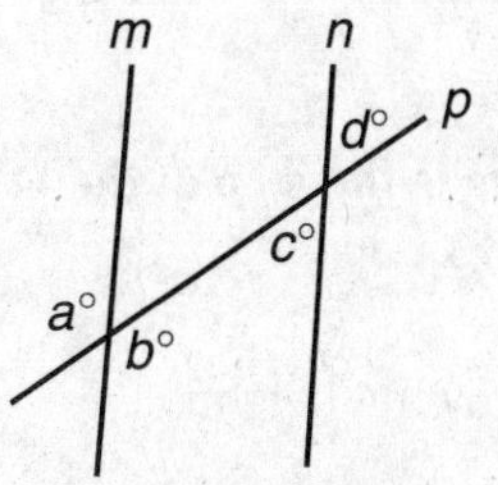 $m \parallel n$ What is the value of $a + b + c + d$?	**Step 1:** The figure shows two parallel lines, *m* and *n*, crossed by transversal *p*. **Step 2:** Find the sum of the four labeled angles. **Step 3:** Two of the labeled angles are acute, and two are obtuse. Every pair of acute + obtuse angles formed by a transversal crossing parallel lines equals 180°. Two acute + obtuse angle pairs = 2 × 180° = 360°.
(A) 120 (B) 180 (C) 200 (D) 360	Choice **(D)** is correct. **Step 4:** Confirm that the answer includes the correct four angles and that your arithmetic is correct.

Polygons

LEARNING OBJECTIVES

In this section, you will learn how to:

- find the perimeter and area of a triangle and a quadrilateral
- solve for the length of one side of a right triangle if given the other two sides

Triangles

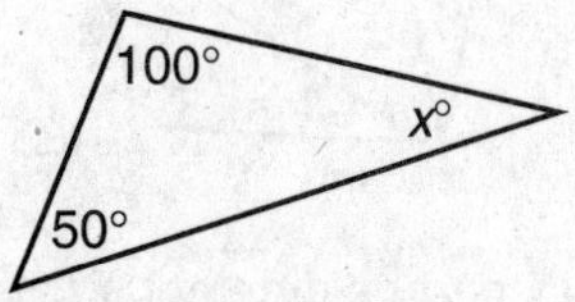

The three **interior angles** of any triangle add up to 180°. In the figure above, $x + 50 + 100 = 180$, so $x = 30$.

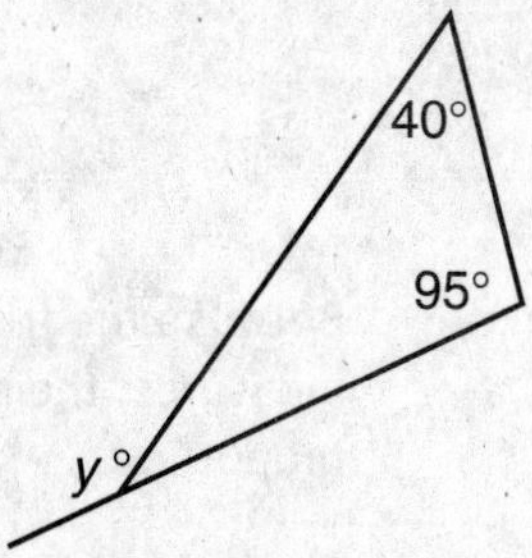

An **exterior angle** of a triangle is equal to the sum of the two interior angles that are not adjacent to that exterior angle. In this figure, the exterior angle labeled *y* is equal to the sum of the remote angles: $y = 40 + 95 = 135$.

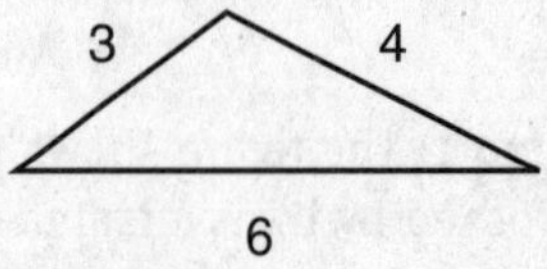

The **perimeter** of a polygon is the sum of the lengths of its sides. The perimeter of a triangle is the sum of the lengths of its three sides. The perimeter of the triangle in the figure above is $3 + 4 + 6 = 13$.

The area of a triangle is equal to half of the base (that is, the length of one leg) multiplied by the height.

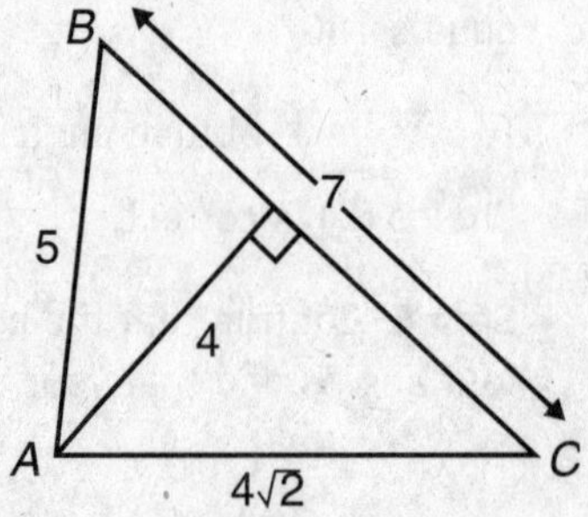

Area of triangle $= \frac{1}{2}$(base)(height)

The **height** is the perpendicular distance between the side that's chosen as the **base** and the opposite **vertex** (opposite corner). In this triangle, 4 is the height when the side with measure 7 (*BC*) is chosen as the base. In the triangle above:

$$\text{Area} = \frac{1}{2}bh = \frac{1}{2}(7)(4) = 14$$

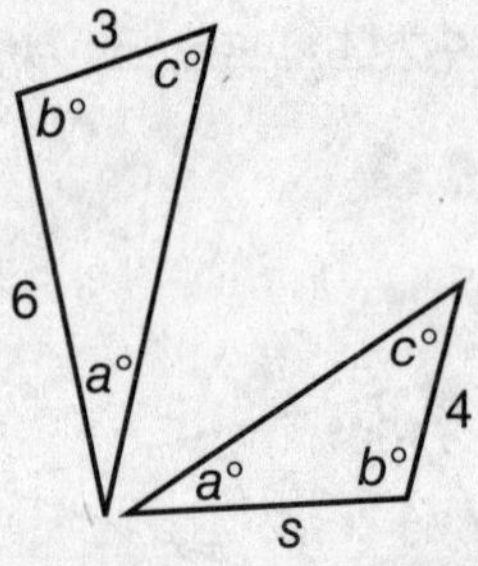

Similar triangles have the same shape, but they may be different sizes. Their corresponding angles are equal, and their corresponding sides are proportional. The triangles shown above are similar because they have the same angles.

Question	Analysis
Find the value of s in the diagram above.	
$\frac{3}{4} = \frac{6}{s}$	Because the triangles have the same angle measures, their sides are proportional. Set up a proportion.
$3s = (4)(6)$ $3s = 24$ $s = 8$	Cross-multiply to solve for s.

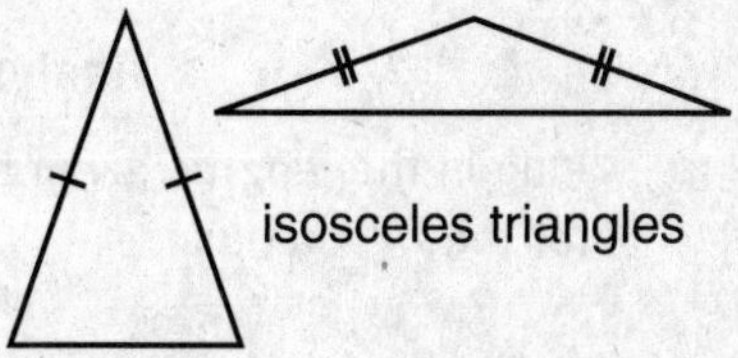

An **isosceles triangle** is a triangle that has two equal sides. Not only are two sides equal, but the angles opposite the equal sides, called **base angles**, are also equal.

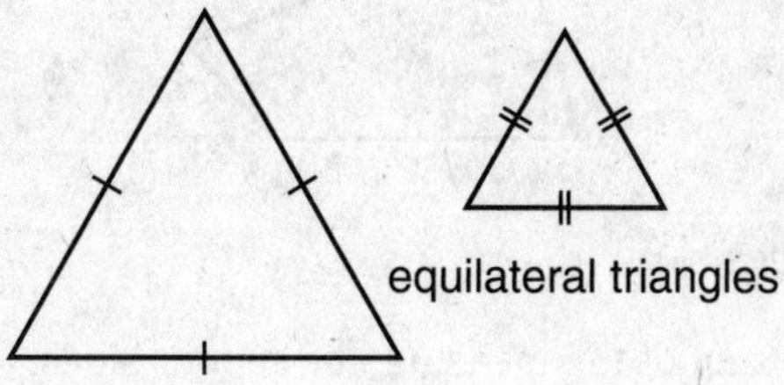

Equilateral triangles are triangles in which all three sides are equal. Since all the sides are equal, all three angles are also equal. All angles in an equilateral triangle measure 60 degrees, regardless of the length of the sides.

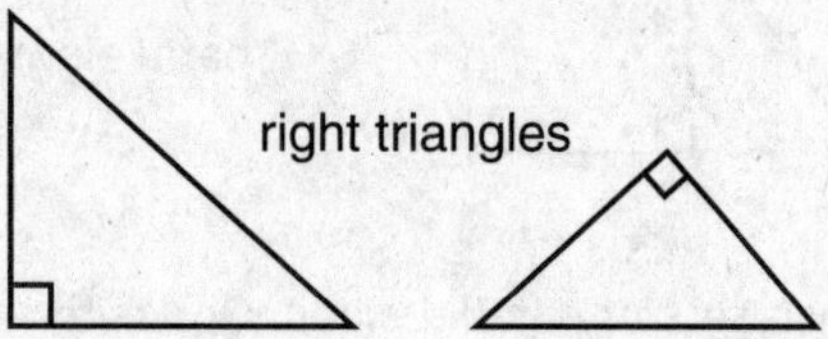

A **right triangle** is a triangle with a right angle. Every right triangle has exactly two acute angles. The sides opposite the acute angles are called the **legs**. The side opposite the right angle is called the **hypotenuse**. Since it's opposite the largest angle, the hypotenuse is the longest side of a right triangle.

If you know two sides of a right triangle, you can solve for the third side using the **Pythagorean theorem**:

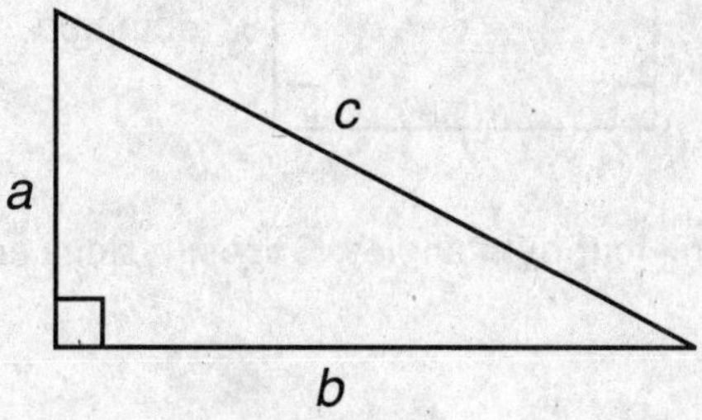

$$a^2 + b^2 = c^2$$

In this equation, *a* and *b* are the perpendicular sides, or legs, of a right triangle, and *c* is the longest side, or hypotenuse, of the triangle.

Question	Analysis
One leg of a right triangle is 2 and the other leg is 3. What is the length of the hypotenuse?	
$a^2 + b^2 = c^2$	Use the Pythagorean theorem.

Question	Analysis
$2^2 + 3^2 = c^2$ $4 + 9 = c^2$ $13 = c^2$ $c = \sqrt{13}$	Plug in the lengths given for the legs and solve for the hypotenuse.

Quadrilaterals

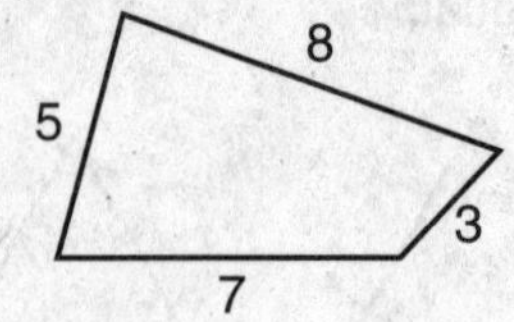

A **quadrilateral** is a figure with four sides.

The perimeter of a quadrilateral is the sum of the measures of all four sides. The perimeter of the quadrilateral in the figure above is $5 + 8 + 3 + 7 = 23$.

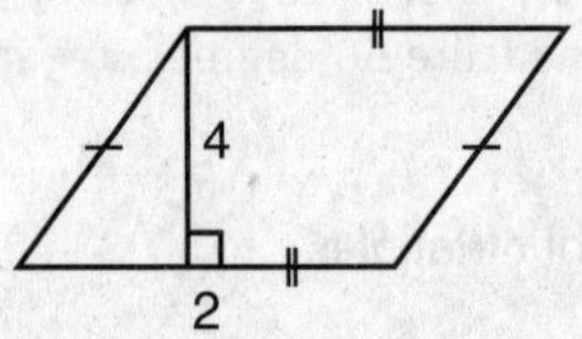

A **parallelogram** is a quadrilateral with two sets of parallel sides. Opposite sides are equal, as are opposite angles. The formula for the area of a parallelogram is

Area = (base)(height)

In the diagram above, the base is 2 and the height is 4, so area $= 2 \times 4 = 8$.

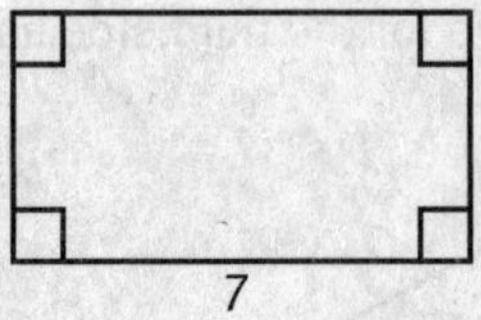

A **rectangle** is a parallelogram containing four right angles. Opposite sides are equal. The formula for the area of a rectangle is

Area = (length)(width)

In the diagram above, the length is 7 and the width is 3, so area $= 7 \times 3 = 21$.

As with any other polygon, the perimeter of a rectangle is all the sides added together. Because a rectangle has two pairs of equal sides, the formula for the perimeter of a rectangle is

Perimeter $= 2(l + w)$

The perimeter of the above rectangle is $2(7 + 3) = 2 \times 10 = 20$.

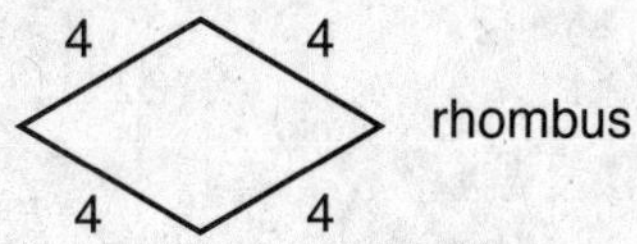

A **rhombus** has four equal sides, with the opposite sides parallel to one another.

The formula for the perimeter of a rhombus is

Perimeter $= 4s$

The perimeter of the above rhombus is $4(4) = 16$.

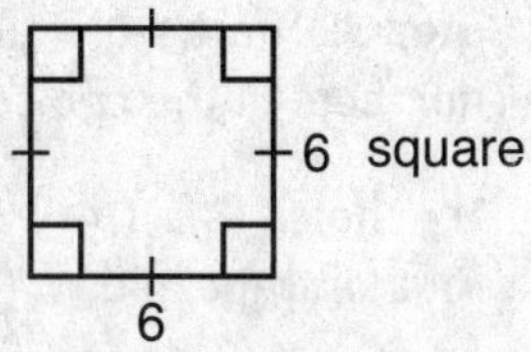

A **square** is a rectangle with four equal sides. Here is the formula for the area of a square:

Area $= (\text{side})^2$

In the diagram above, the length of each side is 6, so area $= 6^2 = 36$.

The formula for the perimeter of a square is

Perimeter $= 4s$

The perimeter of the above square is $4(6) = 24$.

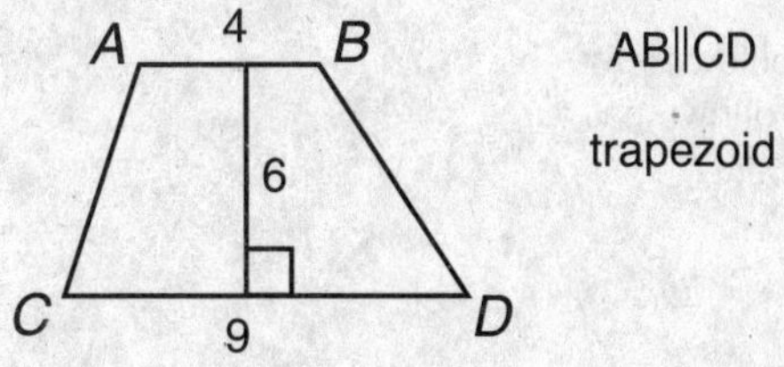

A **trapezoid** is a quadrilateral with one pair of parallel sides. The two parallel sides of a trapezoid are called the **bases**.

The formula for the area of a trapezoid is

Area $= \frac{1}{2}$ (sum of the lengths of the bases)(height)

In the diagram above, the area of the trapezoid is $\frac{1}{2}(4 + 9)(6) = 39$.

Study the example below.

Question	Analysis
The trapezoid below has an area that is equal to greater than twice the area of the square below. Which of the following could be the value of x? 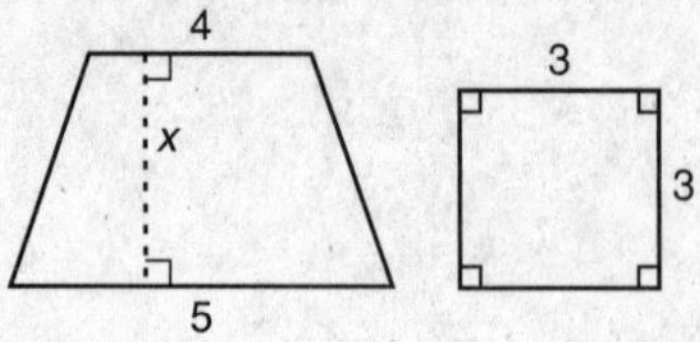 	**Step 1:** You're given two shapes, a square and a trapezoid. Area of square = 9 Area of trapezoid > 2 × Area of square Area of trapezoid > 18 **Step 2:** You're asked for the value of x.
(A) 2 (B) 3 (C) 4 (D) 5	**Step 3:** There's one unknown in the question and numbers in the choices, so use Backsolving. Try choice (B): if $x = 3$, then Area of trapezoid $= \frac{1}{2}(4+5)3 = 13.5$ That's too small. Eliminate (A) and (B). Try choice (C): if $x = 4$, then Area of trapezoid $= \frac{1}{2}(4+5)4 = 18$ That's also too small, since the area of the trapezoid is supposed to be *greater than* 18. Eliminate (C); **(D)** must be correct.
	Step 4: Briefly check that your answer makes sense and move on to the next question.

Now try one on your own.

The value of the area of the trapezoid below is divisible by 5 with no remainder. Which of the following is a possible value of y?

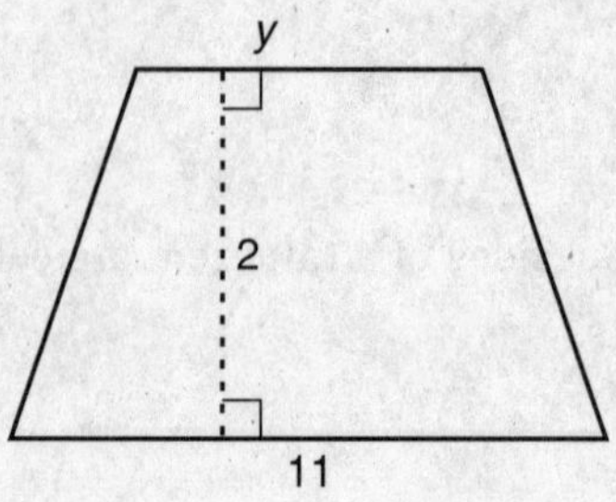

(A) 11
(B) 12
(C) 14
(D) 18

Explanation

Step 1: The area of the trapezoid is a multiple of five. So,

$\frac{1}{2}(y+11)\,2 = (y+11)$ is a multiple of five.

Step 2: You're asked for a possible value of *y*.

Step 3: Start by using Strategic Guessing Using Logic. Multiples of 5 always end in 5 or 0. To produce a number that ends in 5 or 0, *y* would have to end with 4 or 9. Only choice **(C)** fits.

Step 4: Briefly confirm: $\frac{1}{2}(14+11)\,2=(14+11)=25$, which is a multiple of 5.

Circles

LEARNING OBJECTIVE

In this section, you will learn how to:

- determine the circumference and area of a circle

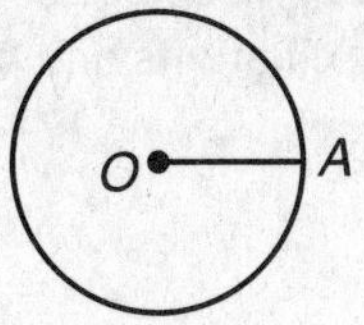

A **circle** is a figure in which each point is an equal distance from its center. In the diagrams in this section, *O* is the center of the circle.

The **radius** of a circle is the straight-line distance from its center to any point on the circle. All radii of a given circle have equal lengths. In the figure above, *OA* is a radius of circle *O*.

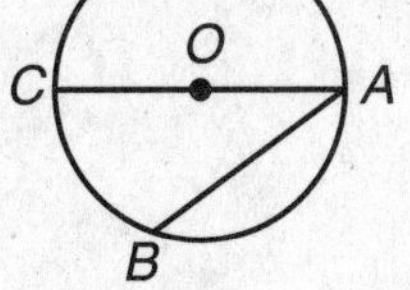

A **chord** is a line segment that connects any two points on a circle. Segments *AB* and *AC* are both chords. The longest chord that can be drawn in a circle is a diameter of that circle.

A **diameter** of a circle is a chord that passes through the circle's center. All diameters are the same length and are equal to twice the radius. In the figure below, *AC* is a diameter of circle *O*.

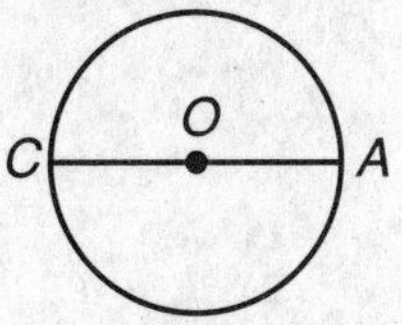

$AC = 6$

The **circumference** of a circle is the distance around it. It is equal to πd or $\mathbf{2\pi r}$. In this example,

$$\text{Circumference} = \pi d = 6\pi$$

The **area** of a circle equals π times the square of the radius, or πr^2. In this example, since *AC* is the diameter, $r = \frac{6}{2} = 3$, and

$$\text{Area} = \pi r^2 = \pi(3^2) = 9\pi$$

Study the following example.

Question	Analysis
In the figure below, the radius of circle A has a length of 1 unit, and the radius of circle B has a length of 2 units. What is the ratio of the area of A to the area of B? 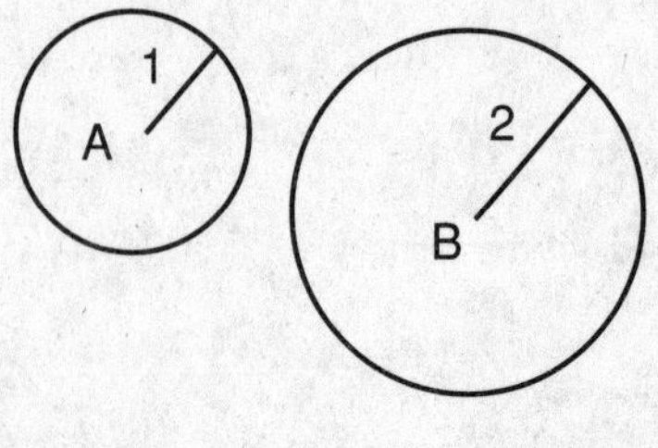 	**Step 1:** The question gives you two circles and their radius measures. **Step 2:** You need the ratio of the area of A to the area of B. The answer choices are all expressed as fractions, so it will look like this: $\frac{\text{Area of A}}{\text{Area of B}}$ **Step 3:** Calculate the area of each. Area of A $= \pi 1^2 = \pi$ Area of B $= \pi 2^2 = 4\pi$ $\frac{\text{Area of A}}{\text{Area of B}} = \frac{\pi}{4\pi} = \frac{1}{4}$
(A) $\frac{1}{4}$ (B) $\frac{1}{2}$ (C) $\frac{1}{\pi}$ (D) $\frac{\pi}{2}$	**Step 4:** Choice **(A)** is correct. Briefly double-check your calculations and confirm.

Now, try this one on your own.

The diameter of the circle below has measure x. The area of the circle is 64π. What is the value of x?

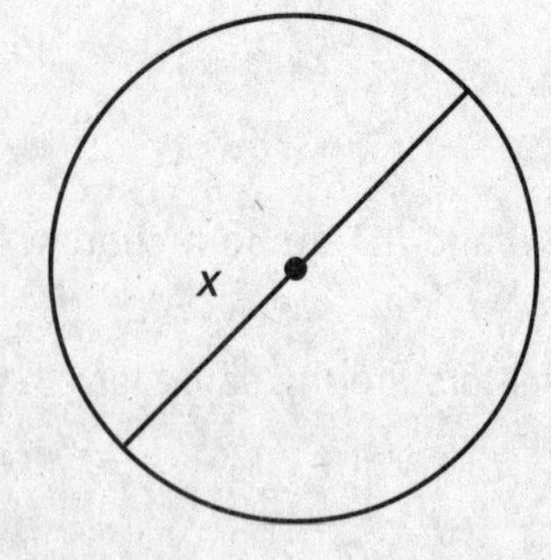

(A) 4
(B) 2π
(C) 8
(D) 16

Explanation

Step 1: The question supplies this information:
$A = 64\pi$
$d = x$

Step 2: You're asked for the measure of the diameter, or x.

Step 3: Work backwards from what you know.

$A = 64\pi = r^2\pi$

$r^2 = 64$

$r = 8$

$x = \text{diameter} = 2r = 16$

Step 4: Choice **(D)** is correct. Briefly double-check your math and confirm.

Solid Geometry

LEARNING OBJECTIVE

In this section, you will learn how to:

- find the volume and surface area of a rectangular solid

These formulas pertain to a rectangular solid box. Length is ℓ, width is w, and height is h.

Volume of a rectangular solid or box $= \ell \times w \times h$
Surface area of a rectangular solid or box $= 2\ell w + 2wh + 2\ell h$

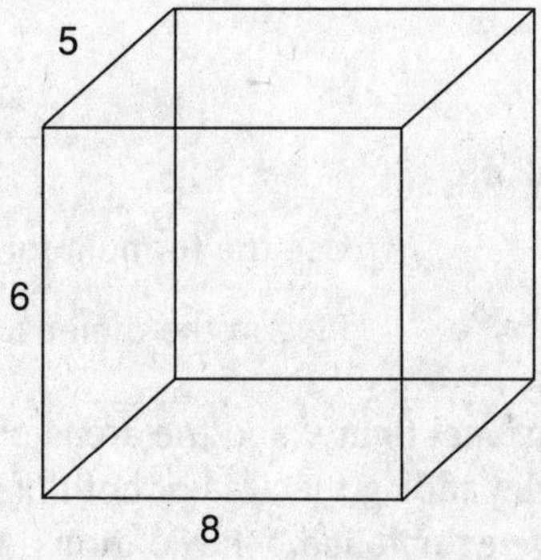

Question	Analysis
Find the volume of the box in the figure above.	
$V = \ell \times w \times h$	Use the formula for the volume of a rectangular solid.
$V = 8 \times 5 \times 6 = 240$	Plug in the dimensions given and solve.

Now try a problem on your own.

Find the surface area of the box above.

Explanation

Step 1: The question provides a diagram of a box.

Step 2: The question asks for the surface area of that box.

Step 3: Use the formula for surface area of a rectangular solid, plug in the values given, and solve:

$SA = 2\ell w + 2wh + 2\ell h$

$SA = (2 \times 8 \times 5) + (2 \times 5 \times 6) + (2 \times 8 \times 6)$

$SA = 80 + 60 + 96 = 236$

Step 4: Confirm that you plugged in the correct values for the correct variables.

To find the volume of a **cylinder**, you find the area of its base (which is a circle) and multiply it by the height.

Volume of a cylinder $= \pi r^2 h$

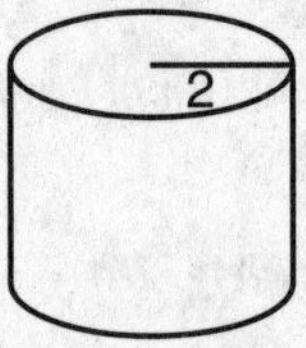

Question	Analysis
Find the volume of the cylinder in the figure above.	
$V = \pi r^2 h$	Use the formula for the volume of a cylinder.
$V = \pi \times 2^2 \times 5 = \pi \times 4 \times 5 = 20\pi$	Plug in the dimensions given and solve.

To find the surface area of any three-dimensional figure, add the areas of each side of the figure. Accordingly, finding the surface area of a cylinder involves adding the area of both of the circular ends to the area of the curved sides. Note that, if the sides of a cylinder were "unrolled," they'd form a rectangle like this:

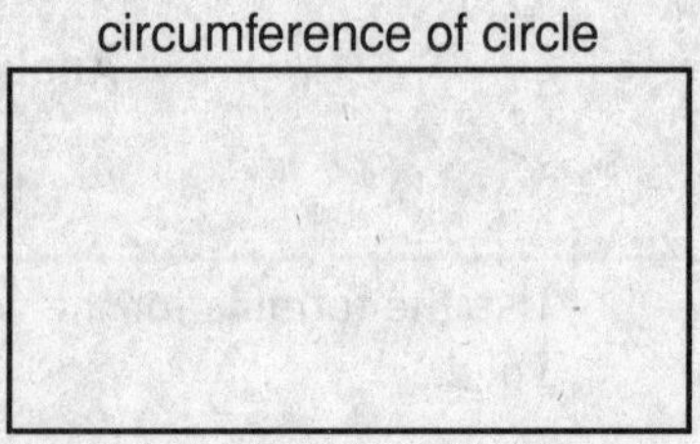

Thus, the formula for the surface area of a cylinder is:

$SA = 2(\pi r^2) + 2\pi rh$

Combined Figures

When you are presented with a diagram that looks like two shapes put together, consider splitting the shapes to make calculations easier. For example, take a look at the next question:

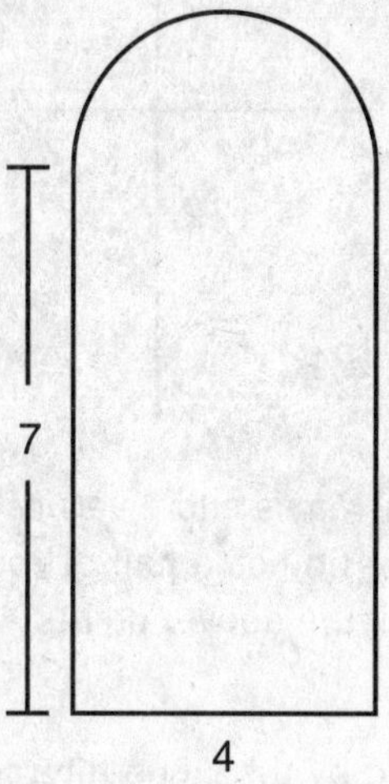

Question	Analysis
Find the area of the figure above.	Note that the figure is half of a circle sitting on top of a rectangle. Split the figure up and work with the two shapes independently.
Area $= l \times w = 7 \times 4 = 28$	Plug the values from the diagram into the area formula for a rectangle.
Area $= \pi r^2 = \pi(2)^2 = 4\pi$	Since 4 is the diameter of the circle, 2 is the radius. Use that information in the area formula for a circle.
$4\pi \div 2 = 2\pi$	Since it is only half of a circle, you need to divide the area by two.
$28 + 2\pi$	Add the area for the rectangle and the area of the half circle together to get your answer.

If you see a complicated figure on the test, try to break it down into more manageable components and solve for each individually.

Coordinate Geometry

LEARNING OBJECTIVES

In this section, you will learn how to:

- choose the equation that correctly represents a given line
- calculate the slope and *y*-intercept of a line if given two points on the line
- find the distance between two given points on a coordinate plane

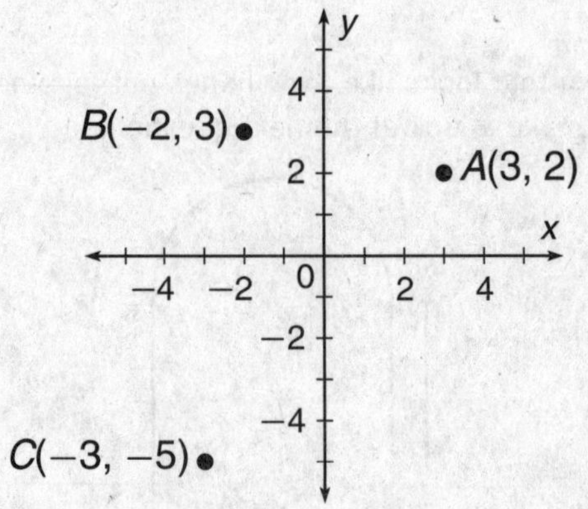

The **coordinate grid** is composed of a horizontal *x*-axis and a vertical *y*-axis. The place where the axes meet is called the **origin**. Every point on the coordinate grid has a pair of **coordinates**, with the first being the *x*-coordinate and the second being the *y*-coordinate. In the figure above, points *A*, *B*, and *C* are at (3, 2), (−2, 3), and (−3, −5), respectively.

A line on the coordinate grid can be represented by the **slope-intercept form of a linear equation**. In this equation, *x* and *y* are the coordinates of a point on the line, *b* is the *y*-intercept, and *m* is the slope. Study the examples below.

Slope-intercept form of a linear equation: $y = mx + b$

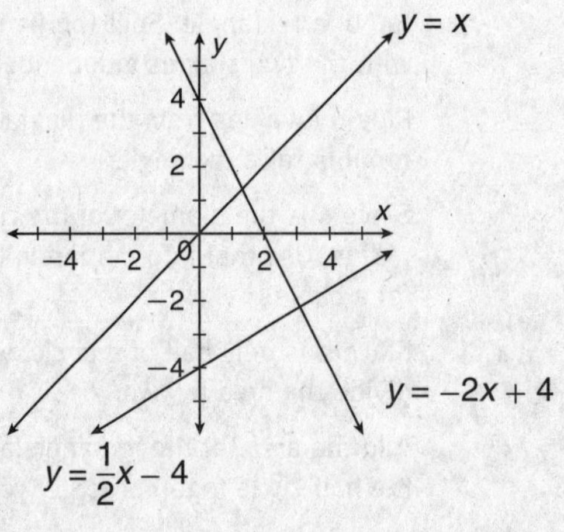

The ***y*-intercept** is the point where the line crosses the *y*-axis. The *x*-coordinate is always zero at the *y*-intercept.

The **slope** defines if a line goes down to the right or up to the right. Slope can be thought of as "rise over run" or "change in *y* over change in *x*."

$$\text{Slope} = \frac{\text{Change in } y}{\text{Change in } x}$$

If a line has a positive slope, then *y* gets larger as *x* gets larger. If a line has a negative slope, then *y* gets smaller as *x* gets larger. The steeper the line, the greater the absolute value of the slope. For example, lines with slopes of −3 or 3 are steeper than lines with slopes of −1 or 1, which in turn are steeper than lines with slopes of $-\frac{1}{3}$ or $\frac{1}{3}$.

Study this example about the coordinate plane.

Question	Analysis
A line on the coordinate grid has a slope of 2 and a y-intercept of -8. What is the correct equation of this line?	
$y = mx + b$	Use the slope-intercept equation for a line.
$y = 2x - 8$	Plug in 2 for m and -8 for b.

Now try this coordinate geometry problem on your own.

The following points lie on a line on the coordinate grid: (0, 2) and (6, 5). What is the slope of this line?

Explanation

Step 1: You are given two points on a coordinate grid.

Step 2: You need to find the slope of the line that contains these two points.

Step 3: Use the formula for the slope of a line:

$$\text{Slope} = \frac{\text{Change in } y}{\text{Change in } x}$$

Plug in the y-coordinates from the second point and the first point, respectively, and the x-coordinates in the same order.

$$\text{Slope} = \frac{5-2}{6-0} = \frac{3}{6} = \frac{1}{2}$$

Step 4: Check your math and your fraction reduction. The line has a slope of $\frac{1}{2}$.

If you know the slope of a line and one point on that line, you can calculate the y-intercept. Continuing the example above, plug the information into the slope-intercept formula and solve for the y-intercept.

Question	Analysis
Find the y-intercept of the line that includes points (0, 2) and (6, 5).	
$y = mx + b$	Use the slope-intercept formula.
$y = \frac{1}{2}x + b$	Plug in the slope, which was calculated above to be $\frac{1}{2}$, for m in the equation.
$2 = \frac{1}{2} \times 0 + b$	Plug in the x- and y-values from one pair of coordinates.
$2 = 0 + b$ $2 = b$	Solve for the variable b.

If you are given two points on a grid and asked to find the distance between them, you can do so using the Pythagorean theorem. Continuing to use the points above, find the distance between the two points on the x-axis and the difference between the two points on the *y*-axis. Then, using those two values as the legs of the right triangle, calculate the hypotenuse, which is the distance between the points.

Question	Analysis
What is the distance between the two points (0, 2) and (6, 5)?	
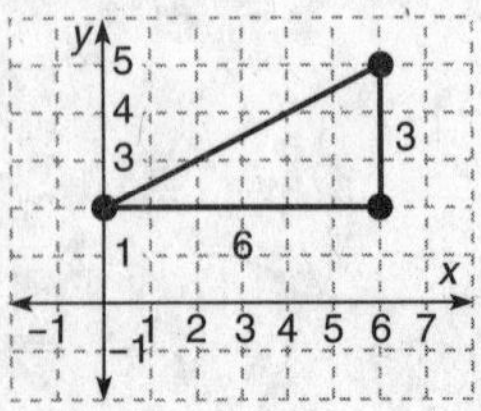	Draw the points on a coordinate plane. Then, draw a third dot that makes a right triangle when all points are connected. The third point will be drawn where the *y*-axis of one point crosses the *x*-axis of the other. The distance along the *x*-axis is 6 and the distance along the *y*-axis is 3.
$\sqrt{6^2+3^2}=\sqrt{36+9}=\sqrt{45}$	Use the Pythagorean theorem to solve for the distance between the two points. The distance between the two points is $\sqrt{\mathbf{45}}$.
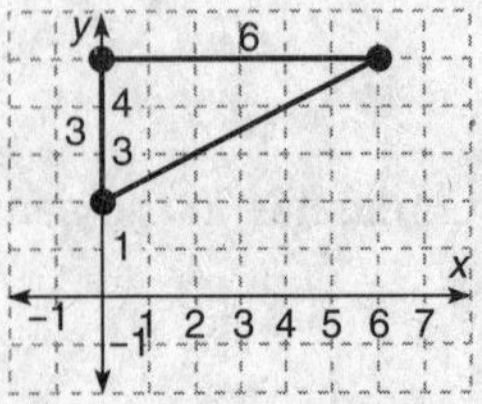	Note: The triangle also could have been drawn with the third point opposite, with the *y*-axis of the lower point intersecting the *x*-axis of the higher point. Either way, the result is the same.

If you are asked to find the distance between two points on the coordinate plane, draw a right triangle to help you use the Pythagorean theorem to solve.

Check out the companion video for this section.

COMPANION VIDEO

kaptest.com/login

MATHEMATICS KNOWLEDGE PRACTICE SET 1

Select the best answer for each question. This practice set contains 15 practice questions, which is the number of Mathematics Knowledge questions you will see if you take the CAT-ASVAB.

1. What is the height of a rectangular box with volume of 140, length of 10, and width of 7?

(A) 2
(B) 17
(C) 70
(D) 9,800

2. If $x = \sqrt{3}$, $y = 2$, and $z = \frac{1}{2}$, then $x^2 - 5yz + y^2 =$

(A) 1
(B) 2
(C) 4
(D) 7

3. Solve for a: $\frac{a}{-2} + 20 < -100$

(A) $a < -240$
(B) $a > 40$
(C) $a > 180$
(D) $a > 240$

4. $(3d - 7) - (5 - 2d) =$

(A) $d - 12$
(B) $5d - 2$
(C) $5d + 12$
(D) $5d - 12$

5. Which of the following are possible values of x, if $x^2 - x - 20 = 0$?

(A) $-5, 4$
(B) $-4, 5$
(C) $-2, 10$
(D) $5, 4$

6. The following two points lie on a line on the coordinate grid: $(-2, 0)$ and $(4, 6)$. Which of the following is the equation of the line?

(A) $x = y + 2$
(B) $y = 3x + 6$
(C) $y = 2x + 1$
(D) $y = x + 2$

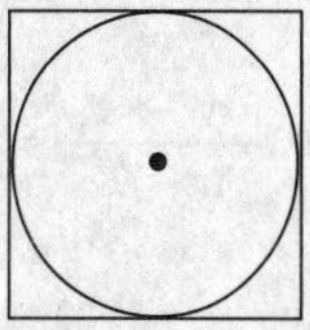

7. If the perimeter of the above square is 36, what is the circumference of the circle?

(A) 6π
(B) 9π
(C) 12π
(D) 18π

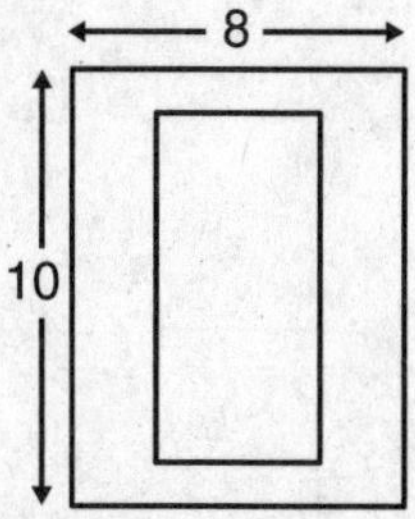

8. What is the area of the frame in the above diagram if the picture inside has a length of 8 and a width of 4?

(A) 16
(B) 24
(C) 48
(D) 56

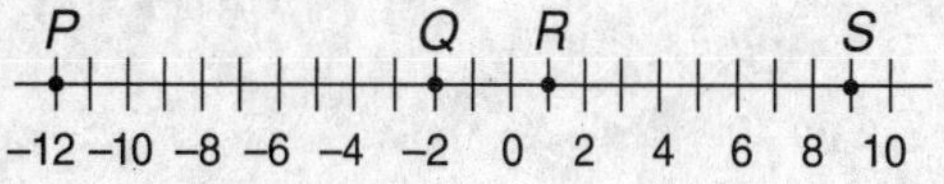

9. In the figure above, what is the distance from the midpoint of $\overline{PQ}$ to the midpoint of $\overline{RS}$?

(A) 12
(B) 14
(C) 16
(D) 18

10. If $s - t = 5$, what is the value of $3s - 3t + 3$?

(A) 11
(B) 12
(C) 15
(D) 18

11. Given the equation $\frac{5xy - y^2}{y - 2} = 21$, if $y = 3$, what is the value of x?

(A) $\frac{1}{2}$
(B) 1
(C) 2
(D) 4

12. At what point do the lines described by the equations $y = 3x - 8$ and $y = -x + 4$ intersect?

(A) (−1, −3)
(B) (−3, −1)
(C) (1, 3)
(D) (3, 1)

13. Given the equation $\frac{rw + 3k}{k + w} = 4$, what is r in terms of w and k?

(A) $r = \frac{k}{w} + 4w$

(B) $r = \frac{k}{w} + 4$

(C) $r = \frac{k}{4w} + w$

(D) $r = \frac{k}{4} + 4w$

14. If $b < 0$ and $\frac{a}{b} > \frac{c}{d}$, which of the following is always true?

(A) $a < \frac{bc}{d}$

(B) $a > \frac{bc}{d}$

(C) $ad > bc$

(D) $a < bcd$

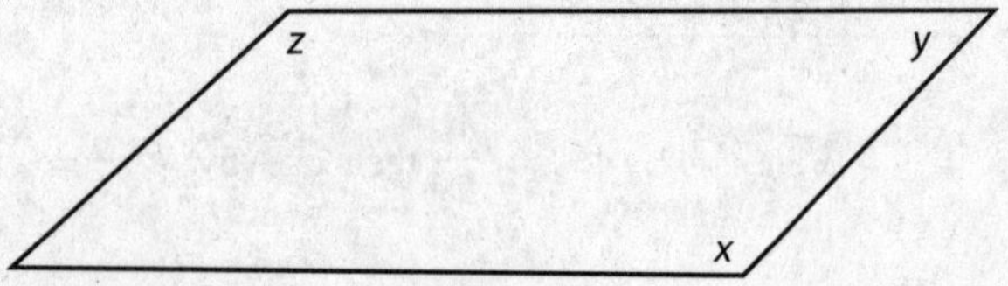

15. In the parallelogram above, if $x = 4y$, what is the value of z?

(A) 36°
(B) 108°
(C) 135°
(D) 144°

MATHEMATICS KNOWLEDGE PRACTICE SET 2

Select the best answer for each question. This practice set contains 15 practice questions, which is the number of Mathematics Knowledge questions you will see if you take the CAT-ASVAB.

1. Which of the following statements cannot be true under any circumstances?

   (A) Triangle *ABC* is an equilateral right triangle.
   (B) Triangle *ABC* is an isosceles right triangle.
   (C) Triangle *ABC* has three acute interior angles.
   (D) Triangle *ABC* has an interior angle greater than 170°.

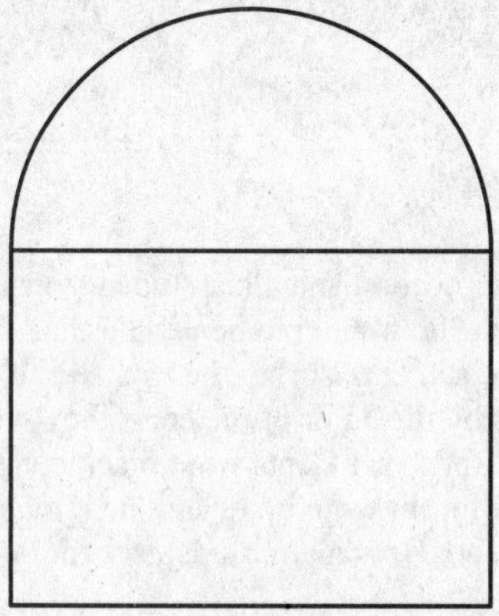

2. In the figure above, what is the area of the semicircle that is on top of the square if the area of the square is 400 cm^2?

   (A) 10π cm^2
   (B) 25π cm^2
   (C) 50π cm^2
   (D) 100π cm^2

3. If $7x + 2y = x - 4y$ and both sides of the equation equal 15, what is the value of x?

   (A) −3
   (B) 1
   (C) 3
   (D) 6

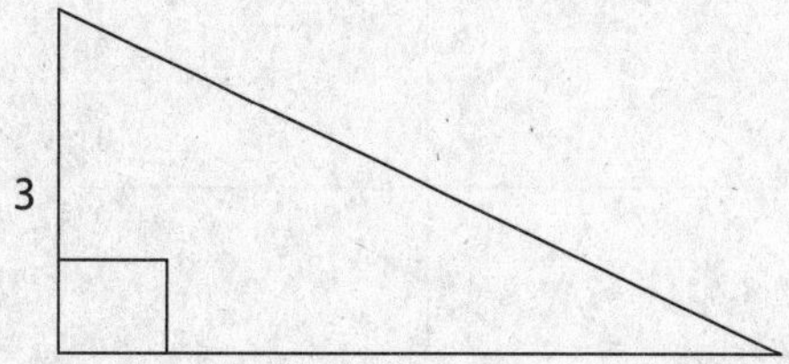

4. If the area of the triangle above is 6, what is the perimeter?

   (A) 9
   (B) 12
   (C) 15
   (D) 18

5. If the perimeter of an equilateral triangle is 24 cm, what is its area?

   (A) $4\sqrt{3}$
   (B) $8\sqrt{3}$
   (C) 24
   (D) $16\sqrt{3}$

6. Solve $\dfrac{(6-3)(3-1)^2}{2+\dfrac{4}{x}} = 5$ for x.

   (A) 2
   (B) 5
   (C) 10
   (D) 20

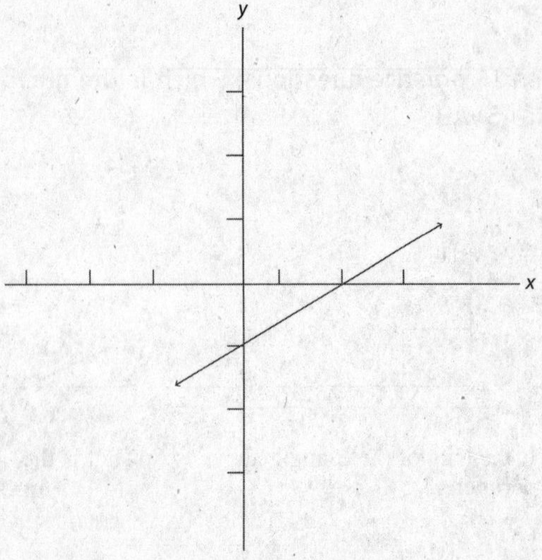

7. If the straight line in the graph were extended to a point where $y = 100$, what would be the value of x at that point?

(A) 100
(B) 102
(C) 200
(D) 202

8. The Greek letter delta, Δ, represents the sensitivity of the price of stock options to price movements of the stock upon which the derivative option is based. If an option has $\Delta = 0.50$, then a \$1.00 increase in the underlying stock should result in a \$0.50 increase in the value of that option. Joseph owns an option on 200 shares of XYZ Corporation. That stock is currently priced at \$45.00 per share. If the price of the stock of XYZ were to decrease by 2%, what would be the expected change in the value of Joseph's option if its $\Delta = 0.40$?

(A) −\$72
(B) −\$36
(C) \$36
(D) \$72

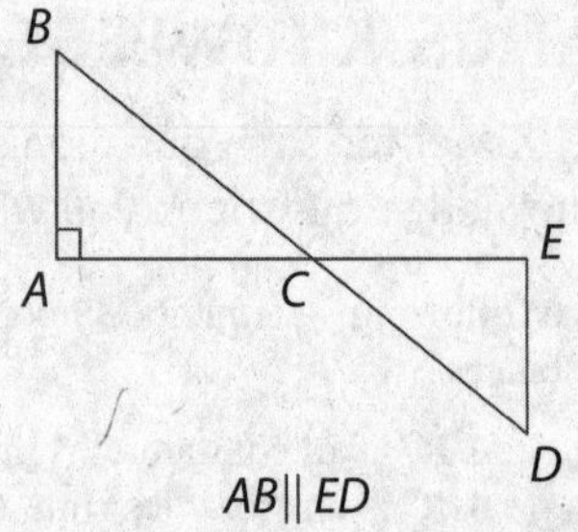

$AB \parallel ED$

9. In the figure above, $ED = 3$ cm, $CD = 5$ cm, and $BA = 4$ cm. What is the perimeter of triangle ABC?

(A) 12 cm
(B) 15 cm
(C) 16 cm
(D) 18 cm

10. A teacher accompanied her students on a trip to the museum. She wanted to be certain that she kept track of all the students during the trip, so she counted the students by the color of the coats they were wearing. One-half of the students wore black coats, one-sixth wore red, one-fourth wore blue, and the remaining two students wore green. How many students went on the trip?

(A) 12
(B) 24
(C) 36
(D) 48

11. Which of the following equations describes a straight line on a standard coordinate plane?

(A) $x = 3$
(B) $y = \frac{1}{x}$
(C) $z = 2x + y$
(D) $y = x^2 + 3$

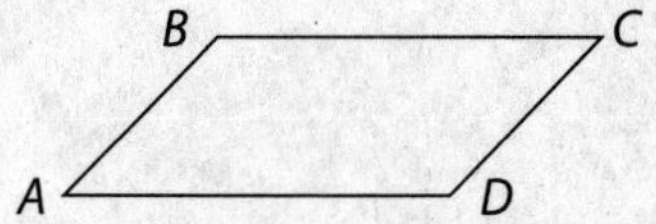

12. Quadrilateral *ABCD* is a parallelogram. Vertex *A* is located at (−5, −2), vertex *B* has coordinates (−3, 3), and the position of vertex *D* is (5, −2). What is the length of diagonal *AC*?

    (A) 10
    (B) 13
    (C) 15
    (D) 20

13. If $x + y = 2$ and $x^2 - xy - 10 - 2y^2 = 0$, what does $x - 2y$ equal?

    (A) 0
    (B) 1
    (C) 2
    (D) 5

14. If each side of a regular hexagon is 6 cm long, what is the area of the hexagon?

    (A) $27\sqrt{3}$ cm^2
    (B) $54\sqrt{3}$ cm^2
    (C) $72\sqrt{3}$ cm^2
    (D) $108\sqrt{3}$ cm^2

15. If *a* is an integer and $2a^2 + a = 28$, what is the value of *a*?

    (A) −4
    (B) 0
    (C) 4
    (D) 7

MATHEMATICS KNOWLEDGE PRACTICE SET 1

Answers and Explanations

1. **A** This question involves a rectangular box, so you are working with a geometric solid. You are given the box's volume and two of its dimensions, and you're asked for the third dimension. The volume of a rectangular solid is found by multiplying the figure's three dimensions, $\ell \times w \times h$. Each of those dimensions has to be smaller than the volume. Therefore, you can rule out (D) right away—it's way too big.

Now you can choose to backsolve or use straightforward math. If you backsolve, start with (B) 17, and see what happens: 10×17 is already 170, which is greater than 140. So 17 is too big, and the answer must be (A). To confirm, check whether $10 \times 7 \times 2$ equals 140. It does, and (A) is the correct answer.

The straightforward math looks like this:

$$\ell \times w \times h = V$$
$$10 \times 7 \times h = 140$$
$$70h = 140$$
$$h = \frac{140}{70}$$
$$h = 2$$

2. **B** You're given values for three variables and asked to solve an equation with those three variables. Plug the values into the equation and solve with arithmetic:

$$x^2 - 5yz + y^2 = (\sqrt{3})^2 - 5(2)\left(\frac{1}{2}\right) + 2^2 = 3 - 5 + 4 = 2.$$

3. **D** This question gives you an inequality and asks you to solve for a variable, a. First, get the only term with an a in it by itself on one side of the inequality by subtracting 20 from both sides: $\frac{a}{-2} < -120$. Now multiply each side by -2, remembering to switch the inequality sign because you are multiplying by a negative number: $a > 240$.

4. **D** This question involves an equation with one variable, and the answer choices are simplified forms of the equation. You can use Picking Numbers or straightforward math to solve.

Try picking 2 for d. Then $(3d - 7) - (5 - 2d) = (3 \times 2 - 7) - (5 - 2 \times 2)$. Use PEMDAS to solve: $(6 - 7) - (5 - 4) = -1 - (1) = -2$. -2 is your target number.

Now plug 2 in for d in each of the answer choices to see which gives you the target number -2: (A) $2 - 12 = -10$; (B) $5(2) - 2 = 8$; (C) $5(2) + 12 = 22$; (D) $5(2) - 12 = -2$. (D) is the correct answer.

If you use algebra to solve, start by eliminating the parentheses by distributing the minus sign over the second set of parentheses: $(3d - 7) - (5 - 2d) = 3d - 7 - 5 - (-2d) = 3d - 7 - 5 + 2d$. Now combine like terms and perform the addition and subtraction: $3d + 2d - 7 - 5 = 5d - 12$.

5. **B** A quadratic expression is set equal to zero and you are asked to find a pair of values that x could equal. Factor the equation by finding a pair of numbers that multiply to -20 and add to -1. Those numbers are -5 and 4, so the equation becomes $(x + 4)(x - 5) = 0$. When multiplication results in zero, at least one factor must be zero. If $x + 4 = 0$, then $x = -4$; if $x - 5 = 0$, then $x = 5$.

You could also use Backsolving here. Because 4 appears in two answer choices, try it first: $4^2 - 4 - 20 = -8 \neq 0$. So x cannot be 4, and you can eliminate (A) and (D). Now try 5: $5^2 - 5 - 20 = 0$. That works! Of the two remaining choices, only (B) has 5 as a value of x, so that is the correct answer.

6. **D** The question gives two points on a line and asks for the equation of the line. Three of the equations in the answer choices are in slope-intercept form: $y = mx + b$. Choice (A) starts with x, which you might recognize as not being the correct form of an equation of a line. Therefore, in order to evaluate (A), you should convert it to the format $y = x - 2$.

Slope is $\frac{y_2 - y_1}{x_2 - x_1} = \frac{6 - 0}{4 - (-2)} = \frac{6}{6} = 1$. The coefficient of x is 1.

Find the y-intercept by plugging one of the coordinate pairs and the slope into the $y = mx + b$ equation: $6 = (1)(4) + b$; $b = 2$.

Plug 1 in for m and 2 in for b in the slope-intercept equation to get the answer. $y = 1x + 2$, which can also be written $y = x + 2$. Choice (D) is the correct answer.

7. **B** This question shows a circle inscribed inside a square. It says that the square's perimeter is 36 and asks for the circle's circumference. A square has four equal sides, and the formula for the circumference of a circle requires knowing either the radius or the diameter of the circle. Your task is to take the information you're given about the square and apply it to the circle.

If the perimeter of the square is 36, then each of the four sides equals $\frac{36}{4} = 9$. The side of the square is the same length as the diameter of the circle, so the diameter of the circle is also 9. Finally, the circumference is $\pi d = 9\pi$.

If it's more efficient for you, you can backsolve this question. Start with (B). If the circle's circumference is 9π, then its diameter is 9. That makes the side of the square also 9, and the square's perimeter is $4 \times 9 = 36$. That matches the information in the question, and you know that (B) is correct.

8. **C** The diagram shows one rectangle inside another and the dimensions of the larger rectangle. The question says this is a picture in a frame and gives the dimensions of the smaller rectangle (the picture). You need to find the area of the frame. The area of the frame is the area of the larger rectangle minus the area of the smaller rectangle, $(8 \times 10) - (4 \times 8) = 80 - 32 = 48$.

9. **A** The diagram gives a number line with four points labeled. The question names two line segments and asks for the distance between their midpoints. First, find the midpoints and then calculate the distance between them. To find the midpoint of a segment on the number line, add the endpoints and divide the sum by 2. The midpoint of PQ is $\frac{(-12)+(-2)}{2} = \frac{-14}{2} = -7$. The midpoint of RS is $\frac{1+9}{2} = \frac{10}{2} = 5$. Finally, the distance between the two points is the positive difference between their coordinates, $5 - (-7) = 12$.

10. **D** You're given an equation with two variables and asked for the value of an expression with the same two variables. You might notice that the variables in the second expression contain the variables in the first expression multiplied by 3. If $s - t = 5$, then if you multiply both sides by 3, you get $3s - 3t = 15$. Thus, $3s - 3t + 3 = 15 + 3 = 18$.

Alternatively, you could pick numbers for s and t so that $s - t = 5$; say $s = 7$ and $t = 2$. Now plug those values into the expression and solve:

$3s - 3t + 3 = 21 - 6 + 3 = 18$.

11. **C** This problem has an equation with two variables, but it provides the value of one of the variables. It could be solved by simplifying the equation to get x in terms of y, but plugging in 3 for y in the equation is a more straightforward way to get the answer.

$$\frac{5x \times 3 - 3^2}{3 - 2} = 21$$

$$\frac{15x - 9}{1} = 21$$

$$15x - 9 = 21$$

$$15x = 30$$

$$x = 2$$

12. **D** The question gives the equations for two straight lines in standard $y = mx + b$ form and asks at what point the lines intersect. While it may be tempting to draw the lines, this question can be answered using algebra. At the point where the two lines intersect, both their x and y values will be equal. The question provides two different equations for y, so start by setting these equal to each other: $3x - 8 = -x + 4$. Add x to both sides: $4x - 8 = 4$. Add 8 to both sides: $4x = 12$, so $x = 3$. Therefore, the x value at the point where the lines intersect is 3. To find the y value for this single point, plug 3 in for x in either equation: $3(3) - 8 = 1$ or $-(3) + 4 = 1$. The proper listing for an ordered pair of coordinates is (x, y), so $(3, 1)$ is correct.

13. **B** The question contains an equation with three variables and asks you to solve for one of the variables in terms of the other. Starting with $\frac{rw + 3k}{k + w} = 4$,

Multiply both sides by $k + w$:

$$\frac{(k + w)(rw + 3k)}{k + w} = 4(k + w)$$

$$rw + 3k = 4k + 4w$$

Subtract $3k$ from both sides:

$$rw = 4k + 4w - 3k$$

Combine like terms:

$$rw = k + 4w$$

Divide both sides by w:

$$r = \frac{k + 4w}{w}$$

$$r = \frac{k}{w} + 4$$

14. **A** The question gives an inequality with four variables, one of which must be a negative number, and asks which of the inequalities in the answer choices must be true. If both sides of the original inequality are multiplied by b, the resulting inequality will have a on one side and $\frac{bc}{d}$ on the other. The question stem states that b is negative. When multiplying or dividing an inequality by a negative number, the inequality sign must be "flipped." Therefore, (A) rather than (B) is correct. Choice (C) could be true if d were positive, but the question does not limit d to positive numbers. Choice (D) is mathematically incorrect; the inequality cannot be algebraically manipulated to establish a relationship between a and the product of b, c, and d.

15. **D** The question consists of a parallelogram with certain angles given as variables and asks you to use that information to calculate a numerical value for one of the variables. Since opposite angles in parallelograms are equal, the measurement of z will be the same as x. Adjacent angles in parallelograms add up to 180°, so $x + y = 180°$. Substituting $4y$ for x, $4y + y = 5y = 180°$, so $y = 36°$ and $x = z = 180° - 36° = 144°$.

MATHEMATICS KNOWLEDGE PRACTICE SET 2

Answers and Explanations

1. **A** The question asks which of the answer choices is never true (is always false). A glance at the choices reveals that the question deals with properties of triangles. An equilateral triangle has three 60° interior angles. A right triangle must have one interior angle of 90°. Therefore, there is no such thing as a "right equilateral triangle." An isosceles triangle could have angles of 45°, 45°, and 90°, so (B) could be a true statement. Since the total of the interior angles of a triangle is 180° and acute angles are less than 90°, there are unlimited combinations of three acute interior angles, so (C) could be true. An equilateral triangle, with each of the three angles equal to 60° is a prominent example. Similarly, a triangle could have one obtuse angle between 170° and 180°, although the other two angles would be very small. For example, a triangle could have interior angles of 172°, 5°, and 3°, so (D) could be a true statement as well.

2. **C** The question states that the two parts of the figure are a semicircle and a square, so you can use the properties of those geometric figures to answer the question. The side of the square is the same as the diameter of the semicircle. The area of a square is the length of a side squared, so $s^2 = 400$ cm^2 and $s = 20$ cm. The area of a circle is πr^2, so the area of a semicircle is $\frac{\pi r^2}{2}$. Since the side of this square is 20 cm, the diameter of the circle is also 20 cm. The radius is half that, or 10 cm. Plug that value into the formula for the area of a semicircle: $\frac{\pi 10^2}{2} = \frac{100\pi}{2} = 50\pi$.

3. **C** The question has one equation with two variables, and both sides of the equation have the same value. You are asked to solve for one of the variables. Merely knowing $x + 2y = x - 4y$ will not be enough to solve the problem, since we can't solve one equation with two variables. Instead, set up two different equations: $7x + 2y = 15$ and $x - 4y = 15$. Multiply the first equation by 2 so that the y coefficient becomes 4:

$$2(7x + 2y) = 2(15)$$

$$14x + 4y = 30$$

Add this to the second equation: $+(x - 4y = 15)$

And the y coefficient becomes 0: $15x = 45$

So $x = 3$. Since this is all that the question asked, there is no need to solve for y.

4. **B** The question provides a value for one leg of a right triangle and the area of that triangle and asks you to calculate the perimeter. The area of a triangle can be calculated using the formula $A = \frac{1}{2}bh$. Since this is a right triangle, the two legs are the base and height, so $6 = \frac{1}{2}b(3)$ and $b = 4$. Use the Pythagorean theorem to find the hypotenuse: $3^2 + 4^2 = h^2$. (This is a "pattern" 3:4:5 right triangle; using that fact is quicker than using the Pythagorean theorem.) The perimeter is $3 + 4 + 5 = 12$.

5. **D** The question states the perimeter of an equilateral triangle and asks you to determine the area. Drawing a quick sketch can help to answer this question.

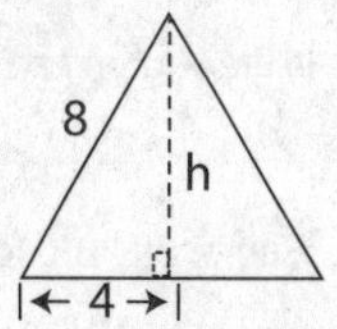

Since the three sides are equal, the length of a side is $\frac{24}{3} = 8$. The dotted line is the height of the triangle. Since the height bisects the base at a right angle (as the sketch shows), it creates right triangles with one leg of 4 cm and a hypotenuse that is 8 cm. The height can be calculated using the Pythagorean theorem: $4^2 + h^2 = 8^2$, so $64 = 16 + h^2$; $h^2 = 48 = 16 \times 3$; $h = 4\sqrt{3}$. The area of a triangle is $\frac{1}{2}$ (base) (height), and $\frac{1}{2}(8)\left(4\sqrt{3}\right) = 16\sqrt{3}$.

6. **C** The question has a complex single variable equation with "stacked" fractions and asks you to solve for the value of the variable. In order to isolate the variable x, you will first need to simplify the equation using PEMDAS.

Parentheses $\quad \dfrac{3 \times 2^2}{2 + \frac{4}{x}} = 5$

Exponents $\quad \dfrac{3 \times 4}{2 + \frac{4}{x}} = 5$

Multiplication $\quad \dfrac{12}{2 + \frac{4}{x}} = 5$

Now continue to solve by multiplying both sides of the equation by the denominator of the left side:

$$\frac{\left(2+\frac{4}{x}\right)\times 12}{2+\frac{4}{x}} = 5\left(2+\frac{4}{x}\right)$$

$$12 = 10 + \frac{20}{x}$$

Multiply both sides by x: $12x = 10x + 20$. So $2x = 20$ and $x = 10$.

7. **D** The graph shows a portion of a straight line and asks what would be the value of the x-coordinate if the line were extended so that $y = 100$. In order to answer this question, you will first need to determine the equation for the line shown on the graph in the standard $y = mx + b$ form. Two good reference points that the line passes through are (2, 0) and (0, −1). The slope, m, is $\frac{0-(-1)}{2-0} = \frac{1}{2}$. Since b is the y-intercept, the second point (or the graph itself) shows that $b = -1$. So the line is defined by the equation $y = \frac{1}{2}x - 1$. Substitute 100 for y: $100 = \frac{1}{2}x - 1$. Multiply both sides by 2: $200 = x - 2$, so $x = 202$.

8. **A** This is a word problem that requires careful attention to the details of the application of an unfamiliar variable with percentage changes. The question asks for the dollar change of Joseph's option based upon a percentage change in the underlying stock. Start by converting the price change of the stock to dollars. Since the change was a decrease, (−2%)($45) = −$0.90. Applying the Δ of 0.40 to this change, option values would decrease by (0.40)($0.90) = $0.36 per share. However, Joseph's option is based on 200 shares of the stock, so his expected change in value would be

−$0.36 × 200 = −$72.00.

9. **C** The question diagram shows two triangles with a common vertex and states that two of the line segments that are sides of the triangles are parallel. Since angle *BAC* is a right angle and *ED* and *BA* are parallel, angle *CED* must also be a right angle. Additionally, angles *BCA* and *ECD* are vertical angles, so they must be equal. Consequently, the two triangles are similar. The question asks you to determine the perimeter of ΔABC. Given that the two triangles are similar and the ratio *BA*:*ED* = 4:3, the length of each side of ΔABC will be $\frac{4}{3}$ the corresponding side of ΔCED. The legs of ΔCED are 3 cm and 4 cm, so the hypotenuse must be 5 cm ($3^2 + 4^2 = 5^2$) and the perimeter is 3 cm + 4 cm + 5 cm = 12 cm. The perimeter of ΔABC is $\frac{4}{3} \times 12$ cm = 16 cm.

10. **B** The problem subdivides a group by a combination of fractions and a whole number and asks you to determine the number of members of the group. The makeup of the group can be expressed as $\frac{1}{2}T + \frac{1}{6}T + \frac{1}{4}T + 2 = T$, where T represents the total number of students. Convert the fractions to their least common denominator (12) so that they can be added:

$\frac{6}{12}T + \frac{2}{12}T + \frac{3}{12}T + 2 = T$. Thus, $2 = \frac{1}{12}T$, so $T = 24$.

Instead of setting up an equation, you could have chosen to backsolve. Start by trying out (C), for instance. If there were 36 students, 18 (one-half) would have worn black coats, 6 (one-sixth) red coats, and 9 (one-fourth) blue coats. Since 18 + 6 + 9 = 33, that would mean that 36 − 33 = 3 students wore green, so 36 is too many. Try (B): 12 + 4 + 6 = 22. Since there are 24 − 22 = 2 students left over wearing green coats, (B) is correct.

11. **A** The question asks you to determine which of the answer choices can be plotted as a straight line on the coordinate plane. Since $x = 3$ does not include y in the equation, you might be tempted to eliminate this answer. However, because y does not appear in the equation, $x = 3$ for *all* values of y. Thus, this equation represents a straight, vertical line crossing the x-axis at 3. You can verify that $y = \frac{1}{x}$ is not a straight line by picking 3 numbers for x such as 1, 2, and 3. The corresponding y values are 1, $\frac{1}{2}$, and $\frac{1}{3}$. Since y decreases by different amounts for the same increase in x, this is not a straight line. Eliminate choice (B). Choice (C) can be eliminated because it contains three variables; the coordinate plane has only two. (This equation might describe a straight line in three dimensions.) Choice (D) has the term x^2, so it is not a linear equation and therefore would not be graphed with a straight line.

12. **B** The question lists the coordinates of three of the four corners of a parallelogram and asks you to find the length of the diagonal that ends at the corner with unknown coordinates. First, determine the coordinates of vertex C. Because the figure is a parallelogram, line segments *AB* and *CD* will be parallel and of equal length. The x-coordinate of B (−3) is 2 greater than the x-coordinate of A (−5), so the x-coordinate of C will be 2 greater than that of D, or 7. Similarly, the y-coordinate of B is 5 greater than that of A, so the coordinates of C are (7, 3). To find the length of a "slanted" line in coordinate geometry, construct a right triangle with the line of unknown length as the hypotenuse, as shown in the diagram below, by dropping a perpendicular vertical line from C to the extension of *AD* to create new point E:

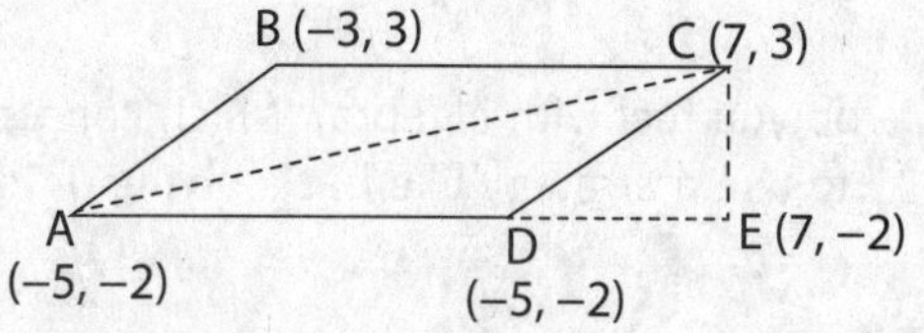

AC is the hypotenuse of right triangle *AEC*. Leg *AE* is 12 units long (7 −(−5)), and leg *CE* is 5 units long (3−(−2)). Applying the Pythagorean theorem, $AC^2 = 12^2 + 5^2 = 144 + 25 = 169$. $AC = \sqrt{169} = 13$.

13. **D** The question provides two equations with two variables and asks for the value of an expression containing both variables. Ideally, you can solve directly for the value of $x - 2y$ rather than having to solve for each variable individually. Since one of the equations is a quadratic, make it easier to factor by adding 10 to each side of $x^2 - xy - 10 - 2y^2 = 0$. Thus $x^2 - xy - 2y^2 = 10$. The x^2 terms can result from the multiplication of two factors, so $(x \pm ?)(x \pm ?) = 10$. The missing terms will have a product of $-2y^2$, so they will be either $-2y$ and y or $-y$ and $2y$. Since their sum must have a coefficient of −1 (the same numerical coefficient as in the middle term of the quadratic), then the individual terms must be y and $-2y$, so $(x + y)(x - 2y) = 10$. The question tells you that $x + y = 2$, so substitute that value to get $2(x - 2y) = 10$, and then $x - 2y = 5$.

14. **B** The question describes a regular hexagon with sides 6 cm long and asks you to determine the area of the hexagon. Like many complex geometry questions, this one can be solved by subdividing the figure into smaller, less complex shapes.

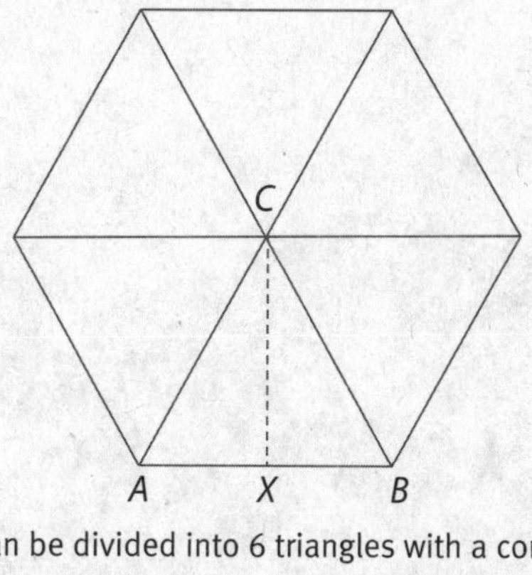

A hexagon can be divided into 6 triangles with a common vertex at the center as shown above. Since a regular hexagon is symmetrical, all 6 of the triangles will be identical (congruent). Therefore, the area of the entire hexagon will be 6 times the area of one of the triangles. There are 6 triangles that share the common vertex *C*, so each of the angles at that point will be $\frac{360°}{6} = 60°$. Since *C* is at the center of the hexagon, it is equidistant from each vertex, so $AC = BC$ and, therefore, $\angle A = \angle B$. Because all three angles total to 180° and $\angle C = 60°$, $\angle A = \angle B = 60°$. Thus, $\triangle ABC$ is an equilateral triangle.

In order to calculate the area of *ABC*, draw an altitude from *C* to *X*. This dotted line bisects and is perpendicular to *AB*. If each side of the hexagon is 6 cm, then $AX = 3$ cm and, because *ABC* is equilateral, $AC = 6$ cm. Use the Pythagorean theorem to determine the length of *CX*: $CX^2 + 3^2 = 6^2$, so $CX = 3\sqrt{3}$. The area of a triangle is $\frac{1}{2}bh$, in this case $\frac{1}{2}(3)(3\sqrt{3}) = 4.5\sqrt{3}$. This is the area of *AXC*; the area of *ABC* is twice that amount or $9\sqrt{3}$, and the total area of the hexagon is $6(9\sqrt{3}) = 54\sqrt{3}$ cm².

15. **A** The question contains a quadratic equation with one variable and also tells you that the variable is an integer. You are asked to find a single solution for the variable's value. Start by rearranging the equation into the standard quadratic form, $2a^2 + a - 28 = 0$. When factored, that will be $(2a \pm ?)(a \pm ?)$. The factors of the coefficient of the first term (2) are 2 and 1. The factors of the third term (−28) could be 1 and −28, −1 and 28, 2 and −14, −2 and 14, 4 and −7, or −4 and 7. You need to find a combination of the two sets of factors and signs that will produce a sum of 1, which is the coefficient of *a* in the second term of the quadratic equation. The correct combination would be $2 \times 4 = 8$ and $1 \times -7 = -7$, since $4 \times -7 = -28$ and $8 + (-7) = 1$. The factored equation is $(2a - 7)(a + 4) = 0$. If $2a - 7 = 0$, then $a = 3.5$, but that is not an integer. If $a + 4 = 0$, then $a = -4$, which is a negative integer. Note that it was not necessary to include the negative factors of the first term $-a$ and $-2a$ since that would have produced factors of $(-2a + 7)$ and $(-a - 4)$, which have the same two solutions, 3.5 and −4.

Alternatively, backsolve to answer this question.
Substitute −4 for *a* to see if the result is 28:
$2(-4)^2 + (-4) = 32 - 4 = 28$. It is not necessary to check the other choices.

Review and Reflect

After this chapter and these practice question sets, how comfortable do you feel with algebra? Which concepts do you need to review? How often were you able to spot the opportunity to use a strategy like Backsolving or Picking Numbers?

Also, how did you do on the items related to geometry—which formulas do you know, and which do you need to brush up on?

Use your thoughts about these questions to guide your review of this chapter.

CHAPTER 8

CONQUERING THE TECHNICAL SUBTESTS

Congratulations! If you've been working through this book in order, then you've reviewed all of the portions of the AFQT, which is crucial for your eligibility to enlist in the armed forces. However, perhaps you don't just want to enlist—you may want to enlist *and* to qualify for a specific job. There are many varieties of specialized job opportunities in the armed forces, which could form the foundation for a lifelong military career or could prepare you for a successful career in civilian life after you've completed your military service. Doing well on the ASVAB's technical subtests, in addition to the AFQT, is important for qualifying for many of those specialized jobs.

The next six chapters will prepare you for the ASVAB technical subtests: General Science, Electronics Information, Auto Information, Shop Information, Mechanical Comprehension, and Assembling Objects. The amount of emphasis you should give to each of these chapters will depend on which job or jobs you desire. Be sure to speak with a recruiter, do your research, and form a sense of which subtests are most important for your goals. That knowledge, combined with your performance on the Diagnostic Test which forms chapter 2 of this book, will tell you which topics you need to study most.

If you aren't sure which career path you are interested in, pay attention to your strengths and the parts of the ASVAB you enjoy. If there is a particular topic that is of interest to you and you would like to learn more about it, you can use that passion to help you decide a potential career path. Conducting online research and speaking to a recruiter can help you determine which paths would help you pursue your interests in the best way. This information can help focus your study time on the most beneficial sections of the test for your ideal path.

THE KAPLAN METHOD FOR THE ASVAB TECHNICAL SUBTESTS

LEARNING OBJECTIVE

In this section, you will learn to:

- apply the Kaplan Method for the Technical Subtests to ASVAB questions

Step 1: Identify what you are being asked for.

Just as with math questions, it's very important to figure out what your task is. Does the right answer represent the value of a variable? The name of a planet? The definition of a unit of measure? The name of a type of tool? How much force is required to move an object? and so on.

Often, a quick glance at the answer choices can help point you in the right direction. Don't read them in depth at this point—save that for Step 4. Simply use them to get a sense of what your task is. If all the answer choices are numbers followed by "mHz," you will be calculating the frequency of a wave and expressing that frequency in megahertz. If all the answer choices are the names of planets, your job is to identify a planet. If it's an Auto Information question and all of the answer choices start with verbs, it's likely you'll be describing how to go about some type of auto repair.

Step 2: Simplify or solve.

This step varies quite a bit depending on the subject matter. Therefore, this section will discuss a variety of approaches depending on the information you are given.

Diagrams

Some technical subtest question may include a diagram. On those questions, ask yourself:

- What does the diagram represent?
- Are there any important relationships depicted in the diagram?
- How does the diagram help me to answer the question?

Numbers, variables, or formulas

Some technical subtest questions resemble math questions. You may be asked to perform calculations, manipulate variables, or apply a formula (especially on Mechanical Comprehension, but on other tests as well). You can often approach these questions using the same tactics you use for the math subtests. Review chapter 5 to remind yourself of these approaches. For each technical subtest question involving math, ask yourself:

- How can I simplify the information I'm given?
- Am I given a formula or expected to recall one out of my own knowledge?
- Can I backsolve, pick numbers, or guess strategically using logic, or should I simply do the math?

Charts, graphs, or tables

Some technical subtest questions may include a graphical representation of data. If you see a chart, graph, table, or other visual arrangement of data, ask yourself:

- How does the graph relate to the information in the question?
- What are the units?
- What do the axes represent?
- Is there a trend or pattern in the data?

Written information

If a technical subtest question includes a few sentences of introductory text, take a moment to paraphrase them to ensure that you understand before you move on to Step 3.

Step 3: Make a prediction.

Just as on Paragraph Comprehension and Word Knowledge, having a prediction is a great strategy for finding the right answer efficiently and not being distracted by tempting wrong answer choices. After all, it's easier to recognize the correct answer if you already have a good idea what it is. Always try to predict unless you're using a strategic math approach such as Backsolving or Strategic Guessing; those strategies do not require making a prediction. As you practice, put a sticky note or your hand over the answer choices as you're working through Step 2, above, to get into the habit of predicting. By Test Day, making a prediction will be second nature, and you will not need to cover the answer choices.

Step 4: Evaluate the choices strategically.

Got a solid prediction? Look for it among the choices.

If you don't have a solid prediction, seek to discard wrong answer choices that do not make sense. You might be able to eliminate choices that:

- contain numbers that are obviously too large or too small
- are expressed using the wrong units
- belong to a different category than that you were asked for. For example, if you are asked to name a planet, "Sirius" would be a wrong answer because it is a star rather than a planet.
- are synonyms: each ASVAB question has only one right answer, so two answer choices that mean the same thing cannot be correct. For example, if you are working on a Shop Information question and two of the answer choices are "Vise Grips" and "locking pliers," cross both those choices off. They both refer to the same tool, so it cannot be the case that one of those answer choices is right and the other is wrong.

Bringing It All Together

Memorize the method below as soon as possible. You don't want to put off learning it. You'll have lots of content review to do in the coming chapters, and you will want the method in place so that you can use what you learn about the subject matter to your advantage on ASVAB questions.

THE KAPLAN METHOD FOR THE ASVAB TECHNICAL SUBTESTS

Step 1: Identify what you are being asked for.

Step 2: Simplify or solve.

Step 3: Make a prediction.

Step 4: Evaluate the choices strategically.

Note: The method above should form your approach to General Science, Electronics Information, Auto Information, Shop Information, and Mechanical Comprehension questions. However, since the Assembling Objects test requires a somewhat different skill set, we will present a different method for the Assembling Objects test in chapter 14.

Check out the companion video for this section.

COMPANION VIDEO

kaptest.com/login

Final Thought Before You Dive In

As you study the technical subtest material in the second half of this book, don't forget to periodically review the math and verbal topics in the first half of this book. Good luck with your preparations for the technical subtests!

CHAPTER 9

GENERAL SCIENCE

KNOW WHAT TO EXPECT

The General Science section of the ASVAB covers a grab bag of topics that you may or may not have studied in high school or elsewhere. This subtest covers a wide variety of material, but you'll never be expected to know too much about a single topic. As you study this chapter, seek first to develop a broad overview of the subject matter, learning the broad outlines of anything that's completely unfamiliar and refreshing your memory about topics that seem familiar. Then, if you have time before Test Day, you can dig deeper into these topics.

ASVAB science topics fall into three broad categories: life science, Earth and space science, and physical science. This chapter is divided into three parts—one for each of those categories—each with its own set of practice questions. As usual, we've included worked examples and practice questions throughout. Also, key terms you may want to memorize are listed in **bold type**.

You'll be answering 25 questions in 11 minutes on the paper-and-pencil version, and you are given 10 minutes for 15 questions on the CAT-ASVAB. That means you need to be able to answer ASVAB science questions in roughly half a minute. To help you work efficiently, use the Kaplan Method covered in chapter 8 on every science question you encounter.

PART I: LIFE SCIENCE

NUTRITION AND HEALTH

LEARNING OBJECTIVES

In this section, you will learn to:

- describe the role of important micro- and macronutrients in the human body and to identify where these nutrients come from
- identify some diseases that can be caused by nutrient deficiencies

Although nutritionists themselves don't always agree about what constitutes a healthy diet, certain facts are clear. The human body requires a combination of protein, carbohydrates, fat, minerals, vitamins, and fiber. Proteins, carbohydrates, and fats (macronutrients) are necessary to provide energy. Minerals and vitamins (micronutrients), along with fiber, are necessary to maintain proper bodily functions.

Proteins are necessary for the body's maintenance, growth, and repair. Animal proteins are contained in meat, fish, eggs, and cheese. Vegetable proteins are found in peas, beans, nuts, and some grains.

Carbohydrates include both starches and sugars. They are major sources of energy for the body. Starches are found in bread, cereal, rice, potatoes, and pasta. Sugars are found in fruits, cane sugar, and beets, as well as processed foods.

Fats also provide energy for the body. There are three types of fats: saturated, monounsaturated, and polyunsaturated. Saturated fats can raise bad cholesterol (LDL), but mono- and polyunsaturated fats can actually decrease levels of bad cholesterol. Diets high in saturated fat can lead to high cholesterol, which can cause heart disease or stroke. Sources of saturated fats include meats, shellfish, eggs, milk, and milk products. Sources of monounsaturated fats include olives and olive oil, almonds, cashews, Brazil nuts, and avocados. Sources of polyunsaturated fats include corn oil, flaxseed oil, pumpkin seed oil, safflower oil, soybean oil, and sunflower oil.

Fiber is an important part of a healthy diet that provides bulk to help the large intestine carry away waste matter. Good sources of dietary fiber include leafy green vegetables, carrots, turnips, peas, beans, and potatoes, as well as raw and cooked fruits and whole-grain foods.

Water is also essential for survival. The body loses approximately four pints of water each day, which must be replenished. Most foods contain water, facilitating proper water maintenance, although it is still necessary to drink water as well! Insufficient water consumption leads to dehydration, which can cause muscle cramps, dizziness, and, if not remedied, even death.

Minerals in small quantities are needed for a balanced diet. Some necessary minerals are iron, zinc, calcium, magnesium, and sodium chloride (salt). Calcium is important for building strong teeth and bones. Iron, on the other hand, is necessary for red blood cell development. Minerals play many different roles in the development and maintenance of a healthy body.

Vitamins, such as vitamins C and D, are organic compounds that are necessary for a wide variety of physiological processes from bone hardness to healthy gums. Fruits and vegetables are rich sources of vitamins. Vitamin D is unique in that one of the best sources comes not from your diet, but from the Sun. Exposure to sunlight allows your body to synthesize its own vitamin D.

A lack of the proper amount of certain necessary nutrients in the diet can lead to **deficiency diseases**. One such disease is iron-deficiency anemia, which may cause weakness, dizziness, and headaches. It is especially common among children, young adults, and pregnant women who do not get enough iron in their diets. Another example of a deficiency disease is scurvy, which is caused by a lack of vitamin C. Though at one time very common among pirates and sailors who did not have access to fresh fruits and vegetables, scurvy is now relatively rare.

The following table lists a selection of nutrients and their sources.

Vitamin/Mineral	Sources	Benefits
Iron	Meat (especially liver), beans, whole grains	Allows red blood cells to transfer oxygen to body tissues
Calcium	Dairy products like milk, yogurt, and cheese; spinach	Bone growth, muscle function
Magnesium	Nuts, whole grains, green leafy vegetables, fortified foods	Bone development, muscle and nerve function, enzyme function
Potassium	Bananas, sweet potatoes, nuts and seeds	Balances fluid levels in the body
Vitamin A	Liver, milk, eggs, carrots, spinach	Vision, immune system, cell growth
Vitamin C	Red and green peppers, citrus, broccoli	Collagen formation, immune system function, antioxidant (helps protect cells from damage)
Vitamin D	Fortified milk, juice, or cereal; body makes majority of vitamin D when exposed to sunlight	Bone strength (by helping the body absorb calcium), muscle and nerve function, immune system

Take a look at how an expert test taker might approach an ASVAB question about health and nutrition.

Question	Analysis
A pregnant woman feels dizzy and has a headache. Which of the following would be the most likely diagnosis? (A) iron-deficiency anemia (B) scurvy (C) cancer (D) diabetes	**Step 1:** You're given three important facts: the woman is pregnant, has a headache, and is feeling dizzy. **Step 2:** What condition that causes headache and dizziness is common in pregnant women? **Step 3:** Those symptoms match iron-deficient anemia, which is especially common among pregnant women. **Step 4:** Select choice **(A)**.

Try one on your own.

If a person has high cholesterol, which one of the following foods might his doctor suggest that he avoid?

(A) olive oil
(B) red meat
(C) almonds
(D) avocados

Explanation

A person with high cholesterol should limit the intake of saturated fat. The only choice here that is high in saturated fat is **(B)**, red meat. The others are filled with healthier fats (mono- and polyunsaturated fats) that could possibly help lower high cholesterol.

HUMAN BODY SYSTEMS AND DISEASES

LEARNING OBJECTIVES

In this section, you will learn to:

- identify the general components and processes of major human body systems, including the musculoskeletal, respiratory, circulatory, nervous, digestive, and reproductive systems
- differentiate between blood types and how they influence blood donation
- describe the roles of the main components of the brain
- describe the causes and effects of common health problems
- differentiate between bacteria and viruses and describe several common illnesses caused by each

The Skeleton and Muscles

The skeleton and muscles are responsible for holding the body together as well as for movement. Without a skeleton you would be just an immobile mass of organs, veins, and skin. Some organisms, namely **arthropods** (including insects, spiders, and crustaceans), have **exoskeletons**, or external skeletons. However, **vertebrate** animals, including humans, have internal skeletons, or **endoskeletons**.

The human skeleton contains both **bone** and **cartilage**. Bones provide the primary support, while cartilage, which is more flexible, is found at the end of all bones, at the joints, in the nose, and in the ears. Bones not only provide structural support for the body and protect vital organs, but also produce blood cells and store minerals such as calcium. **Tendons**, tough fibrous cords of connective tissue, connect muscles to the skeleton. **Ligaments**, another type of connective tissue, connect bones to other bones at joints such as the elbow, knee, fingers, and vertebral column.

Selected Bones

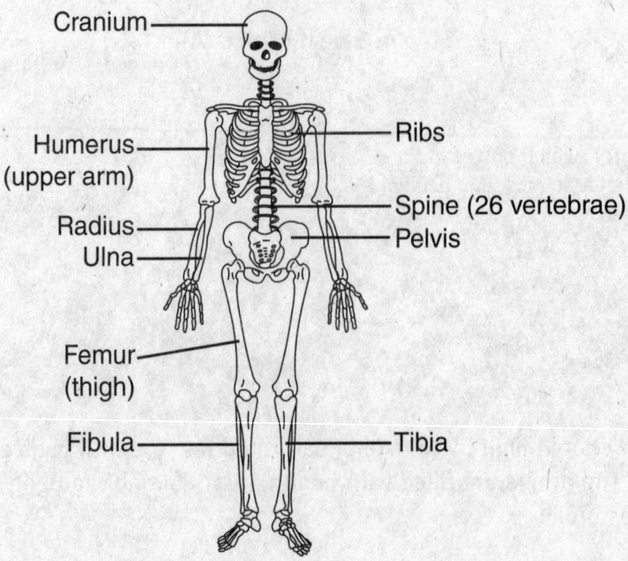

Here's how an expert test taker would approach an ASVAB question about the skeletal system.

Question	Analysis
What type of tissue would connect the humerus to the ulna? (A) bone (B) muscle (C) tendon (D) ligament	**Step 1:** The question asks what type of tissue connects two bones to each other. **Step 2:** There is nothing to simplify. **Step 3:** Make a prediction: ligaments are connected to bone. **Step 4: (D)** is the correct answer.

The Respiratory System

Respiration—the process by which blood cells absorb **oxygen** and eliminate **carbon dioxide** and **water vapor**—is performed by the **respiratory system.**

The Respiratory System

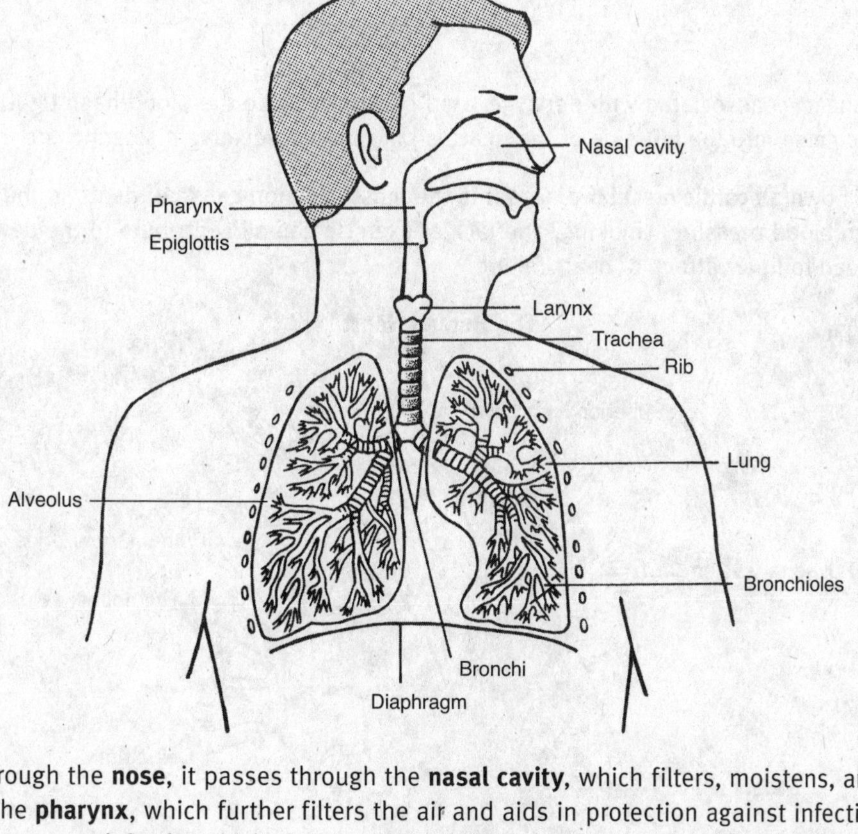

When air enters through the **nose,** it passes through the **nasal cavity,** which filters, moistens, and warms air, and then through the **pharynx,** which further filters the air and aids in protection against infection. The air then passes through the open **epiglottis,** which closes when swallowing to prevent food from going down the airway, and into the **trachea,** which further cleanses the air. The trachea branches into the **left and right bronchi,** which are two tubes that lead to the **lungs**. There the bronchi further subdivide into smaller tubes called **bronchioles.** Each bronchiole ends in a small sac called an **alveolus.** It is in the alveolus that oxygen from the air enters into the bloodstream via tiny blood vessels called **capillaries**. The **diaphragm** is a system of muscles that allows breathing. When the diaphragm causes the lungs to expand, air rushes in to fill the space, in a process called inhalation. When the diaphragm causes the lungs to contract, air is pushed out in an exhalation.

Blood and the Circulatory System

In conjunction with the respiratory system, the **circulatory system** functions to transport oxygen throughout the body while removing carbon dioxide. Additionally, the circulatory system transports nutrients provided by the digestive system and clears away waste by transporting it to the excretory system. The organ that drives the circulatory system is the **heart**. The human heart is a four-chambered pump, with two collecting chambers called **atria** (singular: atrium), and two pumping chambers called **ventricles**. The **right atrium** receives deoxygenated blood from the **venae cavae** (plural of **vena cava**), the two largest veins in the body, and passes it to the **right ventricle**, which pumps the blood to the lungs through the **pulmonary artery**. Blood picks up oxygen in the lungs and returns to the **left atrium** via the **pulmonary vein**. From there it passes to the **left ventricle** and is pumped through the **aorta**, the body's largest artery, into several smaller branching **arteries** that take it through the rest of the body. The heart's **valves** are essential to efficient pumping of the heart. When blood is pumped out of the ventricles, valves close to prevent the blood from flowing backward into the heart after the contraction of the ventricles is complete.

How blood flows from the heart to the body:

right atrium → right ventricle → lungs → left atrium → left ventricle → body

The right side of the heart is associated with deoxygenated blood (because the blood hasn't gotten to the lungs yet), whereas blood coming into the left side of the heart is oxygenated because it's sent there from the lungs.

Heart disease (also known as **cardiovascular disease**) is the most common cause of death in the United States. High cholesterol, high blood pressure, smoking, and lack of exercise can all contribute to the development of heart disease, which can lead to heart attack or heart failure.

The Human Heart

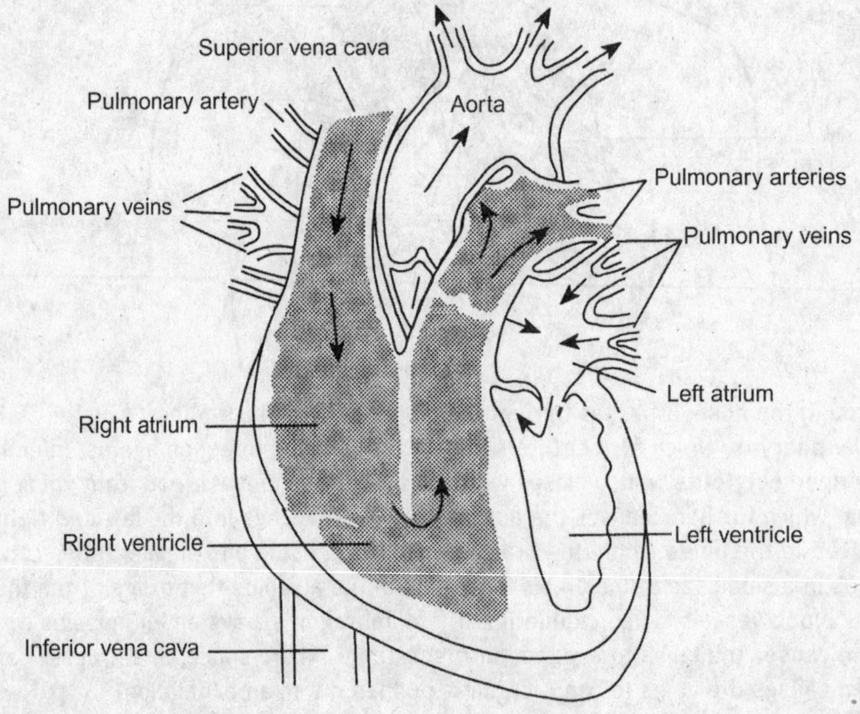

The **arteries** carry blood from the heart to the tissues of the body. They repeatedly branch into smaller arteries (**arterioles**), which supply blood to bodily tissues via the capillaries. Arteries carry blood away from the heart and thus must be thick-walled because they carry oxygenated blood at high blood pressure. Only the pulmonary artery, which carries blood from the heart into the lungs, does not contain oxygenated blood.

Veins, on the other hand, carry blood back to the heart from other parts of the body. Veins are relatively thin-walled, conduct blood at low pressure, and contain many valves to prevent backflow. Veins have no pulse and carry dark red, deoxygenated blood. The lone exception is the pulmonary vein, which carries freshly oxygenated blood from the lungs back into the heart.

Finally, **capillaries** are thin-walled vessels that are very small in diameter. Capillaries, rather than arteries or veins, permit exchange of materials such as oxygen, carbon dioxide, nutrients, and waste between the blood and the body's cells through **diffusion**.

The Capillary System

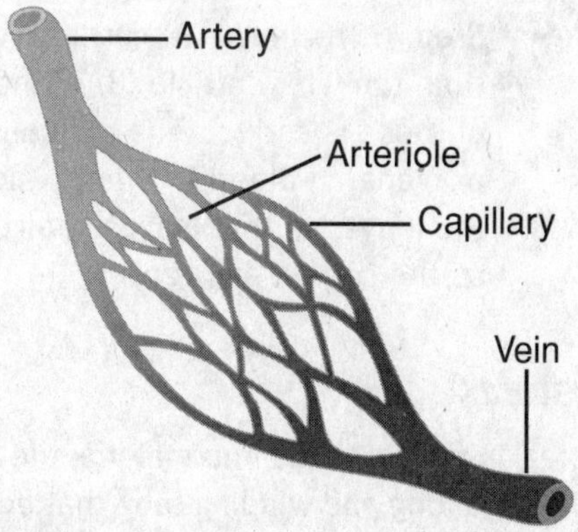

Hypertension, also known as high blood pressure, can cause damage to blood vessels as well as other parts of the body like the kidneys. Limiting salt intake, maintaining a healthy weight, and exercising can help prevent or manage hypertension.

Blood consists of **cells** suspended in **plasma**, the liquid component of blood. There are three types of cells found in blood: **red blood cells**, which are the oxygen-carrying cells; **white blood cells**, which fight infection by destroying foreign organisms; and **platelets**, which are cell fragments that allow blood to clot. All blood cells are created in the **bone marrow**, which is located in the center of bones.

Each of these blood cell types can be measured in the blood as an indicator of overall health. When white blood cell levels are higher than normal, that indicates that the body is fighting off some sort of infection by either bacteria or a virus.

It is also important to note that blood comes in four different types: A, B, AB, and O, which can be further designated as either negative or positive. One combination may be written as A+ or A positive, for example. The letter designation is determined by the type of molecules (**antigens**) found on the outside of the red blood cells. The positive or negative designation is assigned based on whether or not cells have a third type of antigen called the **Rh factor**. A person who has blood that is Rh-factor negative cannot receive blood with a positive type; however, a person with positive type blood can receive donor blood that is Rh-negative. Type O negative is the **universal donor**, which means that type O negative blood can be given to anybody. Type AB positive is the **universal recipient**, which means that someone with this type of blood can receive any other type of blood.

Here's how an expert test taker would think through an ASVAB question about the circulatory system.

Question	Analysis
Which of the following does not contain oxygen-rich blood?	**Step 1:** The question asks which listed item does not contain oxygen-rich blood. **Step 2:** Possible answer choices will contain three choices that *do* contain oxygen-rich blood and one (the correct answer) that *does not.* **Step 3:** A vague question like this can be hard to predict, because it doesn't tell you if you are looking for a blood vessel or a part of the heart. Either way, you should start thinking about what you know about blood flow in the heart and to the body.
(A) aorta (B) left ventricle (C) pulmonary vein (D) right atrium	**Step 4:** The aorta (A) carries oxygenated blood away from the left ventricle (B) of the heart, so neither of those is correct. The pulmonary vein (C) carries oxygenated blood to the left side of the heart from the lungs, so that leaves choice **(D)**, right atrium, as the correct answer.

The Digestive and Excretory Systems

The digestive system is responsible for breaking down foods into material the body can use for energy and building body tissues. The digestive tract is essentially a long and winding tube that begins at the mouth and ends at the anus.

The process of digestion progresses as follows:

- In the mouth, the teeth and the tongue aid in mechanical digestion (chewing), while the enzyme **salivary amylase**, contained in the saliva, begins to break down starch.
- From the mouth, the chewed food moves into the **esophagus**. Contractions push the food down through the esophagus and into the stomach.
- In the **stomach**, food is mixed with **gastric acids** and **pepsin**, which help break down **protein**.
- Most digestion takes place in the **small intestine**. The small intestine is very long, about 23 feet on average. Food is broken down completely by enzymes produced in the walls of the small intestine, in the pancreas, and in the liver. The acids produced by the **pancreas** contain **lipase**, which converts fat to glycerol and fatty acids; **pancreatic amylase**, which breaks down complex carbohydrates into simple sugars; and **trypsin**, which converts polypeptides (the molecules that compose proteins) into amino acids. **Bile**, which is produced by the **liver**, aids in digestion by emulsifying fat (physically separating it into individual molecules). All these digested substances, except for the fatty acids and glycerol, are then absorbed in the small intestine through capillaries that carry the blood into the liver and then throughout the rest of the body.
- In the **large intestine**, also known as the **colon**, water and minerals remaining in the waste matter are absorbed back into the body.
- Chemical waste, such as urea, excess salts, minerals, and water, are filtered from the blood by the **kidneys** and secreted into the urine. Urine is transported to the bladder from the kidneys through the **ureters**.
- In the **rectum**, solid waste matter is stored. Liquid waste (urine) is stored in the **bladder**.
- Solid waste matter is periodically released through the **anus**, and urine is released through the **urethra**.

The Digestive System

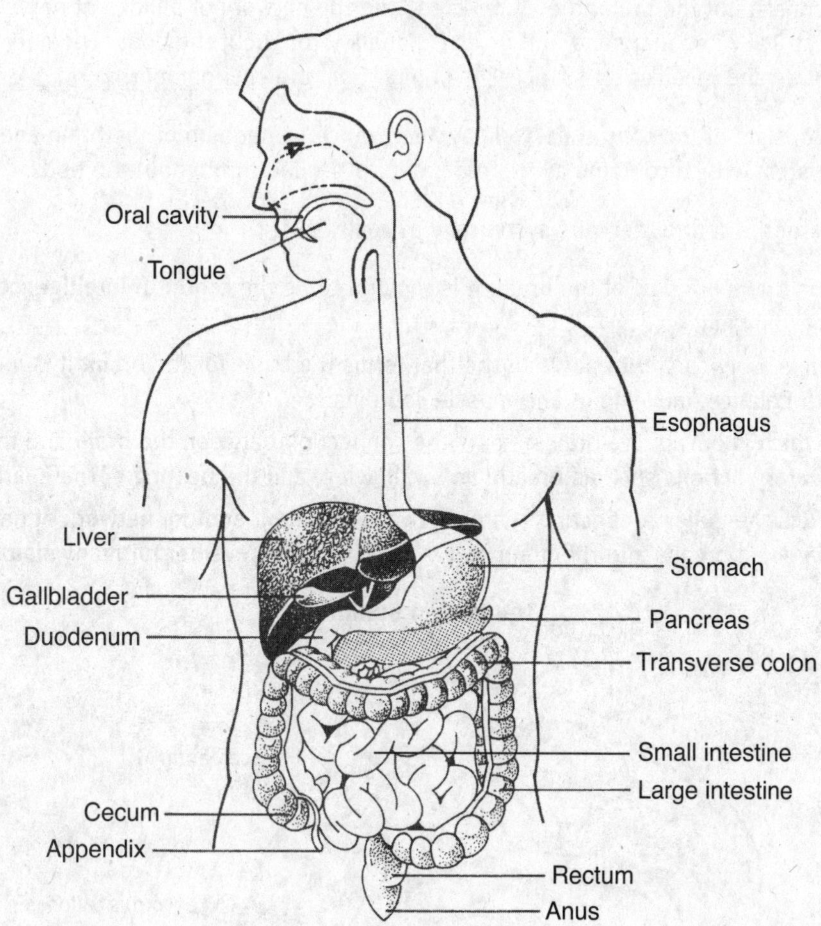

Study the example below to see how an expert test taker might approach a question about the digestive system.

Question	Analysis
Where does food go after the step where proteins begin to be digested?	**Step 1:** The question is asking about the next step in digestion after digestion of proteins begins. **Step 2:** This is essentially two questions in one. First, identify where proteins are digested. Second, recall where food goes after that step. **Step 3:** Recall that protein digestion begins in the stomach with the help of stomach acid and pepsin. Predict where food goes after being in the stomach: the small intestine.
(A) esophagus (B) stomach (C) small intestine (D) large intestine	**Step 4:** Choice **(C)** is correct.

The Nervous System

The nervous system consists of the **brain**, the **spinal cord**, and the network of billions of nerve cells called **neurons**, which behave like electrical wires that send and receive signals throughout the body. The nervous system controls the functions of the body and receives and processes stimuli from the environment.

The nervous system consists of the **central nervous system**, which is made up of the brain and spinal cord, and the **peripheral nervous system**, which contains all the other neurons found throughout the body.

The main components of the central nervous system are as follows:

- The **cerebrum** is the major part of the brain. It is thought to be the **center of intelligence**, responsible for hearing, seeing, thinking, etc.
- The **cerebellum** is a big cluster of nerve tissue that forms the basis for the brain. It is most closely associated with balance, movement, and muscle coordination.
- The **medulla**, which is part of the brainstem, is the connection between the brain and the spinal cord. It controls **involuntary actions** such as breathing, swallowing, and the beating of the heart.
- The **spinal cord** is the major connecting center between the brain and the network of nerves. It **carries impulses** between all organs and the brain and is also the control center for many **simple reflexes**.

The Human Brain

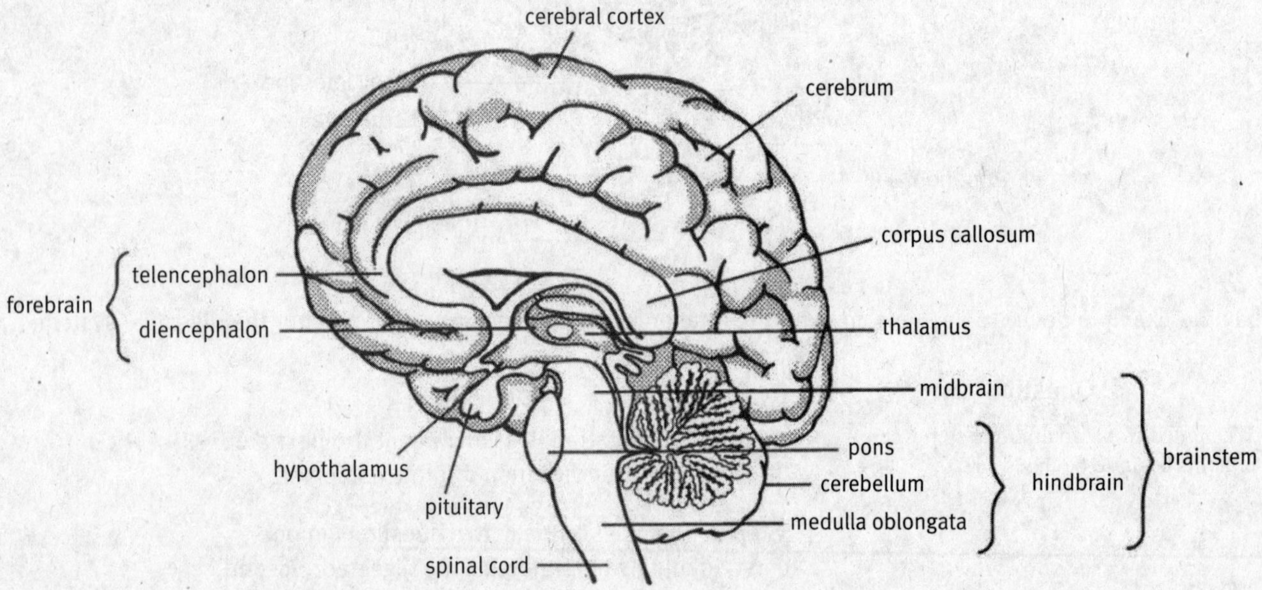

The peripheral nervous system can be subdivided into:

- The **somatic nervous system**, which consists of nerve fibers that send sensory information to the central nervous system and control **voluntary actions**.
- The **autonomic nervous system**, which regulates **involuntary activity** in the heart, stomach, and intestines.

The example below walks you through the expert approach to an ASVAB question about the nervous system.

Question	Analysis
Which of the following parts of the brain, if damaged, would likely cause a person to have poor balance?	**Step 1:** The question asks for a part of the brain that would affect balance if it were damaged. **Step 2:** There's nothing to simplify. **Step 3:** Make a prediction based on what you know about the brain: the cerebellum is closely related to balance.
(A) medulla (B) spinal cord (C) cerebellum (D) cerebrum	**Step 4:** Choose **(C)**. The medulla and spinal cord are associated with more basic, involuntary functions like breathing and reflexes, whereas the cerebrum handles higher functions like our intelligence, so none of these choices fit.

The Reproductive System

Human reproduction occurs when a female's egg is fertilized by a male's sperm. During female **ovulation**, which occurs approximately every 28 days, an egg (**ovum**) is released from one of the ovaries and begins to travel through the **oviduct (fallopian tube)** and into the **uterus**. At the same time, the **endometrial lining** of the uterus becomes prepared for implantation.

Male and Female Reproductive Systems

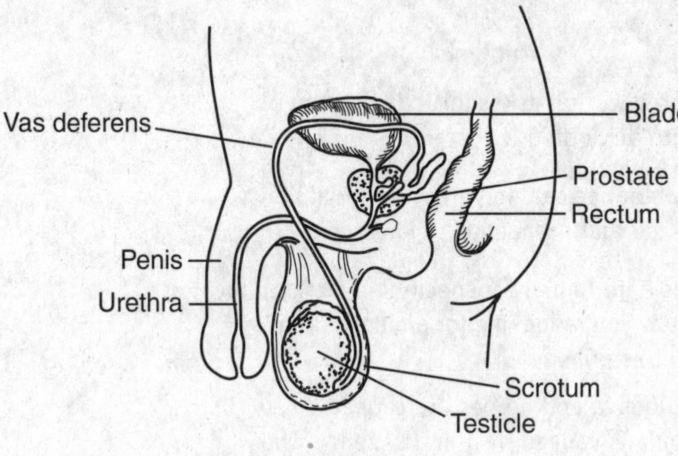

Male: side view

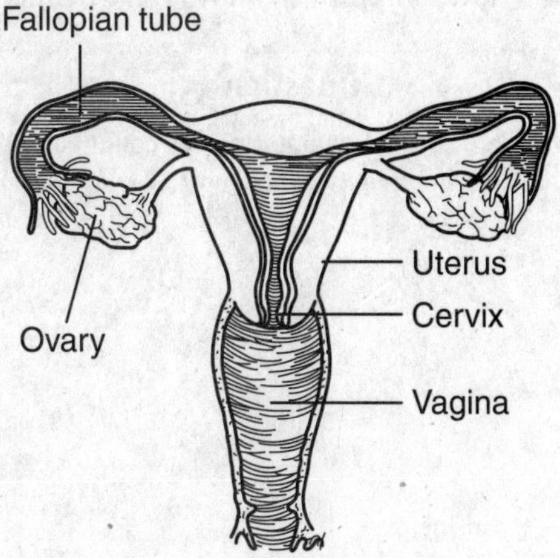

Female: front view

During intercourse, the **penis** ejaculates more than 250 million **sperm**, produced in the **testes**, into the **vagina**. Some of the sperm make their way to the uterus, where they may encounter an egg to fertilize. If the sperm unites with the ovum, a fertilized egg (**zygote**) is formed, which may implant in the uterus and eventually develop into a **fetus**. During pregnancy and after childbirth, **prolactin**, a hormone secreted by the pituitary, activates the production of breast milk (**lactation**).

If the ovum fails to become fertilized, the uterine lining sloughs off during **menstruation**. From puberty to menopause, this **menstrual cycle** repeats monthly except during pregnancy.

Human Pathogens

Some diseases, such as deficiency disorders, hypertension, and heart disease, are caused by diet and/or lifestyle factors. Another major cause of human disease are **pathogens**—disease-causing agents—such as bacteria and viruses. **Bacteria**, which are single-celled organisms, are responsible for diseases such as strep throat, staph infections, and pneumonia; these illnesses may be treated with antibiotic medications.

On the other hand, **viruses** are not technically living things because they are only able to replicate inside a host's cells. Viral illnesses in humans include the common cold and flu, AIDS (acquired immunodeficiency syndrome), and herpes. These illnesses cannot be treated with antibiotics, but may be treated with specially-designed antiviral drugs.

Both bacteria and viruses may be spread from person to person in several different ways. Some bacteria and viruses (including the cold and flu) may be passed through the air, wherein an infected person coughs or sneezes and another individual inhales the pathogen. Some viruses, like HIV (human immunodeficiency virus), can only be transmitted through contact with infected body fluids, as in sexual intercourse or intravenous drug use. Others, like herpes, can be spread by skin-to-skin contact. Every type of virus and bacteria has a unique profile when it comes to how it is transmitted and what type of cells it infects.

Luckily, many diseases caused by viruses (and a few caused by bacteria) can be prevented through **vaccination.** Vaccination, also called **immunization**, prevents many diseases that, not long ago, would have been very severe if not fatal, including smallpox, polio, and the measles. When a person receives a vaccine, a small amount of deactivated, weakened, or partial pathogen is injected into the body, causing the immune system to react; if the body is exposed to that pathogen in the future, the immune system will have a quick response to it, protecting the person from infection.

Here's how to approach an ASVAB question about human health on Test Day.

Question	Analysis
Which of the following illnesses could be treated with an antibiotic?	**Step 1:** The question asks which illness can be treated with an antibiotic. **Step 2:** Antibiotics can only treat bacterial infections, so that's what you're looking for. **Step 3:** It's hard to make a specific prediction, but you do know you're looking for an illness that's caused by bacteria.
(A) influenza (B) anemia (C) strep throat (D) herpes	**Step 4:** Influenza and herpes are caused by a virus, while anemia is caused by iron deficiency. The correct answer is **(C)**, strep throat, which is caused by bacteria.

Now that you've gotten an overview of the human body systems and causes of disease, try a few questions:

1. Blood pressure is generally highest in which of the following?

   (A) veins
   (B) capillaries
   (C) arteries
   (D) lungs

2. White blood cells are produced in

   (A) the heart
   (B) the superior vena cava
   (C) the lymph nodes
   (D) the bones

3. Processing of signals from the heart and regulation of cardiac rhythm occurs in

   (A) the cerebrum
   (B) the medulla, or brain stem
   (C) the cerebellum
   (D) the spinal cord

Explanations

1. (C) arteries Arteries pump blood away from the heart to the rest of the body, so the blood pressure is highest in those vessels. Some veins, (A), have such low blood pressure that they need valves to keep the blood from flowing the wrong way! Capillaries, (B), are tiny vessels located far from the heart, so their blood pressure is relatively low as well.

2. (D) the bones Reading the question stem carefully, home in on the key word, "produced." Try to make a prediction before looking at your answer choices, if possible. Although white blood cells are found in the blood, and thus pass through the vena cava and heart, they are produced in the center of bones, the marrow, so look for an answer like "marrow" or "bone."

3. (B) the medulla, or brain stem Make a prediction. Regulation of the heart is an autonomic (involuntary) process. It occurs in the brain stem (medulla).

GENETICS

LEARNING OBJECTIVES

In this section, you will learn to:

- describe how genetics determine the physical characteristics of an organism
- distinguish between genotype and phenotype
- recognize several health problems that are genetically inherited

Genetics is the study of heredity, the process by which characteristics are passed from parents to offspring. The basic laws of genetics have been understood since the late eighteenth century, when they were first discovered by Gregor Mendel. What Mendel discovered was that in sexual reproduction, individual heredity traits separate in the reproductive cells, so that reproductive cells, known as **gametes**, have half as many **chromosomes** (large strings of hereditary units) as normal cells. Normal body cells are called **diploid**, and gametes are called **haploid**. (Think *DIploid = Double* and *HAploid = HAlf* to help you remember it.)

In human reproduction, the female gamete (ovum) combines with the male gamete (sperm), each of which contains 23 unpaired chromosomes (haploid), to produce a zygote, which contains 23 pairs of chromosomes, or a total of 46 (diploid). **Meiosis** is the process by which gametes are created. Sexual reproduction by meiosis and fertilization results in a great deal of variation among offspring.

DNA (**deoxyribonucleic acid**) is the molecule that contains genetic information. It is "written" with the **genetic code**, a combination of **nucleotides** that bind together in a specific pattern that can be "read" by the cell to instruct it how to grow and behave. DNA is shaped in a **double helix**, which looks like a ladder that is twisted along its axis.

DNA Molecule

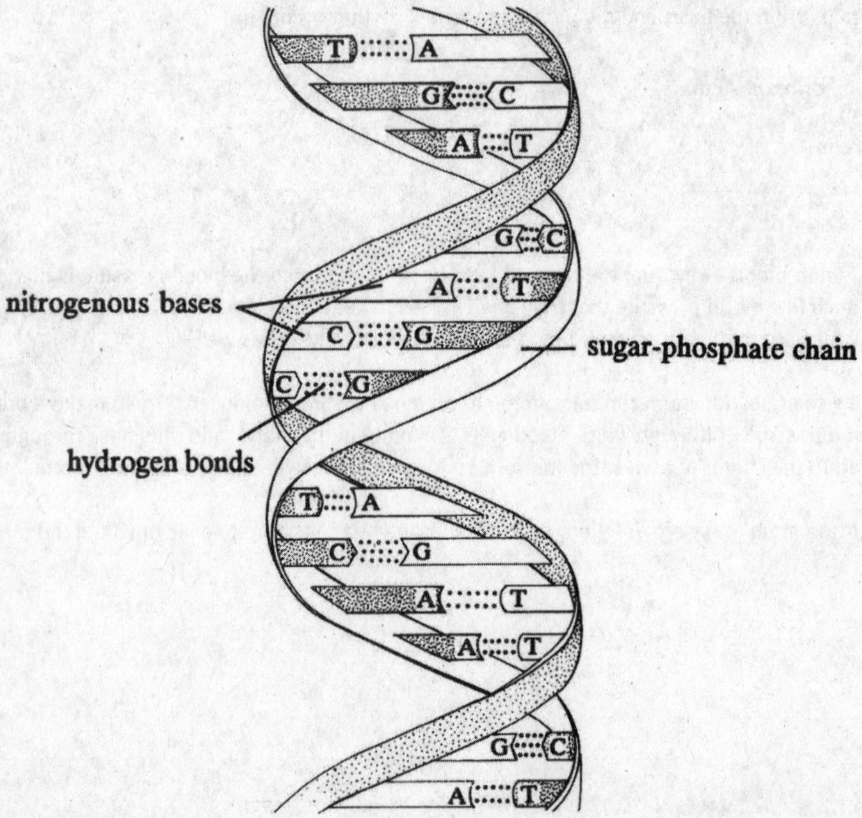

A **gene** is defined as the unit of inheritance contained within the DNA of an individual. A gene may come in several varieties, known as **alleles**. For example, in the gene that determines eye color, a person may have the allele for brown eyes or green eyes. Each of us has two alleles for every gene, one inherited from each parent. These alleles may or may not be alike. If the alleles are alike, that person is **homozygous** for that particular gene. If the alleles are different, the person is **heterozygous** for that particular gene. The sex of babies is determined by the genes contained on the sex chromosome. In females, the two sex chromosomes are alike and are designated as XX. In males, the sex chromosomes are heterozygous and designated as XY.

Genetic traits are inherited independently of one another, and when different traits are paired up during the fertilization of an egg, often one trait is **dominant** and the other is **recessive**. A recessive trait is only expressed if the offspring has two copies of that trait. A dominant trait, on the other hand, will be expressed even if only one copy (paired with a recessive trait) is present. For example, Huntington's disease, which causes degeneration of nerve cells and loss of muscle control, is passed from parent to offspring as a dominant trait, so only one copy is necessary for the offspring to inherit the disorder. On the other hand, cystic fibrosis, which affects the lungs, is caused by a recessive trait, which must be inherited from both parents in order to cause the disease. Other disorders, such as some types of cancer, Down syndrome, and color blindness, are also caused by genetic traits.

A person's **genotype** is their genetic makeup, including both dominant and recessive alleles. A person's **phenotype** is simply how their genes express themselves in physical characteristics. Take eye color, for example. If a woman has brown eyes, then brown eyes are part of her phenotype.

Study the following example.

Question	Analysis
Suppose that black fur in rabbits is a dominant trait and that white fur in rabbits is a recessive trait. A black rabbit and a white rabbit mate and produce a white kit (that is, baby rabbit). Which of the following inferences is supported by this information?	**Step 1:** You're being asked for a deduction based on a set of facts. Two rabbits—one black and one white—have a white kit. **Step 2:** If white is a recessive trait, the white kit must have inherited a white-fur allele from each of its parents. The white rabbit parent must have two white-fur alleles, and one of those alleles was passed on to the kit. The black rabbit parent must have a black-fur allele and a white-fur allele, if it passed a white-fur allele along to the kit. **Step 3:** One of those insights will show up in the answer choices.
(A) The parents' next kit will be black. (B) The white rabbit has a recessive black-fur allele. (C) The black rabbit has a recessive white-fur allele. (D) The kit is an albino.	**Step 4:** Choice **(C)** matches one of the deductions made earlier and is correct. You do not have enough information to infer anything about other kits the parents might have, so choice (A) is wrong. Choice (B) contradicts the information in the question stem: you're told that black is dominant. There is no support for the claim (D) that the kit displays albinism.

CELLULAR STRUCTURES AND FUNCTIONS

LEARNING OBJECTIVES

In this section, you will learn to:

- describe the broad outlines of cell theory
- identify different types of cells
- identify the structures of eukaryotic cells
- understand cell growth and cell processes

Cell theory states that (1) all living things are composed of cells, (2) cells are the basic units and structure of living things, and (3) new cells are produced from existing cells.

Cells are classified into two categories based on the absence or presence of a nucleus: **prokaryotic** and **eukaryotic**. Prokaryotic cells are characterized by not having a nucleus; bacteria are one example.

Structure of a Prokaryotic Cell

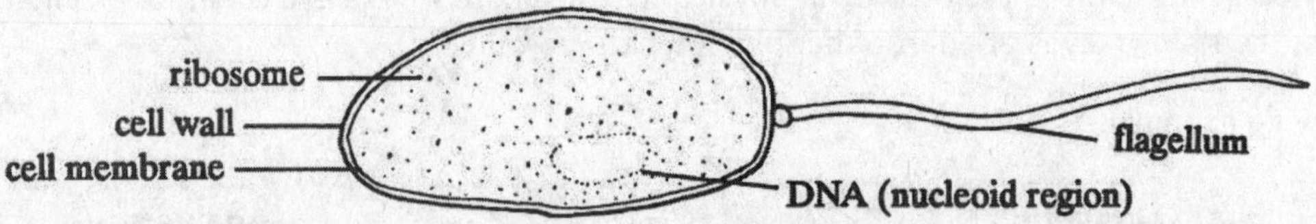

Plants, animals, fungi, and protists are made up of eukaryotic cells and characterized by having a nucleus and a more complex structure than a prokaryotic cell. The **nucleus** of a eukaryotic cell contains the genetic material of the cell. Outside the nucleus lies the **cytoplasm**, a substance which surrounds the other cell structures. The cytoplasm contains many other **organelles** (cell parts with specific functions). These include:

- **ribosomes**, which produce proteins
- **mitochondria** (singular: *mitochondrion*), which produce energy
- **endoplasmic reticulum**, which is involved in the synthesis of proteins and fats
- **Golgi apparatus**, which "packages" proteins for use
- **lysosomes**, which help the cell manage waste
- **centrosomes**, which can be important in guiding the cell's reproduction

Structure of a Eukaryotic Cell

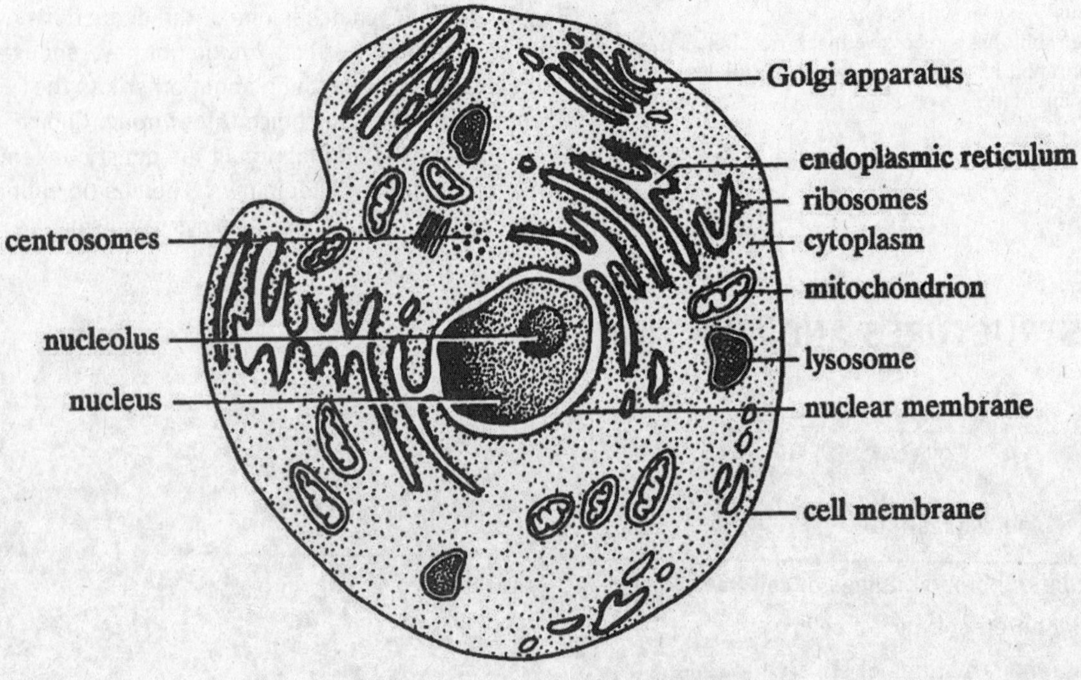

Plant cells also have a somewhat rigid **cell wall** surrounding the membrane. The cell wall provides structure and support for cells.

Some plant cells produce their own energy through the process of photosynthesis. **Photosynthesis** is the process by which sunlight, carbon dioxide, and water react to make sugar and oxygen. It serves as a source of energy for the cells and takes place in plant cells.

Animal cells are surrounded by a semipermeable membrane which allows for the diffusion of water and oxygen from inside the cell to the outside of the cell and vice versa. These cells cannot produce their own energy and rely on consuming outside sources to provide them with the tools to make energy through cellular respiration. **Cellular respiration** is the process by which the mitochondria process sugar and oxygen to produce energy, water, and carbon dioxide. Cellular respiration serves as the energy source for animal cells. If no oxygen is present, cellular respiration will result in fermentation, where either lactic acid or alcohol is produced instead of sugar.

Cell division is the process where genetic material is replicated in the nucleus. Cell division begins in interphase, where DNA replication occurs. This results in the replication of the chromosomes in the nucleus. **Chromosomes** are tightly coiled threads of DNA composed of twin strands called **chromatids**. Interphase is the longest part of cell division and is divided into periods of cell growth and DNA replication. The cell grows in size to accommodate the increase in chromosomes. Following interphase is prophase. During prophase, chromatids begin to pair up with their sister chromatids. This leads into metaphase, where the sister chromatids move to opposite poles of the cell. During the next phase, anaphase, the chromatids begin to pull apart into two separate poles. The cell becomes elongated during this phase, which makes it very easy to identify. During telophase, the two new nuclei become completely separated. The final phase in cell division is cytokinesis, where the cytoplasm and cell membranes complete their separation and two daughter cells are formed.

Normally, cell reproduction is closely regulated by genetic signals in the cell that tell it when to stop reproducing; **cancer** occurs when the signals are mutated and cells can grow without limit. Factors such as smoking, sun exposure, and genetic mutations can cause damage to cells and may lead to cancer. The most common type of cancer is skin cancer, caused by exposure to UV rays in sunlight.

Study the example below to see how an expert test taker would approach an ASVAB question about cellular functions.

Question	Analysis
Cellular fermentation takes place when	**Step 1:** The question is asking the circumstances during which cellular fermentation takes place.
	Step 2: There is nothing to simplify here.
	Step 3: Cellular fermentation occurs when no oxygen is present. When oxygen is not present, the cell is in anaerobic respiration and produces either alcohol or lactic acid instead of ATP, a molecule that stores and transfers energy.
(A) no oxygen is present during cellular respiration (B) oxygen is present during cellular respiration (C) lactic acid is present during cellular respiration (D) yeast is present during cellular respiration	**Step 4:** Based on the prediction above, select answer choice **(A)**.

Try the problem below:

In which organelle does DNA replication take place?

(A) mitochondria
(B) ribosome
(C) nucleus
(D) endoplasmic reticulum

Explanation

Answer choice **(C)** is correct. The nucleus contains all genetic material. DNA replication takes place in the nucleus during interphase. Interphase is the longest phase of mitosis.

ECOLOGY

LEARNING OBJECTIVES

In this section, you will learn to:

- understand the relationship between organisms and their physical environment
- understand the relationships between living organisms

Ecology is the study of the interrelationships between organisms and their physical surroundings. Just as biologists classify organisms according to terminology that goes from the general to the specific, ecologists employ a similar set of terminology:

Biosphere: The zone of planet Earth where life naturally occurs, including land, water, and air, extending from the deep crust to the lower atmosphere.

Biome: A major life zone of interrelated species bound together by similar climate, vegetation, and animal life.

Ecosystem: A system made up of a community of animals, plants, and other organisms as well as the abiotic (non-living) aspects of its environment.

Community: The collection of all ecologically connected species in an area.

Population: A group of organisms of the same species living in the same region.

An ecosystem can be large or small, and can include both pristine and highly developed areas. An ecosystem contains a community, and this community may contain many populations of organisms. The various populations within a community fall into one of several roles in the food chain.

Producers (mainly plant life): Also known as **autotrophs**, they make their own food via photosynthesis.

Decomposers (bacteria and fungi): Also known as **saprotrophs**, they break down organic matter and release minerals back into the soil.

Scavengers (many insects and certain vertebrates, such as vultures and jackals): These animals exhibit characteristics of decomposers by consuming refuse and decaying organic matter, especially **carrion**, or decaying flesh. Similar organisms called **detritivores** consume the small pieces of decaying organic matter called detritus that are too small for most scavengers to want.

Consumers (most animals): Also known as **heterotrophs**, refers to animals that consume other organisms to survive. Consumers are divided into three types:

Primary consumers: Also known as **herbivores**, they subsist on producers, such as plants. Examples include grasshoppers, deer, cows, and rabbits.

Secondary consumers: Also known as **carnivores** or predators, they subsist mainly on primary consumers. Examples of secondary consumers include birds of prey (such as owls and falcons), foxes, and snakes. Some secondary consumers are also **omnivores**, meaning they consume producers and consumers as well. Examples include chickens, rats, some lizards, and sea otters.

Tertiary consumers: Also known as **top carnivores**, they are capable of eating secondary consumers. Many tertiary consumers are also **omnivores**. Examples include lions, wolves, and sharks, as well as human beings.

The diagram below shows roughly the relationships between these groups in an ecosystem:

Hierarchy of Consumers

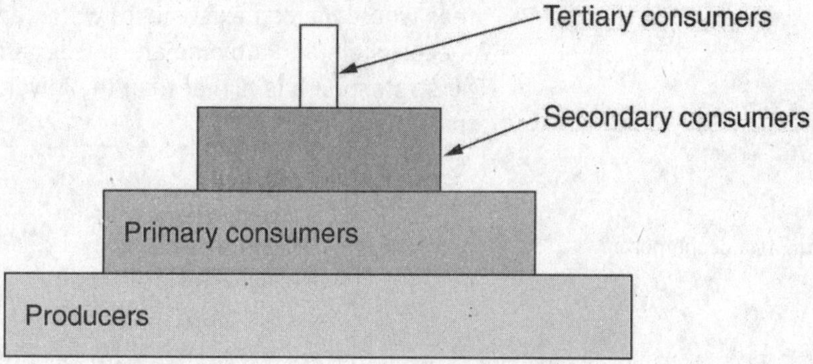

Food chains are a basic way to see the levels in an ecosystem. However, food webs can show more complex relationships that exist amongst the levels. In the example below, the organism to which an arrow points might eat the organism on the other end of that arrow.

A Food Web

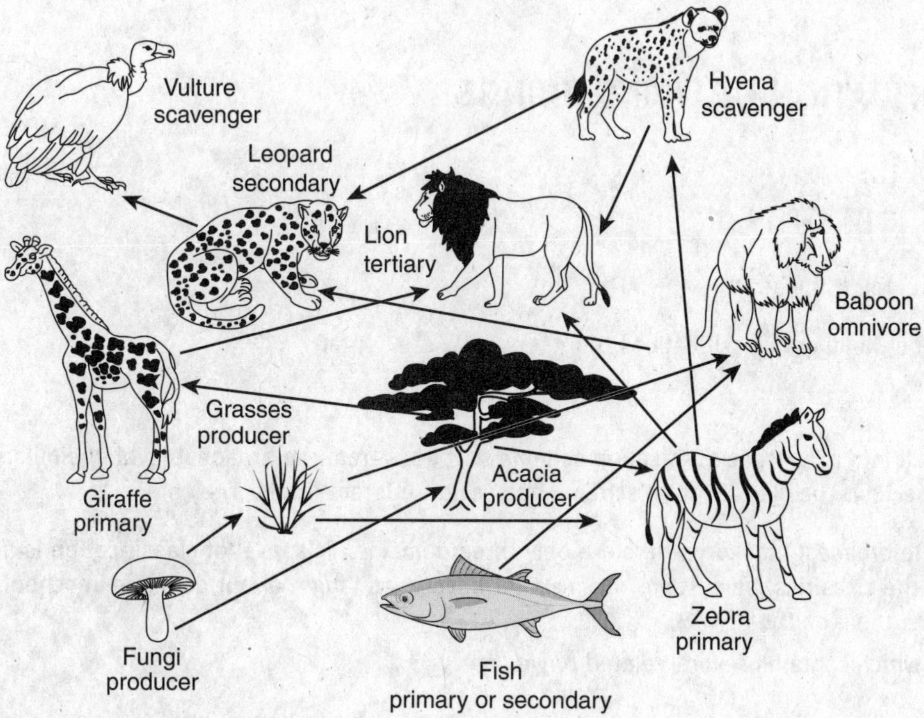

Study this example of a solid approach to an ecology question.

Question	Analysis
Which of the following is the term for the zone that extends from the Earth's crust through the lower atmosphere and which encompasses all life on Earth?	**Step 1:** The question is asking for the name of the zone containing all life on Earth.
	Step 2: Not much to simplify here.
	Step 3: The question describes the biosphere.
(A) biome (B) biosphere (C) ecosystem (D) stratosphere	**Step 4:** Based on the prediction above, select answer choice **(B)**. The biosphere begins in the deep crust of the Earth and includes all areas where life can exist—land, water, and air. An ecosystem and a biome are smaller areas. The stratosphere is higher than the lower atmosphere.

Try problem below:

Which of the following is a decomposer?

(A) vulture
(B) fly
(C) oak tree
(D) bacteria

Explanation

Choose **(D)**. Bacteria are considered a decomposer because they break down dead matter and release it back to the soil as minerals. Flies and vultures would be scavengers, and a tree would be an autotroph.

THE CLASSIFICATION OF LIVING THINGS

LEARNING OBJECTIVE

In this section, you will learn to:

- describe the major classifications of life and how they differ

All living things fall into a careful classification scheme that goes from the broadest level of similarity (**domain**) to the narrowest (**species**). The classification scheme has seven different levels. They are:

Domain: the broadest category; there are only three domains. This level of classification is a fairly recent change to the classification system. You may not have been taught about domains in school, and you may or may not see them on the ASVAB.

Kingdom: which contains several related *phyla*.

Phylum: which contains several related *classes*.

Class: which contains several related *orders*.

Order: which contains several related *families*.

Family: which contains several related *genera*.

Genus: which contains several related *species*.

Species: which contains organisms so similar that they can only reproduce with one another to create viable fertile offspring.

For example, this is the classification of human beings:

Domain — Eukaryota
Kingdom — Animalia
Phylum — Chordata
Class — Mammalia
Order — Primates
Family — Hominidae
Genus — Homo
Species — Sapiens

> Remember: **Domain Kingdom Phylum Class Order Family Genus Species.**
> Here's a mnemonic to help you: **Dear King Philip Came Over For Good Soup.**
> (Remember, Domain is a fairly recent addition to the classification system, so you may or may not see it on the ASVAB.)

The three domains are:

Eukaryota: All living things whose cells have nuclei are in this domain. Almost all multi-celled organisms (including plants, animals, and fungi) are in this domain.

Bacteria and **Archaea**: Both of these domains contain single-celled organisms whose cells do not have nuclei. Living things in the two domains are distinguishable by metabolic and chemical differences.

The five or six kingdoms are:

Kingdom	Description
Monera (sometimes broken into two kingdoms, one of which belongs to domain Bacteria and one of which belongs to domain Archaea)	Includes bacteria, cyanobacteria (blue green algae), and primitive pathogens. Considered the most primitive kingdom, it represents prokaryotic (as opposed to eukaryotic) life forms—that is, the cells of Moneran organisms do not have distinct nuclei.
Protista	Protista are the simplest eukaryotes (that is, their cells have nuclei). Includes protozoa, unicellular and multicellular algae, and slime and water slime molds. Ancestor organisms to plants, animals, and fungi; many can move around by means of flagella. Some are also photosynthetic.

Kingdom	Description
Fungi	Includes mushrooms, bread molds, and yeasts. Fungi lack the ability to photosynthesize; they are called decomposers, breaking down and feeding on dead protoplasm (extracellular digestion).
Plantae	Have the ability to photosynthesize, so they are called producers. There are four major phyla: Bryophyta, or mosses; Tracheophyta, which have vascular systems; gymnosperms; and angiosperms.
Animalia	Produce energy by consuming other organisms, so they are called consumers. Can be either vertebrates (which belong to phylum Chordata) or invertebrates such as mollusks, arthropods, sponges, coelenterates, worms, etc. Human beings belong to the kingdom Animalia.

There are many, many phyla (the plural of *phylum*), classes, orders, families, genera (the plural of *genus*), and species. *If* you have additional time before Test Day *and* you have mastered all of the other subject tests *and* maximizing your General Science score is crucial to your career goals, you may choose to use Internet sources to learn more about some of the largest phyla, classes, etc. Otherwise, that study is unlikely to be the best use of your time.

Study how an expert test taker would approach a general science question about the classification of living things:

Question	Analysis
Mammals are part of the kingdom	**Step 1:** The question is asking you to determine which kingdom mammals belong to.
	Step 2: There is nothing to simplify here.
	Step 3: Mammals are animals; they belong to kingdom Animalia.
(A) Animalia (B) Mammalia (C) Protista (D) Monera	**Step 4:** Based on the prediction above, you would select answer choice **(A)**.

Try one on your own.

Not including domains, the correct order of the categories of taxonomy, from most specific to most general, is:

(A) Species, Genus, Family, Order, Class, Phylum, Kingdom
(B) Kingdom, Phylum, Class, Order, Family, Genus, Species
(C) Kingdom, Phylum, Order, Class, Family, Genus, Species
(D) Species, Genus, Order, Family, Class, Phylum, Kingdom

Explanation

"Species" is the most specific category in taxonomy and "Kingdom" is the most general (not including domains), so choice **(A)** is correct. Choice (B) lists the categories from largest to smallest. Choice (C) almost does the same but confuses Order and Class. Choice (D) is incorrect because it switches the places of "Order" and "Family."

LIFE SCIENCE PRACTICE QUESTIONS

1. Which of the following describes the proper pathway of blood through the heart?
   (A) vena cava → right atrium → right ventricle → pulmonary artery → pulmonary vein → left atrium → left ventricle → aorta
   (B) vena cava → right atrium → right ventricle → pulmonary vein → pulmonary artery → left atrium → left ventricle → aorta
   (C) vena cava → right atrium → left atrium → pulmonary artery → pulmonary vein → left atrium → left ventricle → aorta
   (D) vena cava → right atrium → left atrium → pulmonary vein → pulmonary artery → left atrium → left ventricle → aorta

2. Most human digestion takes place in the
   (A) esophagus
   (B) stomach
   (C) small intestine
   (D) large intestine

3. Which blood type can be donated to anyone?
   (A) A positive
   (B) B negative
   (C) O negative
   (D) AB positive

4. A typical human gamete contains
   (A) 2 chromosomes
   (B) 23 chromosomes
   (C) 46 chromosomes
   (D) 92 chromosomes

5. Which of the following is an example of a primary consumer?
   (A) moss
   (B) mushroom
   (C) jackal
   (D) deer

6. How many domains are recognized in taxonomy?
   (A) 3
   (B) 5
   (C) 7
   (D) 9

7. Which of the following foods are a good source of fiber?
   (A) oils
   (B) fruits and vegetables
   (C) dairy products
   (D) meat and seafood

8. Which components of a prokaryotic cell produce proteins?
   (A) ribosomes
   (B) mitochondria
   (C) lysosomes
   (D) centrosomes

9. A zygote is a
   (A) diseased cell
   (B) mutated male reproductive cell
   (C) female reproductive cell
   (D) fertilized egg

10. Which of the following describes how the process of immunization takes place?
    (A) A highly active form of a pathogen is injected, causing the formation of strong antibodies that will fight off the injected pathogen and remain in the blood stream to fight off future infections.
    (B) Synthetic antibodies are injected into the bloodstream that will persist in the body and fight off future infections.
    (C) A small amount of a deactivated pathogen is injected, causing the immune system to react so that if an active form of the pathogen is encountered in the future, the immune system will respond quickly.
    (D) Chemicals that can destroy a pathogen's ability to reproduce are injected into the bloodstream.

Answers and Explanations

1. **A** To make a prediction on this question, think about what kind of blood (oxygenated or deoxygenated) flows into and out of the heart.

2. **C** The small intestine is the largest digestive organ and does the most work in breaking down food into materials the body can use.

3. **C** Type O negative blood, also known as the "universal donor" type, can be donated to anyone.

4. **B** A typical human gamete contains half the number of chromosomes as a normal cell, or 23.

5. **D** A deer is an example of a primary consumer, that is, an animal that consumes only vegetation.

6. **A** There are three domains in taxonomy: Eukaryota, Bacteria, and Archaea.

7. **B** Sources of dietary fiber include fruits and vegetables, whole grains, and legumes.

8. **A** The primary function of ribosomes is to produce proteins. Mitochondria produce energy, lysosomes help a cell manage waste, and centrosomes are important in a cell's reproduction.

9. **D** When a sperm unites with an ovum, the product is a fertilized egg, which is called a zygote.

10. **C** Immunization uses a deactivated form of a bacterial or viral pathogen. Though inactive, this pathogen triggers the formation of antibodies that attack it, thus "training" the immune system to respond quickly should the active pathogen be introduced in the future.

PART II: EARTH AND SPACE SCIENCE

Earth and space science is the study of the Earth and the universe around it. For purposes of the ASVAB, it's helpful to know a few facts about our planet and the solar system in which it travels.

GEOLOGY

LEARNING OBJECTIVES

In this section, you will learn to:

- identify the layers of the Earth and types of rocks
- describe the general principles of plate tectonics
- describe the composition of the Earth

Geology is the science that deals with the history and composition of the Earth and its life, especially as recorded in rocks.

Structure of Earth

In part by studying rocks, scientists have been able to determine that the Earth is made up of three layers. The outermost layer, or **crust**, comprises roughly one percent of the total volume of the Earth. It varies in thickness from 10 kilometers to as much as 100 kilometers. Beneath the crust lies the **mantle**, which comprises more than 75 percent of the Earth's volume. Roughly 3,000 kilometers thick, the mantle contains mostly iron, magnesium, and calcium, and is much hotter and denser than the Earth's surface because temperature and pressure inside the Earth increase with depth.

At the center of the Earth lies the **core**, which is nearly twice as dense as the mantle because its composition is metallic (iron-nickel alloy) rather than stony. The Earth's core contains two distinct parts: a 2,200-kilometer-thick liquid **outer core** and a 1,300-kilometer-radius solid **inner core**.

Structure of the Earth

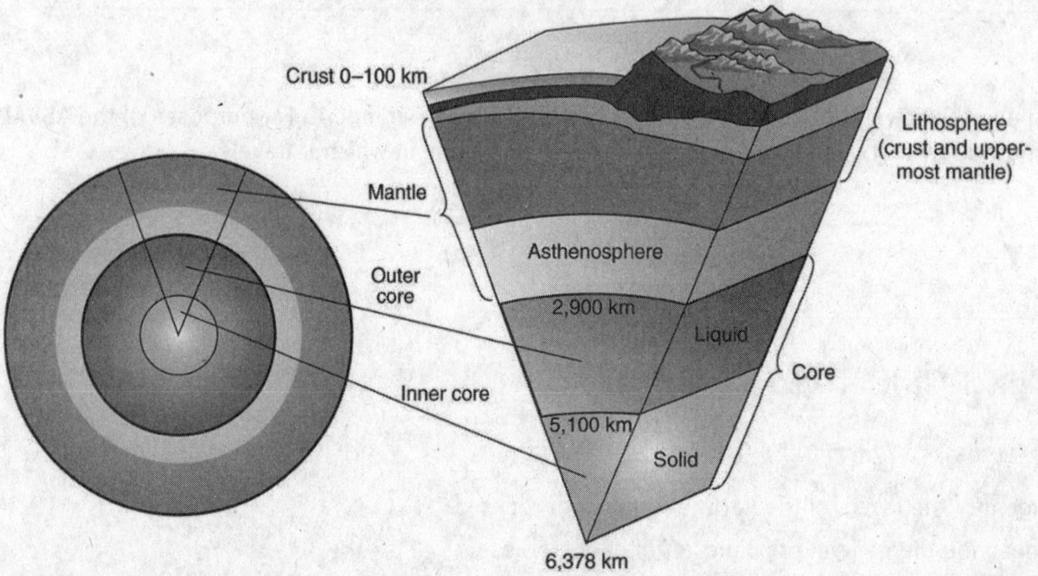

Plate Tectonics

As noted, the interior of the Earth is quite hot, somewhere between 3,000°C and 4,000°C. This heat is generally prevented from escaping thanks to the solid rock in the Earth's upper mantle and crust. The crust and the rigid upper part of the mantle (the **lithosphere**) consist of approximately 30 separate pieces called **plates.** These plates move very slowly upon the more movable mantle beneath (the **asthenosphere**), and this has caused the continental landmasses to drift slowly apart over the course of hundreds of millions of years.

Along the edges of these plates are **fault lines**. (Fault lines are simply places where the plates slide relative to each other.) When plates slide relative to each other along fault lines, earthquakes can occur. When an earthquake occurs, scientists use the **Richter scale** or the **moment magnitude scale** to measure its intensity. The Richter scale is perhaps more familiar to most people. The Richter scale begins at 1, with each larger integer representing a magnitude that is about 10 times greater than the preceding step.

Types of Rocks

The Earth's rocks fall into three categories, based upon how they are formed.

Rock type	How it's formed	Examples
Igneous	Formed from the hardening of molten rock, or magma, which is called lava when it reaches the surface of the Earth	Granite, pumice, basalt, obsidian
Sedimentary	Formed by the sedimentation, or gradual depositing, of small bits of rock, clay, and other materials. Over time this deposited material becomes cemented together. Most fossils are found in sedimentary rocks.	Shale, sandstone, gypsum, dolomite, coal
Metamorphic	Formed when existing rock material is altered through temperature, pressure, or chemical processes	Marble, slate, gneiss, quartzite

Geologic Time Scale

Most of what is known about the history of our planet has been learned by studying the fossil record found in sedimentary rock. By studying rocks, we now know that the Earth is approximately 4.6 billion years old, and that for most of that time, very few fossil traces were left. This is why the period from 4.6 billion years to 570 million years ago is called the **Precambrian eon**, meaning the period before the fossil record began. It turns out, however, that early geologists who studied the Precambrian eon were unable to recognize early, primitive fossils, and in fact life first appeared on Earth as early as 3.5 billion years ago!

Here's how a well-prepared test taker would approach a question about geology on Test Day.

Question	Analysis
Studies of the Precambrian eon have shown that life first appeared on Earth approximately	**Step 1:** The Precambrian eon allowed scientists to narrow down the time frame for life formation.
	Step 2: The Precambrian eon ranges from 4.6 billion to 570 million years ago.
	Step 3: The correct number should be within the range of the Precambrian eon.
(A) 570 million years ago (B) 3.5 billion years ago (C) 4.6 billion years ago (D) 5 billion years ago	**Step 4:** Based on the prediction above, select answer choice **(B)**.

Now try the problem below.

Which of the following is an igneous rock?

(A) marble
(B) gypsum
(C) pumice
(D) coal

Explanation

Choice **(C)** is correct. Pumice comes from volcanic rock when it cools very quickly on the surface. The small air bubbles formed in it are visible to the naked eye. The bubbles give it its lightweight feel and increase its usefulness in many everyday products.

CYCLES IN EARTH SCIENCE

LEARNING OBJECTIVES

In this section, you will learn to:

- identify the stages of the water cycle
- identify the stages of the carbon cycle

There are several biogeochemical cycles. They all work in similar fashion in that the element or compound is released in the air and returned to the ground. The two basic cycles this section will focus on are the water cycle and the carbon cycle.

The **water cycle** is also known as the **hydrologic cycle**. It involves the movement of water in all states of matter (solid, liquid, and gas) through the atmosphere and back to the Earth. As water from the surface of the ocean and from other bodies of water **evaporates** (becomes a gas), it rises into the atmosphere. Water also evaporates into the atmosphere from the leaves of plants; this process is known as **transpiration**. This water vapor **condenses** to form clouds. When the clouds become too heavy, they release **precipitation** in the form of a solid (snow and ice) or a liquid (rain). Some of the water that accumulates on the surface will travel down in the form of snowmelt or surface **runoff** and return to the ocean via rivers and streams. The rest of the water is absorbed into the Earth's surface and travels to the water table (a process known as **infiltration**). The water table serves as a reservoir and can be drawn upon through the digging of wells.

The Water Cycle

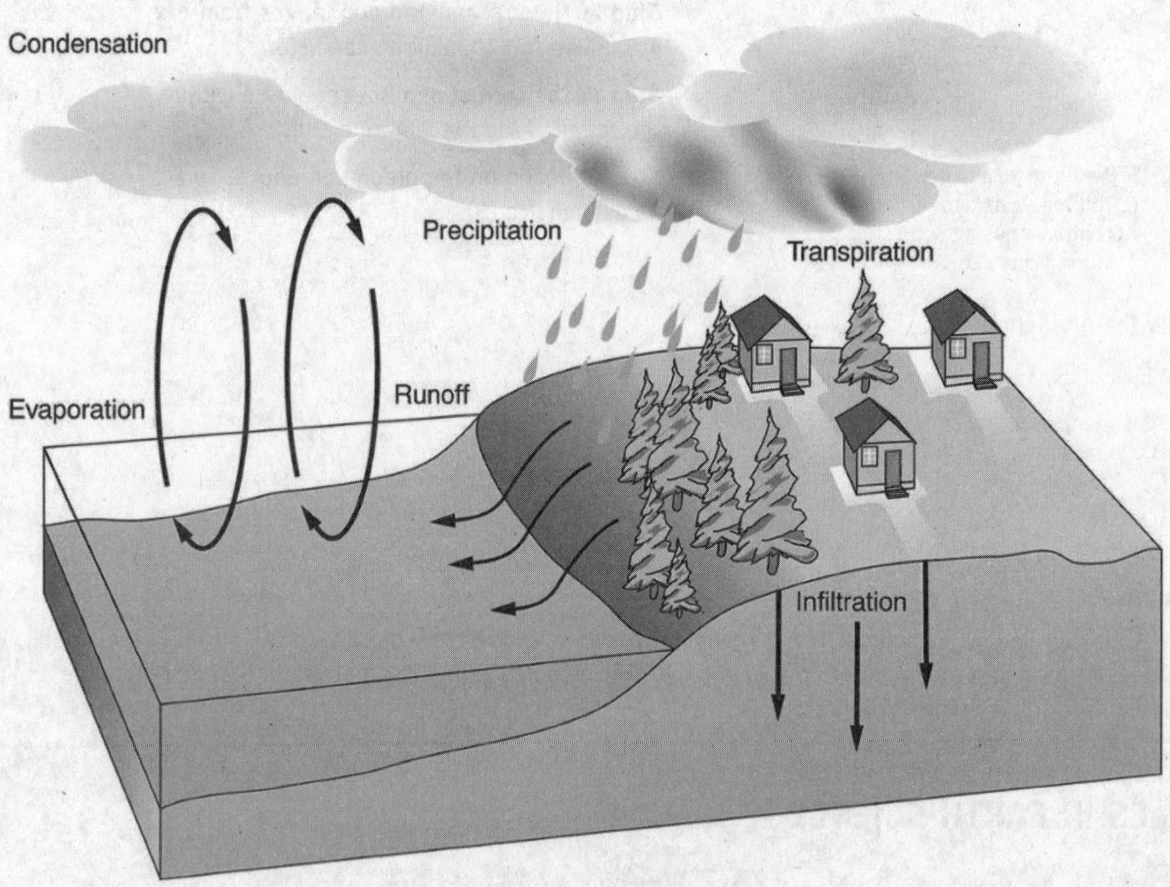

Carbon is one of the most common elements on our planet. The **carbon cycle** helps maintain Earth's ecosystem. Carbon gas exchange is important in maintaining a breathable atmosphere. Carbon is released through human **emissions** (manufacturing) and **respiration**. Respiration results in the release of carbon dioxide waste products into the atmosphere; an animal's breathing is one example. Respiration release is due to human release as well as release from decomposing plant and animal life. Carbon gathers in the atmosphere and is reabsorbed through plants (land and ocean) and soil.

The Carbon Cycle

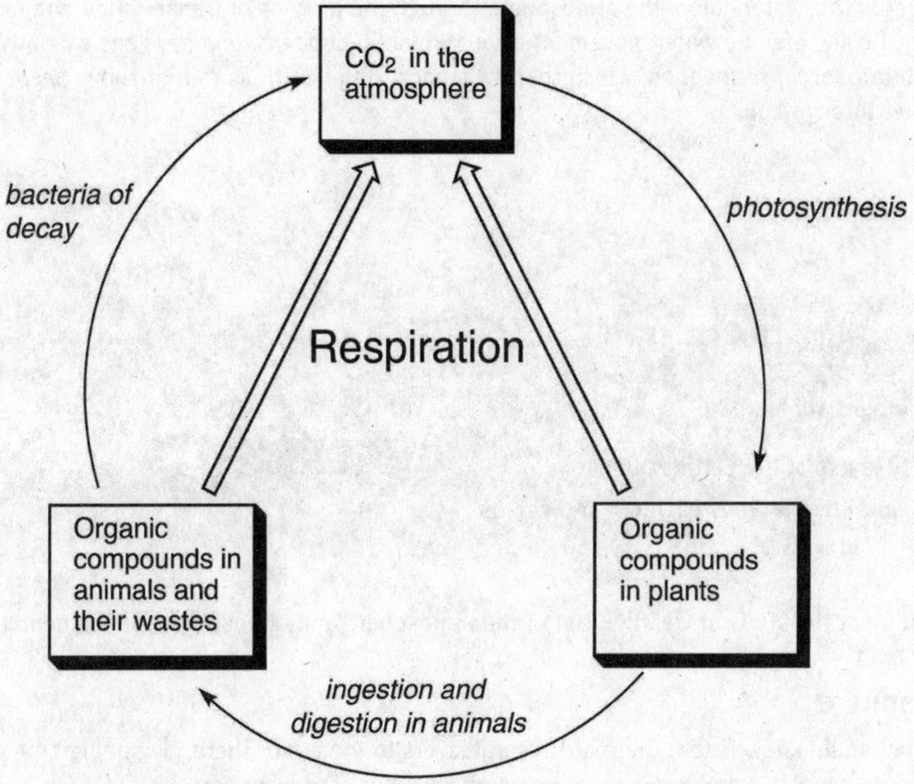

Study this example of an ideal Test Day approach to this question about the carbon cycle.

Question	Analysis
Which of the following is a way in which carbon is released into the atmosphere?	**Step 1:** Think of the ways that carbon is released into the atmosphere.
	Step 2: It is released through respiration, decomposition, and manufacturing.
	Step 3: The correct answer should discuss one of those possibilities.
(A) pine trees engaging in photosynthesis (B) sediment settling in the ocean (C) dead algae decomposing (D) new plants growing in a forest	**Step 4:** Based on the prediction above, select answer choice **(C)**.

Now try one on your own.

During the water cycle, water enters the atmosphere as a gas through

(A) transpiration
(B) condensation
(C) precipitation
(D) seepage

Explanation

The correct answer is **(A)**. Water enters the atmosphere through the process of transpiration and evaporation. *Transpiration* specifically refers to water movement through plants. *Condensation* happens when water vapor condenses together to form *precipitation*, which then falls back to the Earth as rain or snow. *Seepage* is the movement of water through soil.

METEOROLOGY

LEARNING OBJECTIVES

In this section, you will learn to:

- identify layers of the Earth's atmosphere
- describe different types of fronts and clouds

Meteorology is not just the study of weather, but of the atmosphere and atmospheric phenomena in general.

Earth's Atmosphere

The first thing you should know is that there are several layers to the atmosphere, beginning here on the surface of the Earth and continuing up several thousand kilometers above us. The layers are:

Troposphere: The troposphere is the lowest level of the atmosphere, where all weather takes place; it is a region of rising and falling packets of air. Depending on the latitude and the season, it can range from 6 to 17 kilometers thick. Most of the air surrounding the Earth, which is roughly 79 percent nitrogen and 21 percent oxygen, is found in the troposphere.

Stratosphere: Above the troposphere is the stratosphere, where airflow is mostly horizontal. The thin **ozone layer** in the upper stratosphere has a high concentration of ozone, a particularly reactive form of oxygen. This layer is primarily responsible for absorbing the ultraviolet radiation from the Sun. As you enter the stratosphere, the temperature is about −60°C.

Mesosphere: Above the stratosphere is the mesosphere, which extends to about 90 kilometers above the Earth. As you enter the mesosphere, the temperature starts to drop again, to as low as −90°C. This is where we see "falling stars," meteoroids that fall to the Earth and burn up in the atmosphere.

Thermosphere: Beyond the mesosphere, temperatures actually increase with altitude in the thermosphere because there is little matter to deflect solar radiation. Temperatures as high as 2,000°C have been recorded in the thermosphere.

Earth's Atmosphere

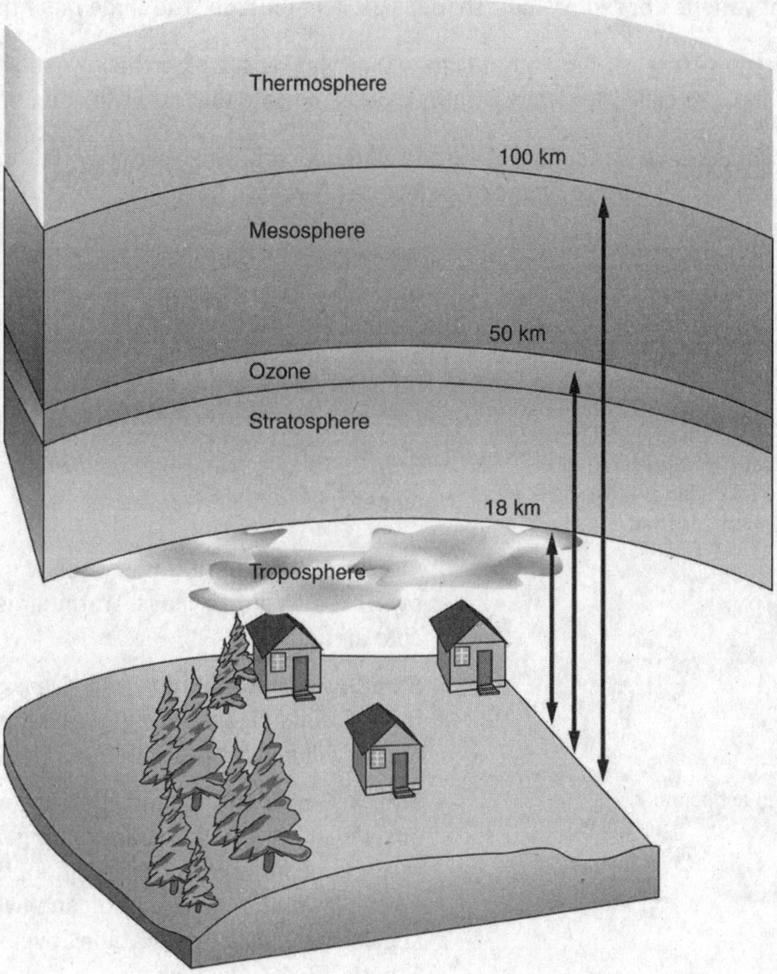

Fronts

Differences in air pressure cause wind and the movement of air masses of different temperatures toward each other. When a warm air mass overtakes a cold air mass, you have a **warm front**. As the warm air advances, it rides over the cold air ahead of it, which is heavier. As the warm air rises, the water vapor in it condenses into clouds that can produce rain, snow, sleet, or freezing rain—often all four.

When a cold air mass overtakes a warm air mass, you have a **cold front**. Most cold fronts are preceded by a line of precipitation as they roar across an area. However, some cold fronts produce very little or no precipitation as they move. The only sign that a front has moved through your area is a sudden change in winds and temperature.

Sometimes two air masses meet and neither is displaced. Instead, the two fronts push against each other in a stalemate. This is called a **stationary front**. Stationary fronts often cause cloudy, wet weather that can last a week or more.

Clouds

Clouds come in different varieties based on their shape, size, and altitude. The three main types are:

Stratus clouds are low-hanging, broad, flat clouds that blanket the sky. The lowest of low clouds, when they occur on the ground, are called fog. Dark stratus clouds indicate that rain will soon occur.

Cumulus clouds are massive clouds that are puffy, like popcorn, with relatively flat bottoms and rounded tops. When cumulus clouds darken, you can expect heavy rain.

Cirrus clouds are the thin, wispy clouds that occur much higher in the atmosphere, at elevations of 20,000 feet or more.

Study the following example, which illustrates an expert approach to a meteorology question.

Question	Analysis
The weather report indicates a warm air front is moving in. Which of the following is most likely to occur as a result of that warm air front?	**Step 1:** Think of what happens when a warm front comes in.
	Step 2: A warm front forms when warm air overtakes a cold air mass. Warm air is heavier than cold air and rises.
	Step 3: When warm, heavy air rises, clouds will form. Clouds mean precipitation, so the answer choice will involve precipitation.
(A) a sudden change in temperature (B) freezing rain (C) hurricanes (D) weather stalemate	**Step 4:** Based on the prediction above, select answer choice **(B)**. Warm air fronts generally result in some form of precipitation, though not a full-blown storm like a hurricane (which is a specialized storm that develops over warm ocean water), (C). A sudden change in temperature, (A), is the result of a cold air front and a stalemate, (D), is caused by a stationary front.

Try one on your own.

The sky is filled with dark, puffy clouds, and there is a thunderstorm approaching. What type of clouds are you seeing?

(A) cirrus
(B) stratus
(C) cumulus
(D) stratocumulus

Explanation

The correct answer is **(C)**. Cumulus clouds are often referred to as thunderheads. They have high fluffy peaks that are perfect for generating static electricity and producing thunderstorms. A few scattered cumulus clouds are usually not a threat of a storm, but if several groups of them converge, they could produce heavy rain, thunder, and lightning.

OUR SOLAR SYSTEM

LEARNING OBJECTIVES

In this section, you will learn to:

- understand the structure of the solar system
- understand the relationship of the Earth to other planets and solar bodies
- understand how Earth's position relative to the Sun produces Earth's seasons

Our solar system consists of one star, which we call the Sun, eight planets and all their moons, several thousand minor planets called asteroids, and an equally large number of comets.

The Sun

Our **Sun** is classified as a **G2V** star, or **yellow dwarf**. G2 stars are approximately 6,000°C at the surface, are yellow, and contain many neutrally charged metals such as iron, magnesium, and calcium. The "V" indicates that the Sun is a dwarf star, or fairly small by the standards of stars. The Sun's age is calculated to be around 4.7 billion years, which is only slightly older than the Earth itself.

Although a dwarf in comparison to other stars, the Sun contains almost 99.9 percent of the mass of our solar system. Like all stars, the Sun is a gigantic ball of superheated plasma, kept hot by atomic reactions emanating from its center. The temperature at the core of our Sun is thought to be about 15,000,000°C; temperatures at the surface range between 4,000° and 15,000°C. The diameter of the Sun is about 1.4 million kilometers, or more than 100 times that of Earth, and its surface area is approximately 12,000 times that of Earth.

The Planets and Other Phenomena

The four planets closest to the Sun—**Mercury**, **Venus**, **Earth**, and **Mars**—are called **terrestrial planets**, meaning they are similar to our own planet in composition, with inner metal cores and surfaces of rock. Earth and Mars are the only terrestrial planets that have moons of their own, although the moons of Mars are much smaller than our own Moon. Earth is the largest of the terrestrial planets.

The four planets beyond Mars—**Jupiter**, **Saturn**, **Uranus**, and **Neptune**—are referred to as the **outer planets**. They also all have rings, most notably so on Saturn. The rings of Saturn, and most likely the other planets as well, are composed mostly of ice crystals. **Pluto**, once considered the ninth planet, is no longer categorized by scientists as a true planet. You may have learned a mnemonic (or memory trick) for the planets in school; here's one that allows for the omission of Pluto: *My Very Educated Mother Just Served Us Nachos*. (In the diagram that follows, Pluto is included, but remember that it is no longer considered a planet. Also, note that the diagram that follows is not drawn to scale.)

The Solar System

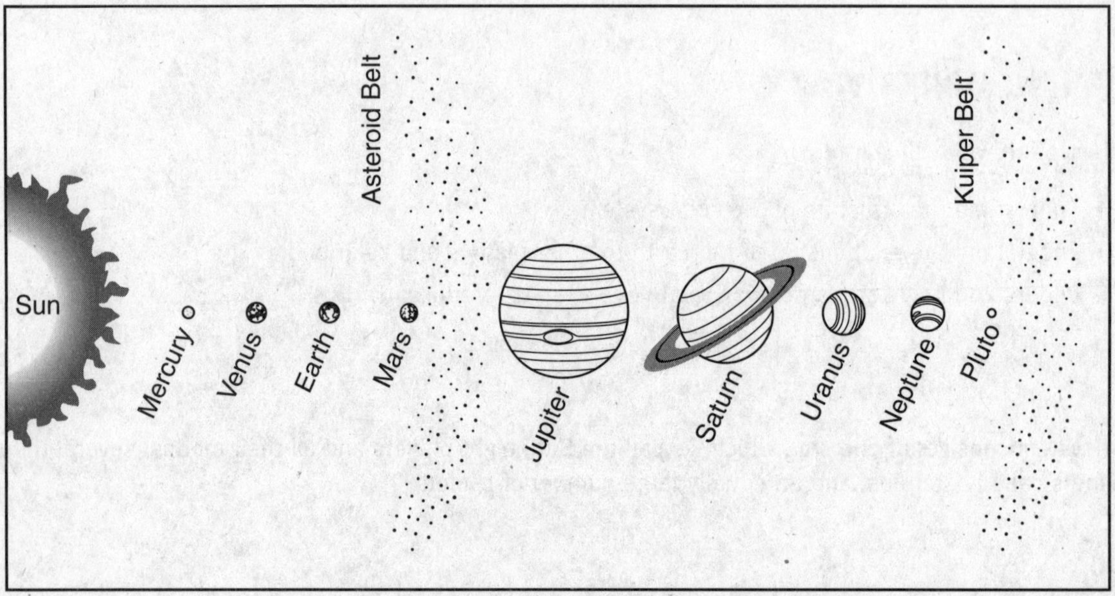

In addition, the solar system contains thousands of small bodies such as **asteroids** and **comets**. Smaller fragments of asteroids and particles shed by comets are known as **meteoroids**, and when they fall into the Earth's gravitational field, they are seen as "falling stars," called **meteors**, as they burn up in the Earth's mesosphere. Those meteoroids that make it to the Earth's surface are called **meteorites**. A belt of asteroids lies between Mars and Jupiter. The **Kuiper Belt**, a much larger collection of asteroids and other objects left over from the formation of the solar system, lies beyond the known planets.

Comets are sometimes called "dirty snowballs" or "icy mudballs." They are a mixture of ices (both water and frozen gases) and dust that for some reason didn't get incorporated into planets when the solar system was formed. Comets are invisible except when they are near the Sun. When they are near the Sun and active, comets have highly visible **tails**, up to several hundred million kilometers long, composed of plasma and laced with rays and streamers caused by interactions with the solar wind.

By far the most important body in the solar system aside from the Sun, as it relates to life on Earth, is our own **Moon**. Because of the gravitational pull that exists between the Moon and Earth, we have **tides**. High tides occur twice a day, when the Moon is at the points closest to and farthest from the affected mass of water. It is suspected that life would never have evolved on land without the constant ebbing and flowing of the oceanic tides on coastal areas.

Earth orbits the Sun once per year. Because the Earth is somewhat tilted on its axis, the northern pole is tilted toward the Sun and the southern pole is tilted away from it for part of the year. That portion of the year is summer in the northern hemisphere and winter in the southern hemisphere. For part of the year that situation is reversed: that period is winter in the northern hemisphere and summer in the southern hemisphere. In between summer and winter, the Earth is tilted so that neither the northern pole nor the southern pole is pointing toward the Sun. These periods are spring and fall. Study the diagram that follows.

Earth's Seasons

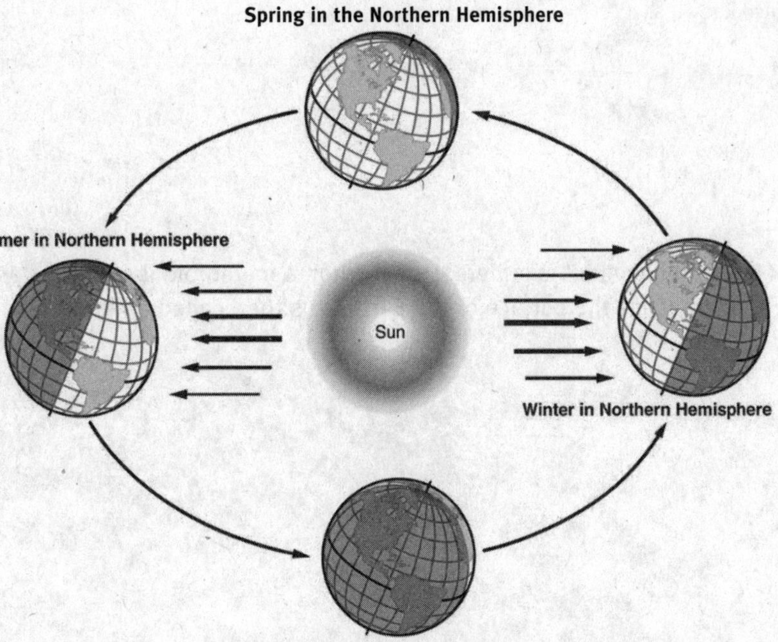

Imagine how terrified ancient humans must have been when their view of the Sun or the Moon became partially or completely obscured during an eclipse. Today we understand how eclipses occur and can predict them well in advance.

A solar eclipse occurs when the Moon passes directly between the Earth and the Sun during daylight hours. The relatively small shadow cast by the Moon onto the surface of the Earth can either partially or totally obscure the Sun. Because the Moon is moving in its orbit around the Earth and the Earth is rotating on its axis, solar eclipses are brief and move across the face of the Earth rapidly.

A lunar eclipse occurs when the Earth passes directly between the Sun and the Moon and casts a shadow on the Moon. Therefore, a lunar eclipse can be seen from any place on Earth where the Moon is above the horizon. Like solar eclipses, lunar eclipses can be either partial or total. Because the Earth has a much larger shadow than does the Moon, lunar eclipses last much longer than their solar counterparts.

Study this example of how an expert test taker would approach an ASVAB question about the solar system.

Question	Analysis
Which of the following is a characteristic of terrestrial planets?	**Step 1:** The Earth is a terrestrial planet. What are its major characteristics?
	Step 2: The Earth has a solid, rocky surface. There are many layers to it that include a solid inner core.
	Step 3: The answer choice will list one of those characteristics.
(A) an inner core made of ice crystals (B) a surface temperature similar to that of Earth (C) an inner core of metal (D) a surface composed of gases	**Step 4:** Based on the prediction above, select answer choice **(C)**. Terrestrial planets have an inner core of metal and a rocky surface. Outer planets have inner cores of ice crystals.

Try this one on your own.

A "falling star" is also known as a

(A) meteor
(B) asteroid
(C) comet
(D) meteorite

Explanation

The correct answer is **(A)**. A "falling star" is a meteor seen when a meteoroid is burning up in the Earth's atmosphere. If the meteroid reaches the surface of the Earth, it is then called a meteorite.

EARTH AND SPACE SCIENCE PRACTICE QUESTIONS

1. The ozone layer is found in the
   (A) troposphere
   (B) stratosphere
   (C) mesosphere
   (D) thermosphere

2. What is the layer that is located immediately beneath Earth's crust?
   (A) inner core
   (B) plates
   (C) mantle
   (D) outer core

3. The clouds that occur at the highest altitude are called
   (A) cirrus
   (B) cumulus
   (C) nimbus
   (D) stratus

4. The Kuiper Belt is _________?
   (A) a group of meteors that orbit around the Earth
   (B) a collection of asteroids and other objects left over from the formation of the solar system
   (C) a layer of Earth's atmosphere
   (D) a grouping of asteroids that orbit the Sun between Mars and Jupiter

5. Which of the following is most responsible for the oceanic tides?
   (A) the gravitational pull of the Sun on the Earth
   (B) the gravitational pull of the Moon on the Earth
   (C) the heat of the Sun
   (D) the magnetic pull of the poles

6. Oxygen makes up approximately _________ percent of Earth's atmosphere.
   (A) 10
   (B) 21
   (C) 78
   (D) 90

7. The Richter scale is used to measure the intensity of _________.
   (A) hurricanes
   (B) blizzards
   (C) earthquakes
   (D) tornadoes

8. The sequence of the movement of water in the hydrologic cycle is
   (A) evaporation → transpiration → condensation → runoff/infiltration → precipitation
   (B) runoff → condensation → infiltration → evaporation/transpiration → precipitation
   (C) precipitation → evaporation/transpiration → condensation → runoff/infiltration
   (D) precipitation → runoff/infiltration → evaporation/transpiration → condensation

9. Granite is an example of what type of rock?
   (A) compound
   (B) igneous
   (C) metamorphic
   (D) sedimentary

10. Earth is somewhat shielded from harmful ultraviolet radiation by the
    (A) Van Allen belt
    (B) thermosphere
    (C) ozone layer
    (D) troposphere

Answers and Explanations

1. **B** The ozone layer is found in the upper stratosphere.

2. **C** The mantle lies beneath Earth's crust.

3. **A** Cirrus clouds are found at the highest altitude of all clouds.

4. **B** There is a group of asteroids that orbit the Sun between Mars and Jupiter, but the Kuiper Belt is located beyond the outermost known planets.

5. **B** The oceanic tides are caused by the gravitational pull of the Moon on the Earth.

6. **B** Oxygen accounts for approximately 21% of Earth's atmosphere. Nitrogen makes up about 78% of the atmosphere.

7. **C** The Richter scale, like the moment magnitude scale, is a logarithmic measure of the intensity of earthquakes. An increase of 1 unit on the Richter scale represents an increase by a factor of 10 in the intensity of an earthquake.

8. **D** Precipitation from the sky falls to the ground, where it either becomes surface runoff or infiltrates into the ground. Evaporation from bodies of water or transpiration from plants causes water vapor to rise into the sky, where it condenses to form clouds and the cycle begins again.

9. **B** Granite and other rocks formed by the hardening of molten magma are classified as igneous rocks.

10. **C** The ozone layer, which is between the stratosphere and mesosphere, is primarily responsible for absorbing much of the ultraviolet radiation from the Sun.

PART III: PHYSICAL SCIENCE

Scientists would know very little about our solar system without an understanding of physics and chemistry, collectively known as the physical sciences. This section will cover those two disciplines as well as the system of measurement used in both.

MEASUREMENT

LEARNING OBJECTIVES

In this section, you will learn to:

- employ the common units and basic conventions of the metric system
- relate the major temperature scales (Fahrenheit, Celsius, and Kelvin)

Scientists don't use British measurement units such as ounces, miles, and gallons; instead, they use the **metric system**. The metric system has been used throughout most of this chapter, but it's time for a more detailed examination. The key idea of the metric system is to designate one base unit for every kind of measurement, and then make bigger or smaller units by adding prefixes to it (groups of letters added to the beginnings of words, like *anti–* or *pro–*). For example, the base unit for **length** in metric is called the **meter (m)**, which is just over a yard (about 39.4 inches).

To measure small things like firearm ammunition or large things like the distance between cities, a prefix is added. *Milli–* means $\frac{1}{1,000}$, so a **millimeter (mm)** is $\frac{1}{1,000}$ the size of a meter. A 9 mm handgun like the Beretta M9 (the US military's standard sidearm) fires rounds that are .009 m in bullet diameter. A **centimeter (cm)** is 10 times as big as a millimeter, and there are about 2.54 cm in one inch. A **kilometer (km)** is 1,000 meters. Metric measurements typically have two- or three-letter symbols (like mm, cm, and km), which are usually the first letter of the prefix followed by the first letter of the base unit.

Prefix	Symbol	Value relative to base unit
mega	M	10^6 or 1,000,000
kilo	k	10^3 or 1,000
hecto	h	10^2 or 100
deka	da	10^1 or 10
base (no prefix)	—	1
deci	d	10^{-1} or $\frac{1}{10}$ or 0.1
centi	c	10^{-2} or $\frac{1}{100}$ or 0.01
milli	m	10^{-3} or $\frac{1}{1{,}000}$ or 0.001
micro	μ (the Greek letter mu, to avoid confusion with "mega")	10^{-6} or $\frac{1}{1{,}000{,}000}$ or 0.000001

The most common prefixes (and the most likely to come up on the ASVAB) are *milli–*, *centi–*, *kilo–*, and sometimes *mega–*. The other prefixes on the table are included for the sake of completion, but it's not important to memorize them, nor other prefixes beyond those listed here. Know both the prefix names as well as their symbols, as the symbol (for example, km for kilometer) is more likely to come up in a calculation question than the name.

Applying the prefix scheme to other measurements, **mass** is measured in a base unit of **grams (g)**. Therefore, large objects can be measured in **kilograms (kg)**, and very small things can be measured in **milligrams (mg)**; for example, the amount of certain minerals and nutrients per serving as listed on a food package's nutritional information tend to be in milligrams. There are approximately 28.3 grams in an ounce, and a mass of one kilogram will have a weight of approximately 2.2 pounds (at the surface of the Earth).

Volume is the measurement of three-dimensional space. A cube (square box) which is one centimeter on a side can be called just that: a **cubic centimeter (cc)**; but it can also be called a **milliliter (mL)**. The definition of the milliliter automatically implies the **liter (L)**: the prefix means a milliliter is one-thousandth of a liter, which is the same as saying a liter is equal to one thousand milliliters. A liter is equal to slightly more than a quart in liquid measure, or about 33.8 ounces, with about 3.79 liters in a gallon.

Time is measured in **seconds (s)** and metric prefixes are used when smaller units are required (as in milliseconds). However, **minutes (min)** and **hours (h)** are also common.

Finally, the metric system equivalent of temperature is the **Celsius scale**, also known as **degrees centigrade**. According to the **Fahrenheit scale**, which is more familiar to most people in the United States, water freezes at 32°F and boils at 212°F. On the Celsius scale, water freezes at 0°C and boils at 100°C. The general equations for converting from the Fahrenheit scale to the Celsius scale, or vice versa, are as follows:

$$F^\circ = \frac{9}{5}C^\circ + 32$$

$$C^\circ = \frac{5}{9}(F^\circ - 32)$$

Finally, there is one other temperature scale commonly used by scientists, known as the **Kelvin scale**, or the **absolute zero scale**. Absolute zero is the temperature at which matter has no heat and its molecules are completely still; in theory, absolute zero is the lowest temperature possible. On the Kelvin scale, absolute zero is set at 0 K, which is equal to −273°C. Otherwise, the Kelvin scale uses the same increments as degrees Celsius, so that water freezes at 273 K and boils at 373 K. Note that there is no degree symbol when writing out temperatures in the Kelvin scale.

The unit types discussed here are not exhaustive. Different base units come up in different areas of science, but the beauty of metric standards is that the value of each prefix is a constant. Metric prefixes will be applied in new contexts elsewhere in this chapter, and again in chapter 10: Electronics Information.

Question	Analysis
In a chemistry lab, the experimental procedure says to add 50 mL of room temperature water to a beaker. How many liters are in 50 mL?	**Step 1:** The question asks for the value of 50 mL converted to L units.
	Step 2: The prefix *milli–* has a value of 0.001 of the base unit. Since L are 1,000 times larger than mL, the equivalent number in L will be 1,000 times smaller than the number in mL. The decimal will move by three steps. $50 \text{ mL} = \frac{50}{1{,}000} \text{ L}$ 0.05 L is equivalent to 50 mL.
	Step 3: The solution is 0.05 L.
(A) 50,000 L (B) 5 L (C) 0.5 L (D) 0.05 L	**Step 4:** Answer choice **(D)** is correct.

Now you try one:

In an experiment, you must measure out 0.13 kg of sodium chloride. However, your instruments are only set to work with gram units. Convert 0.13 kg into grams.

(A) 1.3 g
(B) 13 g
(C) 100 g
(D) 130 g

Explanation

Choice **(D)** is correct. Since there are 1,000 g in 1 kg, and this is a conversion to the smaller unit value, simply move the decimal place three places to the right (once for every 0 in 1,000).

PHYSICS

LEARNING OBJECTIVES

In this section, you will learn to:

- calculate values for velocity, acceleration, and momentum based on motion formulas
- describe the relation between work, energy, and force
- apply Newton's laws to predict the behavior of physical objects in different situations
- understand visible light as just one part of a spectrum of different kinds of radiant energy
- understand the basic wave behavior of light as it applies to interactions with mirrors and lenses
- understand and identify all instances of heat transference as either conduction, convection, or radiation
- understand the basic origin and behavior of magnetic phenomena

Physics is the science dealing with the properties, changes, and interactions of matter and energy. There are many branches of physics, including mechanics, thermodynamics, magnetism, optics, and electricity. This review will cover only the physics that might appear on the General Science section of the ASVAB.

MASS vs. WEIGHT

In physics, words that are often used imprecisely in everyday speech have very strict definitions. For example, the words **mass** and **weight** are often used interchangeably, when in fact they have very different meanings. Mass is defined as the amount of matter that something has, whereas weight is defined as the force exerted on an object's mass by gravity. A person can become nearly weightless in deep space, far from the nearest gravity-producing objects, but that person still has all her mass.

Motion

Velocity is the rate at which an object changes position. Change in position is called **displacement** and velocity is defined as the total displacement per unit time. It can be calculated as **velocity = displacement of an object ÷ time.**

$$\vec{v} = \frac{\vec{d}}{t}$$

In physics, velocity is called a **vector quantity**, meaning it is fully described by both a magnitude *and* a direction. For example, a car traveling west that covers fifteen meters in two seconds would be described as having a velocity of 7.5 m/s (that is, miles per second) west. Displacement is also a vector, and both symbols often have a little arrow above them to signify this. Time is not a vector.

Momentum is a measure of the quantity of motion of an object. It corresponds to how difficult it is for a moving object to stop. The formula definition of momentum is **Momentum = mass × velocity**. In the symbolic version, momentum is represented by the letter *p*, since *m* is already in use for mass.

$$\vec{p} = m\vec{v}$$

This relationship means, for example, that a semitrailer truck moving at 5 km/h has more momentum than a person walking at the same speed, and also that you have more momentum when running than when walking. Momentum is also a vector quantity, so that two objects moving towards each other have opposite directions of momentum which will partially or completely cancel out if they collide.

Acceleration is the rate of change of velocity. **Acceleration = change in velocity ÷ change in time.** The **Δ (delta) symbol** represents change.

$$\vec{a} = \frac{\Delta\vec{v}}{t}$$

You can see acceleration in a stopped vehicle when the light turns green and the driver depresses the gas pedal. The movement of the speedometer needle shows acceleration, as the car's velocity is increasing moment by moment until it plateaus at cruising speed. Acceleration is also a vector quantity.

Question	Analysis
A sports car hits the brakes and changes its velocity from 65 m/s to 40 m/s in five seconds. What is its average rate of acceleration?	**Step 1:** The question asks for the acceleration of the car. The change in velocity and time of acceleration are required.
	Step 2: Change in velocity is the difference between the final and initial velocity. This goes into the acceleration formula. Velocity has decreased, so acceleration should be negative. $\Delta\vec{v} = 40 - 65 = -25$ m/s $a = -\frac{25}{5} = -5$ m/s^2
	Step 3: The acceleration is –5 m/s^2, or the car decelerates at a rate of 5 m/s^2. The prediction has a negative value to denote a direction opposite motion.
(A) –5 m/s (B) 5 m/s (C) 8 m/s (D) 13 m/s	**Step 4:** Select choice **(A)**.

Now try one on your own.

A bicycle initially at rest at the top of a hill accelerates as its rider coasts down. An accelerometer (which measures acceleration) records a fairly constant acceleration of 6.5 m/s^2. If the rider reaches the halfway point 5.2 s into the trip, what was the velocity of the bicycle and rider at this time?

(A) 33.8 m/s, uphill
(B) 1.25 m/s, uphill
(C) 33.8 m/s, downhill
(D) 1.25 m/s, downhill

Explanation

Answer choice **(C)** is correct. The change in velocity can be calculated by rearranging the acceleration equation, then written as $\Delta v = at$. Since the bicycle and rider are initially at rest, the change in velocity is equivalent to the final velocity (after 5.2 s). This also means that both acceleration and the velocity any time after movement has begun are in the downhill direction. Thus, $v2 = (6.5\ \text{m/s}^2)(5.2\ \text{s}) = 33.8\ \text{m/s}$, downhill.

Forces and Energy

Force is the push or pull that causes an object to change its speed or direction of motion.

Weight is just one example of a force; in this case, the force is due to gravity. A unit of force is called a **newton (N)**, which is the force required to impart an acceleration of one meter per second squared to a mass of one kilogram.

Work is performed on an object when there is an applied force that is along the same line of movement. **Work** = **force** × **displacement**, where the directions of force and movement are parallel.

$$W = \vec{F}\vec{d}$$

A unit of work is called a **newton-meter** or **joule (J)**. Performing work uses up **energy**, also measured in joules, which is equal to the amount of work performed. The reason we have to regularly consume food is that we are constantly using up energy: when we move, when our heart pumps blood, when our lungs inhale and exhale, when we generate warmth to maintain our body temperature, and so on. The nutritional information on food packages sometimes lists the food energy per serving in **kilojoules (kJ)** in addition to the traditional British unit of kilocalories (usually referred to, confusingly, as "calories" in everyday speech).

Power is the rate at which work is performed, or energy is converted. It's defined as the amount of work done or energy converted per unit of time and can be calculated as **Power** = **work ÷ time**, or **Power** = **(force** × **distance) ÷ time.**

$$P = \frac{W}{t}$$

The main unit of power is the **watt (W)**, where one watt is defined as one joule per second. Be sure not to mix up the symbol for the Watt unit with the symbol for work in the formula, nor the formula symbol for mass with the symbol for the meter unit. Units go with a number. The letter symbols in each formula stand for an unknown measurement value which will be replaced by a number during calculation.

Study this example of an expert approach to an ASVAB question about measurement in science.

Question	Analysis
Once outside the Earth's atmosphere, a space shuttle's propulsion system effects a net force of 10,000 N over a distance of 50 m. The potential energy of the craft is unchanged during this time. What increase in kinetic energy will result?	**Step 1:** The question asks for the increase in energy. Because energy and work are both measured in joules, the answer will be in joules.
	Step 2: Since work performed results in a conversion of energy from one form to another, the amount of increased kinetic energy (energy of movement) has to be the same as the amount of work done. The formula for work is known. $W = F \times d$ $= (10{,}000\text{ N}) \times (50\text{ m})$ $= 500{,}000\text{ J}$
	Step 3: The increase in kinetic energy will be equal to the amount of work done, or 500,000 J.
(A) 0 J (B) 0.005 J (C) 200 J (D) 500,000 J	**Step 4:** Select choice **(D)**.

Now try this question.

A hydroelectric dam captures the energy of falling water in order to provide electrical power to the grid. The water turns a turbine a total distance of 2.5 m every 10 s. The average force the water applies to the turbine is 10,000 N. What is the average rate of hydroelectric power generation of that turbine?

(A) 25 W
(B) 2,500 W
(C) 25,000 W
(D) 250,000 W

Explanation

Choice **(B)** is correct. The work done by the falling water is equal to

$W = 10{,}000\text{ N} \times 2.5\text{ m} = 25{,}000\text{ J}$

Power is the rate at which work is done, or at which energy is used, converted, or delivered.

$P = 25{,}000\text{ J} / 10\text{ s} = 2{,}500\text{ W}$

Newton's Laws

Sir Isaac Newton was an English mathematician and physicist. In the seventeenth century, he came up with some of our most important formulas for understanding the properties of motion and gravity.

Newton's first law of motion. *An object at rest tends to stay at rest, and an object in motion tends to stay in motion at a constant speed in a straight line (constant velocity), unless acted upon by an unbalanced force.* An example of an unbalanced force—one that keeps objects in motion from staying in motion on Earth—is friction, the force that resists relative motion between two bodies in contact. This law is also known as the **law of inertia**, with **inertia** referring to the tendency of all matter to resist changes in its motion.

Newton's second law of motion. *When dealing with an object for which all existing forces are not balanced, the acceleration of that object, as produced by the net force, is in the same direction as the net force and directly proportional to the magnitude of the net force, and is inversely proportional to the object's mass.* Expressed mathematically, **acceleration = net force ÷ mass**. When using this formula, the units for each of these measures must be m/s^2, N, and kg, respectively.

$$\vec{a} = \frac{\vec{F}}{m}$$

The greater the mass of an object, the greater the force needed to overcome its inertia. This law actually encompasses the first law, as the special case of zero net force results in zero acceleration, i.e., no change in motion. Sometimes this law is written in the equivalent form, **net force = mass × acceleration**.

$$\vec{F} = m\vec{a}$$

Newton's third law of motion. *For every action, there is an equal and opposite reaction.* In other words, when an object exerts a force on another object, the second object exerts a force of the same magnitude but in the opposite direction on the first object. For example, consider what happens when a gun is fired. A bullet fires and the gun recoils. The recoil is the result of action-reaction force pairs. As the gases from the gunpowder explosion expand, the gun pushes the bullet forward and the bullet pushes the gun backward. The acceleration of the recoiling gun is, however, smaller than the acceleration of the bullet, because acceleration is inversely proportional to mass, and the bullet, as a rule, has a smaller mass than the gun, and is more easily accelerated.

In addition to those three laws of motion, Newton also developed **Newton's law of universal gravitation**. *All objects in the universe attract each other with an equal force that varies directly as a product of their masses, and inversely as a square of their distance from each other. This force is known as gravity.* Newton's law of universal gravitation is expressed by the following equation (where G is a constant with a value of 6.67×10^{-11} and r is the distance between the two objects' centers of mass):

$$\vec{F_g} = \frac{Gm_1m_2}{r^2}$$

Take, for example, the gravitational force between the Sun and the Earth. The following consequences follow from the law of universal gravitation:

- If the mass of the Earth were doubled, the force on the Earth would double.
- If the mass of the Sun were doubled, the force on the Earth would double.
- If the Earth were twice as far away from the Sun, the force on the Earth would be a factor of four smaller.
- The force exerted on the Earth by the Sun is equal and opposite to the force exerted on the Sun by the Earth (Newton's third law).

At the surface of the Earth, the acceleration due to gravity is 9.8 m/s^2. This bears remembering, and can be applied to the formula for Newton's second law to determine the weight (gravitational force) of any object given its mass.

Question	Analysis
A 600 g squid propels itself by firing a jet of water with a force of 21 N. The water jet is pointed in an easterly direction. What will be the acceleration of the squid?	**Step 1:** The question asks for acceleration and provides mass and force, which means Newton's laws are being used.
	Step 2: The squid is firing a jet towards the east; however, the question asks about the acceleration of the squid, not the water jet. From Newton's third law, the water being pushed out in a jet to the east is simultaneously pushing on the squid to the west with the same force, which means acceleration will be westward. The mass of the squid must be converted to kg, and then, along with the given force, used to find magnitude of acceleration. Force on squid (equal and opposite) = 21 N, west Mass of squid $= \frac{600}{1{,}000} = 0.6$ kg $a = \frac{21}{0.6} = 35\ m/s^2$, west
	Step 3: Acceleration is 35 m/s^2, west. Both the calculated magnitude and the reasoned direction must be included in the prediction.
(A) 0.035 m/s^2, west (B) 35 m/s^2, east (C) 35 m/s^2, west (D) 28.6 m/s^2, east	**Step 4:** Choice **(C)** matches the prediction.

Now try one on your own:

A man pushes a 15 kg shopping cart across a rough surface, applying a force of 15 N in the direction of motion. A friction force resists the movement of the wheels across the surface with a magnitude of 15 N opposite to the direction of motion. What will be the effect on the motion of the shopping cart?

(A) The cart will slow down at a rate of 1 m/s^2.
(B) The cart will continue its uniform velocity.
(C) The cart will speed up at a rate of 1 m/s^2.
(D) The cart will speed up at a rate of 2 m/s^2.

Explanation

The correct answer is **(B).** Since the two forces acting on the cart are equal in magnitude but opposite in sign (direction), the net force will be equal to zero. Net force = +15 N − 15 N = 0 N. By Newton's second law, with no net force there can be no acceleration, which means, by definition, the velocity will not change.

Energy

As mentioned earlier in this chapter, **energy** can be defined as the **capacity to do work**. Many ASVAB energy questions deal with **mechanical energy**, which may be either **kinetic** or **potential**. **Kinetic energy** is the energy possessed by a moving object. **Potential energy** is the energy stored in an object as a result of its position, shape, or state.

According to the law of conservation of energy, energy can neither be created nor destroyed. Instead, it changes from one form to another. For example, if a rock is poised right at the edge of a cliff, the rock has potential energy relative to the ground at the bottom of the cliff. If the rock is dislodged and falls freely, that potential energy is converted completely to kinetic energy at the instant just before the rock hits the ground.

Sound and light energy travel in waves (although it gets complicated in the case of light). So let's take a look at the properties of waves.

Sound Waves

Sound waves are produced when an object vibrates, disturbing the medium around it, creating an outward ripple in all directions. These ripples (waves) can travel through air, liquids, and solids, but they cannot be transmitted through a **vacuum**, or empty space. Sound waves transmitted through air do not travel as fast as those transmitted through water, and those transmitted through water do not travel as fast as those transmitted through metal or wood.

The pitch of sound is directly related to the **frequency** (rate of vibration) of the sound waves. Sound waves with a high frequency (high rate of vibration) produce a high pitch. Frequency is usually measured in **hertz (Hz)**, defined as the number of repetitions per second. Sound waves with a very high pitch (high number of vibrations per second) are inaudible to humans, although they can be heard by dogs and other creatures. The typical audible hearing range for a human is from about 20 Hz to 20 kHz. Sometimes, a human (or a dog, for that matter) will perceive a sound as being a different frequency than the actual frequency of the sound. This is due to the **Doppler effect**. The Doppler effect occurs when either the source of the sound waves, the listener, or both, are moving closer together (pitch frequency sounds higher than it is) or farther apart (pitch frequency sounds lower than it is). A perfect example of the Doppler effect is the way the sound of a police or ambulance siren seems to change when it zooms by.

The Electromagnetic Spectrum

It's important to realize that visible light makes up only one small part—the visible part—of the electromagnetic spectrum. The **electromagnetic spectrum** covers all the different wavelengths and frequencies of radiation. Visible light waves fall in the middle of the electromagnetic spectrum. Starting with lowest frequency (which corresponds to the longest wavelength), the electromagnetic spectrum goes from **radio waves** to **microwaves** to **infrared waves** to **visible light** to **ultraviolet light** to **X-rays** and finally to **gamma rays**, the most active radiant energy known to exist.

Visible light breaks down into different colors as well, based upon the frequency of the waves. Red has the lowest frequency, which is why wavelengths just below the frequency of visible light are called infrared; likewise, violet has the highest frequency, so wavelengths just above the frequency of visible light are called ultraviolet.

Optics

As noted above, light, as well as the entire electromagnetic spectrum, possesses properties of waves. However, unlike sound waves, which are mechanical, light waves are electromagnetic and can travel through empty space. They also travel at much higher speeds than do sound waves. The speed of light in a vacuum is 299,792,458 meters per second (or roughly 300 million meters per second or 186,000 miles per second).

Refraction

It should be noted, however, that the effective speed of light can vary depending on the material the light waves are passing through; for example, light passes more slowly through water or glass than through a vacuum. The ratio by which light is slowed down is called the **refractive index** of that medium. For instance, the refractive index of a diamond is 2.4, which means that light travels 2.4 times faster when passing through a vacuum than when traveling through a diamond.

The change in speeds causes light to bend when passing from one medium to another (like a vehicle changing direction when hitting a slick patch of road). This bending is what's called **refraction**, and light bends at a greater angle when the change in the index of refraction is greater.

Reflection

Any wave, including light, that bounces off a flat, smooth barrier follows the **law of reflection**, which states that the **angle of incidence** is equal to the **angle of reflection** as measured from a line normal (at a 90° angle) to the barrier. In the case of light, this barrier is often a mirror.

The Law of Reflection

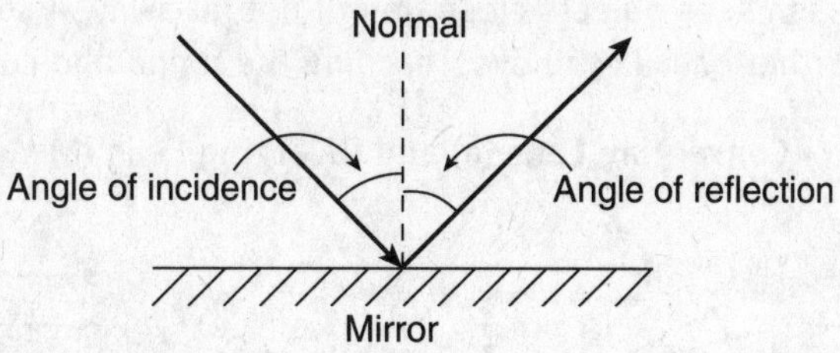

Concave and Convex Mirrors and Lenses

Mirrors and lenses either can be flat, or they can be concave (curved in, like a cave) or convex (curved out, like a swelling). Because of the law of reflection, a **concave mirror** is also called a **converging mirror**, because the angles of incidence of rays of light parallel to the normal all converge upon a point.

Converging Mirror

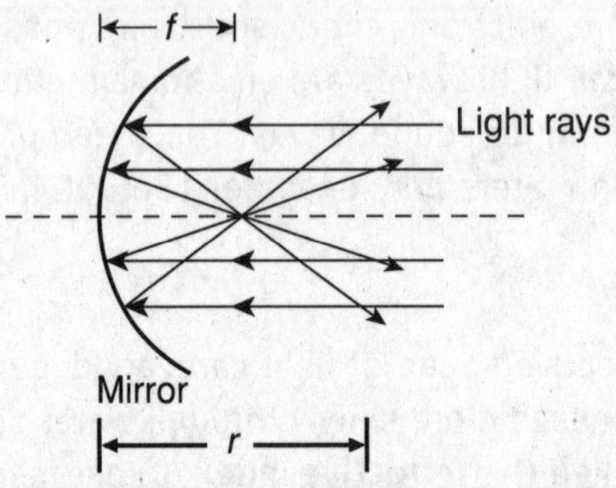

If you draw the mirror away from the source of the image, the image falls out of focus as the image source nears the mirror's **focal point**, the point where the mirror's angles of incidence converge. As the image source keeps moving beyond the focal point, the image reappears in the mirror, only upside down.

A **convex mirror**, on the other hand, is known as a **diverging mirror** because it diverges the light waves that strike it.

Lenses, unlike mirrors, operate on the principle of refraction. A **convex lens**—one that is thicker in the middle than on the edges—is also called a **converging lens** because it converges parallel waves that pass through it. This type of lens is used in reading glasses to correct farsightedness, as well as in magnifying glasses, cameras, telescopes, and microscopes.

A **concave lens**—one that is thicker on the edges than it is in the middle—is also known as a **diverging lens** because it diverges the light waves that pass through it. In nearsightedness, light waves converge before they meet the retina. A nearsighted person sees objects close up but not far away. A concave lens placed before the eye bends light so that it converges further back in the eye, reaching the retina and correcting nearsightedness.

Converging Lens (a) and Diverging Lens (b)

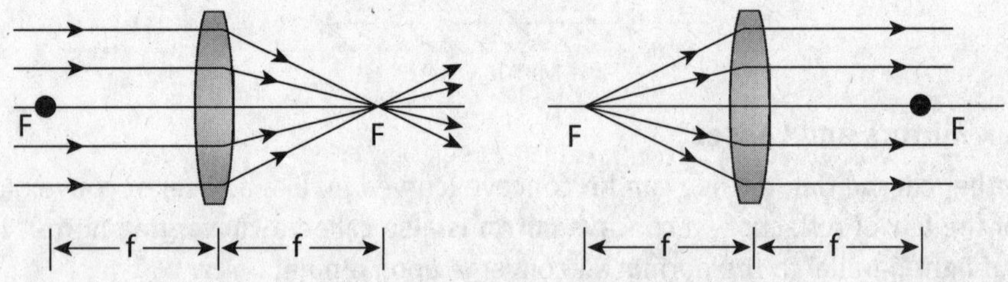

(a)

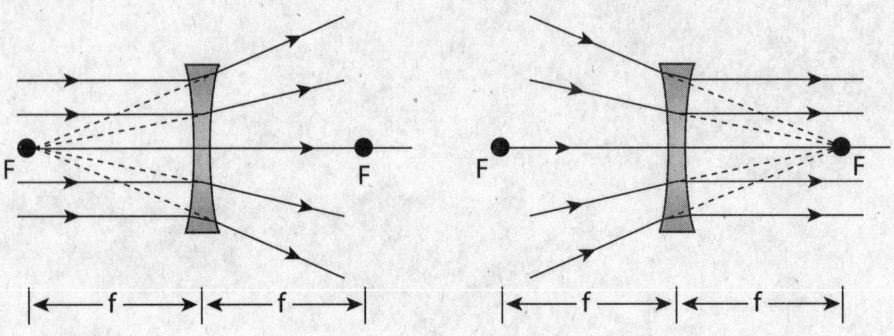

(b)

Study the approach an expert test taker would use on an ASVAB question about optics.

Question	Analysis
Through which medium will light travel the fastest?	**Step 1:** The question asks for the fastest medium for light to travel through.
	Step 2: Light is slowed down by different materials. Materials with high indices of refraction tend to be denser, and slow down light the most. The least dense material should provide light the greatest speed.
	Step 3: The answer choice with the least dense material will be correct.
(A) a vacuum (B) air (C) diamond (D) pure hydrogen	**Step 4:** Of the available choices, empty space, or **(A)** a vacuum, is the least dense material (or lack of material) possible. Light should travel fastest through a vacuum. In fact, the "speed of light" is defined for a vacuum.

Try the question below.

"Waves" of heat can be seen above a hot stove element or flame, as the background appears to be distorted or out of focus just at that spot. This is an example of

(A) diffraction
(B) refraction
(C) reflection
(D) convergence

Explanation

Choice **(B)** is correct. The background can only appear distorted if the light rays are following a different path through the heated area than other light rays moving through the room. Since the light is still traveling through the space, rather than being bounced away, this is refraction, not reflection. It occurs because the air near the heat source is of a different density, resulting in different indices of refraction in different patches of air.

Heat

There are three means by which heat energy may be transferred from one object to another: **conduction**, **convection**, and **radiation.** Heat energy is always transferred from warmer to cooler environments.

Conduction is the simplest method of heat transfer. It is accomplished by direct contact, such as placing your finger on a hot iron, which we recommend that you not try at home. Metals are generally good **conductors** of heat. Other materials, such as wood, Styrofoam, and plastic, are poor conductors of heat, which makes them good **insulators**.

Convection transfers heat by the actual movement of hot particles of a fluid. The hot air rising from a bonfire is an example of convection. Ocean currents and wind are caused by convection movements caused by temperature differences.

Radiation occurs when electromagnetic waves transmit heat. The heat we get from the Sun travels through space as radiation.

Magnetism

Simple magnets have two poles: a **north pole** and a **south pole**. Much as it happens in a Hollywood romance, opposites attract. If you try to bring together two north poles of a magnet—or two south poles, for that matter—they will repel one another, and you can feel their repulsive force. If, on the other hand, you move the north pole of one magnet toward the south pole of another magnet, they will attract each other.

Because the Earth itself is magnetized and has a North Pole and a South Pole, a **magnetic compass**, which contains a small, lightweight magnet balanced on a nearly frictionless, nonmagnetic surface, can be used to tell direction. The magnet, which is generally called a **needle**, has one end marked with an arrow and often the letter "N." This end of the needle is the magnet's south pole, which constantly orients itself to point toward the Earth's North Pole, allowing the person reading the compass to gain bearings from that direction.

Study the example below.

Question	Analysis
The arrival of a warm front, a mass of hot air moving into an area, is an example of	**Step 1:** The question asks for a phenomenon to be identified as one of three methods of heat transfer or as a magnetic effect.
	Step 2: The methods of heat transfer are identified by how heat energy physically travels. If actual hot physical molecules bring heat with them, it's convection.
	Step 3: The question stem specifies that the warm air itself is moving into the area. By definition, that makes this an example of convection.
(A) conduction (B) convection (C) radiation (D) the Earth's magnetic field at work	**Step 4:** Choice **(B)** matches the prediction.

Now you try one:

Wooden spoons are sometimes used in kitchens instead of metal ones when stirring hot pots of soup. What's a reasonable explanation for this?

(A) Wood does not conduct heat as well as metal, so the person stirring won't get burned.
(B) Wood draws in convection heat so that it does not flow into the pot.
(C) Wood is magnetic and diverts heat.
(D) Wood is a good conductor of radiation.

Explanation

Choice **(A)** is correct. Of the methods of heat transfer, only conduction would allow heat to move from a hot liquid into a solid spoon. It makes sense that a less conductive material would be purposely chosen to minimize the amount of heat that is able to travel from the soup, through the spoon, to the cook's fingers. The other choices, on the other hand, suggest a misunderstanding of how the methods of heat transfer work.

CHEMISTRY

LEARNING OBJECTIVES

In this section, you will learn to:

- understand what differentiates one element from another
- find information using the periodic table
- differentiate compound substances from pure elements, and understand the ways in which they are built up from simpler elements and/or compounds
- understand the basic definitions of acids and bases and their relation to the pH scale
- describe phase transitions and the defining features of the solid, liquid, and gaseous states of matter
- understand the basic definition and process of a chemical reaction, and differentiate this from physical change

Elements and the Periodic Table

Chemistry is the science dealing with the composition and properties of **matter,** and with the reactions by which matter is produced from or converted into other types of matter. Matter is anything that has mass and occupies space, and any form of matter has certain chemical properties based upon its molecular composition. To start, here are some basic terms:

Element: A pure type of matter that cannot be separated into different types of matter by ordinary chemical methods. All matter is composed of elements, and all the known elements are listed on the Periodic Table of Elements (see the figure on the next page).

Atom: The smallest component of an element that still retains the properties of the element. An atom may combine with similar particles of other elements to produce compounds. Atoms consist of a complex arrangement of electrons in motion about a positively charged nucleus containing protons and (except for hydrogen) neutrons.

Proton: A subatomic particle found in the atom's nucleus that carries a positive electric charge.

Neutron: A subatomic particle found in the atom's nucleus that does not have an electric charge and is therefore neutral.

Electron: A subatomic particle that orbits the nucleus of an atom. An electron carries a negative charge and has a miniscule mass compared to the other subatomic particles (both neutrons and protons are more than 1,800 times as massive as an electron). Ordinarily, an atom has the same number of negative electrons around the nucleus as the number of positive protons in the nucleus.

Structure of an Atom

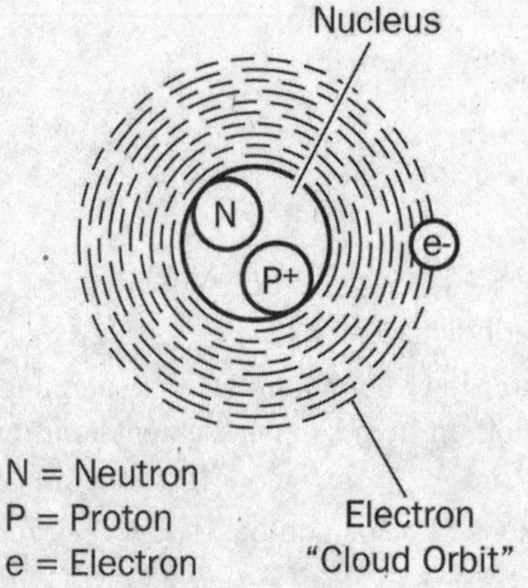

Molecule: The smallest multi-atom particle of an element or compound that can exist in the free state and still retain the characteristics of the element or compound. The molecules of elements consist of two or more similar atoms; the molecules of compounds consist of two or more different atoms.

Through decades of research, scientists have organized all the known elements into a structure called the periodic table, which conveys multiple pieces of information about each element.

Periodic Table of Elements

1 IA																	18 VIIIA
1 H 1.0	2 IIA											13 IIIA	14 IVA	15 VA	16 VIA	17 VIIA	2 He 4.0
3 Li 6.9	4 Be 9.0											5 B 10.8	6 C 12.0	7 N 14.0	8 O 16.0	9 F 19.0	10 Ne 20.2
11 Na 23.0	12 Mg 24.3	3 IIIB	4 IVB	5 VB	6 VIB	7 VIIB	8	9 VIIIB	10	11 IB	12 IIB	13 Al 27.0	14 Si 28.1	15 P 31.0	16 S 32.1	17 Cl 35.5	18 Ar 39.9
19 K 39.1	20 Ca 40.1	21 Sc 45.0	22 Ti 47.9	23 V 50.9	24 Cr 52.0	25 Mn 54.9	26 Fe 55.8	27 Co 58.9	28 Ni 58.7	29 Cu 63.5	30 Zn 65.4	31 Ga 69.7	32 Ge 72.6	33 As 74.9	34 Se 79.0	35 Br 79.9	36 Kr 83.8
37 Rb 85.5	38 Sr 87.6	39 Y 88.9	40 Zr 91.2	41 Nb 92.9	42 Mo 96.0	43 Tc (98)	44 Ru 101.0	45 Rh 102.9	46 Pd 106.4	47 Ag 107.9	48 Cd 112.4	49 In 114.8	50 Sn 118.7	51 Sb 121.8	52 Te 127.6	53 I 126.9	54 Xe 131.3
55 Cs 132.9	56 Ba 137.3	*	72 Hf 178.5	73 Ta 180.9	74 W 183.8	75 Re 186.2	76 Os 190.2	77 Ir 192.2	78 Pt 195.1	79 Au 197.0	80 Hg 200.6	81 Tl 204.4	82 Pb 207.2	83 Bi 209.0	84 Po (209)	85 At (210)	86 Rn (222)
87 Fr (223)	88 Ra (226)	†	104 Rf (267)	105 Db (268)	106 Sg (269)	107 Bh (270)	108 Hs (269)	109 Mt (278)	110 Ds (281)	111 Rg (281)	112 Cn (285)	113 Nh (286)	114 Fl (289)	115 Mc (288)	116 Lv (293)	117 Ts (294)	118 Og (294)
	*	57 La 138.9	58 Ce 140.1	59 Pr 140.9	60 Nd 144.2	61 Pm (145)	62 Sm 150.4	63 Eu 152.0	64 Gd 157.3	65 Tb 158.9	66 Dy 162.5	67 Ho 164.9	68 Er 167.3	69 Tm 168.9	70 Yb 173.1	71 Lu 175.0	
	†	89 Ac (227)	90 Th 232.0	91 Pa 231.0	92 U 238.0	93 Np (237)	94 Pu (244)	95 Am (243)	96 Cm (247)	97 Bk (247)	98 Cf (251)	99 Es (252)	100 Fm (257)	101 Md (258)	102 No (259)	103 Lr (262)	

In order to be able to read the Periodic Table of the Elements, there are a few things you need to understand. For starters, elements are arranged in order of increasing **atomic number**, from left to right and from top to bottom. An element's atomic number represents the number of its protons (and also the number of its electrons, since those two numbers are the same). The different rows of elements are called **periods**. The periods correspond to the **shells**, or the different orbits that electrons occupy around atoms. As a rule, electrons will occupy the lowest shell they can, and move on to higher shells only after lower shells are occupied (although this rule is sometimes violated) so as to minimize the energy of the atom. Every element in the top row (the first period) has one shell for its electrons. Every element in the second row (the second period) has two shells for its electrons, and so on. At this time, the maximum number of shells is seven.

The Periodic Table has a special name for its columns, too. The elements in a column are called a **group**. The elements in a group have the same number of electrons in their outer shell. It is the group that an element occupies, much more than the period, that determines its chemical properties. This is because the number of electrons needed to complete an element's outer shell shapes the way in which it reacts with other elements to form molecules. For instance, every element in the first column (Group I) has one electron in its outer shell. The elements in this group are called **alkali metals**, which describe soft, silvery metals that react strongly with water. The further down the group you go, the more violent this reaction is.

By the same token, every element on the second column (Group II) has two electrons in the outer shell. As you keep counting the columns, you'll know how many electrons are in the outer shell. Note that the far right-hand column is composed of a group called the **noble gases**, sometimes referred to as inert gases, because these elements generally don't react with other elements since their outer shell is completely filled.

Atomic Structure of Lithium and Sodium

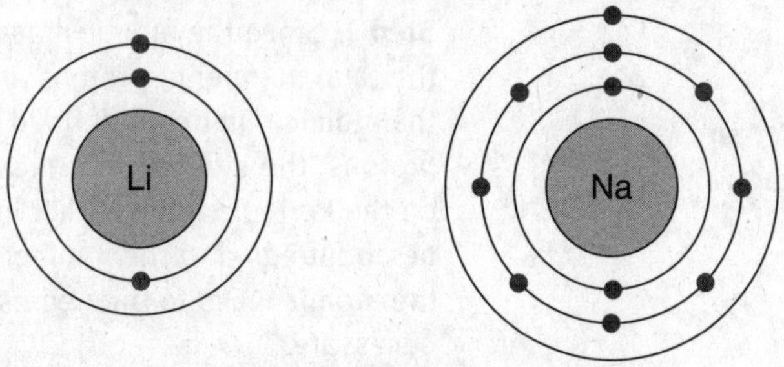

Notice that Li has two electron shells, so it is in the second row of the periodic table; Na has three shells, so it is in the third row. Both elements have one electron in their outer shell and are, therefore, in Group I.

Besides these important chemical properties, an element's **atomic mass** is also related to its position on the Periodic Table. The atomic mass listed in an element box represents the average mass of a single atom. Why average? Because some elements can vary in their number of neutrons, so individual atoms of an element may be found in a couple of slightly different sizes. These different sizes are referred to as **isotopes**.

A pretty good estimate of atomic mass comes from adding the total number of neutrons and protons together. Each proton and neutron has a mass very close to what's called an **atomic mass unit (amu)**. The majority of hydrogen atoms have only one proton, so its atomic mass should be very close to one, and is, at 1.007 amu.

Now take a look at an individual element box within the Periodic Table. Here's chlorine.

Chlorine

17
Cl
35.5

Chlorine's **atomic number** is 17, which is the number of protons in the nucleus of a chlorine atom and also the number of electrons orbiting its shells. Its **atomic symbol**, Cl, is fairly close to its name, unlike the symbols of some other common elements, such as Fe (iron), Au (gold), Ag (silver), W (tungsten), or Na (sodium), symbols that derive from the elements' Latin names, rather than their familiar English names.

Finally, the atomic mass of chlorine is listed as **35.5**, due to about half of chlorine atoms being close to 35 amu and the other half closer to 36 amu. Since all chlorine atoms have 17 protons, this means chlorine atoms are almost as likely to have either 18 or 19 neutrons.

Question	Analysis
How many neutrons does a single sulfur (S) atom have?	**Step 1:** The question asks for the number of neutrons in sulfur, which are found in the nucleus of an atom with the protons.
	Step 2: Since the atomic mass is about equal to the total number of protons and neutrons, and the atomic number is defined by the number of protons, the difference in these values (available on the Periodic Table earlier in this section) should be about equal to the number of neutrons. Round the atomic mass to the nearest whole number as necessary. For sulfur: Atomic # (protons) $= 16$ Atomic mass $= 32.1$ Number of protons and neutrons $= 32$ Number of neutrons $= 32 - 16 = 16$
	Step 3: It isn't possible for an atom to have something other than a whole number of neutrons, so round to the nearest whole number. There are exactly 16 neutrons.
(A) 16 (B) 16.1 (C) 32 (D) 48	**Step 4:** Select choice **(A)**.

Now try one on your own:

How many protons, neutrons, and electrons do argon (Ar) atoms have?

(A) 18, 18, 18
(B) 18, 22, 18
(C) 18, 21.9, 18
(D) 18, 22, 22

Explanation

Answer choice **(B)** is correct. Argon's atomic mass is about 40, so there are 18 protons and 22 neutrons, and, in neutral atoms, there are always the same number of electrons as protons.

Compounds

Unstable elements readily form into **compounds** with properties very distinct from the elements from which they are composed. For instance, sodium and chlorine, both of which are extremely unstable and noxious as elements, combine to form the stable and edible compound sodium chloride (NaCl), more commonly known as table salt. Table salt is called an **ionic compound** because each chlorine atom borrows an electron from each sodium atom and the atoms stick closely together to form a very tightly bound crystalline structure when salt is in solid form. When it is placed in solution, however, such as when table salt is poured in water, the atoms dissociate into sodium ions and chloride ions. An **ion** is an electrically charged atom; in this case, each of the sodium atoms is positively charged because it has lent an electron to its corresponding chloride atom, and each of the chloride atoms is negatively charged for the same reason.

Table sugar is an example of a **covalent compound**, which means, among other things, that it does not ionize when dissolved in water. In a covalent compound, the atoms in the molecule have covalent bonds; that is, they share electrons in pairs, so that each atom provides half the electrons, and the pair is held tightly by both atoms. For this reason, they will not separate as ions do.

Acids and Bases

An **acid** is a substance that gives up positively charged hydrogen ions (H^+) when dissolved in water. Acids corrode metals and generally have a sour taste (though not all acids are safe to drink). Some common acidic solutions are vinegar (which has acetic acid) and lemon juice (which has citric acid). More potent acids include hydrochloric acid, nitric acid, and sulfuric acid (commonly used in lead batteries), all of which are extremely corrosive and must be handled with the utmost care.

A **base** is a substance that gives up negatively charged hydroxyl ions (OH^-) when dissolved in water. Basic substances are also referred to as **alkaline**. Bases typically taste bitter.

Some common basic substances found in a kitchen include baking soda (sodium bicarbonate) and liquid soap (which often includes potassium hydroxide). More potent bases include sodium hydroxide (also known as lye) and sodium hypochlorite (the most common active ingredient in bleach), and these are just as corrosive and dangerous as comparable acids. When acids and bases react (and they react together rather powerfully), the substances neutralize each other and turn into water and a salt.

The **pH** of a solution is a number from 0 to 14 that indicates how basic or acidic that solution is. According to the pH scale, solutions with a pH less than 7 are acidic, with the degree of acidity increasing tenfold with each declining number. For instance, black coffee has a pH of 5; vinegar, which is 100 times as acidic as coffee, has a pH of 3; and battery acid, which is 100 times as acidic as vinegar, has a pH of 1. A solution with a pH of 7 is neutral. Pure water has a pH of 7.

Common Substances on the pH Scale

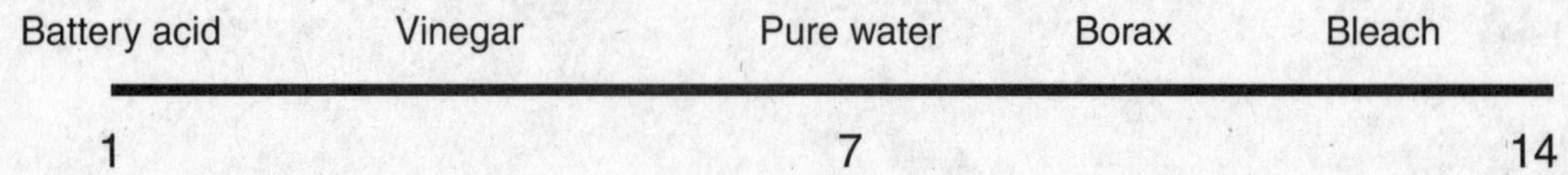

Solutions with a pH greater than 7 are basic, with the degree of alkalinity once again increasing tenfold with each increasing number. Baking soda has a pH that is a little over 8. Borax, commonly used in cleaning solutions and detergents, is about 10 times as basic, at just over 9 on the pH scale, while household bleach, over 12 pH, is 1,000 times as basic as borax, and 10,000 times as basic as baking soda. Human blood, just above pH 7, is just slightly basic, with about one-tenth the alkalinity of baking soda.

Study this example of an expert test taker's approach to a question about acids and bases.

Question	Analysis
In a chemistry lab, you are given four solutions and measure their pH values. Which solution contains the greatest number of hydrogen ions?	**Step 1:** The question asks for the highest concentration of hydrogen ions.
	Step 2: Acids produce hydrogen ions, and more acidic substances produce them in greater concentrations. Lower pH values indicate more acidic substances, which means the most hydrogen ions.
(A) the pH 2.5 solution (B) the pH 3.7 solution (C) the pH 7.1 solution (D) the pH 9 solution	**Step 4:** Answer choice **(A)** has the lowest pH.

Now try this question about acids and bases.

You find a mystery substance in the lab and note the following qualities: it's bitter to the taste, it releases OH^- ions in solution, and it reacts with an acid to form a salt and water. Which of the following could the mystery substance be?

(A) vinegar
(B) soap
(C) lemon juice
(D) water

Explanation

Choice **(B)** is correct. The properties of the mystery substance match those of a base. Of the choices given, the only basic substance is soap.

Physical Change

Matter may undergo either a **physical change** or a **chemical change**. The form, size, and shape of matter may be altered in a physical change, but the molecules remain unchanged. Building a sand castle or bending a piece of plastic into a new shape are both physical changes. The matter has been rearranged, but it is still made of up of the same kinds of atoms and molecules. In other words, its chemical identity has remained intact.

One of the most important kinds of physical change is what's known as a **phase transition**. A phase transition is when matter switches from one **state of matter**, like solid, liquid, or gas, to another.

The **solid state** exists at lower temperatures relative to the liquid and gaseous states (although the specific temperature ranges depend on the element or compound). In the solid state, atoms or molecules are packed very close together and do not move freely. Solids maintain a constant volume and shape. A good example is an ice cube, consisting of solid-state water molecules.

The **liquid state** exists at a higher temperature range relative to the solid state equivalent of the same element or compound. In the liquid state, molecules flow freely around each other. Liquids have a constant volume, but do not maintain their shape. They will either take on the shape of their container or, let free, will spread out into a puddle. Water from a faucet comes out in the liquid state.

The **gaseous state** occurs at a higher temperature range relative to the previous two states of matter. In the gaseous state, molecules flow even more freely from each other, and will also spread out as far as they can. Gases do not maintain either a constant volume or shape. A small amount of gas will spread to fill a large volume if space is available, or can be highly compressed into smaller spaces (oxygen tanks include a large amount of compressed oxygen gas). Steam or vapor are the common terms for water in a gas state.

Under certain circumstances, it's possible for solids and gases to transition between each other directly, without transforming to the liquid state first. The specific terms for each type of phase transition are summarized below.

Phase Changes

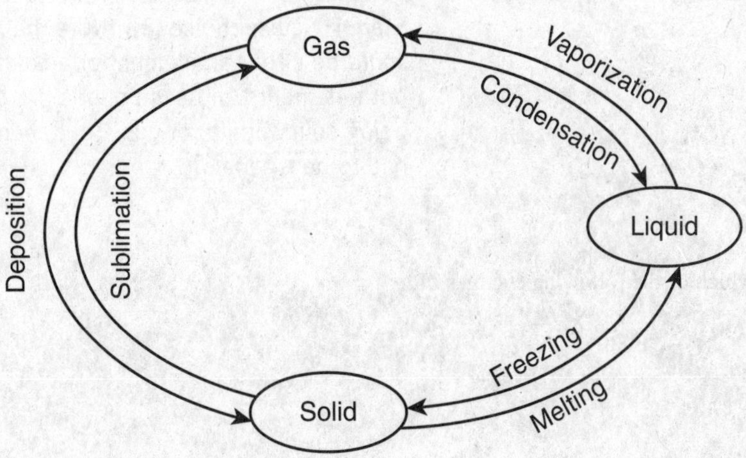

Chemical Change

In a chemical change, molecules of new matter are formed that are different from the original molecules of matter. When elements and compounds undergo a chemical change, the process is called a **chemical reaction**. During a chemical reaction, atoms are rearranged into new combinations, resulting in different kinds of molecules. The molecules and atoms that enter the reaction are called **reactants** and the molecules and atoms that result from the reaction are called **products**.

For example, water is also known as H_2O because a single water molecule is made of two hydrogen atoms and one oxygen atom. One kind of chemical reaction will cause many water molecules to break apart and the atoms to rearrange into hydrogen molecules (H_2) and oxygen molecules (O_2). This is how the reactant, water, is turned into the products, pure hydrogen gas and oxygen gas. Other common examples of chemical reactions include the formation of rust on iron, and the conversion of wood into charcoal, carbon dioxide, and steam in a fire.

To illustrate the profound effects of chemical change, consider one more example: the reaction of chlorine gas and sodium metal. Both substances, in their pure forms, are extremely hazardous to humans. Sodium metal cannot be touched with bare hands; it reacts with any moisture, giving off heat and sparks. It's stored in oil when not in use. Chlorine gas was used as a devastating chemical weapon in the First World War. Yet, these two dangerous substances can undergo a chemical reaction that results in a product of sodium chloride: common table salt, found in any kitchen. That's the power of chemical change.

Here's how the well-prepared test taker would approach a question about states of matter.

Question	Analysis
A sample of matter with an amorphous shape but constant volume has to be	**Step 1:** The question asks for the type or phase of matter with a constant volume but non-fixed shape.
	Step 2: A solid has a fixed shape. A liquid will take the shape of its container, or more exactly, find the lowest possible level.
	Step 3: The only thing that logically *must be* true of the matter sample based on the qualities described in the question is that the substance must be in the liquid phase.
(A) in the solid phase (B) in the liquid phase (C) a pure element (D) a compound	**Step 4:** Select choice **(B)**. (Note that the sample *could be* either an element or a compound, since both elements and compounds are capable of being solid, liquid, or gas. Thus, neither (C) nor (D) has to be true.)

Now try one on your own.

Energy is needed for which of the following changes of state?

(A) solid to liquid
(B) gas to liquid
(C) liquid to solid
(D) gas to solid

Explanation

Choice **(A)** is correct. Liquids occur at higher temperatures than solids, and gases occur at higher temperatures still. The melting of a solid into a liquid is the only answer choice listed that involves movement to a phase that occurs at higher temperatures, requiring a net input of energy to allow the molecules the freer movement that occurs in the liquid phase.

PHYSICAL SCIENCE PRACTICE QUESTIONS

1. What is 77°F in degrees Celsius?

    (A) 25°C
    (B) 32°C
    (C) 37°C
    (D) 5°C

2. Which of the following best explains the recoil action of a shooting gun?

    (A) Newton's first law of motion
    (B) Newton's second law of motion
    (C) Newton's third law of motion
    (D) Newton's law of universal gravitation

3. The weight of a pendulum clock transforms potential energy to usable mechanical work. Assuming no energy loss, approximately how much work will be done if a 200 g weight descends 0.5 m?

    (A) 0.1 J
    (B) 1 J
    (C) 100 J
    (D) 1000 J

4. Which of the following waves on the electromagnetic spectrum has the highest frequency?

    (A) microwaves
    (B) X-rays
    (C) visible light
    (D) radio waves

5. Pure water has a pH of

    (A) 1
    (B) 7
    (C) 0
    (D) 14

6. During a game of baseball, two players both attempt to catch the same ball. The 50 kg Player A is traveling west at 12 m/s, while the 60 kg Player B is traveling east along the same line at 10 m/s. What will happen when they collide?

    (A) Player A will continue forward at reduced velocity, knocking back Player B.
    (B) Player B will continue forward at reduced velocity, knocking back Player A.
    (C) Player B will continue forward without any decrease in velocity, knocking back Player A.
    (D) Player A and Player B will both come to a complete stop.

7. A truck was driven $2 \times 10^5 \times 10^5$ meters. At the beginning of the trip, the fuel tank contained 40,000 mL of fuel; at the end of the trip, it still had 15,000 mL of fuel. What was the average fuel consumption, measured in liters per kilometer, for the trip?

    (A) 0.0125 L/km
    (B) 0.125 L/km
    (C) 8.0 L/km
    (D) 80 L/km

8. The apparent sound of a train's horn changes from a higher to a lower pitch as the train approaches and then passes a stationary observer. This phenomenon is called

    (A) a Newtonian shift
    (B) sound wave compression
    (C) the Doppler effect
    (D) an auditory aberration

9. If an object is viewed through a converging lens, when will the object appear to be upside down?

    (A) always
    (B) never
    (C) when the observer is closer to the lens than the focal length
    (D) when the observer is farther from the lens than the focal length

10. If a block of solid frozen carbon dioxide ("dry ice") is left in a container that is open to the air, some frozen carbon dioxide will become carbon dioxide gas without becoming a liquid first. This phase change is called __________.

    (A) sublimation
    (B) evaporation
    (C) vaporization
    (D) gasification

Answers and Explanations

1. **A** To convert the temperature from degrees Fahrenheit to degrees Celsius, first subtract 32, and then multiply by $\frac{5}{9}$.

$$77 - 32 = 45$$

$$45 \times \frac{5}{9} = 25$$

2. **C** The recoil action of a shooting gun is explained by Newton's third law of motion: for every action, there is an equal and opposite reaction.

3. **B** Using Newton's second law, and the fact that acceleration near the surface of the Earth is 9.8 m/s^2, the force of gravity can be calculated. The mass must be converted to kg units first: $m = 200\text{ g} = 0.2\text{ kg}$. Then the force is $(0.2\text{ kg}) \times (9.8\text{ m/s}^2) = 1.96\text{ N}$. This force acts over a distance of 0.5 m, so $W = (1.96\text{ N}) \times (0.5\text{ m}) = 0.98\text{ J}$. Answer choice (B) is the closest match.

4. **B** Of the wave states listed, X-rays have the highest frequency in the electromagnetic spectrum (gamma rays have an even higher frequency).

5. **B** Pure water has a pH of 7, making it a neutral solution.

6. **D** Determine each player's momentum. Player A has a momentum of $50\text{ kg} \times 12\text{ m/s} = 600\text{ kg} \cdot \text{m/s}$ in a westward direction. Player B has a momentum of $60\text{ kg} \times 10\text{ m/s} = 600\text{ kg} \cdot \text{m/s}$ in an eastward direction. Since the two players' momentums are equal in magnitude but opposite in direction, the total momentum is zero ($P_{total} = (+600) + (-600) = 0\text{ kg} \cdot \text{m/s}$). To put it another way, since neither player has a greater quantity of motion than the other, neither one can "win" in a collision. If neither is able to overcome the other, both coming to a stop is the only option that makes sense.

7. **B** First convert the quantities given to the units of measurement of the answer. Since *kilo–* is the prefix for 1,000, the truck traveled $2 \times \frac{10^5}{10^3} = 2 \times 10^2 = 200 \times 10^2 = 200$ kilometers (km). The amount of fuel consumed was 40,000 – 15,000 = 25,000 mL. The prefix *milli–* means one thousandth, so that converts to 25 liters (L). Since the desired units are liters/kilometer, divide 25 L by 200 km to get 0.125 L/km.

8. **C** This is a classic example of the Doppler effect, which is caused by the relative motion of a sound source and receptor.

9. **D** Light waves converge at a distance equal to the focal length of a lens. Beyond that distance, the waves from the top of the observed object are below those from the bottom of the object, making the object appear to be inverted.

10. **A** When a liquid changes to a gas, that process is evaporation under natural conditions and vaporization if heat is applied. The transformation directly from solid to gas is called sublimation.

PART IV: GENERAL SCIENCE PRACTICE QUESTIONS

GENERAL SCIENCE PRACTICE SET 1

Choose the best answer for each multiple-choice question. This question set has 15 practice questions, which is the number of General Science questions you will see if you take the CAT-ASVAB.

1. Which of the following best describes the difference between eukaryotic cells and prokaryotic cells?
   (A) Prokaryotic cells do not have nuclei.
   (B) Prokaryotic cells are only found in plants.
   (C) Eukaryotic cells are only found in plants.
   (D) Eukaryotic cells do not have nuclei.

2. Which of the following is a terrestrial planet?
   (A) Mars
   (B) Jupiter
   (C) Saturn
   (D) Neptune

3. When a skydiver jumps from a plane, she eventually reaches a maximum falling velocity of −56 m/s. After opening up her parachute, her falling velocity decreases to about −11 m/s. What effect on her motion does the skydiver's parachute have?
   (A) The parachute does not change the skydiver's velocity.
   (B) The parachute applies a negative acceleration on the skydiver.
   (C) The parachute applies a positive acceleration on the skydiver.
   (D) The parachute applies first a negative acceleration, then a positive acceleration on the skydiver.

4. A freight elevator is used to lift loads of 12,000 N a vertical distance of 250 m. The elevator takes 10 s to make this trip. Assuming no energy loss, what is the average power expenditure of the elevator, when working?
   (A) 300,000 W
   (B) 3,000,000 W
   (C) 12,000,000 W
   (D) 30,000,000 W

5. The air in the troposphere is made up primarily of
   (A) oxygen
   (B) helium
   (C) carbon dioxide
   (D) nitrogen

6. The cell membrane that surrounds an animal cell is a
   (A) non-permeable structure
   (B) semi-permeable structure
   (C) fully permeable structure
   (D) cell wall

7. In which level of the atmosphere does all weather take place?
   (A) troposphere
   (B) stratosphere
   (C) mesosphere
   (D) thermosphere

8. Which of the following is NOT true about the element hydrogen (H)?
   (A) It is the only element with no neutrons.
   (B) It is the lightest element.
   (C) It is a noble (inert) gas and rarely reacts.
   (D) It sometimes behaves like an alkali metal.

9. Which of the following activities results in a chemical change?
   (A) mixing salt and sugar
   (B) drying wet clothes
   (C) boiling a pot of water
   (D) burning pieces of wood

10. Air flow in the stratosphere is primarily

    (A) horizontal
    (B) vertical
    (C) clockwise
    (D) counterclockwise

11. Iron is a mineral that is vital to red blood cell development. Foods that are good sources of iron include __________.

    (A) meats, beans, and whole grains
    (B) nuts and green leafy vegetables
    (C) bananas, sweet potatoes, nuts, and seeds
    (D) milk, yogurt, cheese, and spinach

12. If a pencil is placed in a clear glass of water and viewed from the side, the pencil will appear to bend sharply at the top of the water. This is due to a phenomenon known as __________.

    (A) elusivity
    (B) reflection
    (C) refraction
    (D) rotation

13. Red blood cells are produced in __________.

    (A) the liver
    (B) kidneys
    (C) bones
    (D) the spleen

14. Which of the following is NOT an example of phase change?

    (A) fuel burning in a combustion engine
    (B) dry ice changing directly from a solid to a gas
    (C) sleet melting on a sidewalk
    (D) morning dew appearing on a lawn

15. HIV (human immunodeficiency virus) is transmitted __________.

    (A) by touch
    (B) through the air
    (C) by bodily fluids
    (D) by proximity to an infected person

GENERAL SCIENCE PRACTICE SET 2

Choose the best answer for each multiple-choice question. This question set has 15 practice questions, which is the number of General Science questions you will see if you take the CAT-ASVAB.

1. The Earth experiences seasonal patterns of weather because __________.
   (A) its distance from the Sun varies at different times of the year
   (B) solar wind activity varies
   (C) its orientation to the Sun is tilted on its axis
   (D) the Moon's position relative to the Earth varies

2. Body functions such as heart rate and digestion that take place without having to think about them are regulated by the __________.
   (A) somatic nervous system
   (B) cerebrum
   (C) neurons
   (D) autonomic nervous system

3. Which factor is most important in determining how atoms of an element will chemically react with other atoms?
   (A) the number of electrons in the outer shell
   (B) the number of neutrons in the nucleus
   (C) the chemical symbol of the element
   (D) the atomic mass

4. Our Sun is classified as what type of star?
   (A) yellow dwarf
   (B) white dwarf
   (C) yellow giant
   (D) white giant

5. Shale and sandstone are types of __________ rocks.
   (A) igneous
   (B) sedimentary
   (C) metamorphic
   (D) compound

6. A DNA (deoxyribonucleic acid) molecule is shaped like a __________.
   (A) sphere
   (B) double helix
   (C) oblate spheroid
   (D) triple helix

7. A net force of 15 N is applied to an object that starts at rest and accelerates to a velocity of 12 m/s in 4 seconds. What is the mass of the object?
   (A) 3 kg
   (B) 5 kg
   (C) 12 kg
   (D) 15 kg

8. A lunar eclipse is caused by __________.
   (A) an anomaly in the Moon's orbit
   (B) solar flares
   (C) the Moon passing between the Earth and the Sun
   (D) the Earth passing between the Sun and the Moon

9. Earthquakes occur due to movement of plates in the Earth's __________.
   (A) asthenosphere
   (B) lithosphere
   (C) outer core
   (D) lower mantle

10. Which of the following organisms has an exoskeleton?
   (A) human being
   (B) spider
   (C) amoeba
   (D) eagle

11. Our lungs convey __________ to blood cells and receive __________ and __________ from blood cells.

(A) oxygen, carbon dioxide, water
(B) oxygen, carbon dioxide, nitrogen
(C) carbon dioxide, oxygen, water
(D) nitrogen, carbon dioxide, oxygen

12. A fireplace is sealed behind a clear glass screen to prevent sparks and ashes from getting out. Nevertheless, a person sitting in front of the fireplace will be warmed by the heat of the fire primarily because some heat is still transferred by __________.

(A) convection
(B) radiation
(C) conduction
(D) osmosis

13. Sodium (Na), which is in Group 1 of the Periodic Table of the Elements, and chlorine (Cl), which is in Group 17, will bond together to form sodium chloride (NaCl). How will this compound behave when dissolved in pure water?

(A) The molecules will group together to form large crystals.
(B) The molecules will separate into Na and Cl atoms.
(C) The molecules will separate into negatively charged sodium ions and positively charged chlorine ions.
(D) The molecules will separate into positively charged sodium ions and negatively charged chlorine ions.

14. Meteorology is the study of __________.

(A) meteors and meteorites
(B) weather
(C) the atmosphere and atmospheric phenomena
(D) the solar system

15. Human eggs are produced in the female's __________.

(A) uterus
(B) fallopian tubes
(C) cervix
(D) ovaries

GENERAL SCIENCE PRACTICE SET 1

Answers and Explanations

1. **A** Prokaryotic cells, such as bacteria, do not have nuclei. Their genetic material is contained within the main part of the cell body. Eukaryotic cells do contain nuclei where all genetic material is contained. Eukaryotic cells are plant, animal, fungi, and protist cells.

2. **A** Terrestrial planets are those closest in composition to the Earth. Not surprisingly, they are also the inner planets and those closest in proximity to the Earth. Terrestrial planets have an inner core of metal and rocky surfaces similar to the Earth's, though their atmospheric characteristics can be vastly different from those of the Earth. Jupiter, Saturn, and Neptune are outer planets.

3. **C** The direction of velocity of the skydiver is negative for the entire time period discussed in the question, but due to the release of the parachute, the later velocity is smaller in magnitude. Since the parachute acts against the direction of motion of the skydiver (as established by the fact that the magnitude of the velocity is decreased instead of increased), the acceleration must be positive, opposite in sign to the negative velocity. Alternately, this question can be reasoned out as follows: the skydiver is falling and has a negative velocity, therefore down is negative; parachutes slow down skydivers by pulling them in an upwards direction, therefore the parachute provides a positive acceleration.

4. **A** The applied force of the elevator has to counteract the weight of the load in order to ensure a safe, uniform velocity up. So the work done can be calculated as $W = (12{,}000 \text{ N}) \times (250 \text{ m}) = 3{,}000{,}000 \text{ J}$. Power is energy expended per unit time, so $P = (3{,}000{,}000 \text{ J}) \div (10 \text{ s}) = 300{,}000 \text{ W}$.

5. **D** The air in the troposphere is approximately 78% nitrogen and 21% oxygen. This is the layer of air closest to the Earth's surface.

6. **B** Animal cells have semi-permeable membranes which allow for osmosis. Osmosis is essential for maintaining homeostasis within the cell and preventing it from shrinking or bursting.

7. **A** The troposphere is the lowest level of the atmosphere and where all weather affecting Earth takes place. This is the level where clouds form and from where precipitation falls.

8. **C** The question asks for the statement about hydrogen which is not true. Both answer choices (A) and (B) are true statements. (The atomic number of 1 means there is one proton, and that one proton accounts for the atomic mass, which is also 1). The question can now be answered by identifying just one of the remaining statements as definitely true or false. Since hydrogen is definitely not a noble gas (helium is the lightest noble gas, with atomic number two), (C) is the correct answer. Alternatively, (D) can be identified as the last true statement (hydrogen has a single electron in its outer shell, like alkali metals), which by process of elimination means that (C) is the correct answer.

9. **D** Mixing salt and sugar is a physical change. When mixed together, the salt and sugar retain their molecular makeup. Drying wet clothes is also a physical change. When clothes are dried, water molecules evaporate, becoming a gas. The clothes and water molecules remain chemically unchanged. When boiling a pot of water, once again, liquid water molecules move into the gas phase, but there is no change in the chemical composition of the molecules. Finally, burning wood is a chemical change, because the molecules in the wood are converted into new molecules that are different than the original.

10. **A** Air flow in the stratosphere is primarily horizontal, in contrast to air flow in the troposphere, which has a strong vertical component.

11. **A** Nuts and green leafy vegetables are a good source of magnesium. Bananas, sweet potatoes, nuts, and seeds contain potassium. Milk, yogurt, cheese, and spinach provide calcium. Meats, beans, and whole grains are all sources of iron.

12. **C** The speed of light is slower in water than in air causing light to bend at the interface between air and water. While this may look like what could be called an optical illusion, elusivity is not a phenomenon related to light. Reflection would be due to light bouncing off a surface such as a mirror. Rotation is irrelevant to this question.

13. **C** Red blood cells are produced by the marrow inside human bones. The liver has many functions, most of which are related to the digestive system. Kidneys, in addition to their role in purifying blood, produce the hormone that stimulates red blood cell production in the bone marrow. The spleen stores and purifies blood.

14. **A** Burning is a chemical change, not merely a change of a physical state. Although most materials do not transition directly from a solid to a gas, dry ice does, and the phase change is called sublimation. Sleet melting is an example of a solid liquefying (melting). Morning dew is the result of water vapor in the air condensing on the grass.

15. **C** HIV can only be transmitted by an infected person's bodily fluids, such as blood, saliva, or semen.

GENERAL SCIENCE PRACTICE SET 2

Answers and Explanations

1. **C** Because of the tilt of the Earth, during the summer months the Sun's rays are more directly overhead and the days are longer; hence summer is the warmest season. Because of the tilt, summer occurs at "opposite" times of the year in the northern and southern hemispheres. The distance between the Earth and Sun does vary, but the effect on climate is much smaller than the tilt. The Earth is actually farthest from the Sun during the months that those who live in the northern hemisphere call summer. Solar winds are electrically charged particles that, while they may be disruptive to electronic communications, have little effect on climate. The Moon's position affects tides, not climate.

2. **D** The somatic nervous system controls voluntary actions and also sends sensory information to the brain. The cerebrum is the portion of the brain that is considered the center of intelligence. Neurons are individual nerve cells.

3. **A** Chemical reactions typically involve either the transfer or sharing of electrons among atoms. Neutrons are in the nucleus, which usually remains unchanged in chemical reaction (although the nucleus can be changed in a nuclear reaction). The chemical symbol of an element has absolutely no bearing on the element's properties. The atomic mass is made up almost totally of protons and neutrons, so chemical reactions have little to do with atomic mass.

4. **A** Our Sun is technically classified as a G2V star, more commonly referred to as a yellow dwarf because it is small relative to many other stars.

5. **B** Sedimentary rocks are formed by the gradual deposit of sediments over a very long period of time. The sediments could be sand, which would form sandstone, or clay that could eventually become shale. Igneous rocks originate as molten magma, and metamorphic rocks are formed from other rocks altered by temperature, pressure, or chemical processes. Compound is not a recognized type of rock.

6. **B** The two biopolymer strands of DNA coil around each other to form a double helix structure which, when the covalent bonds between the strands are included, resembles a twisted ladder that bears no resemblance to a sphere or oblate spheroid. Since there are only two main strands, a triple helix configuration would not be possible.

7. **B** Use Newton's second law of motion, $F = ma$, to solve this problem. The object reached a velocity of 12 m/s in 4 seconds, so the acceleration was $\frac{12}{4} = 3$ m/sec^2. $15 = m(3)$, so $m = \frac{15}{3} = 5$ kg.

8. **D** When the Earth is exactly in line between the Sun and the Moon the Earth's shadow is cast over the Moon, resulting in a lunar eclipse. Eclipses have nothing to do with an anomaly of the Moon's orbit or solar flares. When the Moon is between the Earth and the Sun, the Moon's shadow falls on a portion of the Earth, resulting in a solar eclipse.

9. **B** The lithosphere refers to the Earth's crust and upper mantle, the two layers closest to the surface. In order towards the center of the Earth from the lithosphere are the lower mantle, asthenosphere, and outer core.

10. **B** The prefix *exo–* means outside, as in "external." Human beings and eagles have endoskeletons inside their bodies. An amoeba is a single-celled organism that does not have a skeleton. A spider's outer shell serves as its skeleton.

11. **A** Blood carries the oxygen that cells need to function to the cells and picks up the carbon dioxide and water that they generate, so the lungs have to support that process by supplying oxygen to the blood and carrying away carbon dioxide and water.

12. **B** Heat is transferred by radiation when electromagnetic waves transfer heat, just as we are warmed by the Sun. Convection transfers heat by the actual movement of warmed matter, quite often air. If the fireplace were not sealed behind glass, heat would likely be transferred to a room by convection. Conduction of heat occurs when two bodies of different temperature directly touch. While the air touching the screen will warm up and spread a very slight amount of heat into the space in which the fireplace is located, that heat transfer would be very small compared to the transfer by radiation. Osmosis, the passage of a fluid through a membrane, is unrelated.

13. **D** Elements on the left side of the Periodic Table, such as those in Group 1, have weak bonds with the electrons in their outer shells and will give them up relatively easily. Elements in Group 17 are on the right side of the table and have a strong attraction for electrons. Therefore, compounds of atoms from these two groups will form ionic, rather than covalent, bonds. When compounds with ionic bonds are dissolved in water, they typically separate into charged ions. Since the Na atoms have given up an electron, they will become positively charged ions. The Cl atoms will retain an extra electron and become negatively charged ions.

14. **C** Watching meteorologists on TV presenting weather forecasts, one could assume that meteorology deals exclusively with weather. However, the science is much more broadly based, encompassing the atmosphere as a whole. Although choice (A) might be tempting, meteorology is unrelated to meteors.

15. **D** The Latin word for eggs is "ova," so it makes sense that eggs would be produced in ovaries. From the ovaries, eggs travel through the fallopian tubes where, if they are fertilized, they may implant in the uterus. The cervix is located at the bottom of the uterus.

Review and Reflect

As you look back over your work in this chapter, think about these questions:

- Which concepts are you more comfortable with, and which seem less familiar?
- As you worked practice questions in this chapter, were you careful to make predictions whenever possible?
- If you missed practice questions that called for calculations, was it because you didn't understand or remember the concept being tested or because you made a misstep while doing the math?
- Did you pay careful attention to the units required by a problem?
- Which science topics should you review again before Test Day?

Want more practice with General Science? Try more questions in the Qbank online.

QBANK

kaptest.com/login

CHAPTER 10

ELECTRONICS INFORMATION

KNOW WHAT TO EXPECT

The discovery of electricity and its use in transferring energy has been the foundation of modern technology. It's hard to imagine a world without automobiles, computers, or even flashlights. Though they have vastly different functions, all electrical and electronic devices operate due to the movement of subatomic particles called *electrons* (hence the word *electronics*).

In this chapter, you will learn the basic principles of electricity and electronic components. By the end of the chapter, you will be able to identify some major circuit components and their functions, perform calculations to determine circuit properties, and understand some of the safety devices used in modern household wiring.

On the CAT-ASVAB, you will have 10 minutes to complete 15 Electronics Information (EI) questions. If you take the paper-and-pencil ASVAB, you will have 9 minutes to complete 20 EI questions.

ELECTRON FLOW THEORY

LEARNING OBJECTIVES

In this section, you will learn to:

- understand electricity as the flow of electrons through conducting materials
- understand the meaning of voltage, current, and resistance, and how they relate to different aspects of electrical flow

A solid understanding of electricity begins at the atomic level. Matter is composed of **atoms**, which are the smallest particles that elements can be broken into and still retain the properties of that element. An atom is itself broken into even smaller components, called subatomic particles.

Two subatomic particles are found within the nucleus: **protons** and **neutrons**. Protons have a positive charge, and neutrons are neutrally charged. The nucleus is the heaviest part of the atom and accounts for the majority of the atom's mass. **Electrons**, the third type of subatomic particle, are in motion around the nucleus and have a negative charge. For a neutral atom, there is one electron for each proton that resides in the nucleus. From static shock to a lightning strike, all electricity is the movement of electrons.

The electrons occupy various energy levels around the nucleus, known as **shells**, and each time one becomes full, a new one is begun. The outer shell of an atom is known as its **valence shell**, and the number of electrons that reside in the valence shell are what determines whether an element is a **conductor**, a **semiconductor**, or an **insulator**. In many elements, the valence shell holds a maximum of eight electrons.

Boron, a Conductor; Silicon, a Semiconductor; and Phosphorus, an Insulator

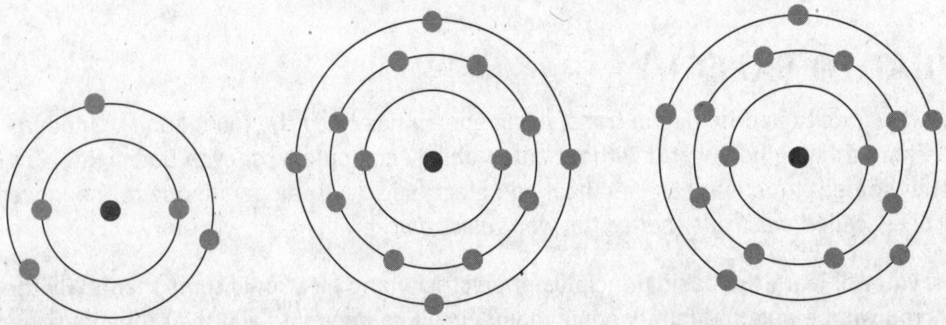

Boron: 5p 6n 5e
3 valence electrons

Silicon: 14p 14n 14e
4 valence electrons

Phosphorus: 15p 15n 15e
5 valence electrons

A **conductor** is an element that allows electrons to flow freely. What all conductors have in common is one or more mobile valence electrons per atom that are free to move from one atom to another, as each valence shell has more empty spots than electrons. An **insulator** has a valence shell that is more than half full or is completely full. It does not conduct electricity much at all, because its electrons are all tightly bound and will not leave their packed valence shells. A **semiconductor**, with a valence shell that is exactly half full, is neither a good conductor nor a good insulator, but it has some remarkable properties that make it very useful for making electronic components.

A convenient way of visualizing how electricity works is to imagine the electrons as drops of water flowing through a pipe. Such a flow is due to applied pressure on one end, forcing the water the opposite way. Likewise, when a conductor is connected across the terminals of the battery, the negative terminal of the battery applies a repulsive pressure on the electrons and pushes them toward the positive terminal. (A basic rule of electricity is that like charges repel each other, while opposite charges attract each other.) The resulting flow of electrons through a conductor is an electric current.

Current

The rate of flow of electrons through a conductor is known as **current**. Just as the gush of water through a pipe has a current, so does the flow of electricity. When you turn on a faucet, it's easy to distinguish between a trickle and a roaring flood by the very different amounts of water flowing out each moment. Electrical current, similarly, is measured by the amount of **charge** flowing past per unit of time. Current is measured in **amperes**, or amps for short, and an ampere is defined as one **coulomb (C)**, the basic unit of electrical charge, flowing past a given point in one second. The symbol used for amperes is **A**. If five coulombs of charge were to pass by a point every second, this would be a current of five amps (5 A).

Try this EI question.

What would be the current if 18 coulombs passed by in three seconds?

(A) 3 A
(B) 6 A
(C) 18 A
(D) 54 A

Explanation

The correct answer is **(B)**. Since 18 C of charge pass by in 3 s (that is, seconds), there must be 6 C passing by per 1 s. By definition, that's a current of 6 A.

One coulomb, incidentally, is the amount of negative charge in 6.25×10^{18} or 6,250,000,000,000,000,000 electrons. It takes this many electrons flowing past a point in a conductor every second for a modest one ampere of current to be flowing. Current can be measured using a device called an **ammeter**.

Voltage

In order to make electrons move through a conductor, there must be a force that causes them to move. Water will move from a high-pressure area in a pipe toward a lower-pressure area, but if the pressure is equal at both ends, no movement takes place. Electricity is no different. There must be electrical "pressure" applied to a conductor to cause electrons to move.

For an electron, the "pressure" is caused by electromagnetic repulsion from a large negative charge, like that found at the negative terminal of a battery. This negative terminal has a huge excess of electrons, all pushing at each other. When a piece of conducting wire connects the positive and negative terminals of a battery (known as "completing the circuit"), electrons are pushed from the negative terminal into the wire, themselves pushing along the conducting electrons already there, and so on down the line, so that every electron will eventually reach the positive terminal, where there is a shortage of electrons and an excess of attractive protons pulling the electrons in. This will continue until the excess of electrons at the negative terminal is exhausted, and the battery is dead.

Electrical pressure is known as **voltage**, and it is measured in volts (symbolized by the letter **V**). The higher the voltage that is applied to a conductor, the greater the electrical pressure the electrons will experience, and the higher the current (rate of flow of electrons) will tend to be. Voltage and current are therefore **directly proportional**.

To be more exact, voltage is really the difference in electric pressure between two points, such that electrons will tend to be pushed from areas of greater **electric potential** (the proper term for this pressure-like quantity) to areas of lesser electric potential (lower pressure). For this reason, voltage is also known as **electrical potential difference**. It's also referred to sometimes as **electromotive force**. A technician would measure voltage using a **voltmeter**.

Resistance

Opposition to the flow of current is known as **resistance**. Resistance is measured in **ohms**, and one ohm is defined as the amount of resistance that will allow one ampere of current to flow if one volt of electrical pressure is placed on a conductor. The symbol for the ohm is Ω (the Greek letter omega), and resistance can be measured using an **ohmmeter**.

As resistance in a conductor increases, current flow decreases. This means that current and resistance are **inversely proportional**. The smaller the resistance a material has, the better a conductor it is. But voltage and current, remember, are **directly proportional**. Any time voltage increases, current will also increase. So in materials with high resistance (that is, in poorer conductors), a higher voltage must be used to get the same current.

All conductors have a certain amount of resistance, with some having more than others. Three materials that are most often used as conductors are silver, copper, and aluminum. Of these three, silver is the best conductor, as it exhibits the lowest resistance. Unfortunately, silver is relatively expensive, so the next best choice is copper. Copper has only a slightly higher resistance than silver, but is much less expensive. Most electrical cable and wire is currently made from copper. Aluminum has a higher resistance than copper, and exhibits some other characteristics that make it even less than desirable for use in electrical applications. While aluminum was used extensively in residential wiring at one time, it is now used only in a few select applications.

Exceptions and Additional Notes

The analogy of electron flow to the flow of water through a pipe is helpful, but like all analogies, there are places where it breaks down. For example, when a pipe is broken, water will spray out, but when a conducting wire in an electronic device is broken, current flow simply stops along that path. That's because electricity travels *through* the conducting metal of a wire, not inside a hollow pipe like water does. With no conducting material, electrons cannot move forward, and air molecules are very poor conductors. Except when the gap is very small (spark plugs are designed this way) or the amount of voltage extremely large (as occurs with lightning), an electric current will not flow through air.

Another interesting quirk in the study of electricity is what's known as **conventional current**, which is defined by the (imaginary) flow of positive charge and is opposite in direction to actual electron flow. This strange concept came about because of the difficulty in actually observing electricity at a microscopic level.

Seeing and feeling the direction that water is moving in is easy enough, but electricity is another story. When scientists first started studying electricity seriously, not enough was known about atoms yet, and they realized that either a flow of negative electrons in one direction or a flow of positive protons in the opposite direction could explain the movement of charges discovered to be at the heart of electricity.

Scientists took a guess that it was the protons that moved, and defined the direction of current based on that. Later they discovered that protons stay in the nucleus and don't move, but current direction had already been defined this way. As a result, if a circuit has a conventional current that is moving in a clockwise direction, this means that, in reality, electrons are moving around the circuit in a counter-clockwise direction. Nowadays, we can refer to either conventional current or electron flow, but it's necessary to remember to switch the direction when going from one to the other.

It is straightforward to think of electrical current as a flow of electrons, but an individual electron does not necessarily traverse an entire circuit. Often this is more like a relay race as subsequent electrons carry the charge onward. Some references may, therefore, refer to the flow of a charge.

This chapter hasn't discussed circuits yet, but you still have enough information to understand the worked example below. In the diagram, V stands for a source of voltage, such as a battery cell, and L1 and L2 are lamps connected to that voltage source. An unbroken line means an unbroken connection, and a broken line means a broken connection.

Question	Analysis
In the wiring diagram pictured, which of the lamps will be lit up?	**Step 1:** The question asks which lamp will light up—that is, which will receive a flow of current.
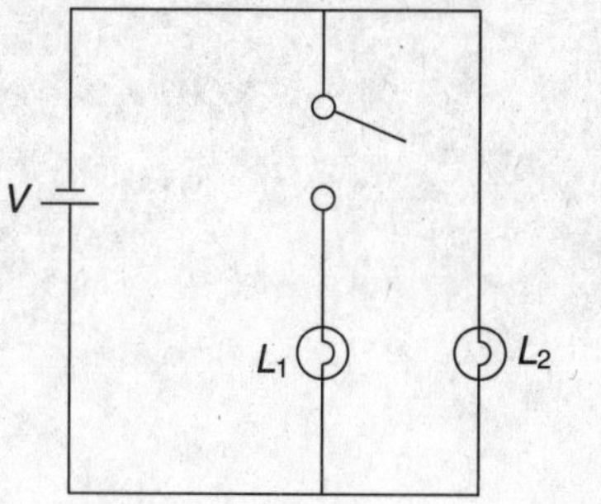	**Step 2:** The open switch (the break in the middle line) prevents current from flowing across the path containing lamp L1. That's because current cannot flow into a dead end. However, the path from the battery cell, through L2, and back to the opposite terminal of the battery does make a complete circuit. L1 should not have a current passing through it, so will not be lit up. L2 should have a current passing through it, so will be expected to light up.
	Step 3: Since there are no gaps in its path, L2 should be lit up, but L1 should not.
(A) L1 only will be lit. (B) L2 only will be lit. (C) L1 and L2 will both be lit. (D) Neither lamp will be lit.	**Step 4:** Choose option **(B)**.

Now try one on your own.

Certain conducting materials decrease in resistance as the wires heat up. As resistance decreases, which of the following would be expected to happen?

(A) The amount of current should increase.
(B) The amount of current should decrease.
(C) The direction of current should switch.
(D) The current should remain constant.

Explanation

The answer is **(A)**. Resistance and current are inversely proportional, so when resistance decreases, current increases. Remember that resistance is so called because it resists current. If the resistance is less, it makes sense that the amount of current is able to increase, similar to how the rate of traffic flow is greater on the highway where there are fewer traffic lights.

CIRCUITS

LEARNING OBJECTIVES

In this section, you will learn to:

- effectively use Ohm's law to perform circuit calculations
- effectively use the power formula in an electrical context
- recognize the features of a basic circuit and give definitions of each
- simplify complex circuits

In order for electric current to flow, there must be a **circuit** for it to follow. A circuit is a complete loop or path that electricity can follow. There are three essential components of an electrical circuit: a **voltage source**, a **load**, and **conductors** to connect the load to the voltage source. When these three components are connected so that current can flow, we have a **closed (completed) circuit.**

Simple Closed Circuit

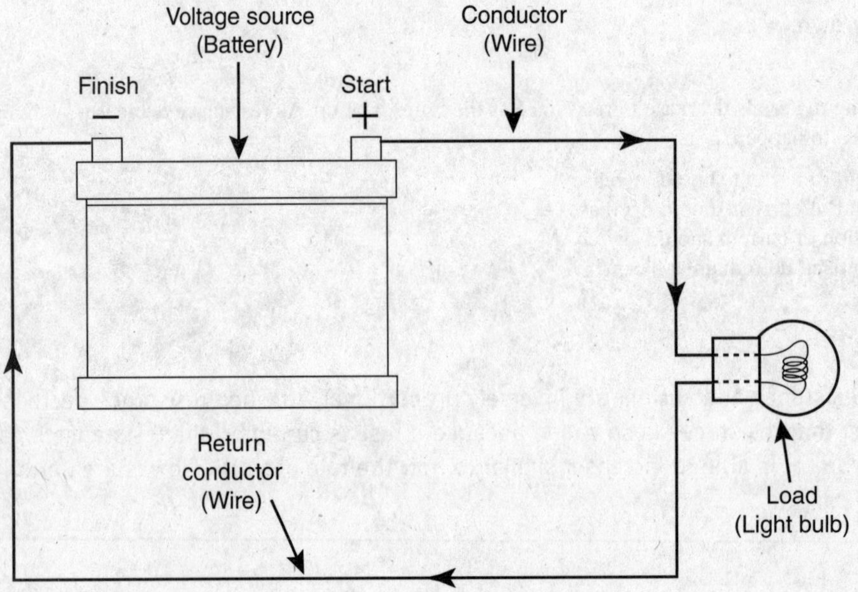

A **load** is basically a source of resistance that converts electrical energy into some other energy form. For instance, a light bulb is a load. It has resistance and converts electrical energy into light energy and heat. Other examples of loads are electric motors, heating elements, and solenoids.

If a load's resistance were to decrease, the current flowing in the circuit would increase. However, if the circuit were broken, as in a wire (conductor) being disconnected, no current would flow. This would be called an **open circuit**.

When current is flowing, each electron is carrying the voltage the length of the conductor: pushing the electrons ahead of them to keep moving forward, while being pushed themselves by those behind. When the line stops, it stops everywhere, thus when there is no clear path to their destination, as in the gap that occurs in an open circuit, electron flow stops entirely and the current is zero.

Ohm's Law

The relationship between voltage, current, and resistance through a wire is summed up in **Ohm's law**. Ohm's law states that voltage (potential difference) in volts is equal to the current in amperes multiplied by the resistance in ohms. This is represented by the formula:

$$V = IR$$

V represents *voltage*, also known as potential difference, while I represents *current*, and R represents *resistance*. If the *I* symbol seems odd, think of current (defined as the amount of charge flowing past a point per unit time) as the *intensity* of charge flow. The other two are easier to remember.

The voltage source is the beginning and end of the circuit. The negative side repels electrons, forcing them to move through the circuit. The positive side helps by attracting electrons, and an electron emitted from the negative side will eventually traverse the complete circuit before arriving at the positive side. Remember that voltage is essentially electric pressure on a charged object, so the voltage source is ultimately responsible for electric flow in a circuit by pushing the electrons forward.

To put it another way, a non-zero voltage will produce a certain amount of current in a closed circuit for a given resistance, as shown by Ohm's law above. Batteries, which have positive and negative terminals, are voltage sources (for example, a 12 V battery). A household electrical socket also provides voltage, which ultimately comes from the power plant. They are also referred to as power sources, because the electric current they create carries energy.

The conductor is just the wire that connects the load(s) and voltage source. For the sake of simplicity, the wire is often considered to have zero resistance. Although this is not true, the resistance in a conducting wire is often small enough compared to any given load that it can be ignored for most calculations. At zero resistance, Ohm's law requires the voltage across such a conducting wire to also be zero, meaning the electrical pressure does not change from one side to another.

Remember, a load is any component with resistance that converts electrical energy into some other energy form. In accordance with Ohm's law, were a load's resistance to decrease, the current flowing in the circuit would increase. However, if the circuit were broken, as in a wire (conductor) being disconnected, no current would flow. This would be called an open circuit. This, too, can be explained by Ohm's law, as the break or gap in the circuit is like a load with infinite resistance (recall from the previous section that air is a terrible conductor).

Series Circuits

An electrical circuit that has only one path for current to flow is known as a **series circuit**. A break (opening) at any point in the circuit will cause current to stop flowing in all parts of the circuit. The simplest possible circuit, one voltage source connected to one load by conductors, is an example of a series circuit.

Series Circuit

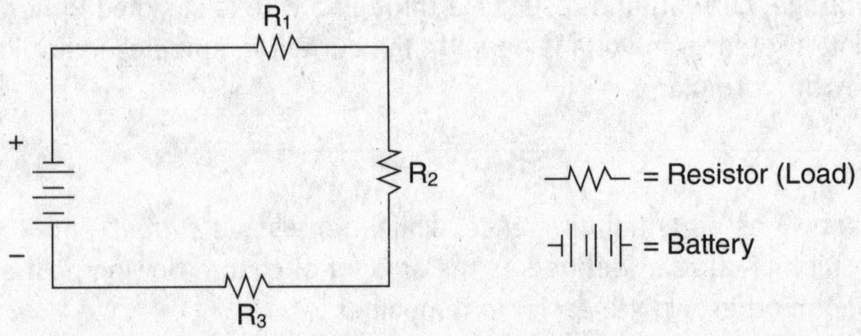

Since there is only one path for current to follow in a series circuit, the current flow will be the same in all parts of the circuit. It wouldn't make sense, for example, if one million electrons were leaving the negative terminal of a battery each second, but two million were arriving at the positive terminal at the other end. It would be like a plane taking off from New York and then landing in Miami with twice as many passengers. Thus, the current passing through any single load in a series circuit is the same.

$$I_1 = I_2 = I_3 = \ldots$$

On the other hand, what happens if multiple loads were found in a series circuit? If two or more loads are added to a series circuit, their resistances add together, creating what's called the total or effective resistance across the circuit.

$$R_{tot} = R_1 + R_2 + R_3 \ldots$$

Voltage, which refers specifically to a *difference* in electric potential ("electric pressure") at two different points (like from the negative to the positive terminals of a voltage source), can also be calculated across a single load. This is also known as the voltage drop across a load, and the total of all voltage drops across each load in a series circuit is equal to the total voltage of the complete circuit, which is equal to the voltage of the voltage source itself.

$$V_{tot} = V_1 + V_2 + V_3 \ldots$$

Try applying these ideas to solve the following problem.

Three different loads with different rated resistances are wired in series. Which of the following will be the same through all of them?

(A) the voltage drop
(B) the resistance
(C) the current passing through
(D) the heat given off

Explanation

Choice **(C)** is correct. The current passing through one point in a series circuit has to be the same in other parts of the circuit, as there is nowhere else for the current to go.

Parallel Circuits

It is much more common for loads to be wired in parallel. With this arrangement, each load is wired in a separate path. If any one of these paths were to have a break or gap, current flow would still continue through the other paths, so there would still remain a closed circuit.

Parallel Circuit

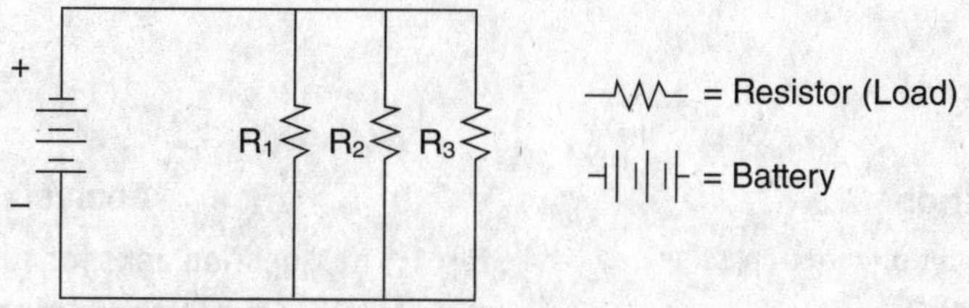

Parallel circuits are the exact opposite of series circuits in that *voltage is the same throughout each parallel branch of the circuit*, while current flow varies. Since voltage is a measure of the difference in electric potential between two points, the voltage across alternate paths connecting the same two points in a circuit must also be equal.

$$V_1 = V_2 = V_3 = \ldots$$

On the other hand, current flow becomes divided when there are parallel paths. It's commonly misstated that electricity follows the path of least resistance. This is not actually true. Electricity actually follows every possible path, though a larger proportion of electrons will take a path of low resistance while a smaller proportion will take a path with higher resistance. For a parallel circuit, therefore, the total current through the circuit is equal to the sum of currents through each closed path.

$$I_{tot} = I_1 + I_2 + I_3 \ldots$$

What about the effective resistance of a circuit when multiple loads are wired in parallel? It's natural to instinctively think that adding more loads is bound to add more resistance, but this is not the case. This is a little less intuitive at first glance, so consider an analogy.

There's a big sale on at a popular retail store. One of the aisles is choked with customers trying to grab the best deal. The jostling customers, abandoned shopping carts, and crates of items waiting to be shelved all create resistance, making it difficult to move forward and slowing the rate at which people move through the aisle. In this analogy, the people are electrons, and the rate at which they move is the current value.

Suddenly, a store clerk clears away a pile of merchandise that was blocking an adjacent aisle. This aisle, too, has a lot of junk in the way. It may even have higher resistance than the original aisle, as it has more stuff. Even so, some shoppers opt for that cramped but less crowded aisle. The total number of shoppers passing through the store each moment is now greater, with the addition of that parallel path.

Likewise, adding a load in parallel, even if that load's resistance is high, decreases the total or effective resistance of a circuit, because even though it has resistance, it provides an additional path while doing nothing to constrain the existing one. The formula for adding resistances in parallel is a slightly more complicated one, but it's very useful.

$$\frac{1}{R_{tot}} = \frac{1}{R_1} + \frac{1}{R_2} + \frac{1}{R_3} + \ldots$$

Study the example below.

Question	Analysis
What is the approximate amount of current passing through switch S when it is closed?	**Step 1:** The question asks for current through the whole circuit. Effective resistance of the circuit and total voltage are needed to calculate this.
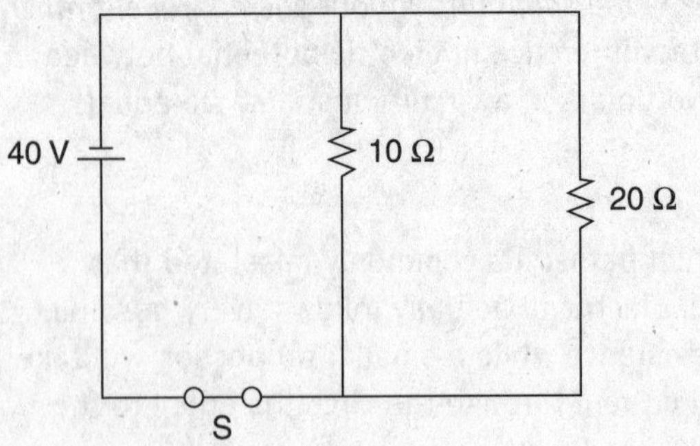	**Step 2:** There are two loads in parallel, so the parallel resistances formula is needed. Given the effective resistance (R_{eff}), Ohm's law can be used to determine the current drawn. The reciprocals of the resistance values are $R_1 = \frac{1}{10}$ and $R_2 = \frac{1}{20}$. Find the least common denominator and add: $R_1 + R_2 = \frac{2}{20} + \frac{1}{20} = \frac{3}{20}$ $R_{eff} = \frac{20}{3}\ \Omega$ or $6.67\ \Omega$ $I = \frac{40}{6.67} \approx 6$ A
	Step 3: The predicted value is 6 A.
(A) 0 A (B) 2 A (C) 4 A (D) 6 A	**Step 4:** Select answer choice **(D)**.

Now try one on your own.

Which of the following would not increase the rate of direct current flow, *I*, through a conductor?

(A) decreasing effective resistance via a variable resistor
(B) increasing voltage
(C) adding a resistor to the circuit in parallel
(D) adding a resistor to the circuit in series

Explanation

Answer choice **(D)** is correct. Using the process of elimination, (A) would decrease the overall resistance of the circuit, which, according to Ohm's law, would increase current given a constant voltage source. Choice (B) would also increase current flow if resistance remained constant. Choice (C) would also decrease the overall resistance of the circuit and increase current. However, adding a resistor in series would increase effective resistance, which means a decrease in current drawn.

Series-Parallel Circuits

The most popular arrangement is the series-parallel circuit. A series-parallel circuit has some components, such as an on/off switch, wired in series with a number of loads that are connected in parallel.

Series-Parallel Circuit

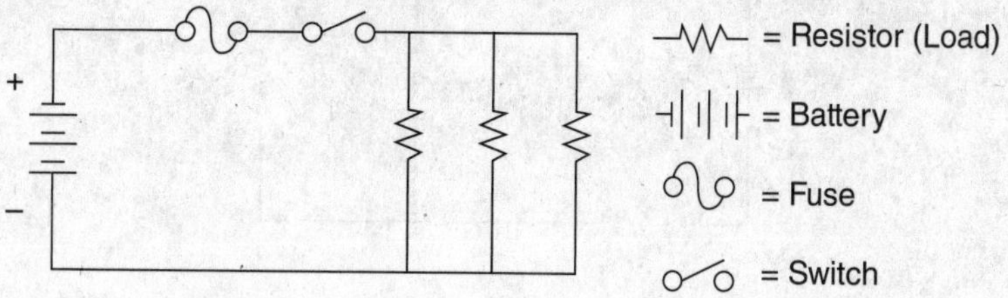

Most residential wiring circuits are series-parallel. Wall outlets are wired in parallel, but are all fed from a circuit breaker (shown in the diagram as a fuse) that is wired in series. Switching off the circuit breaker will turn off power to all of the outlets. However, if the circuit breaker is on, voltage is provided to all of the outlets whether there are loads plugged into them or not.

Determining the total effective resistance across a series-parallel circuit is a step-by-step process. Use the appropriate formula for each group of resistors, either in series or parallel, to simplify the circuit to fewer and fewer calculated effective resistances until a single value for resistance has been found.

To determine the voltage drop or amount of current flowing through any particular load, combine Ohm's law with the rules for current and voltage in series and parallel wiring to solve for a given variable of any particular load.

Question	Analysis
In the diagram shown, all three resistors have a resistance of 4 Ω. What is the effective resistance of the circuit?	**Step 1:** The question asks for the effective resistance of a series-parallel circuit.
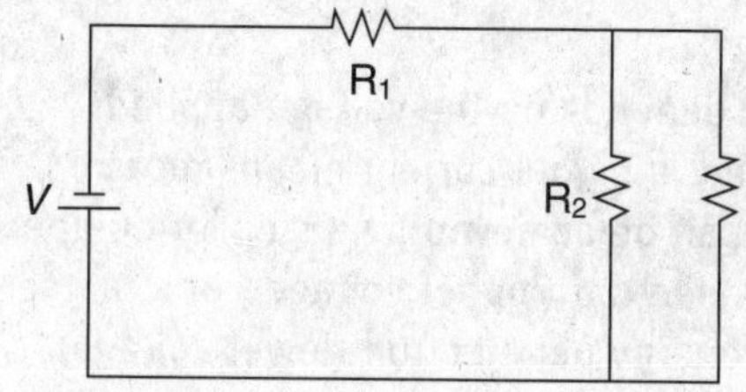	**Step 2:** The two parallel loads can be combined into a single effective resistance. The effective resistance of the last two loads can be added to the first load's resistance since they're in series with the first load. $\frac{1}{R_{23}} = \frac{1}{4} + \frac{1}{4} = \frac{2}{4} = \frac{1}{2}$ $R_{23} = 2\ \Omega$ $R_{tot} = 4 + 2 = 6\ \Omega$
	Step 3: The prediction is 6 Ω.
(A) $\frac{4}{3}\ \Omega$ (B) 4 Ω (C) 6 Ω (D) 12 Ω	**Step 4:** Select choice **(C)**.

Now try the question below on your own.

What is the amount of current passing through the 3 Ω resistor in the circuit pictured?

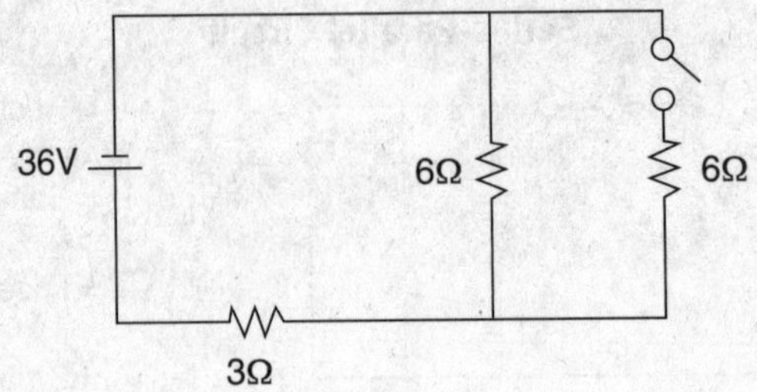

(A) 3 A
(B) 4 A
(C) 6 A
(D) 12 A

Explanation

The answer is **(B)**. Since a switch is open on the branch of one of the parallel 6 Ω resistors, it can be ignored. That means this circuit will behave like a simple series circuit. The two connected resistors sum to an equivalent resistance of 3 + 6 = 9 Ω. Using Ohm's law, the current flowing through the 3 Ω resistor (and one of the two 6 Ω resistors) can be calculated as $I = \frac{V}{R} = \frac{36}{9} = 4$ A.

Electrical Power

Electrical power is a term that refers to the actual rate at which energy is provided to and consumed by an electric circuit. Power is expressed in watts (joules per second) and can be calculated by multiplying the voltage (in volts) applied to the circuit by the current (in amperes) that flows in the circuit. This is represented by the formula:

$$P = IV$$

Each electron flowing through a circuit or wire has energy. How much energy depends on the voltage applied. Electrons moving through a large voltage have more energy. On the other hand, a higher current means more electrons passing by each moment. So an equivalent rate of energy delivery can be achieved by a smaller number of electrons each moment (lower current) with a larger amount of energy per electron (higher voltage), or a larger number of electrons each moment (higher current) with a smaller amount of energy per electron (lower voltage).

This formula applies not only to the rate of energy delivery (power found in long-distance power lines) but to the rate of energy production (power provided by a generating plant) and to the rate of energy consumption (power usage of an appliance or load). The word *consumption* is actually a little bit misleading, as energy is never used up but just changes form. A forty-watt (40 W) light bulb doesn't really *use up* 40 joules every second. It transforms 40 joules of electrical energy to 40 joules of light and heat energy each second.

Question	Analysis
An electric generator provides a power supply to a cabin at the North American household standard of 120 V. If the effective resistance of the cabin's wiring is 30 Ω, what is the power usage of the generator?	**Step 1:** The question asks for the power of the generator.
	Step 2: Both current and voltage provided by the generator are needed, but only voltage and resistance are given. However, these can be used to find current using Ohm's law. Now, power can be calculated. $I = \frac{120V}{30\,\Omega} = 4\text{A}$ $P = (4\text{ A}) \times (120\text{ V}) = 480\text{ W}$
	Step 3: The predicted answer is 480 W.
(A) 4 W (B) 150 W (C) 480 W (D) 3,600 W	**Step 4:** Select choice **(C)**.

Now try a question on your own.

The power usage of a coffee pot is 1,200 W while running, and is powered by a standard North American 120 V electrical supply. What is the resistance of the coffee pot?

(A) 10 Ω
(B) 12 Ω
(C) 120 Ω
(D) 1,200 Ω

Explanation

Answer choice **(B)** is correct. Using the power law and solving for current, $I = \frac{1{,}200}{120} = 10$ A. Then using the current and voltage values in Ohm's law and solving for resistance, $R = \frac{120}{10} = 12\ \Omega$.

Standard Electrical Units and the Metric System

When using Ohm's law and other electrical formulas, it is important to keep in mind that the quantities used must be expressed in ohms, amperes, volts, and watts. If any of these quantities are given in other units, these should be converted before making calculations with them.

Any of these base units can be modified with metric prefixes. For example, smaller generating plants typically measure their power output in megawatts (MW), while cardiac doctors may measure tiny electrical signals in the human heart on the scale of milliamps (mA).

For more detail on metric system prefix values, please refer to the measurement information at the beginning of the Physical Science section of chapter 9: General Science.

STRUCTURE OF ELECTRICAL AND ELECTRONIC SYSTEMS

LEARNING OBJECTIVES

In this section, you will learn to:

- explain the safety mechanisms that are used in modern household electric systems
- recognize the basic electrical components in a circuit via their symbols
- explain the basic functions of several electrical components and their uses and benefits

Thus far, you've learned the basic operational and mathematical concepts behind an electric circuit. In this next section, you will continue to build on these skills. The section will include a discussion of different types of current, some safety considerations, and components that give circuits special qualities and enable them to perform specific tasks.

AC *vs.* DC

There are two types of current. They are **direct current** (DC) and **alternating current** (AC). DC is current that flows in only one direction in a conductor. This is the type of current that is delivered by a battery. Many electronic devices, like cell phones and laptops, work on DC. AC is current that changes direction (moves back and forth) many times in a second as it moves in a conductor. Household electrical outlets deliver alternating current. AC provides a more efficient way to transport electricity from the power stations to your home or business.

Here in North America, AC is always delivered at 60 cycles per second (hertz). One cycle would be voltage starting at zero, increasing to a maximum level in one direction, then decreasing to zero, increasing to a maximum level in the reverse direction, then returning to zero again. The unit hertz (Hz) describes the frequency of the signal and is a measure of how many complete cycles occur in 1 second. The figure below shows a 60 Hz AC signal.

60 Hz AC signal

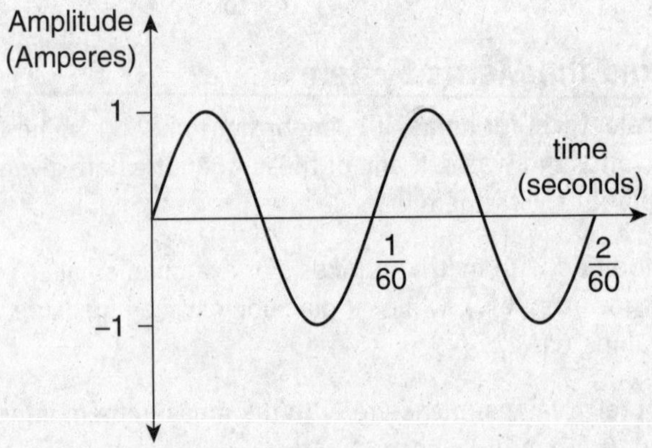

This is the frequency equation:

$$f = \frac{1}{T}$$

In the equation above, f = frequency in Hz and T = the amount of time for one cycle in seconds. You can use that equation to solve for either frequency or period when given one of them. Thus, with a frequency of 60 Hz, it can be determined that one complete cycle occurs in $\frac{1}{60}$ of a second, since:

$$T = \frac{1}{f}$$

So, if power stations deliver AC and most electronic devices work on DC, where and how does the conversion take place? Most electric devices come with built-in AC to DC conversion systems. Later in this section, we will introduce an electric component, the diode, that makes this conversion possible.

Ground

Grounding electrical devices and residential wiring is an important safety factor. **Ground** represents a place of lowest potential in a circuit. Since the potential difference (voltage) is largest between any point in the circuit and the ground point, any "stray" electricity will follow this path since resistance is low here. This is important to prevent shock due to external influences such as lightning, or due to internal circuit failure where conducting wires are compromised.

In residential wiring, ground is a common connection throughout the wiring system that protects against electrical shock. All of the wiring grounds are connected to an earth ground (such as a copper rod driven into the ground or buried conduit) and this is used to guide electrical current away from panels and equipment, should an internal short circuit take place.

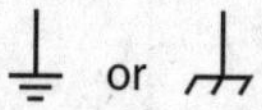

Ground Symbol

Important Electric and Electronic Components

The properties of a particular electrical circuit will change depending on the types of components present. In this section, several of the major ones are introduced and discussed.

Resistors

Not all components within an electrical circuit require the same voltage. There are also times when it is useful to be able to raise or limit current in a circuit in order to control certain functions (e.g., volume in a stereo). A component that can be used to limit current and/or voltage is a resistor. A resistor does what its name suggests: it creates a specific amount of resistance that generates a voltage drop when current passes through it.

As you learned previously, resistance can influence current, via Ohm's law. An increase in R will therefore cause a decrease in the current.

Resistor Symbols

Resistor Symbol

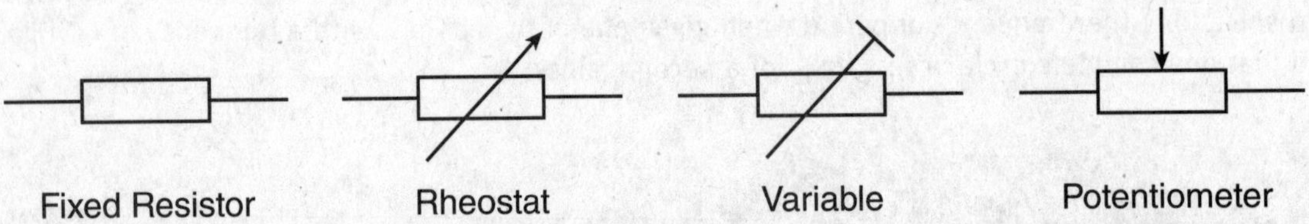

There are two major types of resistors used today: the **fixed resistor** and the **variable resistor.** Fixed resistors, as their name implies, have a fixed resistance. Variable resistors (**rheostats** or **potentiometers**), have resistances that can be changed. Potentiometers are useful in changing the voltage drop across a component in a circuit. In electronic applications, they can be used to help adjust the volume on a TV, radio, or stereo. Rheostats are useful in changing the current in a circuit and can be used as light dimmers or to control the speed of small motorized devices.

Fixed resistors use a color band system to enable us to determine their resistance in ohms. The first color gives the first number. The second color gives the second number, and the third number gives the multiplier or the number of 0s behind the numbers.

Fixed Resistor Color Bands

Color Name	1st digit 1st stripe	2nd digit 2nd stripe	Multiplier 3rd stripe
Black	0	0	×1
Brown	1	1	×10
Red	2	2	×100
Orange	3	3	×1,000
Yellow	4	4	×10,000
Green	5	5	×100,000
Blue	6	6	×1,000,000
Violet	7	7	-
Grey	8	8	-
White	9	9	-

So, given a resistor with a red, violet, and yellow band, what is the value of the resistance? According to the chart, the red represents the number 2, the violet represents the number 7, and the yellow represents 4 zeros. Putting it all together, the indicated resistance is 270,000 ohms.

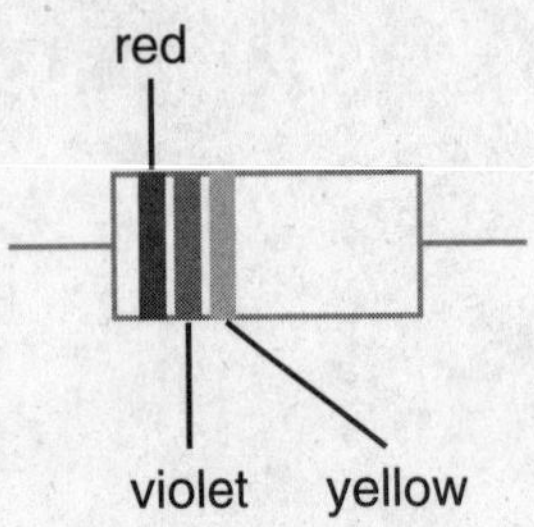

Given a simple series circuit with unknown resistance, how could you go about determining the value of resistance in the circuit? This can be done experimentally in one of two ways: first, by measuring resistance directly using an ohmmeter. To use an ohmmeter correctly, remove the device with unknown resistance from the circuit or disconnect the power source from the circuit (an ohmmeter has its own power source). The resistance can then be measured by attaching the ohmmeter leads across both ends of the device.

Alternatively, if no ohmmeter is available, an ammeter connected in series with the resistor and a voltmeter connected in parallel with the resistor as shown below will give *I* and *V*. Via Ohm's law, resistance is equal to *V* divided by *I*.

Using an Ammeter and Voltmeter to Measure Resistance

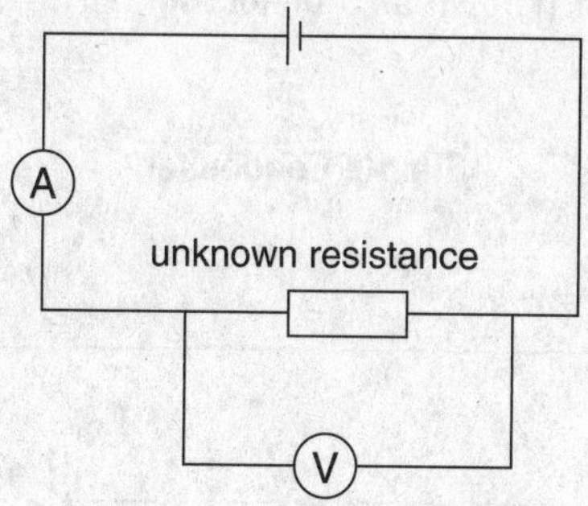

Fuses and Circuit Breakers

An electric ground is one essential safety component in a circuit. It helps prevent shock by providing an alternate path for electricity to flow, away from the device. Fuses and circuit breakers are also important safety components in a circuit. When current in a circuit increases too much, wires can overheat, leading to fires.

A **short circuit** occurs when a load is bypassed with a conductor for some reason. Since current and resistance are inversely proportional, the flow of current will increase. This could be caused by something as simple as the insulation on the wires leading to and from the load becoming frayed and allowing the wires to come into direct contact with each other. If a short circuit occurs in part of a series circuit, the net effect is merely to remove the load that was bypassed as a result of the short circuit. If, however, the short circuit occurs at *any* load in parallel circuits or bypasses *all* the loads in series circuits, the resistance of the circuit as a whole is reduced to near zero and a dangerous surge in current could occur. For this reason, it is a good safety precaution to protect circuits with fuses or circuit breakers.

Fuses are thin wires that melt when current exceeds a prescribed amount, thereby preventing any further electricity flow. This prevents any potential damage to the electric device. One disadvantage of using a fuse is that when it has melted, or "blown," it has to be replaced before the circuit will work again. Fuses with different current ratings can be used depending on the device specifications.

Fuse Symbol

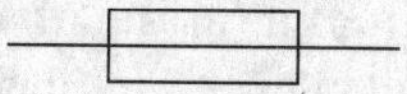

Circuit breakers serve the same function as fuses, but have the advantage of being able to be reused multiple times. However, they respond more slowly to increases in current than fuses do, and are more expensive to install. There are several types of circuit breakers, but the underlying principle behind their function remains the same. One class of circuit breakers consists of a bimetallic strip that bends away from its contact in a circuit when too much current is flowing. This makes a break in the circuit and prevents further current flow. Another class of circuit breakers uses an electromagnet to cause a breach in the circuit. When the current rises to a certain value, the ferrous material is sufficiently magnetized to cause the circuit to open, thereby inhibiting further current flow. Later in this chapter we will explore the relationship between electricity and magnetism.

Circuit Breaker Symbol

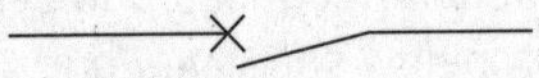

Capacitors

Capacitors (also known as **condensers**) are electrical storage units. They are constructed using two metal conducting plates with a very thin insulator (known as a **dielectric**) between them. Air can also serve as a dielectric.

Simple Condenser

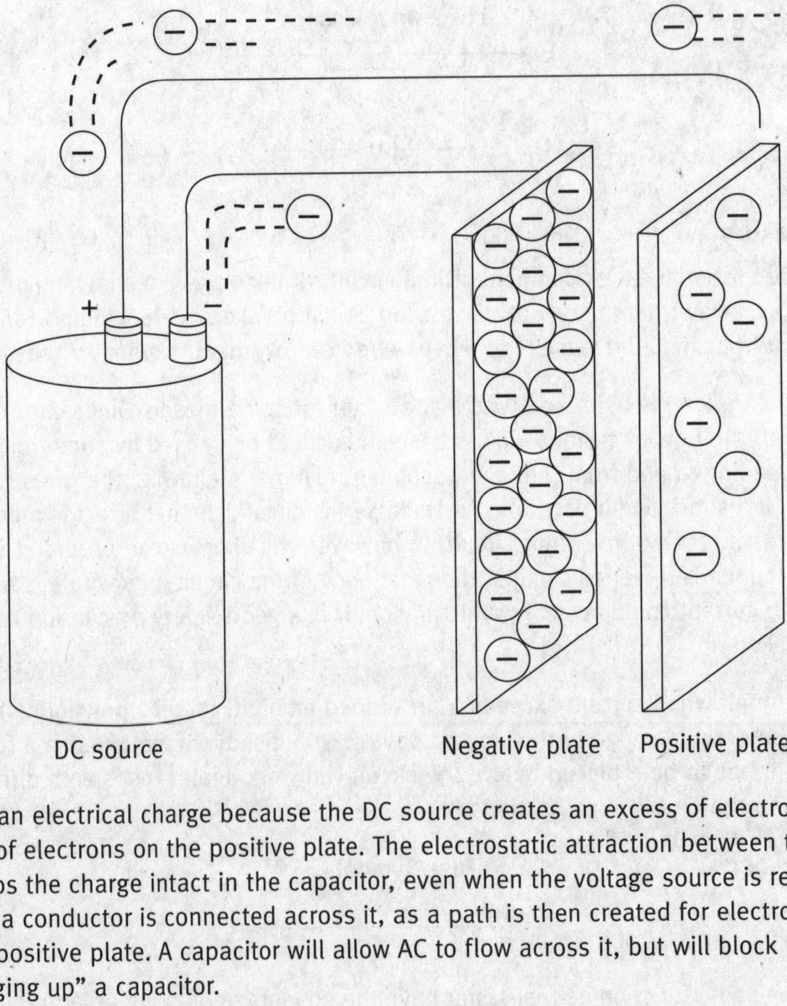

A capacitor can store an electrical charge because the DC source creates an excess of electrons on the negative plate and a shortage of electrons on the positive plate. The electrostatic attraction between the positive and negative charges keeps the charge intact in the capacitor, even when the voltage source is removed. The capacitor will discharge itself if a conductor is connected across it, as a path is then created for electrons to flow from the negative plate to the positive plate. A capacitor will allow AC to flow across it, but will block DC. This is why DC is most useful for "charging up" a capacitor.

A capacitor's opposition to the flow of current is known as **capacitive reactance**, and this is measured in ohms. Capacitive reactance is inversely proportional to the frequency of the AC signal. In other words, the higher the frequency, the less opposition there is to the flow of AC across the capacitor.

General Capacitor Symbol

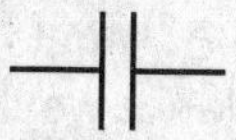

Capacitance, or the ability of the capacitor to store charge, is represented by the symbol ***C***, and its unit of measurement is the **farad**. A farad is sufficient capacitance to store one coulomb of electrons with an electrical potential of one volt applied.

Question

In the wiring diagram pictured, the fuse is rated for 0.1 A. When the switch is closed, what will happen to the lamp in the circuit?

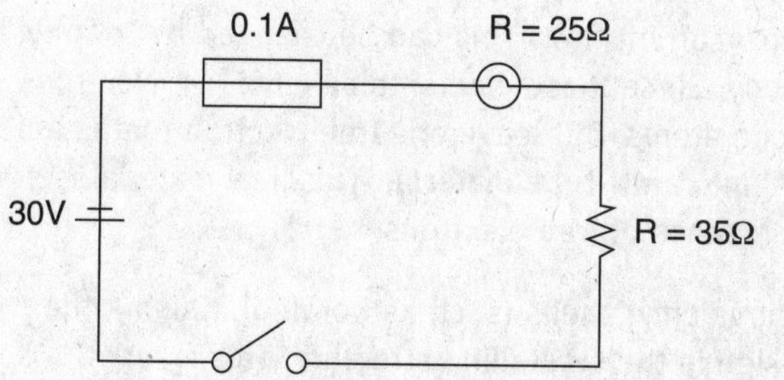

(A) The lamp will stay lit.
(B) The fuse will blow, preventing the lamp from staying lit.
(C) The fuse will stay intact, but the lamp will go off.
(D) The lamp will blink continually.

Analysis

Step 1: The question asks what will happen to the lamp in the circuit when the switch is closed; i.e., will it light or not?

Step 2: Normally, when the switch is closed, current will flow and the lamp will come on. However, this circuit has a fuse, so its impact on the circuit needs to be considered. The fuse is rated at 0.1 A. This means that it will only allow currents to pass that are 0.1 A or less. The current flowing in the current needs to be determined.

Via Ohm's law, $I = \frac{V}{R}$. Based on the circuit, which is a simple series, $V = 30\text{V}$, and $R_{tot} = 25\ \Omega + 35\ \Omega = 60\ \Omega$. Thus, $I = \frac{30\text{ V}}{60\ \Omega} = 0.5\text{ A}$.

Step 3: Since the current in the circuit is larger than the fuse's 0.1 A rating, the light may come on briefly, but the fuse will blow, breaking the circuit and thus the light will go off.

Step 4: Choose option **(B).**

Now try a question on your own.

A fuse rated at 0.5 A blows in a circuit when the switch is closed. If the total resistance in the circuit is 100 Ω, which of the following represents the smallest the voltage could have been?

(A) 10 V
(B) 25 V
(C) 60 V
(D) 120 V

Explanation

Answer choice **(C)** is correct. If the fuse is rated at 0.5 A, it will blow when current exceeds this amount. Via Ohm's law, a 50 V power source in a circuit with 100 Ω resistance would create a current of 0.5 A. Therefore, any voltage above 50 V, such as 60 V, would blow the fuse. Choice (D) would lead to a blown fuse as well, but the question asks for the smallest the voltage could have been.

Semiconductors

The term **semiconductor** refers to an element that has four electrons in its valence shell. Since the bonds between these four electrons and the nucleus are somewhat strong, these elements are neither good conductors nor good insulators. Elements that are widely recognized as semiconductors are silicon and germanium.

In their pure form, silicon and germanium are not particularly useful. However, when impurities are added to their crystalline structure in a process called **doping**, a whole new world of possibilities springs forth. The crystalline structure of pure silicon is very stable. The four valence electrons in each silicon atom bond with all the valence electrons in the atoms around it, so no free electrons exist to allow current flow. This can be changed by "doping" the silicon's crystal structure with phosphorus, arsenic, or antimony. Since these elements all have five electrons in their valence shell, they will bond themselves to the other silicon atoms, but leave one free electron that is able to migrate throughout the crystal. This changes the silicon crystal into an **N-type material**. This new material is still electrically neutral, but is able to conduct electricity due to the presence of free electrons.

If silicon is doped with elements that have three electrons in their valence shells (such as boron or indium), the result is a "hole" being left where an electron would normally reside in the crystalline structure. This creates repeating regions with an overall positive charge. As the free electrons move into these spaces, they create "holes" in the spots they previously occupied. With such doping, the silicon is now known as a **P-type material**.

Diodes

P-type and N-type materials can do little by themselves; however, when the two are joined, a **diode** is formed. A diode is an electrical one-way valve that results from interactions at the junction of P-type and N-type material. Current can pass easily in one direction, but is blocked in the opposite direction.

Since current can only flow in one direction, the orientation of the battery in a circuit is important. If current is able to flow, this is called **forward-biased**. In order for the circuit to be forward-biased, the P-type material (also known as the **anode**) should be connected to the battery's positive terminal and the N-type material (also known as the **cathode**) should be connected to the battery's negative terminal.

The excess of electrons at the battery's negative terminal repels the free electrons in the N-type material and sends them toward the junction. The electrons in the N-type material then cross over the junction to fill the "holes" in the P-type material, and current flows through the diode. Reversing the diode's connection to the battery creates a new set of conditions. The N-type material's free electrons move away from the junction, since they are attracted to the opposite charge on the battery's positive terminal. Since the electrons have moved away from the junction, no electron transfer takes place there and current flow stops. The diode is now in a **reverse-biased** condition.

The most common application for diodes is **rectification**. Rectification is the conversion of AC into DC. Since a diode only allows current to flow in one direction, it will only pass either the upper or lower half of the AC waves. This is known as half-wave rectification. Four diodes can be connected in a diamond configuration to create a full-wave rectifier, which is the foundation for most DC power supplies. Diodes are also used for rectification in automotive alternators, where generated AC is converted into DC to be used in the vehicle's electrical system.

Half Wave Rectification (Single Diode)

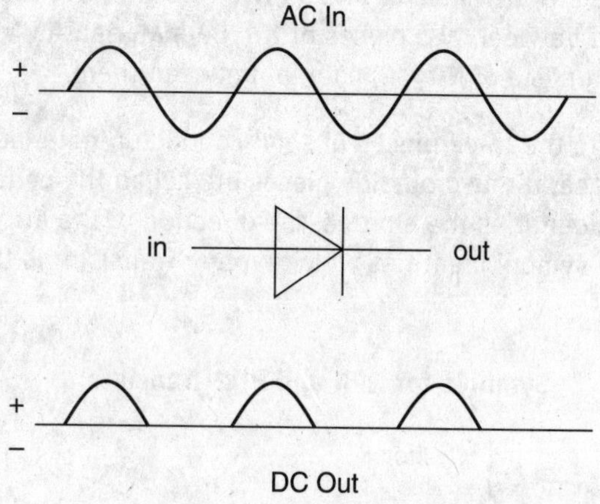

Full Wave Rectification (4 Diodes in Diamond Formation)

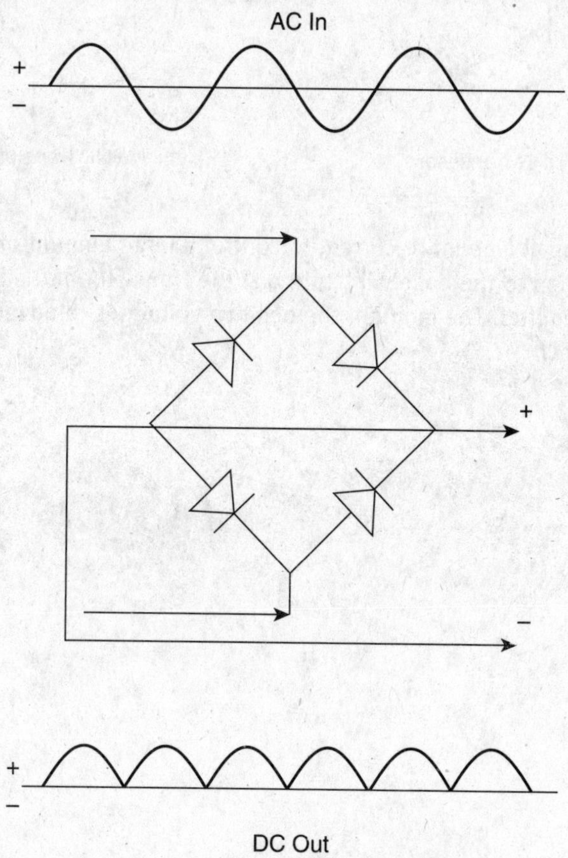

Transistors

The **transistor** is a component that has revolutionized the construction of electronic devices such as computers, calculators and radios. A transistor is versatile, as it can be used as an electrical switch, an amplifier, or a current regulator. Since there are no moving parts in a transistor and it is made from solid silicon, it is known as a **solid-state device**, which makes it very reliable. It is also very compact in its design, as thousands of transistors can fit into a single integrated circuit. An integrated circuit is one that contains all necessary circuit components in one single silicon chip. Transistors are the bedrock of logic operations that a computer uses to process information.

There are two types of transistors: an **NPN transistor** and a **PNP transistor**. An NPN transistor is made up of a thin piece of P-type material sandwiched between two pieces of N-type material. A PNP transistor is the opposite: two pieces of P-type material that have a piece of N-type material between them.

Inside the transistor, wires connect to the three pieces of semi-conductor material. Each piece is given a name. The middle piece is always called the **base**; the two outside pieces are called the **collector** and the **emitter**. The symbol for the transistor has an arrow that identifies the emitter. The direction of the arrow tells us what type of transistor it is. The key to remembering which symbol identifies which type of transistor is that the arrow always points in the direction of the N-type material.

Symbols for NPN and PNP Transistors

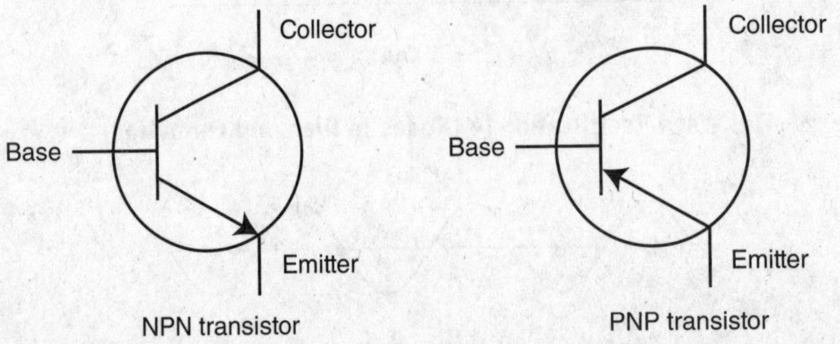

The transistor works by using a small amount of current to control a large amount of current. In the case of an NPN transistor, a positive voltage applied to the base will "turn on" the transistor and allow a relatively large current to flow from the collector to the emitter. The moment the positive voltage is removed from the base, the transistor turns off and stops the current flow.

Current Flow in an NPN Transistor

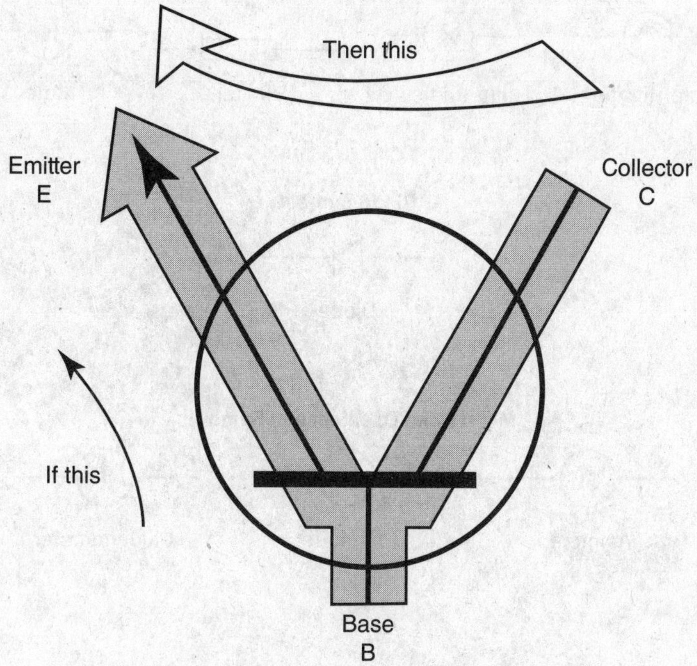

PNP transistors work opposite to NPN transistors. A PNP transistor requires a negative voltage at the base to turn it on, and the current then flows from the emitter to the collector.

Here are additional electronic component symbols that you may come across when you take the ASVAB. Some of them, including transformers, inductors, motors, and generators, will be discussed in the next section.

Wires and Connection Symbols

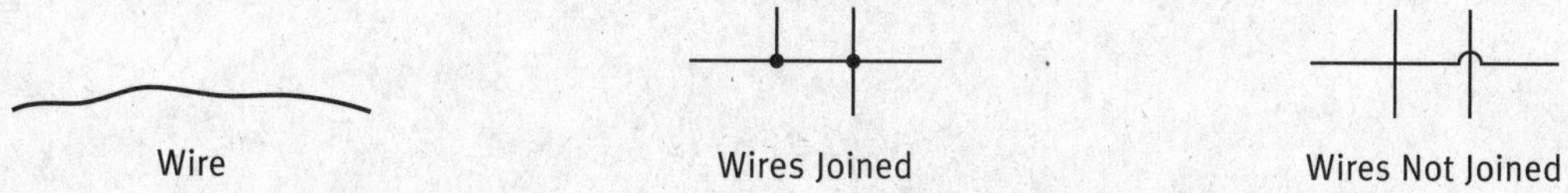

Power Supply Symbols

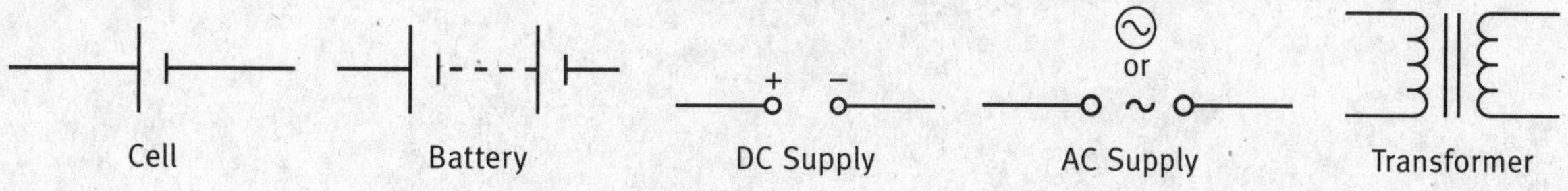

Switch Symbols

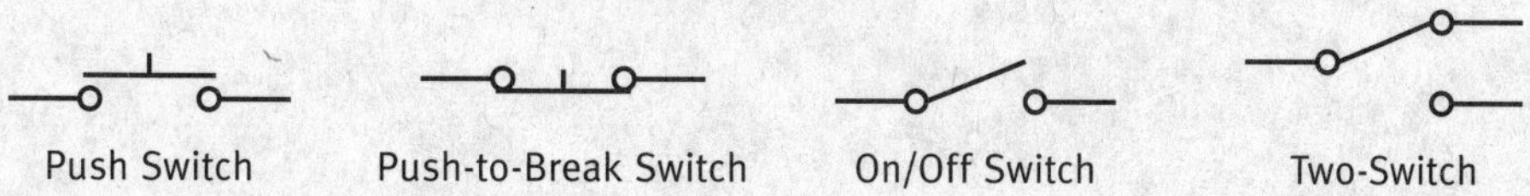

Output Device Symbols

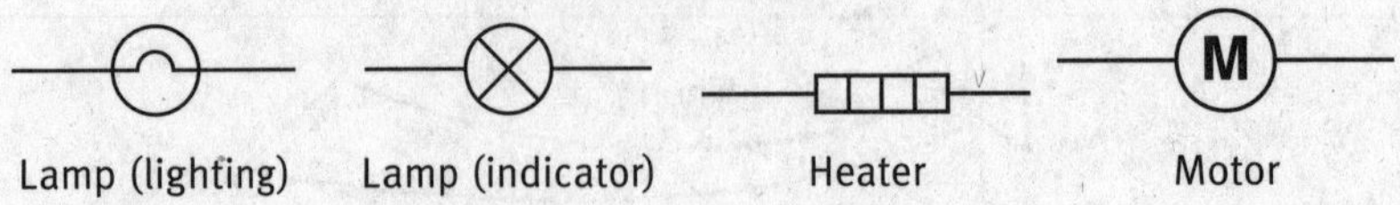

Diode Symbol

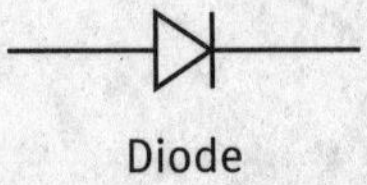

Meters and Oscilloscope Symbols

Question	Analysis
Which of the following electrical components can be used to convert the given input into the output shown?	**Step 1:** The question asks what circuit component is capable of receiving the input signal and producing the output shown.
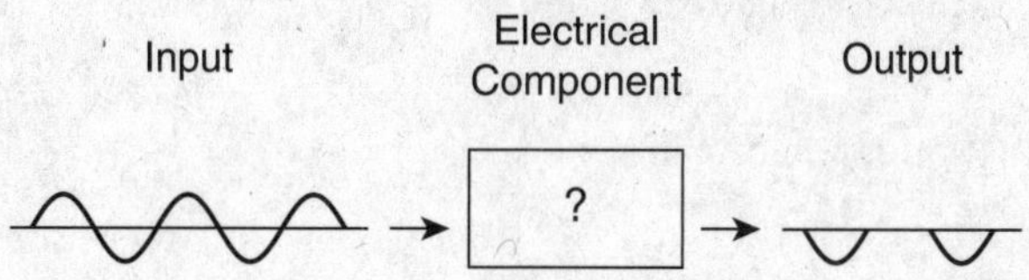	**Step 2:** The image shows AC as the input being converted into DC as the output.
	Step 3: A diode outputs only DC current. Any AC current that is passed through a diode will be converted to DC. Thus, only the positive or negative values of the AC signal will be output. None of the other electrical components listed converts AC into DC.
(A) capacitor (B) fuse (C) resistor (D) diode	**Step 4:** Choose option **(D)**.

Now try a question on your own.

A capacitor is used to store charge. When a given capacitor is "charged up" and then discharged, it will output which of the following:

(A) direct current
(B) alternating current
(C) capacitance
(D) reactance

Explanation

Answer choice **(A)** is correct. A capacitor stores charge when a DC signal is applied across it. When the capacitor is discharged, it can only output DC.

ELECTRICITY AND MAGNETISM

LEARNING OBJECTIVES

In this section, you will learn to:

- recognize the basic components in a circuit that function by using the magnetic properties of electricity
- explain the basic function of each of the electrical components in this section and their use/benefit

Electricity and **magnetism** are closely linked. Moving charges produce **magnetic fields**, and the moving electrons in a circuit are no exception. From circuit breakers to transformers, from motors to generators, from microphones to door bells, the interplay between electricity and magnetism has proven useful in a range of applications.

A current-carrying wire generates its own magnetic field. The strength of the field depends on the amount of current flowing in the wires. When the wires are wrapped around a ferrous material and a current is passed through the wires, the resulting magnetic field magnetizes the iron core, producing an electromagnet. This induced magnetism is the basic principle behind many of the aforementioned electrical devices.

Inductors

The magnetic field formed by a current-carrying wire can be made stronger by winding the wire into a coil. If the coil is wound onto a ferrous (iron) core, the magnetic field around the coil becomes even stronger, since magnetic lines of force travel more easily through iron than through air. When current first flows through a coil, the magnetic field builds relatively slowly. This is because the expanding magnetic field generates a voltage in the coil that opposes the original current flow, known as **counter-electromotive force**, or **counter-emf**. When the current is cut off, the magnetic field then collapses, and this collapsing magnetic field generates a voltage in the coil that keeps the current flowing. This resistance to the change in current flow in a circuit is known as self-induction and is a property exhibited by electrical components known as **inductors**. Inductors resist change in current flow. If current is increasing, the inductor opposes the increase by generating a voltage that moves against the applied current. If current decreases, the inductor uses the magnetic energy in the coil to oppose the decrease and to keep the current flowing.

Inductor Symbol

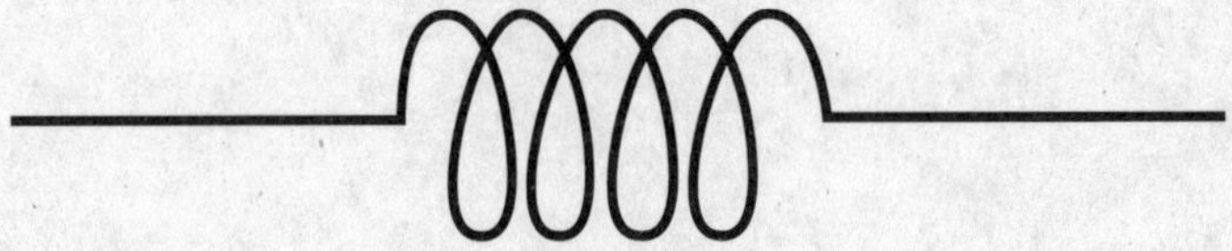

Induction is measured in a unit known as **henries**, and the symbol used to represent induction is ***L***. Inductors work exactly opposite to capacitors, in the sense that they allow DC to pass easily, but resist the flow of AC. This resistance to current flow is known as inductive reactance and is measured in ohms. It increases with increasing frequency of the AC signal.

Transformers

A transformer is used to increase or decrease the voltage in a circuit. A transformer uses the properties of an inductor to accomplish this. Alternating current flowing in wires wrapped around an iron core magnetizes the core, and in turn produces a changing magnetic field in the core. This changing magnetic field generates a voltage in a neighboring coil of wire. Depending on the number of turns of the wire in the primary versus the secondary coil, and the proximity of the coils, a smaller or larger voltage can be induced in the secondary coil. The primary coil is the one that is connected to the source. The secondary coil is the one in which an electric current is induced. A larger number of secondary coils means a larger voltage. The closer the secondary coils are to the primary coils, the more efficient the transformer is in producing a voltage in the secondary coils.

Transformers are especially useful to transmit electricity from power plants to residences and businesses. It is more energy efficient to transmit low current, high voltage electricity. However, by the time it arrives at homes, it must be reduced or "stepped down" to the standard 120–240 volts that most appliances in our homes use.

Basic Electrical Motors and Generators

A current-carrying wire positioned in a permanent magnetic field will experience a force (push or pull). This force can be harnessed to produce useful mechanical energy. This is the principle by which a simple motor works. The mechanical energy can be used to move the tires on a car, spin the blades of a fan, or lift heavy objects, just to name a few examples. The generated force on the current-carrying wire is directly proportional to the amount of current flowing, the length of the wire, and the strength of the magnetic field. Increasing any of these will increase the force on the wire and thus the capacity to do work. Motors today are more complex than this, and many use electromagnets in place of the current-carrying wire alone, but the underlying principle remains the same. A generator, on the other hand, is simply a motor in reverse. Moving a wire in a permanent magnetic field induces a current in the wire if it is connected in a complete circuit. Thus, a generator takes mechanical energy and converts it into electrical energy.

Here's how an expert test taker would approach a question about the structure of electrical systems.

Question	Analysis
Given the following circuit, with a switch that has been closed for a long time, what will happen to the lamp immediately after the switch is opened?	**Step 1:** The question asks what will happen to the lamp in the circuit when the switch is opened. Will it stay on or turn off? Burn less brightly or more brightly?
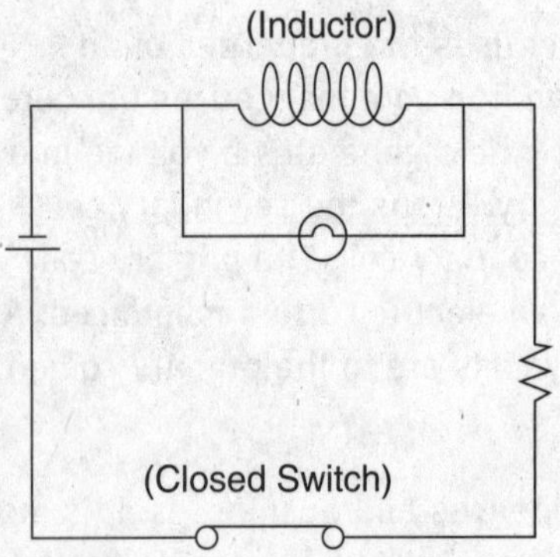	**Step 2:** Normally, in a simple series circuit, when the switch is opened, the lamp will turn off immediately. However, the presence of an inductor in the circuit must be accounted for. An inductor allows DC to pass easily with little resistance, and so most of the current will pass through the inductor and less through the lamp. Thus initially, the lamp will not be glowing very brightly. Once the switch is opened, the current generated by the voltage source will stop, but since the inductor resists this change, it will briefly produce a voltage that keeps current flowing in the circuit.
	Step 3: Therefore, immediately after the switch is opened, the lamp will still stay lit briefly and will burn more brightly, since all the voltage generated by the inductor will be dropped across it.
(A) The lamp will stay lit for a long time. (B) The lamp will burn more dimly. (C) The lamp will immediately turn off. (D) The lamp will briefly burn more brightly and then turn off.	**Step 4:** Choose option **(D)**.

Now try a question on your own.

Which of the following circuit components, used when electricity travels from power plants to residences, can change the value of the input voltage, in a process known as "stepping-up" or "stepping-down"?

(A) generator
(B) motor
(C) transformer
(D) inductor

Explanation

Answer choice **(C)** is correct. A transformer can take low current, high voltage electricity from a power plant and reduce the voltage, or step it down, before it enters a home. In general, a transformer can increase or decrease the source voltage depending on the application.

ELECTRONICS INFORMATION PRACTICE SET 1

Select the best answer for each question. This practice set contains 15 questions, which is the number of Electronics Information questions you will see if you take the CAT-ASVAB.

1. How much current will flow in a circuit that has 60 mV applied to a 15 KΩ resistance?

(A) 0.004 mA
(B) 0.9 A
(C) 4.0 A
(D) 900 A

2. What voltage is required for 30 A to flow through a 60 kΩ resistance?

(A) 1,800 V
(B) 1.8 kV
(C) 18 kV
(D) 1.8 MV

3. What is the current in the 20-ohm resistor?

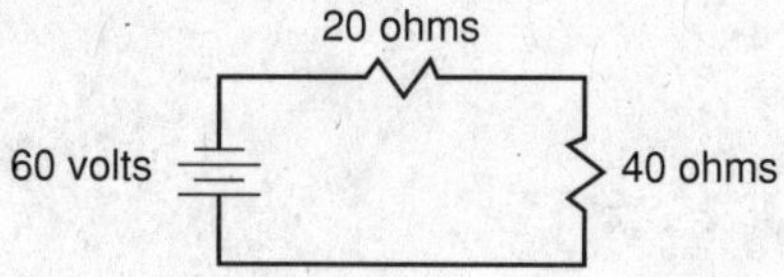

(A) $\frac{1}{2}$ amp
(B) 1 amp
(C) 30 amps
(D) 100 amps

4. Given the image below, which shell is the valence shell in a copper atom?

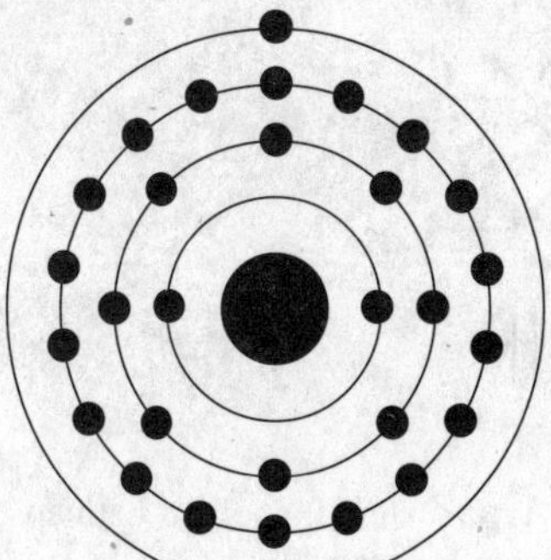

Copper atom (29 electrons)

(A) the first
(B) the second
(C) the third
(D) the fourth

5. Which is the collector in the transistor symbol below?

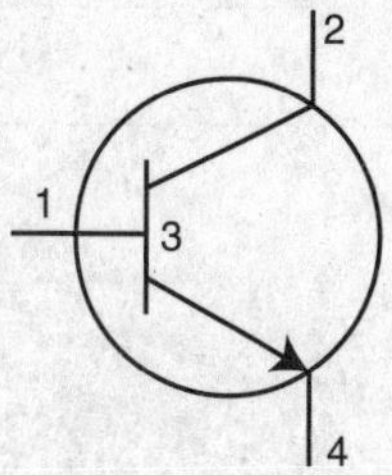

(A) 1
(B) 2
(C) 3
(D) 4

6. Current can be expressed in

(A) kilohms
(B) kilovolts
(C) milliamperes
(D) millivolts

7. A circuit with flowing current in it can be assumed to be
   (A) open
   (B) closed
   (C) short
   (D) dead

8. Which of the following symbols represents a polarized capacitor?
   (A) 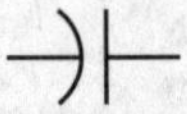
   (B) 
   (C) 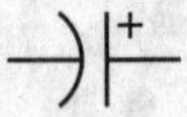

   (D) 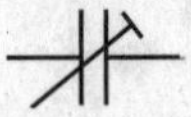

9. When a P-type material is joined with an N-type material, what is created?
   (A) a resistor
   (B) a diode
   (C) an inductor
   (D) a capacitor

10. Which switch, if flipped, would not close the circuit in the diagram shown?

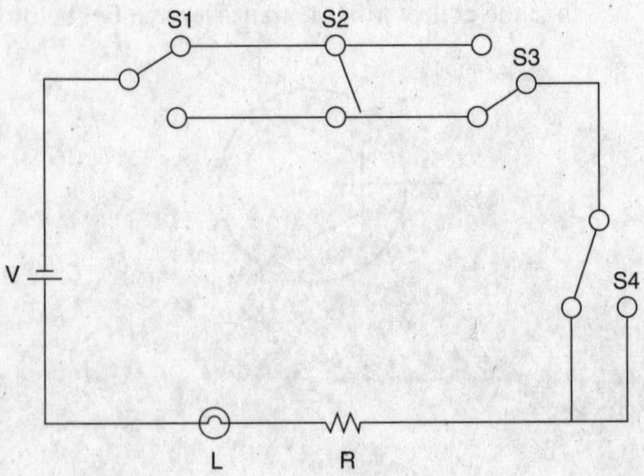

   (A) S1
   (B) S2
   (C) S3
   (D) S4

11. Given a circuit with a single resistor, adding two more resistors to the circuit, both with the same resistance value as the original resistor, one in parallel with the original resistor, and one in series with them both, will cause the overall resistance to
   (A) definitely increase
   (B) definitely decrease
   (C) stay the same
   (D) either increase or decrease depending on wiring order

12. Which of the following would not increase the rate of direct current flow, *I*, through a conductor?
   (A) decreasing effective resistance via a variable resistor
   (B) increasing voltage
   (C) adding an inductor to the circuit (in series)
   (D) adding a resistor to the circuit (in parallel)

13. A circuit includes three 2 Ω resistors wired in parallel. What is their effective resistance?
   (A) $\frac{1}{2}\ \Omega$
   (B) $\frac{2}{3}\ \Omega$
   (C) $\frac{3}{2}\ \Omega$
   (D) $6\ \Omega$

14. Why is copper favored over other conducting materials like aluminum and silver?
   (A) Copper has the lowest resistance.
   (B) Copper has the highest resistance.
   (C) Copper is inexpensive.
   (D) Copper has low resistance relative to its cost.

15. A 6 V battery is used to provide power to a flashlight whose 10 W light bulb is the only load in the circuit. What is the amount of current flow through the flashlight?
   (A) 0.6 A
   (B) 1.67 A
   (C) 6 A
   (D) 60 A

ELECTRONICS INFORMATION PRACTICE SET 2

Select the best answer for each question. This practice set contains 15 questions, which is the number of Electronics Information questions you will see if you take the CAT-ASVAB.

1. A 2.5 V battery is used to power a cordless phone. An ammeter shows the current flow is 30 A through the phone when it is on. How much power does the phone use?

    (A) 0.067 W
    (B) 12 W
    (C) 32.5 W
    (D) 75 W

2. A shop vac is plugged into a 120 V generator. Given a total current flow of 15 A, what must the total resistance of the vacuum cleaner be?

    (A) 8 Ω
    (B) 15 Ω
    (C) 120 Ω
    (D) 1,800 Ω

3. Two identical resistors are wired in parallel in a circuit. The voltage drop across these resistors is 18 volts, and the total current in the circuit is 6 amps. What is the resistance of either resistor?

    (A) 1.5 Ω
    (B) 3 Ω
    (C) 6 Ω
    (D) 9 Ω

4. Which of the following is one of the reasons neighborhoods are connected to the power grid via electrical substations instead of being wired into long-distance power lines directly?

    (A) The substations use special components to connect aluminum power lines to copper household wiring.
    (B) The substation transforms direct current to alternating current for household use.
    (C) The substation transforms alternating current to direct current for household use.
    (D) The substation transforms the efficient high-voltage power coming off the lines to a safer, lower value for household use.

5. A fixed resistor has a color band pattern of black, orange, orange. What is its resistance?

    (A) 33 ohms
    (B) 3,000 ohms
    (C) 30,000 ohms
    (D) 33,000 ohms

6. Given the following DC circuit, which has been open for some time, what will happen after the switch is closed?

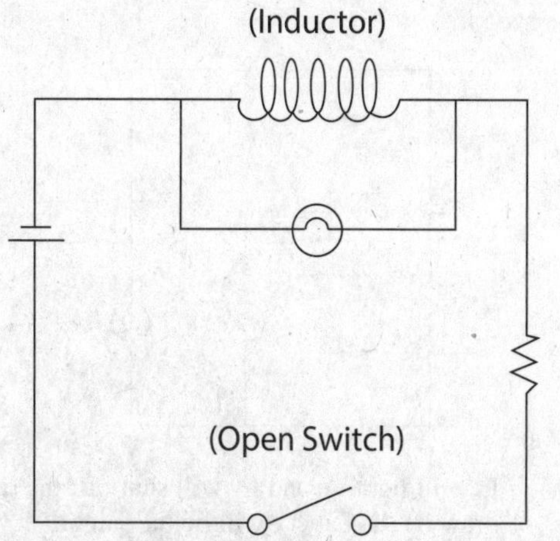

    (A) The total current flow will be smaller when the lamp first lights, then increase dramatically.
    (B) The lamp will light up brightly and then burn out.
    (C) The countercurrent of the inductor will cause the closed switch to spontaneously open again.
    (D) The power draw on the battery will be higher at first, then lower as the resistance decreases.

7. A European traveler brings his hair dryer to the United States. The specifications on the small appliance say it draws 10 amps when on a standard household circuit, which is 220 volts in Europe. What will it draw if connected directly to a U.S. power source of 110 volts?

    (A) 5 amps
    (B) 10 amps
    (C) 15 amps
    (D) 20 amps

8. A portable air compressor converts electric energy to the potential energy of pressurized air. A mechanics student calculates the total energy of the pressurized air at about 800,000 J. If it took 400 seconds to fully pressurize the machine, what must be the electric power rating of the compressor?

    (A) 400 W
    (B) 2,000 W
    (C) 800,000 W
    (D) 320,000,000 W

9. Given a lamp wired in a simple AC household circuit, which of the following will cause the light to dim the most?
   (A) A capacitor wired in parallel with the lamp
   (B) An inductor wired in parallel with the lamp
   (C) A capacitor wired in series with the lamp
   (D) An inductor wired in series with the lamp

10. In the circuit diagram pictured, where the two lamps have identical loads, what will be the effect of closing the switch?

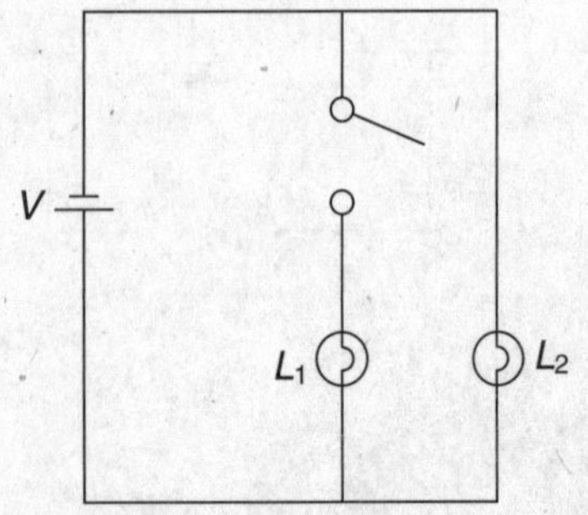

   (A) L_1 will light up and L_2 will shut off; the power draw on the battery will be the same.
   (B) L_1 will light up and L_2 will dim; the power draw on the battery will double.
   (C) L_1 will light up and L_2 will remain at the same brightness; the power draw on the battery will stay the same.
   (D) L_1 will light up and L_2 will remain at the same brightness; the power draw on the battery will double.

11. A simple parallel circuit with resistances of 2.5 Ω and and 5.0 Ω, respectively, will have
   (A) equal amounts of current through each resistor
   (B) twice as much current in the 2.5 Ω resistor
   (C) twice as much current in the 5.0 Ω resistor
   (D) 100 percent of the current in the 2.5 Ω resistor

12. Plugs are often differently shaped in different countries to prevent foreign electronics from drawing on an electric grid they weren't designed for. What would happen if a device designed for a 150-volt source were forced into a plug providing 240 volts?
   (A) The device would function normally, as current and resistance are constant within a device.
   (B) The device might be weaker due to lower current.
   (C) The device might burn out due to higher current.
   (D) The device would function normally but waste more energy.

13. Which one of the following circuit components will least affect the flow of an alternating current?

   (A) 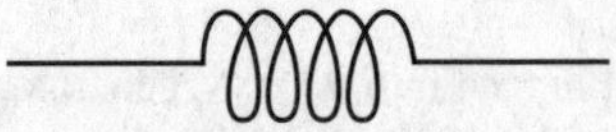

   (B) 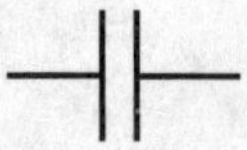

   (C) 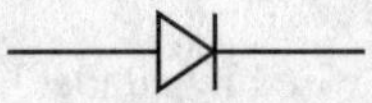

   (D) 

14. In the wiring diagram below, the fuse is rated for 0.4 A. The lamps and resistor have the resistances given, and the fuse has no resistance. What will happen when the switch is closed?

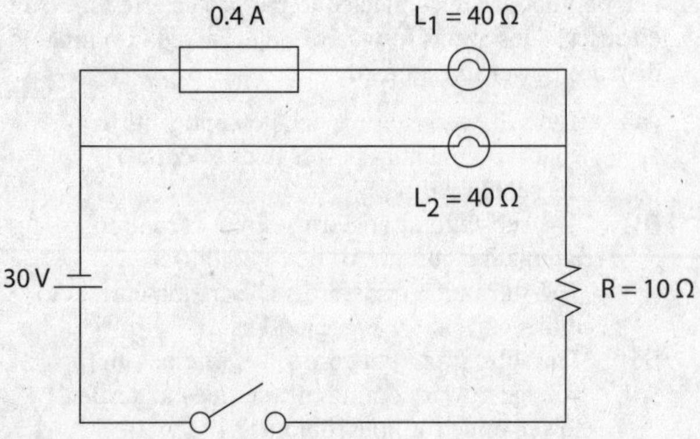

   (A) Both lamps will stay lit.
   (B) Both lamps will light briefly and go out.
   (C) L_1 only will stay lit.
   (D) L_2 only will stay lit.

15. A technician wants to measure the total current in the circuit described in the wiring diagram shown. If she attaches ammeters just after the 10 Ω and 20 Ω resistors, the total current will be equal to

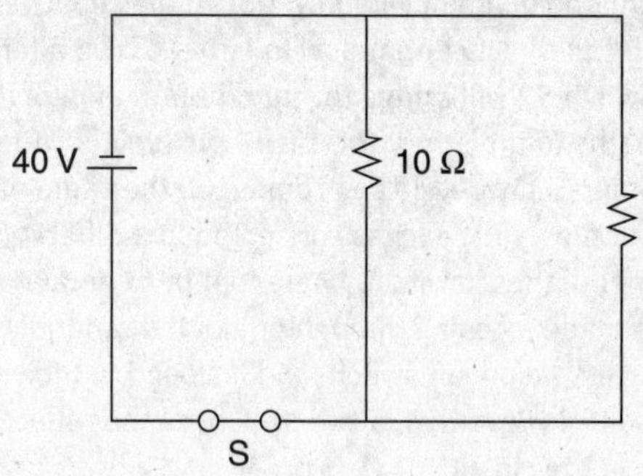

(A) the sum of the values on each ammeter
(B) the value on either ammeter, since they'll show the same value
(C) the difference in the values on the two ammeters
(D) the average of the values on the two ammeters

ELECTRONICS INFORMATION PRACTICE SET 1

Answers and Explanations

1. **A** To solve this problem, the first step is to convert 60 mV to V. You can quickly make this conversion by moving the decimal point three places to the left, which yields 0.060 V. Next, convert 15 kΩ to Ω by moving the decimal point three places to the right, which yields 15,000. Since current is the unknown, the formula needed to finish the calculation is $I = E \div R$. Dividing 0.060 by 15,000 gives 0.000004 amperes. This can be converted to 0.004 mA.

2. **D** Like the last problem, this one requires you to first convert the given measurements. Take the resistance and move the decimal. 60 kΩ becomes 60,000 Ω. Using our formula for calculating voltage, $E = I \times R$, we get 30 A × 60,000 = 1,800,000 V, or 1.8 MV.

3. **B** The resistors in a series circuit add up, so one 40-ohm resistor in series with a 20-ohm resistor is equal to a 60-ohm resistor. To find current flow through a circuit, divide the total circuit resistance (60 ohms) by the voltage across the circuit (60 volts) to get 1 amp.

4. **D** The valence shell is the outer shell of an atom, which in a copper atom is the fourth shell. A copper atom has 29 electrons, so it would fill the first three shells completely and have one electron left to begin a fourth shell.

5. **B** In the symbol, the number 2 indicates the collector.

6. **C** The ampere is the basic unit of current, so current can be expressed in milliamperes.

7. **B** A closed circuit has continuity and will allow current to flow in it.

8. **C** The choices represent four different capacitor symbols: a fixed capacitor (A), variable capacitor (B), polarized capacitor (C), and trimmer capacitor (D).

9. **B** When a P-type material is joined with an N-type material, a diode is created.

10. **D** Examine the diagram carefully. Notice that switches S1, S3, and S4 all can direct current flow to either of two alternative paths. Switch S2 will bridge the current from one of the top parallel paths to the other. The circuit pictured is currently open. The S1 and S3 two-switches connect to the same parallel paths, but presently, each is connected to a different branch. If either of these switches were to flip, a complete circuit would be made, passing through one or the other of these paths. S2 is a simple on/off switch, and if closed, it too would close the circuit. However, S4, a two-switch, cannot affect the current break in the circuit.

11. **A** This question has no numbers, which can make it tougher to grasp. It can be reasoned out if your understanding of parallel and series resistors is very strong. Otherwise, your best bet in a situation like this might be to just choose some simple numbers and do the math. The question has three identical resistors in a series-parallel circuit. Choose an easy number—three 1 Ω resistors would work just fine, and you can determine the correct answer choice by seeing if the effective resistance is greater than, less than, or equal to 1 Ω. The effective resistance of two 1 Ω resistors in parallel is $\frac{1}{2}$ Ω. Add to this an additional 1 Ω resistor in series, and the final effective resistance is $1\frac{1}{2}$ Ω, an increase. Time permitting, the calculation can be repeated with a different resistance value for the three resistors, just to ensure that the resistance still increases.

12. **C** Each answer choice would increase the rate of current flow, except one. Answer choices (A) and (D) would both decrease the overall resistance of the circuit, which, according to Ohm's law, would increase current. Answer choice (B) would also increase current flow if resistance was not changed. However, for a direct current, an inductor in the circuit has little effect. It will resist changes in current flow but not tend to either decrease or increase it.

13. **B** The effective resistance of several resistors wired in parallel can be determined by taking the reciprocal of the sum of the reciprocals of each individual resistance value. Since each resistor has a resistance of 2 Ω, you get $\frac{1}{2}+\frac{1}{2}+\frac{1}{2}=\frac{3}{2}$. Since the reciprocal of this sum is needed in parallel resistor calculations, it's best to stick with fractions at least until the end. The reciprocal of $\frac{3}{2}$ is simply the fraction created by switching the numerator and denominator values, that is, $\frac{2}{3}$ Ω.

14. **D** Copper isn't the best conductor. Silver has a lower resistance and is thus better at conducting electricity than copper is. But though silver is a somewhat better conductor, it is far, far more expensive. Copper is a very good (but not the best) conductor, while also being very affordable. Aluminum is inexpensive but is a worse conductor than copper. Copper is also safer than aluminum, which, due to some of its heating and expanding properties, is more likely to cause electrical fires in household wiring setups.

15. **B** Rearranging the power formula, $I = P \div V = 10 \div 6 = 1.67$ A (rounded to the nearest hundredth).

ELECTRONICS INFORMATION PRACTICE SET 2

Answers and Explanations

1. **D** The relevant formula here is $P = I \times V$, and both the current and voltage are given, so $P = 30 \times 2.5 = 75$ W.

2. **A** The relevant formula here, $V = I \times R$, needs to be rearranged in order to calculate resistance. $R = V \div I = 120 \div 15 = 8\ \Omega$.

3. **C** These two resistors are in parallel and can be treated as a unit. Since we're given the voltage drop across them both (the voltage drop either in total or across either resistor is the same when they're wired in parallel), as well as the total current through them both, we can calculate the total effective resistance of the two resistors. $R = V \div I = 18 \div 6 = 3\ \Omega$. The question asks for the resistance of either resistor, though, so there's one more step. The formula for the total effective resistance of these two resistors in parallel is $\frac{1}{R_{tot}} = \frac{1}{R_1} + \frac{1}{R_2}$. However, since we know both resistors are the same, and we know the total resistance, the formula can be written as $\frac{1}{3} = \frac{1}{R} + \frac{1}{R}$. R is the resistance of a single resistor and is exactly what we're trying to figure out. Multiply both sides of the equation by R to get $\frac{R}{3} = 1+1$, or $\frac{R}{3} = 2$. Then multiply both sides by 3 to get $R = 6\ \Omega$.

Alternatively, you might recall that in the special case of multiple identical resistors in parallel, you can use the shortcut of dividing the individual resistance by the total number of resistors. In order to have a total effective resistance of 3 Ω then, there must be two 6 Ω resistors ($6 \div 2 = 3$).

4. **D** Household wiring and long-distance high-voltage power lines are connected via step-down transformers in the local substations. The other statements are not true. The current is AC in both power lines and household wiring, and power lines do not always use aluminum wiring.

5. **B** The first and second color determine digits, and the last one determines how many zeroes to add. Black is zero, then orange (in either of the first two positions) is three, and then orange (in the final position) is three zeroes. A zero at the beginning of the number has no effect, as 03000 is just written as 3,000. Answer choice (C) would have been correct if the pattern were orange, black, and then orange.

6. **A** Inductors resist changes in current and for this reason act as resistors in alternating current circuits. However, this property is also important in a direct current circuit as seen here. An open circuit has no current, but the moment the switch is closed, current will begin to flow. The inductor, resisting the sudden change in voltage, will resist current flow at first. The lamp, since it is parallel to the resisting inductor, will light up immediately without difficulty. However, the inductor will soon "get used" to the current flow, and the resistance of its path will drop significantly. As more current flows through the easy inductor path, the total current (and therefore energy use) of the circuit will increase significantly as the overall resistance drops. (D) is exactly the opposite of what will happen. Something similar to (B) could happen only if the voltage source were limited in the amount of current it could provide and the small fraction of current passing through the lamp were not enough to light the lamp once the inductor dropped its resistance, or if the total current draw caused a fuse or breaker to cut off the current. However, in neither of these cases would the lamp itself "burn out."

7. **A** The hair dryer has the same resistance no matter what circuit it's attached to. You can calculate the resistance from the specifications given and then use that same resistance to determine current under a lower voltage. $R = V \div I = 220 \div 10 = 22$ ohms. Plugging this 22-ohm hair dryer into an American socket gives you the current as $I = V \div R = 110 \div 22 = 5$ amps, which is answer choice (A).

Another way to answer this question is by reasoning it out from the proportionality given by Ohm's law: $V = I \times R$. Given a constant resistance, voltage and current are constant, so if the voltage is halved from 220 to 110 volts, the current must also be halved from 10 to 5 amps.

8. **B** This question about power takes you back to the more general definition of power, as the rate of energy flow over time. You do not need the electrical power formula, $P = I \times V$. Instead, divide the total energy, in joules, by the total time, in seconds, to determine the total power, in watts. The information in the question is already given in the correct units, so calculate $800{,}000 \div 400 = 2{,}000$ W.

9. **D** You can dismiss (A) and (B), since for loads in parallel, the voltage drop is the same across all paths; the brightness of the light would not change if the lamp were wired in parallel with either a capacitor or an inductor. Loads in series, on the other hand, affect the overall resistance of the circuit and therefore the total power draw on the battery. Recall that capacitors and inductors can either let current flow freely or act as resistors, depending on the type of current. In an AC

circuit, a capacitor lets current flow freely while an inductor resists current flow. An inductor placed in series with the lamp will therefore act as a resistor over which a voltage drop occurs, and the lamp will be correspondingly dimmer than if the voltage drop of the circuit occurred entirely over the lamp itself.

10. **D** Before the switch is closed, there is only one path through the circuit, through L_2, so it is lit while the other lamp is dark. Closing the switch will provide a second possible path, and as a result, both lamps will now be lit. However, it's also necessary to determine what will happen to the brightness of L_2 itself. The general rule is that lights in parallel will not change their brightness as more paths are added, because the voltage drop across each path remains the same and therefore the amount of current passing through each lamp also stays the same. However, the fact that there are now two lamps brightly lit instead of one does mean that the battery will drain twice as fast because the total power draw of the circuit will be doubled when the switch closes.

11. **B** $I = V \div R$, and since the voltage across parallel loads is always equal, any difference in current will be a result of the different resistances. You can see that current and resistance are inversely proportional, meaning any change to one will have the opposite effect on the other. But this question may be a little easier to figure out by just making up a value for voltage, say 5 V. Given this value, then, the currents through each load will be $I = 5 \div 5 = 1$ A, and $I = 5 \div 2.5 = 2$ A. The 2.5 Ω resistor has twice as much current as does the 5.0 Ω resistor.

12. **C** Since the device's resistance is constant, the formula $V = I \times R$ shows that if voltage increases, current increases proportionally. To put it another way, the greater the voltage, the greater the "pressure" pushing those electrons forward. This has the effect of producing a higher current, assuming the resistance is the same. Only answer choice (C) mentions the higher current that must result from this situation. And yes, you should definitely only plug your device into a socket that fits so you don't burn it out or start a fire.

13. **B** This is the general symbol for a capacitor, which acts as a resistor to direct current but does little to disrupt alternating current. It's the opposite of an inductor, pictured in answer choice (A), which acts as a resistor to alternating current. Choice (C) is the diode symbol, which allows current flow in only one direction. Since alternating current constantly changes direction, half of the current flow would be disrupted by the process known as rectification. Finally, choice (D) displays an open switch, which will not allow either AC or DC power to flow.

14. **D** The total current in the circuit can be determined from the voltage source and total resistance of the circuit. This is a tricky series-parallel circuit that has to be taken in two steps. The total resistance of the parallel loads L_1 and L_2 can be calculated without the full parallel-resistors formula by using the rule that equal resistors in parallel have an effective resistance equal to their individual resistance divided by the number of parallel loads. The effective resistance of L_{12} then is $40 \div 2 = 20$ Ω. This is in series with R, so the total resistance of the whole circuit is $20 + 10 = 30$ Ω. Now, calculate the current draw of the circuit as $I = V \div R = 30 \div 30 = 1$ A. Once again, the fact that L_1 and L_2 have an equal amount of resistance makes things easier. Since the two paths are equivalent, each will take half the current, meaning 0.5 A passes through either one. This is greater than the maximum current allowed by the fuse, so it will burn out, cutting the current flow through L_1 and leaving only L_2 lit.

15. **A** Current will separate when reaching parallel paths. Since the two resistors are different, they'll draw different amounts of current, so the statement in (B) is false. However, regardless of whether they're the same or different, it's the sum of the two currents that will give you the total current. This technician would have been better off placing her ammeter just after the battery or perhaps near the switch, however, as doing so would have allowed her to get the total current with a single measurement.

Review and Reflect

Look back over your work on these practice sets. If you struggled with some of the questions, ask yourself why.

Did you misunderstand a formula or simply make errors in calculations? If it's the latter, be sure to slow down and double-check your math when you do Electronics Information questions. If you need to review the formulas discussed in this chapter, consider making flashcards to help you remember them.

Did you have trouble remembering some of the technical terms introduced in this chapter? If so, consider making flashcards to help you memorize the meaning of those terms.

Want more practice with Electronics Information? Try more questions in the Qbank online.

QBANK

kaptest.com/login

CHAPTER 11

AUTOMOTIVE INFORMATION

KNOW WHAT TO EXPECT

On the CAT-ASVAB, you are given 7 minutes to answer 10 Auto Information (AI) questions. On the paper-and-pencil test, the Auto Information and Shop Information subtests are combined into one, with 11 minutes to answer 25 questions about both topics.

The required pace can be daunting. However, we're confident that by the time you finish this section, you will have a solid understanding of all of the automotive systems and will be able to use Kaplan's proven methods to conquer AI questions. Take your time with this chapter, and remember that you can't possibly know everything after one read-through. Review this chapter multiple times if a high AI score is important to your career goals.

Also remember that there are many, many minute details involved in the makeup of automotive systems. The test maker cannot possibly ask about all of them. Your best approach to studying is to develop a thorough understanding of automotive systems, their major parts, and how those systems work together. This high-level understanding will likely equip you to make solid inferences about details when you need to.

WHAT'S IN THIS CHAPTER

The modern automobile is a technological marvel. Thousands of parts are assembled to form the means that many of us use for our daily transportation. Space-age materials and advanced computer control systems are used extensively in its construction, but in spite of all this, the basic systems within the automobile still function the same as they always have.

Despite its size, any automobile is still only made up of many smaller **parts.** Parts are the smallest pieces that are used in the construction of the automobile. Parts include such items as bolts, screws, spark plugs, and belts. The parts of the vehicle cannot be disassembled any further, so they form the foundation for the construction of the vehicle. **Systems** are combinations of parts that work together to perform a particular function.

This chapter will be organized around four major types of systems:

- **Engine systems** generate power to drive the vehicle's wheels and various accessories. The foundation of engine systems is engine theory, which defines the mechanics and components involved in making the engine run. Engine systems also include the cooling system, which acts to remove excess heat from the engine. Finally, engine systems include the lubrication system, which circulates motor oil through the engine to reduce friction and make the engine run smoothly.

- **Combustion systems** comprise the fuel system, which ensures that correct amounts of air and fuel are available for efficient combustion in the engine; the ignition system, which generates and times the spark that initiates combustion; and the exhaust system, which forms a "pipeline" for waste gases to be removed from the engine and then dissipated to the open atmosphere.
- **Electrical and control systems** are becoming more and more significant in modern vehicles. These include the electrical system—which includes starting, charging, lighting, and accessory systems—as well as the computer system, which controls all aspects of vehicle operation, including engine, drivetrain, brakes, and suspension.
- **Chassis systems** are made up of the drivetrain system, which transmits power from the engine to the vehicle's drive wheels, the suspension and steering systems, which control the vehicle's ride quality and handling, and the brake system, which stops the vehicle safely and predictably.

This chapter will conclude with a set of practice questions.

ENGINE SYSTEMS

LEARNING OBJECTIVES

In this section, you will learn to:

- identify the basic function and the major components of the internal combustion engine
- differentiate between the different types and configurations of internal combustion engines
- identify the basic function and the major components of the cooling system
- identify the basic function and the major components of the lubrication system

Basic Engine Theory

The type of engine used in automobiles is known as an **internal combustion engine**. Combustion is the rapid burning of an air-fuel mixture. Internal combustion means exactly that: fuel is burned *internally* and the resulting heat is used directly to power an engine.

Typical fuels used with internal combustion engines are gasoline, diesel fuel, propane, natural gas, and alcohol. Gasoline and diesel fuel are by far the most common.

Three things must be present before combustion can take place: air, fuel, and a heat source that can be used to ignite the air-fuel mixture. If any one of those three elements is missing, combustion stops and the engine will not run. Internal combustion engines convert the **chemical energy** in the fuel and air into **heat energy**, and this heat energy is then converted into **mechanical energy**. (See chapter 9: General Science and chapter 13: Mechanical Comprehension for more information about types of energy.)

Components

There are common components in all internal combustion engines. These components include:

- **Engine block**: forms the framework for the engine cylinders and reciprocating assembly.
- **Piston**: a cylindrically shaped object with a solid crown (top) that moves up and down in the engine's cylinders. Hot gases produced from the combustion of the air-fuel mixture push on the piston to do the actual work.
- **Cylinder**: forms a guide for the piston to move in; allows the piston to move up and down as the engine completes its cycle.
- **Piston rings**: seal the piston to the cylinder and prevent combustion gases from leaking past. Oil rings prevent oil from the engine crankcase from making its way into the combustion chamber.
- **Wrist pin**: connects the piston to the connecting rod, and forms a pivot point for the small end of the connecting rod to move on.
- **Connecting rod**: connects the piston/wrist pin assembly to the engine's crankshaft. The large end of the connecting rod attaches to the crankshaft on the connecting rod journal.
- **Crankshaft**: converts the linear (straight line) motion of the piston into rotary motion, which can then be used to power a vehicle or drive an accessory.
- **Cylinder head**: located above the piston, it houses the combustion chamber, the intake and exhaust valves, and the intake and exhaust ports.
- **Combustion chamber**: located in the cylinder head directly above the piston, it is where the actual combustion of the air-fuel mixture takes place.
- **Intake valve**: allows the air-fuel mixture to be drawn into the combustion chamber. When closed, it must seal the combustion chamber from the intake port.
- **Exhaust valve**: allows waste gases to be removed from the combustion chamber. When closed, it must seal the combustion chamber from the exhaust port.
- **Camshaft**: responsible for the opening and closing of the engine's intake and exhaust valves. The camshaft turns at one-half the speed of the engine's crankshaft.

Basic Engine Components

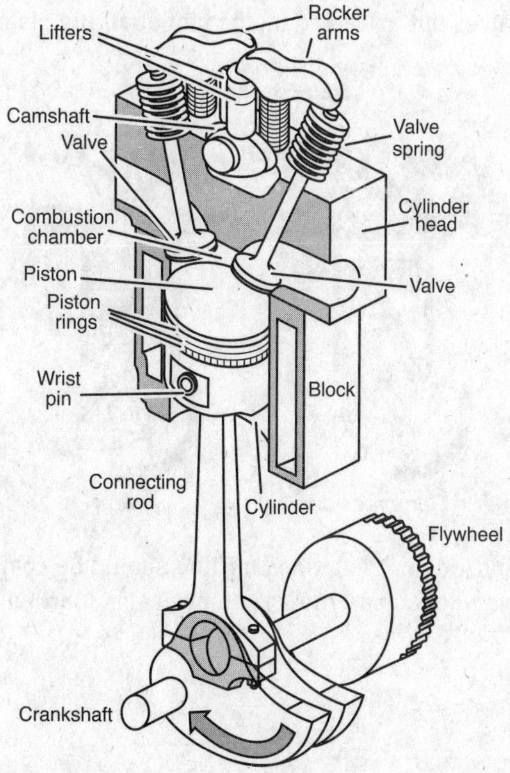

The conversion of linear (straight line) motion into rotary motion as achieved by the piston-connecting rod-crankshaft combination is very similar to the leg of a bicycle rider. As the rider's upper leg moves up and down (piston), her lower leg (connecting rod), the bicycle pedals, and the sprocket (crankshaft) convert that straight line motion into rotary motion that then can be used to drive the wheels of the bicycle.

Most internal combustion engines are built to utilize a **four-stroke cycle**. This means that it takes four strokes of the piston to complete one cycle of events. A stroke of the piston is defined as the piston movement from the top of its travel in the cylinder (**top dead center** or **TDC**) to the bottom of its travel (**bottom dead center** or **BDC**) and back again to TDC.

The four-stroke cycle begins with the **intake stroke**. With the piston at TDC, the intake valve begins to open. As the piston moves downward in the cylinder, it forms a low-pressure area (or vacuum). Because the intake valve is open to the atmosphere, air is sucked into the combustion chamber due to the vacuum in the cylinder during the intake stroke. As the air is traveling through the intake system, fuel is injected into the air stream before it enters the combustion chamber. This allows the cylinder to fill with a fresh mixture of air and fuel. Once the piston reaches BDC, the intake valve is almost closed again and the engine is ready to begin the second stroke of the cycle, the compression stroke.

During the **compression stroke**, the piston starts moving upward in the cylinder while both of the engine's valves are closed. With the combustion chamber then sealed, the continued upward motion of the piston toward TDC compresses the air-fuel mixture in the cylinder, as the air-fuel mixture is forced into a progressively smaller and smaller space. The air-fuel mixture in the combustion chamber becomes progressively hotter as the particles of fuel get closer together due to the compression, making the air-fuel mixture easier to ignite and ultimately increasing the power of the next stroke, the power stroke.

Just before the piston reaches TDC of the compression stroke, the spark plug fires and ignites the air-fuel mixture, starting a flame that then travels across the combustion chamber. This flame further heats the gases in the combustion chamber and the resulting rapid expansion of these gases pushes on the piston as it passes TDC and then continues downward on the third stroke of the cycle, the **power stroke.** During the power stroke, both the intake and exhaust valves remain closed to ensure that the pressure generated from the combustion of the air-fuel mixture is not blown out past the valves but extracted by the motion of the piston toward BDC.

Four-Stroke Cycle

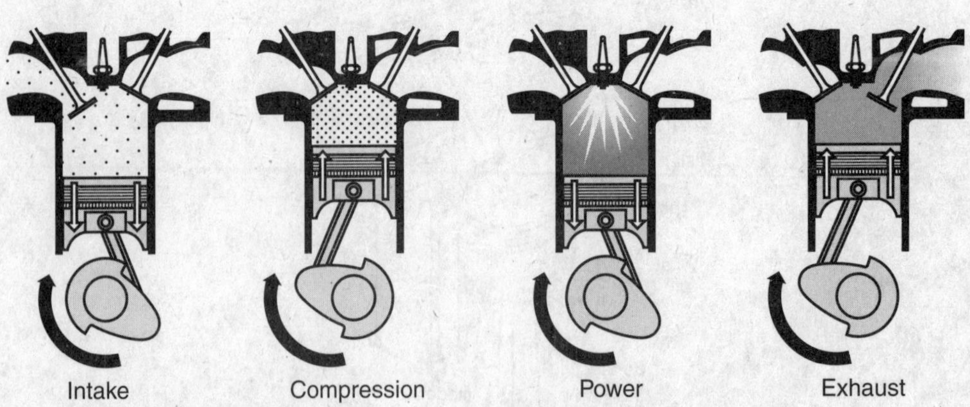

Before the piston reaches BDC, combustion of the air-fuel mixture should be completed. Just before the piston reaches BDC, the exhaust valve starts to open and the engine begins its fourth and final stroke, the **exhaust stroke.**

The gases in the combustion chamber are now spent and must be purged from the engine before the next cycle can begin. As the piston begins its upward movement, it pushes exhaust gases past the open exhaust valve and into the engine's exhaust system, where the exhaust gases are eventually sent out to the open atmosphere. The piston continues its travel toward TDC, and at that point will have completed one complete cycle of events. This cycle then starts over as the intake stroke begins again.

Cylinder Arrangement

Automotive engines can be built in a number of different configurations, and these can most easily be classified in terms of the engine's **cylinder arrangement**, which is the position of the various cylinders relative to each other. The most common numbers of cylinders used in automobile engines are four, six, and eight, but there are also designs that utilize three, five, ten, twelve, and even sixteen cylinders.

The simplest cylinder arrangement is known as the **inline** design, in which all of the engine's cylinders are lined up in a row. This is a practical design for four- and six-cylinder engines. Inline four-cylinder engines are very popular in front-wheel drive cars with transverse (sideways) mounted engines. Inline engines are most often found in small- to medium-sized vehicles. Another arrangement that has been utilized by engine designers is the **horizontally opposed** or **flat** design. This arrangement has all of the cylinders lying in a horizontal plane, with half of the cylinders facing away from the other half and the crankshaft located between them. Some refer to this design as a "boxer" engine because the pistons move back and forth like a boxer throwing punches. The one major advantage to this design is that the engine's center of gravity is much lower, so it is easier to build a more stable, better-handling vehicle.

The last and most popular design for six- and eight-cylinder engines is the V-type engine. These engines have one crankshaft connected to pistons on both sides of the V-shaped engine. The two rows of cylinders are located 60 to 90 degrees away from each other, and each row has one cylinder head, for a total of two heads for the engine. A primary advantage of this configuration is the inherent physical balance of the engine's moving components, resulting in a smooth-running engine. Another advantage is the decrease in height and length relative to the in-line engine. The V-type engine improves the aerodynamic flow outside the vehicle; because the engine is smaller, the front of the vehicle can be lower. If an eight-cylinder engine is needed, this configuration creates the power required while also making the engine more compact.

Camshaft Location

Another classification used to describe automotive engines is **camshaft location**. The camshaft is responsible for the opening and closing of the engine's valves, and is driven by the crankshaft through a **timing chain** or **timing belt**. Some modern engines have the camshaft located in the engine block and the intake and exhaust valves located in the cylinder head. Engines built in this manner would be known as **overhead valve** or **OHV** engines because the valves are arranged above the piston and the combustion chamber. In OHV engines, a lever called a rocker arm, which is actuated by the camshaft through a pushrod, operates each valve. The OHV arrangement can be used with either V or flat configuration engines.

Engine designers often locate the camshaft above the valves and are then able to eliminate the pushrods that are necessary for the OHV configuration. The valve operating mechanism itself becomes simpler and lighter. Less mass in the valve train means that higher engine speeds can be attained. This design is known as the **overhead cam** or **OHC**. If only one overhead camshaft is used to operate the intake and exhaust valves of a cylinder for an OHC engine, that arrangement is known as **single overhead cam** or SOHC. In a V-type engine with two cylinder heads, there would be two camshafts, with one installed above each cylinder head.

One final step to make even higher engine speeds possible is to go to a **double overhead cam** or **DOHC** arrangement. This puts two camshafts into each cylinder head, and makes it so one cam operates the exhaust valves in that head, and the other operates all the intake valves. With this arrangement, it is possible to eliminate the rocker arms that are often used in SOHC engines and have the camshafts operate the valves directly through a follower.

Multiple-Valve Cylinder Heads

Engines may also be classified according to the number of valves used for each cylinder. The least expensive and most common arrangement is to use a **two-valve cylinder head**, with one intake valve and one exhaust valve per cylinder. To improve airflow through the engine and, thus, engine performance, it is possible to design a multiple (more than two) valve cylinder head. Manufacturers have designed engines that use three valves per cylinder, but now it is much more common to use a **four-valve** cylinder head. This arrangement has two intake valves and two exhaust valves for each cylinder. The four-valve cylinder head design has several advantages over the two- or three-valve designs. The four-valve design allows higher engine operating speeds and more complete combustion of the air-fuel mixture, which helps improve engine power and efficiency.

Firing Order

The order that the cylinders fire in is known as the **firing order**. A common firing order for four-cylinder engines is 1-3-4-2. This means that once the first cylinder fires, the crankshaft will make one half turn, and then the third will fire and so on. Cylinders 1 and 3 will fire on the first turn of the crankshaft, and cylinders 4 and 2 will fire on the second turn of the crankshaft. In the case of a V-8 engine with a firing order of 1-8-4-3-6-5-7-2, cylinders 1, 8, 4, and 3 will fire on the first turn of the crankshaft, and 6, 5, 7, and 2 will fire during the second turn.

Many engines have their firing order cast directly into the intake manifold for easy reference. For those that don't, it is necessary to refer to the engine's service information.

Diesel Engines

Another variation of the internal combustion engine is the diesel. **Diesel engines**—also known as compression ignition engines—are much simpler and more reliable than gasoline engines, mostly because they do not incorporate a spark-ignition system. Instead of using a spark to initiate combustion, diesel engines use a much higher compression ratio to generate sufficient **heat of compression** to ignite the air-fuel mixture. The compression ratio for a diesel engine can range anywhere from 16:1 to 22:1.

Another major difference between a diesel engine and most gasoline engines is that all diesel engines must inject their fuel directly into the combustion chamber or into a portion of the combustion chamber called the pre-combustion chamber in order for combustion to take place. The diesel engine must inject its fuel into the combustion chamber because it needs the heat of the highly compressed air in the combustion chamber to ignite the fuel, whereas all gasoline engines, including direct injected ones, rely upon the spark-ignition system to initiate combustion.

Diesel Engine

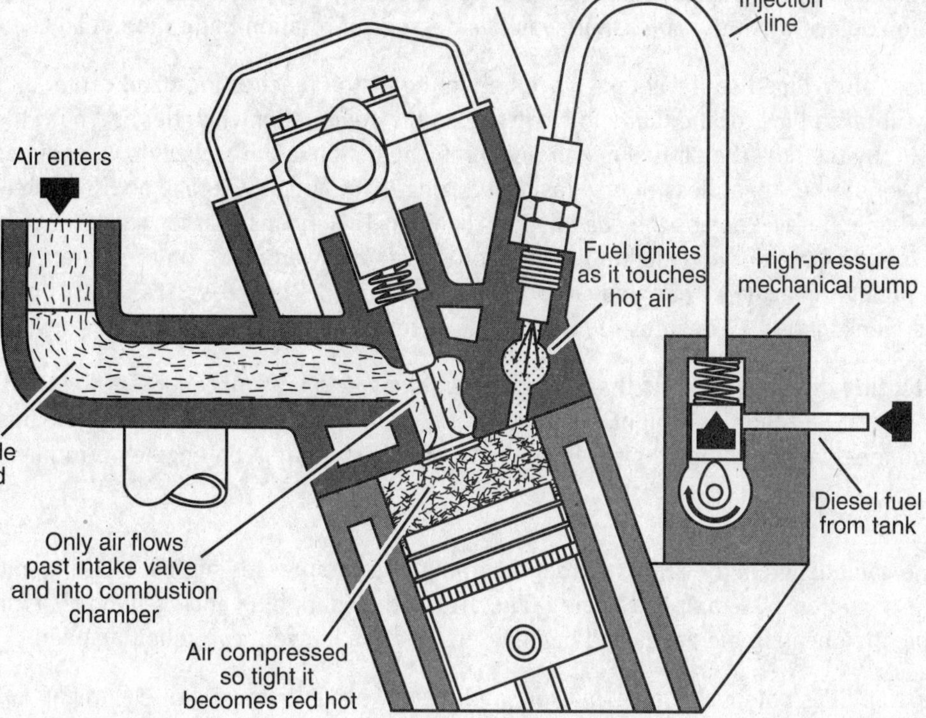

Engine Operating Conditions

Air-fuel mixture

For an engine to run efficiently, it is necessary to mix air and fuel in the correct amounts. Combustion suffers if there is too much fuel and not enough air, or if there is too much air and not enough fuel. The ideal ratio of air to fuel is known as the **stoichiometric ratio**, and it is the responsibility of the engine's fuel system to maintain that balance.

Air-fuel ratio is a comparison of the mass of the air relative to the mass of the fuel that has been mixed with it. The stoichiometric air-fuel ratio for a gasoline engine is 14.7:1. This means that 14.7 pounds of air is combined with 1 pound of fuel to create an ideal air-fuel mix. When fuel is mixed with air by the fuel system, particles of fuel are **atomized** and directed into the air stream that is entering the engine.

Too much air and not enough fuel would be described as a **lean** mixture. Lean mixtures burn relatively slowly because there is greater space between the fuel molecules and it takes more time for a flame to jump from particle to particle. Lean mixtures also burn much hotter, and thus can cause serious engine damage. A typical lean air-fuel ratio might be 17:1.

In contrast, a **rich** air-fuel mixture has too much fuel and not enough air. Rich mixtures burn quicker because of the small distances between fuel particles, and they also burn much cooler. A rich air-fuel mixture can cause spark plug fouling and black exhaust smoke, but is certainly less threatening to an engine than lean mixtures. A typical rich air-fuel mixture might be 10:1.

Ignition timing

Another major factor in making an engine run efficiently is **ignition timing**. Ignition timing is the point in time during the combustion cycle that a spark is generated at the spark plug. This is described relative to the position of the engine's crankshaft. For instance, an ignition timing of five degrees **before top dead center (BTDC)** would mean that the spark took place when the crankshaft was five degrees of rotation before top dead center.

The spark is most often timed so it will take place before the piston reaches top dead center on the compression stroke, because it takes time for the flame to move across the combustion chamber and burn the air-fuel mixture contained within. By the time the air-fuel mixture is burnt, the piston is on its way down again and is ready to make an effective power stroke. Regardless of how fast the engine is turning, it will take approximately the same amount of time for the flame to travel across the combustion chamber. This means that for higher engine speeds, the flame must be started earlier in order to generate the most effective downward push on the piston. This is known as **advancing the timing**. An example of an advanced timing would be 25 degrees BTDC. In this case, the spark takes place when the crankshaft is 25 degrees of rotation before top dead center.

In contrast, **retarding the timing** means that the spark is adjusted to take place later in the combustion cycle. Certain engine operating conditions might call for a retarded timing as part of normal engine operation. Ignition timing that is unnecessarily retarded, however, will have an adverse effect on engine performance.

Combustion

Gasoline engine combustion is the rapid, thorough burning of a compressed air-fuel mixture, initiated by a spark from the engine's ignition system. It is a flame that starts at the spark plug and then moves rapidly across the combustion chamber, heating the gases and building pressure in an even, controlled manner.

Normal combustion does not involve an explosion, and it is always initiated by an electric arc at the spark plug. If combustion is started by something other than an electric arc at the spark plug, this is an abnormal condition known as **pre-ignition**. Pre-ignition takes place when the air-fuel mixture is ignited prematurely (before the time the spark plug would normally fire), typically by a hot spot in the combustion chamber, such as a glowing spark plug electrode or a hot piece of carbon.

Detonation is when an air-fuel mixture explodes, rather than burns. It can often take place when an engine's air-fuel mixture is lean. This is because lean mixtures burn very slowly, and as the flame moves across the combustion chamber, the unburned gases become heated due to the advancing flame and the increasing compression in the cylinder. If the temperature of these unburned gases rises sufficiently, they will "auto ignite" and the flame front from this explosion will collide with the original flame front. The pressure spike from this abnormal combustion results in a brutal shock to the engine assembly, and can be heard as a "knock." Severe engine damage can result from unchecked detonation.

Cooling System

There are two major types of cooling systems in modern automotive engines. The first type is **air-cooling,** where air is circulated over cooling fins on the outside of the engine to remove excess heat. The second type is **water-cooling**. A water-cooled engine uses a liquid coolant to pick up excess heat and then rejects that heat through a radiator. In order to ensure optimized emissions and efficiency, cylinder temperatures need to quickly reach operating temperature and remain consistent throughout a wide range of ambient temperatures. Therefore, virtually all newly designed automotive engines use liquid cooling, due to its greater flexibility in coolant circuit design, its heat transfer ability, and its high heat capacity.

Coolant

The most critical component of the cooling system is the coolant itself. Engine coolant is normally made up of a 50/50 mix of **antifreeze** and water. Frozen coolant can lead to serious engine damage (even a possible cracked block and/or cylinder head), so it is important that the coolant be freeze-protected. The most common type of antifreeze is ethylene glycol. A 50/50 mix of ethylene glycol and water will not freeze until its temperature reaches −34° Fahrenheit. This same 50/50 mix will also raise the boiling point of the coolant, which is important in hot weather, because it makes the coolant that much more efficient at transferring heat.

Components

The major components of a water-cooling system include:

- **Water pump**: responsible for moving coolant through the cooling system in order to transfer and control heat.
- **Water jacket**: hollow sections in the engine block and cylinder head that allow coolant to be transferred through them. These are the areas that the coolant must absorb heat from.
- **Thermostat**: controls engine temperature by allowing coolant to flow into the radiator when the coolant temperature rises above a certain level.
- **Bypass tube**: allows coolant to flow back into the water pump from the cylinder head when the thermostat is closed.
- **Radiator hoses**: flexible hoses that allow hot coolant to flow between the engine and the vehicle's radiator.
- **Radiator**: responsible for transferring heat from the coolant to the outside air.
- **Radiator cap**: responsible for maintaining pressure in the system, and allowing coolant to transfer between the coolant reservoir and the radiator.
- **Coolant recovery bottle**: forms a reservoir for coolant to flow in and out of the cooling system as the engine increases and decreases in temperature.

Operation

Coolant is circulated through the engine by a belt- or timing-chain-driven water pump. The engine crankshaft drives the belt, so the water pump actually uses engine power to operate it. The water pump takes coolant in and pushes it into the engine block. The coolant then makes its way upward into the cylinder head, and then returns to the inlet of the water pump through the bypass tube. As long as the thermostat is closed (engine is below operating temperature), the coolant will continue to circulate in this manner.

When the thermostat opens, hot coolant moves past the thermostat into the upper radiator hose, and then the coolant enters the radiator itself. One of the methods used to raise the boiling point of the water has already been mentioned: using coolants made up of a 50/50 mix of ethylene glycol and water. The other method that can be utilized is raising the pressure in the cooling system. The **radiator cap** is responsible for this, and most are designed to maintain anywhere from 9 to 16 pounds per square inch (psi) of pressure in the cooling system. For every 1 psi of pressure that is placed on the cooling system, the boiling point of the coolant is raised approximately 3°F. A 15-psi radiator cap would then raise the boiling point of the coolant to between 212°F and 260°F.

Radiator caps incorporate two separate valves: a pressure valve and a vacuum valve. As the engine warms up, the coolant in the cooling system will tend to expand. This expansion will raise the pressure in the cooling system, but when the pressure reaches the radiator cap's rating, the pressure valve in the radiator cap will lift and allow some coolant to flow from the radiator into the coolant recovery bottle. This flow will continue as the cooling system pressure exceeds the rating of the radiator cap. The coolant level in the recovery bottle will rise until the engine reaches operating temperature.

When the engine is shut down and begins to cool, the coolant will contract and create a low-pressure area inside the cooling system. If left unchecked, this low-pressure area would cause the radiator hoses to collapse and have an adverse effect on the cooling system at engine startup.

Maintenance

Cooling systems require maintenance in order to keep them operating at peak efficiency. Again, the coolant itself is absolutely critical and making sure to have it replaced every two to three years is a good maintenance practice. The strength of the coolant should be kept at a 50/50 mix of antifreeze and water. This will ensure that the coolant has sufficient freeze protection and corrosion resistance.

It is also important to check that the belt driving the water pump is in good condition, and that there are no leaks in the system. Hoses and the radiator should also be inspected for visible signs of damage or wear. Most often, a leak will be external and can be observed as it drips from hose connections or faulty cooling system components.

Comparison of Cold and Warm Engine Conditions on Automotive Cooling Systems

Engine cold, thermostat closed

Coolant circulates through bypass to pump and back to engine but does not go through radiator.

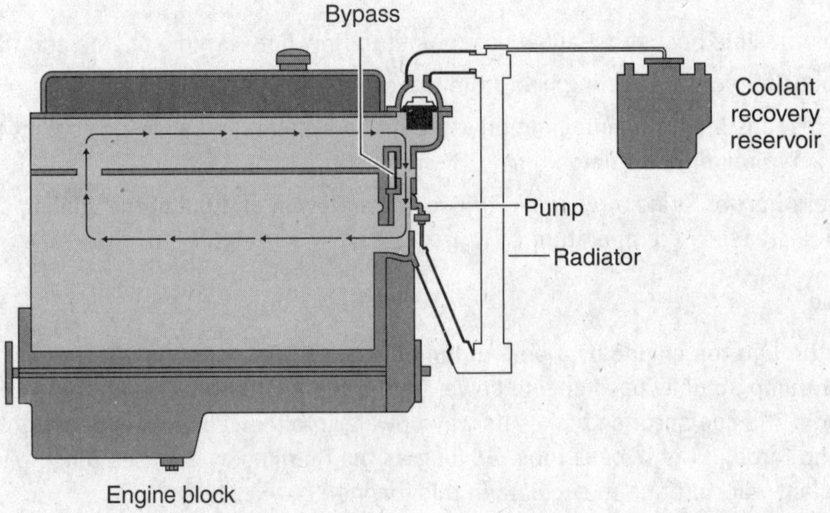

Engine warm, thermostat open

Coolant circulates through radiator.

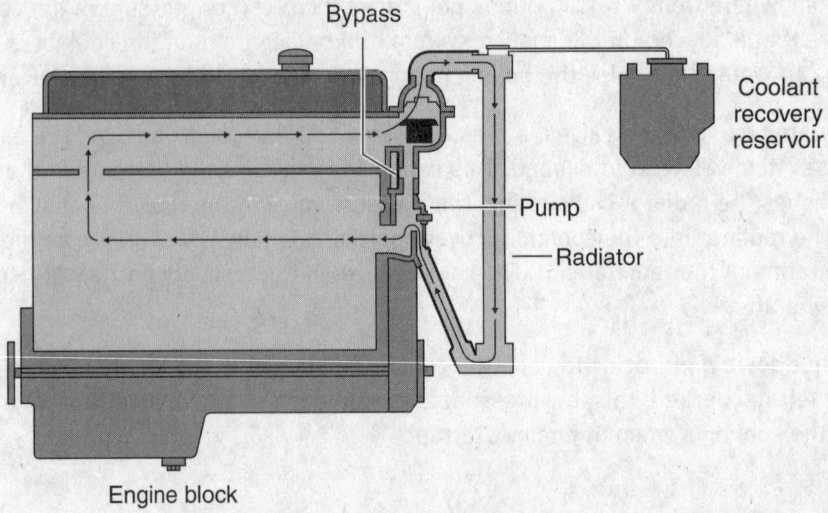

Lubrication System

Another system critical to engine operation is the **lubrication system**. Without lubrication, the internal parts of the engine would very quickly develop enough friction to stop (seize) the engine completely. The lubrication system is responsible for the following functions:

- **Lubricates:** It puts an oil film between moving parts to reduce friction and smooth engine operation.
- **Cools:** It puts motor oil in contact with hot engine parts (such as the underside of the piston) and transfers heat to the oil pan, or the engine oil cooler if applicable.
- **Seals:** Motor oil acts as a sealer between the piston, the piston rings, and the engine cylinder walls. This helps seal combustion gases in the combustion chamber and makes the engine run more efficiently.
- **Cleans:** Additives in the motor oil cause contaminants to be suspended in the oil, so they can be filtered out by the engine's oil filter.
- **Quiets:** Motor oil damps engine noise and makes the engine run more quietly.

Engine oil

Engine oil is a major component of the lubrication system. The engine oil is the lubricant to the moving parts of the engine. The oil must be carefully engineered to provide peak performance under the toughest engine operating conditions, so choosing and using high quality engine oil is extremely important for the life of an engine. Engine oil is made up of two main components: **base oil** and an **additive package**.

One of the most important properties of engine oil is **viscosity**. Viscosity is resistance to flow, and is expressed as a number that is directly proportional to the thickness of the oil. The Society of Automotive Engineers (or SAE) is responsible for developing the standards concerning engine oil viscosity. Oil with a viscosity rating of SAE 5 would have a relatively low viscosity (low resistance to flow), whereas SAE 50 oil would have a high viscosity.

Another important engine oil rating system is the API quality rating. API is the American Petroleum Institute, and it is responsible for setting standards for engine oil quality. For a gasoline engine, this rating would have a prefix of "S," and then the next letter would identify the specific quality standard that the engine oil meets. The first gasoline engine oils produced had a quality rating of SA, but greater demands on engine oils led to ratings that currently exist at the SN, SM, SL and SJ levels. Today's engines would not last very long if they were operated with an SA motor oil.

Engine oil with a "C" prefix for its quality rating would be suited for diesel engine use (e.g., a CJ-4, CI-4 plus, CI-4, and CH-4 rating). Engine oil that had both an "S" and a "C" rating would be suited for either gasoline or diesel engine use. An example of this would be engine oil with a rating of SJ/CD.

Components

The primary components of the lubrication system are as follows:

- **Oil pan**: forms the reservoir for the engine oil at the bottom of the engine.
- **Oil pickup tube and screen**: immersed in engine oil, this filters out large solids and directs oil into the oil pump.
- **Oil pump**: responsible for pumping the oil through the engine oil galleries. It is normally driven by the engine's camshaft.
- **Pressure relief valve**: prevents excessive pressure from building in the lubrication system.
- **Oil filter**: filters oil from the oil pump before it is sent to the various parts of the engine.
- **Oil galleries**: passages or "drillings" in the engine assembly that transport oil to critical components.

Operation

There are two main designs for lubrication systems: **dry sump** and **wet sump lubrication**. The dry sump system is very complex and typically only used on racing vehicles and, therefore, is rarely encountered. Virtually all mass-produced commercial and automotive engines use the wet sump lubrication system design. In the wet sump system design, the engine's oil reservoir is located at the bottom of the engine in the oil pan. All oil used in the engine drains to the oil pan due to gravity. There, it has an opportunity to cool. The oil pump draws oil from the oil pan into its inlet through the pickup tube and screen. The engine oil moves through the oil pump, and then is transferred under pressure to the engine's oil filter. The oil filter removes particulates from the oil, and sends the clean oil on to the main oil galleries.

When the engine is cold, the oil has a higher resistance to flow and thus requires a good deal more energy to pump it through the system. This also causes the engine's oil pressure to rise. If the oil pressure rises above the pressure relief valve's setting, the relief valve will open and allow some of the oil to drain back into the oil pan before it is pumped into the oil filter. This prevents engine oil pressure from rising to the point where it can damage components such as the oil filter, bearings, and engine seals.

The most critical area in the engine in terms of lubrication requirements is the engine's reciprocating assembly. The pistons, connecting rods, and crankshaft all require oil to lower friction and to remove heat. The largest oil galleries in the engine are used to move oil to these areas, and galleries are drilled directly through the crankshaft to provide oil under pressure to the connecting rod bearings. Oil that leaks off the connecting rod bearings is thrown onto the cylinder walls, where it helps lubricate and cool the pistons.

Other oil galleries take oil from the main oil gallery and direct it to the engine's valve train. This is where the camshaft, lifters, push rods, rocker arms, and valve stems get their lubrication. Other parts that may receive lubrication from this oil would be the camshaft drive, or timing chain, and associated gear drives for the oil pump, ignition distributor, etc.

Oil that has circulated through the engine eventually drains back into the oil pan. The oil pan is located at the bottom of the engine and thus sits low in the engine compartment, where cooler air typically resides. This cool outside air helps absorbs heat from the oil pan and helps cool the engine oil. For heavy-duty or hot-weather operation, a liquid-to-air heat exchanger (similar to the radiator in the cooling system) can be used to further cool the engine oil. Once cooled, the oil is ready to be picked up by the oil pump and circulated back into the lubrication system.

Maintenance

The key to lubrication system maintenance is to change the engine oil and filter on a regular basis, and to use top-quality engine oil and filters. This is the cheapest and most effective maintenance that can be performed to minimize wear on the engine.

Questions on the ASVAB may ask you to recall or interpret how the various systems of a vehicle operate. Let's take a look at a sample question from the area of engine systems, which encompasses engine theory, the cooling system, and the lubrication system.

Question	Analysis
An internal combustion engine converts chemical energy of a fuel into ________ energy to power a vehicle.	**Step 1:** This question asks about the conversion of energy in an internal combustion engine. **Step 2:** The answer relates to Engine Theory. Otherwise, there isn't much to simplify here. **Step 3:** Recall that an internal combustion engine converts chemical energy of a fuel into heat energy by burning the fuel and then converts the heat energy into mechanical energy that is used to produce useful work.
(A) electrical (B) mechanical (C) work (D) sonic	**Step 4:** The correct answer is **(B)**. Answer choice (A) is incorrect because electrical energy is used to release the chemical energy but is not the result of the release of the energy. Choice (C) is incorrect because work is the result of the mechanical energy produced by the engine. And (D) is incorrect because even though sonic energy (sound) is produced in the energy conversion process, sonic energy is not the source of the engine's power.

Now try a question on your own.

Engine coolant is normally made of a 50/50 mixture of

(A) water and soap solution
(B) salt and alcohol
(C) water and ethylene glycol
(D) antifreeze and saltwater

Explanation

Engine coolant is composed of 50% water and 50% ethylene glycol (also known as antifreeze). The correct answer is **(C)**. You would never want to introduce a corrosive material like salt into an automotive engine coolant; therefore, choices (B) and (D) are incorrect. Answer choice (A) is incorrect because water with a small amount of soap solution added as a "wetting agent" could function as an engine coolant, but not in a mixture of 50/50. Soap does not increase the temperature performance of the coolant as needed in automotive applications.

COMBUSTION SYSTEMS

LEARNING OBJECTIVES

In this section, you will learn to:

- understand the fundamentals of combustion within an internal combustion engine
- identify the basic function and the major components of the fuel system
- identify the basic function and the major components of the ignition system
- identify the basic function and the major components of the exhaust system

Fuel System

The **fuel system** is responsible for maintaining the correct air-fuel mixture for efficient engine operation. If the air-fuel mixture is not adjusted correctly, drivability, emission control, fuel economy, and engine service life can suffer.

Up until about 35 years ago, it was most common for the air-fuel mixture to be determined through mechanical means. This was accomplished through the use of a **carburetor**, which was very reliable but incapable of providing the precision required in modern fuel systems. Due mostly to the demands of emission control regulations, the carburetor has been made obsolete in favor of **electronic fuel injection**. This system is controlled by an onboard computer and incorporates a feedback function that gives it flexibility to adjust to changing engine conditions very quickly.

Components

The following are the major components of an electronic fuel-injection system:

- **Electric fuel pump**: located in the vehicle's fuel tank; supplies fuel under pressure to the fuel injectors.
- **Fuel filter**: filters contaminants from the fuel before it reaches the fuel rail.
- **Fuel rail**: a manifold that supplies fuel under pressure to the inlets of all the engine's fuel injectors.
- **Fuel pressure regulator**: regulates pressure in the fuel rail according to intake manifold vacuum. Excess fuel is bled to the fuel return line, where it is sent back to the fuel tank.
- **Fuel injector**: sprays fuel into the intake air stream as it receives electrical signals from the powertrain control module (PCM). The location of the injector will determine the specific type of fuel injection system being used on the engine.
- **Powertrain control module (PCM)**: another name for the vehicle's central computer. Responsible for control of all functions associated with the engine and transmission.
- **Intake manifold**: distributes air to the intake ports on the cylinder heads.
- **Intake air filter**: removes airborne contaminants that could damage internal engine parts. All air entering the engine passes through the air filter.
- **Throttle body/throttle plate**: connected to the throttle pedal; controls engine speed and output torque.

Fuel injection system designs

When fuel injection systems first came into common use, they were designed to utilize either one or two injectors mounted in a **throttle body** that took the place of the carburetor. This is known as a **throttle body injection system** or **TBI**. While this system proved to be reliable, it was not capable of providing a high enough level of fuel control to meet emission control requirements. Most engines are now built with either **multiport fuel injection**, which has an injector for each engine cylinder, or **direct injection**, where the fuel is directly injected into the combustion chamber.

With multiport fuel injection, the injectors are located in the intake manifold, with their spray directed toward the intake valves. As air enters the intake manifold, it will flow all the way to the cylinder head before fuel is injected into it. This allows for much better air-fuel mixing and prevents fuel droplets from falling out on the intake manifold runners. With direct injection, a high-pressure fuel injector is located to spray highly pressurized fuel directly into the combustion chamber. Therefore, only air flows over the intake valves and enters the combustion chamber. Fuel can then be sprayed directly into the combustion chamber at an optimized point in the compression stroke. Also, the fuel acts as a coolant to the hot compressed air in the combustion chamber, which allows the engine to be designed with even greater levels of compression, which further improves power and efficiency.

Fuel Injection Types

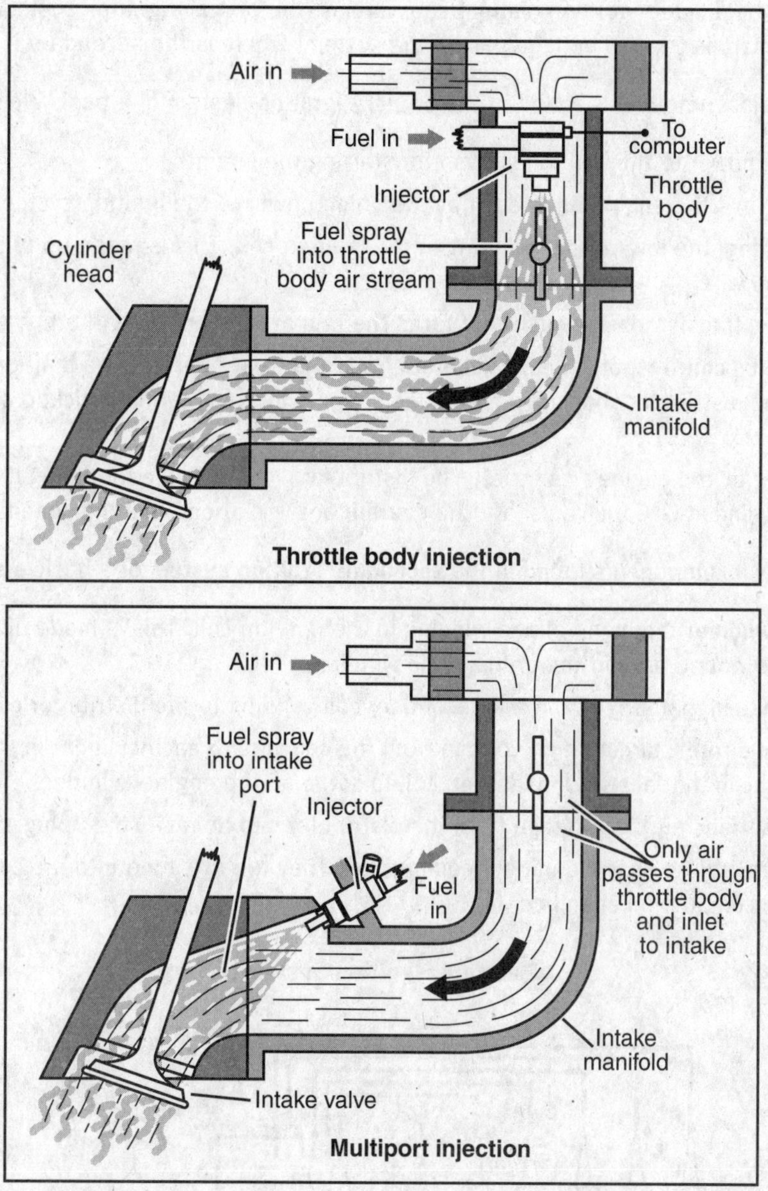

Maintenance

The fuel-injection system requires very little maintenance. The filters (both air and fuel filters) must be changed periodically, based on the manufacturer's recommendations. Manufacturers may also require periodic cleaning of the injection system. Aside from that, fuel injection systems are built to give reliable service with a minimum of maintenance.

Ignition System

One of the most critical vehicle systems is the **ignition system.** Without it, combustion cannot take place, so the engine will not run. The ignition system must generate high-voltage sparks at the correct time in order to make the engine run smoothly and efficiently.

Components

The ignition system can be divided into two distinct subsystems: the primary ignition system and the secondary ignition system. The primary is the low-voltage part of the system, whereas the secondary is high voltage.

The following are the major components found in the **primary ignition system** of a basic electronic ignition system:

- **Battery**: supplies power to the ignition system for starting the engine.
- **Ignition switch**: turns the engine on and off by switching power to the ignition system.
- **Primary coil winding**: the low-voltage winding in the ignition coil. This is made up of several hundred turns of relatively heavy wire.
- **Ignition module**: a transistorized switch that turns the primary current on and off.
- **Reluctor and pickup coil**: responsible for generating a signal that operates the ignition module. The reluctor is mounted on the distributor shaft, and generates a signal in the stationary pickup coil as it rotates with the distributor.
- **Distributor**: driven by the engine's camshaft, the distributor is responsible for timing the spark and distributing it to the correct cylinder. Like the camshaft, the distributor turns at one-half the speed of the engine.

The following are the major components found in the **secondary ignition system** of a basic electronic ignition system.

- **Secondary coil winding**: the high-voltage winding in the ignition coil. This is made up of several thousand turns of fine wire, wound around the primary coil winding.
- **Coil wire**: transmits high voltage from the secondary coil winding to the distributor cap.
- **Distributor cap and rotor**: directs high voltage from the coil wire to each cylinder in the firing order. This is a switching mechanism that allows one ignition coil to serve all the engine cylinders.
- **Spark plug wires**: transmit high voltage from the distributor cap to each spark plug.
- **Spark plugs**: generate the spark to initiate combustion. They are threaded into the cylinder head and protrude into the combustion chamber.

Electronic Ignition Systems

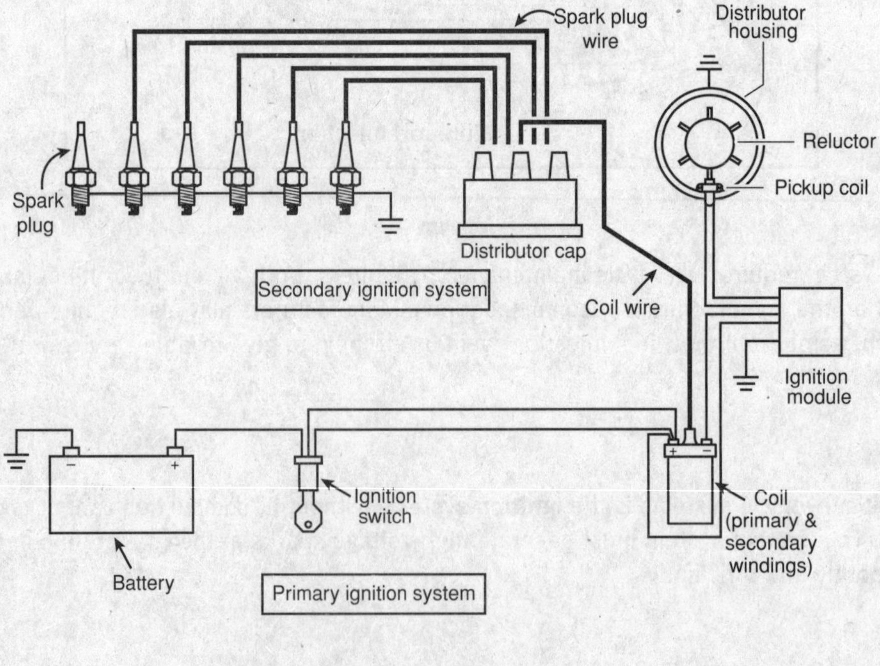

Primary ignition operation

When the driver turns the ignition switch to the "run" position, current is sent from the battery, through the ignition switch, and on to the primary coil winding. From the coil winding, that same current flows through the ignition module and back to the battery through the vehicle ground circuit. Within the ignition system as a whole, the low-voltage (primary) system is responsible for the control of the high-voltage (secondary) system. Whatever happens in the secondary system is in response to the events in the primary system.

The ignition system works because of a phenomenon known as **electromagnetic induction**. If a magnetic field moves across a stationary wire, voltage is induced in the wire. Another dimension of this principle is that electric current passing through a wire will produce a magnetic field around that wire. This means that electric current can be used to generate a magnetic field, and a magnetic field can be used to generate electric current.

Secondary ignition operation

The collapse of the magnetic field in the primary coil winding induces a high voltage in the secondary winding. These secondary system voltages typically range from 20,000 to 40,000 volts, and modern systems can go even higher. This huge increase in voltage in the ignition coil is achieved because of the large number of turns of wire in the secondary winding relative to the primary. The magnetic field in the primary winding cuts across a much larger number of turns of wire in the secondary, creating a **step-up** effect in the coil.

The high-voltage current in the secondary winding must be directed to the correct cylinder in the firing order. This current is directed through the **coil wire** to the center tower of the **distributor cap**. From this point, it flows to the **rotor**, which is mounted on top of the distributor shaft and rotates with the distributor. As the engine turns, the rotor will direct the current to the appropriate cylinder and send it through the **spark plug wire** to the **spark plug**. The electric arc at the spark plug is used to initiate combustion.

Current ignition system design

Since engine performance relies heavily on accurate ignition timing, engineers look carefully at how to eliminate more moving parts from the ignition system. This led to the elimination of the distributor in a design named the **distributorless ignition system** or **DIS**. DIS systems use one ignition coil to operate two spark plugs, so a V-8 engine would require four separate coils instead of one. The increase in the number of coils allows each coil more time to build up its electric charge and, therefore, deliver a more powerful spark when the time comes to discharge. However, this system still utilizes spark plug wires, which degrade over time and eventually need replacement for the DIS system to function properly.

The most recent ignition system designs have eliminated the spark plug wires altogether. These systems are known as **coil-on-plug** ignitions. Individual ignition coils are mounted directly onto the spark plugs, and the discharge of the coils is controlled by the vehicle computer system. While the incorporation of computer control has increased ignition system complexity, the elimination of moving and high-maintenance components has led to very accurate control of spark timing, which is critical for optimum engine efficiency and emissions, as well as increased reliability of the ignition system.

Exhaust Systems

The exhaust system is responsible for removing waste gases from the engine. It must do this in a way that allows these gases to flow freely, muffles the sound of the exhaust, and keeps the gases and heat away from the vehicle cabin and its occupants.

Components

Typical components found in an exhaust system include the following:

- **Exhaust manifolds**: attached directly to the exhaust ports on the cylinder head. The majority of the exhaust heat and noise is focused on the exhaust manifolds. These are often made from cast iron for durability under high heat conditions.
- **Catalytic converter**: responsible for converting the toxic components of engine exhaust into relatively harmless compounds such as carbon dioxide and water.
- **Muffler**: incorporates an expansion chamber and sound absorbing material to diminish loud exhaust noises.
- **Tailpipe**: the exit point for exhaust gases as they enter the open atmosphere. The tailpipe normally exits at the rear of the vehicle.

Exhaust System Components

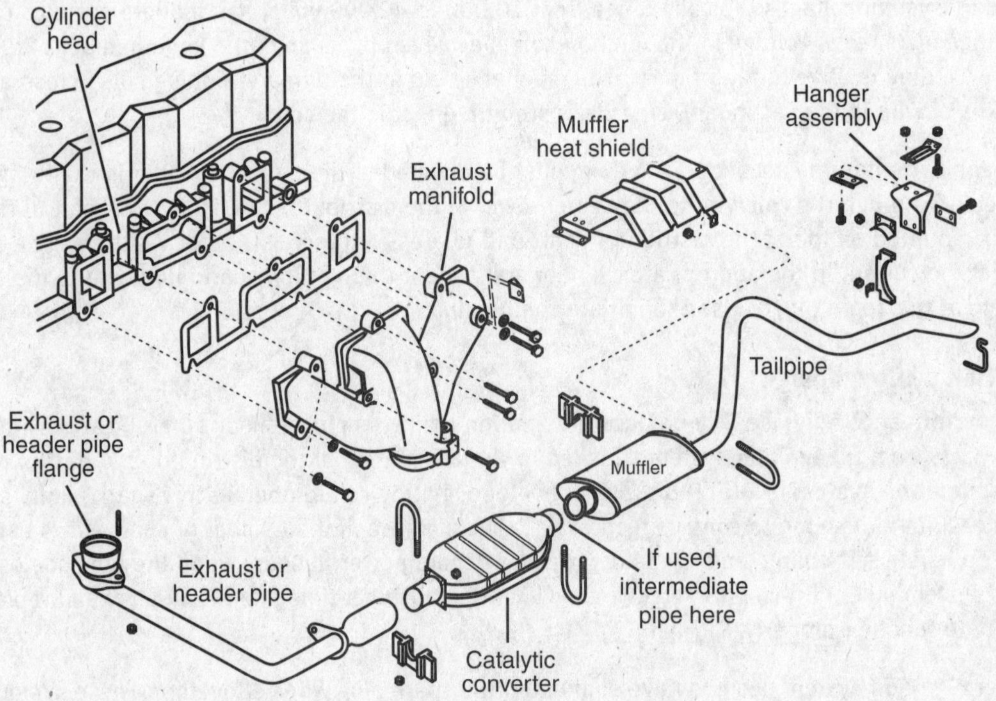

Operation

As gases flow from the exhaust ports of the engine, the **exhaust manifold** collects them. On a V-type engine there would be two exhaust manifolds, one for each cylinder head. These manifolds feed the gases into steel exhaust pipes, which connect the major components of the exhaust system.

The exhaust gases are then sent into the **catalytic converter**. High heat is developed in the catalytic converter as it reduces the toxic components of the exhaust to gases that are less toxic. Some vehicles use two, and even three, catalytic converters in an effort to meet emission control regulations.

After leaving the catalytic converter, exhaust gases are directed into the **muffler**. The muffler has expansion chambers built into it that absorb the loud sounds generated by the engine's combustion. The muffler is normally located toward the rear of the car, somewhere after the catalytic converter and before the tailpipe.

Even after passing through the catalytic converter, engine exhaust is highly toxic. The most dangerous of the toxic gases emitted by the engine's exhaust is carbon monoxide, an odorless and colorless deadly gas. It is extremely important that these gases be routed in such a way that they do not come into contact with the driver or passengers in the vehicle. This usually involves sending these gases to the rear of the vehicle where they are dissipated to the open air by the **tailpipe**.

To increase engine efficiency and power, the exhaust system can be tuned to match the design of the engine. "Tuned" exhaust manifolds are often known as **header pipes**. The idea is to allow exhaust gases to flow more freely and allow the engine to breathe better. Increased performance is usually gained at the expense of durability, as header pipes are much more fragile than cast-iron exhaust manifolds.

Maintenance

The system should be inspected from time to time for physical damage; there should be no restrictions and no leaks. The system should also be tight in terms of its mounting brackets and clamps. Any damage should be repaired immediately.

The catalytic converter will enjoy a long service life if the vehicle's engine is kept in good running order. An engine that burns oil or has ignition or fuel system problems will damage the catalytic converter.

Questions on the ASVAB may ask you to recall the components or functions of the various systems of a vehicle. Questions may be situational, asking you to interpret the situation at hand and to make an assessment based on your foundational knowledge of automotive systems. Let's take a look at a sample question from the area of combustion systems, which covers the fuel system, the ignition system, and the exhaust system.

Question	Analysis
Multiport fuel injection prevents	**Step 1:** This question asks you to recall one of the benefits of multiport fuel injection. **Step 2:** There is nothing to simplify, but remember what you know about multiport fuel injection. **Step 3:** One of the features of multipoint fuel injection is the ability to precisely atomize the fuel into the intake air stream within the intake manifold runners. This fine atomization enables the fuel to be suspended in the intake air stream as fine particles and prevents larger fuel droplets from falling out onto the walls of the intake manifold runners.
(A) fuel droplets from falling out on the exhaust valves (B) fuel droplets from falling out on the intake manifold runners (C) the overheating of the engine (D) engine fires	**Step 4:** The correct answer is **(B)**. Answer choice (A) is incorrect because injected fuel does not flow over the exhaust valve. Choices (C) and (D) are incorrect because, generally, multiport fuel injection does not directly affect engine overheating or engine fires.

Now give this one a try on your own.

A major component of the primary ignition system is the

(A) reluctor and damper coil
(B) inductor
(C) primary resistor
(D) reluctor and pickup coil

Explanation

The correct answer is **(D)**. The reluctor and pickup coil are responsible for generating a trigger signal that operates the ignition module. Answer choices (A), (B), and (C) are not major components of the primary ignition subsystem, and therefore are not correct.

ELECTRICAL AND CONTROL SYSTEMS

LEARNING OBJECTIVES

In this section, you will learn to:

- identify the basic function and the major components of the electrical system
- understand the basic function and the major components of the computer system

Electrical System

The vehicle's electrical system is growing in importance with each new vehicle model year. Many systems that were once driven mechanically are now being redesigned to operate electrically.

Subsystems

The major subsystems of the electrical system include the following:

- **Battery**: stores electrical energy in chemical form. Provides direct current (DC) for engine starting and accessory operation.
- **Starting system**: responsible for cranking the engine to get it started. The battery and starter motor are the major components of the starting system.
- **Charging system**: responsible for supplying electrical current to charge the battery, as well as for vehicle operation. The major component of the charging system is the alternator.
- **Lighting system**: headlights, marker lights, brake lights, tail lights, etc.
- **Accessories**: include the rear-window defogger, windshield wipers, stereo, blower motors, and all other electrically powered accessories.

Battery

The battery is the foundation for the entire electrical system. It provides electrical current for starting the engine, provides current to the electrical system when the load exceeds the output of the alternator, and acts as an electrical "shock absorber," preventing voltage spikes when there is excessive current in the electrical system.

An automobile battery is made up of lead plates immersed in an electrolyte made up of sulfuric acid and water. This is why this type of battery is known as a **lead-acid battery**. As the battery discharges, the sulfuric acid in the electrolyte is reduced to water, and the lead plates become lead sulfate. Charging the battery restores the chemical composition of the lead plates and the electrolyte. Care must be taken when working with an automotive battery because of the great amount of electrical energy and the highly corrosive sulfuric acid stored in it.

Starting system

In order to start the engine, an electrically operated starter motor is utilized. When the ignition switch is moved to the "start" position, an electrical current is sent to the starter solenoid, which engages the starter drive gear onto the engine's ring gear (located on the flywheel). When the drive is engaged, the solenoid connects the battery to the starter motor, which then turns the engine at sufficient speed to start it.

Charging system

Once the engine is running, the charging system provides electrical current to recharge the battery and power the vehicle's electrical system. The main component of the charging system is the **alternator**. The alternator is belt-driven by the engine's crankshaft, and converts mechanical energy into electrical energy. The alternator produces alternating current (AC), which is then **rectified** by an internal set of diodes known as the **rectifier bridge**. The rectifier bridge converts AC to direct current (DC), which can be used to power the vehicle's electrical system.

A **voltage regulator** controls the output of the alternator. Normal system voltage during engine operation is around 14.5 volts. Turning on the headlights, heater motor, and other accessories on the vehicle increase the **load** on the electrical system and system voltage drops. The voltage regulator senses this decrease in system voltage and responds by increasing the alternator's output to compensate for it. As long as the alternator's output is able to match the load on the electrical system, the system voltage will remain close to 14.5 volts.

Lighting system

There are many lights built into a vehicle's lighting system. **Headlights** illuminate the road ahead of the car, **taillights** mark the rear of the car for other drivers, and **interior lights** help the driver see the instrument panel and other areas inside the car when necessary. The various lights are controlled by the driver through **switches** that turn the electrical current to the lighting circuits on and off.

Fuses or circuit breakers protect the vehicle's lighting circuits. If, for some reason, the circuit should draw more current than it has been designed for, the fuse will "blow" and cut off current flow. This protects the wires in the circuits from overheating and may even prevent an electrical fire.

Computer System

Computer control systems work much like the human nervous system. **Sensors** provide data to the **computer**, which processes this information and sends signals to **actuators** to control vehicle functions. In the case of the human body, the eyes and ears send signals to the brain, which processes this information and sends signals to the various muscles to control body movement.

Components

The major components of the automotive computer system include the following:

- **Sensors**: generate signals based on rotational speed, temperature, pressure, and relative position. The sensors are the "eyes and ears" of the computer system. An example of a sensor would be an oxygen sensor, which is responsible for sending data to the computer concerning the oxygen content of the exhaust gases. The computer uses this information to determine whether the engine's air-fuel mixture is rich or lean.
- **Computer**: processes data from the sensors based on a preprogrammed strategy (software) and then generates outputs to control vehicle functions. The computer is the "brain" of the computer system. In today's vehicles, the computer is often referred to as the **powertrain control module (PCM)** or the **engine control unit (ECU)**.
- **Actuators**: receive output signals from the computer and control vehicle functions. An example of an actuator would be a fuel injector, because it responds to signals from the computer to inject a specific amount of fuel into the intake air stream.

Automotive technicians communicate with the ECU using a **scan tool**, which allows them to get information from the computer concerning the operation of the various vehicle systems. The scan tool is connected to the vehicle's computer system through a **diagnostic data link** that is typically located near the driver's seat.

Now, let's look at how you will use your understanding of electrical and control systems on Test Day.

Question	Analysis
The alternator produces ___________ current.	**Step 1:** This question asks you to recall the type of current produced by the alternator. **Step 2:** There is nothing to simplify here. **Step 3:** The rotating magnetic field of the alternator rotor produces an alternating current (AC) in the alternator stator windings. The AC produced by the alternator must be rectified into a direct current (DC) in order to be compatible with the battery and the rest of the vehicle's electrical system. The rectifier bridge is the component that performs the rectification of AC into DC.
(A) fast (B) slow (C) direct (D) alternating	**Step 4:** Only answer choice **(D)** addresses the nature of the current produced by the alternator. So it is correct.

Now try another question on your own.

The PCM (or ECU) is the ___________ of the computer system.

(A) actuator
(B) sensor
(C) data link
(D) brain

Explanation

The correct answer is **(D)**. The PCM (or ECU) is the processor that performs the analysis of the sensor signals and sends control signals to the actuators which makes the PCM (or ECU) the brains of the computer system. Answer choices (A), (B), and (C) are supporting subcomponents of the computer system that do not perform any analysis. Therefore, they do not answer the question.

CHASSIS SYSTEMS

LEARNING OBJECTIVES

In this section, you will learn to:

- identify the basic function and the major components of the drivetrain system
- recognize the basic function and the major components of the suspension and steering systems
- understand the basic function and the major components of the brake system

Drivetrain System

The engine can produce power to move the vehicle, but this energy must be processed and transmitted effectively in order to accelerate the vehicle smoothly and quickly. Transmitting power from the engine to the wheels is the responsibility of the vehicle's drivetrain.

An important part of any drivetrain is the **transmission**. The transmission is the device that matches engine speed to a desired speed of the vehicle. The two main types of transmissions are the **automatic** and the **manual**. Drivers who like to have more control over the operation of the vehicle will often choose a manual transmission, because the driver is responsible for shifting the gears up and down. A car with an automatic transmission is much easier to operate because it does not use a driver-operated clutch and does all of the gear shifting automatically.

While rear-wheel drive was the most common design in years past, front-wheel drive has become very popular, especially in small- to medium-sized cars.

Components

The typical components found in a front-wheel drive vehicle with a manual transmission include the following:

- **Clutch**: transmits torque from the engine to the transmission. The clutch can be released and engaged to allow smooth starts and transmission shifting.
- **Transaxle**: has several "gears" that are selected according to the speed of the vehicle and how fast the driver would like to accelerate. The transaxle is a combination of the **transmission** and the **drive axle**, which are normally two separate components in a rear-wheel drive vehicle.
- **Half shaft**: a short drive shaft that transmits power from the transaxle to the drive wheels. Used mostly in front-wheel drive vehicles, the half shafts must be able to apply power to the drive wheels while allowing them to move up and down and turn the vehicle. There are two half shafts, one for each drive wheel.
- **Constant-velocity (CV) joints**: can transmit power through very steep angles, and are located at each end of a half shaft. The inboard CV joint is located at the transaxle, while the outboard CV joint is located behind the vehicle wheel.

Rear- and Front-Wheel Drive

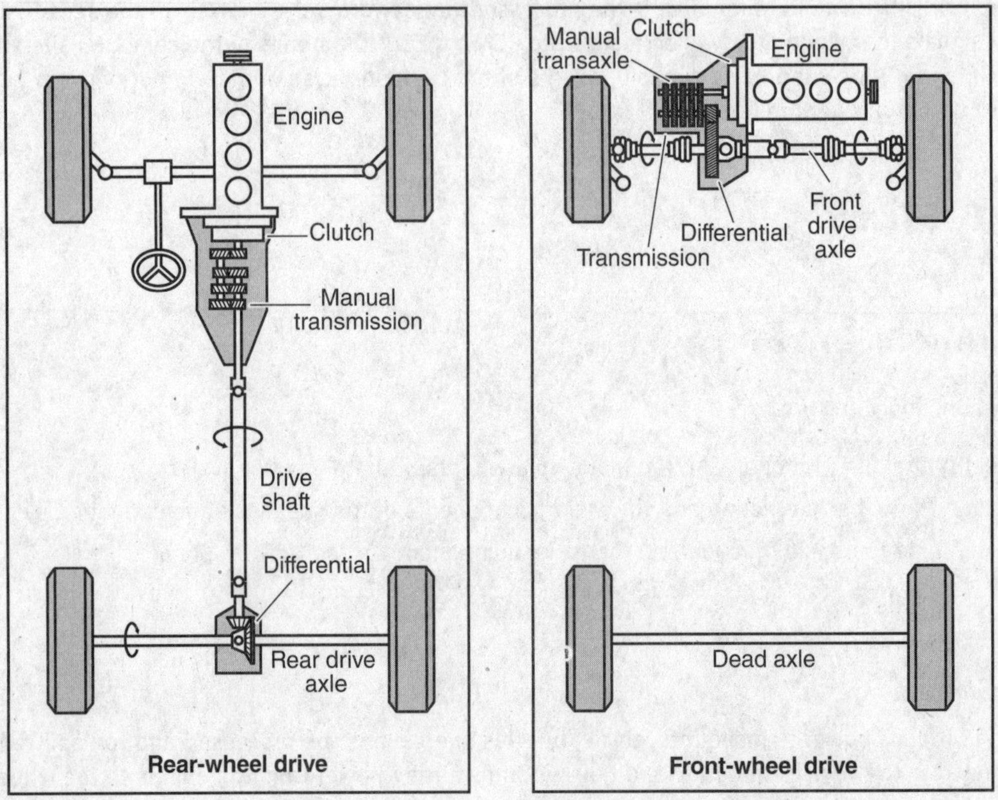

Other drivetrain components that may be found in other vehicle designs include the following:

- **Transmission**: used in rear- and four-wheel drive vehicles, the transmission is responsible for matching engine speed to the desired speed of the vehicle.
- **Drive shaft**: a longer version of the half shaft, the drive shaft transmits torque from the transmission to the drive axle. The drive shaft is also required to allow up and down movement of the vehicle's wheels, but not to the same extent as the half shaft in a front-wheel drive.
- **Universal joints**: located at each end of the drive shaft, the universal joint allows the shaft to operate at an angle with the component that it is driving. Also known as **U-joints**.
- **Drive axle**: transmits engine power through a 90-degree angle, and splits that power between the two drive wheels. The wheels are attached to the ends of the drive axle. The drive axle incorporates a **differential**, which allows the left and right wheels to turn at different speeds as the vehicle goes around a corner.
- **Transfer case**: located between the transmission and the drive axles on a four-wheel drive vehicle. The transfer case splits the engine's power between the front and rear drive axle.

Gear selection

Transmissions are built with several forward ratios that are based on the specific gears that are used. Typical transmissions are therefore said to have "speeds" or "gears." For instance, a four-speed transmission would have four forward gear ratios and one reverse ratio. A driver would typically start from a stop using **low (first) gear**. In this gear, the transmission would increase the **torque** from the engine sufficiently to be able to start the vehicle from a standstill. However, this would limit the vehicle's speed once it did start rolling, so the driver would then select **second gear** and so on as the vehicle continued to accelerate.

High (fourth) gear would be used to allow the vehicle to cruise on the highway. Engine speed would be moderate to low in this gear, allowing for maximum fuel economy and quiet running. To accelerate quickly, however, the driver would **downshift** to a lower gear in order to increase engine speed and power. Proper gear selection is important when operating a vehicle with a manual transmission. Modern transmissions now have as many as eight forward speeds.

Automatic transmissions

Automatic transmissions are much more complex than their manual counterparts, yet are much easier to operate. The reason that many drivers prefer them is that automatic transmissions do not use a driver-operated clutch.

Instead of using a clutch, automatics transmit engine torque from the engine to the transmission through a **torque converter**. The torque converter uses fluid to transmit power, and allows for a certain amount of slippage when the vehicle is stopped.

Automatic transmissions do all of the gear selection for the driver. This is done in the transmission using hydraulic pressure to select the proper **planetary gear sets**. Virtually all modern automatic transmissions are controlled electronically by the vehicle's powertrain control module.

Another modern automatic transmission is the **continuously variable transmission** or **CVT**. A CVT transmission does not utilize specific gear sets to determine the transmission ratio (or gears). Instead, it relies on two opposing sets of cones, one for input and one for output, with a chain or belt running between the two cones. The effective ratio of the transmission depends on where the chain (or belt) is riding on the two cones. Because the transmission's effective ratio is not limited to specific gears as in a typical transmission, the ratio can be varied continuously between the limits of the diameters of the two cones.

Torque Converter

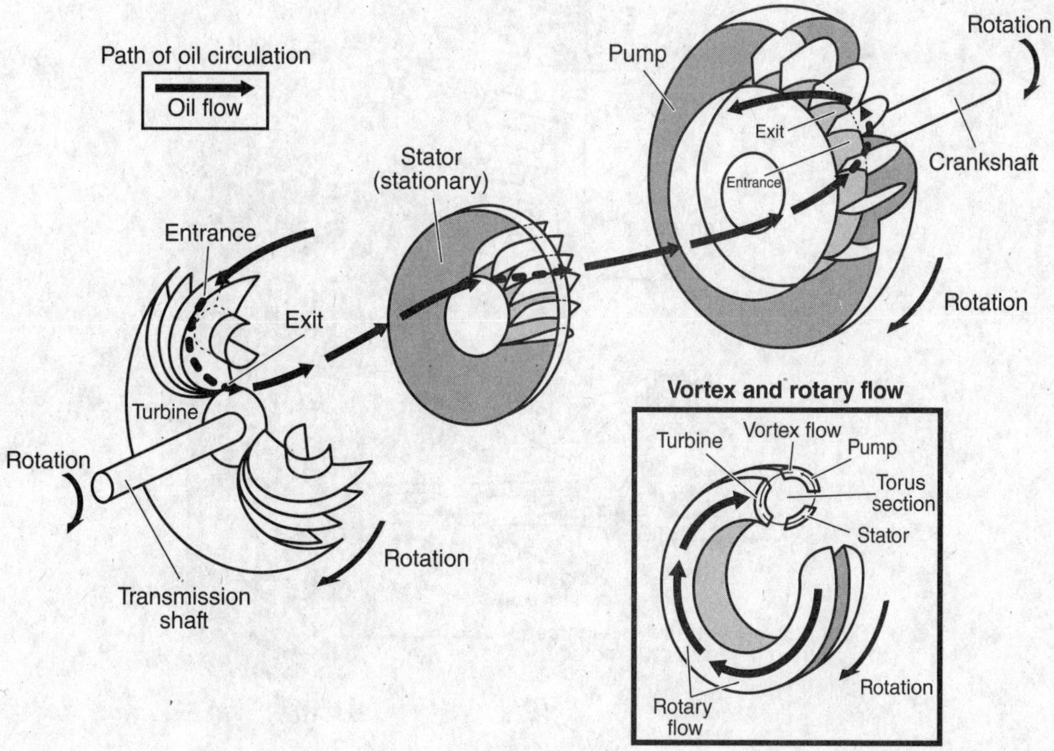

Suspension and Steering System

The newest suspension and steering systems have incorporated computer control in an effort to improve vehicle performance. While the sophistication of computer control has increased the overall complexity of these systems, the basic principles remain the same.

Components

The major components of the long-short arm suspension system include the following:

- **Springs**: hold the vehicle's chassis up and allow the wheels to move up and down in relation to it.
- **Shock absorbers**: absorb the energy released by the up-and-down movement of the vehicle wheels.
- **Control arms (A-arms)**: the long-short arm suspension system uses an upper and lower control arm to maintain the vertical orientation of the steering knuckle as the wheel moves up and down.
- **Steering knuckle**: connects to the upper and lower control arms through the use of ball-joints. The wheel hub mounts on the spindle, which is part of the steering knuckle.
- **Ball joints**: ball-and-socket assemblies that allow the steering knuckle to turn and move up and down simultaneously.
- **Steering linkage**: connects the steering wheel to the steering knuckle.
- **Wheel hub**: forms the mounting point for the vehicle's tire assembly.
- **Tire**: makes contact with the road and provides a "footprint" to aid in vehicle stability and handling.

Long-Short Arm Suspension

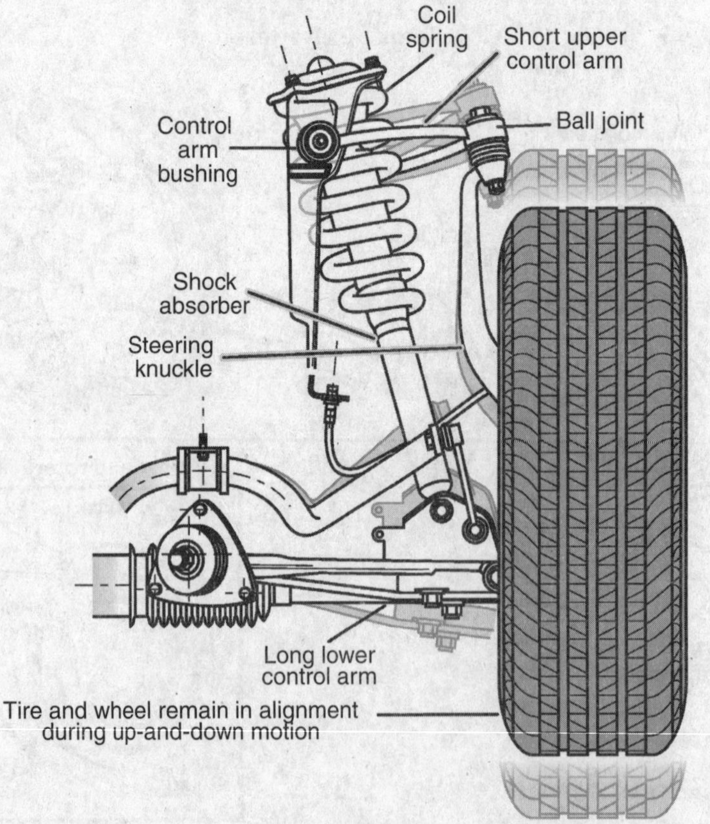

Construction and operation

In order to make a vehicle ride smoothly, the wheels must be able to move up and down while the vehicle chassis stays steady. This is the job of the vehicle **springs**; they must be able to allow this movement of the wheels without transmitting road shock. However, the springs absorb energy as this happens and must release this energy without bouncing the vehicle after encountering a bump.

The **shock absorber** is installed between the chassis and a control arm to help dissipate or absorb the energy stored in the spring as it is either compressed (or **jounced**) or extended (or **rebounded**). Without properly functioning shock absorbers, the vehicle will continue to bounce up and down after the vehicle has hit a bump in the road. The shock absorber uses a piston and hydraulic oil to absorb the excess mechanical energy from the suspension system and then releases that same energy in the form of heat.

The **upper and lower control arms** form the supports for the steering knuckle. The control arms are attached to the chassis through the use of **control arm bushings**. The spring is placed between the chassis and one of the control arms. The spring can be installed on either the upper or lower control arm in a long-short arm suspension. The control arm with the spring on it becomes the load-bearing arm as the vehicle weight is supported through it on the way to the steering knuckle and tire assembly.

An **upper ball joint** and a **lower ball joint** are used to attach the steering knuckle to the control arms. Ball joints are built similar to a human hip joint, with a ball and stud rotating in a socket to allow a wide arc of movement. As the control arms move up and down, the ball joints allow the steering knuckle to move freely with them. They also allow the steering knuckle to turn left and right as it is moved by the steering linkage.

The **steering linkage** forms the connection between the steering wheel and the steering knuckle. There are two main designs in common use, the **linkage steering** and the **rack and pinion**. Long-short arm suspensions most commonly use linkage steering to connect the steering column to the **pitman arm**, **center link**, and **idler arm**. **Tie rods** are then used to make the final connection to the steering knuckles. Smaller vehicles most commonly use the rack-and-pinion steering system because it is the most compact of the two major steering systems.

Linkage Steering

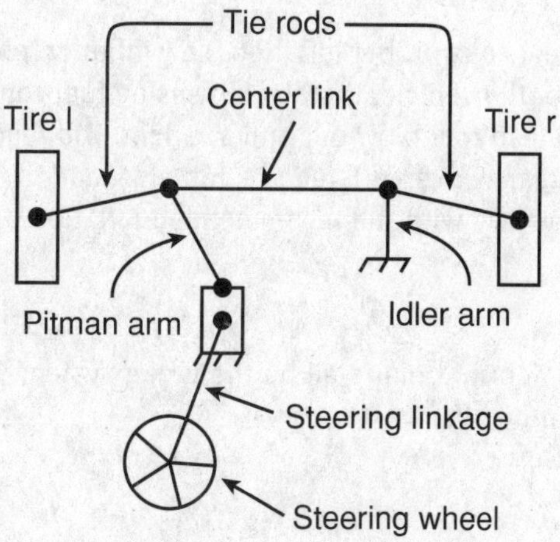

Rack and Pinion Steering

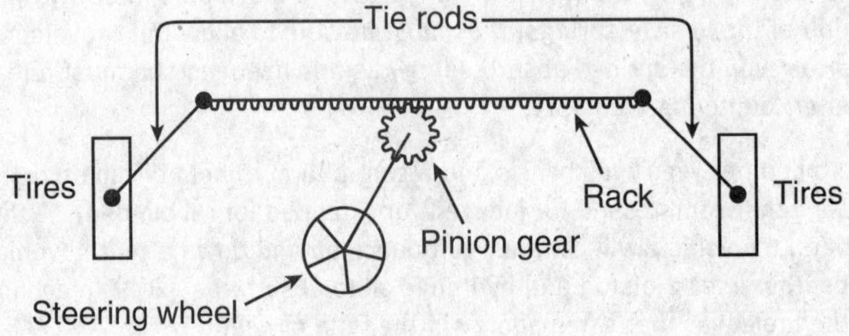

Tires

Tires are one of the most critical components of the suspension and steering system. They support the weight of the vehicle and make final contact with the road surface. The most common tire design is the **radial tire**, which is known for its stable footprint and low rolling resistance.

Tire construction begins with the **beads**. The bead is a circular piece of high strength material like steel wire that is encased in rubber. The bead forms the mounting point for the tire on the **rim**. **Body plies** form the main body of the tire and run from bead to bead. All other parts of the tire attach to the body plies, including the **liner** (sealed surface inside the tire), the **sidewalls**, and the **tread**.

Belts are used between the plies and the tread to help stabilize the tire's footprint (where the tire makes contact with the road). A stable footprint means better traction under all road conditions, and makes the vehicle handle and brake better.

Maintenance

The primary maintenance that should be performed on suspension and steering systems is lubrication of the ball joints, tie rod ends, and other components. This should be done with a hand-powered grease gun, and the amount of grease should be metered carefully.

The most important part of tire maintenance is proper inflation. Low inflation pressures can be very dangerous as well as increase the rolling resistance of the vehicle, thereby increasing fuel consumption. An underinflated tire can seriously damage the tire, which could lead to a blow out and accident. Too much tire pressure, or overinflation, can lead to excessive tire wear. An extremely over-inflated tire can also lead to a possible blow out and an accident. Automotive manufacturers label each vehicle with the recommended inflation pressures for the front and rear tires.

Brake Systems

Of all the systems on a car, the most important may well be the brake system. It is one thing to not be able to move forward; it is another thing again to not be able to stop.

Components

The major components found in any brake system include the following:

- **Brake pedal**: the mechanical connection between the driver's foot and the master cylinder.
- **Master cylinder**: located in the engine compartment just in front of the driver, the master cylinder generates the fluid pressure to operate the brake assemblies at the wheels.

- **Fluid reservoir**: provides fluid to the brake circuits. The fluid reservoir is located on top of the master cylinder.
- **Brake lines**: transmit fluid pressure from the master cylinder to the brake assemblies. These can be steel lines that run along the vehicle chassis or flex hoses that are used to connect the steel lines to the brake assemblies.

Assemblies

Brakes work by converting the energy of the vehicle's motion into heat energy through friction. The two major types of brake assemblies are as follows:

- **Drum brakes**: expanding shoes make contact with a rotating drum to create friction.
- **Disc brakes**: brake pads on either side of a rotating disc are "pinched" together to slow the vehicle.

Brake systems are **hydraulically operated**. A pumping piston, located in the **master cylinder**, is operated by the **brake pedal** and puts pressure on the system's **brake fluid**. Since fluids cannot be compressed, the brake fluid travels through the **brake lines** and moves the pistons in the **brake assemblies** to operate the brakes. The harder the driver presses on the brake pedal, the more fluid pressure is developed and more braking power is generated.

Dual Hydraulic Brake Systems

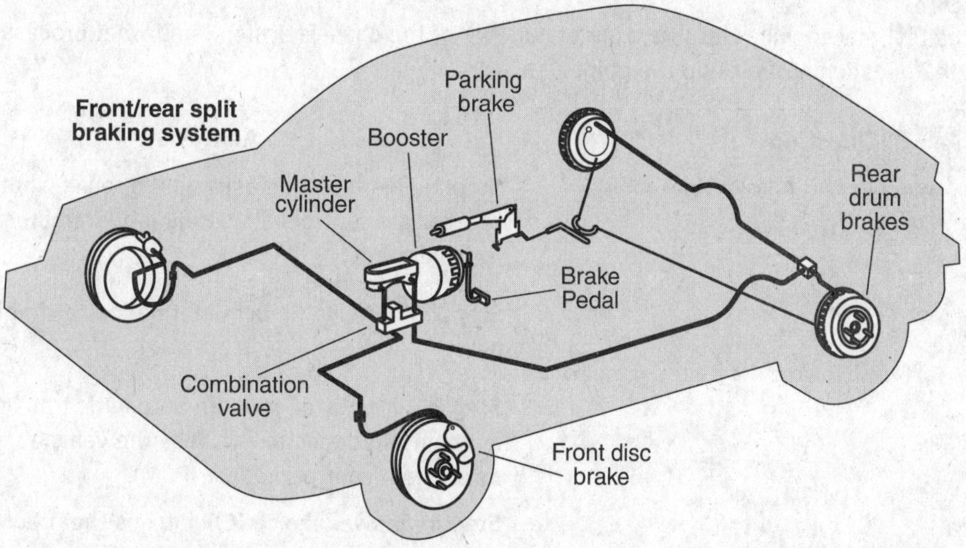

Drum brakes

The parts that develop the friction in a drum brake are the **brake shoes**. There are two brake shoes in any drum brake assembly. As hydraulic pressure is applied to the wheel cylinder, which is located between the brake shoes, the pistons inside it push apart and act against them. The brake shoes then push outward against the inner surface of the brake drum, and the resulting friction between the brake shoes and the brake drum slows the vehicle.

Disc brakes

Disc brakes are much more powerful than drum brakes because of disc brakes' superior ability to reject heat energy. The pistons used to actuate disc brakes are much larger, and can apply more force to the brake pads. Also, brake rotors used in disc brakes are more easily cooled, as they are better exposed to the cool air under the car and are often designed with air passages through them to enhance heat rejection.

The piston that operates a disc brake is housed in a **brake caliper**. This caliper "floats" laterally as the brake is applied and released, and makes it possible for one piston to operate the pads on both sides of the rotor. This makes the caliper operate similar to a C-clamp. As the screw is tightened on one side, the entire assembly is drawn together and pinches tightly on the item being clamped; in this case, the brake rotor.

Power brakes

Braking pressure is usually boosted using a **brake booster**. The brake booster is located between the brake pedal and the master cylinder. Engine intake manifold vacuum is utilized to generate greater force on the master cylinder, so higher hydraulic pressures can be generated in the brake system for the same amount of pedal force provided by the driver. Some brake booster systems use fluid pressure from the power steering system to increase braking power.

Antilock brakes

As in every other vehicle system, computer control is being incorporated into the brake system. Most cars now come equipped with **antilock brakes (ABS)**, which prevent wheel lock under hard braking conditions. The ABS system uses speed sensors attached to each wheel to tell a computer (the ABS computer) the relative speeds of each wheel. If the ABS computer detects a difference in wheel speed more than a preset amount, the computer uses pumps and valves in the ABS system to adjust the brake pressure for the affected wheel or wheels. This gives more control to the driver in slippery conditions and allows the vehicle to stop more predictably and safely.

Let's see how the test maker will assess your understanding of the chassis system, and what process you should use to address the question and pick up the points.

Question	Analysis
Virtually all modern automatic transmissions are controlled by	**Step 1:** This question asks you to recall what component controls almost all modern automatic transmissions. **Step 2:** There isn't much information to simplify here. **Step 3:** Virtually all modern automatic transmissions are controlled electronically by the vehicle's powertrain control module (PCM).
(A) the driver (B) oil pressure (C) the PCM (D) the engine	**Step 4:** Answer choice **(C)** matches the prediction.

Now try a similar question on your own.

Brakes slow a vehicle by converting motion energy into _______ energy.

(A) sound
(B) electrical
(C) fluid
(D) heat

Explanation

The correct answer is **(D)**. Brakes convert a vehicle's motion energy into heat energy through friction.

AUTO INFORMATION PRACTICE SET 1

Select the best answer for each question. This question set has 10 practice questions, which is the number of Auto Information questions you will see if you take the CAT-ASVAB.

1. How many revolutions of the crankshaft does it take to complete one cycle of events in a four-stroke cycle engine?

    (A) one-half
    (B) one
    (C) two
    (D) four

2. This image depicts what stroke in the four-stroke cycle?

    (A) intake stroke
    (B) compression stroke
    (C) power stroke
    (D) exhaust stroke

3. A 12:1 air-fuel mixture would be a _____ mixture.

    (A) rich
    (B) lean
    (C) stoichiometric
    (D) none of the above

4. Diesel engines utilize a

    (A) high compression ratio
    (B) spark ignition system
    (C) low compression ratio
    (D) carburetor

5. The antifreeze in the engine's coolant will

    (A) lower the boiling point of the coolant
    (B) raise the boiling point of the coolant
    (C) raise the freezing point of the coolant
    (D) lower the corrosion resistance of the coolant

6. Identify this engine component.

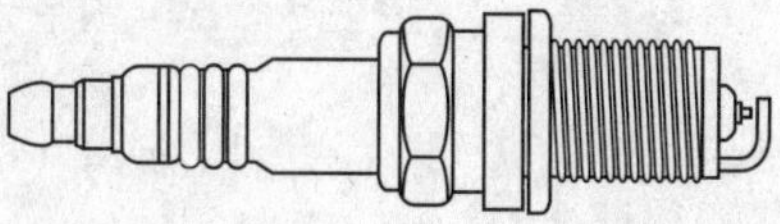

    (A) spark plug
    (B) fuel injector
    (C) wrist pin
    (D) valve spring

7. Electronic fuel injectors are controlled by the

    (A) powertrain control module
    (B) fuel pressure regulator
    (C) electric fuel pump
    (D) ignition switch

8. Where does combustion take place in the engine?

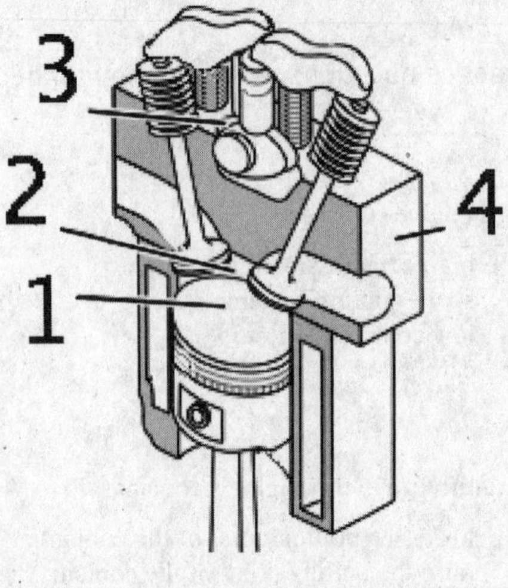

(A) 1
(B) 2
(C) 3
(D) 4

9. The PCM receives signals from the vehicle's
(A) sensors
(B) diagnostic data link
(C) actuators
(D) scan tool

10. Oil galleries in the crank shaft provide pressurized oil to
(A) the pistons
(B) the push rods
(C) the valves
(D) the connecting rod bearings

AUTO INFORMATION PRACTICE SET 2

Select the best answer for each question. This question set has 10 practice questions, which is the number of Auto Information questions you will see if you take the CAT-ASVAB.

1. How do shock absorbers dissipate suspension system energy?
   (A) by converting motion energy into electric power
   (B) by converting motion energy into potential energy
   (C) by converting motion energy into heat energy
   (D) by neutralizing chemical energy

2. The water-cooled engine component that rejects heat is the
   (A) water pump
   (B) thermostat
   (C) radiator
   (D) antifreeze

3. Multiport fuel injectors are located in
   (A) the throttle body
   (B) the combustion chamber
   (C) the intake manifold
   (D) the flame front

4. Which of the following is developed in the catalytic converter?
   (A) toxic gas
   (B) high heat
   (C) hydrocarbon gas
   (D) carbon trioxide

5. Coil-on-plug ignition systems eliminate the need for
   (A) secondary coil winding
   (B) spark plug wires
   (C) spark plugs
   (D) primary coil winding

6. The main component of an engine's charging system is the __________.
   (A) power charger
   (B) battery
   (C) rectifier
   (D) alternator

7. Automatic transmissions do not require a clutch because a __________ transmits power from the engine to the transmission.
   (A) half-shaft
   (B) torque converter
   (C) drive axle
   (D) constant-velocity joint

8. Rack and pinion gears are found in which of the following applications?
   (A) steering system
   (B) manual transmission
   (C) lubrication system
   (D) automatic transmission

9. The most important property that enables a hydraulic brake system to transmit pressure from the master cylinder to all individual brakes is __________.
   (A) viscosity
   (B) incompressibility
   (C) acidity
   (D) adaptability

10. Which of the following is the least expensive and most effective maintenance procedure to maintain maximum engine life?
    (A) lubricating all fittings
    (B) regularly changing automatic transmission fluid
    (C) regularly changing the oil and filter
    (D) properly maintaining the thermostat

AUTO INFORMATION PRACTICE SET 1

Answers and Explanations

1. **C** It takes two revolutions of the crankshaft to complete each four-stroke cycle.

2. **C** This image depicts the power stroke, which is the third stroke in the four-stroke cycle. The power stroke generates the engine's power.

3. **A** A 12:1 ratio of air to fuel would be a rich mixture, fast-burning and possibly producing black smoke.

4. **A** Diesel engines operate with a high compression ratio. The compression ratio for a diesel engine can range anywhere from 16:1 to 22:1.

5. **B** Antifreeze raises the boiling point of your coolant, thereby rendering it much more efficient at transferring heat.

6. **A** This is a spark plug. Spark plugs are threaded into the cylinder head, where they protrude into the combustion chamber and generate the spark to initiate combustion.

7. **A** Of the items given, only the powertrain control module is in charge of firing and regulating electronic fuel injectors.

8. **B** The combustion chamber, indicated in the diagram by the number 2, is where the actual combustion of the air-fuel mixture takes place. It is located in the cylinder head directly above the piston (1 in the diagram).

9. **A** As the "brain" of the car, it is the job of the PCM to gather signals from the car's sensors and generate outputs to control the vehicle's functions. The diagnostic data link, choice (B), connects a scan tool to the PCM; the scan tool, (D), is what technicians use to read the PCM.

10. **D** The oil galleries are drilled directly through the crank shaft in order to provide pressurized oil to the connecting rod bearings. The pistons, push rods, and valves do not receive pressurized oil from the crank shaft oil galleries, so (A), (B), and (C) are incorrect.

AUTO INFORMATION PRACTICE SET 2

Answers and Explanations

1. **C** The shock absorber provides damping to the suspension system. The damping effect converts motion energy into heat, which is then dissipated though the shock absorber body into the atmosphere. The shock absorber does not generate electrical power or potential energy, or neutralize chemical energy. Therefore, choices (A), (B), and (D) are incorrect.

2. **C** The radiator is the engine component designed to reject heat in a water-cooled engine. The other answer choices are incorrect because antifreeze is a subcomponent of the coolant, the water pump circulates the coolant, and the thermostat regulates the coolant temperature.

3. **C** Throttle body fuel injectors are located in the throttle body, so (A) is incorrect. Direct injection fuel injectors are located in the combustion chamber, so (B) is incorrect. The flame front is the burning portion of the fuel air mixture in the combustion chamber, so (D) is incorrect. Multiport fuel injectors are located in the intake manifold, so (C) is correct.

4. **B** The function of the catalytic converter is to further reduce unburned hydrocarbon fuel in the exhaust gas by decomposing the unburned hydrocarbon fuel in the exhaust gas into carbon dioxide, water vapor, and heat. As a result, high heat is developed in the catalytic converter. Answer choices (A), (C), and (D) are incorrect because they represent components not developed in the catalytic converter.

5. **B** Coil-on-plug ignition systems eliminate the need for spark plug wires because the ignition coil is mounted directly over the spark plugs. Answer choices (A), (C), and (D) are still necessary components in a coil-on-plug ignition system, and therefore are not correct.

6. **D** The alternator converts mechanical energy into electrical energy needed to operate the vehicle's systems and to charge the battery as required. Power charger, (A), is a generic term, not a specific component. The battery, (B), stores electrical energy and the rectifier, (C), converts alternating current from the alternator to direct current.

7. **B** A torque converter converts the rotational power of the engine to fluid power. A half-shaft, (A), connects the transaxle to the drive wheels in a front wheel drive vehicle. The drive axle, (C), sends power from the differential to the drive wheels. Constant-velocity joints, (D), are located between the transaxle and front drive wheels.

8. **A** Rack and pinion gears convert rotational motion from the steering wheel to the linear motion needed to change the direction of the wheels. This type of gear arrangement is not used in any of the other systems listed.

9. **B** If brake fluid were compressible, the fluid displacement created by the master cylinder would be diminished at the individual brake piston assemblies and braking power would be severely diminished. Viscosity, (A), refers to how easily a fluid flows. While this could have some effect on brake performance, it is not the most important property. Acidity, (C), is irrelevant to transmitting pressure and fluid movement. Adaptability, (D), is not a relevant property.

10. **C** Proper lubrication of internal moving parts is essential to long engine life. Excessive wear on internal engine parts can occur when the engine oil becomes dirty or experiences degradation of its lubricating properties. Internal combustion engines are lubricated by circulating oil internally, not by lubricating fittings, (A). Changing automatic transmission fluid, (B), does not need to be done frequently and is important to the life of the transmission more than the engine. Thermostats, (D), generally don't require maintenance; if they fail to function properly they are usually replaced.

Review and Reflect

Look back over your work on the practice questions. If you got some questions wrong, think about why.

- Are there specific systems you need to review? If so, review this chapter, and then try to draw diagrams of those systems on your own.
- Did you get tripped up over the names of parts? If so, review the chapter, focusing on the words in bold. Also, consider making those words into flashcards.
- How did you do overall? Do you need to review this chapter comprehensively?

Want more practice with Automotive Information? Try more questions in the Qbank online.

QBANK

kaptest.com/login

CHAPTER 12

SHOP INFORMATION

KNOW WHAT TO EXPECT

In order to work effectively in any industrial or technical environment, a technician must be able to identify tools correctly and use them safely. Hand tools are the foundation of industry; very little work would get done without them. Even the largest and most complex piece of equipment would not run for very long if hand tools were not available for its maintenance and repair.

The ASVAB Shop Information (SI) test will assess your familiarity with common tools and their uses. You are given 6 minutes to answer 10 Shop Information questions on the CAT-ASVAB. On the paper-and-pencil test, the Auto Information and Shop Information subtests are combined into one, with 11 minutes to answer 25 questions about both topics.

MEASURING TOOLS

LEARNING OBJECTIVES

In this section, you will learn to:

- identify tools used to measure distances
- determine when a micrometer is needed
- understand the use of calipers
- differentiate between types of spirit levels
- determine the use of a steel square

A critical skill for any technician is the ability to make accurate measurements.

Tape Measures and Steel Rules

A **steel rule** or a **tape measure** can be used to determine distances. These measure in fractions of an inch as low as $\frac{1}{32}$, but are ineffective when more accuracy is required. This is because the scale becomes too difficult to read with smaller measurements. Steel rules are also useful for the layout of straight lines.

Steel Rule

Micrometers

When accuracy to the thousandths of an inch is required, a **micrometer** is used. The most common type of micrometer is the **outside micrometer**. The outside micrometer is made to measure the thickness of flat objects or the outside diameter of cylindrically shaped objects. It is built similarly to a C-clamp (see the discussion of C-clamps later in this chapter). The **spindle** of the micrometer is rotated in and out to adjust the distance between it and the **anvil**. All measurements are taken between the opposing faces of the spindle and the anvil.

Outside Micrometer

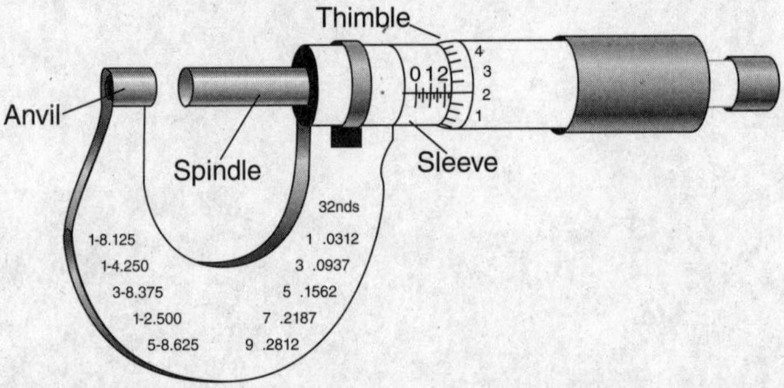

The **thimble** of the micrometer is attached directly to the spindle. The **sleeve** of the micrometer is stationary, and has markings on it indicating how far the spindle has been moved relative to the anvil. As the thimble moves outward, it uncovers the graduations on the sleeve. The position of the thimble relative to the sleeve is what reveals the micrometer reading.

Forty turns of the thimble will move the spindle exactly 1 inch. This means that each turn of the thimble moves the spindle $\frac{1"}{40}$ or 0.025". The graduations on the sleeve are also marked every 0.025" to keep track of how many times the thimble has been turned. Each time the thimble completes one turn, another mark on the sleeve is uncovered. Four complete turns of the thimble move the spindle $\frac{1}{10}$ of an inch or 0.100", so this is marked with a large "1" on the sleeve. The mark at 0.200" is a "2," and so on.

The outside of the thimble has 25 evenly spaced graduations on it. Since a full turn of the thimble moves the spindle 0.025", this means that one of the graduations on the thimble is the equivalent of 0.001". Reading a micrometer is simply a matter of adding the measurement on the sleeve to the number on the thimble that aligns with the longitudinal line on the sleeve.

Example of Reading a Micrometer

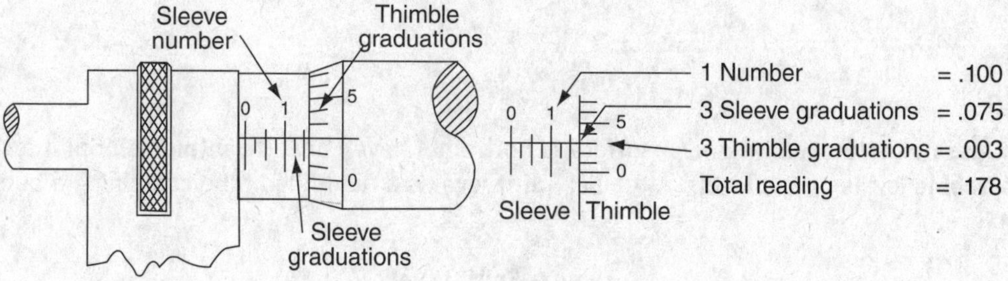

Calipers

A **caliper** is a two-legged instrument that is used to measure the distance between two sides of an object. It can also be used to transfer a measurement from one object to another. Calipers that measure the external size of an object are called **outside calipers**, while those that measure the internal size of an object are called **inside calipers**.

Outside Caliper

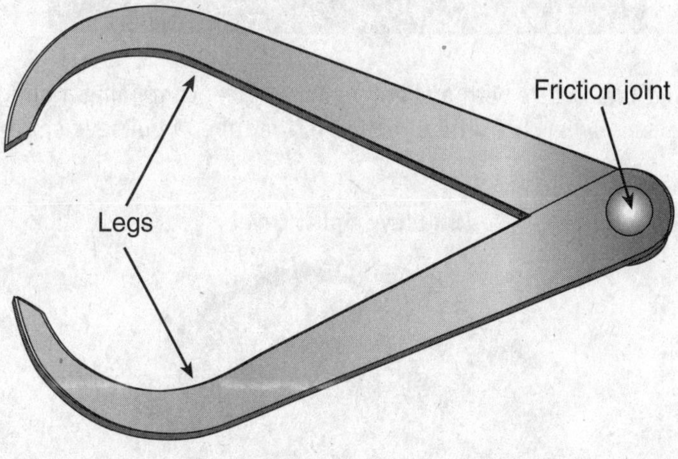

Inside Caliper

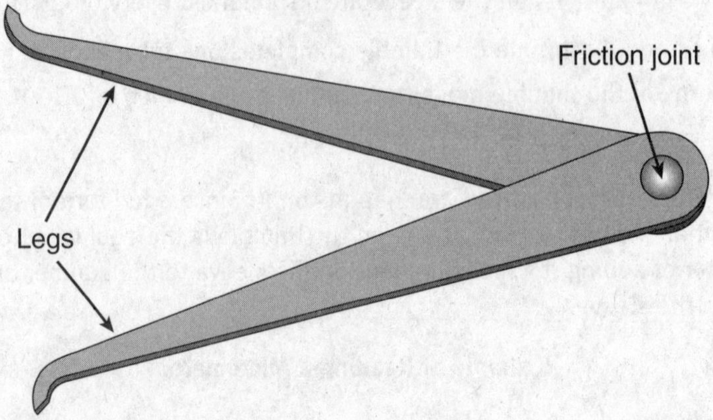

Spirit Levels

A **spirit level** is used to determine whether a surface is horizontal (level) or vertical (plumb). Spirit levels are also known as **bubble levels** or just **levels**. A **tubular spirit level** is a fluid-filled tube containing a bubble that is centered when level.

Tubular Spirit Level

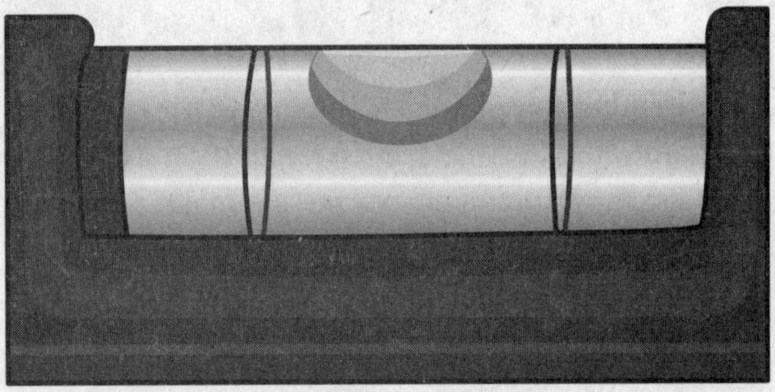

A **bullseye spirit level** is a fluid-filled circle with a slightly convex face, containing a bubble that is centered when level. While a tubular spirit level only levels in the direction of the tube, a bullseye spirit level can level a surface across a plane.

Bullseye Spirit Level

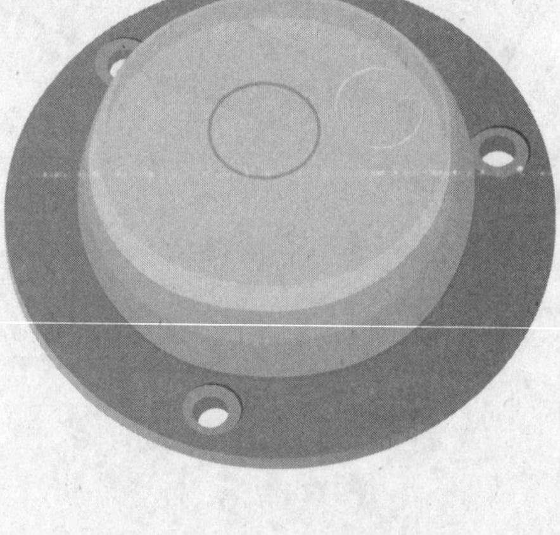

Steel Squares

A **steel square** is used to measure or lay out angles. A steel square is also known as a **carpenter's square** or **framing square**, because carpenters commonly use it to frame stairs and rafters. Its two arms meet at a 90-degree angle. The long arm is known as the blade (or body) and the short arm is known as the tongue.

Steel Square

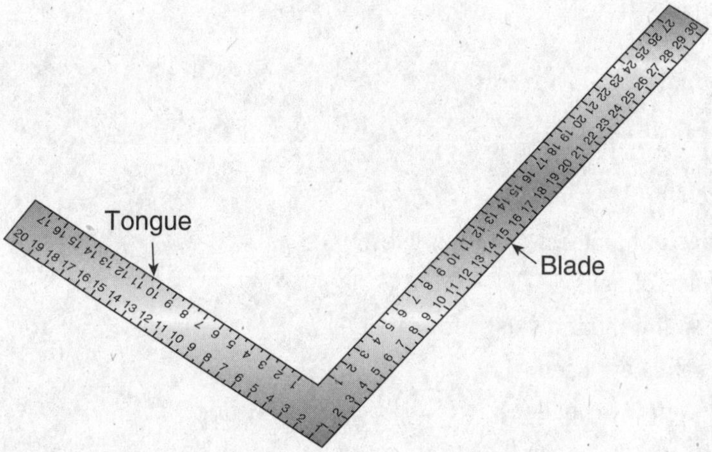

Questions on the ASVAB may ask you to choose the most appropriate tool for a job. Look at this example:

Question	Analysis
You need to measure the length of a piece of wood within $\frac{1}{40}$". Which of the following would be the best tool to select for this job?	**Step 1:** This question asks which tool would be best for measuring distance within $\frac{1}{40}$". **Step 2:** There is nothing to simplify. **Step 3:** You can narrow your prediction to tools used for measuring distance. In particular, the tool must be capable of making very accurate measurements of distance.
(A) steel rule (B) measuring tape (C) micrometer (D) spirit level	**Step 4:** Choice **(C)** matches that prediction. Choice (D) can be eliminated because it is not used to measure distance, and choices (A) and (B) can be eliminated because they cannot perform measurements of distance below $\frac{1}{32}$".

Now give this one a try on your own:

You want to be sure that the joint you are installing is at a flush right angle. Which of the following tools could be used for that?

(A) steel rule
(B) spirit level
(C) steel square
(D) caliper

Explanation

The key term here is "right angle." The only one of these instruments that allows you to measure a right angle is a steel square, choice **(C)**.

STRIKING TOOLS

LEARNING OBJECTIVES

In this section, you will learn to:

- differentiate types of hammers and their uses
- understand the use of nails
- understand rivets and the tools used to install them
- identify how punches are used
- determine when a drift is needed

Hammers

Striking objects in order to remove or install them is the job of the **hammer**. Almost every technician uses a hammer of some kind, whether it is used to drive a nail or to loosen pieces of an assembly. Hammers come in many designs and sizes according to their intended use.

The type most often used by metal workers and mechanics is known as a **ball-peen hammer**. This hammer is designed with a regular striking face, like most hammers, but also has a rounded end that can be used for shaping metal and making gaskets.

Ball-Peen Hammer

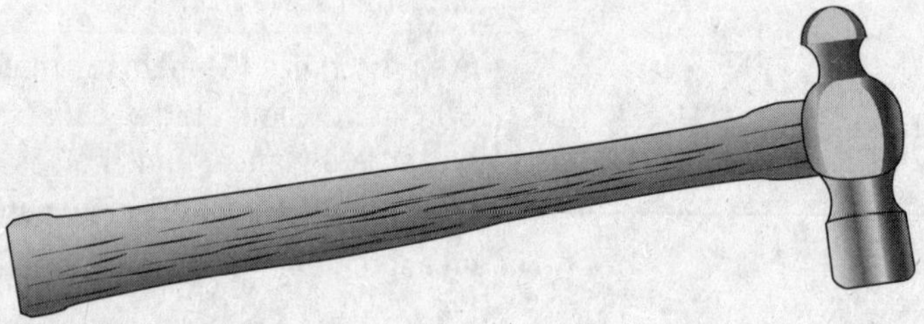

Mechanics will often use a **rubber mallet** to prevent damage to the parts they are striking. Rubber mallets are not made for maximum impact; they are designed to install or remove delicate parts such as hub caps while preventing damage to their surface. A carpenter would use a **wooden mallet** to achieve the same effect.

Rubber and Wooden Mallet

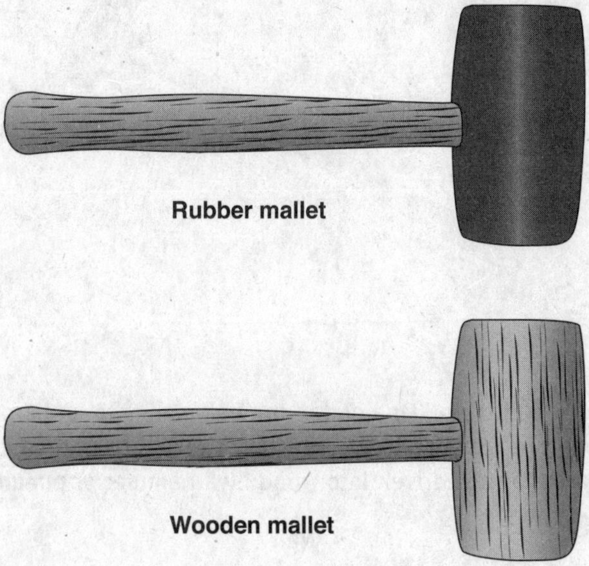

Carpenters often will use a **claw hammer,** which serves a dual purpose. The hammer head has two ends; one to drive nails and the other to remove nails. Claw hammers come in a variety of sizes, and these are determined by the weight of the hammer head. A general-purpose claw hammer would have a 13 oz head, while a rough-framing hammer, typically used for framing wooden houses, might be anywhere from 16 to 20 oz.

Claw Hammer

Heavy jobs, such as driving fence posts or breaking down drywall, require the use of a **sledge hammer**. This is a long-handle hammer with a large steel head that typically requires both hands to operate.

Sledge Hammer

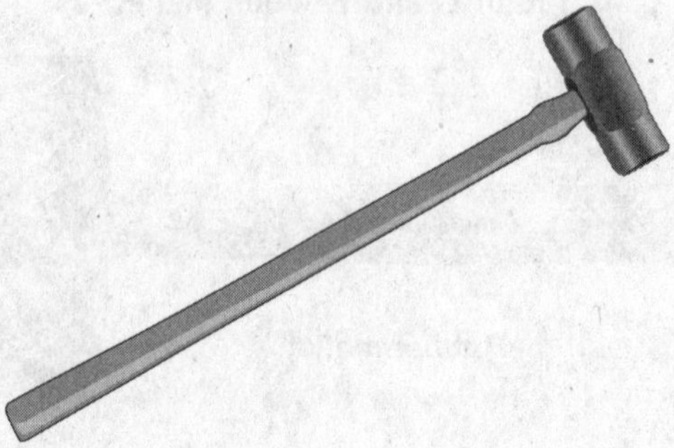

Nails

A **nail** is a pin-shaped fastener made of metal. One end of the nail comes to a sharp point, while the other end usually has a flat head. However, there are also headless nails. The length of the nail body between the head and point is called the shank. Nails are typically driven into wood by a hammer or **pneumatic nail gun**.

Nail

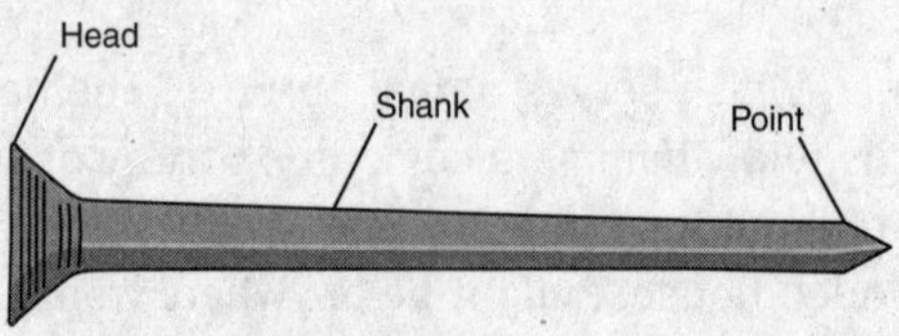

Rivets

Rivets are metal fasteners that can be used to assemble parts. A rivet is simply a pin with a head at one end. A rivet is installed in a hole (the same diameter as the rivet) that is drilled through two pieces that are to be assembled. With the two pieces tightly clamped, the head of the rivet is placed on a hard surface, while the other end is formed into a head using a hammer or special riveting tool. This creates an assembly that is semipermanent, as the rivet must be drilled out to remove it.

Installation of a Regular Rivet

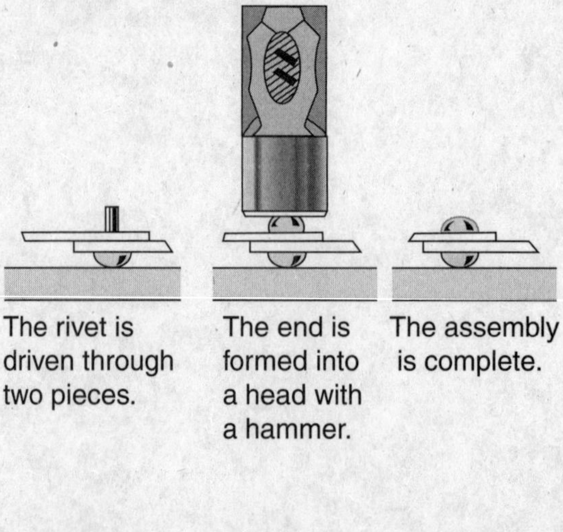

Chisels, Punches, and Drifts

Hammers are often used in conjunction with a chisel, a punch, or a drift. A **chisel** normally has a long, sharp edge and is used for cutting. A **punch** is narrow and is used for driving small fasteners and making layout marks. A **drift** is used for striking an object where it is important that the hammer itself not come in direct contact with the work.

The most common chisel is the **cold chisel**, which has a straight, sharp edge for cutting off bolt heads or separating two pieces of an assembly. Cold chisels get dull from time to time and must be sharpened on a bench grinder.

Punches are made in a number of different designs, but the most common ones are the **pin punch** and the **center punch**. Pin punches are straight and cylindrical in shape. Pin punches come in various sizes, normally starting as small as $\frac{1}{16}$" and going up to $\frac{1}{4}$" in diameter. The pin punch is used to drive pins out of holes, and to follow the pin through the hole as it forces the pin out.

Pin Punch

Center punches are used to make small indentations that serve as starting marks for drilling operations. Making a small indentation with a center punch can help the drill bit stay on target long enough to get a hole started. Attempting to drill a hole in metal without first marking it with a center punch can allow the drill bit to "walk" across the work and completely miss the original target.

Center Punch

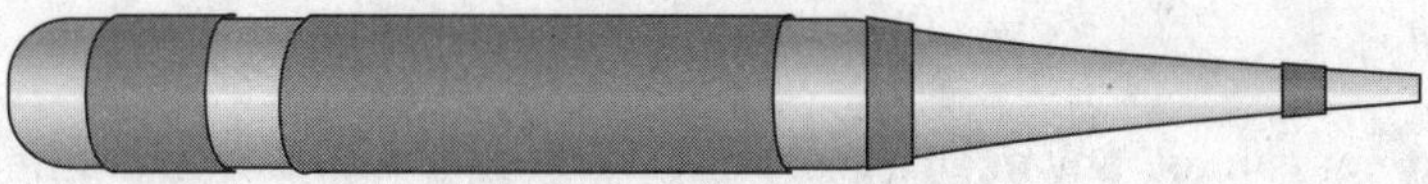

When using a hammer to drive parts in or out of an assembly, it is easy to damage the parts if they are struck directly by the hammer. The head of a ball-peen hammer is made from forged steel, and thus can easily damage parts that are made from softer materials. Placing a **drift** against the object and then striking the drift with a hammer prevents damage to the part that is being driven. Drifts are often made from soft metals such as mild steel, brass, and even aluminum.

Brass Drift

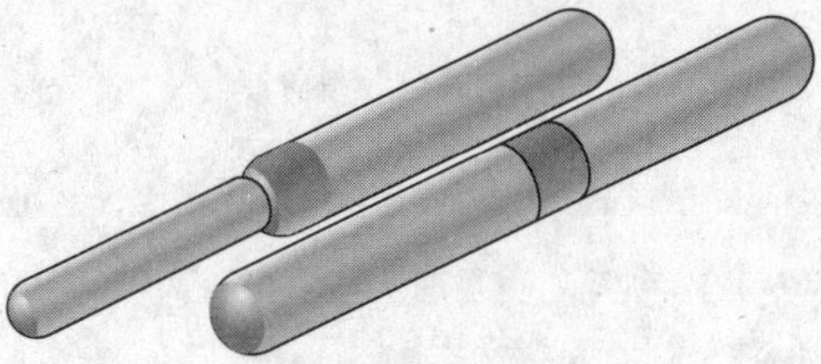

Here is a sample ASVAB question that asks you to differentiate between striking tools:

Question	Analysis
Which of the following hammers would be most appropriate to install a hubcap?	**Step 1:** This question asks which hammer would be most appropriate to install a delicate part. **Step 2:** Not much info to simplify. **Step 3:** You can narrow your prediction to hammers best used for delicate jobs like installing a hubcap.
(A) claw hammer (B) pneumatic hammer (C) ball-peen hammer (D) rubber mallet	**Step 4:** Choice **(D)** matches that prediction. Choices (A) and (C) can be eliminated, because their striking heads are too hard. Choice (B) can be eliminated because it exerts too much force.

Now try one on your own.

You hammered a nail into the wrong part of a board and you want to remove it. With which of the following hammers could you do that?

(A) ball-peen hammer
(B) wooden mallet
(C) rubber mallet
(D) claw hammer

Explanation

Choices (A), (B), and (C) do not contain any device that could remove a nail. However, the claw on a claw hammer, **(D),** is designed especially to remove nails. That is the tool you would need.

TURNING TOOLS

LEARNING OBJECTIVES

In this section, you will learn to:

- identify different types of screwdrivers and their coordinating fasteners
- distinguish between different types of wrenches and the benefits of each
- understand how wrenches and sockets are sized
- describe what type of wrench to use for a particular project
- understand the concept of torque as it applies to hand tools
- understand when and why nuts are used
- identify the types of fasteners used with wrenches
- understand the system of labeling common fasteners

Screwdrivers

Screwdrivers, which come in many different sizes and styles, are one of the most common tools used to install and remove fasteners. The oldest screwdriver design is the **flat tip** type, which is basically a flat blade made to turn a screw with a single slot across the top of it.

Flat Tip Screwdriver

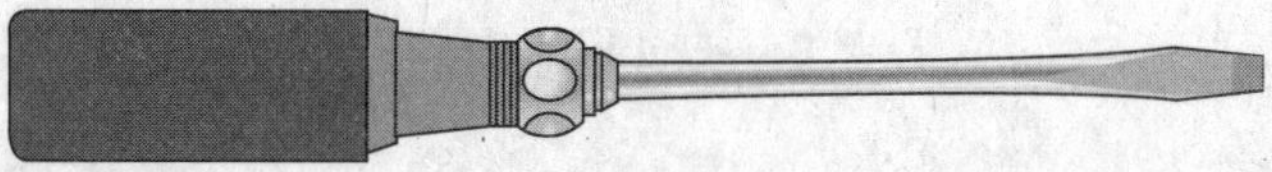

The flat tip screwdriver is far less popular now in light of some newer designs, including the **Phillips**, **Robertson**, and **Torx** screwdrivers. A Phillips screwdriver is recognizable by its tip that looks like a plus sign. Robertson screwdrivers have a square tip, and Torx screwdrivers have a characteristic six-pointed star-shaped tip. All of these newer designs grip the fastener, making the screw easier to remove and install. Since the screwdriver makes better contact with the screw, it is also possible to fasten it more tightly.

Phillips Tip Screwdriver

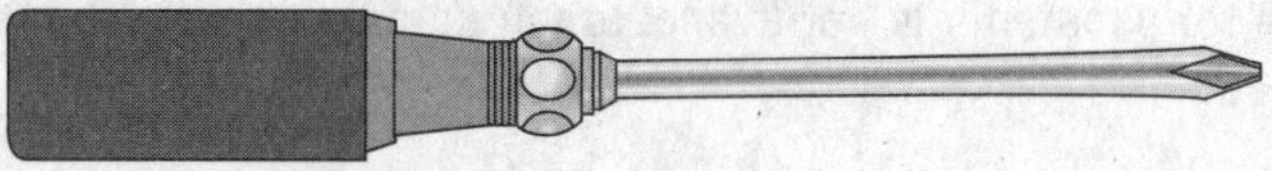

Robertson Screwdriver

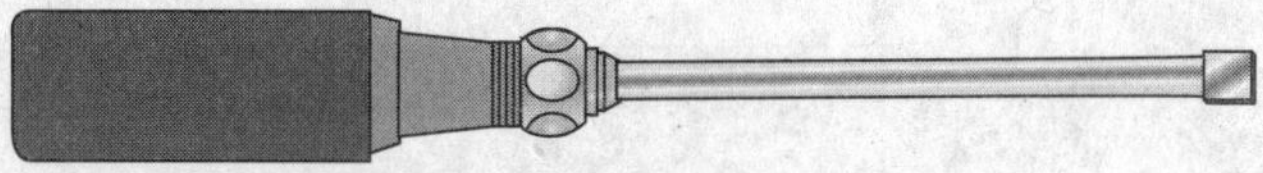

Torx Screwdriver

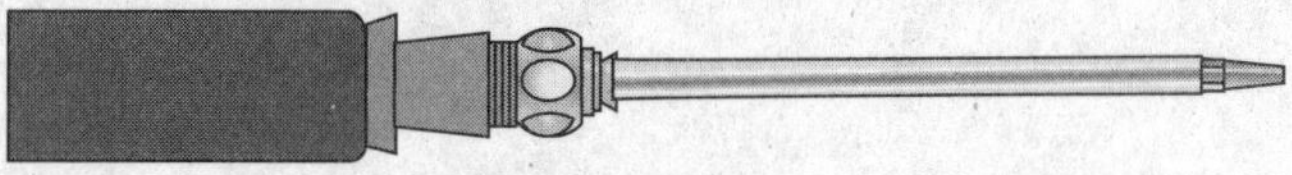

Screwdrivers are used with only one type of fastener—**screws**, which are **threaded** fasteners that vary in size and shape depending on the project. The threads on screws are designed to tighten when turned clockwise. Read more about how fasteners are classified later in this chapter in the "Nuts and Bolts" section. **Washers,** disk-shaped rings, may be used alongside screws to protect the work surface and distribute the force caused by tightening the screw.

Wrenches

With all wrenches, the longer the wrench, the more leverage and the more twisting force (**torque**) it can apply to tighten or loosen a fastener. There are two basic types of wrenches; the **open-end** and the **box-end.** The open-end wrench is made for speed. Since the end is open, it is easy to slide the wrench on and off a fastener, such as a cap screw. To loosen tight fasteners, it is a good idea to use a box-end wrench. The box end wraps completely around the head of a bolt, and therefore makes greater surface contact, distributing the force more evenly.

Open-End Wrench

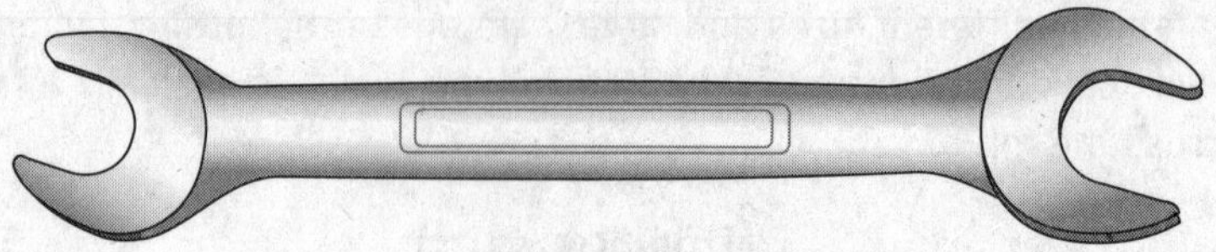

Box-End Wrench

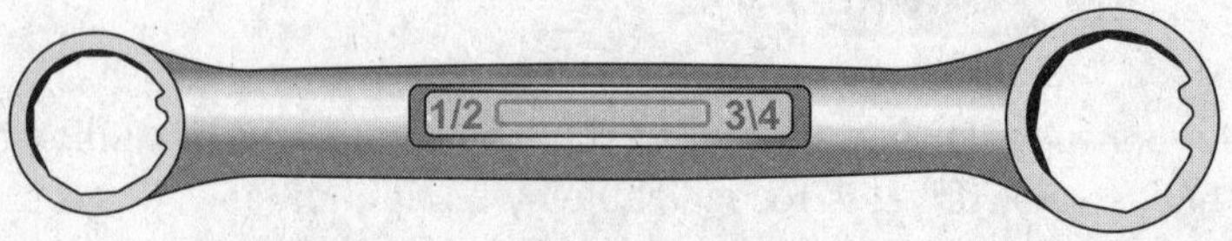

The most common wrench arrangement is the **combination wrench**. This design has an open end and a box end on opposite ends of a wrench. Both ends are made to fit the same size fastener, but the technician can loosen the bolt with the box end, and then finish removing the bolt more quickly using the open end.

When the specific size wrench for a fastener is not available, an **adjustable wrench** (sometimes referred to as a **Crescent® wrench**) can be used.

Adjustable Wrench

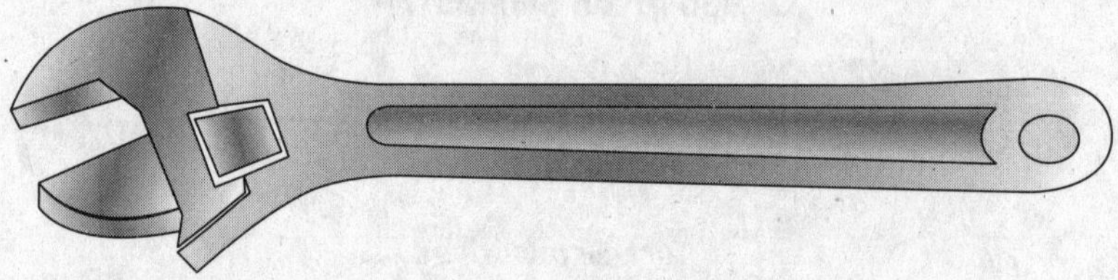

Sockets

An alternative to using wrenches to loosen fasteners is the use of **sockets**. Like wrenches, sockets come in two designs. **Six-point** is a stronger design because of its greater wall thickness, and is usually the mechanic's first choice in the smaller socket drive sizes. However, **twelve-point** is a useful design for certain applications, as it is easier to align with the bolt head in tight spaces.

Six-Point and Twelve-Point Sockets

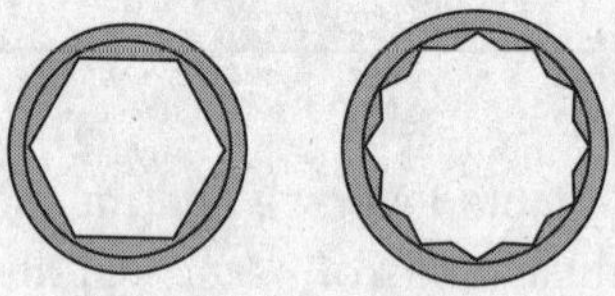

To determine the size of a socket required for a job, simply measure the distance between two parallel sides of the bolt head. If the bolt head measures $\frac{9}{16}$" across two parallel sides, then a $\frac{9}{16}$" socket is required to loosen it.

Sockets come in a variety of **drive sizes**, which are determined by the size of the opening that attaches to the **drive tool**. For instance, if the square end of a socket measures $\frac{3}{8}$" across, then that would be a $\frac{3}{8}$" drive socket. The most popular drive sizes are $\frac{1}{4}$", $\frac{3}{8}$", $\frac{1}{2}$", and $\frac{3}{4}$". Larger drive sizes are available for the very large fasteners used in heavy industry. Take a look at the diagram below.

Common Socket Drive Size Range

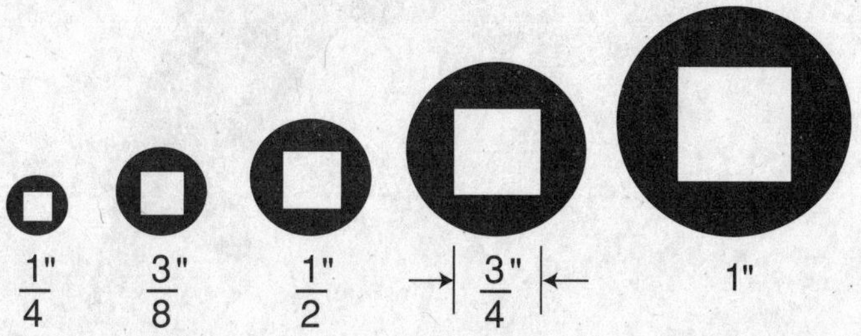

Sockets are very versatile, because they can be used with a variety of drive tools. The most common drive tool for sockets is the **ratchet,** which turns the fastener in only one direction as the handle is moved back and forth through a narrow arc. Ratchets are reversible, so they can be set to tighten or loosen a fastener. They can also be more useful than open-end/box-end wrenches in tight quarters (especially when paired with an extension bar) because you do not need to replace the tool with each turn—just move the handle back and forth.

Ratchet

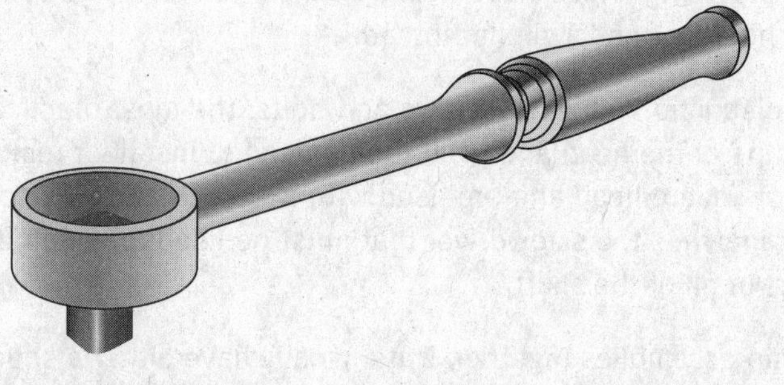

Sockets can also be used with **pneumatic** (compressed air) power tools, such as an **air impact wrench.** The air impact wrench can remove fasteners quickly by applying tremendous amounts of torque (twisting force) and using a hammering action that vibrates fasteners loose. It is important to remember that only **impact sockets** should be used with an air impact wrench.

Air Wrench

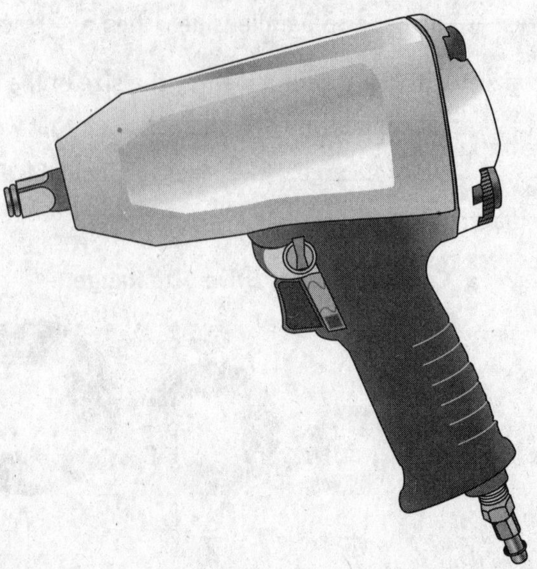

Nuts and Bolts

Some of the most common types of tools to be tested on the ASVAB are fasteners (screws and bolts). It pays to review this wide range of hardware.

Wrenches of all types are used with a variety of **threaded** fasteners called bolts and nuts. A bolt has **external threads,** whereas a nut has **internal threads**. Bolts typically have a hexagonal or square head, which is held in place with a wrench while the nut is tightened (also with a wrench) to fasten the assembly. A threaded bolt can only be inserted into a nut or hole that has a similar thread.

While there is not a clear-cut distinction between screws and bolts, the most practical way to distinguish between the two is based upon the shape of the head and what tool is used to install or remove the fastener. While bolts have a hexagonal (six-sided) or square head and are used with a wrench and nut, screws typically have a round head with an indentation that matches the screwdriver that must be used to install or remove the screw. Both fasteners are threaded on part or all of the shaft.

Nuts thread onto bolts to clamp assemblies together. Nuts usually have either a square head or a hexagonal head, and can be locked into position using several different methods.

Wing nuts make it possible to disassemble a component by hand. The two "wings" attached to the nut make it easy to tighten and loosen without the aid of hand tools.

Wing Nut

A **castellated nut** uses a **cotter pin** to lock it into place. The cotter pin passes through a hole in the bolt or stud that the nut is threaded on, engaging the cutouts in the nut.

Castellated Nut

Cotter Pin

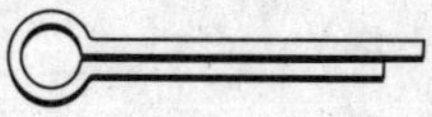

Lock nuts have a nylon insert incorporated into its threads that provides enough interference to prevent the nut from loosening, thus locking it in place.

Lock Nuts

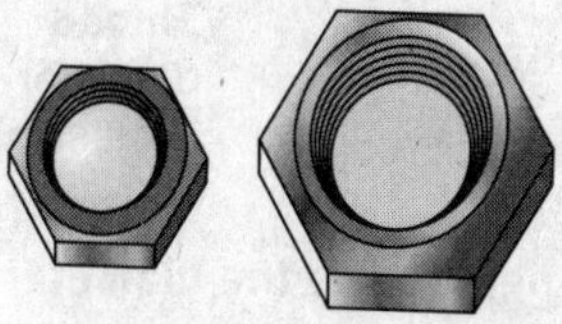

The type of thread that is placed on a fastener will vary according to the diameter of the fastener and the intended strength of the finished product. Threads are identified by their **pitch**, and this is measured using a **thread pitch gauge**.

Fractional-measurement fasteners (measured in fractions of inches) use threads that are identified by the number of threads per inch. There are two basic thread classifications within this group: **Unified National Coarse (UNC)**, and **Unified National Fine (UNF)**. A UNC or coarse thread would have relatively few threads per inch, where a UNF or fine thread would have a larger number of threads per inch.

A bolt that is $\frac{3}{8}$" in diameter could, therefore, have two possible thread pitches. If it were a UNC bolt, it would have 16 threads per inch, whereas if it were a UNF bolt, it would have 24 threads per inch.

Bolt Designation Numbers

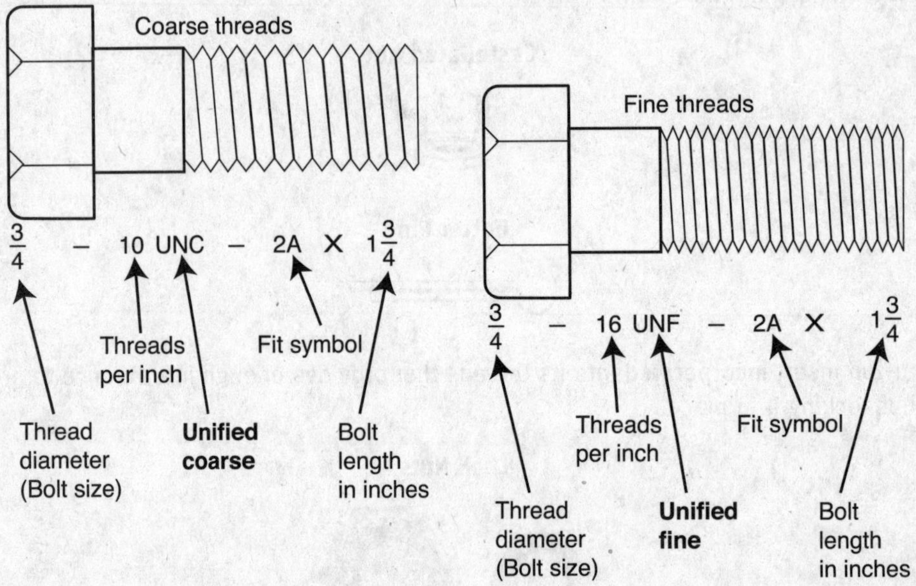

Two other important measurements of a fastener include the **diameter** and the **length.** The diameter is the distance across the unthreaded portion of the bolt. This would give an indication of the size of hole that the fastener is made to be installed in. The length of the bolt is the distance between the underside of the bolt head and the end of the bolt. Note that the bolt head does not count toward the length of the bolt.

Fastener Diameter and Length

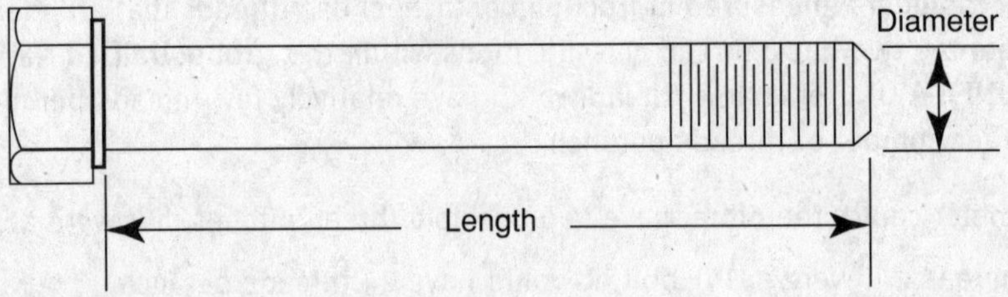

Let's check out one way fasteners could be tested on the ASVAB:

Question	Analysis
You are assembling a project with bolts that you would like to be able to remove later, even if you do not have a wrench with you. What would be the best type of nut to use in this assembly?	**Step 1:** You're asked to determine the best type of nut to use in the given project. **Step 2:** The key phrase is "even if you do not have a wrench with you." Think about what type of nut can be removed easily with just your hands. **Step 3:** Make a prediction: a wing nut can be removed using only your hands.
(A) lock nut (B) standard nut (C) wing nut (D) castellated nut	**Step 4:** Choose answer **(C).**

A Test Day question may ask you about the correct size tool to use. Take a look at this question:

You have a bolt that needs to be removed, but an $\frac{11}{16}$" open-end wrench is just slightly too big to fit properly. Which of the following sizes should be tried next?

(A) $\frac{3}{4}$"

(B) $\frac{5}{8}$"

(C) $\frac{3}{8}$"

(D) $\frac{5}{16}$"

Explanation

The key phrase to focus on in this question is that the $\frac{11}{16}$" wrench is just slightly too big. So, you are looking for a wrench that is a bit smaller. It can be tricky to tell which fraction is smaller, but a quick way to do that is to find a common denominator for each fraction. (Review the fraction rules in chapter 6: Arithmetic Reasoning if you need a quick reminder.) Convert all the fractions to have 16 as the denominator so they are easier to compare. Choice (A) is larger than $\frac{11}{16}$", so that can be eliminated right away. Choices (C) and (D) are significantly smaller than $\frac{11}{16}$", so they also do not fit what we are looking for. Choice **(B)**, on the other hand, is the size of a wrench just smaller than an $\frac{11}{16}$" wrench ($\frac{5}{8}$" is equal to $\frac{10}{16}$"), so that is the correct answer.

FASTENING TOOLS

LEARNING OBJECTIVES

In this section, you will learn to:

- identify where and why snap rings are used
- differentiate ways to join metals
- determine when and how to solder
- determine when and how to weld

Ring Fasteners

Retaining rings (or **snap rings**) are used to prevent end-movement of cylindrical parts in bores or parts mounted on shafts. **External snap rings** are installed in grooves on shafts, whereas **internal snap rings** install in grooves inside a bore. Snap rings are installed and removed using snap-ring pliers.

Internal Snap Rings and External Snap Rings

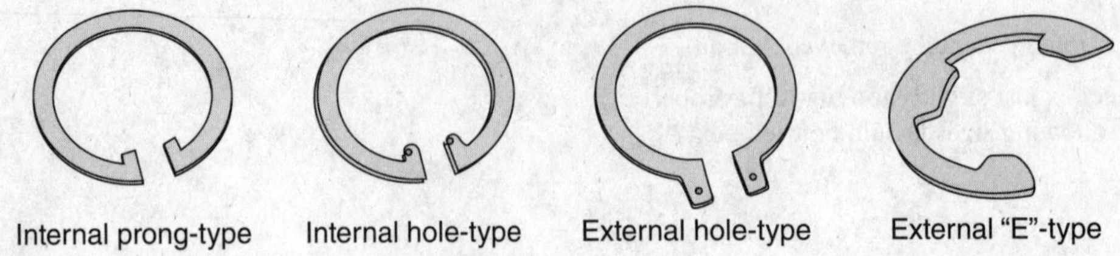

Soldering Tools

Soldering is a process that joins metals by bonding a metal alloy to their surfaces. It is a low-temperature process that can be performed with simple tools and inexpensive materials.

Most solder is an alloy of lead and tin. The percentages of each metal in the solder will vary depending on the desired properties of the solder, e.g., its melting point. Higher percentages of lead will result in a lower melting point.

The most critical part of soldering is the cleaning of the surfaces to be joined. Any oxides or other contaminants on the surface to be soldered will prevent a solid connection from being made. The best way to prepare a surface for soldering is to use a **flux** that will clean the surfaces with a chemical action. Electrical connections require a **rosin flux**, and solders made for this will have the flux contained in the core of the solder. This is known as **rosin-core solder**.

There are a number of tools that can be used to generate the necessary heat to melt the solder, but in most cases a **soldering iron** is used. Most soldering irons are electrically powered, and can draw anywhere from 25 to 100 watts. Low-power irons would be used to solder electrical connections, whereas higher powered ones would be used in sheet metal work.

Soldering Iron

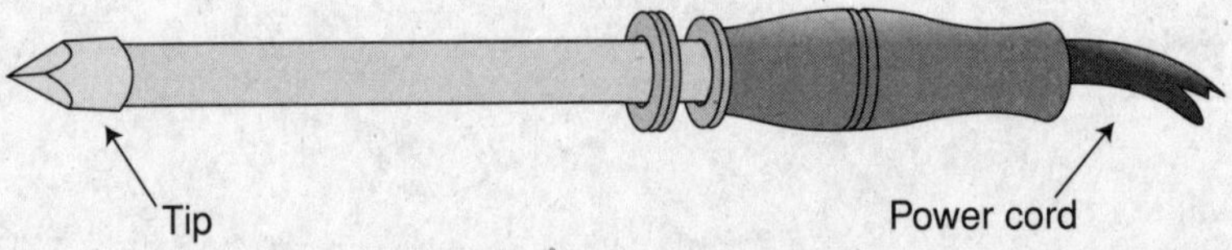

A tool that is often used for soldering electrical connections is the **soldering gun**. A soldering gun has a two-step trigger that allows the technician to quickly select a low or high heat setting. The main advantage to a soldering gun is the very rapid warm-up cycle. A soldering gun also has a light built into it for illuminating the work.

Soldering Gun

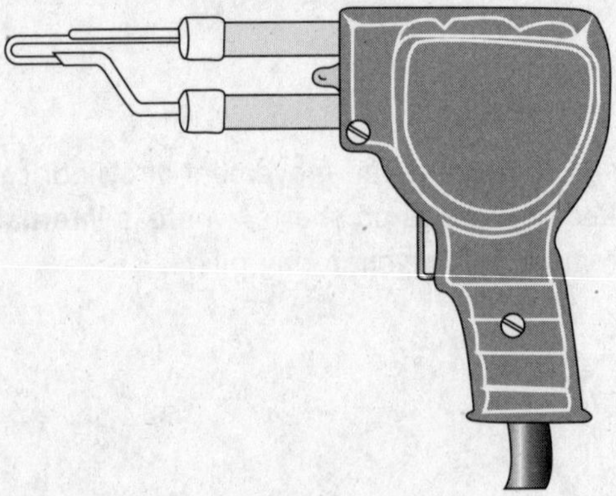

When soldering, start by mechanically cleaning the materials to be joined. This may involve the use of a stainless steel brush or even sandpaper. Heat the surfaces gently while applying flux to them, and then add solder once the flux has thoroughly cleaned the material. **Tin** the surfaces by spreading the solder thinly over the surface, making sure to thoroughly heat any area that is to be tinned.

Before soldering an electrical connection, it is always a good idea to make a solid mechanical connection first. This will involve stripping insulation from wires and twisting them together. Keep in mind that the entire connection needs to be heated before applying solder to it. If only the solder is heated and it then is "melted" onto relatively cool wires, a "cold" solder joint is created that can exhibit unusual electrical properties.

Welding Tools

When joining metals, the best way to achieve a high-strength joint is to use a **welding** process. Welding is very different from soldering in that it involves melting the base metal of the objects to be joined, a process that requires very high temperatures.

There are two major types of welding processes: **oxyacetylene welding** and **electric-arc welding.** Oxyacetylene welding involves the use of a torch that is fueled with oxygen and acetylene. Burning these two gases together creates an extremely hot flame, hot enough to melt steel and other ferrous (iron-based) materials. **Filler rod** is melted along with the base metal to produce a finished weld.

There are many specialized processes that fall under the electric-arc welding classification. The simplest and most easily recognized type is known as **stick welding.** Stick welding involves the use of an electric-arc welding machine and two cables: one that attaches to the work being welded through a ground clamp, and the other going to an electrode that is sometimes referred to as a **stinger**. The stinger is held by the welder and is used to hold the **welding rod.**

Stick Welder

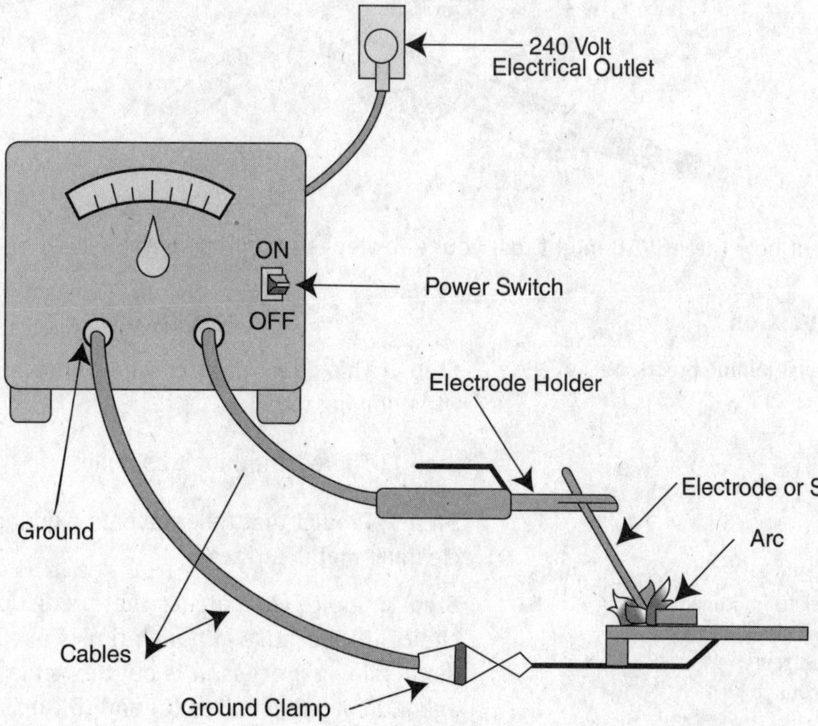

When the welder touches the welding rod on the work piece, an electric arc is formed. This electric arc generates a tremendous amount of heat and accomplishes three things. First, the heat melts the base metal of the material being welded. Second, the heat melts the electrode and deposits this metal in the weld as filler. Lastly, the **flux** on the outside of the electrodes burns and the generated gases form a "shield" around the weld. This gaseous shield prevents air from reacting with the hot metal and therefore weakening the weld.

As the welder continues the weld, the electrode burns and becomes shorter. The means that the welder must continue moving the stinger closer to the work until the electrode has been consumed. Once the electrode is used up, the welder breaks off or stops the arc, and a new electrode is placed in the stinger to continue the weld.

If more heat is needed to perform the weld, the welder can increase the amount of electric current that is supplied by the welding machine. Generally speaking, the larger the pieces that are being welded or the larger the electrode that is being used, the greater the amount of current required to get the weld done.

The electric arc that is generated during this process gives off a very high intensity light. The ultraviolet light is so intense that it can "sunburn" exposed skin and burn the retinas of unprotected eyes. Welders must, therefore, cover all exposed skin with protective clothing (preferably leather) and wear face shields (helmets) with light filters. The filters are usually very dark, so the welder can view the welding process itself but nothing else.

Newer welding helmets have electronic filters that sense the ultraviolet light from an electric arc and switch from clear to shaded in a split second. This allows the welder to leave the helmet down while lining up the next weld, saving the time consumed by flipping the helmet up and down.

Another electric-arc welding process that continues to gain popularity is **MIG (metal inert gas) welding**. MIG welding is also known as **wire-feed welding** because the electrode used for the weld process is a wire that is automatically fed from a spool. MIG uses a bottled inert gas (such as argon) to shield the weld, and requires a relatively sophisticated welding machine.

MIG Gun

Let's look at an example of how the ASVAB might test your knowledge of joining metal:

Question	Analysis
The purpose of flux in metal joining processes is to	**Step 1:** This question asks what function flux serves when joining metal. **Step 2:** There is nothing to simplify. **Step 3:** Predict that flux prevents oxidation by cleaning metal surfaces.
(A) clean metal surfaces to be joined (B) physically bond two metal surfaces (C) transfer electrical current (D) transfer electrical voltage	**Step 4:** Choice **(A)** matches that prediction. Eliminate choice (B), because although flux is used in the metal joining process, it is not the actual bonding agent. Eliminate choices (C) and (D), because flux is not intended to transfer electricity.

Try another example about fasteners on your own:

Using which of the following tools absolutely requires the use of a protective face mask?

(A) electric drill
(B) stick welder
(C) ratchet
(D) hand saw

Explanation

Which tool requires the use of a face mask? That would be the stick welder, **(B)**, which produces high-intensity light that could burn the skin and retinas if a protective face mask is not used. While it might be a good idea to wear a protective mask with the other tools, depending on the particular job, it is not always necessary.

GRIPPING TOOLS

LEARNING OBJECTIVES

In this section, you will learn to:

- identify the different types of pliers
- understand how different types of pliers are used
- describe the importance of clamps and vises for carpentry projects

Pliers

Repair operations often call for objects to be gripped, twisted, bent, or turned. Sometimes these objects are irregularly shaped or otherwise difficult to hold. The best tool to use for gripping objects like this is pliers.

The most common type of pliers is the **combination slip-joint**. These are adjustable at the joint of the two handles of the pliers. With two different positions to choose from, these pliers can grip objects in a wide range of sizes. Sometimes, this design also incorporates a wire cutter for increased versatility.

Combination Slip-Joint Pliers

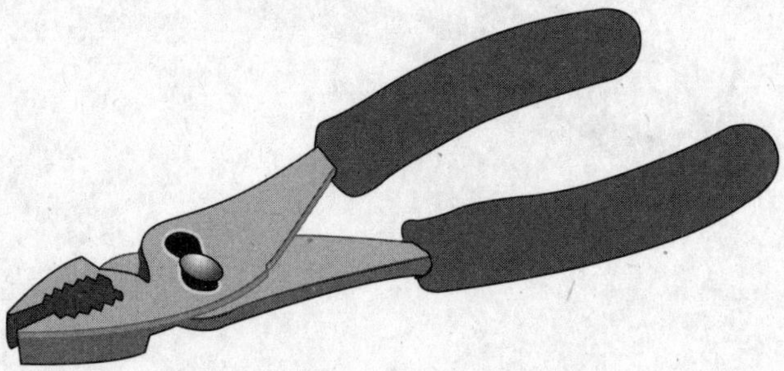

When large diameter objects must be gripped or twisted, a technician would use **adjustable joint pliers** to get the job done. These are adjustable over a large range of sizes, as they have multiple "arc-joints" that the pliers can be set into. The handles are also very long, which gives very good leverage and makes for maximum gripping power. These pliers are also commonly known as **water pump pliers** or **Channellock® pliers**.

Channellock Pliers

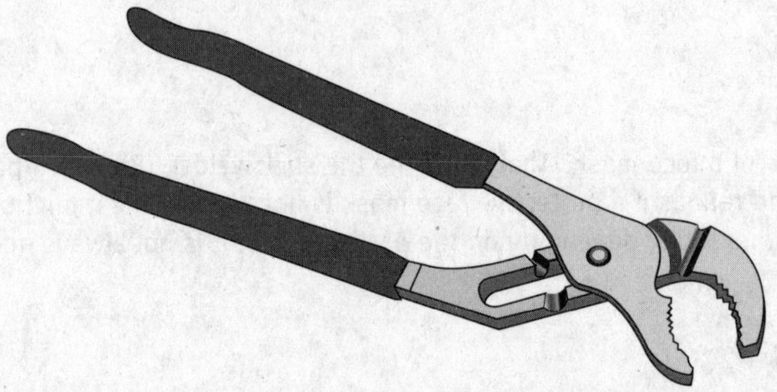

Lineman pliers are used for cutting and bending heavy gauge wire. These are not size-adjustable, but are made for maximum leverage at the jaws to make the cutting process easier. This type of pliers would most often be used by electricians, but all trades utilize them at one time or another.

Lineman Pliers

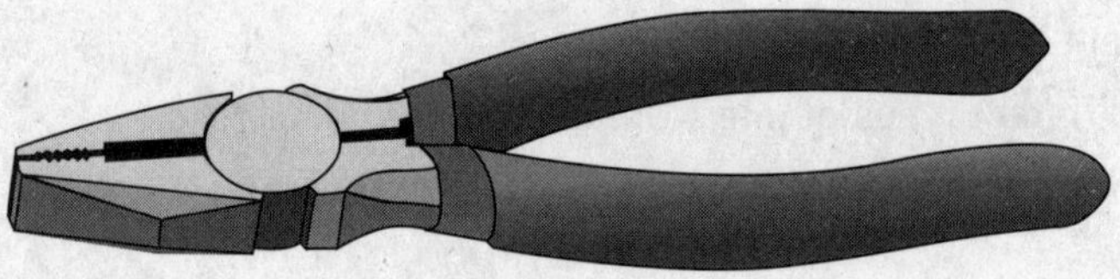

Diagonal cutters are pliers that are made exclusively for cutting. The two jaws are set at an angle (diagonally) to make it easier to cut wires straight across. Diagonal cutters are normally used for cutting wire and small cables.

Diagonal Cutters

For holding small objects in tight places, a pair of **needle nose pliers** could be used. These have very long, pointed jaws for maximum reach. Needle nose pliers are often used for intricate jobs like soldering circuit boards and small components, and they will most often have a wire cutter built into them at the base of the jaws.

Needle Nose Pliers

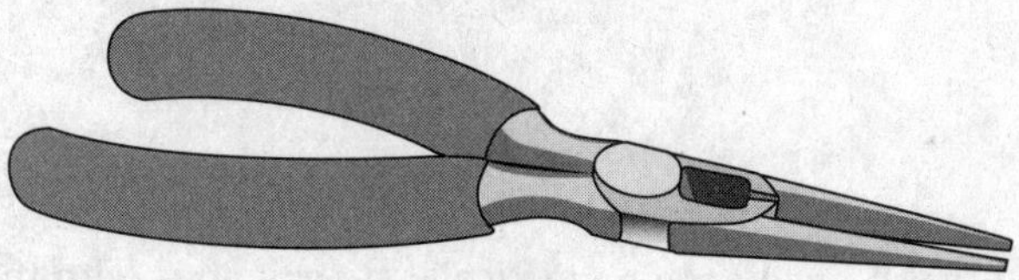

Locking pliers are used by technicians of all trades. Most people know them as **Vise-Grip**® pliers. They are adjustable, made in a large variety of jaw designs, and will lock tightly in place for holding or clamping objects together. Locking pliers often have wire cutters built into them as well for maximum versatility.

Vise-Grip Pliers

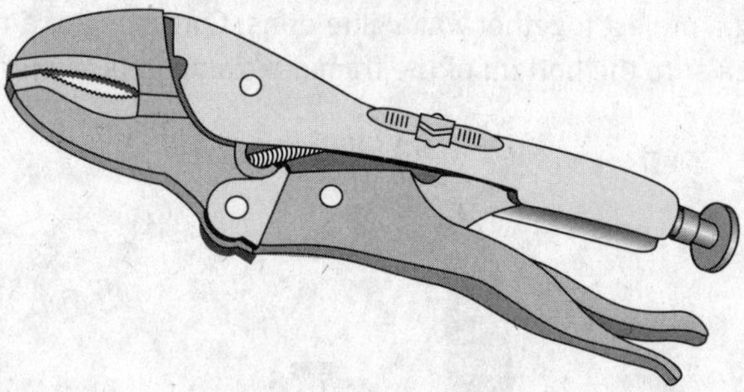

Questions on the ASVAB may give you a hypothetical job and ask what tool would be most useful. Look at this example:

Question	Analysis
You are completing a job that requires gripping several different sizes of pipes. Which of the following would be the best tool to select for this job?	**Step 1:** This question asks which tool would be best for gripping a wide range of pipe sizes **Step 2:** There is nothing to simplify. **Step 3:** You can narrow your prediction to types of pliers, since that is the tool of choice for jobs that require gripping. Your prediction can be even more specific if you recall that adjustable joint pliers provide the most versatility because they can be adjusted over a wide range of sizes.
(A) adjustable joint pliers (B) needle nose pliers (C) lineman pliers (D) adjustable wrench	**Step 4:** Choice **(A)** matches that prediction. Choices (B) and (C) can be eliminated because they are not adjustable, and choice (D) can be eliminated because it is designed for tightening and removing fasteners rather than gripping pipes.

Here is another question, this time dealing with terminology. Try this one on your own:

What type of wrench is commonly known as a Crescent wrench?

(A) open-end wrench
(B) box-end wrench
(C) adjustable wrench
(D) combination wrench

Explanation

Crescent is such a popular brand that it has become practically synonymous with the adjustable wrench; **(C)** is the correct answer. Remember, a combination wrench has an open-end wrench on one end and a box-end wrench on the other.

Clamps

A clamp is a tool that applies pressure to prevent movement between different pieces of a project. For example, a clamp would be useful for holding a project together while glue dries. Consisting of a metal frame (commonly C-shaped) with a flat-edged screw perpendicular to the bottom of the frame, a clamp gets tighter as the screw is turned.

C-Clamp

Vises

Vises are similar to clamps in that they are designed to hold wood or other materials in place. A key difference, however, is that vises are affixed to the workstation and are used to hold a material in place while you saw, sand, drill, or otherwise work on the material. Typically, one side **(jaw)** of the vise is fixed, and the other is moved by turning a handle.

Vise

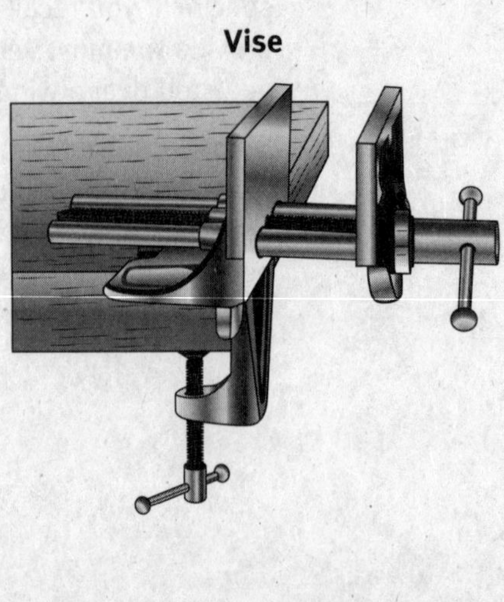

CUTTING TOOLS

LEARNING OBJECTIVES

In this section, you will learn to:

- identify different types of saws
- explain how saw blades differ based on the material they are designed to cut
- select the appropriate hand saw for a particular project
- understand the benefits of using powered saws for certain projects
- understand how power drills and their accessories are used

Manual (Hand) Saws

Cutting a material like wood requires the use of a **saw**. Saws are used in many different trades, but are typically utilized by carpenters.

The **crosscut saw** is designed to do what its name suggests: cut across the grain of the wood. Crosscut saw teeth are unique in that they cut like a knife. This is in contrast to a **rip saw**, which is made to cut with the grain of the wood and whose teeth are shaped like chisels. A rip saw's teeth are also set (bent) alternately from side to side to cut a relatively wide **kerf** (slot). This style of teeth allows the rip saw to cut a straight line even if the grain of the wood is curved.

Crosscut Saw

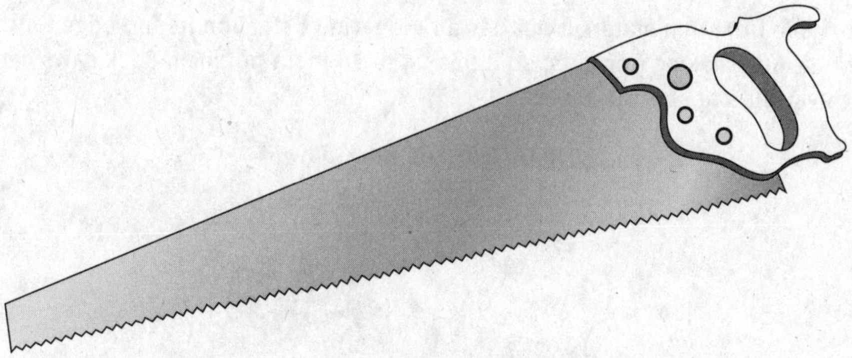

Rip Saw

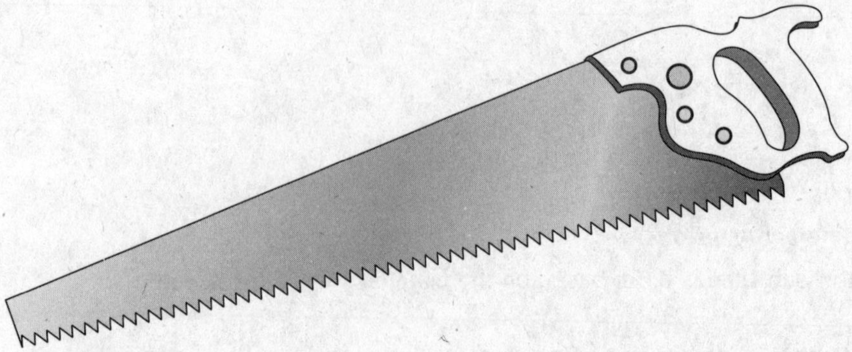

A **coping saw** is used to make fine, curving cuts. This saw uses a thin, flexible blade that is held tight on a wide frame. The blade can be rotated in the frame for further flexibility, making it easier to make difficult cuts on larger pieces of material.

Coping Saw

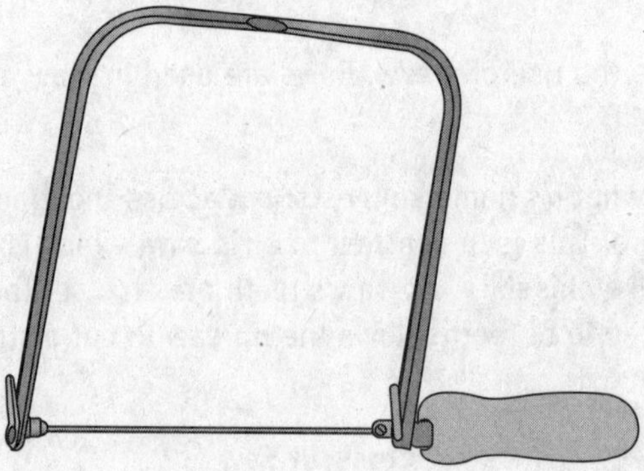

The **back saw** is also made from thin material, but has a rigid strip of steel on its top edge for reinforcement. A back saw is normally used for making fine cuts, so it has 14 to 16 teeth per inch. Back saws can be used with a **miter box** for making even cuts at specific angles.

Miter Box and Back Saw

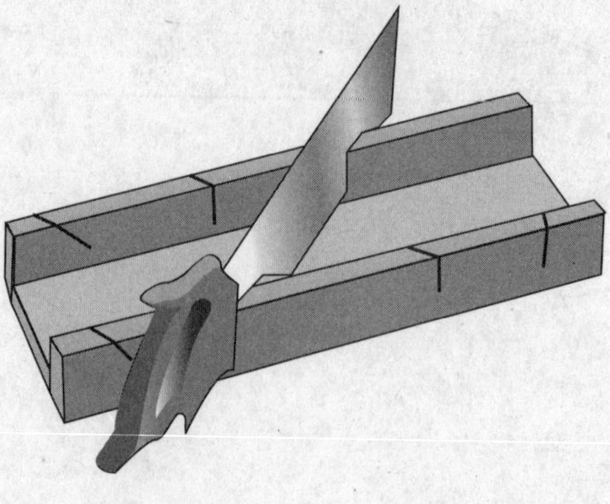

The one saw that can be found in a mechanic's toolbox is a **hacksaw**. Hacksaws are used for cutting metals such as steel, aluminum, or copper. The blades in a hacksaw are replaceable, and it is important to choose the right blade for the material that is going to be cut.

Hacksaw

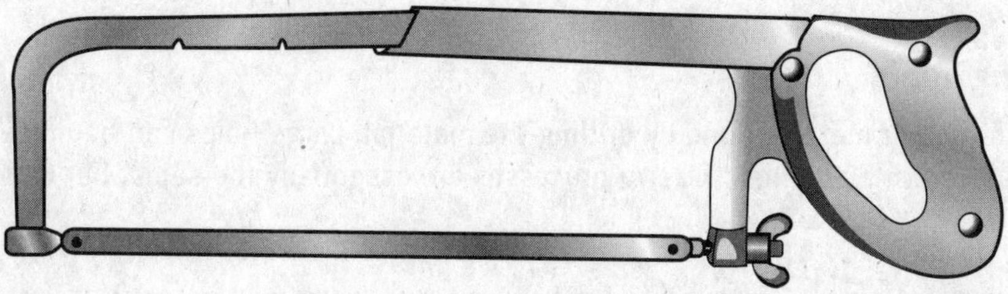

When selecting a hacksaw blade, keep in mind that larger numbers of teeth per inch help when cutting thinner materials. Also, remember that when installing a new blade, the blade must be oriented so that the teeth point away from the handle. This makes the hacksaw cut on the forward stroke, so let up on the downward pressure when pulling the hacksaw back to prevent breaking the blade.

Powered Saws

For jobs that require faster cuts or that involve especially strong materials, powered saws may be used in lieu of hand saws. A common type of powered saw is a **circular saw**, which—as the name suggests—has a circular blade that rotates quickly to make a cut. There are two very common types of circular saw. A **miter saw** is a self-contained tool that sits atop a tool bench and is useful for making crosscuts in lumber, especially when the cut needs to be at a specific angle. Don't get this confused with a **table saw**, which also has a circular blade; a table saw has the blade actually embedded into the table or bench itself. Another key difference lies in how the saws are used. To use a miter saw, clamp or otherwise brace the material to be cut in order to keep it steady, and then lower the rotating circular saw to make the cut. To use a table saw, carefully slide the material to be cut toward the rotating blade.

Miter Saw

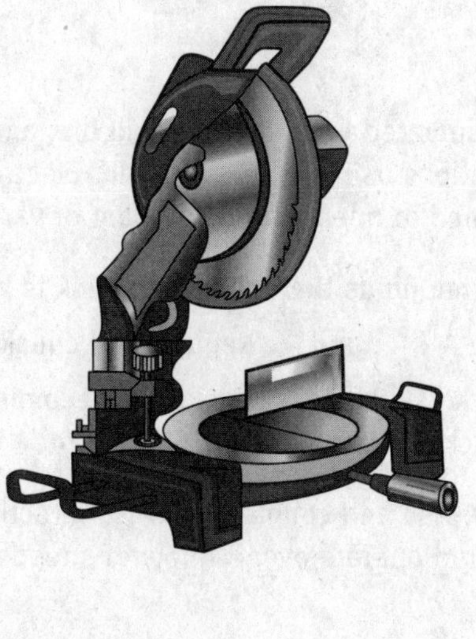

Another common type of power saw is a **band saw**, which is so called because its blade is one continuous band of metal that revolves very quickly on two spinning wheels to make a continuous cut in a piece of wood or metal. Band saws are used for creating lumber out of timber and can also be useful for cutting straight or curved lines in a material. When using a saw like this for cutting metal, it is vital to use a coolant wash to keep the material cool and debris-free.

Drilling and Boring Tools

Making small holes in wood or metal is done by **drilling** the material. Large holes can be made in wood or soft metals using a process known as **boring**. The two processes are essentially the same, but they use different tools to get the job done.

Drill bits are used for drilling holes. A carpenter would use drill bits up to $\frac{1}{4}$" in diameter, but a mechanic would commonly use drill bits as large as $\frac{1}{2}$". If holes larger than this must be made, special tools would be used to bore the hole.

The vast majority of drill bits are made to cut while rotating in a clockwise direction. These are known as **right-hand** drill bits. **Left-hand** drill bits are made to cut in the opposite direction. The most common practical application for a left-hand drill bit is the removal of broken bolts from threaded holes.

Hole saws can also be used for boring large holes. Hole saws are not adjustable, so each one is only capable of drilling one size of a hole.

Hole Saw

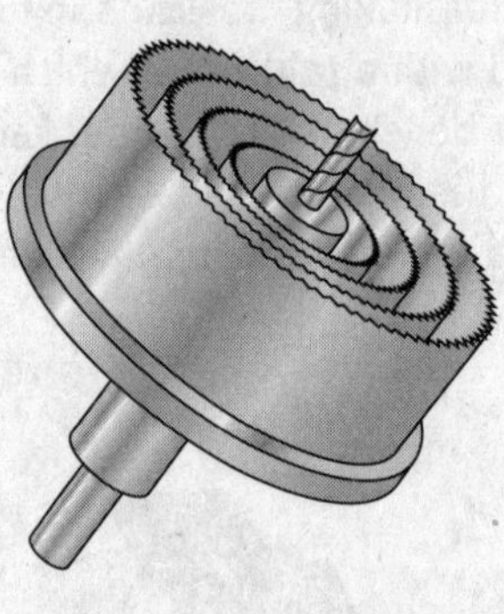

Today, it is rare for anyone to use hand-operated tools to drive drill bits and other boring tools. Instead, it is common practice to perform these operations using an **electric drill**. Electric drills can be identified by chuck size, reversibility, and whether they are designed to operate at a constant or variable speed.

The **chuck** of an electric drill is the part that holds the drill bit. A chuck is identified by the largest diameter bit that will fit in it. Common chuck sizes include $\frac{1}{4}$", $\frac{3}{8}$", and $\frac{1}{2}$". Regular drill chucks can be tightened and loosened using a **chuck key**. Since it is easy to lose a chuck key, **keyless chucks** are becoming more popular. A keyless chuck makes it possible to tighten and loosen the chuck by hand.

Some drills can operate in both the clockwise and counterclockwise directions. These are known as **reversible drills**. **Variable speed drills** are designed to operate over a range of speeds that can be determined by the position of the trigger.

As battery technology advances, it is getting more common for people to use **cordless drills**. Longer times between battery recharges and portability are just two of the reasons why these have become so popular. Many of these cordless drills come with a wide range of attachments (called drivers or bits) and can double as powered screwdrivers and socket wrenches as well.

The most important safety consideration when using a drill is to use a sharp bit. Dull bits require more time and more pressure placed on the drill to get the job done. With smaller bits, this could result in bit breakage and damage to the surface of the material being drilled. Drill bits can be sharpened on a bench grinder using a special attachment.

Question	Analysis
You are attempting to cut a rounded shape from a square piece of wood. What is the best saw to choose?	**Step 1:** This question asks which tool would be best for cutting a round shape from a square. **Step 2:** Not much info to simplify. **Step 3:** You can narrow your prediction by thinking about the type of saw that can cut a round, rather than straight, line.
(A) coping saw (B) crosscut saw (C) hack saw (D) rip saw	**Step 4:** Choice **(A)** matches that prediction. The remaining saws are designed for straight cuts.

Now try this question about saws.

Which of these would a mechanic be most likely to use?

(A) coping saw
(B) crosscut saw
(C) hacksaw
(D) rip saw

Explanation

Of these four choices, only choice **(C)**, hacksaw, is commonly used to cut materials other than wood, so that is the saw most likely to be in a mechanic's toolbox.

FINISHING TOOLS

LEARNING OBJECTIVES

In this section, you will learn to:

- identify tools used to plane, smooth, and shape
- determine when and how to use a plane
- understand the uses of chisels
- determine when to use a file
- identify when to use a rasp

Planes

When working with wood, there are many occasions when it is necessary to shave off a small amount of material to make a piece fit properly or to make a surface smooth. One tool that can be used for this purpose is a **plane.**

There are several different types of planes, but the one that would be best suited for general purpose work is the **jack plane**. The jack plane has a smooth lower surface (plane bottom) and an adjustable-depth blade that protrudes slightly below this surface (through the mouth). This blade is set at an angle, similar to the blade in a disposable razor.

Jack Plane

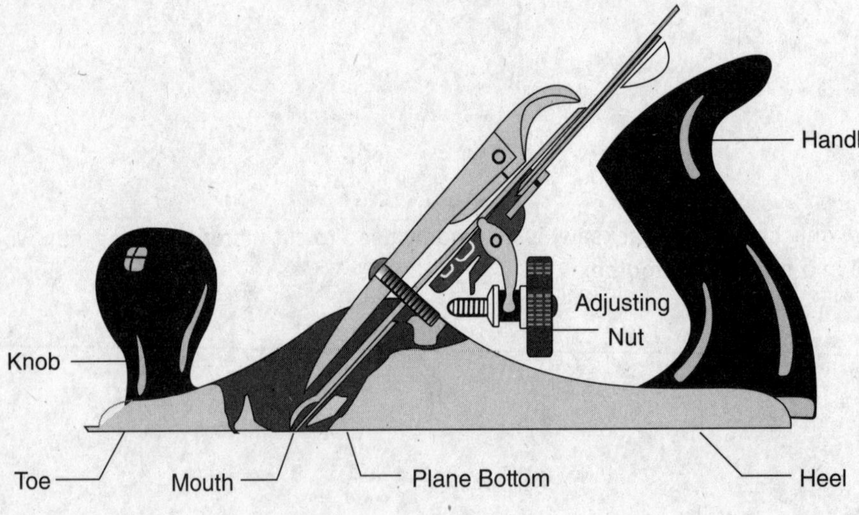

When using a jack plane, the carpenter grasps the plane with both hands and moves it evenly across the surface of the work, shaving a thin layer of wood. If a heavier cut is desired, the blade can be adjusted to protrude farther from the lower surface of the plane.

Wood Chisels

Wood chisels may also be used to shape or smooth a wood surface. Wood chisels come in a variety of widths, which can range from $\frac{1}{8}$" to 2". A wood chisel is normally hand-operated, but when making deep cuts, a soft-faced mallet should be used to lightly tap the chisel through the work.

Wood Chisels

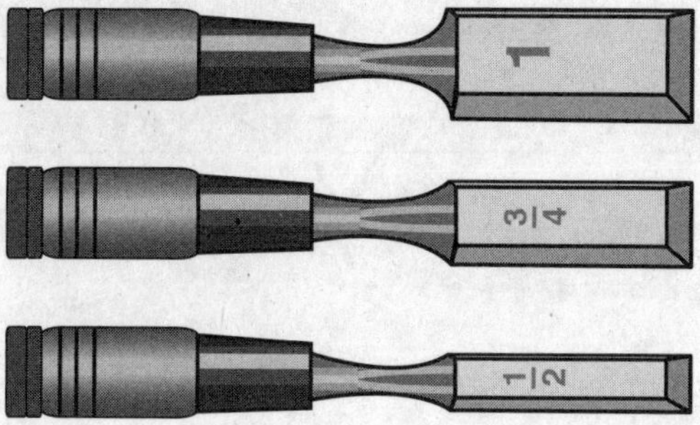

Files and Rasps

Files are used to smooth, polish, and shape materials. Files are made from hardened steel and consist of diagonal rows of teeth. These teeth can be arranged as either a single row of parallel teeth (single-cut) or one row of teeth crisscrossing with another row (double-cut). Files are made in flat, round, half-round, and triangular designs.

Most files do not come with handles. It is important to have a handle ready to attach to a file when it is put to use, as this will prevent the end of the file from being driven into the technician's hand. Remember that a file is made to cut only on the forward stroke, so it should be lifted slightly from the work when it is pulled back.

Flat File, Half-Round File, and Triangular File

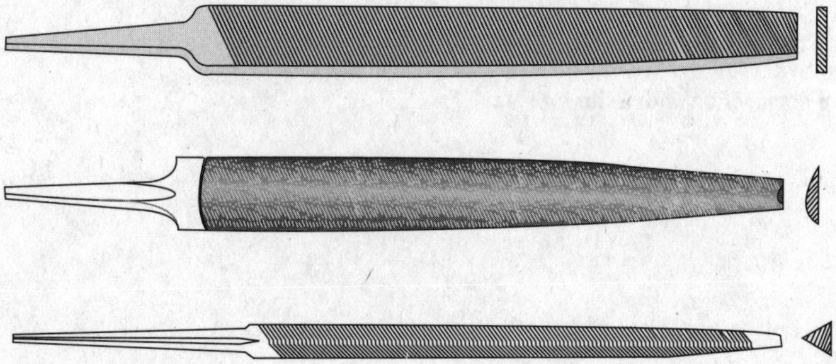

Rasps are basically very coarse files that are used to trim, shape, and smooth materials. While files are used for fine finishes, rasps are used for coarse work. Round rasps are useful for cleaning up holes, whereas a flat rasp would be used to smooth flat surfaces. The cutting teeth on a rasp are coarse enough to clear sawdust easily, so the rasp is always ready for the next stroke.

Rasps

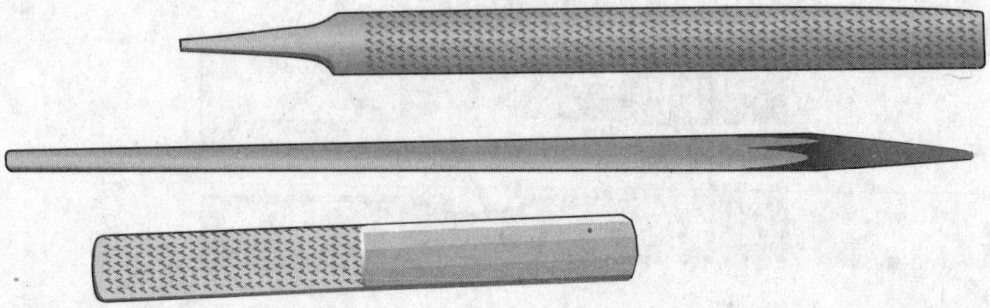

The ASVAB may ask you which tools can be used together on a project. Here is an example:

Question	Analysis
Which of the following finishing tools may be used with a mallet? (A) jack plane (B) wood chisel (C) file (D) rasp	**Step 1:** This question asks which tool would be most appropriate to use with a mallet. **Step 2:** There is nothing to simplify. **Step 3:** Predict the tool that would most likely be struck by a mallet during normal operation. **Step 4:** Only choice **(B)** is commonly struck by a mallet, to make deep cuts.

Try this example on your own.

What must be used in conjunction with a file for safe operation?

(A) mallet
(B) file
(C) chisel
(D) handle

Explanation

The correct answer here is **(D)**, handle. Many files are sold without handles attached, but it is important to use a handle whenever you use a file, to prevent the pointed end of the file from injuring the palm of your hand.

SHOP INFORMATION PRACTICE SET 1

Select the best answer for each question. This practice set has 10 practice questions, which is the number of Shop Information questions you will see if you take the CAT-ASVAB.

1. In order to tighten a bolt as much as possible, a ________ wrench should be used.
   - (A) long-handled
   - (B) short-handled
   - (C) heavy
   - (D) reticulated

2. What is the tool shown below known as?

   - (A) water pump pliers only
   - (B) retaining ring pliers only
   - (C) Channellock pliers only
   - (D) both water pump pliers and Channellock pliers

3. A combination wrench has
   - (A) two different size box ends
   - (B) two different size open ends
   - (C) a box end and an open end of the same size
   - (D) an open end and a ratchet end

4. The blade in a hacksaw should be installed with its teeth pointed
   - (A) toward the handle only
   - (B) away from the handle only
   - (C) either toward or away from the handle
   - (D) neither toward nor away from the handle

5. How is the length of a bolt measured?
   - (A) by measuring the length from end to end
   - (B) by measuring the length of the threaded portion only
   - (C) by measuring the length from the underside of the head to the end of the threads
   - (D) none of the above

6. A bolt with a hexagonal head could be loosened using all of the following EXCEPT
   - (A) an open-end wrench
   - (B) a box-end wrench
   - (C) a 12-point socket and ratchet
   - (D) a Torx screwdriver

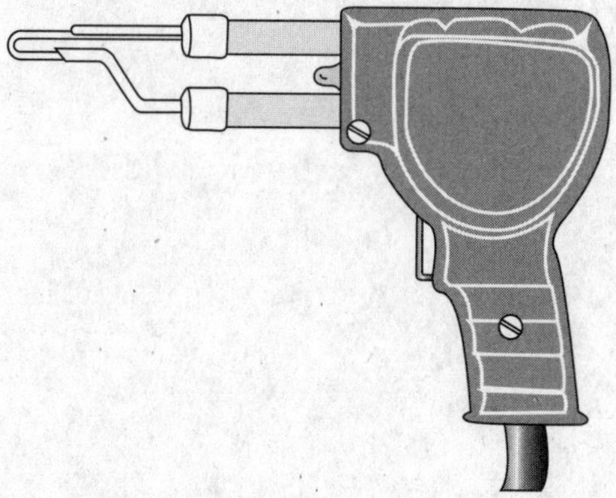

7. The tool shown above is used for
   - (A) installing pop rivets
   - (B) stapling asphalt shingles
   - (C) crimping electrical terminals
   - (D) soldering electrical connections

8. "Tinning" is a process that is related to
   - (A) carpentry
   - (B) roofing
   - (C) soldering
   - (D) welding

9. The following screw head would be used with which type of screwdriver?

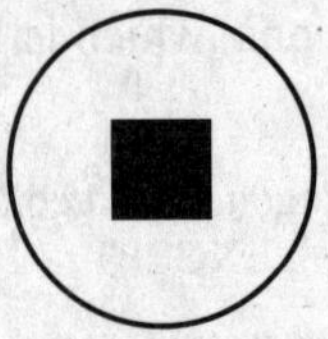

(A) flat tip
(B) Phillips
(C) Robertson
(D) Torx

10. What force is responsible for allowing wrenches to tighten or loosen fasteners?

(A) acceleration
(B) torque
(C) tension
(D) friction

SHOP INFORMATION PRACTICE SET 2

Select the best answer for each question. This practice set has 10 practice questions, which is the number of Shop Information questions you will see if you take the CAT-ASVAB.

1. The best choice to loosen a tight fastener is the ________ wrench.

   (A) open-end
   (B) adjustable
   (C) box-end
   (D) Allen

2. To determine the correct socket size needed for a given bolt head, measure the distance between two ________ of the bolt head.

   (A) adjacent sides
   (B) parallel threads
   (C) opposite points
   (D) parallel sides

3. What is a benefit of using a ball-peen hammer rather than a claw hammer?

   (A) It is more suitable for hammering on metal because of its generally stronger head.
   (B) It is more delicate and can thus be used on fragile surfaces.
   (C) It can remove nails as well as hammer them in.
   (D) It provides more leverage.

4. You have a bolt that needs to be removed, but a $\frac{5}{16}$" open-end wrench is just slightly too small to fit properly. Which of the following sizes should be tried next?

   (A) $\frac{3}{4}$"
   (B) $\frac{3}{8}$"
   (C) $\frac{5}{8}$"
   (D) $\frac{11}{16}$"

5. What is the most accurate name of the tool shown below?

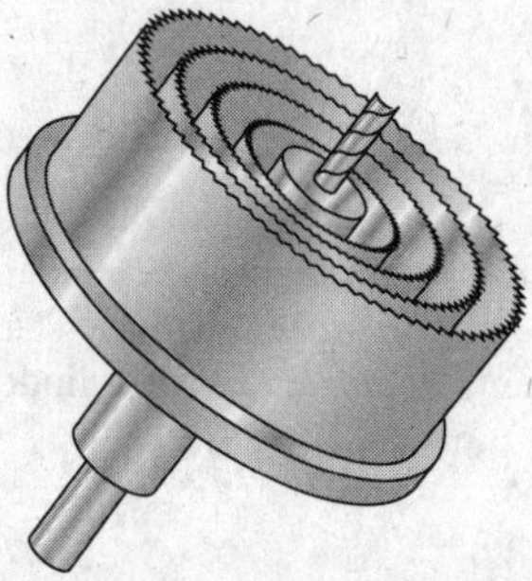

   (A) hole saw
   (B) router
   (C) borer
   (D) smoother

6. Kerry wants to create a jigsaw puzzle by gluing a photograph to a thin piece of wood and then cutting the laminated picture into many odd-shaped pieces, which can then be reassembled into the original picture. Unfortunately, she does not have an electric jigsaw at her disposal. What would be her best choice of a saw type to perform this task?

   (A) hacksaw
   (B) crosscut saw
   (C) coping saw
   (D) rip saw

7. Dirk wants to cut a piece of thick metal into two smaller pieces using his hacksaw. Which tool would be best to hold the metal in place when he cuts?

   (A) vise
   (B) C-clamp
   (C) miter box
   (D) adjustable joint pliers

8. Retaining ring fasteners are primarily used to __________.

   (A) connect two odd-shaped pieces together
   (B) provide a tight seal for a cylinder
   (C) suspend gears that don't have shafts attached
   (D) prevent end movement of cylindrical parts

9. The part of a nail between the head and the point is called the __________.

   (A) cubit
   (B) shank
   (C) body
   (D) span

10. What is a function of the thimble of an outside micrometer?

    (A) to move the spindle so that it touches the object being measured
    (B) to balance the instrument
    (C) to move the sleeve into the proper position
    (D) to keep the measuring graduations clean

SHOP INFORMATION PRACTICE SET 1

Answers and Explanations

1. **A** The tightest connection can be achieved by applying the most torque to the bolt. Since torque is the product of force and the length of the lever arm, a long-handled wrench would apply more torque to the bolt than a short-handled wrench if the same force were applied. Although a heavy wrench might be a long wrench, the weight of the wrench is not a factor. Reticulated, meaning resembling a network, is not applicable.

2. **D** When large-diameter objects must be gripped or twisted, a technician would use adjustable joint pliers to get the job done. These are adjustable over a large range of sizes, as they have multiple "arc-joints" that the pliers can be set into. The handles are also very long, which gives very good leverage and makes for maximum gripping power. These pliers are also commonly known as *water pump pliers* or *Channellock pliers*. Since choice (D) allows for the tools in both choices (A) and (C), it is the correct answer.

3. **C** The box end of a combination wrench is typically used to apply maximum torque, while the open end is used to turn the fastener more quickly.

4. **B** When installing a new hacksaw blade, orient it so the teeth point away from the handle. This makes the hacksaw cut on the forward stroke and also makes it easy to pull the saw back in order to begin the next forward stroke.

5. **C** One important measurement of a fastener is the length. The length of the bolt is the distance between the underside of the bolt head and the end of the bolt. Note that the bolt head does not count toward the length of the bolt. Therefore, the best answer is (C).

6. **D** The two basic types of wrenches—the open end and the box end—are both acceptable for loosening a hexagonal bolt. The open-end wrench is easy to slide on and off a fastener. The downside to the open-end wrench is that it only makes contact with two sides of a six-sided (hex) bolt head. The box-end wraps completely around the head of a bolt, and therefore makes greater surface contact. Box-end wrenches normally come in a 12-point configuration, but some are made as 6-point. Used with a ratchet, the correct size socket is a good choice for loosening fasteners such as hexagonal bolts. A Torx screwdriver, despite its name, generally does not exert as much torque as a wrench; it also requires a bolt that is specially configured to match its shape.

7. **D** The tool shown is a soldering gun. It is used for soldering electrical connections.

8. **C** If two pieces to be joined by soldering are large, it may pay to tin the surfaces before attempting to join them. Heating the surfaces gently while applying flux to them, and then adding solder once the flux has thoroughly cleaned the material will accomplish this. Since soldering is the process involved, the correct answer is (C).

9. **C** The Robertson screwdriver grips the fastener better than a flat tip screwdriver, so the screw is much easier to remove and install.

10. **B** Whenever you see the words "force" and "wrench" in the same sentence, you should immediately think about *torque*, (B). You can think of torque as twisting motion, which is exactly what allows a fastener to be tightened or loosened. Don't get torque confused with *tension*, (C), which instead is a pulling force associated with a weight on a string or wire.

SHOP INFORMATION PRACTICE SET 2

Answers and Explanations

1. **C** The box-end wrench is closed, which allows for more surface area contact of the wrench with the fastener, compared to the contact offered by the open-end and adjustable wrenches; therefore, answer choices (A) and (B) are incorrect. Allen wrenches are typically intended for low-torque applications; therefore, answer choice (D) is also incorrect.

2. **D** The size of a socket for a given bolt head is determined by the distance between two parallel sides of the bolt head.

3. **A** A ball-peen hammer has one main benefit: it generally has a stronger head made of steel. This matches choice (A). Choice (B) more closely describes a mallet, certainly not a steel hammer, and choice (C) describes a benefit of a claw hammer, so that is the opposite of what you're looking for. Finally, it makes little sense to talk about (D) leverage when discussing striking tools.

4. **B** If a wrench is just slightly too small, you will want to try the next size larger. In this case, $\frac{3}{8}$" is the next size up. This is the wrench you would want to try next, so choice (B) is the correct answer.

5. **A** The *hole saw* rotates about its shaft and the saw teeth cut a circular hole perpendicular to the shaft. A *router*, (B), has a single blade that is typically used to cut a channel or to shape material. *Borer*, (C), is a generic term that applies to a variety of tools that can produce holes. Similarly, *smoother*, (D), is a catch-all term, and the hole saw shown would not help to smooth anything.

6. **C** A coping saw has a thin blade that can be rotated within its frame in order to make fine curving cuts. The *hacksaw*, (A), though similar in appearance to a coping saw, has a thicker fixed blade and is typically used for cutting metal. Each of the *crosscut saw*, (B), and *rip saw*, (D), has a large, fixed blade. The crosscut saw is used to cut across the grain of wood, while the rip saw is used to cut with the grain of wood.

7. **A** Since a vise is affixed to a workstation, that tool would hold the metal stationary while Dirk is cutting and both of his hands would be free. A *C-clamp*, (B), is better suited for holding two pieces of material together, as would be required to attach one to the other. A *miter box*, (C), is typically used with a back saw to make cuts at specific angles. It would be difficult, if not impossible, for Dirk to hold the metal stationary with adjustable *joint pliers*, (D).

8. **D** The protrusions on retaining rings (also called snap rings) fit into grooves so that they prevent a cylindrical shaft from sliding out of position. Retaining rings do not connect pieces, nor are they ordinarily used with odd shapes, so (A) is incorrect. They do not provide any type of seal, so (B) is incorrect. Suspending a gear with no shaft, (C), seems to be a useless function.

9. **B** *Cubit*, (A), is a measurement of length. While *body*, (C), and *span*, (D), may seem to be logical answers, the proper term is *shank*.

10. **A** Turning the thimble moves the spindle into contact with the object being measured. While the thimble is also used to read the measurement, it has nothing to do with keeping the graduations clean. There is no need to "balance" a micrometer to use it. The sleeve surrounds the spindle and does not move.

Review and Reflect

How did you do on this practice set? As you review your work, think about these questions:

- Are there certain categories of tools you should review?
- Do you know the names of tools but need more review about when to use them?
- Would it help you to make flashcards out of the bold words in this chapter?

Want more practice with Shop Information? Try more questions in the Qbank online.

QBANK

kaptest.com/login

CHAPTER 13

MECHANICAL COMPREHENSION

KNOW WHAT TO EXPECT

In order to understand how machines work, it is important to have a good grasp of **applied physics**. The study of applied physics is the study of the practical application of the laws of physics. Engineering, architecture, and heavy equipment operation depend upon applied physics. This chapter will build on concepts first discussed in the Physical Science portion of chapter 9: General Science to explain how various common mechanical devices operate.

On the CAT-ASVAB, you will have 22 minutes to answer 15 questions in the Mechanical Comprehension subtest. If you are taking the paper-and-pencil test, you will have 19 minutes to answer 25 questions.

A REVIEW OF THE PHYSICS OF MECHANICAL DEVICES

LEARNING OBJECTIVES

In this section, you will learn to:

- recognize a mechanical device as any machine based on force and motion laws
- define *mass* and *force*
- apply Newton's laws of motion to questions about mechanical devices

Mechanics is the area of physics that encompasses the laws of motion, energy, and forces. Velocity, momentum, and Newton's second law relating force, mass, and acceleration are all mechanical concepts. All mechanical devices, including the simple machines discussed in this chapter, are based on the application of force in order to achieve a movement or change in position of a mass.

Each of the physical quantities reviewed in this section were first introduced in the Physical Science section of chapter 9: General Science. This brief review will focus only on the concepts from that chapter which most directly underlie mechanical operation.

Mass and Force

All matter has mass. **Mass** is a measure of the total quantity of matter in an object. Generally speaking, larger objects tend to have greater mass, but some materials contain more mass than others per unit volume (are denser), so bigger doesn't always mean more massive. Unlike the force of weight, which will be discussed in more detail a little later in the chapter, mass is not a vector quantity, since it has magnitude but no direction.

For the purposes of mechanics, the most important thing to know about mass is that it corresponds to how much force is required to achieve a particular acceleration. In other words, the more mass something has, the more difficult it is to change its motion. The term **inertia** is sometimes used to refer to an object's resistance to changes in its motion, but inertia is just a property of mass.

A **force** is a push or pull. Forces are everywhere in the world. Some are fairly obvious, such as a tractor pulling a plow, a baseball being hit into the stands, or a person shoving his way past others in a crowded store. Other forces are often taken for granted. Earth's gravity applies a force (weight) that always pulls down, and it is what keeps objects and people on the ground instead of floating around in midair. All forces are vector quantities and therefore act in a single direction. The term **net force** refers to the total force acting on an object.

Without a force being applied to them, objects would not move (or, if they were already in motion, they would remain in motion without stopping). Objects with a great deal of mass (such as a freight train) require a large force to alter their motion. The goal of all mechanical technology is to most effectively apply forces to the movement of varying mass quantities.

Newton's Laws of Motion

All mechanical devices take Newton's laws for granted in their operation. Newton's first law describes the unchanging state of motion of an object when it experiences no net force. A heavy crate suspended by a pulley is only able to avoid crashing to the ground because the attached ropes fully counteract the force of gravity pulling the mass downwards.

Newton's second law can be summarized by the formula

$$F = ma$$

where m is measured in kilograms (kg), a is in meters per second squared (m/s^2) and F is in newtons (N). This law expresses the linear relationships between either the mass of the moved object or the desired acceleration, with the force required to achieve that movement.

One of the most important historical roles of machines has been in multiplying the force a human or work animal provides in order to move very massive objects that would be otherwise immobile. Ancient structures including Stonehenge and the Egyptian pyramids are evidence of a long human tradition of mechanical aids to labor.

Newton's third law of motion states that for every action there is an equal and opposite reaction. This law applies to all applications of force, including those mediated by simple machines. It's possible to pull on a rope and pulley to lift a weight only because the force of tension is carried through a rope by a daisy chain of action-reaction force pairs from one section of rope to the next.

Question	Analysis
Simple machines and other mechanical devices	**Step 1:** The question asks which statement applies to mechanical devices.
	Step 2: Think about the traits that simple machines and other mechanical devices have in common: they are used to transfer force, and they obey Newton's laws.
	Step 3: That's as close as you can get to a prediction with a general question like this, so start evaluating answer choices.
(A) are technologies used to get around certain laws of physics (B) are exclusively based on momentum (C) operate according to the rules of force and mass (D) transmit and multiply force in order to increase the amount of energy available	**Step 4:** Mechanical devices use the laws of physics; they don't circumvent them, so (A) is not correct. Momentum is important, but it is not the only basis for any and all machines, so eliminate (B). Choice **(C)** is true. Select it. Choice (D) sounds *almost* correct, but how does a device increase the energy available? By burning fuel? By creating energy from nothing? Simple machines like levers don't themselves burn up fuel. And although force can be multiplied with mechanical advantage (more on that later), energy cannot be created out of nothing.

Now try one on your own.

According to Newton's laws, an object being lifted by a pulley will accelerate

(A) even if the applied force from the rope to the object is a little less than the object's weight, because mechanical advantage makes up the difference
(B) only if the net force on the object is zero and the object is already in motion
(C) if the applied force from the rope to the object is greater than the object's weight
(D) if the object's mass is greater than the applied force

Explanation

Choice **(C)** is correct. There must be a non-zero net force for any object to accelerate. Objects in motion where the net force is zero will continue moving at a constant velocity, but will not accelerate. Both (C) and (D) talk about a difference between two values, but answer choice (D) compares the applied force to the object's mass rather than weight. Since these quantities have different physical meanings, it doesn't make sense to say one is greater than another.

COMMON TYPES OF FORCES

LEARNING OBJECTIVES

In this section, you will learn to:

- distinguish and understand the related terms of mass, weight, and gravity, and apply a formula to calculate one value given the others
- understand friction as a responsive force and apply the formulas for calculating kinetic and static friction
- understand tension as a force found in a cable that stretches it tight
- understand hydraulic pressure as it relates to an applied force, and as a transmitter or multiplier of force
- apply the formula for torque as a measure of how effectively a force can cause an object to rotate

Gravity and Weight

As seen previously, **weight** is different from mass. Mass is an intrinsic property of matter, whereas weight is the force exerted on an object due to gravity. The downward pull that the Earth exerts on bodies outside of itself is a specific case of gravity at work. Weight can be easily measured using a spring scale like that found in many people's bathrooms, and the direction of weight for all objects on Earth is the same: straight down toward the center of the Earth.

The general formula for **Newton's Law of Universal Gravitation** was given in chapter 9: General Science. It can be used to explain why the same person will weigh less on planets of different sizes and masses (for example, when astronauts walked on the Moon, their weight was only $\frac{1}{6}$ of what it was while on Earth). However, unless you are told otherwise, it is safe to assume that for the purposes of everyday mechanical work the location is here on Earth.

For bodies free-falling near the Earth's surface, the **acceleration due to gravity (*g*)** is calculated to be about 9.8 m/s^2, ignoring air resistance. (Note that "near the Earth's surface" really means anywhere on Earth, including on top of Mount Everest or parachuting from a plane thousands of meters in the air. The value of g doesn't change significantly at the top of a stair ladder.)

Applying Newton's second law, $F = ma$, for the specific case of weight, the more general symbol for force, F, is replaced with the more specific force of weight, W, and acceleration, a, is replaced with the constant value of acceleration due to gravity, g. This results in the modified version of the formula

$$W = mg$$

where W is weight in newtons (N), m is mass in kilograms (kg), and g is acceleration due to gravity (either m/s^2 or N/kg). To determine an object's weight in N, therefore, simply multiply its mass in kg by 9.8.

Study the example that follows, which does not have answer choices, to help you practice predicting an answer to this type of question.

Question	Analysis
How many newtons of force due to gravity will act on an 80 kg man while in a plane flying over Denver, Colorado?	The question asks for the amount of force due to gravity.
	The man is not an astronaut (or if he is an astronaut, he's not currently in outer space or on the Moon), so his weight will be pretty much the same as any other 80 kg man on Earth. This weight will be basically unchanged whether he's at sea level, in the United States' highest-altitude large city, or in a plane flying above that same city.
	The phrase "newtons of force due to gravity" is just another way of saying weight. Weight $W = (80\text{ kg})(9.8\text{ m/s}^2) = 784\text{ N}$. The answer is **784 N.**

Friction

There are several responsive forces which act to resist movement. When you move an object across a surface, the rubbing of the two surfaces results in **kinetic friction**, which always acts in a direction opposite to the motion of the object. Because it always opposes motion, kinetic friction will eventually slow any moving object to a stop. Ice is polished to minimize kinetic friction, but unless it is constantly being pushed, a hockey puck will eventually slide to a stop in even the most slippery of rinks.

When an object is not moving, a force that attempts to slide it across a surface will encounter a different type of friction. Attempting to push a stationary box across a concrete floor will require a person to overcome the **static friction** that opposes any attempted movement.

Both kinds of friction are based on two factors: the **coefficient of friction, μ** (the Greek letter mu), which represents how much two materials resist sliding against each other; and the **normal force, F_N**, which represents the equal and opposite force a surface exerts when an object presses against it. The formulas for kinetic and static friction, respectively, are

$$F_f = \mu_k F_N$$

$$F_f = \mu_s F_N$$

where F_f stands for the force of friction and μ_k and μ_s stand for the coefficients of kinetic and static friction. The coefficients of friction are constant for any two surfaces in contact. The coefficients of friction are dimensionless quantities, with no units of their own. Therefore, whatever force unit is used to measure F_N (N or lbs, for example), will also be used to measure F_f.

Slippery surfaces like ice or wet asphalt have low coefficients of both kinetic and static friction with most objects. Friction is directly proportional to these coefficients. That's why it's important for drivers of cars to understand that weather conditions can make braking to a stop (which relies on the friction of the wheels with the road surface) less effective.

Free Body Diagram of Forces Acting on a Sliding Object

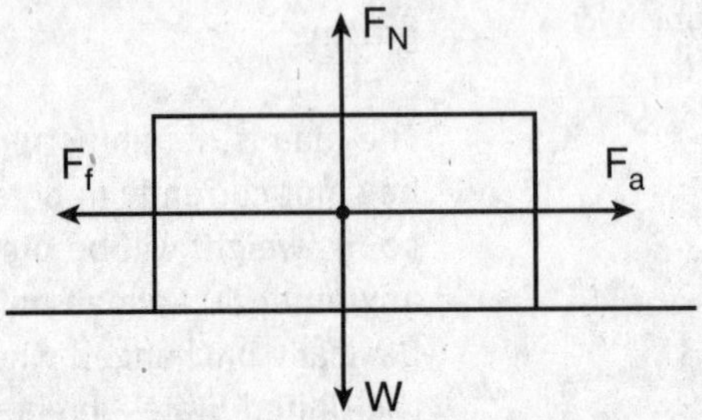

For any given surface, the coefficient of static friction is higher than the coefficient of kinetic friction. This means that more force is required to initially get a stationary object moving than to keep it moving afterwards. Once the large static friction force has been overcome and the object is in motion, the smaller coefficient of kinetic friction takes over, and the resistant force of friction decreases. This is why moving heavy objects usually involves one extra big push at the beginning, but then a somewhat smaller force for the rest of the trip.

One more important difference between static and kinetic friction is that static friction only arises in response to an attempt to move an object along a surface. Like the normal force that holds up objects with weight, it responds to an applied force by matching and counteracting it. The formula is actually used to determine the maximum of a variable value for static friction, beyond which it breaks and an object will begin to move.

Both kinds of friction forces are directly proportional to the normal force of the surface. As an experiment, try laying your palm flat and sliding your hand along a nearby surface like a table or desk. Feel the resistance of the movement due to kinetic friction. Now, with your palm still flat, try pushing harder against the same surface, while still trying to slide your hand along it.

Is it more difficult? If you push very hard against the surface, you may find you are not even able to overcome the force of static friction and slide your hand without decreasing the normal force first. When a person leans against a wall, she allows a portion of the weight to push against the wall, which pushes back with an equal and opposite normal force. This normal force contributes to a corresponding static friction force that prevents the person from sliding down the wall to the floor.

Most frequently, the normal force arises on flat horizontal surfaces as a response to the weight of an object pressing it down. As a result, the normal force is equal to an object's weight much of the time, so that the forces of friction are equal to the coefficients of friction multiplied by the object's weight. This assumption is false only on non-horizontal surfaces (for example, slopes), or when additional forces are acting to push an object down or hold it up.

Question	Analysis
A 15 kg object rests on a concrete floor, and the coefficients of static and kinetic friction are 0.3 and 0.2, respectively. If a continuous force of 35 N is applied to the object in a direction parallel with the floor, what will happen?	**Step 1:** The question asks for a prediction as to whether an object will move given an applied force and taking into account friction.
	Step 2: Comparing the applied force to the frictional forces should make it obvious whether the object will move or not. The weight of the object will be equal to the normal force holding the object up. Then the force of friction can be determined. $F_N = W = (15\ \text{kg})(9.8\ \text{m/s}^2) = 147\ \text{N}$ $F_f = \mu_k F_N = (147\ \text{N})(0.2) = 29.4\ \text{N}$ $F_f = \mu s F_N = (147\ \text{N})(0.3) = 44.1\ \text{N}$
	Step 3: The applied force of 35 N is less than the maximum value for the force of static friction. Since the object is already at rest, and the applied force is not enough to "break" the force of static friction, the net force will be zero and the object will stay put. The lesser force of kinetic friction is irrelevant since the object will not slide to begin with.
(A) The object will be bounced back due to friction. (B) The object will remain at rest as the applied force is unable to overcome the static friction. (C) The object will initially move but then stop as kinetic friction overcomes the applied force. (D) The object will accelerate continuously.	**Step 4:** Answer choice **(B)** is correct. (The key is that an object at rest must experience a large enough applied force to overcome static friction before kinetic friction becomes relevant.)

Now you try one:

Four objects of the same material and dimensions, but different weights, are placed on four different tabletops. When a force is applied that causes the boxes to move, which one generates the greatest amount of kinetic friction?

(A)
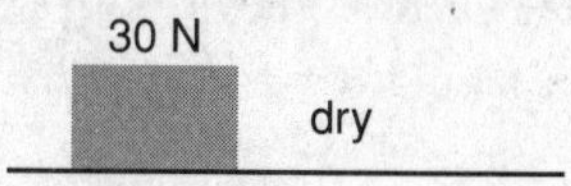

(B)
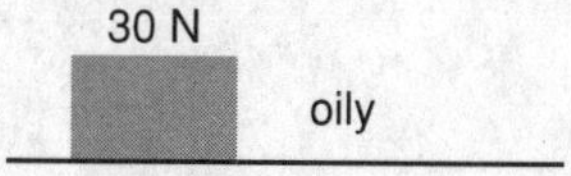

(C)
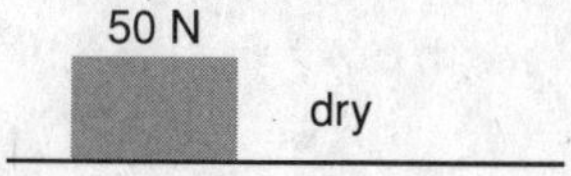

(D)
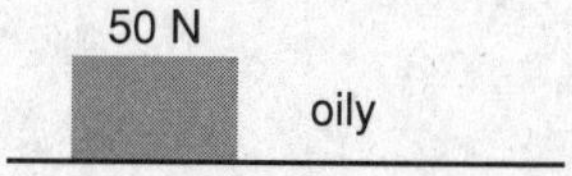

Explanation

Answer choice **(C)** is correct. According to the formula for kinetic friction, the box with the greater weight will have the greater normal force and thus greater frictional force compared to one of less weight. The oily surface will have a lower coefficient of friction than the dry surface. Therefore, the heaviest box on the driest surface will experience the most friction.

Another responsive force comes from air resistance, sometimes known as air friction but more commonly known as **drag**. Drag opposes movement not across a solid surface but through a fluid, like water or air. The two main variables that affect drag are cross-sectional area and speed. High-performance cars, boats, and aircraft are designed to be sleek and pointy, to minimize the area of the side that is plowing through the air or water. Parachutes work by intentionally increasing area for a skydiver, so that drag is increased and the person can be slowed down to a safe speed before hitting the ground.

Tension

When cables are used to pull an object, an applied force exerted on one end results in the attached object receiving the same pulling force from the other end. When force is transmitted through rope or other pullable materials, the internal stretch force of the material is called **tension**. Tension can be measured in N or lbs, or any other force unit measurement.

As a simplification, tension is often assumed to be equal in all parts of the material. This assumes, however, that the mass of the material under tension is either zero or negligible. In reality, tension is greatest where the force is being applied. To understand why this is, imagine a 100-pound weight suspended from a 10-pound chain. The bottommost link in the chain has an important job: it's holding up the 100-pound weight. The tension at this point in the chain, therefore, is 100 pounds.

But the chain link just above it also has an important job. It's holding up the link that's holding up the 100-pound weight, which means it must have a tension equal to 100 pounds, plus the weight of one chain link. If you continue to the top of the chain, where the applied force is holding it in place, the topmost link is responsible for not only the 100-pound weight, but for another 10 pounds from the chain itself. At this section of the chain, the tension is 110 pounds.

The takeaway from this is that, barring imperfections or weaknesses in the material under tension, the most likely place for a rope or cable to snap is not near the object it is holding or pulling, but near the place where the rope or cable is itself being pulled. When a winch (a mechanical device used to wind a rope) draws in a heavy-duty cable and the tension exceeds the material's tensile strength, it will likely snap very close to the winch, rather than close to the load.

Question	Analysis
A concrete block is suspended by two steel cables, spaced equidistant from the edges. What best describes the value of T1? 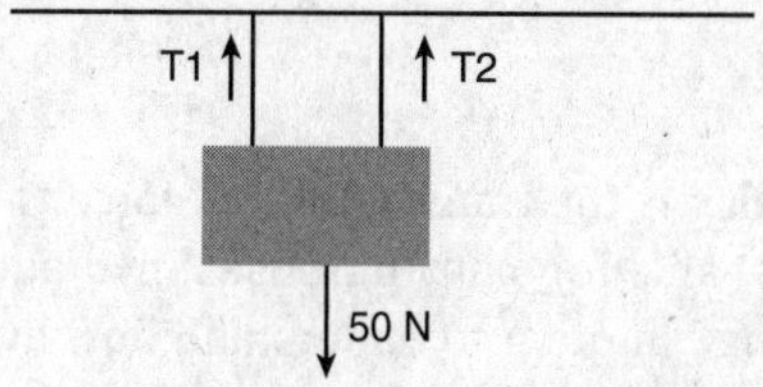 	**Step 1:** The question asks for the tension of one of two cables from which a weight is suspended.
	Step 2: The system is in static equilibrium. Since the block is remaining in one place without accelerating, the net force is zero and the downward force due to gravity must be exactly balanced by the tension in the two cables. Each cable must contribute half of the upward force (tension). $T1 + T2 = 50$ N $T1 = T2 = 25$ N
	Step 3: The tension in the first cable must be equal to half of the downward force on the block.
(A) 5 N (B) 25 N (C) 50 N (D) 100 N	**Step 4:** Choice **(B)** is correct.

Now try one on your own.

Using a cable attached at one end, a force is applied to a slab of metal. At which point along the cable is the cable most likely to snap?

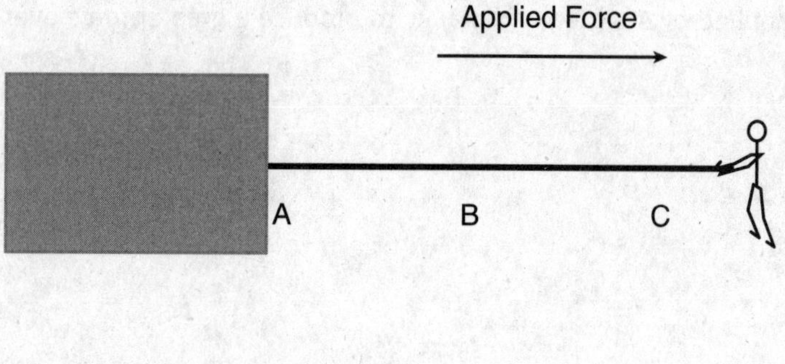

(A) A
(B) B
(C) C
(D) It is equally likely to break at any point along the rope.

Explanation

Choice **(C)** is correct. The force applied to the cable will result in acceleration of the metal slab. However, the metal slab by itself requires less force than the slab and cable combination. Thus, more force is applied at the end of the cable where the force is applied compared to what is actually acting on the metal slab itself. This corresponds to a greater tension near point C compared to points B and A. If the cable were to snap, it would likely happen where there is higher tension.

HYDRAULIC PRESSURE

Fluid power can also be used to gain mechanical advantage. **Hydraulics** is the transmission of force through the use of liquids. While the classic simple machines are effective and reliable for many purposes, hydraulics is a little more versatile at redirecting and multiplying forces in complex systems. In order to understand how hydraulics works, it is first necessary to have a grasp of the concept of **pressure**. If force is applied evenly over a certain area, then pressure is applied to that area. Pressure is calculated using the formula

$$P = \frac{F}{A}$$

where F is force in pounds (lb), A is area in square inches (in^2), and P is pressure in pounds per square inch (psi). While other applications of pressure in physics and chemistry use standard units of newtons and square meters, the above units are more common in hydraulics. Keep in mind that F is divided by A. A force of 100 pounds applied over an area of 10 square inches results in a pressure of 10 psi being developed.

Using the formula for pressure, you can derive $F = PA$, which is useful for calculating the amount of force that is developed when pressure P is applied to area A.

Even when extremely high pressure is applied to a liquid, the volume of the liquid will decrease only a very small amount. This property of near-incompressibility makes liquids very effective at transmitting force, as no force (or very little) is used up in compressing the liquids into a smaller space. Liquids also tend to take the shape of their container, filling up any available space (given enough of the liquid). According to **Pascal's law** (discovered by French mathematician and physicist Blaise Pascal in the seventeenth century), the pressure in any part of an enclosed fluid is the same and points in all directions.

Putting it all together, applying force to one end of a hydraulic system will increase the pressure everywhere, which, when applied to an area of the same size, produces the same force at a different location and in a different direction. Changing the area over which force acts can also change the magnitude of the force. In a hydraulic system, a smaller force is applied over a greater distance to produce a greater force over a smaller distance.

Hydraulic Lift

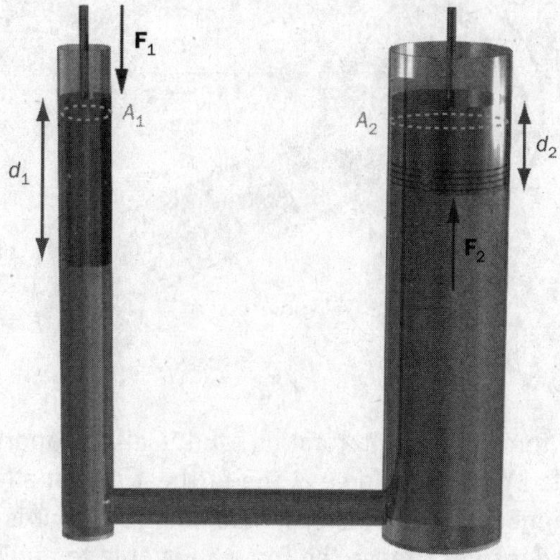

Try this problem about pressure.

Pressure is equal to force acting over

(A) a distance
(B) a period of time
(C) a volume
(D) an area

Explanation

Choice **(D)** is correct. Pressure is equal to force acting over an area.

Torque

Torque measures how much a force is able to cause an object to rotate. Torque, like pressure, cannot be properly called a force itself. The formula for torque is

$$\tau = rF$$

where torque (τ, the Greek letter tau) is the product of an applied force (*F*) perpendicular to the lever arm of the rotated object, and the length of the lever arm (*r*), measured from the pivot point (center of rotation) or fulcrum, to the point where the force is being is applied. Force is normally measured in either N or lb and the length of the lever arm in m or ft. Torque may therefore be measured in standard N-m (newton-meters) or the ft-lb (foot-pound) unit.

However, remember this length measurement is perpendicular to the force applied in rotating the object, not parallel, so torque is not equivalent to work, whose N-m units (more often expressed as joules, J) represent a force applied through a distance. (Nor is torque equivalent to energy, also measured in J.)

Torque Applied via a Wrench

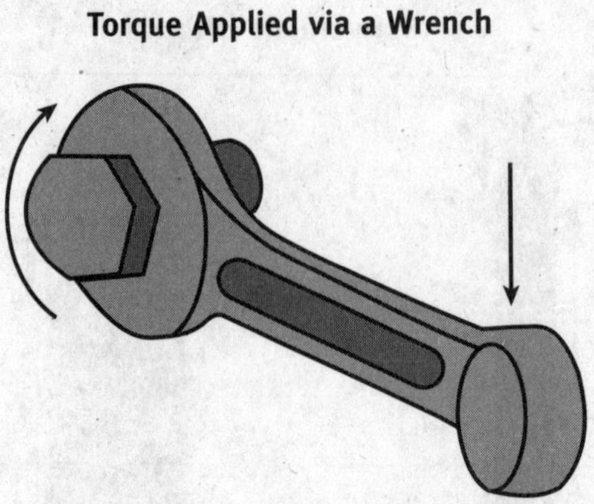

Torque is directly proportional to the force applied to rotation, and is also proportional to the distance from the center of rotation where the force acts. Applying a force at the farthest end of a lever arm is advisable in getting the greatest possible torque. When a mechanic applies his maximum force but is unable to turn a difficult bolt, a longer wrench may do the trick. This doesn't increase the force he is able to apply, but the use of a longer lever arm for the same force does result in increased torque and therefore, hopefully, better success in loosening the bolt. Unlike work, an object does not have to be successfully rotated before a torque is said to have been applied.

Question	Analysis
How much force is needed to apply 50 ft-lb of torque to a bolt using a 2 ft long wrench?	**Step 1:** The question asks for the amount of force applied to a lever arm for a certain torque.
	Step 2: The torque formula relates length, force, and torque. To solve this question, isolate force and plug in values. Based on the units in this question, force is measured in pounds (since pounds multiplied by feet is the only way to get foot-pound units for torque). $\tau = rF$ $F = \tau/r = (50 \text{ ft-lb})/(2 \text{ ft}) = 25 \text{ lbs}$
	Step 3: The force must be 25 lbs, not 25 N.
(A) 100 lbs (B) 50 N (C) 25 lbs (D) 25 N	**Step 4:** There is only one perfect match: choice **(C)**.

Now try a question about torque on your own.

Which of the following applied forces will result in the greatest torque?

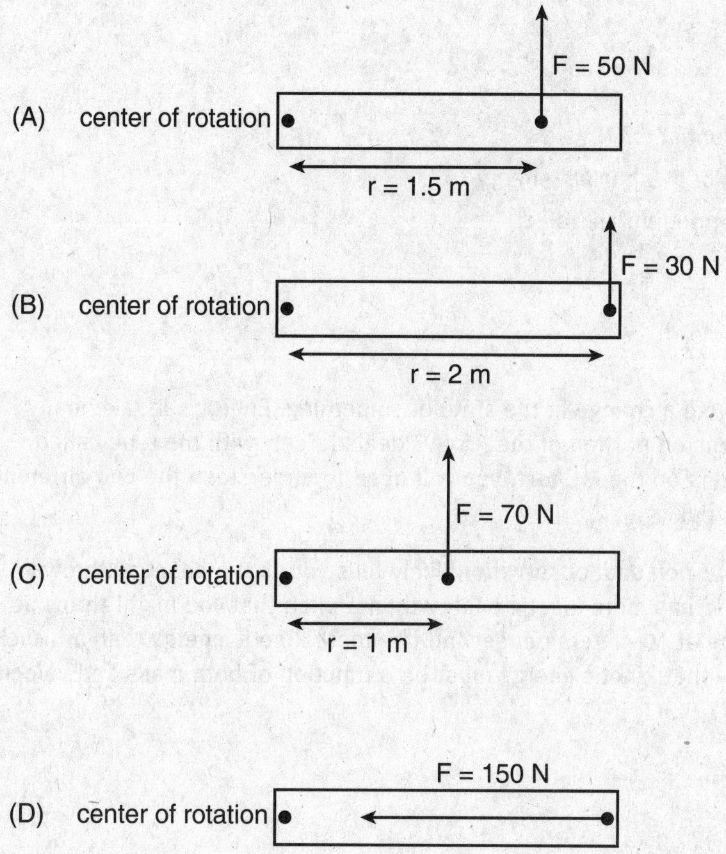

Explanation

Choice **(A)** is correct. The product of the force and the length of the lever arm, where the two quantities are perpendicular to each other, produces torque. For (A), $\tau = (1.5\text{ m})(50\text{ N}) = 75$ N-m, compared to 60 N-m in (B) and 70 N-m in (C). (D) provides no torque at all. Although there is a large applied force, it is pushing along the lever arm toward the fulcrum, and cannot result in any rotation.

ENERGY, WORK, AND POWER

LEARNING OBJECTIVES

In this section, you will learn to:

- compute values for kinetic and potential energy
- apply the principle of conservation of mechanical energy
- calculate the amount of work performed on an object

Energy

In a broad sense, energy is the capacity to cause a change in the state of something. Energy can take many different forms, but the Mechanical Comprehension portion of the ASVAB deals largely with the type called **mechanical energy**. In order to answer questions on the MC test, you will need to understand the two different types of mechanical energy: kinetic and potential.

Kinetic energy is the energy of movement. Your personal observation likely tells you that a baseball thrown by a major league pitcher at 40 meters per second has more kinetic energy than a pitch that you might throw at 25 meters per second. Similarly, a car traveling at 10 meters per second has more kinetic energy than a bicycle traveling at the same speed. So you could say that kinetic energy must be a function of both mass and velocity, and you would be correct.

The formula to calculate kinetic energy (KE) is:

$$KE = \frac{1}{2}mv^2$$

where m is mass in kilograms (kg), v is speed in meters per second $\left(\frac{\text{m}}{\text{s}}\right)$, and KE is kinetic energy in a standard unit called joules (J), which has the units $\left(\frac{\text{kg} \times \text{m}^2}{\text{s}^2}\right)$. Note that KE is a linear function of mass but the velocity component is squared. Here is how the formula could be applied:

Question	Analysis
Truck A has a mass of 10,000 kg and is traveling at a speed of 30 m/s. Truck B has a mass of 5,000 kg and a speed of 15 m/s. What is the ratio of the kinetic energy of Truck A to that of Truck B?	**Step 1:** The question asks for the ratio of the *KE*s of the two trucks, not the actual values.
	Step 2: The values needed to calculate *KE* for both trucks are given, so no further simplification is necessary.
	Step 3: Use relationships rather than actually calculating. Both the mass and speed of A are 2 times those of B. The masses by themselves cause the *KE* of A to be greater by a factor of 2. Since *KE* is a function of v^2, the *KE* of A is 2^2 times that of B due to speed. In total, A's *KE* is $4 \times 2 = 8$ times that of Truck B.
(A) 1:8 (B) 2:1 (C) 4:1 (D) 8:1	**Step 4:** Select choice **(D)**. Double check that the question asked for the *KE* ratio of A:B and that calculations are correct.

Try this problem on your own. This question does not have answer choices.

If an automobile with a mass of 1,500 kg is traveling at 72 kilometers per hour, what is its kinetic energy in joules?

Explanation

Note that the units of joules $\left(\frac{\text{kg} \times \text{m}^2}{\text{s}^2}\right)$ are not the same as the units in the problem. To convert, first restate the speed as 72,000 meters per hour. Since there are 60 seconds in a minute and 60 minutes in an hour, there are 3,600 seconds in an hour. Now you can set up the calculations:

$$V = \frac{\left(72{,}000\frac{\text{m}}{\text{hr}}\right)}{\left(3{,}600\frac{\text{s}}{\text{hr}}\right)} = 20 \text{ m/s}$$

$$KE = \frac{1}{2} \times (1{,}500 \text{ kg}) \times \left(20\frac{\text{m}}{\text{s}}\right)^2 = \frac{1}{2} \times 1{,}500 \text{ kg} \times 400\frac{\text{m}^2}{\text{s}^2} = 300{,}000 \text{ J}.$$

Potential energy (PE), as the name implies, is energy that has the potential to be converted to kinetic energy. Many ASVAB questions use one particular type of *PE*, **gravitational potential energy**. If you hold an object in your hand raised above the floor and then let it fall, its potential energy is fully converted to *KE* by the time it hits the floor. The higher the object is raised, (*h*), the greater the potential energy of the object. Since heavy objects and light objects fall at the same rate, a heavier object released from the same height would have greater *KE* when it hits the floor and therefore greater *PE* when held above the floor. The force acting on the dropped object is, of course, gravity. If you were to drop an object from the same height on the Moon, the speed when it hit the ground would be less than on Earth because the pull of gravity is much lower. Therefore, the object's *PE* would be less on the Moon.

The formula to calculate gravitational potential energy follows the logic above.

$$PE = mgh$$

where *m* is mass in kilograms (kg), *g* is **acceleration due to gravity** $\left(9.8\frac{\text{m}}{\text{s}^2}\right)$, *h* is the height of the object in meters (m), and *PE* is gravitational potential energy measured in joules (J).

On the ASVAB, it does not matter whether the object is dropped or rolled down a ramp because minor factors such as friction will be ignored on energy problems; the velocity, *KE*, and *PE* will be the same in either case, even though it will take longer for the object on the ramp to reach the ground.

Study the following example problem, which does not have answer choices, to help you practice making a prediction on a question of this type.

Question	Analysis
What is the potential energy of a 0.5 kg ball held at the top of a building 20 meters above the ground?	The question asks you to calculate potential energy.
	The question provides two of the variables you need to calculate PE, mass (*m*) and height (*h*). It would be a good idea to memorize that $g = 9.8\ \text{m/s}^2$ or at least that it is approximately 10. $PE = mgh = 0.5 \times 9.8 \times 20 = 98$ joules (J)
	The answer is **98 J**.

A falling object has its potential energy (*PE*) converted into kinetic energy (*KE*) as it falls. Since the ASVAB ignores such minor factors as resistance on energy problems, problems in the MC section will adhere to the **principle of conservation of mechanical energy**. This rule states that the total mechanical energy ($PE + KE$) remains constant as long as no other forces are applied. After the object is dropped but before it hits the ground, it will have some *KE* and some *PE*, totaling up to the initial *PE*. At the instant it hits the ground, all of the initial *PE* will have been converted to *KE*. See how this principle is used to solve the following problem:

Question	Analysis
A 1.63 kg object is dropped from a height of 2.5 meters. What will its velocity be at the moment it hits the ground?	**Step 1:** You are asked to calculate the velocity when the object hits the ground.
	Step 2: You are given the mass and height of the object but nothing about velocity. There is nothing to simplify.
	Step 3: The initial *PE* equals the final *KE*, so $mgh = \left(\frac{1}{2}mv^2\right)$. Notice that *m* appears on both sides of the equation so it cancels out: $gh = \left(\frac{1}{2}\right)v^2$. $9.8 \times 2.5 = \left(\frac{1}{2}\right)v^2$. Multiply both sides by 2 for ease of calculation to get $9.8 \times 5 = 49 = v^2$ $v = 7$
(A) 7 m/s (B) 14 m/s (C) 28 m/s (D) 49 m/s	**Step 4:** Select choice **(A)**. By setting $PE = KE$ you followed the principle of conservation of mechanical energy and the variable *v* is the answer to the right question.

If an object is thrown straight up into the air rather than being dropped, when the object reaches its highest point just before it begins to descend, its $KE = 0$ and the potential energy it has at this point equals the kinetic energy it had when it was first thrown. Apply this concept to the problem below:

Question	Analysis
Joel throws a ball straight up in the air at a velocity of 30 m/s. When he releases the ball, it is 2 meters above the ground. What is the ball's approximate maximum height above the ground?	**Step 1:** You are given an initial velocity, the additional fact that the ball is released 2 meters above the ground, and the knowledge that the answer will be an approximation.
	Step 2: The question asks for the maximum height above the ground rather than above the point of release.
	Step 3: Set $PE = KE$: $mgh = \left(\frac{1}{2}\right)mv^2$. Cancel out *m* and plug in *v*, and $g = 10$ since the question asks for an *approximate* value. $10h = \left(\frac{1}{2}\right)30^2 = 450$ $h = 45$ Add 2 meters to get the total height, 47. Since you used 10 for *g* rather than 9.8, the true value will be greater than 47.
(A) 40 meters (B) 45 meters (C) 48 meters (D) 60 meters	**Step 4:** Only (C) and (D) are greater than 47, but (D) is clearly too high. **(C)** is correct. Check the math and that the question requires the additional 2 meters.

Work

Work is accomplished when force is applied to move an object. Like the formulas for energy, the formula to calculate work coincides with what you can observe in the real world. The more force that has to be exerted to move an object a certain distance, the more work is done. Similarly, the farther an object is moved by exerting force, the more the work. These facts are summarized by the formula:

$$W = Fd$$

where *W* is work in joules (J), *F* is force in newtons (N), and *d* is distance in meters (m).

If a force of 100 N is applied to a car, and the car is moved 10 meters, then a total of $100 \times 10 = 1{,}000$ joules of work has been done.

Force *F* Pushing a Car over Distance *d*

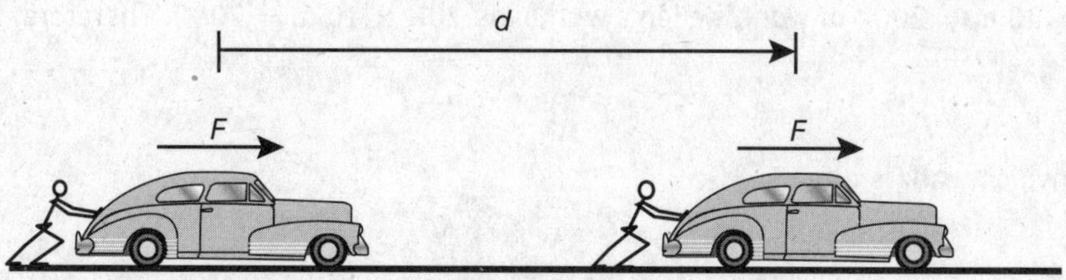

Work is done if force *F* is applied over distance *d*

If force is applied to the car, but the car does not move, no work has been done. This can be seen in the formula $W = Fd$, where zero distance (d) results in work (W) being zero as well. No matter how much force is applied, no movement results in zero work being accomplished.

Note that work and kinetic energy are both measured in joules. This is not a coincidence, as any work that is done to accelerate an object at rest to velocity v will be converted into the kinetic energy of that object. This principle is known as the **work-energy theorem**.

Question	Analysis
A force of 10 N is applied to a stationary object with 1 kg mass for a sufficient time to attain a velocity of 4 m/s. How much work was done to accomplish this?	The question has values for F, m, and v and asks how much work was done to attain that velocity.
	Work can't be calculated directly because the distance isn't known. However, $W = KE$, per the work-energy theorem. $W = \left(\frac{1}{2}\right) mv^2$
	Plug the known values into the equation and solve. $W = \left(\frac{1}{2}\right) \times 1 \times 4^2$
	Answer: **8 joules**

Apply the information in this section to answer the problem below.

Jane pushes a 20 kg box 20 m in 20 seconds by pushing with a constant force of 10 N. Kellen pushes a 40 kg box 10 m in 15 seconds by pushing with a constant force of 20 N. Which person did more work?

(A) Jane
(B) Kellen
(C) They both did the same amount of work.
(D) It cannot be determined from the information given.

Explanation

The data regarding the mass of the boxes and the time it took each person to move a box are irrelevant. Jane accomplished 10N × 20 m = 200 J of work; Kellen's work was 20N × 10 m = 200 J. Therefore, **(C)** is correct.

Power

Power is the **rate** at which work is done:

$$P = \frac{W}{t}$$

Consider the example of a woman pushing a small car over a distance of 100 meters. If the woman pushes with a force of 100 N to move the car, then using $W = F \times d$, you can see that 100 N × 100 m = 10,000 J of work has been done. If she accomplishes the task at a constant rate over the course of 500 seconds, then the power applied was $\frac{10{,}000}{500} = 20\frac{\text{J}}{\text{s}}$. If another woman can do the same amount of work in 250 seconds, then that woman has twice as much power, or $40\frac{\text{J}}{\text{s}}$.

One common unit used to express power is the watt, which equals $1\frac{\text{J}}{\text{s}}$ or $1\frac{\text{N} \times \text{m}}{\text{s}}$. The watt is named after James Watt, an engineer who died in 1819 and whose work on the steam engine helped make the Industrial Revolution possible. The unit of power that many people in the United States are more familiar with is **horsepower,** since this is the unit that is used to rate internal combustion engines. Horsepower uses a different system of measurement: it is defined as 550 foot-pounds per second of work being done. One horsepower is the equivalent of 746 watts. Study the example below, which is presented without answer choices, in order to help you practice predicting on a question of this type.

Question	Analysis
Jane pushes a 20 kg box 20 m in 20 seconds by pushing with a constant force of 10 N. How much power did she exert?	You are asked to calculate how much power Jane exerted.
	You already know that Jane did 200 J of work. Since she accomplished that in 20 seconds, the power she applied was **10 watts.**

SIMPLE MACHINES

LEARNING OBJECTIVES

In this section, you will learn to:

- identify simple machines and their uses
- calculate the mechanical advantage of specific simple machines
- determine forces in simple machine problems

Simple machines were invented centuries ago to reduce the force or effort needed to perform a variety of tasks; heavy objects need to be lifted, parts need to be pushed together or pulled apart, and large machines need to be moved. Often, it is difficult or impossible for a human being to apply enough force to get the job done.

In order to increase the available force to a much larger one, mechanical advantage is used. **Mechanical advantage** is defined as "the advantage gained by the use of a mechanism in transmitting force." Using the proper equipment, it is possible to increase the force applied severalfold. However, while mechanical advantage can be used to amplify force, this is accomplished by exerting a force over a greater distance. In other words, the work output can never exceed the work input. The good news, however, is that simple machines can make hard work easier by breaking it down into smaller, more manageable pieces.

The **efficiency** of a machine describes how much of the power inputted into the machine is turned into movement or force. A machine that turned 100 percent of the power applied to it into outputted force would have 100 percent efficiency. In the real world, however, no machine is 100 percent efficient, because all machines are affected by friction and most are affected by worn-down and/or imperfectly fitted parts.

Inclined Plane

One of the simplest ways of managing lifting tasks is through the use of the **inclined plane**. Lifting a 100 kg (about 220 pound) box is a difficult, if not impossible, task for one person. However, the use of an inclined plane (or ramp) can make this a manageable chore.

Pushing a Box Up a Ramp or Lifting It to Height *h*

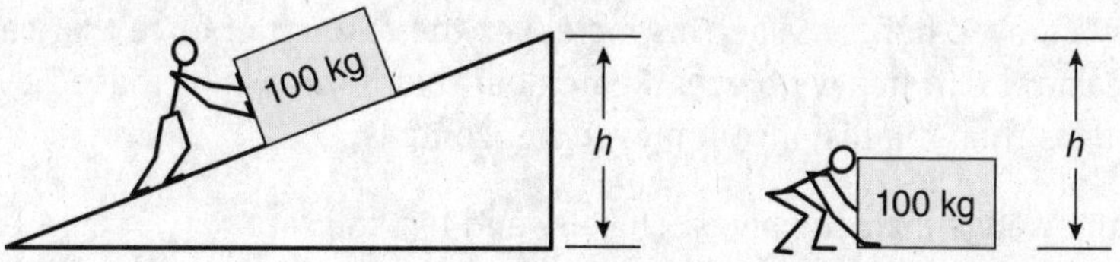

Remember that the work that is done to lift the box will be the same as the final gravitational potential energy of the box ($PE = mgh$). Therefore, the same amount of work will be done whether the box is pushed up the ramp (assuming no friction), or if it is lifted directly. The difference with the ramp is that a much smaller force can be applied to raise the box. This force must be applied over a longer distance, but it will give some mechanical advantage to make the task easier. Study the following problem to see how much the mechanical advantage of inclined planes can reduce the force needed to raise a heavy object.

Question	Analysis
Jose has to push an object that weighs 1,000 N up an inclined ramp to a height 2 m above its current location. If the ramp is 20 m long (measured up the slope), how much force will Jose have to use if the effects of friction are ignored?	The question asks you to calculate force and ignore friction.
	When Jose has completed the task, the object will have 2 m $\times$ 1,000 N $=$ 2,000 J greater potential energy than when he started, so he will have to accomplish 2,000 J of work. Since $W = f \times d$ and the ramp is 20 m in length, he will only have to exert $\frac{2,000}{2} = 100$ N.
	The answer is **100 N**.

Wedge

The **wedge** is a variation of the inclined plane. While the inclined plane stays stationary, the wedge is designed to move. It can be used for many purposes, including lifting, splitting, and tightening. The mechanical advantage that can be gained with a wedge is determined by the ratio of its length to its height.

Wedge

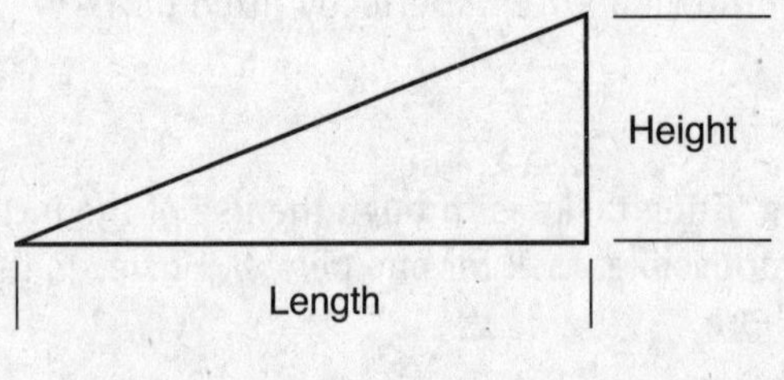

Making the wedge longer relative to its height (that is, with a smaller slope) will decrease the amount of lift that takes place as the wedge is moved horizontally. This increases the amount of force that can be generated by the wedge and makes it easier to lift heavy objects. Remember that if force is increased by the wedge, then the distance it will lift will be less than the distance it moved horizontally.

Common applications of the wedge include knives, chisels, and log splitters.

Levers

Another simple machine that can be used to amplify force or distance, as well as change direction, is the **lever.** There are three basic types of levers, so this machine lends itself to many different applications.

The **first-class lever** is used to increase force or distance, and to change the direction of the force. An example of a first-class lever is a child's teeter-totter.

First-Class Lever

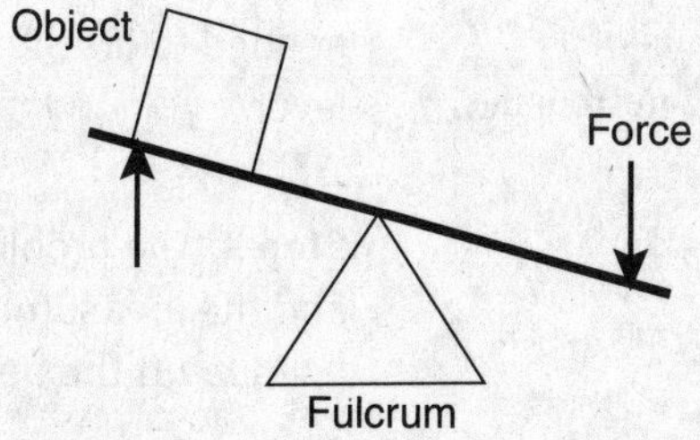

The **fulcrum** is the point upon which the lever pivots. When force is applied to one end of the lever, the movement at the other end is in the opposite direction. The position of the fulcrum will define whether any mechanical advantage is gained. If the fulcrum is closer to the object being lifted, less force is required to do the work. However, the end that the effort is applied to must move farther to get the job done.

The opposite effect takes place when the fulcrum is moved closer to where the effort is applied. Now the distance the object can be moved is amplified, but the force required is increased.

The mechanical advantage gained by the use of a lever is simply the ratio of the distance of the applied force from the fulcrum to the distance of the object from the fulcrum. Thus, a lever where the force is applied 3 meters from the fulcrum and the object being acted upon is 1 meter from the fulcrum will have a mechanical advantage of 3:1. The force applied will move 3 times as far as the object but the force on the object will be 3 times as great as the force applied.

A **second-class lever** is used to increase force on the object in the same direction as the force applied. This type of lever requires a smaller force to lift a larger load, but the force must be exerted over a greater distance. An example of a second-class lever is a wheelbarrow.

Second-Class Lever

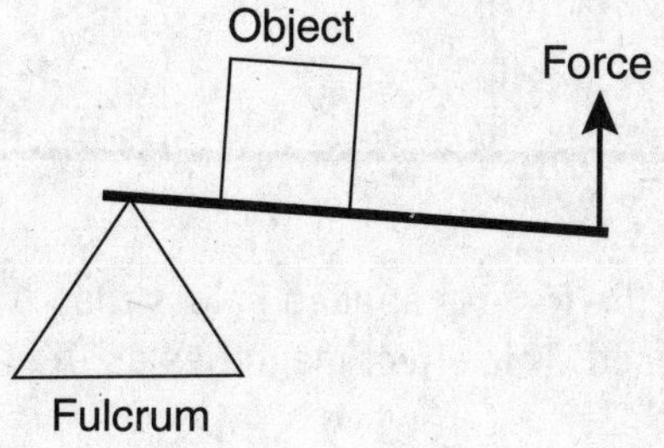

The mechanical advantage of the second-class lever increases as the object is moved closer to the fulcrum. This can also be accomplished by increasing the length of the lever. The following example illustrates the mechanical advantage of a second class lever:

Question	Analysis
Which of the following forces would be the least needed to lift a 10 N object placed 0.5 m from the fulcrum of a second-class lever if the lifting force is exerted 2 m from the fulcrum?	**Step 1:** The question asks for the minimum force needed to lift the object given the dimensions of the lever.
	Step 2: The problem provides the weight of an object and the measurements of a second-class lever that is used to lift the object.
	Step 3: The mechanical advantage of the lever is $\frac{2}{0.5} = 4$, so a force of $\frac{1}{4}$ the weight of the object or 2.5 N is needed. However, this amount of force would balance the downward and upward forces, so the correct answer must be > 2.5 N.
(A) 0.5 N (B) 2 N (C) 3 N (D) 10 N	**Step 4:** Choose option **(C)**. Verify that this is the *minimum* force needed from among these choices (that's why (D) is incorrect), and double check the dimensions and calculations.

The **third-class lever** is also used to increase distance traveled by the object in the same direction as the force applied. It is similar to the second-class lever in that the fulcrum is at one end. The object, however, is at the other end of the lever, with the force being applied somewhere in between the fulcrum and the object. Examples of third-class levers would be a fishing pole and a catapult.

Third-Class Lever

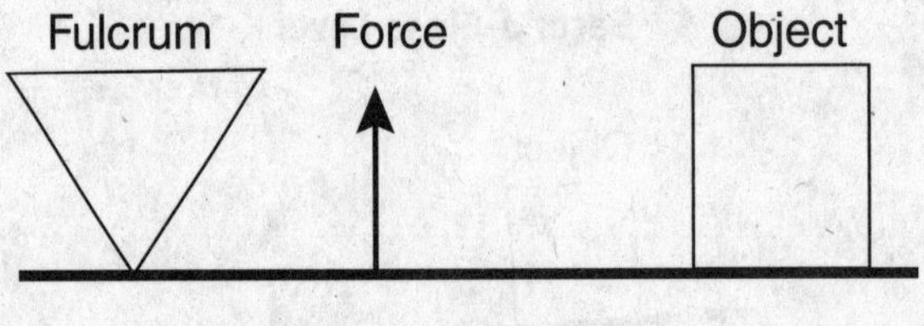

Since the distance moved at the point where the force is applied is less than the distance traveled by the load with a third-class lever, the force needed is increased. This effect magnifies as the force is applied closer to the fulcrum.

Pulleys

Applying force to an object is sometimes accomplished through the use of a **pulley**. A pulley is a system that is used to change the direction, but not the amount, of a force.

Person Pulling a Rope Through a Pulley to Lift an Object

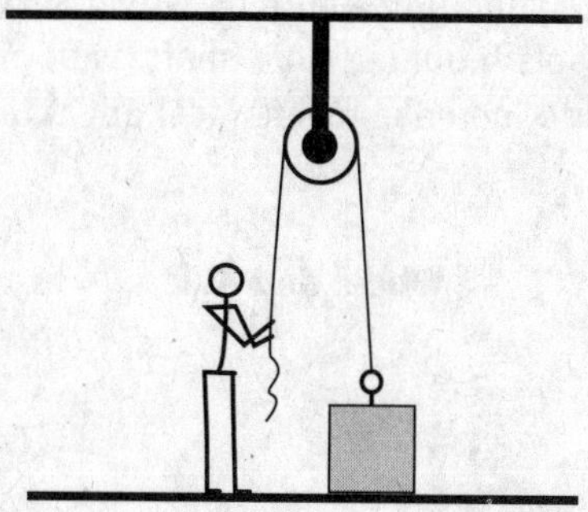

A pulley system consists of a rope, belt, or chain looped over a wheel. It can be useful in certain situations because it is often easier to pull with a downward force than an upward force.

Block and Tackle

It is relatively uncommon for one pulley to be used by itself. Often, two or more pulleys will be used in an arrangement known as a **block and tackle** to increase lifting force.

Block and Tackle

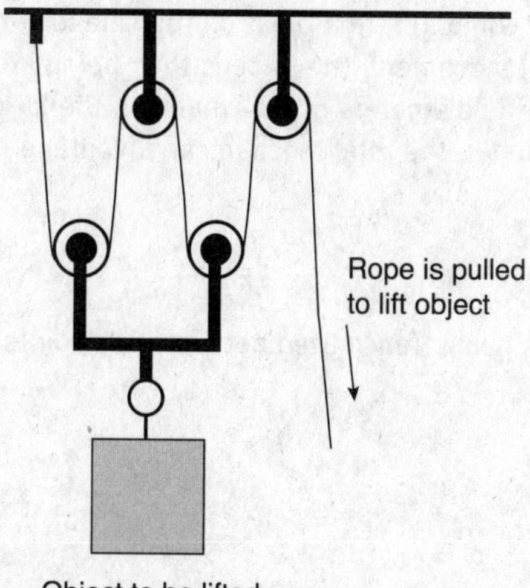

The block and tackle shown here has four pulleys; two are attached to a stationary object (such as the ceiling or wall of a building) and the other two are attached to the lower block (the load). As the rope is pulled, the lower block moves toward the stationary pulleys. Note that in order to lift the load 1 foot, it is necessary to pull the rope a total of 4 feet. This is because each of the four rope links must shorten by 1 foot to get the lower block to move 1 foot. Neglecting friction, this gives a total mechanical advantage of 4:1, so if 4 pounds of force are required to lift a load, only 1 pound needs to be applied to the rope. Note that you can determine the mechanical advantage ratio of a block and tackle system by counting the number of pulleys.

Wheel and Axle

The **wheel and axle** mechanism is not just a simple free-turning wheel such as the front wheel of a typical bicycle. Instead, this type of machine uses two wheels mounted on a shaft, with one wheel having a larger diameter than the other. The wheel and axle can be used to increase mechanical advantage. An example of a wheel and axle is the steering wheel of a car.

Wheel and Axle

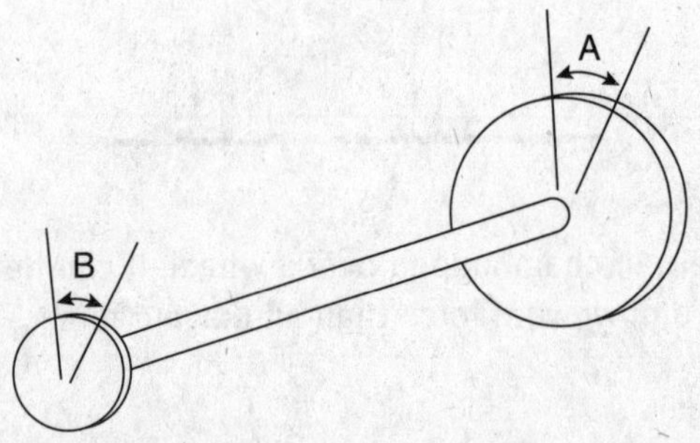

Movement at B is smaller than movement at A—force is increased at B.

The large-diameter wheel will usually be the one that has the effort applied to it. Large movements of the large wheel result in small movements of the small wheel. The larger wheel has a greater circumference, so a point on the edge of it travels further per turn compared to a point on the smaller wheel. Since a large movement (or distance) is being translated into a small movement, the force is then being increased. The amount of mechanical advantage is determined by the ratio of the diameters of the wheels. If the large wheel is 20 inches in diameter, and the small wheel is 10 inches in diameter, the total mechanical advantage is 2:1.

Gears and Gear Ratios

Gears can be used anywhere that force is being transmitted between two points, and are excellent for gaining mechanical advantage.

Meshed Gears

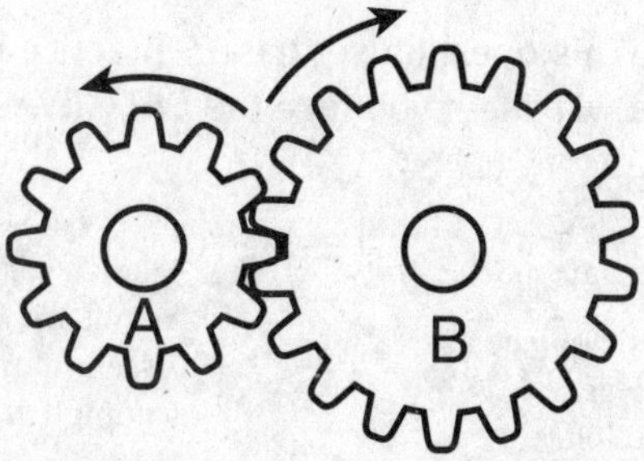

Gears can be used to change rotational speed and torque. The **gear ratio** is determined by comparing the number of teeth on the large gear to the number on the small gear. In the figure, the large gear has 18 teeth and the small gear has 12, so the gear ratio is 3:2.

If a small gear drives a large gear, a **speed reduction** takes place. The large gear will turn more slowly than the small gear and the speed of the output will be slower in proportion to the gear ratio. However, an associated **torque multiplication** (increase) will take place, and the torque output of the large gear will be greater than what was applied to the small gear, in proportion to the gear ratio. A speed decrease means that torque will increase. A mathematician would describe these two properties as **inversely proportional**.

Check your understanding of gears with this problem:

Question	Analysis
Gear A in the figure above turns counter-clockwise at the rate of 60 revolutions per minute (RPM). In what direction and at what rate will Gear B turn?	**Step 1:** The question asks both the direction and the rotational speed of Gear B.
	Step 2: The relationship of the gears, the number of teeth on each, and the RPM and direction of Gear A are given. No simplification is needed.
	Step 3: The two gears move in opposite directions, so eliminate answer choices (C) and (D). Since Gear B is larger and has more teeth, it must turn more slowly.
(A) clockwise at 40 RPM (B) clockwise at 90 RPM (C) counter-clockwise at 40 RPM (D) counter-clockwise at 90 RPM	**Step 4:** The correct answer is **(A)**. The gear ratio is 3:2 and the ratio of the rate of revolutions is 2:3, so the answer checks correctly.

MECHANICAL COMPREHENSION PRACTICE SET 1

Select the best answer for each question. This question set has 15 practice questions, which is the number of Mechanical Comprehension questions you will see if you take the CAT-ASVAB.

1. The work required to lift an object is the same as

    (A) the change in potential energy of the object
    (B) the weight of the object divided by gravity
    (C) the force required to overcome friction
    (D) half of the work required to put the object down

2. A constant force of 2 N is applied to a mass of 6 kg that is initially at rest for 3 seconds. Assuming there is no friction or any other force applied to the mass, what is the velocity of the mass after 3 seconds?

    (A) 0.5 m/s
    (B) 1.0 m/s
    (C) 6.0 m/s
    (D) 12.0 m/s

3. Approximately how much force is needed to lift the weight below?

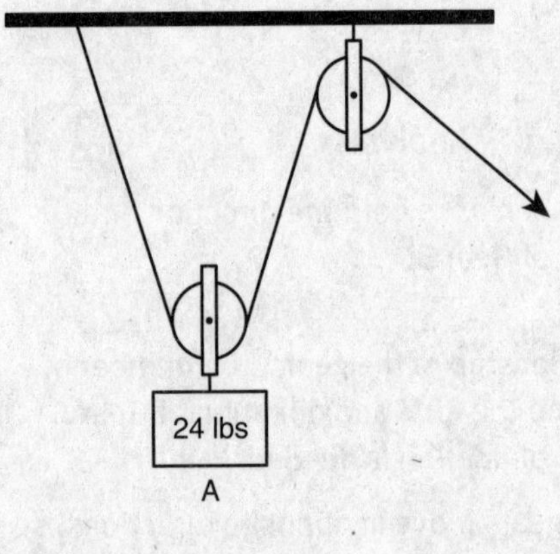

    (A) 6 lbs
    (B) 12 lbs
    (C) 24 lbs
    (D) 48 lbs

4. If a ball is dropped from a height of 5 m, what will be its approximate speed when it hits the ground?

    (A) cannot be determined because the mass is unknown.
    (B) cannot be determined because the time it takes to reach the ground is unknown.
    (C) 5 m/s
    (D) 10 m/s

5. Adam's car is stuck in the mud, so he recruits some friends to help push it out. If they apply 2,000 N of force to push the car 5 m, how much work did they accomplish?

    (A) 400 J
    (B) 1,000 J
    (C) 2,000 J
    (D) 10,000 J

6. A machine in an assembly plant lifts parts weighing 10 N each from the floor to a height of 2 m. If this machine must lift batches of 20 parts at a time in 2 seconds, what amount of power is required?

    (A) 20 watts
    (B) 100 watts
    (C) 200 watts
    (D) 400 watts

7. A man who weighs 900 N wants to compress some loose soil in his garden, so he puts a thick plate that is $\frac{1}{2}$ m wide and $\frac{3}{4}$ m long on the ground and stands on the plate. Assuming that the plate distributes his weight evenly, how much pressure is applied to the soil?

    (A) $900\ \frac{N}{m^2}$
    (B) $1{,}600\ \frac{N}{m^2}$
    (C) $1{,}800\ \frac{N}{m^2}$
    (D) $2{,}400\ \frac{N}{m^2}$

8. Gear A has 15 teeth and gear B has 10. If gear A makes 14 revolutions, how many will gear B make?

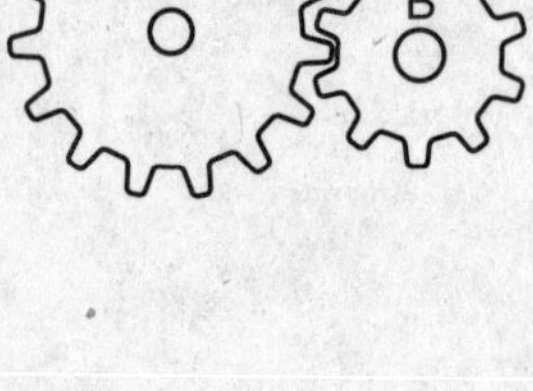

    (A) 7
    (B) 14
    (C) 21
    (D) 28

9. Two wheels are secured to a common axle. If the larger wheel has a radius of $\frac{1}{2}$ m and the smaller wheel has a diameter of $\frac{1}{10}$ m, what is the mechanical advantage for a force applied to the larger wheel?

(A) 5 to 1
(B) 10 to 1
(C) 20 to 1
(D) 40 to 1

10. A basic kitchen knife, like the one below, is based on which of the following simple machines?

(A) wedge
(B) first-class lever
(C) pulley
(D) second-class lever

11. How much force is needed to apply 50 ft-lb of torque to a bolt using a 2 ft long wrench?

(A) 100 pounds
(B) 50 pounds
(C) 40 pounds
(D) 25 pounds

12. If a 10 N force applied to an object produces an acceleration of 10 m/s^2, assuming friction is NOT negligible, what could the mass of the object have been?

(A) less than 1 kg
(B) equal to 1 kg
(C) between 10 kg and 1 kg
(D) greater than 10 kg

13. The force of friction between two dry surfaces moving past each other DOES NOT depend on

(A) the normal force
(B) the nature of the surfaces in contact with each other
(C) the coefficient of kinetic friction
(D) the speed at which the surfaces are moving past each other

14. The gravitational acceleration on Earth, g, is 9.8 m/s^2. The gravitational acceleration on a neighboring planet, Xendor, is 3 times as much. What is the approximate weight of a 100 kg person on Xendor? (Hint: you can approximate the gravitational acceleration on Earth as 10 m/s^2 to make calculations easier.)

(A) 100 N
(B) 300 N
(C) 1,000 N
(D) 3,000 N

15. Assuming no friction, if 80 J of work was done to move an object 2 m, what was the net force that was applied?

(A) 4 N
(B) 16 N
(C) 40 N
(D) 160 N

MECHANICAL COMPREHENSION PRACTICE SET 2

Select the best answer for each question. This question set has 15 practice questions, which is the number of Mechanical Comprehension questions you will see if you take the CAT-ASVAB.

1. A man pushes against an object at rest, but fails to get it moving. Which of the following must be true about the relationship between the applied force and the force of friction?

   (A) The applied force is less than the force of static friction.
   (B) The applied force is equal to the force of kinetic friction.
   (C) The applied force is equal to the force of static friction.
   (D) The applied force is greater than the force of static friction.

2. A ball with a mass of 10 kg starts rolling down a hill 20 m high. What is its approximate velocity when it hits the bottom, ignoring resistance?

   (A) 10 m/s
   (B) 15 m/s
   (C) 20 m/s
   (D) 40 m/s

3. A 500 g block made from a uniform material is suspended from the ceiling by two equal length ropes equidistant from the center of the block, as depicted in the figure below. Which of the following best approximates the tension in rope 1 (*T1*)?

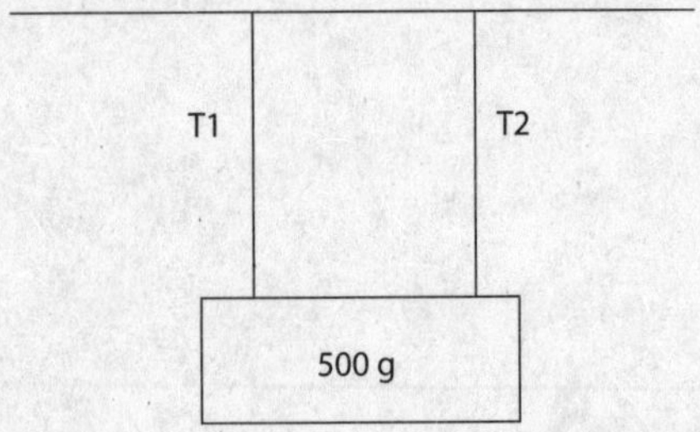

   (A) 2.5 N
   (B) 5.0 N
   (C) 50 N
   (D) 250 N

4. A hydraulic lift is set up as depicted below. Which of the following must be true?

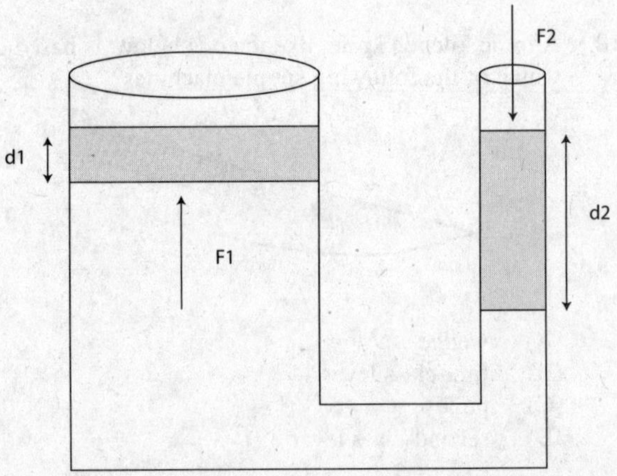

   (A) The work done by Force 2 is greater than the work done by Force 1, since Force 2 is applied over a longer distance.
   (B) To do the same amount of work, Force 2 must be smaller than Force 1, but applied over a larger distance.
   (C) Force 1 must be greater than Force 2, because more work must be done by Force 1.
   (D) The pressure in the left side of the lift must be greater than the pressure in the right side of the lift, since Force 1 is larger.

5. Two weights, one with a mass of 10 kg and another with a mass of 20 kg, are placed on opposite ends of a 9 m rod. To achieve rotational equilibrium, what distance from the 10 kg weight must the fulcrum be placed?

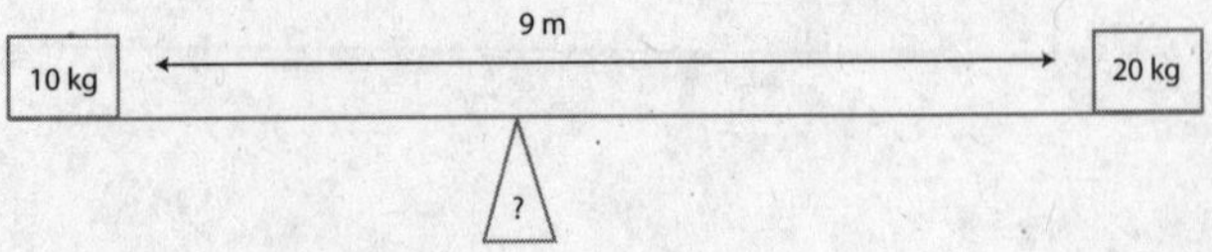

   (A) 3.0 m
   (B) 4.5 m
   (C) 6.0 m
   (D) 8.0 m

6. What is the kinetic energy of a 98 N object moving at 100 m/s?

(A) 10,000 J
(B) 50,000 J
(C) 100,000 J
(D) 500,000 J

7. A man pushes a 50 kg block a distance of 10 meters in 20 seconds with a force of 100 N. How much power did he exert?

(A) 25 W
(B) 50 W
(C) 100 W
(D) 200 W

8. Based on the diagram below, approximately what minimum force must be applied to the rope to lift the object?

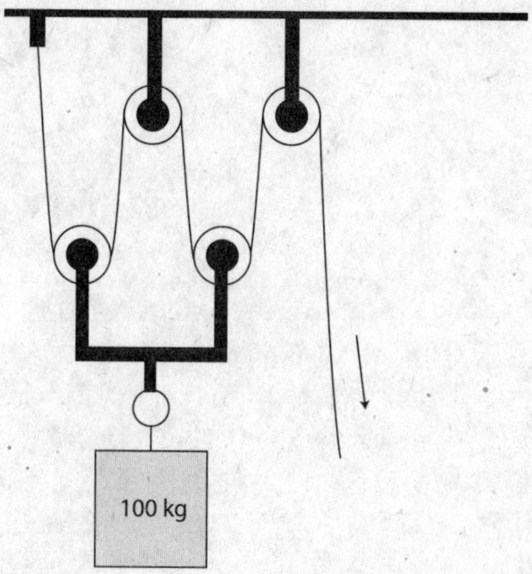

(A) 25 N
(B) 50 N
(C) 100 N
(D) 250 N

9. Snowshoes are used to reduce the pressure exerted by the feet when walking on top of deep snow. A 150 lb woman puts on two rectangular snowshoes with dimensions 10 inches by 15 inches each. What pressure does the woman exert on the snow while both shoes are equally bearing her full weight?

(A) 0.5 psi
(B) 1.0 psi
(C) 5.0 psi
(D) 10.0 psi

10. A 40 kg block is sitting on a table and it is also pushed down by a 100 N force. What is the magnitude of the normal force provided by the table if the object remains at rest? (Use 10 m/s^2 for gravity.)

(A) 140 N
(B) 300 N
(C) 400 N
(D) 500 N

11. A ball with a mass of 5 kg is thrown from a height of 10 meters at an angle of 30° above the horizontal with an initial velocity of 10 m/s. Consider any resistance to be negligible. What is the approximate total kinetic energy when the ball hits the ground?

(A) 250 J
(B) 300 J
(C) 500 J
(D) 750 J

12. Which of the answer choices best describes the topics that are included in the area of physics known as "mechanics?"

(A) the proper use of tools and measuring devices
(B) machines that transmit or use energy
(C) the laws of motion, energy, and forces
(D) the laws of trajectories, energy, and forces

13. Once sufficient force is applied to a stationary object to overcome static friction, if the same force is maintained, the object will accelerate. This is true because ___________.

(A) the force that is applied accumulates
(B) the coefficient of kinetic friction is greater than the coefficient of static friction
(C) the coefficient of static friction is greater than the coefficient of kinetic friction
(D) the object's momentum is cumulative

14. Playground seesaws are constructed in the same way as first-class levers. A boy wants to play on the seesaw with his little brother. The big brother weighs 100 pounds; the little brother weighs 75 pounds and sits 8 feet from the center of the seesaw. In order for the two brothers to be balanced, how far from the center of the seesaw should the big brother sit?

(A) 6.0 feet
(B) 8.0 feet
(C) 10.0 feet
(D) 12.5 feet

15. A hydraulic lift has a 2 cm diameter piston that exerts a downward force of 20 N on the incompressible fluid in a closed system. If the piston that is at the other end has a diameter of 20 cm, how much upward force can it exert?

(A) 20 N
(B) 200 N
(C) 1,000 N
(D) 2,000 N

MECHANICAL COMPREHENSION PRACTICE SET 1

Answers and Explanations

1. **A** The work-energy theorem states that work performed on an object will increase the object's mechanical energy by the amount of work applied to the object. In this case the increase in mechanical energy was the additional potential energy resulting from the object being lifted higher. Therefore the final *PE* equals the work.

2. **B** Since $F = ma$, then $a = \frac{F}{m} = \frac{2}{6} = \frac{1\text{ m}}{3\text{ s}^2}$. Change in velocity = acceleration × time. Therefore: $\frac{1\text{ m}}{3\text{ s}^2} \times 3\text{s} = 1\frac{\text{m}}{\text{s}}$.

3. **B** This block and tackle has 2 pulleys, so the mechanical advantage is 2:1, because the weight will only move half the distance that the rope is pulled. As a result, the force applied will be half the weight, or 12 lbs.

4. **D** Although the mass and time are unknown, they are not needed to solve the problem. When the ball is 5 m above the ground all its energy is *PE*, but when it hits the ground all its energy is *KE*. Since energy is conserved, set *PE* = *KE* or $mgh = \frac{1}{2}mv^2$. The mass cancels out and, since the question asks for the approximate speed, $g = 10$ will suffice. $2 \times 10 \times 5 = v^2$ and $100 = v^2$, so $v = 10$.

5. **D** Work = Force × Distance or 2,000 × 5 = 10,000 J.

6. **C** To calculate power, start by determining the work performed. Since there are 20 parts that weigh (exert a downward force) 10 N each that are lifted 2 m, work = 20 × 10 × 2 = 400 J. The machine accomplishes this work in 2 seconds. Power is $\frac{\text{work}}{\text{time}}$, or 200 watts.

7. **D** Pressure is force divided by the area over which it is applied, so

$$P = \frac{900}{\left(\frac{1}{2}\right) \times \left(\frac{3}{4}\right)} = \frac{900}{\frac{3}{8}} = 900 \times \frac{8}{3} = 2{,}400\frac{\text{N}}{\text{m}^2}.$$

8. **C** This could be solved with two different approaches. Gear A makes 14 revolutions and has 15 teeth, so 14 × 15 = 210 teeth would pass any point. Since gear B has 10 teeth, it would need to make 21 revolutions to match the teeth of gear A. Alternatively, the gear ratio is 3:2, so the proportion $\left(\frac{3}{2}\right) = \left(\frac{x}{14}\right)$ could be used to obtain the same answer, 21.

9. **B** The mechanical advantage of a wheel and axle is equal to the ratios of the radii of the two wheels. The problem states that the smaller wheel's diameter is $\frac{1}{10}$ m, so its radius is $\frac{1}{20}$ m and the ratio of the two radii is 10:1.

10. **A** The cross-section of a kitchen knife is very small on the cutting side and gets thicker toward the top, just as a wedge does.

11. **D** Torque (τ) is determined by multiplying a force (*F*) by the lever arm length (*r*) that the force is acting through or $\tau = rF$. In this case, to achieve an applied torque of 50 ft-lb, with a 2 ft lever arm, where $F = \frac{\tau}{r}, \frac{50\text{ ft-lb}}{2\text{ ft}} = 25$ lb of force is needed.

12. **A** Based on Newton's second law, *F* = ma, where F = force in newtons, *m* = mass in kilograms, and *a* = acceleration in m/s^2. Therefore, rearranging the equation for mass, you get: $m = \frac{F}{a}$. Since you know that friction is not negligible, you also know that the resultant force will be less than 10 N (see diagram below). Therefore, the maximum mass possible in a frictionless system is calculated by dividing 10 N by 10 m/s^2, which gives 1 kg. Knowing that friction will diminish the applied force, the resulting mass has to be less than 1 kg, which matches answer choice (A).

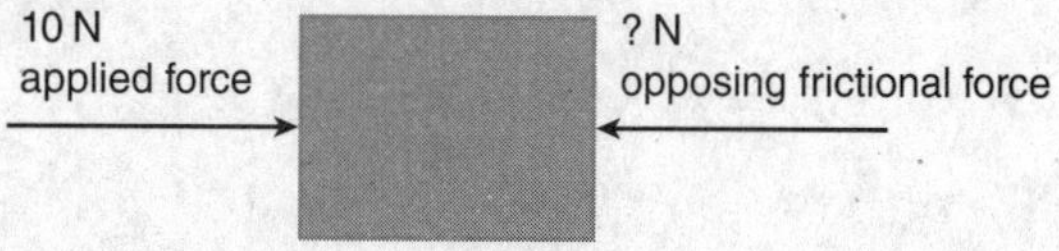

Resultant Force = 10 N – force of friction

Thus, the resultant force will be less than 10 N.

13. **D** Kinetic friction is exhibited when surfaces/objects move past one another. The normal force, which is a factor of the weight of the object, will affect friction (for example: move your hand across a table lightly and then again with more downward force; increased force leads to increased friction). The nature of the surfaces in contact with each other will also affect the friction that develops between two surfaces (for example: moving your hand across a smooth desk is a lot easier than moving it across rough concrete). Since the objects are moving, use the formula $F_f = \mu_k F_N$. So, a change in μ_k would change the force of friction. The speed at which dry surfaces move past each other, however, does not impact the frictional forces (though more heat may be generated).

14. **D** Weight = *gm*, where *g* = gravitational acceleration (m/s^2) and *m* = mass in kg. On Xendor, the gravitational acceleration is 3 times as much as on Earth, giving a new *g* of approximately 30 m/s^2. To calculate weight on Xendor, multiply the new gravitational acceleration by the person's mass: Weight = 30 $m/s^2 \times$ 100 kg = 3,000 N.

15. **C** Recall that $W = Fd$, where *W* is work in joules, *F* is force in newtons and d is distance in meters. Therefore, rearranging the equation for force, you get: $F = \frac{W}{d}$. Taking 80 J and dividing by 2 m, you get 40 N.

MECHANICAL COMPREHENSION PRACTICE SET 2

Answers and Explanations

1. **C** The object remains at rest, so the acceleration is 0 and the net force must also be 0. Given that the object is not moving, the frictional force must be static rather than kinetic. The calculated force of static friction is actually the maximum that can be attained without creating movement. In this instance, not enough information is available to determine if this maximum has been reached. However, since the object remains at rest, the applied force must be equal to the force of static friction.

2. **C** Total mechanical energy is conserved as the ball moves down the hill, meaning that the potential energy at the top will be converted to kinetic energy at the bottom. To solve for the velocity at the bottom, the potential energy must be set equal to the kinetic energy. Algebraically, this can be written as $mgh = \frac{1}{2}mv^2$, which simplifies to $gh = \frac{1}{2}v^2$ by canceling out mass. Solving for velocity, $v = \sqrt{2gh}$. Finally, plugging in the values from the question stem and using 10m/s^2 as an approximation for g, $v = \sqrt{(2)(10)(20)} = \sqrt{400}$, so the velocity of the ball when it reaches the bottom of the hill is 20 m/s.

3. **A** The question gives the mass of the object in grams, so the mass must first be converted to kg by dividing by 1,000, giving 0.5 kg. Then the mass must then be converted into a force by multiplying by g (use 10 m/s^2 for simplicity), so the weight of the mass is $0.5(10) = 5$ N. Since the mass is not accelerating, the total tension must be equal and opposite to the weight of the object. Algebraically, this means $T1 + T2 = 5$ N. The tensions are equal because the ropes are of equal length, they are attached to the block in a symmetric fashion, and the mass of the block is uniformly distributed. So $T1 = T2$ and $2T1 = 5$ N, so $T1 = 2.5$ N.

4. **B** Since the pressure is the same throughout a closed hydraulic system and force is equal to pressure times the area over which it is applied, $F1$ must be greater than $F2$. Also, the work (force times distance) must be equal on both sides. Since $F1 > F2$, it follows that $d2 > d1$.

5. **C** Since the system is in rotational equilibrium, the net torque on the system must be 0, so the torques exerted by the left- and right-hand sides of the fulcrum must be equal. The question is asking the distance from the 10 kg weight; let that distance equal x, so the distance between the fulcrum and the 20 kg weight must be $9 - x$. Setting the torques equal to each other: $10xg = 20(9 - x)g$. Gravity is on both sides of the equation, so it will cancel out, giving $10x = 180 - 20x$. Solving for x, $x = 6$ m.

6. **B** The equation for kinetic energy is $\frac{1}{2}mv^2$. Since the weight of the object was given, not the mass, the first step is to calculate the mass. $W = mg$, so the equation can be rearranged to $m = \frac{W}{g}$ in order to calculate the mass. Given the weight of 98 N, it will be more accurate to calculate the mass using $g = 9.8$ m/s^2 rather than the approximation of 10 m/s^2. So the mass is $\frac{98}{9.8} = 10$ kg. Finally, the mass and velocity can be plugged into the equation for kinetic energy: $KE = \frac{1}{2}(10)(100)^2 = 50{,}000$ J.

7. **B** Since power is work per unit of time and work is force times distance, then $P = \frac{Fd}{t}$. Substituting the known values from the question, $P = \frac{(100)(10)}{20} = 50$ W. Note that the mass of the block was not needed to answer this question. Recognizing and discarding extraneous information is a skill that takes practice but is very helpful.

8. **D** The first step is to convert the mass given into a force, which will be its weight. The force of the weight is $mg = 100 \text{ kg} \times 10 \text{ m/s}^2 = 1{,}000$ N. Since there are 4 pulleys, the mechanical advantage is 4:1, meaning that $\frac{1}{4}$ of the force can be applied over 4 times the distance to pull up the mass. Since the weight is 1,000 N, the force applied would be 250 N.

9. **A** Pressure is equal to force divided by the area over which it is applied. The force is equal to the weight of the woman. To calculate the area, multiple the dimensions of the snowshoes, so $10 \times 15 = 150$ square inches. Since she is wearing two shoes, the total area is $2 \times 150 = 300$ square inches. Divide the force by the area: $\frac{150}{300} = 0.5$ psi.

10. **D** The normal force must be equal and opposite to the force exerted by the block on the table in accordance with Newton's third law of motion. Since the mass of the object is given in kg, the weight must first be calculated using the formula $w = mg$ to get a weight of 400 N. There is also a force applied downward on the block of 100 N, so the total force exerted on the table is 500 N. Since the object remains at rest, the normal force must be equal to 500 N to get a net force of 0 and have no acceleration.

11. **D** The fact that the ball was thrown at an angle may seem to complicate the problem. The ball will travel outward and upward until it reaches its maximum altitude, then begin to travel downward and outward. However, since the question does not specify the direction that the ball is traveling when it hits the ground, the problem can be solved by considering that the total kinetic energy when the ball finally reaches the ground will be the sum of the initial kinetic energy and the potential energy that is converted to kinetic energy. Since air resistance is minimal, no kinetic energy will be lost to air resistance. The potential energy will initially be equal to *mgh*. Substituting the known numbers and using the approximation of 10 m/s^2 for *g*, you get $5(10)(10) = 500$ J. To calculate the beginning kinetic energy, use the equation $KE = \frac{1}{2}mv^2$. Substituting the numbers from the question stem, we get $\frac{1}{2}(5)(10)^2$ or 250 J. Summing the two forces gives 750 J.

12. **C** Tools do use the principles of mechanics but measuring devices do not. Similarly, machines that transmit or use energy are governed by these principles, but the study of mechanics is not limited to machines. While trajectories are a *type* of motion, the more broad category of "motion" is the proper answer.

13. **C** Since the coefficient of static friction is greater than that of kinetic friction, once sufficient force is applied to overcome the maximum value of the force of static friction, there will be a net force acting to move the object. Because of Newton's second law, $F = ma$, a non-zero net force always creates a non-zero acceleration. That formula applies to the net force, not any force that "accumulates." As the object's velocity increases, so will its momentum, but this is an effect of the acceleration, not the cause.

14. **A** In order for the brothers to be balanced, the torque (rotational force) exerted by each boy must be equal. The little brother sits 8 feet from the fulcrum and weighs 75 pounds, so the torque he creates is $8 \text{ ft} \times 75 \text{ lb} = 600 \text{ ft-lb}$.
In order for the big brother to generate 600 ft-lb of torque, he should sit 6 feet from the center of the seesaw, since $6 \text{ ft} \times 100 \text{ lb} = 600 \text{ ft-lb}$.

15. **D** Since force equals pressure times the area over which the pressure is applied and the pressure is the same throughout the system, the ratios of the forces will be equal to the ratio of the areas of the two pistons. The radius of the smaller piston is 1 cm and that of the larger piston is 10 cm, so $\frac{\pi(1)^2}{\pi(10)^2} = \frac{20 \text{ N}}{F}$. Cancel the π and cross-multiply to get $F = 2{,}000$ N.

Review and Reflect

How did you do on this practice set? As you review your work, ask yourself some questions:

- Did you remember the equations you learned in the chapter?
- Did you think about how various forces were at work in the practice questions?
- Did you stop to think carefully about what the question is asking for, simplify or solve, and then make a prediction before looking at the choices?

Want more practice with Mechanical Comprehension? Try more questions in the Qbank online.

QBANK

kaptest.com/login

CHAPTER 14

ASSEMBLING OBJECTS

KNOW WHAT TO EXPECT

The ASVAB analyzes your abilities and aptitude to help you discover what military careers would be the best match for you. For example, pilots, mechanics, machine operators, and engineers need to develop good spatial relationship skills—that is, the ability to see how parts will look when assembled into a whole. The **Assembling Objects** (AO) subtest examines your spatial relationship skills, so it gives you an opportunity to show that you are good at putting things together. As always, we recommend checking with your recruiter or career counselor to learn more about which subtests will be important to you in reaching your goals.

The skills required for the AO test are *learnable*. Even test takers who struggle to visualize how shapes fit together can greatly improve their spatial relationship skills by learning strategies and practicing. In this chapter, you'll have the opportunity to do just that.

The CAT-ASVAB test includes 15 AO questions, with a time limit of 17 minutes. Remember, unlike the paper-and-pencil ASVAB, you will not be able to review or change an answer once you submit it. The paper-and-pencil ASVAB includes 25 AO questions, with a 15-minute time limit.

The Assembling Objects subtest contains two types of test items. We're going to call these:

- **Jigsaw-Puzzle-Type Problems**, which test your ability to choose how an object will look when its parts are put together into a larger shape.
- **Connector-Type Problems**, which test your ability to correctly connect two objects with a line at the indicated points.

In both types of AO items, you are offered five boxes. The leftmost or first box contains drawings of unconnected parts. The four answer-choice boxes hold drawings of shapes, only one of which shows the shape that would result when all the parts are correctly assembled.

JIGSAW-PUZZLE-TYPE PROBLEMS

LEARNING OBJECTIVES

In this section, you will learn to:

- identify what your task is on Jigsaw-Puzzle-Type problems
- apply the Kaplan Method for Assembling Objects questions to Jigsaw-Puzzle-Type problems

Study the question below.

Which figure best shows how the objects in the left box will appear if they are fit together?

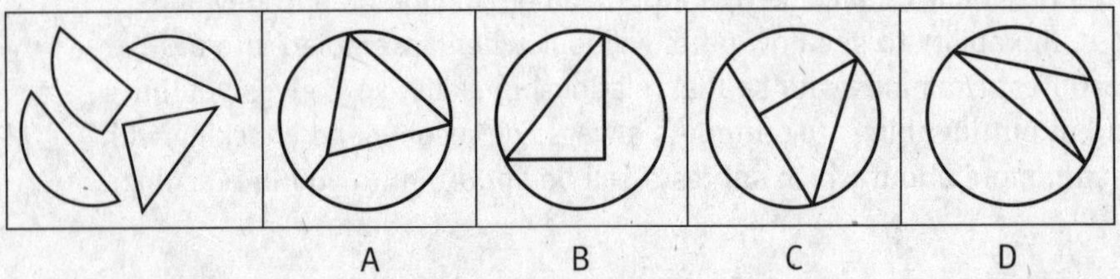

Every AO question has a box on the very left that shows a number of individual parts. Let's call that box on the left the **instructions box.** Each AO question then offers four answer choices. Three of those choices distort the shapes in the instructions box. Only one of them accurately shows how the shapes would fit together into a whole. In this case, **(C)** is correct.

Having trouble visualizing why **(C)** is correct? Study the diagram below.

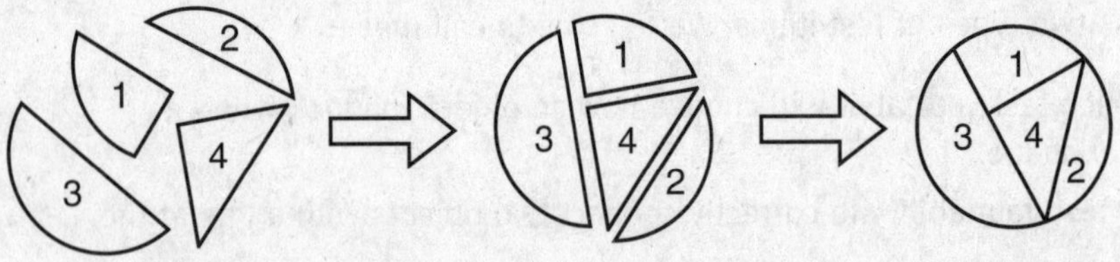

However, you *do not* have to visualize assembling the parts in your head in order to answer AO questions. (That's good news.)

Rather, approach these questions strategically. Did you notice that each of the answer choices in the example above is a circle dominated by a large triangle? That will be helpful. Now, focusing on the circle outline won't be helpful, because it's the same in all four choices. However, the placement and shape of that large triangle varies among the answer choices. Thus, focusing on that prominent triangle will help you eliminate some choices right off the bat. You can call that key shape a **landmark**, in the sense that it helps you find what you're looking for.

What do you notice about the triangle depicted in the instructions box? It's a right triangle, or very close to being a right triangle. How many of the answer choices clearly do *not* have right triangles? (A) and (D). Cross them off. Your landmark just knocked out half the choices.

Only (B) and (C) remain as possible choices. Find another key trait in the instructions box. Here, counting the number of shapes might be helpful. There are four shapes in the instructions box. How many shapes do (B) and (C) have? (B) has only three, so it can't be right. Circle **(C)** and move on to the next question.

Here's a method you'll use on all Assembling Objects questions. Memorize this method now, so that you can work on mastering it in your practice.

THE KAPLAN METHOD FOR ASSEMBLING OBJECTS QUESTIONS

Step 1: Identify a landmark in the instructions box.

Step 2: Eliminate any answer choices that don't display the landmark.

Step 3: Compare other shapes to eliminate remaining choices.

Study the example below.

Question

Which figure best shows how the objects in the left box will appear if they are fit together?

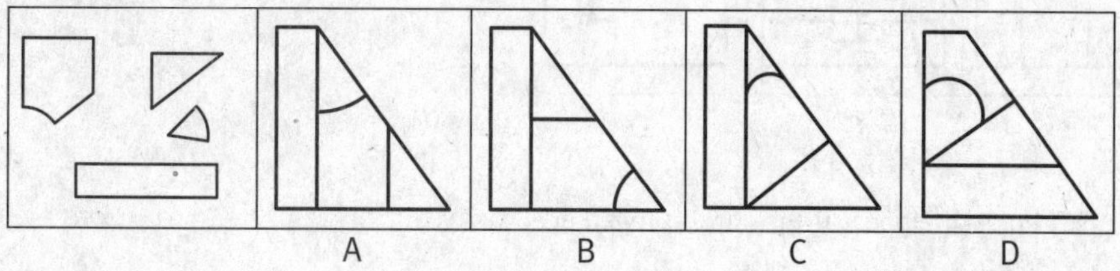

Analysis

Step 1: The easiest landmark here might be the smallest shape, which looks like a small pizza slice.

Step 2: (C) and (D) don't have that small pizza shape, so they're out. (A) has a similar shape, but it's far too big. So (A) is out as well. **(B)** is correct.

Step 3: No need to compare other shapes here.

In **Step 1**, your landmark shape in Jigsaw-Puzzle-Type AO problems might be:

- the largest shape
- the smallest shape
- the shape that seems to differ the most among the answer choices
- the most oddly-shaped piece
- a shape with a recognizable angle or side (such as a right angle, or a steeply curved side)

With practice, you will learn to efficiently spot a useful landmark in the instructions box.

When you get to **Step 3**, here are strategies you can use to eliminate any remaining wrong answer choices:

- Count the number of parts. If you are taking the paper-and-pencil test, you can actually assign each part a number and write it in.
- Find a second landmark. If you used the largest shape in Kaplan Method Steps 1 and 2, perhaps you would use the smallest shape as your second landmark.
- Compare the sizes of the shapes in the instructions box to the remaining choices.

And here are some things to watch out for in wrong answer choices on Jigsaw-Puzzle-Type AO questions:

- You can expect one or more of the shapes in the instructions box to be inverted left-to-right or top-to-bottom in one or more of the wrong answer choices. The correct answer might move the parts around or rotate them, but it won't invert any of them.
- An enlarged or shrunken version of a shape in the instructions box might appear in the wrong answer choices. The correct answer will have the same shapes in the same sizes.
- A shape that has all straight edges in the instructions box may have a curved edge in some wrong answer choices.
- A shape from the instructions box may be carved into two or more smaller shapes in some wrong answer choices.

Try these problems on your own, then read the explanations below.

Which figure best shows how the objects in the left box will appear if they are fit together?

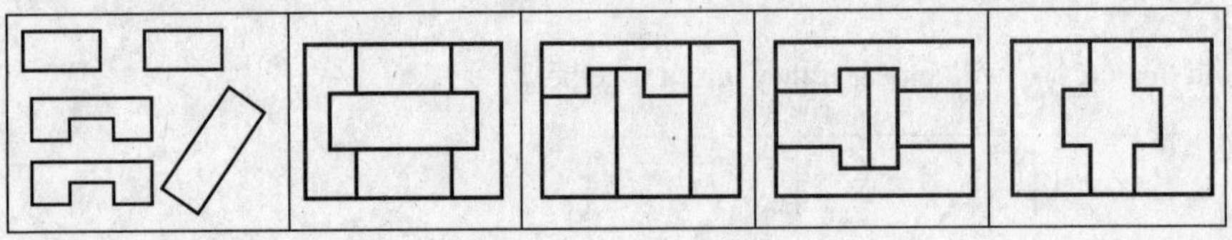

A B C D

Which figure best shows how the objects in the left box will appear if they are fit together?

A B C D

Explanations

Note: It's perfectly fine if you chose different landmarks than the ones below, as long as you were able to work efficiently through the questions.

1. Step 1: The instructions box has two shapes that look like fat square brackets; those are good landmarks.

Step 2: (B) doesn't have two of those shapes; eliminate. (C) has two but they're too long; eliminate. Only (A) and (D) remain.

Step 3: Count the number of pieces in the instruction box: there are five. (D) has three shapes; eliminate.

(A) is correct.

2. Step 1: The instructions box has an odd shape that looks like a pie piece with two branches growing out of the narrow end; that's a good landmark.

Step 2: (A) and (C) have no such piece; eliminate. (B) and (D) remain.

Step 3: The instructions box has four shapes; (B) has five. Eliminate (B).

(D) is correct.

CONNECTOR-TYPE PROBLEMS

LEARNING OBJECTIVES

In this section, you will learn to:

- identify what your task is on Connector-Type problems
- apply the Kaplan Method for Assembling Objects questions to Connector-Type problems

Study the example below.

Which figure best shows how the objects in the left box will touch if the letters for each object are matched?

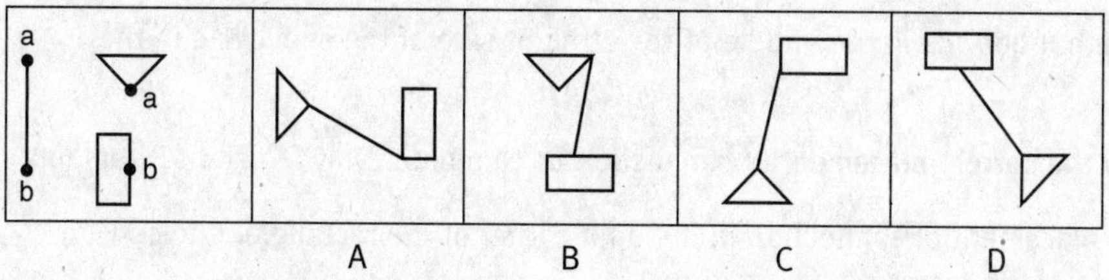

The instructions box of a Connector-Type problem displays two geometric shapes. Some problems have familiar shapes such as circles, squares, rectangles, and triangles, while others have more unusual shapes such as letters, clouds, or hearts. The instructions box will also display a **connector line**. You could think of the connector line as being a string you must tie to the two shapes in precisely the places indicated.

The two shapes and the connector line are labeled with dots and small letters to indicate points where you are to connect the shapes. You will see one dot on each shape, each labeled with a letter. Dots on the ends of the connector line will correspond to these letters. Sometimes the dots will be on the edges of shapes, and sometimes they will be inside the shapes.

You are asked to attach the two shapes to the line at the correct points, based on the diagram in the instructions box. The shapes may be rotated in the solution boxes, but the correct solution cannot be flipped. In the example above, choice **(D)** is correct.

Having trouble seeing it? Study the diagram below.

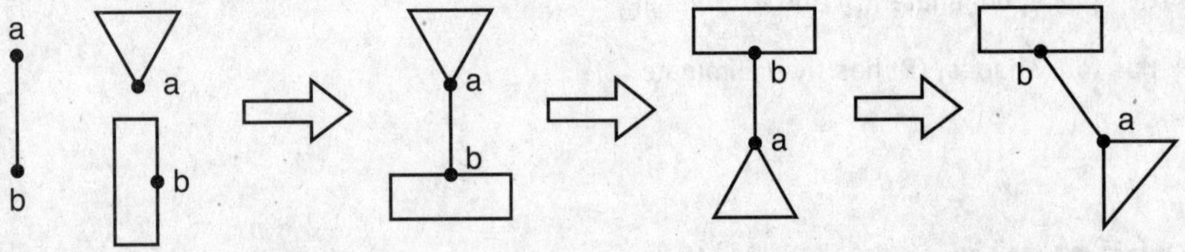

However, just as with Jigsaw-Puzzle-Type questions, you do *not* need to be able to visualize the solution in your head. The Kaplan Method works on Connector-Type problems just as it does on Jigsaw-Puzzle-Type problems. Study the example below.

Question

Which figure best shows how the objects in the left box will touch if the letters for each object are matched?

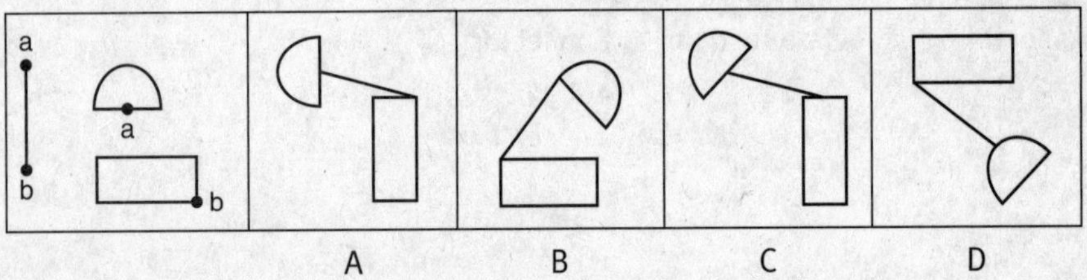

Analysis

Step 1: A good landmark here is the fact that dot *a* is in the middle of the straight side of the semicircle in the instructions box.

Step 2: Choices (B) and (D) don't display the correct placement of connection *a*; eliminate. Only (A) and (C) remain.

Step 3: A second landmark could be the placement of connection *b*: it's on a corner of the rectangle. Choice (C) places it wrongly in the middle of one side of the rectangle. **(A)** is correct.

In **Step 1**, your landmark in Connector-Type AO problems might be:

- whether a dot is placed in the middle of a side or on a corner of a shape
- whether a dot is placed on the edge or inside a shape
- a shape itself

With practice, you will learn to efficiently spot a useful landmark in the instructions box. If you are testing on paper and pencil, you could actually add the labels *a* and *b* to the answer choices in order to help you eliminate choices that don't display your landmark.

When you get to **Step 3**, here are strategies you can use to eliminate any remaining wrong answer choices:

- Find a second landmark. If you used dot *a* as your first landmark, you might use dot *b* as your second landmark.
- If a dot is placed on the corner of a shape in the instructions box, check to see *which* corner is attached to the connector line in the answer choices.
- Check to see if any of the shapes have been flipped or inverted in the answer choices.

Here are some things to watch out for in wrong answer choices on Connector-Type AO questions:

- If the connection point is on the edge of a shape in the instructions box, it may appear inside the shape in some wrong answer choices (or vice versa).
- If the connection point is on the corner of a shape, expect it to either appear on other corners in some of the wrong answer choices *or* appear in the middle of an edge in some wrong answer choices.
- One or more of the shapes in the instructions box may be inverted left-to-right or top-to-bottom in one or more of the wrong answer choices.
- An enlarged or shrunken version of a shape in the instructions box might appear in the wrong answer choices. The correct answer will have the same shapes in the same proportions.

Try it on your own, then review the explanations below.

Which figure best shows how the objects in the left box will touch if the letters for each object are matched?

1.

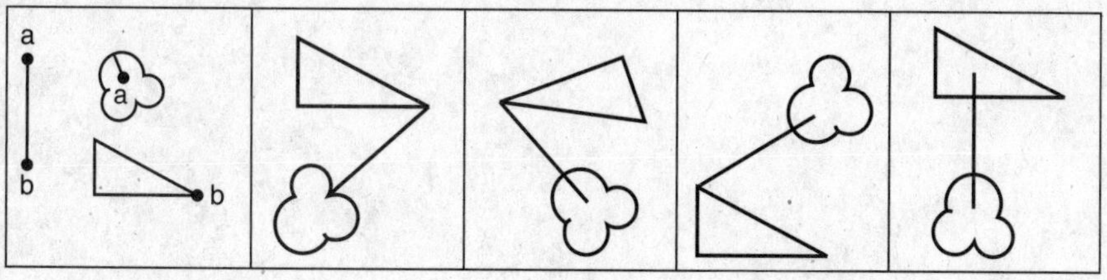

A B C D

Which figure best shows how the objects in the left box will touch if the letters for each object are matched?

2.

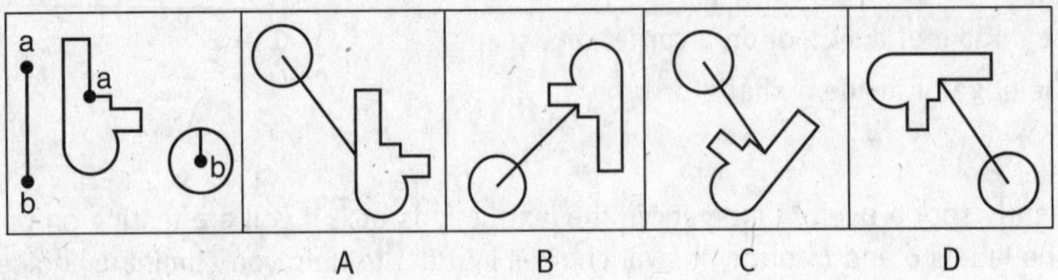

Explanations

Note: Your landmarks may differ. That's fine, as long as you were able to work through the problems efficiently and accurately.

1. Step 1: Dot *b* looks like a good landmark: it's on the narrowest point of the triangle.

Step 2: Choice (C) places the connector on a different corner of the triangle, and choice (D) places it in the middle of the triangle. Eliminate. (A) and (B) remain.

Step 3: A good second landmark would be the placement of dot *a*: it's in the middle of the three-lobed shape. Choice (A) wrongly places it on the edge.

Choice **(B)** is correct.

2. Step 1: The instructions box includes a shape that looks like a hair dryer, and dot *a* is at the point where the trigger of the hair dryer meets the barrel of the hair dryer. That's a good landmark.

Step 2: Choices (A) and (B) misplace the connection point on the hair dryer; eliminate. (C) and (D) remain.

Step 3: The circle with dot *b* is the same in choices (C) and (D), so a different landmark needs to be used. Instead, compare the hair dryers: turns out it's flipped in choice (C).

Choice **(D)** is correct.

ASSEMBLING OBJECTS PRACTICE SET 1

For each question, select the answer that best shows how an object will look when its parts are put together. This practice set has 15 questions, which is the number of Assembling Objects questions you will see if you take the CAT-ASVAB.

1.

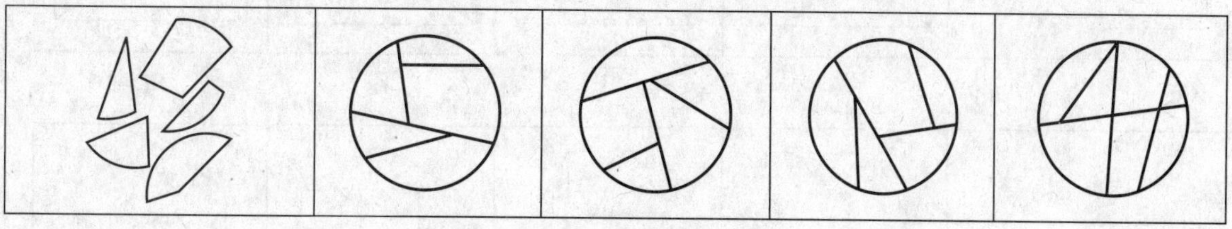

A B C D

2.

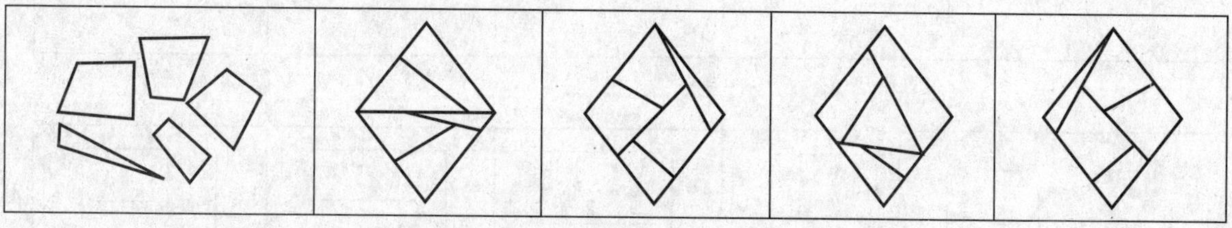

A B C D

3.

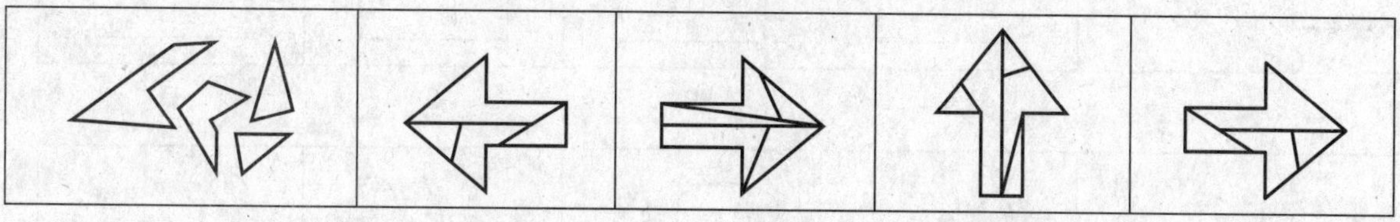

A B C D

4.

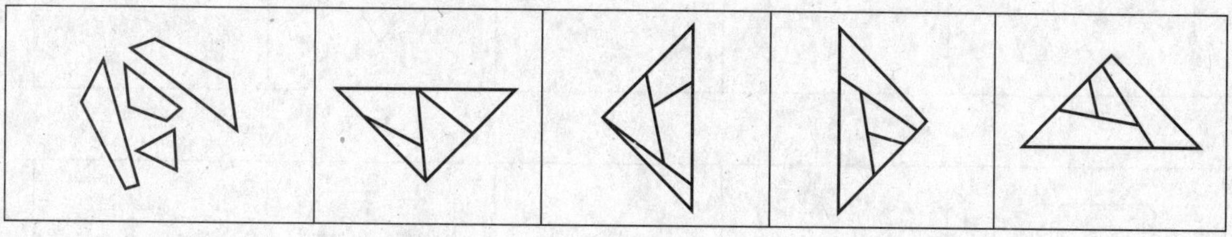

A B C D

5.

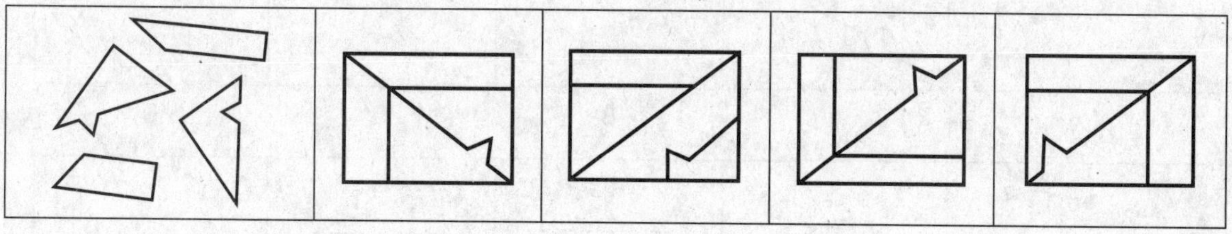

A B C D

6.

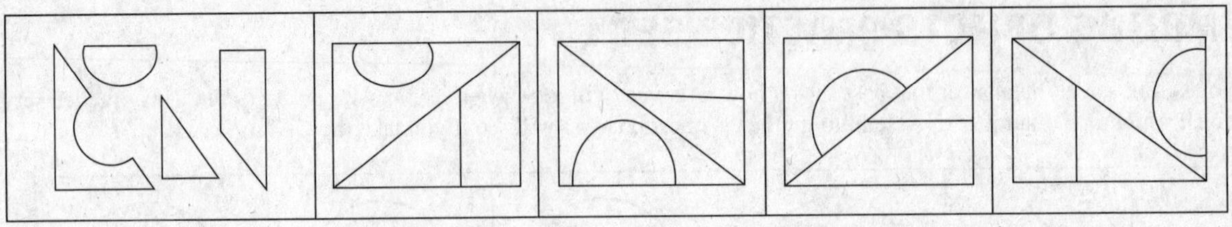

A B C D

7.

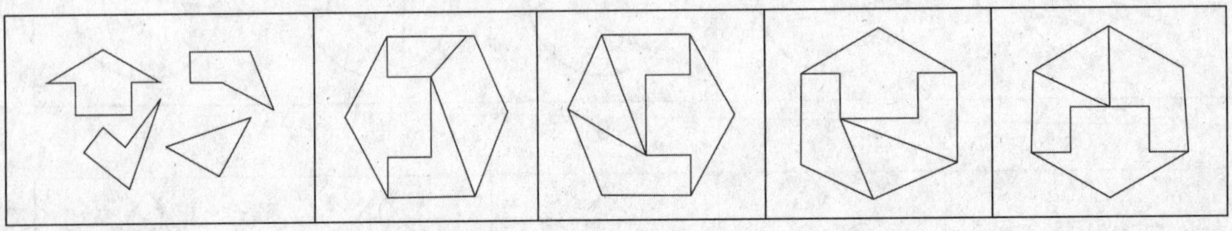

A B C D

8.

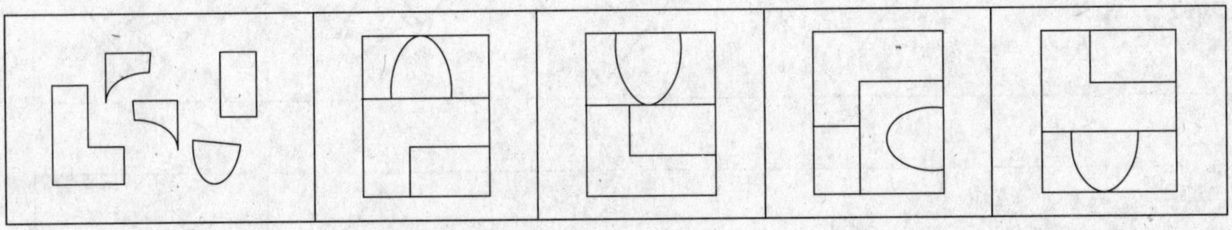

A B C D

9.

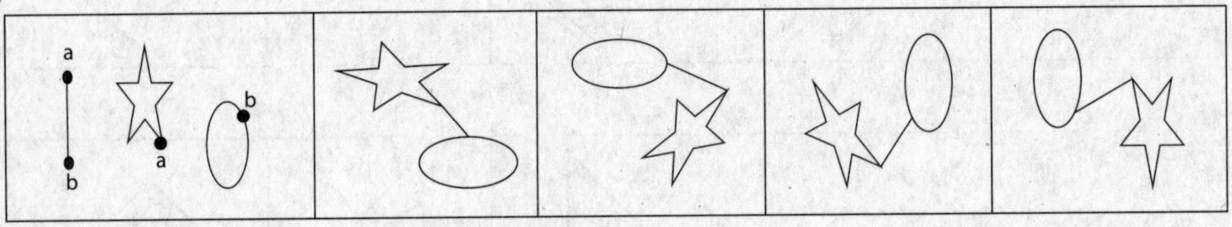

A B C D

10.

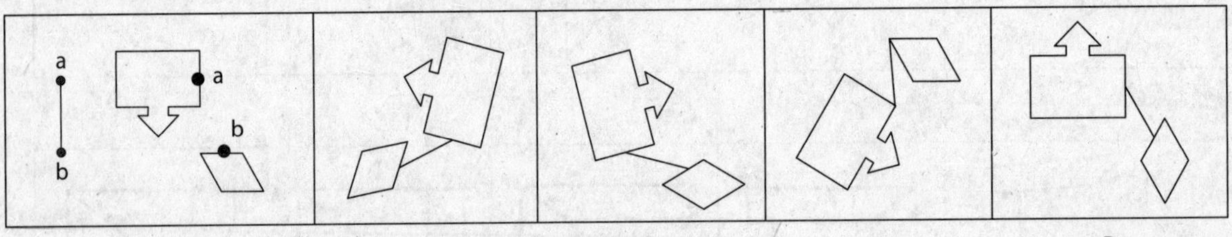

A B C D

11.

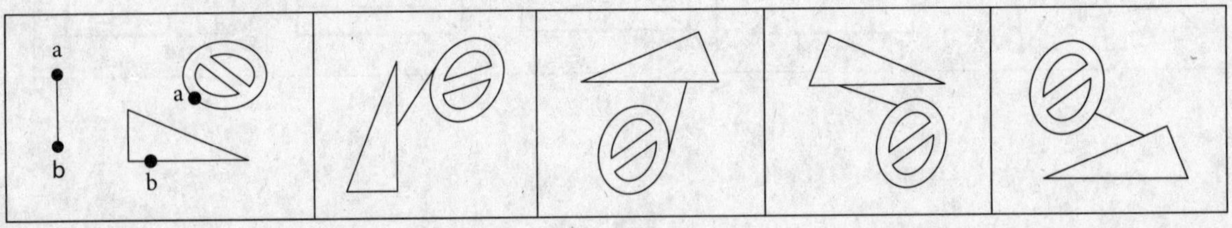

A B C D

12.

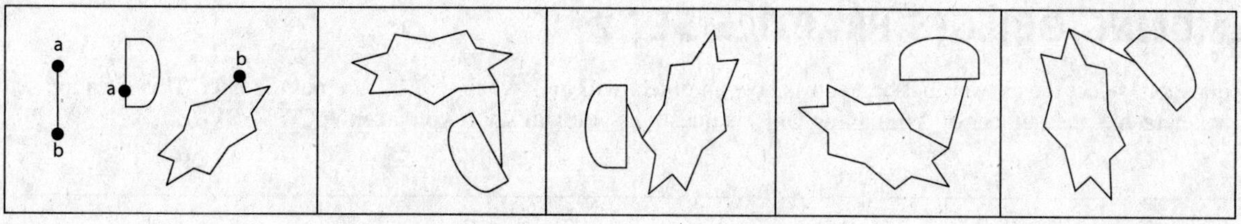

A B C D

13.

A B C D

14.

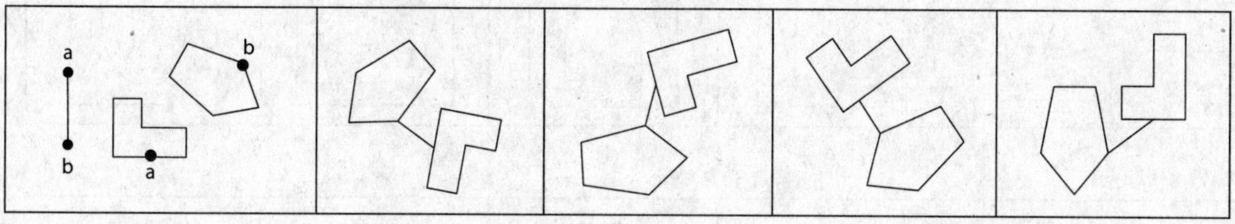

A B C D

15.

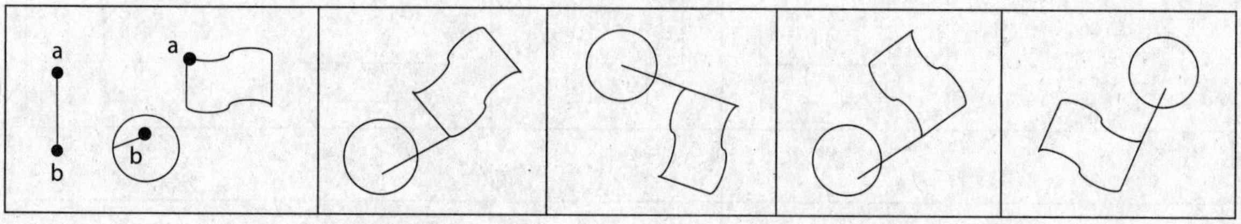

A B C D

ASSEMBLING OBJECTS PRACTICE SET 2

For each question, select the answer that best shows how an object will look when its parts are put together. This question set has 15 questions, which is the number of Assembling Objects questions you will see if you take the CAT-ASVAB.

1.

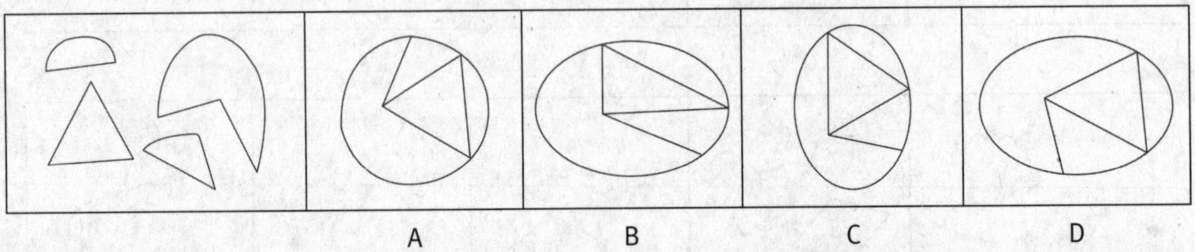

2.

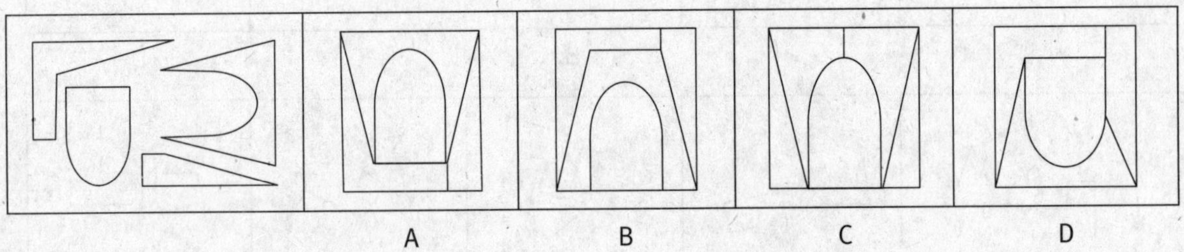

3.

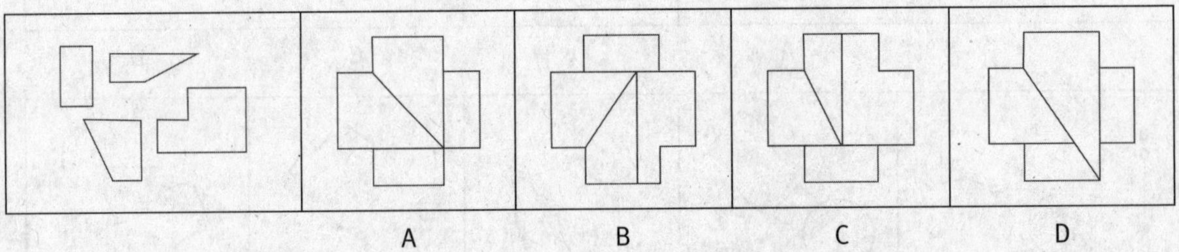

4.

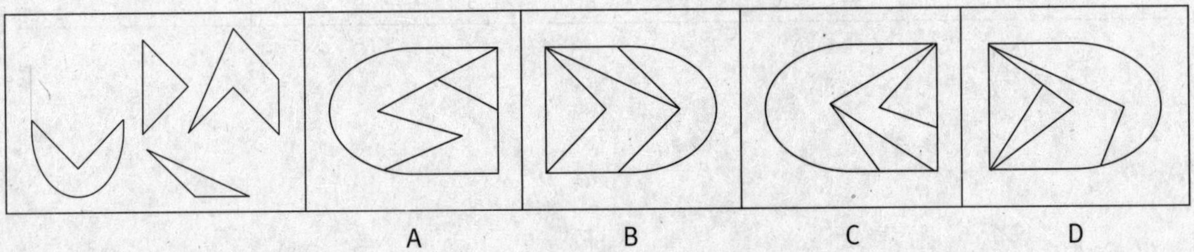

5.

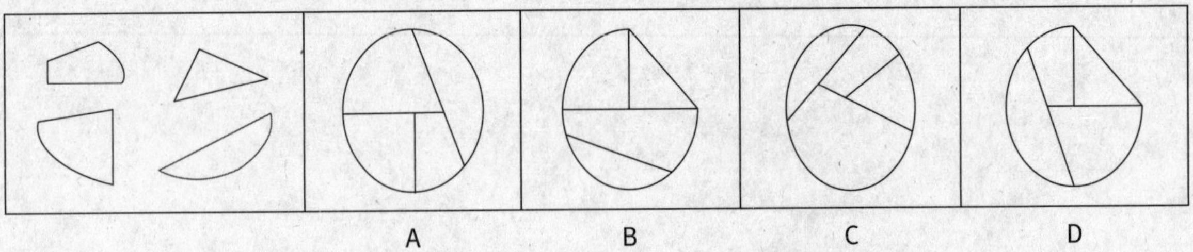

6.

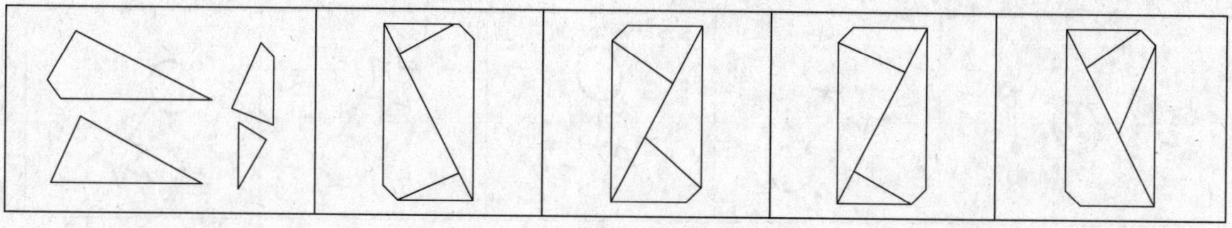

A B C D

7.

A B C D

8.

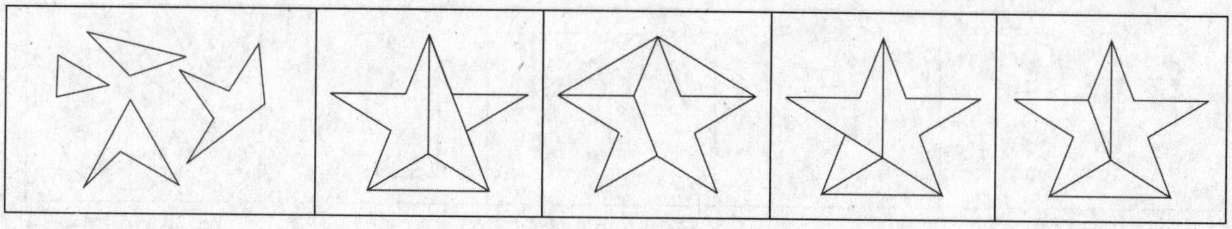

A B C D

9.

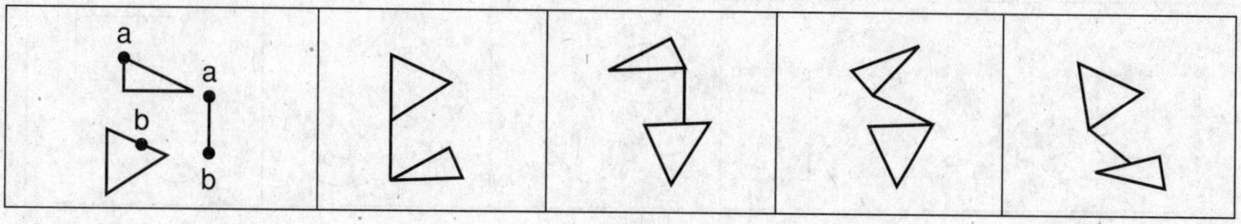

A B C D

10.

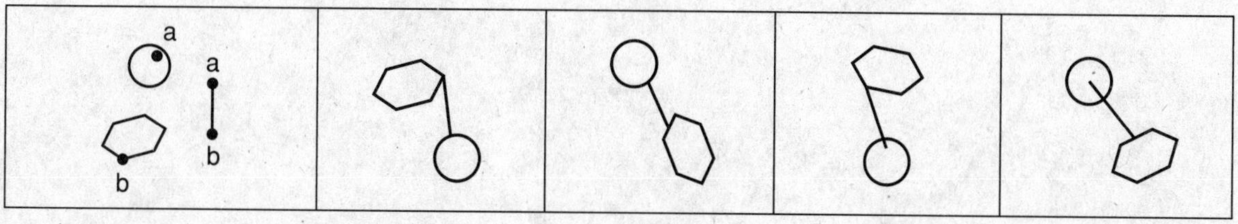

A B C D

11.

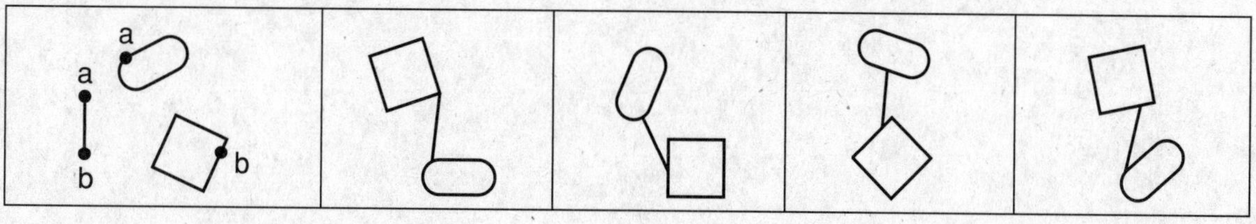

A B C D

12.

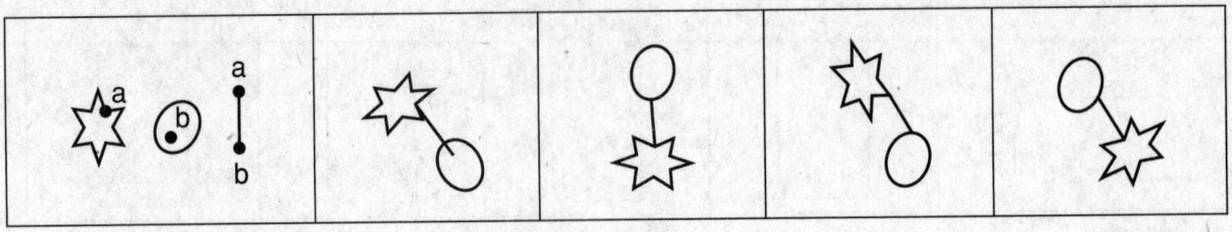

A B C D

13.

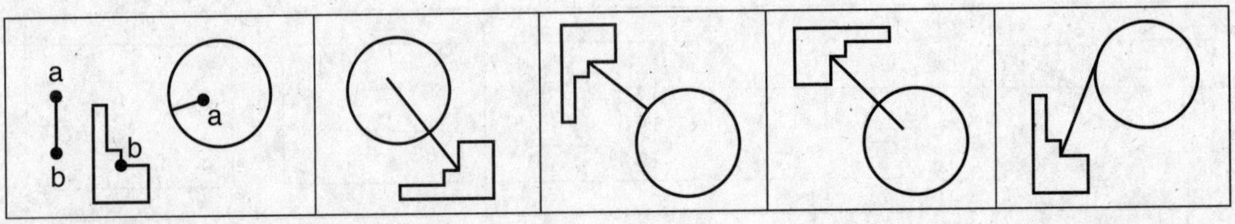

A B C D

14.

A B C D

15.

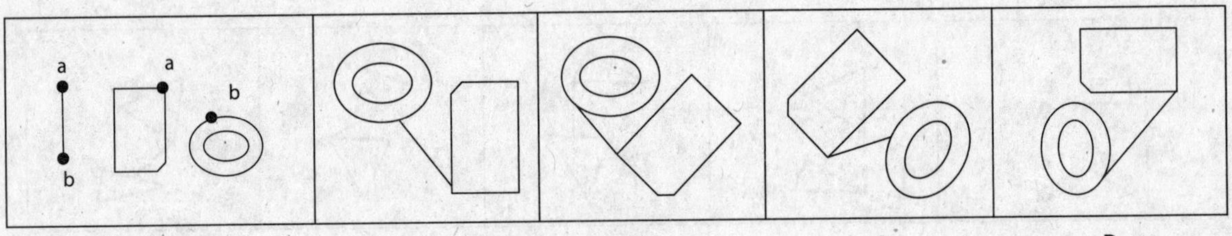

A B C D

ASSEMBLING OBJECTS PRACTICE SET 1

Answers and Explanations

1. **A**

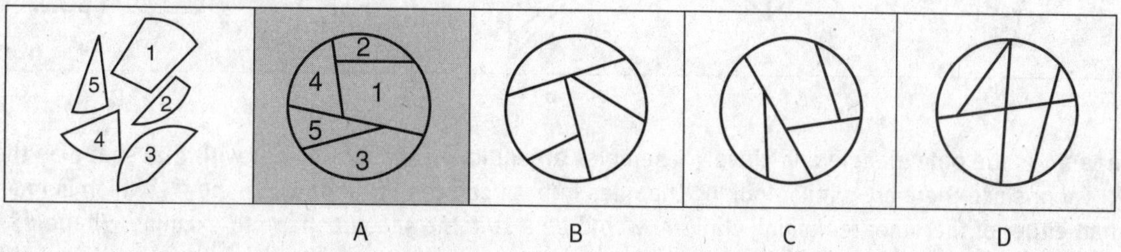

Shape #3 has an unusual shape, so it is a good landmark; it is a fairly long arc of the outer circle defined by two lines that meet in an obtuse angle. This shape appears only in choice (A). The shape at the left side of choice (C) is similar, but the long and short legs are reversed. Since only one answer choice contains this shape, there is no need to select a second shape other than to verify the correct answer.

2. **B**

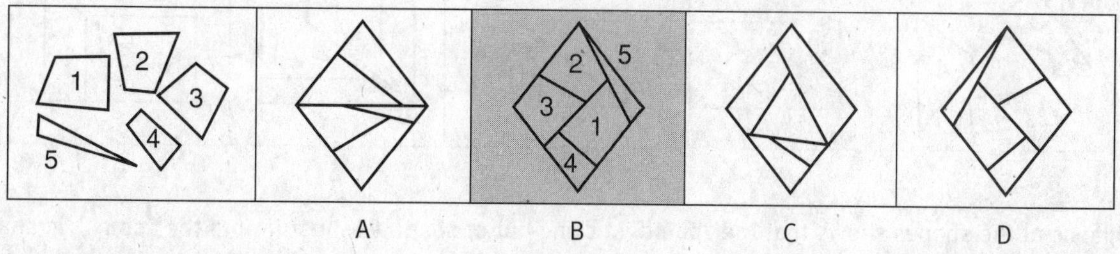

Shape #5 is the most distinct, since it is a triangle and the others are quadrilaterals. This landmark shape can be found only in choice (B). The tall triangle in choice (D) might be tempting, but the longest leg of this triangle is on the right side whereas the longest leg of the correct shape is on the left side. Since the three incorrect choices have been eliminated, there is no need to check a second shape.

3. **D**

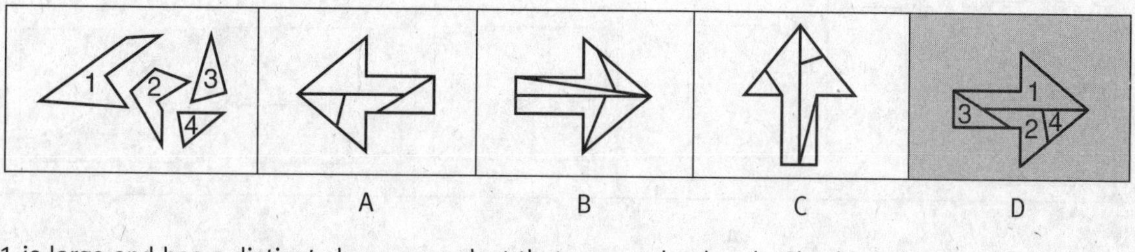

Shape #1 is large and has a distinct shape, so select that as your landmark. Checking the answers, #1 can be found in choice (D). The shape at the top of choice (A) is similar, but it "points" the wrong direction (it is flipped over). Again, there is no need to select a second shape to analyze.

4. **C**

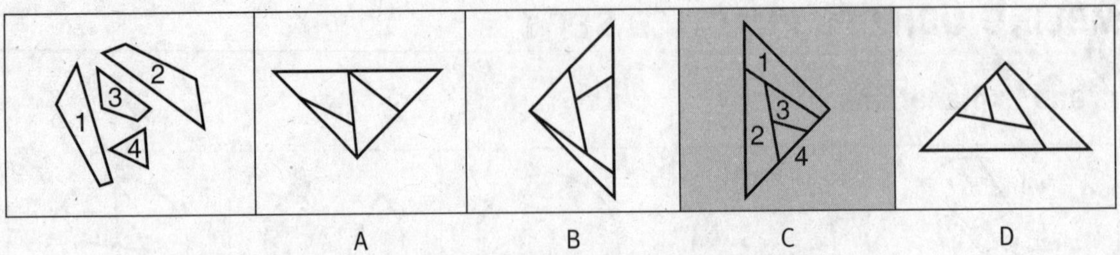

Since shape #4 is the only triangle and closely resembles an equilateral triangle, start with that shape as the landmark. Then, since there are similar-looking triangles in both choices (C) and (D), shape #3, which is much smaller than either of the other remaining choices, would be a suitable second shape to examine. Shape #3 is replicated in choice (C), but the corresponding quadrilateral in choice (D) is a mirror image of the original shape. Remember that shapes can be rotated in the plane to change their placement, but they cannot be rotated about an axis. Therefore, a shape that is inverted top to bottom or left to right will not be part of the correct answer choice.

5. **A**

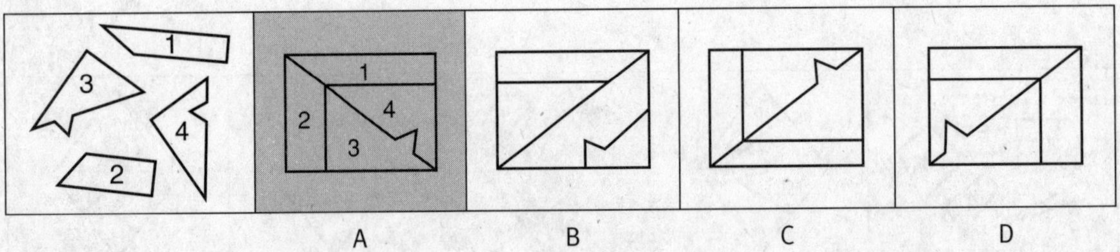

A quick glance at the shapes shows that the "point" sticking out of shape #3 must fit into the "notch" in shape #4 and that if #4 is on top, the notch will be on the lower right side of #4. Thus, the combination of these two shapes can serve as a landmark. The correct combination occurs only in choice (A). Although these two pieces are nested with the notch pointed up in the other choices, none of the alternatives have the "point" in the lower right-hand corner.

6. **C**

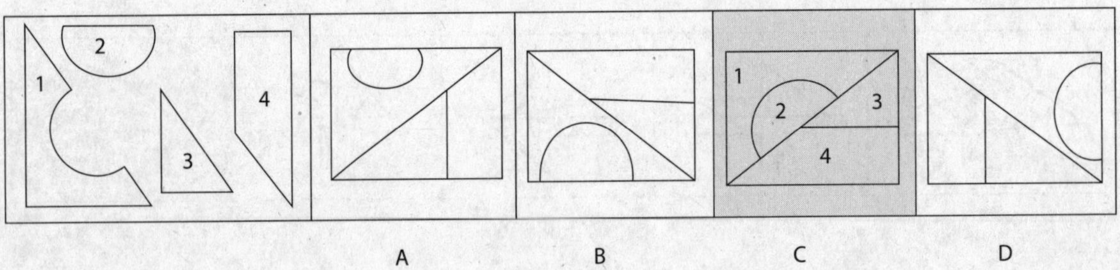

Shape #1 looks unique and should be a good landmark but, unfortunately, only choice (D) does not contain that shape. Shape #4 is the most complex of the remaining shapes. While there is a trapezoid in choice (A), that one is not long enough and it is flipped. The trapezoid in choice (B) is reversed. The trapezoid in (C) is, indeed, shape #4 turned 90° clockwise.

7. **B**

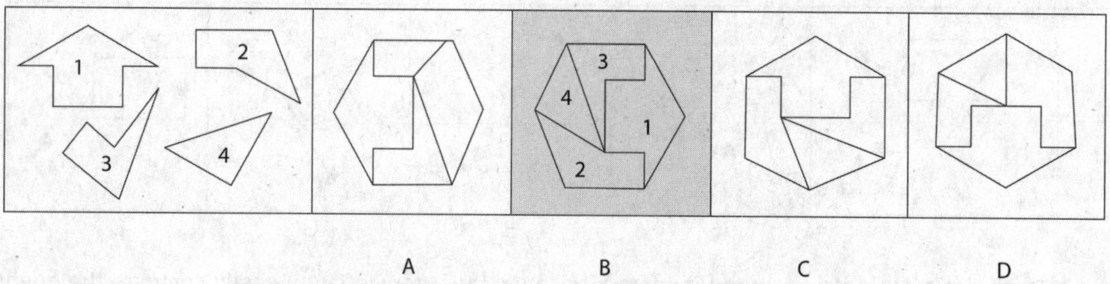

Shape #1 is large and easily identifiable, making it a good landmark. However, it appears in all four answer choices. Shape #3, a "bloated check mark," appears in choices (A) and (B) (it is also in (C), but it is flipped there). Therefore, another shape must be examined. Shape #4 does not appear in choice (A), so choice (B) is correct.

8. **A**

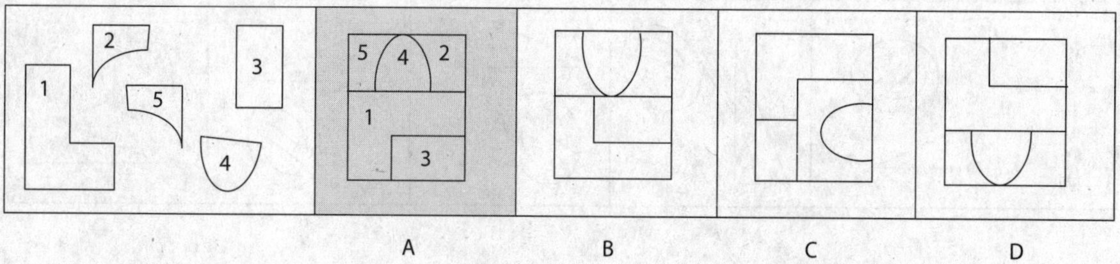

Shape #1 is the largest, so select that as the landmark since it should be easily identifiable in the answer choices. In choices (B) and (D), the "L" is backward; so eliminate those answers. Shapes #2 and #5 are distinct, so they could be examined next. Since neither one appears in choice (C), the correct answer is (A).

9. **C**

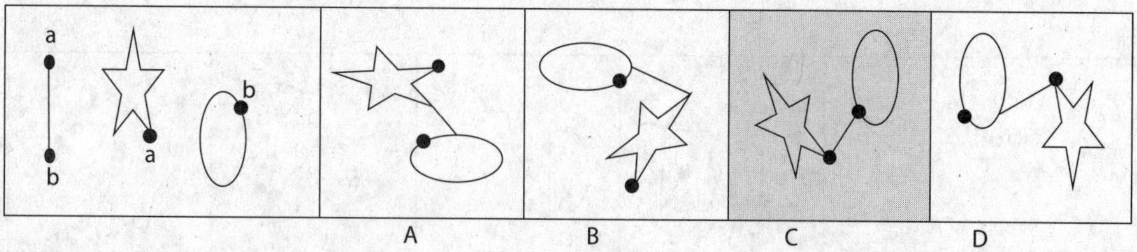

Dot *b* is a good choice for the landmark because it is located near, but not at, one end of the oval. Dot *a*, on the other hand, is at one of 5 points on the star. Using dot *b*, you can eliminate choices (A) and (B) immediately. Choice (D) requires more careful examination. If the initial oval shape were rotated 180° so that the dot were near the bottom, the dot would then be on the left side. However, the connector line ends at the right side of the oval, so (C) is correct.

10. **A**

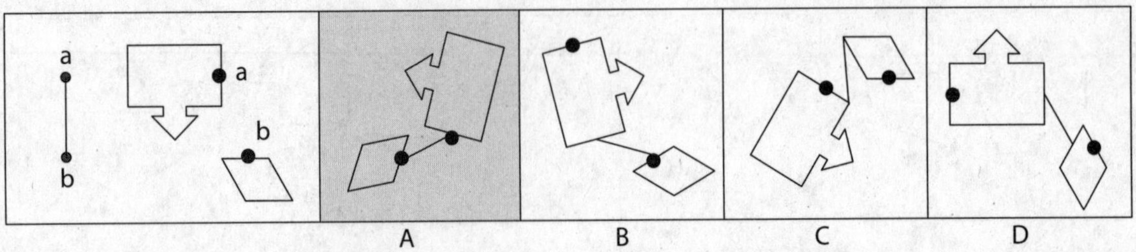

The shape on which dot *a* is placed is a good landmark because the arrow lets you easily compare the orientation of the shape as well as the location of the connector line. The connector line in choice (C) touches the corner of the shape with dot *a* rather than the middle of a short side. If the shape is rotated so that the arrow points up, then dot *a* is on the left, making choices (B) and (D) wrong. Choice (A) is correct.

11. **D**

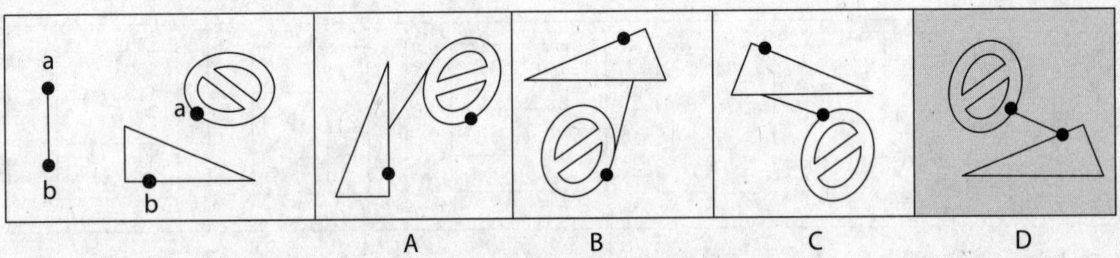

The triangle appears to be a right triangle, and dot *b* is located on the longer leg near the end closest to the short leg, so this should be a good landmark. In choices (B) and (C), the connector line terminates on the hypotenuse of the triangle, so eliminate those answers. In choice (A), the connector line does end on the long leg of the triangle, but at a point that is closer to the acute angle. In choice (D), the dot is at the correct location on the triangle, and a quick double check verifies that the connector line is at the proper place on the other shape; choice (D) is correct.

12. **C**

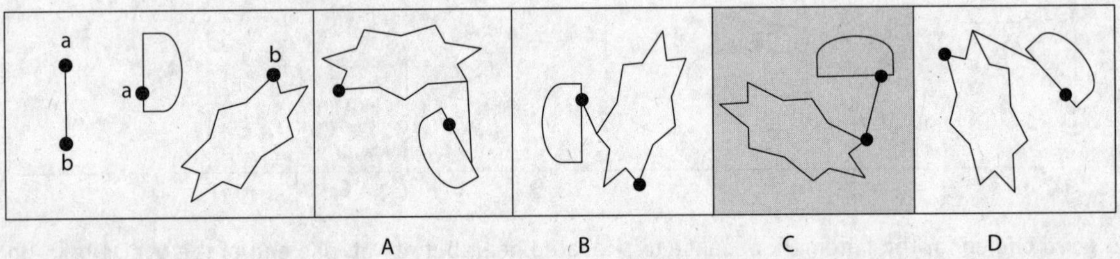

Dot *a*, on the straight edge near the corner of the "D," is easy to spot, so use that for the initial landmark. If the "D" is rotated 180°, the dot moves to the top right, so answer choice (A) is incorrect, but (B) is still in the running. If the "D" is rotated 90° or 45° counterclockwise, as in (C) and (D), respectively, the dot is at the bottom right, so choice (C) is correct for this shape but (D) is eliminated. The flattened, six-pointed star shape has an asymmetry that can be used. If you rotate the object so that its longest axis is horizontal, dot *b* is on either the upper right prong or the lower left prong, depending on the orientation. If the figure in choices (B) and (C) is aligned horizontally, the connector in (B) touches the wrong prong—either the upper left or the lower right, depending on the orientation. The connector in choice (C) touches the upper right prong, making this the correct answer.

13. **C**

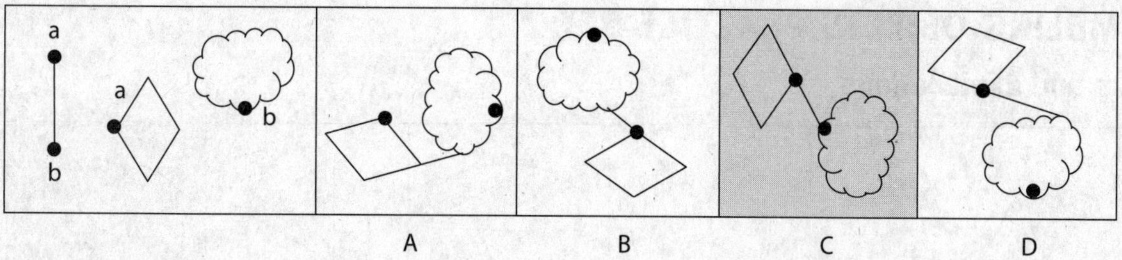

Although the cloud is an intriguing shape, it might be easier to pinpoint the connector line location on the simpler diamond shape, so you could try using dot *a* as the initial landmark. Unfortunately, all the answer choices except (A) correctly place the connector line on the diamond, so it's on to dot *b*. Choice (D) connects at or near the end of the cloud, but dot *b* is in the middle of the "puffier" side of the cloud. In choice (B), the dot appears to be too close to one end of the cloud and, moreover, it connects to the less puffy side. Choice (C) is correct.

14. **A**

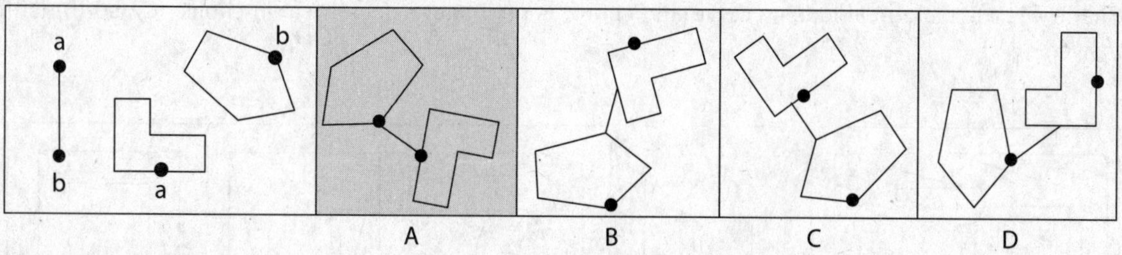

Dot *a*, located approximately midway along the longest side of the reverse "L," should be a good landmark. In choices (B) and (D), the line connects to the shorter side of the shape, so eliminate those answers. In choice (C), the line connects to the proper side but too close to the end. Choice (A) is correct.

15. **C**

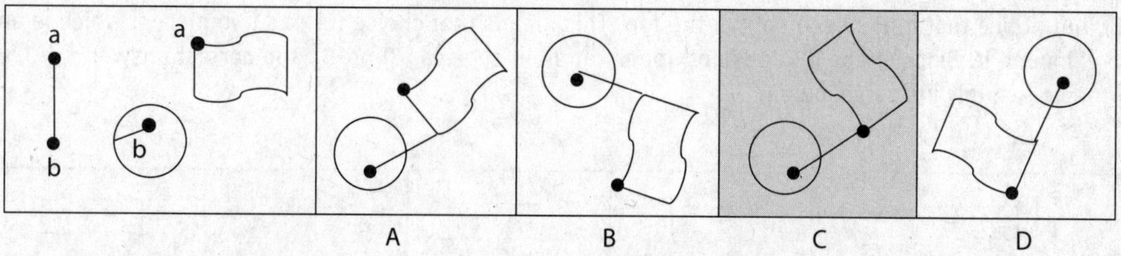

Since its shape is a bit more complex, with a line inside of a circle, use dot *b* for the landmark. Eliminate choice (B) since the connector line terminates in the middle of the circle. The other choices place the end of the connector correctly in the circle. The analysis of dot *a* will be more involved. In choice (A), the banner shape has been rotated approximately 45° counterclockwise, so the correct point is the leftmost corner of the banner rather than the bottom one. Look closely at the banner in choice (D). It could be rotated approximately 150° counterclockwise or 210° clockwise in order to be oriented as shown. However, the curved part closest to the circle would then be convex (curved outward) in the direction of the hole. In choice (D), however, that part of the banner is concave (curved inward), so (C) is correct.

ASSEMBLING OBJECTS PRACTICE SET 2

Answers and Explanations

1. **D**

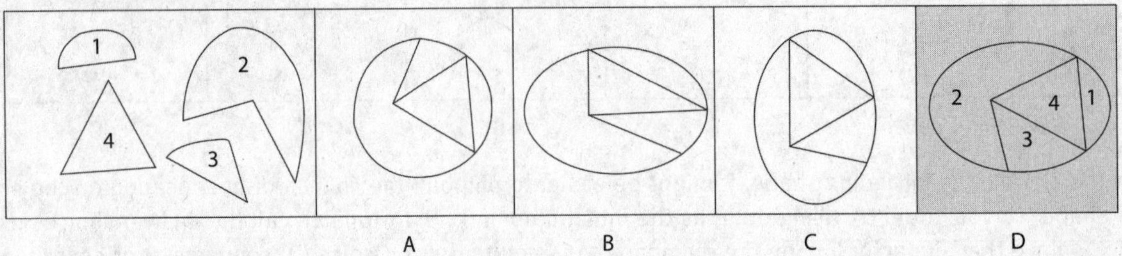

Shape #2, which looks like an oblong about to devour something, is certainly the most distinctive choice for a landmark. In choice (A), the oblong has become a circle, so that answer is out. In choice (B), the long and short "jaws" are reversed. The shape in choice (C) looks suspiciously thin, but pick another shape to test, just to be certain. Shape #1, a symmetrical-looking curve and chord, is no longer symmetrical in choice (C), so (D) is the correct answer.

2. **A**

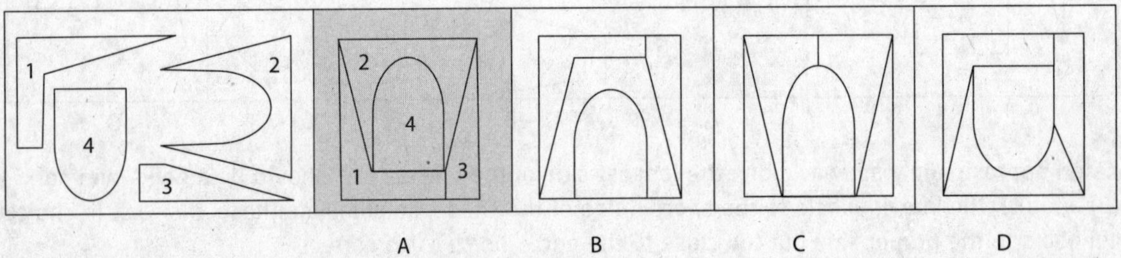

Shape #2 is certainly an oddball, making it a good landmark. At first glance, it might appear that this shape is in choice (C), but notice that it has been split at the top. This means that choice (C) has five pieces, which is another reason to eliminate it. Since shape #2 does not appear at all in choices (B) or (D), the correct answer is (A), with shape #4 nested snugly inside shape #2.

3. **C**

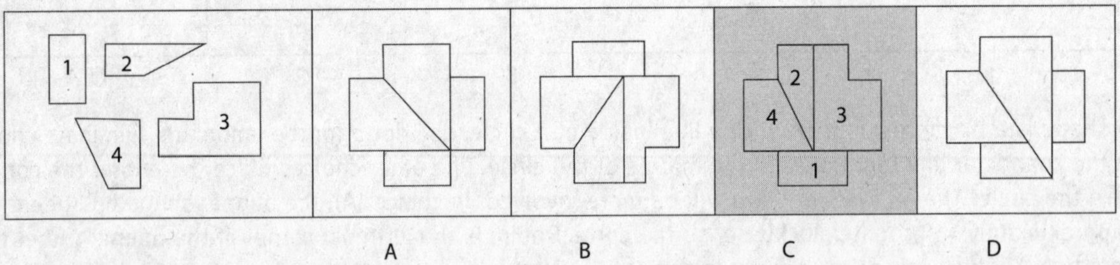

The shapes are all fairly simple in this problem; #2 would be one option to start with as a landmark. That shape clearly is missing from choice (A). Choice (B) might be tempting, but the sloping side is reversed. Choice (D) has a shape similar to #2 pointing down, but it is much too large, so choice (C) is correct. To double-check, note that shape #3 does not appear in choice (D).

4. **B**

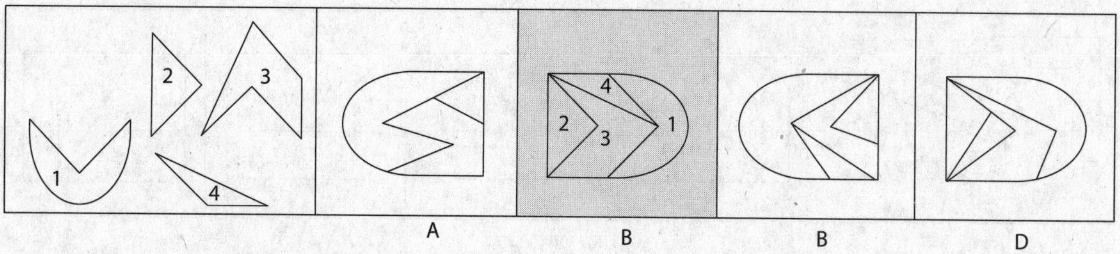

Shape #1, which looks like the dot-gobbler from an old video game, is an easy shape to use as a landmark. Since that shape is absent (or distorted) in choices (A), (C), and (D), eliminate those answers. Shape #1 is found at the right end of choice (B), rotated 90° counterclockwise, so that is the correct answer.

5. **D**

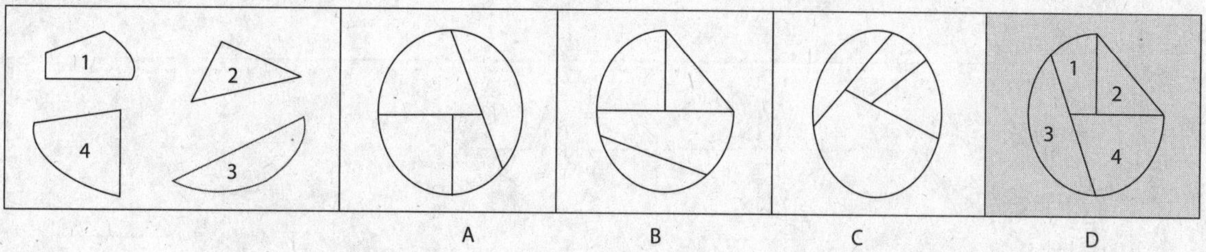

Shape #2 is the only one without a curved side, so that is a good landmark. That shape is not present in choices (A) and (C), so eliminate those and select another shape for a second search. Shape #1 is the most complex remaining shape. Although there is a similarly shaped piece in the lower half of choice (B), that shape is larger than the original, so choice (D) is correct.

6. **C**

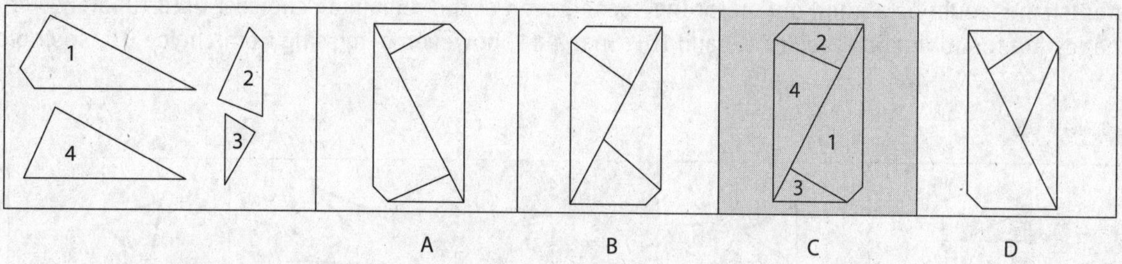

Shape #1 is the largest and most complex, so it is a logical landmark. The notch at the corner of what otherwise would look like a right triangle should be easily identifiable. The notch is missing in choice (B) and is in the wrong corner of the shape in choice (D), so eliminate those two answers. Shape #2 is a readily recognizable choice for the second scan. That shape cannot be found in choice (A), so choice (C) is the correct answer.

7. **A**

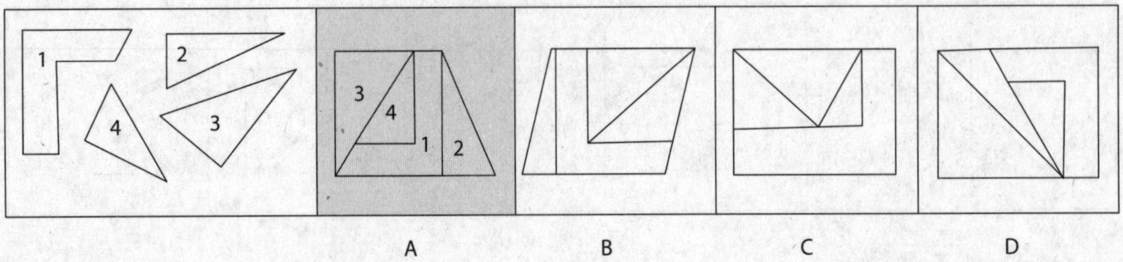

Shape #1 is unique, since the other three shapes are triangles. Use that one as the landmark. In choice (C), the "L" doesn't have the slanted end on the short leg, so eliminate that answer immediately. A careful inspection of choices (B) and (D) reveals that the legs of the "L" are reversed; eliminate those answers. Answer choice (A) is correct.

8. **C**

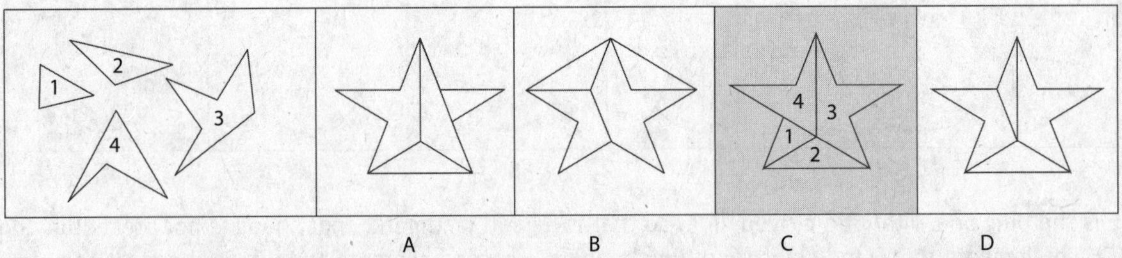

Shape #3 is the most complex of the group and is a good choice for a landmark shape. While the corresponding shape in choice (A) is similar, the long and short sides are reversed. In choice (B), the original six-sided shape has morphed into a seven-sided shape. Shape #3 is present in the remaining answers, so shape #4, another complicated shape, could be a good option for the second scan of the remaining choices. Unfortunately, very similar shapes are found in both choices (C) and (D). Shape #1, however, is missing from choice (D), so choice (C) is correct.

9. **B**

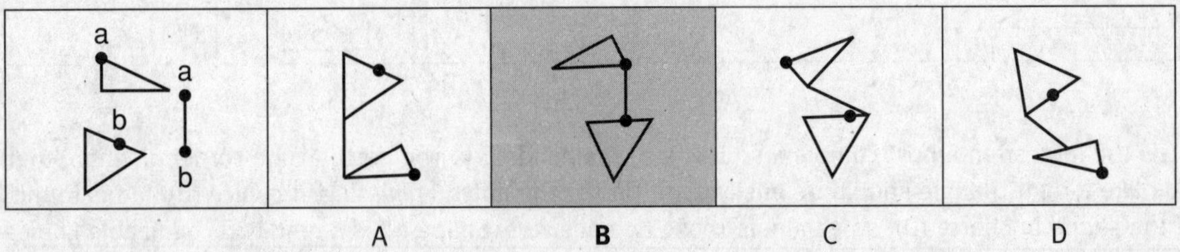

Dot *a*, which is situated at the vertex of the short leg and apparent hypotenuse of the triangle, is a good landmark. The connector line only touches the smaller triangle at that point in choice (B).

10. **C**

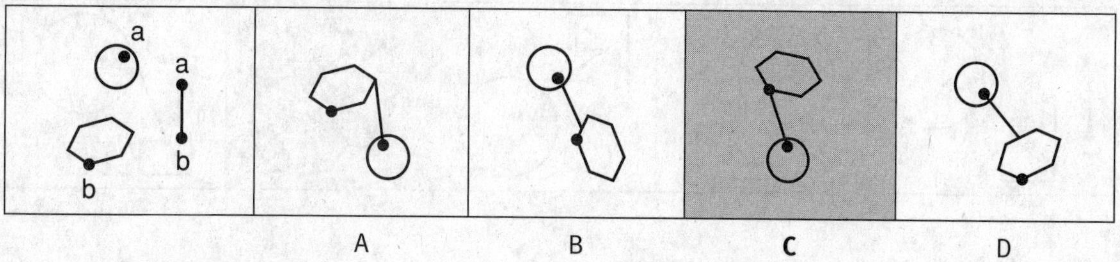

The shape that shows dot *b* is an irregular hexagon; dot *b* is located at the intersection of a long side and a short side of that hexagon. Once again, you don't need a second landmark to solve this one, since the connector line only touches the hexagon at dot *b* in choice (C).

11. **D**

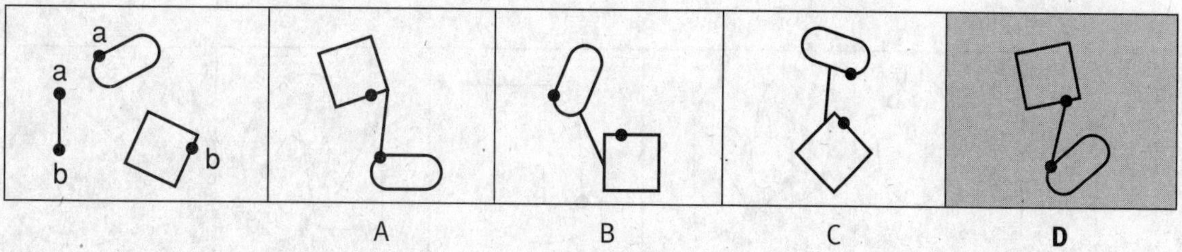

Dot *a* is attached to the "racetrack" shape at the end of the backstretch and can serve as the landmark. The connector line does not meet the "racetrack" at the proper point in choices (B) and (C). Dot *b* is placed very close to the corner of the square. In choice (A), the connector line is at the corner of the square, so choice (D) is correct.

12. **A**

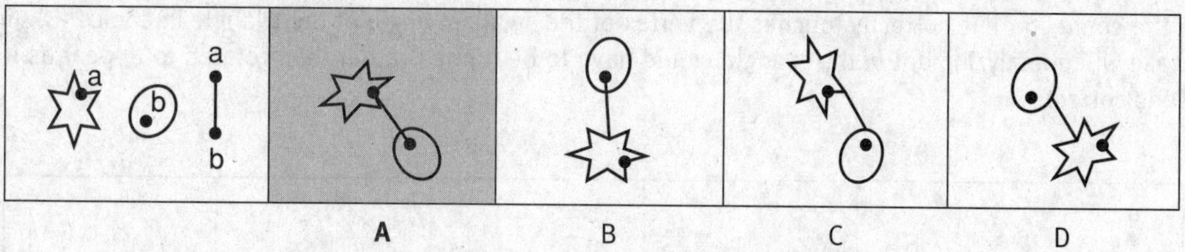

The egg is a nice simple shape, whereas the star is just slightly elongated; analyzing the connection to the star could be more difficult. Therefore, start with dot *b*, located just inside the egg, as the landmark. Since choices (C) and (D) show the connector line touching the outside of the egg, those choices can be eliminated. Now look at the connection to the star. In both (A) and (B), the connector touches the star at an inner vertex. However, in (B) the connection is in the middle of the "longer" or more stretched out side of the star. Therefore, choice (A), which makes the connection on the "narrow" end, is correct.

13. **C**

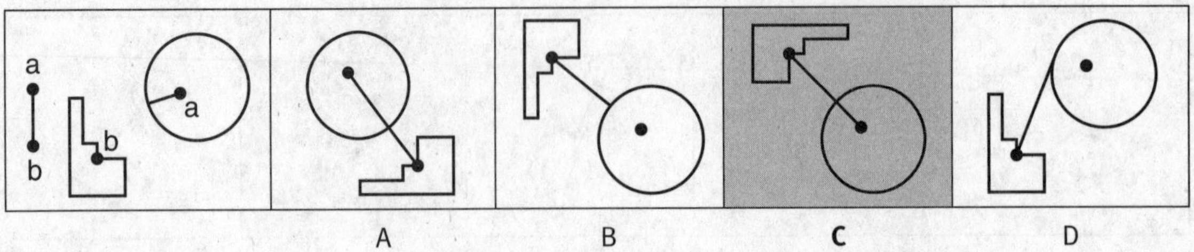

While dot *b* is located on the more complex of the two shapes, dot *a* has the readily identifiable feature of being linked to the outside of the circle by a short line rather than just "floating" inside the circle. Thus, dot *a* is a good starting landmark, and only choice (C) shows this. Had you chosen to use dot *b* as the landmark, you could have eliminated choice (B) because the odd shape is flipped (if it were upside-down, the long "finger" would be on the right, not the left) and then proceeded to answer the question using dot *a*.

14. **D**

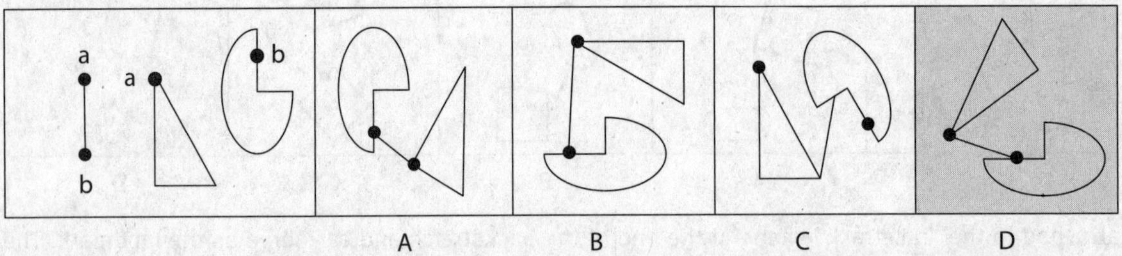

Since it is attached to a more complex shape, use dot *b* as the landmark. In choice (A), the shape would have to be flipped rather than rotated to be oriented in that manner. The shape is proper in choice (C), but the connector line touches the shorter of the two straight lines. Look at dot *a* to decide which of the two remaining choices is correct. Although the connector line correctly touches the vertex of the smallest angle of the triangle in choice (B), just as was the case when analyzing dot *b*, the triangle would have to be flipped rather than rotated to appear as it does. Choice (D) is correct.

15. **A**

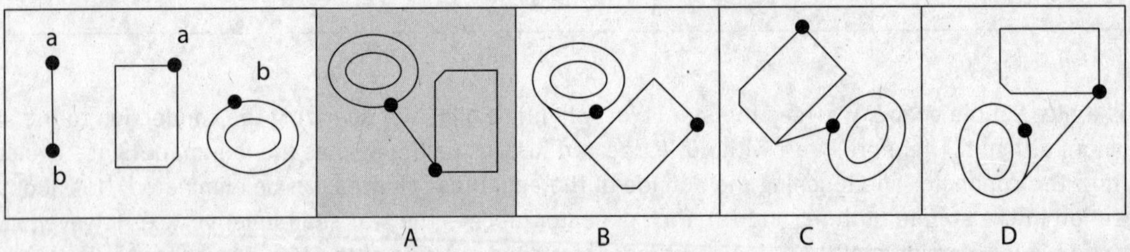

Since one corner of the rectangle is notched, dot *a*, which is at the opposite end of the long side of the rectangle, is a good landmark. In choice (B), the connector line touches the rectangle at the corner that is separated from the notch by the shorter side of the rectangle. Choice (C) reverses the notched and unnotched corners, so eliminate choice (C) as well. Now use dot *b* to eliminate one more wrong answer. In choice (D), the "O" has been rotated to be "upright," so dot *b* is along the upper right side of the double oval. However, the line connects at the lower right. Choice (A) is correct.

Review and Reflect

How did you do? Which questions were more challenging? How could you improve your choice of landmarks? Going over the questions again may prove helpful.

Want more instruction and practice with Assembling Objects? Check out the companion video for this section.

COMPANION VIDEO

kaptest.com/login

CHAPTER 15

ASVAB PRACTICE TESTS

On the following pages, there are two full-length practice tests with complete answers and explanations. The number of questions and timing follow the ASVAB's paper-and-pencil format. Do your best to mimic the Test Day scenario. Take the test in a quiet location, do not use a calculator, and stick to the time frames allotted for each section.

When you're done, review the explanations to determine why the right answers are right and the wrong answers are wrong.

For experience taking the test on a computer, check out the practice tests online.

PRACTICE TESTS

kaptest.com/login

ANSWER SHEET

Part 1: General Science (GS)

1 A B C D	6 A B C D	11 A B C D	16 A B C D	21 A B C D
2 A B C D	7 A B C D	12 A B C D	17 A B C D	22 A B C D
3 A B C D	8 A B C D	13 A B C D	18 A B C D	23 A B C D
4 A B C D	9 A B C D	14 A B C D	19 A B C D	24 A B C D
5 A B C D	10 A B C D	15 A B C D	20 A B C D	25 A B C D

Part 2: Arithmetic Reasoning (AR)

1 A B C D	7 A B C D	13 A B C D	19 A B C D	25 A B C D
2 A B C D	8 A B C D	14 A B C D	20 A B C D	26 A B C D
3 A B C D	9 A B C D	15 A B C D	21 A B C D	27 A B C D
4 A B C D	10 A B C D	16 A B C D	22 A B C D	28 A B C D
5 A B C D	11 A B C D	17 A B C D	23 A B C D	29 A B C D
6 A B C D	12 A B C D	18 A B C D	24 A B C D	30 A B C D

Part 3: Word Knowledge (WK)

1 A B C D	8 A B C D	15 A B C D	22 A B C D	29 A B C D
2 A B C D	9 A B C D	16 A B C D	23 A B C D	30 A B C D
3 A B C D	10 A B C D	17 A B C D	24 A B C D	31 A B C D
4 A B C D	11 A B C D	18 A B C D	25 A B C D	32 A B C D
5 A B C D	12 A B C D	19 A B C D	26 A B C D	33 A B C D
6 A B C D	13 A B C D	20 A B C D	27 A B C D	34 A B C D
7 A B C D	14 A B C D	21 A B C D	28 A B C D	35 A B C D

Part 4: Paragraph Comprehension (PC)

1 A B C D	4 A B C D	7 A B C D	10 A B C D	13 A B C D
2 A B C D	5 A B C D	8 A B C D	11 A B C D	14 A B C D
3 A B C D	6 A B C D	9 A B C D	12 A B C D	15 A B C D

Part 5: Mathematics Knowledge (MK)

1 A B C D	6 A B C D	11 A B C D	16 A B C D	21 A B C D
2 A B C D	7 A B C D	12 A B C D	17 A B C D	22 A B C D
3 A B C D	8 A B C D	13 A B C D	18 A B C D	23 A B C D
4 A B C D	9 A B C D	14 A B C D	19 A B C D	24 A B C D
5 A B C D	10 A B C D	15 A B C D	20 A B C D	25 A B C D

ANSWER SHEET

Part 6: Electronics Information (EI)

1 Ⓐ Ⓑ Ⓒ Ⓓ	5 Ⓐ Ⓑ Ⓒ Ⓓ	9 Ⓐ Ⓑ Ⓒ Ⓓ	13 Ⓐ Ⓑ Ⓒ Ⓓ	17 Ⓐ Ⓑ Ⓒ Ⓓ
2 Ⓐ Ⓑ Ⓒ Ⓓ	6 Ⓐ Ⓑ Ⓒ Ⓓ	10 Ⓐ Ⓑ Ⓒ Ⓓ	14 Ⓐ Ⓑ Ⓒ Ⓓ	18 Ⓐ Ⓑ Ⓒ Ⓓ
3 Ⓐ Ⓑ Ⓒ Ⓓ	7 Ⓐ Ⓑ Ⓒ Ⓓ	11 Ⓐ Ⓑ Ⓒ Ⓓ	15 Ⓐ Ⓑ Ⓒ Ⓓ	19 Ⓐ Ⓑ Ⓒ Ⓓ
4 Ⓐ Ⓑ Ⓒ Ⓓ	8 Ⓐ Ⓑ Ⓒ Ⓓ	12 Ⓐ Ⓑ Ⓒ Ⓓ	16 Ⓐ Ⓑ Ⓒ Ⓓ	20 Ⓐ Ⓑ Ⓒ Ⓓ

Part 7: Auto and Shop Information (AS)

1 Ⓐ Ⓑ Ⓒ Ⓓ	6 Ⓐ Ⓑ Ⓒ Ⓓ	11 Ⓐ Ⓑ Ⓒ Ⓓ	16 Ⓐ Ⓑ Ⓒ Ⓓ	21 Ⓐ Ⓑ Ⓒ Ⓓ
2 Ⓐ Ⓑ Ⓒ Ⓓ	7 Ⓐ Ⓑ Ⓒ Ⓓ	12 Ⓐ Ⓑ Ⓒ Ⓓ	17 Ⓐ Ⓑ Ⓒ Ⓓ	22 Ⓐ Ⓑ Ⓒ Ⓓ
3 Ⓐ Ⓑ Ⓒ Ⓓ	8 Ⓐ Ⓑ Ⓒ Ⓓ	13 Ⓐ Ⓑ Ⓒ Ⓓ	18 Ⓐ Ⓑ Ⓒ Ⓓ	23 Ⓐ Ⓑ Ⓒ Ⓓ
4 Ⓐ Ⓑ Ⓒ Ⓓ	9 Ⓐ Ⓑ Ⓒ Ⓓ	14 Ⓐ Ⓑ Ⓒ Ⓓ	19 Ⓐ Ⓑ Ⓒ Ⓓ	24 Ⓐ Ⓑ Ⓒ Ⓓ
5 Ⓐ Ⓑ Ⓒ Ⓓ	10 Ⓐ Ⓑ Ⓒ Ⓓ	15 Ⓐ Ⓑ Ⓒ Ⓓ	20 Ⓐ Ⓑ Ⓒ Ⓓ	25 Ⓐ Ⓑ Ⓒ Ⓓ

Part 8: Mechanical Comprehension (MC)

1 Ⓐ Ⓑ Ⓒ Ⓓ	6 Ⓐ Ⓑ Ⓒ Ⓓ	11 Ⓐ Ⓑ Ⓒ Ⓓ	16 Ⓐ Ⓑ Ⓒ Ⓓ	21 Ⓐ Ⓑ Ⓒ Ⓓ
2 Ⓐ Ⓑ Ⓒ Ⓓ	7 Ⓐ Ⓑ Ⓒ Ⓓ	12 Ⓐ Ⓑ Ⓒ Ⓓ	17 Ⓐ Ⓑ Ⓒ Ⓓ	22 Ⓐ Ⓑ Ⓒ Ⓓ
3 Ⓐ Ⓑ Ⓒ Ⓓ	8 Ⓐ Ⓑ Ⓒ Ⓓ	13 Ⓐ Ⓑ Ⓒ Ⓓ	18 Ⓐ Ⓑ Ⓒ Ⓓ	23 Ⓐ Ⓑ Ⓒ Ⓓ
4 Ⓐ Ⓑ Ⓒ Ⓓ	9 Ⓐ Ⓑ Ⓒ Ⓓ	14 Ⓐ Ⓑ Ⓒ Ⓓ	19 Ⓐ Ⓑ Ⓒ Ⓓ	24 Ⓐ Ⓑ Ⓒ Ⓓ
5 Ⓐ Ⓑ Ⓒ Ⓓ	10 Ⓐ Ⓑ Ⓒ Ⓓ	15 Ⓐ Ⓑ Ⓒ Ⓓ	20 Ⓐ Ⓑ Ⓒ Ⓓ	25 Ⓐ Ⓑ Ⓒ Ⓓ

Part 9: Assembling Objects (AO)

1 Ⓐ Ⓑ Ⓒ Ⓓ	6 Ⓐ Ⓑ Ⓒ Ⓓ	11 Ⓐ Ⓑ Ⓒ Ⓓ	16 Ⓐ Ⓑ Ⓒ Ⓓ	21 Ⓐ Ⓑ Ⓒ Ⓓ
2 Ⓐ Ⓑ Ⓒ Ⓓ	7 Ⓐ Ⓑ Ⓒ Ⓓ	12 Ⓐ Ⓑ Ⓒ Ⓓ	17 Ⓐ Ⓑ Ⓒ Ⓓ	22 Ⓐ Ⓑ Ⓒ Ⓓ
3 Ⓐ Ⓑ Ⓒ Ⓓ	8 Ⓐ Ⓑ Ⓒ Ⓓ	13 Ⓐ Ⓑ Ⓒ Ⓓ	18 Ⓐ Ⓑ Ⓒ Ⓓ	23 Ⓐ Ⓑ Ⓒ Ⓓ
4 Ⓐ Ⓑ Ⓒ Ⓓ	9 Ⓐ Ⓑ Ⓒ Ⓓ	14 Ⓐ Ⓑ Ⓒ Ⓓ	19 Ⓐ Ⓑ Ⓒ Ⓓ	24 Ⓐ Ⓑ Ⓒ Ⓓ
5 Ⓐ Ⓑ Ⓒ Ⓓ	10 Ⓐ Ⓑ Ⓒ Ⓓ	15 Ⓐ Ⓑ Ⓒ Ⓓ	20 Ⓐ Ⓑ Ⓒ Ⓓ	25 Ⓐ Ⓑ Ⓒ Ⓓ

ASVAB PRACTICE TEST A

Part 1: General Science (GS)

Time: 11 minutes; 25 questions

Directions: In this section, you will be tested on your knowledge of concepts in science generally reviewed in high school. For each question, select the best answer and mark the corresponding oval on your answer sheet.

1. A genus classification contains several related
   (A) species
   (B) families
   (C) orders
   (D) phyla

2. Particles that orbit the nucleus of an atom are
   (A) anions
   (B) electrons
   (C) positrons
   (D) photons

3. As a human breathes, air is first warmed and filtered in the
   (A) trachea
   (B) pharynx
   (C) oral cavity
   (D) nasal cavity

4. In the food chain described below, which is the tertiary consumer?

   dandelion → rabbit → fox → coyote

   (A) dandelion
   (B) rabbit
   (C) fox
   (D) coyote

5. The process by which a solid becomes a gas is known as
   (A) boiling
   (B) sublimation
   (C) melting
   (D) diffusion

6. The gravitational force the Moon exerts on the Earth is
   (A) less than one-half as much as that exerted by the Earth on the Moon
   (B) one-half as much as that exerted by the Earth on the Moon
   (C) the same as that exerted by the Earth on the Moon
   (D) more than that exerted by the Earth on the Moon

7. Fungi are organisms that break down dead matter and return the organic material back into the environment for reuse. They are examples of
   (A) producers
   (B) decomposers
   (C) consumers
   (D) mutualists

8. What is the freezing point of water in Kelvin?
   (A) −273 K
   (B) 0 K
   (C) 100 K
   (D) 273 K

9. Tough elastic tissues found in the joints that connect bones to bones are called
   (A) ligaments
   (B) tendons
   (C) cartilage
   (D) muscles

10. Stress, a poor diet, cigarette smoking, and hereditary factors all contribute to individuals developing
    (A) diarrhea
    (B) high blood pressure
    (C) gall stones
    (D) anemia

11. Which clouds are thin and wispy and occur high in the atmosphere?
    (A) cirrus
    (B) cumulus
    (C) stratus
    (D) stratocumulus

12. Fossils are most likely to be found in which of the following types of rock?
    (A) igneous
    (B) metamorphic
    (C) sedimentary
    (D) volcanic

13. Animals that consume only plants are called

(A) saprophytes
(B) herbivores
(C) carnivores
(D) omnivores

14. What type of star is the Sun?

(A) red dwarf
(B) blue giant
(C) white dwarf
(D) yellow dwarf

15. Which of the following is a group of organisms of the same species living in the same region?

(A) biome
(B) ecosystem
(C) community
(D) population

16. Which one of these planets is NOT an outer planet?

(A) Jupiter
(B) Saturn
(C) Uranus
(D) Earth

17. A boulder that begins to roll down a hill is an example of an energy conversion from

(A) potential to thermal
(B) potential to kinetic
(C) kinetic to thermal
(D) kinetic to potential

18. The lowest layer of the Earth's atmosphere is called the

(A) ionosphere
(B) mesosphere
(C) stratosphere
(D) troposphere

19. According to the law of conservation of energy

(A) energy is the capacity to do work
(B) energy can neither be created nor destroyed
(C) energy is possessed by a moving object
(D) energy is stored in an object as a result of its position, shape, or state

20. As an ambulance passes, its pitch seems to change. This perception is best explained by

(A) convection
(B) Newton's third law
(C) the Doppler effect
(D) momentum

21. A concave lens is also known as a

(A) focal point
(B) converging lens
(C) convex lens
(D) diverging lens

22. The smallest particle of a covalent compound that can exist in a free state and still retain the characteristics of that compound is a(n)

(A) proton
(B) quark
(C) atom
(D) molecule

23. Which of the following is most responsible for oceanic tides?

(A) the orbit of the Earth around its own axis
(B) the magnetic polarity of the Earth
(C) the orbit of the Moon around the Earth
(D) the orbit of the Earth around the Sun

24. In the Periodic Table of the Elements, the elements in a column are referred to as a

(A) period
(B) group
(C) configuration
(D) unit

25. In a chemical change, the molecules that enter the reaction are called

(A) reactants
(B) products
(C) compounds
(D) elements

STOP. IF YOU FINISH BEFORE THE TIME IS UP, YOU MAY CHECK OVER YOUR WORK ON THIS PART ONLY.

Part 2: Arithmetic Reasoning (AR)

Time: 36 minutes; 30 questions

Directions: In this section, you are tested on your ability to use arithmetic. For each question, select the best answer and mark the corresponding oval on your answer sheet.

1. Three apples cost as much as 4 pears. Three pears cost as much as 2 oranges. How many apples cost as much as 72 oranges?

(A) 36
(B) 48
(C) 64
(D) 81

2. If 48 of the 60 seats on a bus were occupied, what percent of the seats were NOT occupied?

(A) 12%
(B) 15%
(C) 20%
(D) 25%

3. Effin sings for $1,000 an hour. Her rate increases by 50% after midnight. If she sings one night from 8:30 p.m. until 1:00 a.m., how much should she be paid?

(A) $4,500
(B) $5,000
(C) $5,500
(D) $6,000

4. Fran has a drawer containing 4 black T-shirts, 3 orange T-shirts, and 5 blue T-shirts. If she picks one T-shirt at random from the drawer, what are the chances that it will NOT be orange?

(A) $\frac{1}{4}$
(B) $\frac{1}{3}$
(C) $\frac{2}{3}$
(D) $\frac{3}{4}$

5. After a 5-hour flight from Newark, Harry arrives in Denver at 2:30 p.m. If the time in Newark is 2 hours later than the time in Denver, what was the time in Newark when Harry began the flight?

(A) 10:30 a.m.
(B) 11:30 a.m.
(C) 3:30 p.m.
(D) 5:30 p.m.

6. A full box of chocolate contains 24 pieces. If Doris starts out with 198 pieces of chocolate, how many pieces will she have left over if she fills as many boxes as she can?

(A) 3
(B) 6
(C) 8
(D) 10

7. If 75% of x is 150, what is the value of x?

(A) 150
(B) 175
(C) 200
(D) 250

8. Rachel's average score after 6 tests is 83. If Rachel earns a score of 97 on the 7th test, what is her new average?

(A) 85
(B) 86
(C) 87
(D) 88

9. A delivery service charges $25.00 per pound for making a delivery. If there is an additional 8% sales tax, what is the cost of delivering an item that weighs $\frac{4}{5}$ of a pound?

(A) $20.00
(B) $21.60
(C) $22.60
(D) $24.00

10. A cake recipe requires $\frac{3}{5}$ of an ounce of vanilla extract. How many cakes can be made using a package containing 60 ounces of vanilla extract?

(A) 48
(B) 80
(C) 96
(D) 100

11. Ed has 100 more dollars than Robert. After Ed spends 20 dollars on groceries, he now has 5 times as much money as Robert. How much money does Robert have?

(A) $16
(B) $20
(C) $24
(D) $30

12. 587 people are traveling by bus for a field trip. If each bus seats 48 people and all of the buses but one are filled to capacity, how many people sit in the unfilled bus?

(A) 37
(B) 36
(C) 12
(D) 11

13. Riley brings 100 cookies to school for her class party. If there are 15 students in the class, including Riley, and there are 25 cookies left after the party, what is the average number of cookies that Riley and her classmates ate?

(A) 1
(B) 4
(C) 5
(D) 6

14. A person 4 feet tall casts a 9-foot shadow at the same time that a nearby tree casts a 21-foot shadow. What is the height of this tree, in feet?

(A) 7
(B) $8\frac{1}{2}$
(C) $9\frac{1}{3}$
(D) 10

15. In a group of 25 students, 16 are female. What percent of the group is female?

(A) 16%
(B) 40%
(C) 60%
(D) 64%

16. Phil is making a 40-kilometer canoe trip. If he travels at 30 kilometers per hour for the first 10 kilometers, and then at 15 kilometers per hour for the rest of the trip, how many more minutes will it take him than if he travels the entire trip at 20 kilometers per hour?

(A) 20
(B) 24
(C) 30
(D) 40

17. In a certain class, 3 out of 24 students are in student organizations. What is the ratio of students in student organizations to students not in student organizations?

(A) $\frac{1}{8}$
(B) $\frac{1}{7}$
(C) $\frac{1}{6}$
(D) $\frac{1}{4}$

18. In a certain baseball league, each team plays 160 games. After playing half of their games, team A has won 60 games and team B has won 49 games. If team A wins half of its remaining games, how many of its remaining games must team B win to have the same number of wins as team A at the end of the season?

(A) 51
(B) 59
(C) 60
(D) 61

19. A vendor bought 10 crates of oranges for a total cost of $80. If he lost 2 of the crates, at what price would he have to sell the remaining crates in order to earn a total profit of 25% of the total cost?

(A) $8.00
(B) $10.00
(C) $12.50
(D) $15.00

20. A machine labels 150 bottles in 20 minutes. At this rate, how many minutes does it take to label 60 bottles?

(A) 2
(B) 4
(C) 6
(D) 8

21. $15 \times (-5) =$

(A) 10
(B) 75
(C) −75
(D) −225

22. Mike has a collection of baseball cards and baseball figurines. If the ratio of cards to figurines is 5:7 and there are 25 cards in his collection, how many figurines does he have?

(A) 20
(B) 35
(C) 40
(D) 45

23. Two hot dogs and a soda cost $3.25. If three hot dogs and a soda cost $4.50, what is the cost of two sodas?

(A) $0.75
(B) $1.25
(C) $1.50
(D) $2.50

24. Danielle drives from her home to the store at an average speed of 40 miles per hour. She returns home along the same route at an average speed of 60 miles per hour. What is her average speed, in miles per hour, for the entire trip?

(A) 48
(B) 50
(C) 52
(D) 55

25. At a certain school the ratio of teachers to students is 1 to 10. Which of the following could be the total number of teachers and students?

(A) 100
(B) 121
(C) 222
(D) 1,011

26. The average of two numbers is equal to twice the positive difference between the two numbers. If the larger number is 35, what is the smaller number?

(A) 9
(B) 15
(C) 21
(D) 27

27. Each of seven runners on a relay team must run a distance of 1.27 kilometers. Approximately what is the total combined number of kilometers run by the team in the race?

(A) 11
(B) 10
(C) 9
(D) 8

28. A scanner can scan 12 photos per minute. At this rate, how many minutes would it take to scan 60 photos?

(A) 2
(B) 3
(C) 5
(D) 7

29. An employee's net pay is equal to gross pay minus total deductions. A certain employee's gross pay is $1,769.23 and her deductions are as follows: FICA, $218.99; Social Security, $107.05; Medicare, $25.03; state tax, $68.65; municipal tax, $42.75. What is the employee's net pay?

(A) $1,306.76
(B) $1,306.66
(C) $1,305.76
(D) $1,305.66

30. A student takes four tests. His grades are 94, 72, 84, and 98. What is this student's average test grade?

(A) 92
(B) 65
(C) 87
(D) 88

STOP. IF YOU FINISH BEFORE THE TIME IS UP, YOU MAY CHECK OVER YOUR WORK ON THIS PART ONLY.

Part 3: Word Knowledge (WK)

Time: 11 minutes; 35 questions

Directions: In this section, you are tested on the meaning of words. Each of the following questions has an underlined word. Select the answer that most nearly means the same as the underlined word and mark the corresponding oval on your answer sheet.

1. Impose most nearly means
   (A) create
   (B) force
   (C) damage
   (D) trade

2. Stunted most nearly means
   (A) halted
   (B) frightened
   (C) aged
   (D) overstated

3. A sturdy home can withstand nearly any disaster.
   (A) huge
   (B) strong
   (C) cold
   (D) cautious

4. He was undeterred in his quest to find her.
   (A) surprised
   (B) persistent
   (C) careless
   (D) brazen

5. Ennui most nearly means
   (A) patient
   (B) gloating
   (C) boredom
   (D) tasteful

6. Mutable most nearly means
   (A) changeable
   (B) silent
   (C) big-hearted
   (D) calm

7. The way they squandered the money was shameful.
   (A) gathered
   (B) stole
   (C) wasted
   (D) owned

8. The handwriting was nearly illegible.
   (A) unreadable
   (B) unethical
   (C) creative
   (D) dangerous

9. Infinite most nearly means
   (A) costly
   (B) unending
   (C) babyish
   (D) daily

10. Longevity most nearly means
    (A) training
    (B) duration
    (C) girth
    (D) lifestyle

11. Tourists always fall for that ruse.
    (A) trick
    (B) display
    (C) itinerary
    (D) backtalk

12. The president proclaimed it a national holiday.
    (A) suggested
    (B) renamed
    (C) announced
    (D) unmade

13. Even the most <u>rudimentary</u> details are crucial to understanding this problem.
    (A) basic
    (B) ecstatic
    (C) illogical
    (D) fancy

14. <u>Rout</u> most nearly means
    (A) careful
    (B) defeat
    (C) blatant
    (D) open

15. His <u>rogue</u> attitude didn't fit the "team" concept.
    (A) aggressive
    (B) sad
    (C) rebellious
    (D) low-brow

16. Her point of view didn't <u>resonate</u> with everyone on the committee.
    (A) create irritation
    (B) display clarity
    (C) evoke agreement
    (D) generate discussion

17. Duplicating steps is far too <u>redundant</u> at this late stage.
    (A) repetitive
    (B) upsetting
    (C) colorful
    (D) wishy-washy

18. Far be it from me to <u>prescribe</u> how to raise one's own children.
    (A) predict
    (B) dictate
    (C) judge
    (D) decide

19. <u>Logical</u> most nearly means
    (A) cognizant
    (B) easy
    (C) rowdy
    (D) sensible

20. <u>Prerequisite</u> most nearly means
    (A) requirement
    (B) evaluation
    (C) glee
    (D) good taste

21. It seems that loud people tend to <u>gravitate</u> toward other loud people.
    (A) be drawn
    (B) be hostile
    (C) be inquisitive
    (D) be competitive

22. The police felt it was best to be <u>cautious</u> when approaching the house.
    (A) excited
    (B) apathetic
    (C) dodgy
    (D) careful

23. <u>Braggart</u> most nearly means
    (A) clown
    (B) leader
    (C) boaster
    (D) deputy

24. Basing his opinions on <u>hearsay</u> doomed the case.
    (A) rumor
    (B) samples
    (C) truth
    (D) belief

25. <u>Taunt</u> most nearly means
    (A) grade
    (B) relate
    (C) ridicule
    (D) party

26. Most of our employees respond well to <u>constructive</u> criticism.
    (A) damaging
    (B) productive
    (C) gentle
    (D) flattering

27. Quantifiable most nearly means
    (A) laughable
    (B) standard
    (C) countable
    (D) breakable

28. The least effective way of dealing with children is yelling.
    (A) tantalizing
    (B) creative
    (C) useful
    (D) positive

29. Apathy most nearly means
    (A) greatness
    (B) laziness
    (C) disinterest
    (D) boredom

30. It left a glaring blemish on his permanent record.
    (A) impact
    (B) defect
    (C) commendation
    (D) compliment

31. It will take a consensus to get this measure passed.
    (A) agreement
    (B) discussion
    (C) infighting
    (D) reprimand

32. No coach can handle dissent from his players for long.
    (A) discussion
    (B) laughter
    (C) laziness
    (D) insubordination

33. Tempo most nearly means
    (A) heartbeat
    (B) safety
    (C) modernity
    (D) speed

34. She can take solace in the fact that it couldn't get much worse.
    (A) protection
    (B) comfort
    (C) glee
    (D) depth

35. Many pitfalls await the inexperienced climber.
    (A) joys
    (B) traps
    (C) ropes
    (D) talents

STOP. IF YOU FINISH BEFORE THE TIME IS UP, YOU MAY CHECK OVER YOUR WORK ON THIS PART ONLY.

Part 4: Paragraph Comprehension (PC)

Time: 13 minutes; 15 questions

Directions: This section contains paragraphs followed by incomplete statements or questions. For each question, read the paragraph and select the answer that best completes the statements or answers the question that follows. Mark the corresponding oval on your answer sheet.

Questions 1 and 2 refer to the following passage.

In modern society, a form of folktale called the *urban legend* has emerged. These stories persist both for their entertainment value and for the transmission of popular values and beliefs. Urban legends are stories many have heard; they are supposed to have really happened, but they cannot be verified. It turns out that the people involved can never be found. Researchers of urban legends call the elusive participant in these supposedly "real-life" events a "FOAF": friend of a friend.

One classic urban legend involves alligators in the sewer systems of major metropolitan areas. According to the story, before alligators were a protected species, people vacationing in Florida purchased baby alligators to take home as souvenirs. After the novelty of having a pet alligator wore off, people would flush their souvenirs down the toilet. The baby alligators found a perfect growing environment in city sewer systems, where to this day they thrive on an ample supply of rats.

1. The passage suggests that the real-life participants of urban legends

   (A) can be very difficult to track down
   (B) are usually known, but only barely, by the teller
   (C) are friends with a large number of people
   (D) are the original tellers of the stories

2. According to the passage, the successful urban legend contains all of the following characteristics EXCEPT

   (A) the capacity to entertain
   (B) messages that conform to popular values
   (C) qualities of a folktale
   (D) a basis in reality

For most students of biology, Charles Darwin is considered the father of evolution. But few realize that he was but one of many theorists who noticed that the genetics of animals showed progression over time. Henri Bergson, for instance, formulated a theory that today is known more commonly as *theistic evolution,* or evolution from God. However, Darwin's ability to articulate his theories in writing and to account for many diverse examples of evolutionary biology means that other theories are often seen as little more than offshoots of the original idea of Darwinian evolution.

3. According to the passage, Darwin's theory persists because

   (A) Bergson's theory was incorrect
   (B) Darwin's theory accounted for many examples of evolution
   (C) genetics is an inexact science
   (D) evolution is not accepted by the mainstream

Local elementary schools have changed considerably over the past 50 years. Where once we had schools in every small town, now students bus for miles to attend larger, more advanced schools. While most parents see this as a positive step for progress and education, some worry about their children losing touch with the simple things around them. A few have even decided to homeschool their children instead of sending them to nearby towns.

4. The author's tone in this passage is

   (A) embittered
   (B) informative
   (C) biased
   (D) ambivalent

Questions 5 and 6 refer to the following passage.

Most life is fundamentally dependent on photosynthetic organisms that store radiant energy from the Sun. In almost all the world's ecosystems and food chains, photosynthetic organisms such as plants and algae are eaten by other organisms, which are then consumed by still others. The existence of organisms that are not dependent on the Sun's light has long been established, but until recently they were regarded as anomalies.

Over the last 20 years, however, research in deep-sea areas has revealed the existence of entire ecosystems in which the primary producers are chemosynthetic bacteria that are dependent on energy from within the Earth itself. Indeed, the growing evidence suggests that these sub-sea ecosystems model the way in which life first came about on this planet.

5. The passage suggests that most life is ultimately dependent on which of the following?

(A) photosynthetic algae
(B) the world's oceans
(C) bacterial microorganisms
(D) light from the Sun

6. Which of the following conclusions about photosynthetic and chemosynthetic organisms is supported by this passage?

(A) Both perform similar functions in different food chains.
(B) Both are known to support communities of higher organisms at great ocean depths.
(C) Sunlight is the basic source of energy for both.
(D) Chemosynthetic organisms are less nourishing than photosynthetic organisms.

Halley's Comet has been known since at least 240 B.C.E. and possibly since 1059 B.C.E. Its most famous appearance was in 1066 C.E. when it appeared right before the Battle of Hastings. It was named after the astronomer Edmund Halley, who calculated its orbit. He determined that the comets seen in 1530 and 1606 were the same object following a 76-year orbit. Unfortunately, Halley died in 1742, never living to see his prediction come true when the comet returned on Christmas Eve 1758.

7. It can be inferred from the passage that the appearance of Halley's Comet in 1758

(A) presaged catastrophic events in world history
(B) could not have been foreseen
(C) shocked the scientific community
(D) confirmed Halley's calculations

Both alligators and crocodiles can be found in southern Florida, particularly in the Everglades National Park. Alligators and crocodiles look similar but there are several physical characteristics that differentiate the two giant reptiles. The most easily observed difference between alligators and crocodiles is the shape of the head. The crocodile's skull and jaws are longer and narrower than the alligator's. When a crocodile closes its mouth, the long teeth remain visible, protruding outside the upper jaw. When an alligator closes its mouth, those long teeth slip into sockets in the upper jaw and disappear. In general, if you can still see a lot of teeth even when the animal's mouth is closed, you are looking at a crocodile.

8. According to the passage, one can distinguish a crocodile from an alligator

(A) only when the animal's mouth is closed
(B) by the location in Florida where the animal is found
(C) by its thick, heavily armored skin
(D) by the narrower snout found on a crocodile

A talent agent analyzed her company's records in an attempt to determine why it was placing so few actors in roles. She attributed the company's poor performance to the fact that often the actors sent to an audition were completely inappropriate for the role.

9. It can be inferred from the passage that the agent believes that

(A) certain actors are inappropriate for certain roles
(B) the actors her company represents are not very good
(C) it is difficult to predict how appropriate an actor will be for a role
(D) her company does not send enough actors to audition for major roles

One of the most commonly used of all poetic meters is iambic pentameter. *Pentameter* refers to the number of feet, or groupings of syllables. The prefix *penta–* means five, so a poem written in pentameter has five feet per line. In iambic pentameter, the type of foot used is an iamb. An *iamb* is a word or words with two syllables with the stress on the second syllable. William Shakespeare is perhaps the most famous writer of iambic pentameter. Shakespeare composed hundreds of poems and plays in this meter and popularized it for the masses.

10. According to the passage, a *foot* is

(A) a grouping of syllables
(B) similar to a meter
(C) an innovation credited to William Shakespeare
(D) a type of poem

It is often said that American involvement in World War I would not have begun in earnest were it not for the German sinking of the passenger ship the *Lusitania* on February 18, 1915. Preferring prior to then to stay neutral in the war, America was until that time offering only financial and tactical support for Britain and France against Germany and Austria-Hungary. The sinking of the huge passenger ship, however, altered public opinion about U.S. involvement and subsequently led to military escalation, truly making the war a matter for the whole world.

11. The main idea of the passage is that
    (A) the Germans sank the *Lusitania* because of America's financial and tactical support for Britain and France
    (B) Austria-Hungary and Germany were allies in World War I
    (C) American involvement in World War I was minimal
    (D) the sinking of the *Lusitania* prompted increased American involvement in World War I

When a movie criticized by most film critics is a popular success, it is often seen as a sign of poor taste on the part of general audiences. But there is little diversity among film critics, and the critics' preferences are often a product of their backgrounds and class prejudices. Thus, their opinions are no more likely to be an unerring guide to quality than those of the average moviegoer.

12. The passage above best supports which of the following conclusions?
    (A) Judgments of film quality by professional film critics are usually wrong.
    (B) Judgments of quality applied to movies are meaningless.
    (C) Film critics usually consider popular movies to be of poor quality.
    (D) When critics and general audiences disagree about a movie's quality, the critics' opinion is not necessarily more valid.

The tomato originated in the New World. It was first domesticated around 700 C.E. by the Aztec and Incan civilizations. In the sixteenth century, European explorers were so appreciative of the tomato that they introduced it to the rest of Europe. Although the French, Spanish, and Italians quickly began to adapt their recipes to use tomatoes, the English considered tomatoes poisonous. This myth traveled to the colonies, where settlers in America continued to avoid the supposedly deadly tomato. It wasn't until the middle 1800s that the tomato began to gain acceptance in the U.S.

13. According to the passage, Americans did not start using the tomato until the mid-nineteenth century because
    (A) it was unavailable in the New World
    (B) they lacked recipes for using it
    (C) they believed it to be poisonous
    (D) it was viewed as a Mediterranean food

Daria has been picking up many extra shifts at the local restaurant. Her school work has also been increasing at an alarming rate. The past few weeks have left Daria feeling very drowsy. She looks forward to heading to bed each night and getting as much sleep as possible.

14. What does the word "drowsy" mean in context?
    (A) drunk
    (B) tired
    (C) happy
    (D) satisfied

The sun was relentless as I started my hike through the woods. Sitting under the large branches of the trees was the only way I was able to escape the oppressive heat. Without the leafy canopy sheltering me throughout my journey, I would have had to cut my hike short.

15. The author implies which of the following about the trees in the forest?
    (A) They provide adequate shade.
    (B) They do not provide adequate shade.
    (C) They impair her ability to enjoy the sunshine.
    (D) They provide a home to many animals.

STOP. IF YOU FINISH BEFORE THE TIME IS UP, YOU MAY CHECK OVER YOUR WORK ON THIS PART ONLY.

Part 5: Mathematics Knowledge (MK)

Time: 24 minutes; 25 questions

Directions: In this section, you will be tested on your knowledge of basic mathematics. For each question, select the best answer and mark the corresponding oval on your answer sheet.

1. For all x, $(3x + 4)(4x - 3) =$

   (A) $7x - 12$
   (B) $12x^2 - 12$
   (C) $12x^2 - 25x - 12$
   (D) $12x^2 + 7x - 12$

2. If a tree grew 5 feet in n years, what was the average rate at which the tree grew, in inches per year?

   (A) $\frac{60}{n}$
   (B) $\frac{5}{12n}$
   (C) $\frac{12n}{5}$
   (D) $\frac{n}{60}$

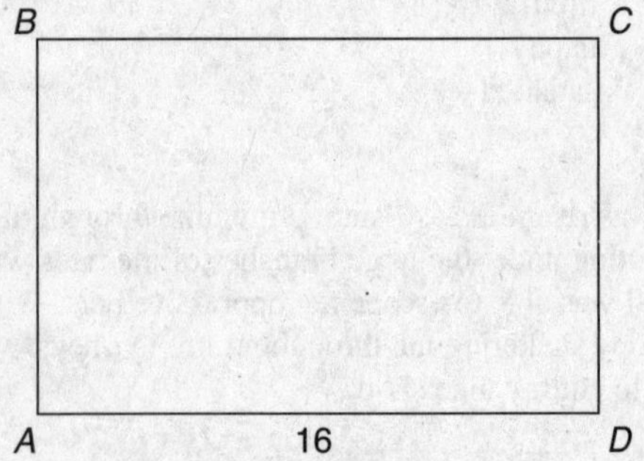

3. In the figure above, if the perimeter of rectangle $ABCD$ is 56, and the length of AD is 16, what is the area of $ABCD$?

   (A) 40
   (B) 64
   (C) 160
   (D) 192

4. A square is a rectangle with four ________ sides.

   (A) equal
   (B) parallel
   (C) unequal
   (D) curvilinear

5. If $a < b$ and $b < c$, which of the following must be true?

   (A) $b + c < 2a$
   (B) $a + b < c$
   (C) $a - b < b - c$
   (D) $a + b < 2c$

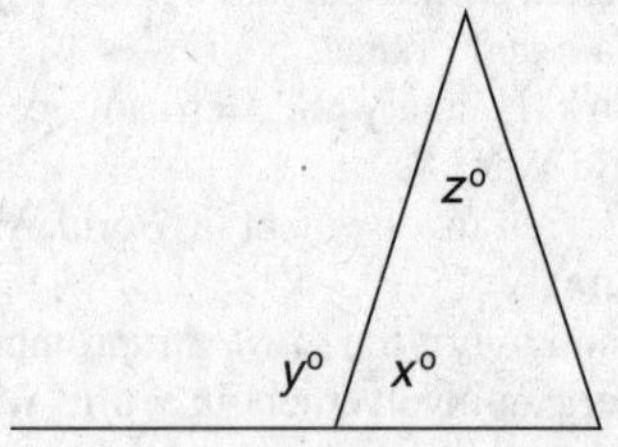

6. In the figure above, $x = 2z$ and $y = 3z$. What is the value of z?

   (A) 24
   (B) 30
   (C) 36
   (D) 54

7. If the minute hand of a properly functioning clock has just moved 45 degrees, how many minutes passed while it did so?

   (A) 6
   (B) 7.5
   (C) 12.5
   (D) 27

8. What is the area of the figure below?

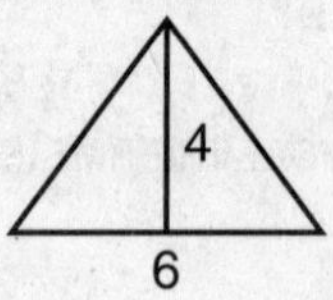

   (A) 4
   (B) 6
   (C) 12
   (D) 24

9. In the figure below, if line p is parallel to line q, what is the value of y?

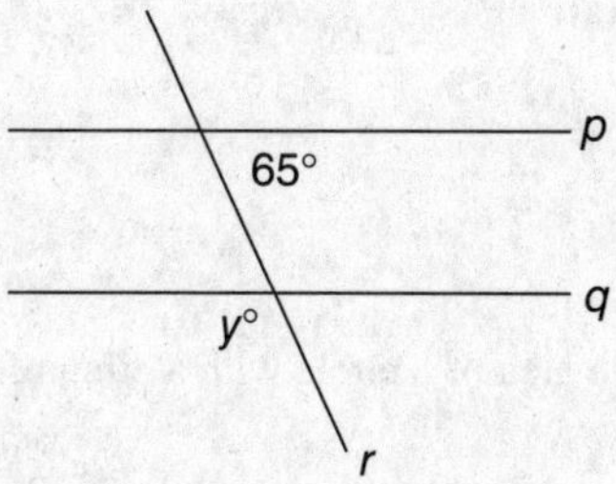

(A) 65
(B) 115
(C) 125
(D) 130

10. 15% of 15% of 200 is

(A) 4.5
(B) 6
(C) 12.5
(D) 45

11. What is the average (arithmetic mean) of $\frac{1}{20}$ and $\frac{1}{30}$?

(A) $\frac{1}{25}$
(B) $\frac{1}{24}$
(C) $\frac{2}{25}$
(D) $\frac{1}{12}$

12. If an angle measures $y°$, what will its supplement measure in terms of y?

(A) $(90 - y)°$
(B) $(90 + y)°$
(C) $(180 - y)°$
(D) $(180 + y)°$

13. Diane painted $\frac{1}{3}$ of her room with $2\frac{1}{2}$ cans of paint. How many more cans of paint will she need to finish painting her room?

(A) $2\frac{1}{2}$
(B) 5
(C) $7\frac{1}{2}$
(D) 10

14. If $13 + a = 25 + b$, then $b - a =$

(A) 12
(B) 8
(C) −8
(D) −12

15. For all x, $3x^2 \times 5x^3 =$

(A) $8x^5$
(B) $8x^6$
(C) $15x^5$
(D) $15x^6$

16. If the sides of a square increase in length by 10%, the area of the square increases by

(A) 10%
(B) 15%
(C) 20%
(D) 21%

17. If $\frac{m}{2} = 15$, then $\frac{m}{3} =$

(A) 7.5
(B) 10
(C) 22.5
(D) 30

18. Which of the following is a factor of $6x^2 - 13x + 6$?

(A) $2x + 2$
(B) $2x + 3$
(C) $3x - 2$
(D) $3x - 3$

19. If the product of 3 and x is equal to 2 less than y, which of the following must be true?

(A) $\frac{3x - 2}{y} = 0$
(B) $3x - y - 2 = 0$
(C) $3x + y - 2 = 0$
(D) $3x - y + 2 = 0$

20. If $x > 1$ and $\frac{a}{b} = 1 - \frac{1}{x}$, then $\frac{b}{a} =$

(A) $\frac{x}{x-1}$

(B) $\frac{x-1}{x}$

(C) $x - 1$

(D) $\frac{1}{x} - 1$

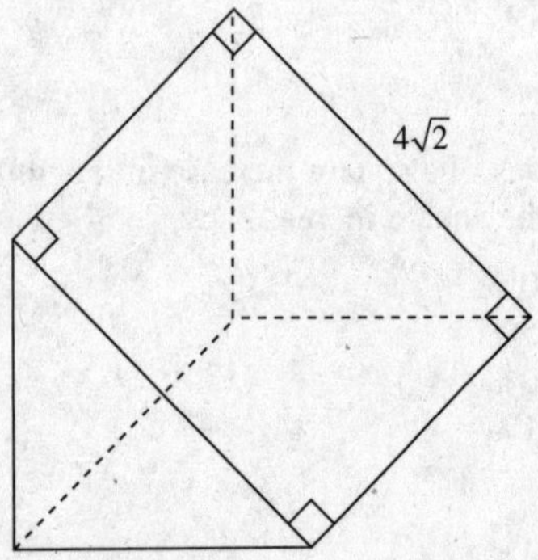

21. If the solid above is half of a cube, then the volume of the solid is

(A) 16
(B) $16\sqrt{2}$
(C) 32
(D) $32\sqrt{2}$

22. When 7.6 is divided by 0.019, the quotient is

(A) 4,000
(B) 400
(C) 40
(D) 4

23. What is the area of a circle with a circumference of 8?

(A) $\frac{4}{\pi}$

(B) $\frac{16}{\pi}$

(C) 4π

(D) 16π

24. 36% of 18 is 18% of what number?

(A) 9
(B) 36
(C) 72
(D) 200

25. $\sqrt{104{,}906}$ is between which two numbers?

(A) 100 and 200
(B) 200 and 300
(C) 300 and 400
(D) 400 and 500

STOP. IF YOU FINISH BEFORE THE TIME IS UP, YOU MAY CHECK OVER YOUR WORK ON THIS PART ONLY.

Part 6: Electronics Information (EI)

Time: 9 minutes; 20 questions

Directions: In this section, you will be tested on your knowledge of electronics basics. For each question, select the best answer and mark the corresponding oval on your answer sheet.

1. Resistance is measured in

(A) amperes
(B) ohms
(C) V
(D) W

2. Protons have a _______ charge, and neutrons are _______ charged.

(A) positive, negatively
(B) positive, neutrally
(C) negative, positively
(D) neutral, negatively

3. What is the term for a current flowing in a conductor which changes direction (moves back and forth) many times in a second?

(A) alternating current
(B) direct current
(C) electromotive force
(D) variable current

4. What type of circuit does this symbol represent?

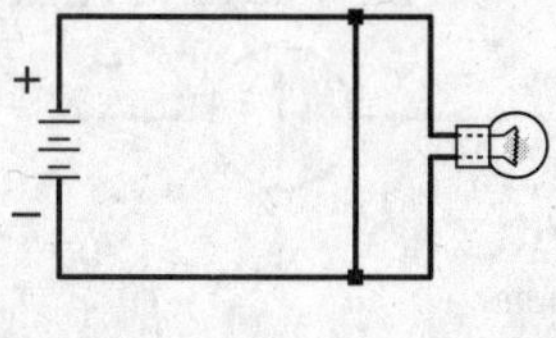

(A) parallel circuit
(B) series circuit
(C) series-parallel circuit
(D) short circuit

5. In the formula for Ohm's law, $V = I \times R$, the I represents

(A) voltage
(B) current
(C) resistance
(D) impedance

6. The P-type material in a diode is also known as the

(A) electrode
(B) cathode
(C) base
(D) anode

7. Introducing impurities into the crystal structure of silicon is also known as

(A) tinning
(B) enhancing
(C) processing
(D) doping

8. Which of the following causes the least resistance to a high frequency alternating current?

(A) capacitor
(B) fixed resistor
(C) variable resistor
(D) inductor

9. A circuit contains four resistors of 2 Ω, 4 Ω, 6 Ω, and 8 Ω in series. What is their effective resistance?

(A) 2 Ω
(B) 5 Ω
(C) 8 Ω
(D) 20 Ω

10. All of the following statements about transistors are true EXCEPT

(A) N-type material has free electrons
(B) NPN transistors require a positive voltage at the base to turn them on
(C) the arrow in the transistor symbol is always placed on the collector
(D) P-type material conducts current through the movement of holes

11. This is the symbol for which type of meter?

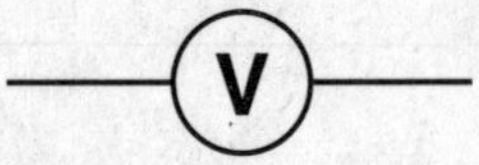

(A) voltmeter
(B) ammeter
(C) ohmmeter
(D) galvanometer

12. Which of the following is NOT an essential component of an electrical circuit?

(A) voltage source
(B) watt
(C) load
(D) conductor

13. What type of switch would be utilized by a doorbell?

(A) 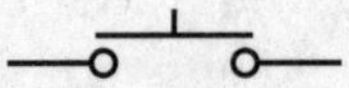

(B) 

(C) 

(D) 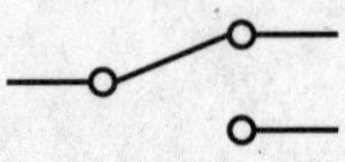

14. Which of the following is the symbol for an AC voltage source?

(A) 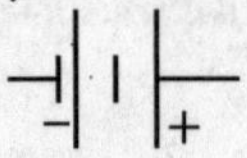

(B) 

(C) 

(D) 

15. Capacitors are also known as

(A) capacitances
(B) dielectrics
(C) condensers
(D) inductors

16. A material that will not conduct electricity is known as

(A) a conductor
(B) a semiconductor
(C) an insulator
(D) a voltage source

17. Which of the following materials is NOT a conductor?

(A) rubber
(B) silver
(C) copper
(D) aluminum

18. A circuit with four 6-ohm resistors wired in series has 12 volts applied to it. What current flows through this circuit?

(A) 5 mA
(B) 50 mA
(C) 500 mA
(D) 5 A

19. An element that is widely recognized as a semiconductor is

(A) silicon
(B) aluminum
(C) helium
(D) wood

20. This is the symbol for which type of meter?

(A) voltmeter
(B) ammeter
(C) ohmmeter
(D) galvanometer

STOP. IF YOU FINISH BEFORE THE TIME IS UP, YOU MAY CHECK OVER YOUR WORK ON THIS PART ONLY.

Part 7: Auto and Shop Information (AS)

Time: 11 minutes; 25 questions

Directions: In this section, you will be tested on your knowledge of automotive and shop basics. For each question, select the best answer and mark the corresponding oval on your answer sheet.

1. Wood can be shaped using all of the following tools EXCEPT

   (A) jack plane
   (B) wood chisel
   (C) flat rasp
   (D) miter box

2. Ride quality is the job of the vehicle's __________ system.

   (A) steering
   (B) suspension
   (C) charging
   (D) brake

3. The cut in the wood left behind by a saw is known as the

   (A) gap
   (B) line
   (C) trace
   (D) kerf

4. A four-wheel drive vehicle uses a __________ to distribute power to the front and rear drive axles.

   (A) transmission
   (B) drive shaft
   (C) torque converter
   (D) transfer case

5. What type of cut is this saw made for?

   (A) a cut with the grain of the wood
   (B) a cut across the grain of the wood
   (C) a cut made in plywood only
   (D) a cut made in steel only

6. This image depicts what stroke in the four-stroke cycle?

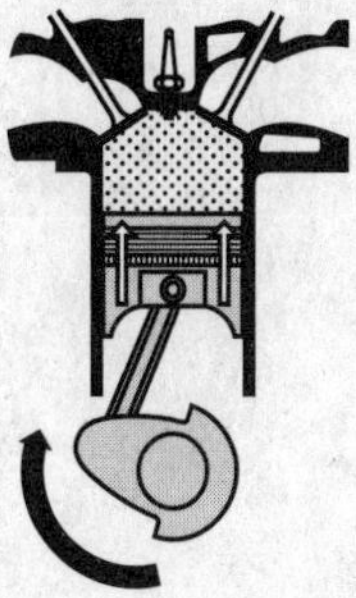

   (A) intake stroke
   (B) power stroke
   (C) compression stroke
   (D) exhaust stroke

7. Thread pitch of a fractional measurement fastener is determined by

   (A) counting the number of threads per inch
   (B) measuring the distance in millimeters between threads
   (C) measuring the distance in inches between threads
   (D) counting the number of threads per millimeter

8. Which would be best to use to loosen tight fasteners?

   (A) only this tool: 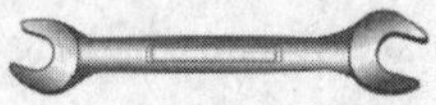

   (B) only this tool: 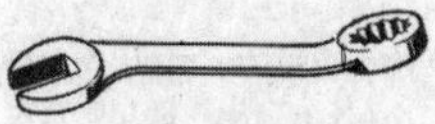

   (C) only this tool: 

   (D) the tools shown in both (B) and (C)

9. All of the following are parts of the vehicle's steering system EXCEPT

   (A) coil spring
   (B) steering knuckle
   (C) tie rod end
   (D) idler arm

10. Which engine component is responsible for the opening and closing of the engine's intake and exhaust valves?

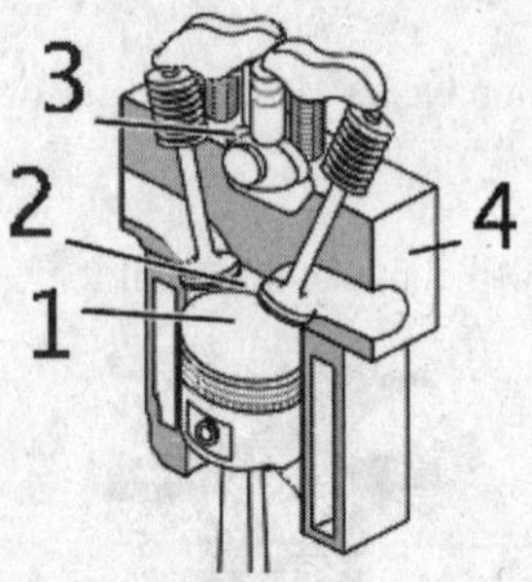

(A) 1
(B) 2
(C) 3
(D) 4

11. On a fuel-injected engine, the fuel pump is normally located

(A) on the engine
(B) on the vehicle frame
(C) near the rear bumper
(D) in the fuel tank

12. UNC fasteners have __________ threads per inch when compared to UNF fasteners of the same diameter.

(A) the same number of
(B) fewer
(C) twice as many
(D) thrice as many

13. In power brake systems, braking pressures are increased using

(A) only the engine intake manifold vacuum
(B) only the fluid pressure from the power steering system
(C) both the engine intake manifold vacuum and fluid pressure from the power steering system
(D) neither the engine intake manifold vacuum nor the fluid pressure from the power steering system

14. All of the following are parts of the ignition primary system EXCEPT the

(A) battery
(B) ignition switch
(C) reluctor and pickup
(D) rotor

15. "Stoichiometric" is a term that describes

(A) the ideal air-fuel mixture in a gasoline engine
(B) the voltage required to generate a spark
(C) the engine's breathing efficiency
(D) theoretical horsepower

16. An oil's resistance to flow is known as its

(A) quality rating
(B) pumpability
(C) lubricity
(D) viscosity

17. All cars are currently being built with __________ electrical systems.

(A) only direct current
(B) only alternating current
(C) both direct and alternating current
(D) neither direct nor alternating current

18. When arc-welding, the ground clamp should be connected to the

(A) floor
(B) welding rod
(C) electrical outlet
(D) work

19. What cylinder design is referred to as a "boxer" engine?

(A) inline only
(B) flat only
(C) horizontally opposed only
(D) both flat and horizontally opposed

20. What are these tools used for?

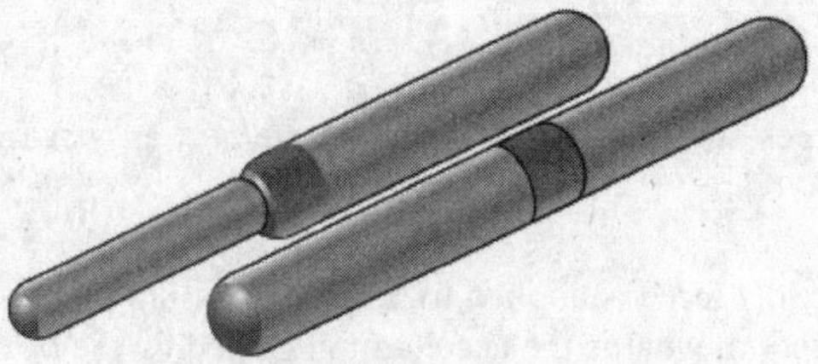

(A) securing a rivet in place where a tight rivet is important
(B) striking an object where it is important that the hammer itself not come in direct contact with the work
(C) driving small fasteners and making layout marks in metalwork
(D) guiding a saw when making a cutting across the grain

21. In a V-type engine with two cylinder heads, which of these designs would have two camshafts, with one installed above each cylinder head?

(A) single overhead cam or SOHC
(B) double overhead cam or DOHC
(C) triple overhead cam or TOHC
(D) overhead value or OHV

22. Air entering an engine's combustion chamber must pass through each of the following EXCEPT

(A) the air filter
(B) the intake manifold
(C) the intake port in the cylinder head
(D) the catalytic converter

23. Which tool would be best suited for loosening a very tight nut?

(A) open-end wrench
(B) 12-point socket and breaker bar
(C) 6-point socket and ratchet
(D) 6-point socket and breaker bar

24. What type of retaining ring is depicted here?

(A) internal prong-type
(B) internal hole-type
(C) external "E"-type
(D) external hole-type

25. As a car's battery discharges its energy, which of the following takes place in the battery's lead plates?

(A) They release water.
(B) They are converted to lead sulfate.
(C) They discharge impurities.
(D) They are converted to electrolytes.

STOP. IF YOU FINISH BEFORE THE TIME IS UP, YOU MAY CHECK OVER YOUR WORK ON THIS PART ONLY.

Part 8: Mechanical Comprehension (MC)

Time: 19 minutes; 25 questions

Directions: In this section, you will be tested on your knowledge of mechanics and basic physics. Select the best answer for each question and mark the corresponding oval on your answer sheet.

1. A car rolls down a hill. It travels progressively faster because
   (A) its potential energy is being converted into kinetic energy
   (B) its kinetic energy is being converted to potential energy
   (C) its energy is maximized
   (D) its energy is minimized

2. When a ball is thrown into the air, it has its maximum potential energy when
   (A) it leaves the hand of the person throwing it
   (B) it enters the hand of the person catching it
   (C) its velocity is zero
   (D) its velocity is greatest

3. Objects that require more force to get them moving have greater
   (A) momentum
   (B) inertia
   (C) torque
   (D) pressure

4. Isaac Newton discovered that every particle in the universe is attracted to every other particle. This law is called Newton's
   (A) first law of motion
   (B) third law of motion
   (C) law of universal gravitation
   (D) second law of motion

5. A 30N force is applied to 5 kg box. Assuming no friction, what is the acceleration gained by the box?
   (A) 0.15 m/s^2
   (B) 0.6 m/s^2
   (C) 1.5 m/s^2
   (D) 6 m/s^2

6. Compared to lifting a heavy box directly from ground level up to the top of a ramp, pushing the box up the ramp (neglecting friction) is doing:
   (A) less work
   (B) more work
   (C) the same amount of work
   (D) no work

7. All of the following statements about work are true EXCEPT
   (A) force is a factor of work
   (B) it is possible to get more work out of a machine than what is put in
   (C) to do work, an object must be moved
   (D) work can be measured in foot-pounds

8. A chisel is illustrated below. A chisel is an example of a(n)

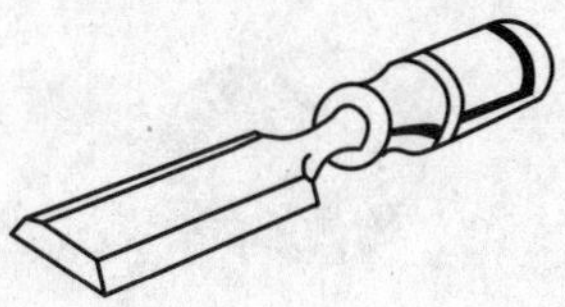

(A) wedge
(B) first-class lever
(C) inclined plane
(D) wheel and axle

9. The difference between a wedge and an inclined plane is

(A) a wedge has a steeper slope than an inclined plane
(B) a wedge is made to move, whereas an inclined plane stays stationary
(C) a wedge stays stationary, whereas an inclined plane is made to move
(D) a wedge has less friction

10. Which post holds up the greater part of the load?

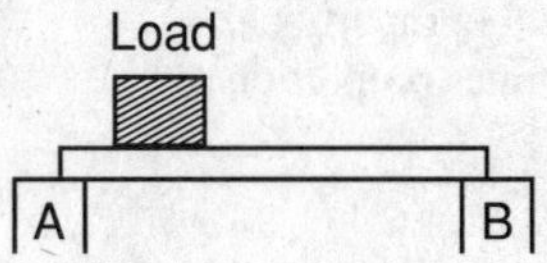

(A) post A
(B) post B
(C) both hold up the load equally
(D) answer cannot be determined based on information provided

11. Work and kinetic energy are both measured in

(A) kilograms
(B) meters per second
(C) joules
(D) newtons

12. The design of a screwdriver is based on which of the following simple machines?

(A) inclined plane
(B) wheel and axle
(C) first-class lever
(D) pulley

13. An increase in force accomplished through mechanical advantage is always accompanied by _______ in distance moved.

(A) an increase
(B) a decrease
(C) no change
(D) an increase at times and a decrease at other times

14. The design of a prybar, as illustrated, is based on which of the following simple machines?

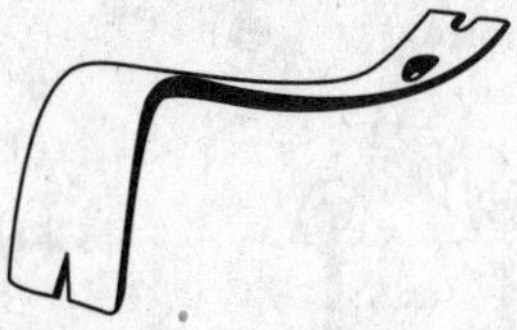

(A) first-class lever
(B) second-class lever
(C) third-class lever
(D) inclined plane

15. One horsepower is equal to how many watts?

(A) 550
(B) 746
(C) 10,000
(D) 100

16. In order to increase force with a first-class lever, the fulcrum should be

(A) moved toward where the effort is applied
(B) kept in the same position
(C) moved toward where the object is moved
(D) removed altogether

17. Which of the following statements is correct?

(A) The work put out by a simple machine can sometimes exceed the work put in.
(B) The work put out by a simple machine always exceeds the work put in.
(C) The work put out by a simple machine can never exceed the work put in.
(D) The work put out by a simple machine is equal to and opposite of the work put in.

18. If a small gear drives a large gear, a ________ takes place.

(A) mechanical advantage
(B) block and tackle
(C) torque multiplication
(D) reduction in rotational speed

19. A block and tackle, as shown below, utilizes

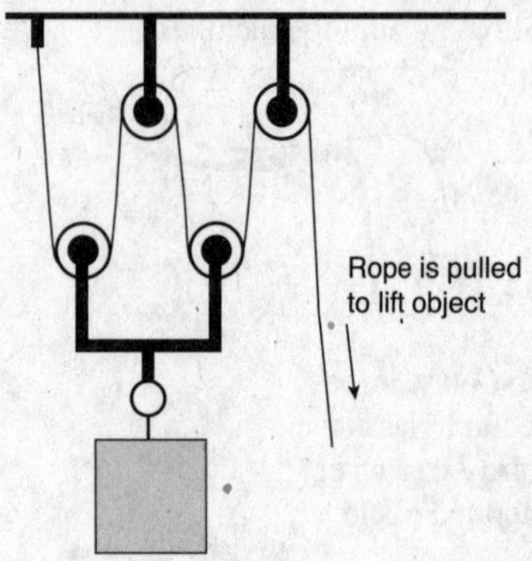

(A) a single pulley
(B) multiple pulleys
(C) only moveable pulleys
(D) only fixed pulleys

20. The force required by a person pulling on a rope using a single pulley to lift an object is equal to

(A) the weight of the object
(B) twice the weight of the object
(C) one half of the weight of the object
(D) the weight of the object divided by the diameter of the pulley

21. The design of a doorknob is based on which of the following simple machines?

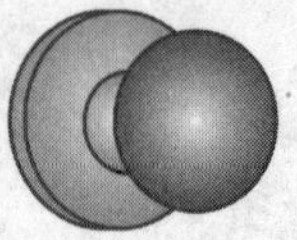

(A) inclined plane
(B) wheel and axle
(C) first-class lever
(D) pulley

22. An astronaut's mass is 30 kg on Earth. What is his mass on the Moon?

(A) the same
(B) lower
(C) higher
(D) cannot be determined

23. In a gear train, torque and speed can be described as

(A) exactly the same
(B) directly proportional
(C) inversely proportional
(D) sometimes proportional

24. An object's potential energy can be calculated using the formula

(A) $PE = \frac{1}{2}m \times v^2$
(B) $PE = m \times v$
(C) $PE = m \times g \times h$
(D) $PE = m \times a$

25. A child applies a 10 N force to a large box. The box however, does not move. How much work has the child done?

(A) 0 J
(B) 5 J
(C) 10 J
(D) 100 J

STOP. IF YOU FINISH BEFORE THE TIME IS UP, YOU MAY CHECK OVER YOUR WORK ON THIS PART ONLY.

Part 9: Assembling Objects (AO)

Time: 15 minutes; 25 questions

Directions: In this section, you will be tested on your ability to determine how an object will look when its parts are put together. For each question, select the best answer and mark the corresponding oval on your answer sheet.

1.

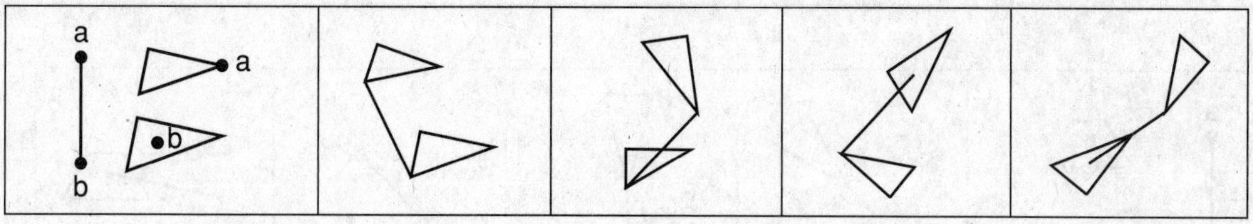

A B C D

2.

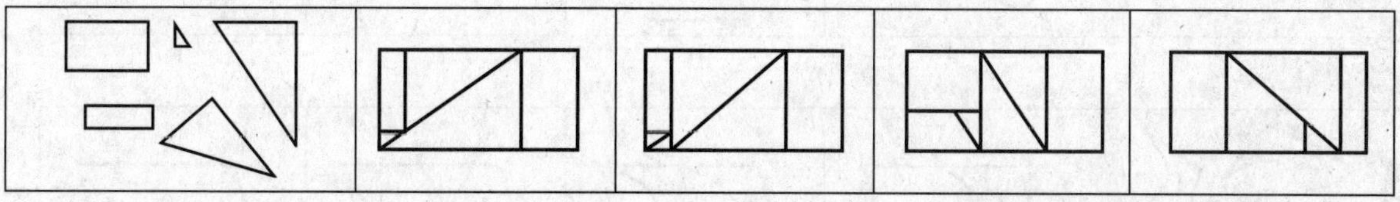

A B C D

3.

A B C D

4.

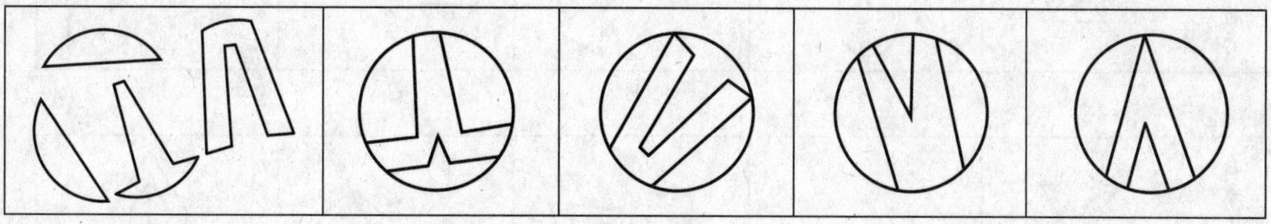

A B C D

5.

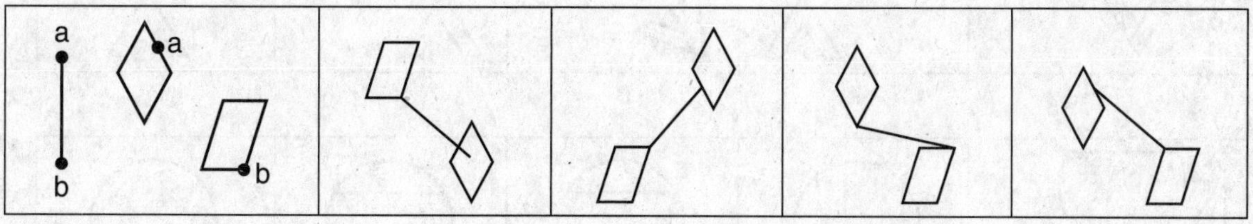

A B C D

6.
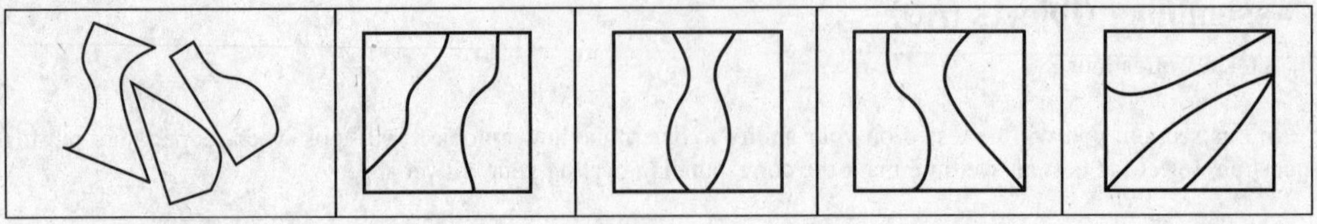

A B C D

7.
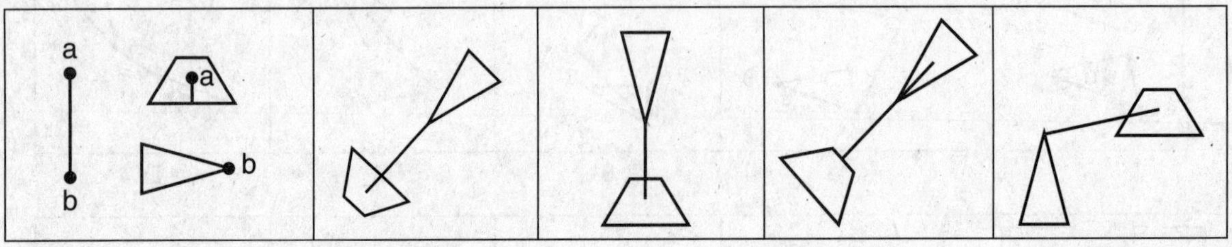

A B C D

8.
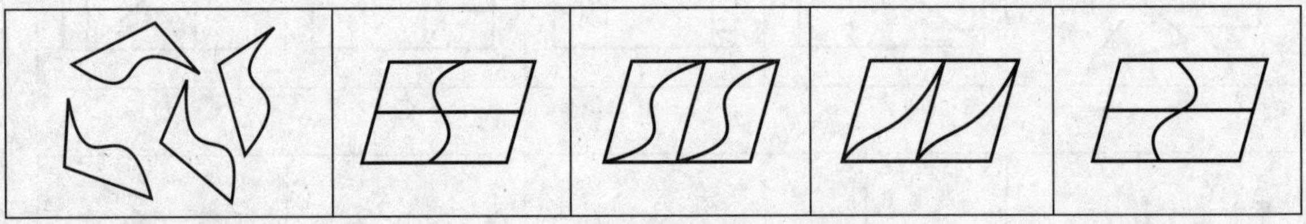

A B C D

9.
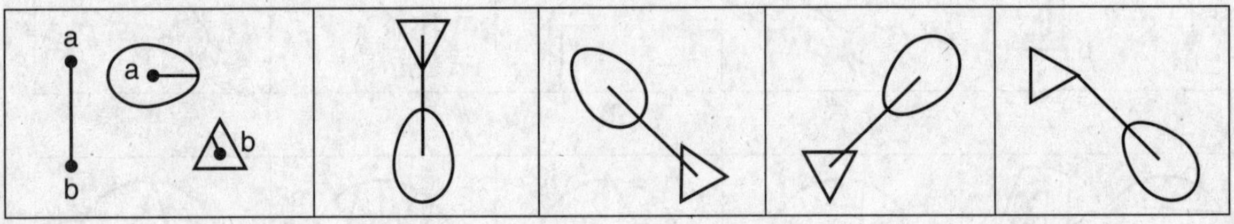

A B C D

10.

A B C D

11.
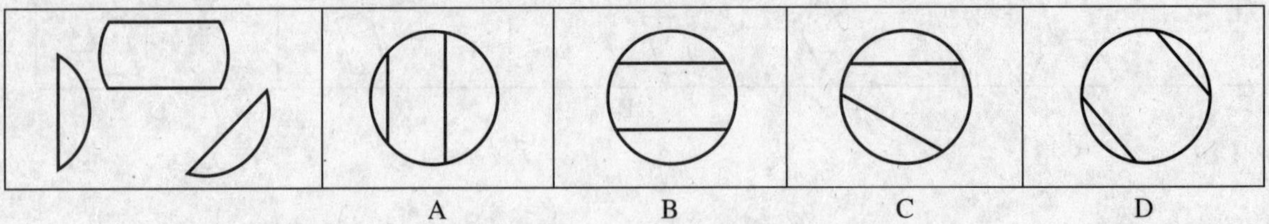

A B C D

12.

A B C D

13.

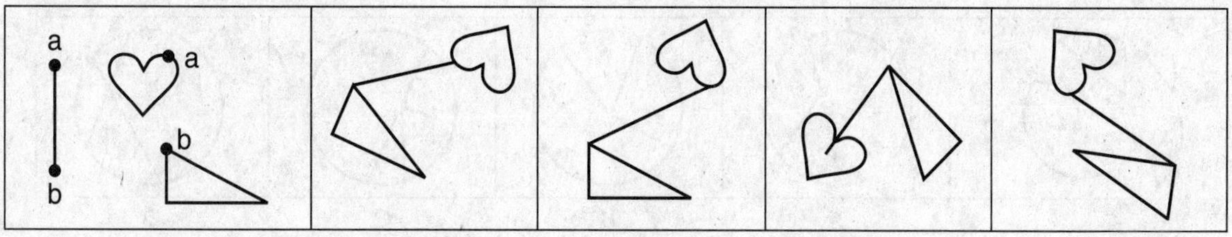

A B C D

14.

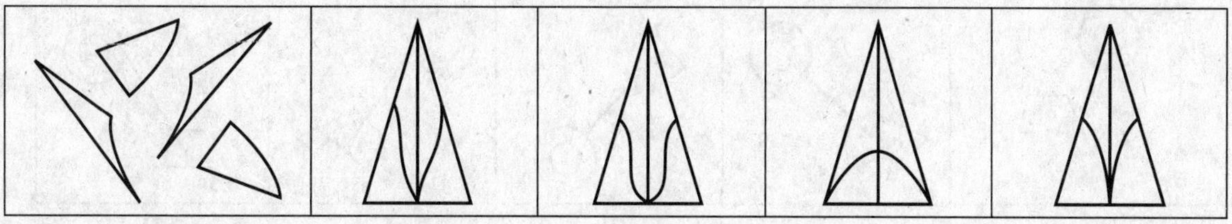

A B C D

15.

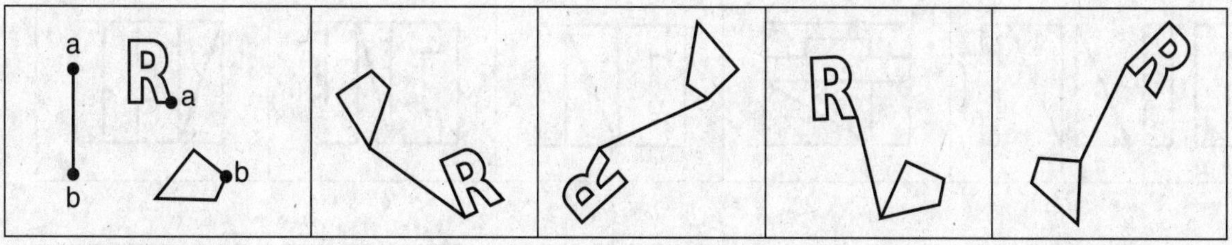

A B C D

16.

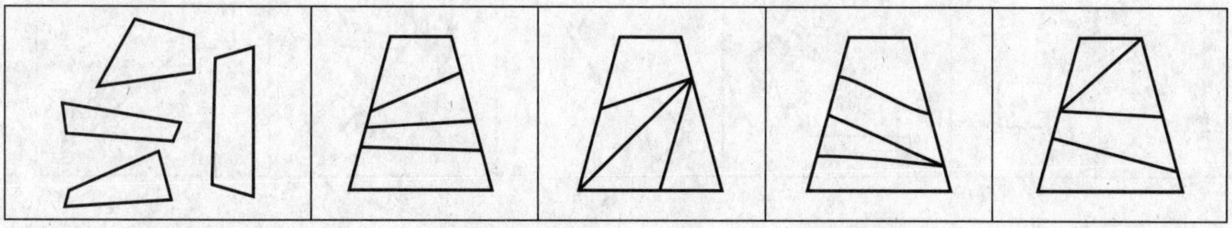

A B C D

17.

A B C D

18.

A B C D

19.

A B C D

20.

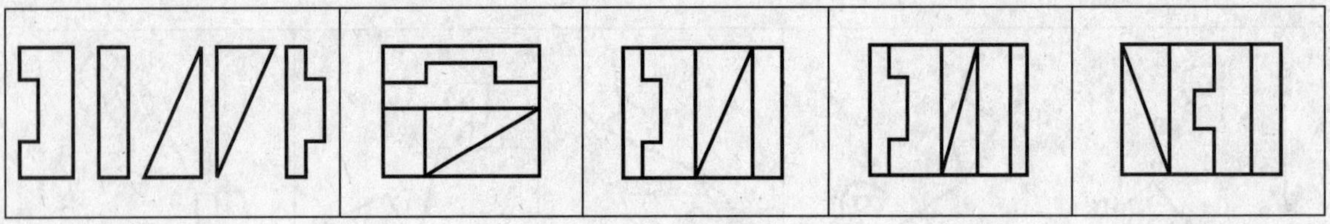

A B C D

21.

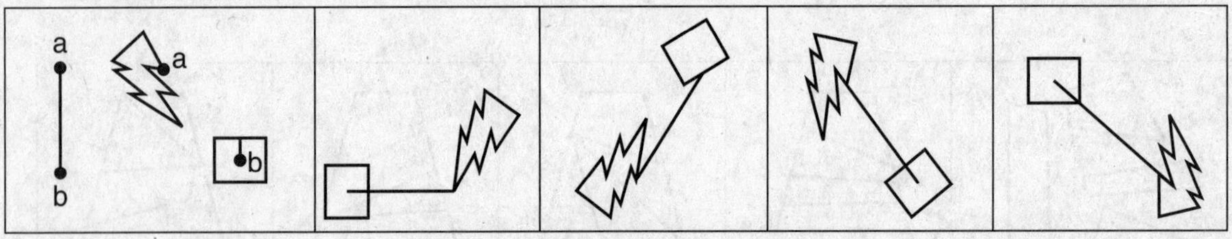

A B C D

22. 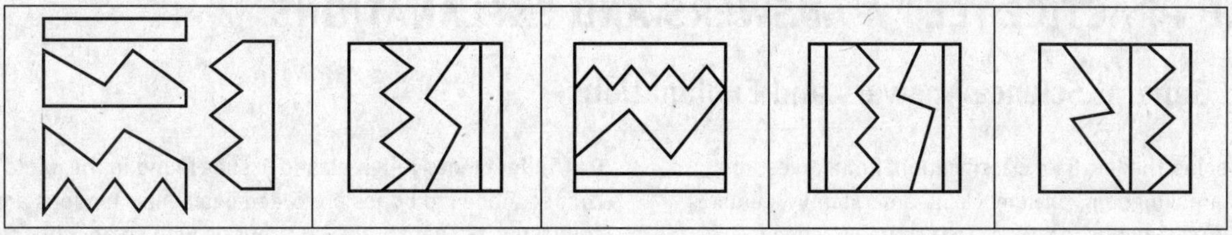

A B C D

23. 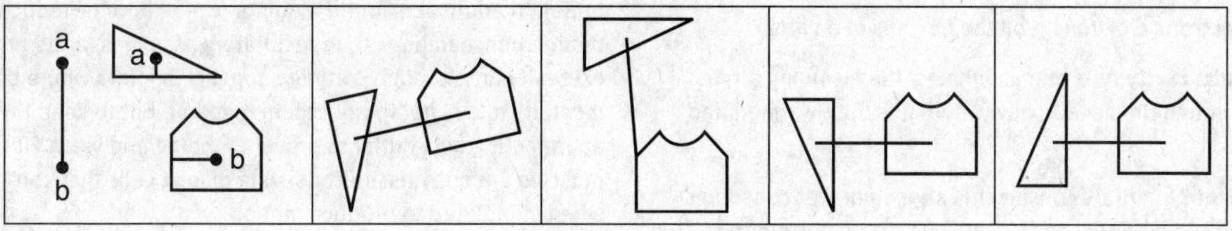

A B C D

24. 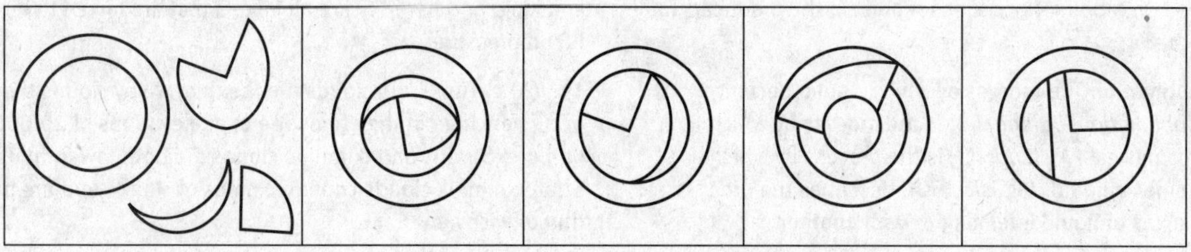

A B C D

25. 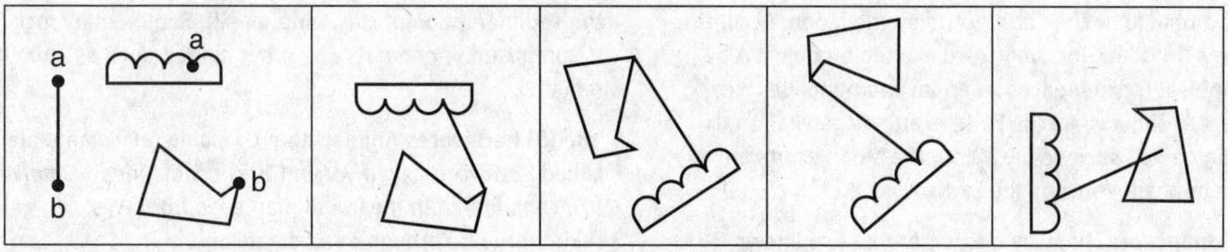

A B C D

STOP. IF YOU FINISH BEFORE THE TIME IS UP, YOU MAY CHECK OVER YOUR WORK ON THIS PART ONLY.

CONGRATULATIONS! YOU HAVE COMPLETED ASVAB PRACTICE TEST A. CHECK YOUR WORK USING THE EXPLANATIONS THAT BEGIN ON THE NEXT PAGE.

ASVAB PRACTICE TEST A ANSWERS AND EXPLANATIONS

Part 1: General Science Answers and Explanations

1. **(A) species** The levels of classification, from largest to smallest, are: kingdom, phylum, class, order, family, genus, species. Thus, a genus is a group that contains related species.

2. **(B) electrons** Electrons orbit the nucleus of an atom.

3. **(D) nasal cavity** As a human inhales, the incoming air first passes through the nasal cavity, where it is both warmed and filtered.

4. **(D) coyote** A tertiary consumer is an animal that consumes secondary consumers, or other carnivores. Coyotes are the tertiary consumer in this food chain. They feed on foxes, which in turn feed on rabbits. Dandelions represent the producers in the food chain.

5. **(B) sublimation** The process by which a solid becomes a gas is *sublimation*. *Boiling*, (A), is the process by which a liquid becomes gas. *Melting*, (C), is the process by which a solid becomes a liquid. *Diffusion*, (D), describes the process by which a gas or liquid intermingles with another.

6. **(C) the same as that exerted by the Earth on the Moon** The gravitational force the Moon exerts on the Earth is the same as the gravitational force the Earth exerts on the Moon. According to Newton's third law, for every force exerted on object A by object B, object B experiences an equal and opposite force from object A. Thus, when the Earth exerts its gravitational force on an object, such as the Moon, the Moon exerts an equal and opposite force on the Earth.

7. **(B) decomposers** Fungi are *decomposers* (also known as saprophytes), returning the organic material from dead matter such as leaves, trees, and animal remains back into the environment. *Producers*, (A), make their own food. *Consumers*, (C), must eat other organisms in order to gain energy. *Mutualism*, (D), is a type of symbiosis whereby both organisms benefit.

8. **(D) 273 K** The freezing point of water on the Kelvin scale occurs at 273 K. Kelvin and Celsius share a similar scale (i.e., a change of one Kelvin is the same as a change of one degree Celsius). However, they differ by 273 units. To convert Celsius to Kelvin, add 273. To convert Kelvin to Celsius, subtract 273. Since water freezes at 0 degrees Celsius, this is equivalent to $0 + 273 = 273$ K.

9. **(A) ligaments** Tough elastic tissues found in the joints that connect bones to bones are called ligaments. Tendons are connective tissues that unite a muscle with some other part, such as a bone. Cartilage is a somewhat elastic tissue (unlike bone). In adults it is found in joints (where it helps reduce friction between bones), in respiratory passages, and in the external ear. Although cartilage appears in joints where bones meet, its role is not to hold them together, but to keep them apart, acting as a buffer to prevent rubbing and wear. Finally, muscles are body tissue consisting of long cells that contract when stimulated to produce motion.

10. **(B) high blood pressure** Stress, a poor diet, cigarette smoking, and heredity are all factors that are linked to high blood pressure.

11. **(A) cirrus** Cirrus clouds are the thin, wispy clouds that occur very high in the atmosphere, at elevations of 20,000 feet or more. Stratus and cumulus clouds are both low-forming. Stratocumulus clouds contain clumps of clouds and are not thin or wispy.

12. **(C) sedimentary** Fossils are most likely to be found in sedimentary rock such as shale, which is rock formed by the sedimentation of clay, mud, or silt. Sedimentary rock is composed of minerals and other matter, such as animal remains.

13. **(B) herbivores** Animals that consume only plants are called *herbivores*. *Carnivores*, (C), consume meat. *Omnivores*, (D), consume both meat and plants. *Saprophytes*, (A), gain their nutrients through dead organisms.

14. **(D) yellow dwarf** The Sun is a fairly young star and therefore has lower temperature and luminosity, resulting in its yellow color. The Sun is also fairly small as stars go and falls under the "dwarf" category.

15. **(D) population** A population is a group of organisms of the same species living in the same region. Biomes, ecosystems, and communities all encompass more than one type of species.

16. **(D) Earth** The four planets beyond Mars—Jupiter, Saturn, Uranus, and Neptune, are referred to as the outer planets. Earth is one of the four planets closest to the Sun and is thus considered an inner planet.

17. **(B) potential to kinetic** A boulder that begins to roll down a hill is an example of an energy conversion from potential energy to kinetic energy. The boulder has maximum potential energy at the top of the hill, and this energy is converted into kinetic energy as the boulder rolls down the hill and picks up speed. At the foot of the hill, all potential energy has been converted into kinetic energy.

18. **(D) troposphere** The lowest layer of the Earth's atmosphere is called the troposphere. This layer extends from the surface to the bottom of the stratosphere. Most weather changes occur here and the temperature generally decreases rapidly with altitude.

19. **(B) energy can neither be created nor destroyed** According to the law of conservation of energy, energy can neither be created nor destroyed. Instead, it changes from one form to another.

20. **(C) the Doppler effect** As an ambulance approaches, the sound waves from its siren are compressed toward the observer. The intervals between waves diminish, which translates into an increase in frequency or pitch. As the ambulance recedes, the sound waves are stretched relative to the observer, causing the siren's pitch to decrease. By the change in pitch of the siren, one can determine if the ambulance is coming nearer or speeding away. This perceived change in pitch is known as the Doppler effect.

21. **(D) diverging lens** A concave lens—one that is thicker on the edges than it is in the middle—is also known as a diverging lens because it diverges the light waves that pass through it.

22. **(D) molecule** A *molecule* is the smallest particle of an element or compound that can exist in a free state and still retain the characteristics of the element or compound. In elements whose atoms form covalent bonds, the molecules consist of two or more similar atoms (e.g., N_2). (Note that atoms of some elements form metallic bonds, and these groups of atoms are not molecules; for example, pure copper (Cu) is made up of atoms that do not form molecules.) The molecules of covalent compounds consist of two or more different atoms (e.g., H_2O). *Protons* and *quarks* are subatomic particles; *atoms* are the building blocks of molecules.

23. **(C) the orbit of the Moon around the Earth** The orbit of the Moon around the Earth, which exerts a gravitational pull on the ocean's waters, is the phenomenon most responsible for oceanic tides.

24. **(B) group** The elements in a column on the Periodic Table are referred to as a group; each member of a group has the same number of electrons in its outer shell. Each row on the Periodic Table is referred to as a period.

25. **(A) reactants** When molecules undergo a chemical change, the process is called a chemical reaction. The molecules that enter the reaction are called reactants and the molecules that result from the reaction are called products.

Part 2: Arithmetic Reasoning Answers and Explanations

1. **(D) 81** The ratios of apples to pears and pears to oranges for a given cost are provided. What's needed to solve this problem, however, is the ratio of apples to oranges. To merge these into one ratio relating all three fruits, you need to express both ratios using the same number of pears. In one ratio it's 4, and in the other it's 3. Since the least common multiple of 3 and 4 is 12, you can multiply both individual ratios so that the number of pears comes to equal 12:

apples:pears = 3:4 = 9:12 (multiply both sides of the ratio by 3)
pears:oranges = 3:2 = 12:8 (multiply both sides of the ratio by 4)
Now you can combine the ratios:
apples:pears:oranges
9:12:8
Now you can also state the ratio of apples to oranges:
apples:oranges = 9:8
Given a value for the number of oranges as 72, you can now set up a proportion:

$\frac{\text{apples}}{\text{oranges}} = \frac{9}{8} = \frac{72}{x}$

Cross-multiply and solve for x:

$\frac{9}{8} = \frac{x}{72}$. Cross-multiply, getting $8x = (9)(72)$. Divide both sides by 8 to get $x = \frac{(9)(72)}{(8)}$. To make the arithmetic easier, divide 72 by 8 first, which equals 9, then multiply this answer by 9, getting $9 \times 9 = 81$.

2. **(C) 20%** If 48 of 60 seats are occupied, then 12 of 60 seats are unoccupied (since $60 - 48 = 12$). What percent of the seats are unoccupied?

$\text{Percent} = \frac{\text{Part}}{\text{Whole}} \times 100\% = \frac{12}{60} \times 100\% =$

$\frac{1}{5} \times 100\% = 20\%$, choice (C).

3. **(B) $5,000** Break this question down into pieces. Before midnight, Effin gets paid $1,000 an hour. If she starts singing at 8:30 p.m., then she sings for $3\frac{1}{2}$ hours before midnight. Her earnings for that time are $3\frac{1}{2} \times \$1{,}000 = \$3{,}500$. After midnight, her rate increases 50%, so her hourly rate after becomes $1,000 + $1,000(50%) = $1,000 + $500 = $1,500. She works one hour after midnight, earning $1,500. Add the amount she earns before midnight and the amount she earns after midnight, so her total pay is $3,500 + $1,500 = $5,000.

4. **(D) $\frac{3}{4}$** The probability of an event occurring is a fraction: the number of possible outcomes in which the event can occur divided by the total number of possible outcomes. Fran is going to pick one shirt and the question asks for the probability that it will not be orange. In other words, you need to find the probability of picking a black or blue shirt. There are 4 black shirts and 5 blue shirts, so altogether there are 9 possible successful outcomes. The total number of shirts to choose from, including the orange shirts, is 12, so there are 12 possible outcomes. The probability of Fran picking a non-orange shirt is $\frac{9}{12}$. Since $\frac{9}{12}$ is not available in the answer choices, reduce the fraction by dividing the top and bottom by 3 to get $\frac{3}{4}$.

5. **(B) 11:30 a.m.** Newark is two hours later than Denver, so if the time in Denver when Harry arrives is 2:30 p.m., then the time in Newark when he arrives is 4:30 p.m. The flight takes 5 hours, so the time he began in Newark is 5 hours earlier than 4:30 p.m., or 11:30 a.m.

6. **(B) 6** How many times does 24 go into 198? $198 \div 24 = 8$ with a remainder of 6, so Doris will have 6 pieces of chocolate left over.

7. **(C) 200** If 75% of x is 150, then $0.75(x) = 150$ or $\frac{3}{4}(x) = 150$. Multiply both sides by $\frac{4}{3}$ and $x = 200$. You could also backsolve. Only choice (C), 200, works. If $x = 200$, then $0.75(x) = 150$ since $0.75(200) = 150$.

8. **(A) 85** The formula for finding the average is:

$\text{Average} = \frac{\text{Sum of the terms}}{\text{Number of terms}}$

Don't just average the old average and the last test score—that would give the last score as much weight as all the other scores combined. The best way to deal with changing averages is to use the sums. Use the old average to figure out the total of the first 6 scores:

$$\text{Sum of first 6 scores} = (83)(6) = 498$$

Then add the 7th score and divide:

$\frac{498 + 97}{7} = \frac{595}{7} = 85.$

9. **(B) $21.60** If the charge for one pound is $25.00, then the charge for $\frac{4}{5}$ of a pound would be $\frac{4}{5} \times \$25.00 = \$20.00$ plus 8% sales tax. 8% of $20.00 = 0.08($20.00) = $1.60. $20.00 + $1.60 = $21.60.

10. **(D) 100** To answer this question, determine how many times $\frac{3}{5}$ goes into 60. It will be more than 60, so right away you can eliminate (A). $60 \div \frac{3}{5} = \frac{60}{1} \times \frac{5}{3}$. This is easier to multiply if you cancel first. 3 goes into 60 twenty times, so replace the 3 with a 1, and replace the 60 with a 20 to get $\frac{20}{1} \times \frac{5}{1} = 20 \times 5 = 100$.

11. **(B) $20** Set up equations based on the information given in the problem. The question asks for the amount of money Robert has.

Ed has 100 more dollars than Robert:

$E = R + 100.$

Ed then spends 20 dollars and has five times as much money as Robert:

$E - 20 = 5R.$

Solve a question like this using algebra if it's a strong area for you. Substitute $R + 100$ for E and solve:

$R + 100 - 20 = 5R$

$R + 80 = 5R$

$80 = 4R$

$R = \$20$

You could also backsolve, starting with choice (B) or (C) for Robert's money. If you started with (C) 24, you would plug 24 into both equations and see if you got the same value for E.

In the first equation:

$E = R + 100$

$E = 24 + 100$

$E = 124$

In the second equation:

$E - 20 = 5R$

$E - 20 = 5(24)$

$E - 20 = 120$

$E = 140$

The numbers don't match, so eliminate (C) and try (B) 20.

In the first equation:

$E = R + 100$

$E = 20 + 100$

$E = 120$

In the second equation:

$E - 20 = 5R$

$E - 20 = 5(20)$

$E - 20 = 120$

$E = 120$

The value for E is the same in both equations, which tells you that 20 is the correct answer.

12. **(D) 11** This question is actually a remainder question in disguise. If 587 people are to be divided among buses that each seat 48 people, divide 587 by 48 to see how many buses would be filled completely. The remainder is the number of people in the unfilled bus. $587 \div 48 = 12$ with a remainder of 11.

13. **(C) 5** Riley brought 100 cookies, and 25 are left over. That means that Riley and her classmates ate 75 cookies.

$$\text{Average} = \frac{\text{Sum of the terms}}{\text{Number of terms}} = \frac{75 \text{ cookies}}{15 \text{ students}}$$

$= 5$ cookies for each student.

14. **(C) $9\frac{1}{3}$** This is a proportion question:

$$\frac{4 \text{ ft object}}{9 \text{ ft shadow}} = \frac{x \text{ ft object}}{21 \text{ ft shadow}}$$

Cross-multiply to solve:

$4 \times 21 = 9x$

$\frac{84}{9} = \frac{9x}{9}$

$x = 9\frac{1}{3}$

15. **(D) 64%** *Percent* times *whole* equals *part*:

$\text{Percent} \times 25 = 16$

$\text{Percent} = \frac{16}{25} \times 100\% = 64\%$

16. **(A) 20** First find how long the trip takes him at each of the two different rates, using the formula

$$\text{Time} = \frac{\text{Distance}}{\text{Rate}}$$

He travels the first 10 km at 30 km per hour, so he takes $\frac{10}{30} = \frac{1}{3}$ hour $= 20$ minutes for this portion of the journey. He travels the remaining 30 km at 15 km per hour, so he takes $\frac{30}{15} = 2$ hours for this portion of the journey.

The whole journey takes him

$2 + \frac{1}{3} = 2\frac{1}{3}$ hours $= 2$ hours 20 minutes.

Now calculate the amount of time it would take to make the same trip at a constant rate of 20 km per hour. If he traveled the whole 40 km at 20 km per hour, it would take $\frac{40}{20} = 2$ hours. Now $2\frac{1}{3}$ hours is more than 2 hours by $\frac{1}{3}$ hour, or 20 minutes.

17. **(B) $\frac{1}{7}$** Since 3 out of 24 students are in student organizations, the remaining $24 - 3 = 21$ students are not in student organizations. Therefore, the ratio of students in organizations to students not in organizations is: $\frac{3}{21} = \frac{1}{7}$.

18. **(A) 51** By midseason, team A had won 60 and team B had won 49 games. Over the remaining 80 games, team A won 40 and lost 40. That means that team A wound up with $60 + 40 = 100$ wins. To win 100 games, team B would have to win $100 - 49 = 51$ games over the remainder of the season.

19. **(C) \$12.50** A vendor bought 10 crates for a total cost of \$80. He wants to make a total profit of 25% over the total cost. So the total sale would have to equal $\$80 + (0.25)(\$80) = \$80 + \$20 = \$100$. He lost 2 crates, so he would need to sell the remaining $10 - 2 = 8$ crates for \$100 to make the required profit. $\frac{\$100}{8 \text{ crates}} = \$12.50$ per crate.

20. **(D) 8** Set up a proportion on questions that give rates:

$$\frac{150 \text{ bottles}}{20 \text{ minutes}} = \frac{60 \text{ bottles}}{x \text{ minutes}}$$

Before you cross-multiply, reduce the first fraction as much as possible to make the calculations easier. Divide the numerator and denominator by 10:

$$\frac{15 \text{ bottles}}{2 \text{ minutes}} = \frac{60 \text{ bottles}}{x \text{ minutes}}$$

Cross-multiply:

$15x = 120$

Divide both sides by 15:

$x = 8$

21. **(C) –75** To figure out whether the product is positive or negative, count the number of negatives given. If there are an odd number of negatives, the product is negative. That eliminates (A) and (B). Multiplying the two numbers gives: $15 \times (-5) = -75$.

22. **(B) 35** The ratio of baseball cards to baseball figurines is $\frac{5}{7}$. You are given the number of cards, so you can set up a proportion: $\frac{5 \text{ cards}}{7 \text{ figurines}} = \frac{25 \text{ cards}}{x \text{ figurines}}$. Cross-multiply to get $5x = 175$ and divide both sides by 5 so $x = 35$.

23. **(C) $1.50** Set up equations to represent the relationships given in the problem:

$2h + 1s = \$3.25$

$3h + 1s = \$4.50$

The difference between the two totals is the price of a single hot dog. Thus, a single hot dog costs: $\$4.50 - \$3.25 = \$1.25$.

Plug the price of a single hot dog into the first equation to get the price of a soda:

$$\begin{aligned} 2(\$1.25) + 1s &= \$3.25 \\ \$2.50 + 1s &= \$3.25 \\ s &= \$0.75 \end{aligned}$$

The question asks for the cost of two sodas:

$2 \times \$0.75 = \$1.50$. Always double-check to be sure you are solving for the value the question asks for. In this case, choice (A) is a trap answer because it is the price of one soda.

24. **(A) 48** When you are given two rates, you cannot simply take the average of the two. Thus, answer choice (B) 50 is a trap answer here. Because it takes more time to travel at a slower speed, the average for the entire trip will be closer to the slower speed. To solve this problem, pick a number for the distance of the entire trip. Since 40 and 60 are the rates, use the least common multiple, 120, for the number of miles. First, find the time spent traveling at each speed:

120 miles at 40 mph = 3 hours

120 miles at 60 mph = 2 hours

Total distance = 240 miles. Total time = 5 hours.

The average for the entire trip will be the total distance divided by the total time.

$$\text{Average speed} = \frac{\text{Total distance}}{\text{Total time}} = \frac{240 \text{ miles}}{5 \text{ hours}} = 48 \text{ mph.}$$

25. **(B) 121** Call the number of teachers n. Then the number of students is $10n$. The total number of teachers and students is $n + 10n = 11n$.

Since you can't have a fraction of a person, n must be an integer, which means the total number of teachers and students must be a multiple of 11. Scan the answer choices and locate (B), 11×11, as the correct answer. None of the other answer choices are multiples of 11.

26. **(C) 21** Use Backsolving for this one, beginning with choice (C). If the smaller number is 21 (and the larger number is 35), does the math from the question work?

The average of 21 and $35 = \frac{21 + 35}{2} = \frac{56}{2} = 28$.

The positive difference between 21 and 35 is 14. Twice the positive difference would be $2 \times 14 = 28$.

Thus, 21 is the smaller number and the correct answer.

You can also solve this question algebraically. Set the smaller number equal to y. Then, the average of the two numbers is: $\frac{35+y}{2}$. Twice the positive difference of the two numbers is: $2(35 - y)$. The average is equal to twice the positive difference, so $\frac{35+y}{2} = 2(35 - y)$. Solve:

$$\begin{aligned} \frac{35 + y}{2} &= 2(35 - y) \\ 35 + y &= 4(35 - y) \\ 35 + y &= 140 - 4y \\ 5y &= 105 \\ y &= 21 \end{aligned}$$

27. **(C) 9** If a question uses the word "approximately" you can approximate the answer. If each of seven runners is running a little less than 1.3 kilometers, the total distance run would be a little less than $7 \times 1.3 = 9.1$, which is approximately 9 kilometers.

28. **(C) 5** Set up a proportion:

$$\frac{12 \text{ photos}}{1 \text{ minute}} = \frac{60 \text{ photos}}{x \text{ minutes}}$$

Now cross-multiply and solve: $12x = 60$, so $x = \frac{60}{12} = 5$.

29. **(A) $1,306.76** This is a pure arithmetic computation question. First, add up all of the deductions:

$218.99 + 107.05 + 25.03 + 68.65 + 42.75 = 462.47$

Now subtract total deductions from gross pay:

$1{,}769.23 - 462.47 = 1{,}306.76$.

30. **(C) 87** This is a standard average problem. Take the sum of all the terms and divide by the number of terms. $94 + 72 + 84 + 98 = 348$. $348 \div 4 = 87$. If dividing 348 by 4 was time consuming for you, you could have used Backsolving. Once you found the sum was 348, you could multiply the answer choices by 4, starting with (B) or (C), to see which answer choice gives 348. $87 \times 4 = 348$.

Part 3: Word Knowledge Answers and Explanations

1. **(B) force** *Impose* as a verb means to "establish by authority." Of the answer choices given, only one implies one person dictating to another. *To force* as a verb can be construed as "to establish by authority." Thus, answer choice (B) is correct. Remembered real-life context might be helpful. If you have ever heard someone described as "an imposing figure," you may remember that person as an authoritative or forceful individual.

2. **(A) halted** *Stunted* means "stopped short or canceled abruptly." Answer choice (A) *halted* also means "stopped short." Remembered real-life context may be helpful here as well. If you have ever heard that someone's "growth was stunted," something prevented that person from growing as much as he otherwise might have.

3. **(B) strong** While a *sturdy* home might be huge, it does not by nature have to be. Something sturdy, however, would be considered *strong*. Choice (B) is the most accurate answer.

4. **(B) persistent** *Undeterred* uses the prefix *un–*, meaning "opposite of," with the root word "deter." To deter something is to stop or hinder it. So something *undeterred* would be something unhindered or *persistent*.

5. **(C) boredom** *Ennui* derives from a French word meaning *boredom* or "listlessness." Choice (C) is correct.

6. **(A) changeable** *Mutable* means "prone to change." Consider the word root *mut*, which appears in *mutant* (something or someone changed from the norm). Choice (A) *changeable* is the best answer.

7. **(C) wasted** To *squander* is to "waste or fritter away" something. Of the choices given, only (C) *wasted* means the same. This is a good question on which to use context to eliminate some of the choices. There's nothing necessarily shameful about (A) *gathering* or (D) *owning money*. *Stealing*, choice (B), is certainly shameful, so at that point you may have to make a 50-50 guess. Or you could use another tool, like remembered real-life context. It is unlikely you have ever heard "squander" on the news in the context of a story about theft.

8. **(A) unreadable** You could use context on this one: handwriting is more likely to be described as *unreadable* than as *unethical* or *dangerous*. A written work might be *creative*, but *handwriting* refers to the way the letters are made and not to the content of the writing.

9. **(B) unending** The word *infinite* means "lasting forever" or *unending*. Break the word into parts: *in–* means *not*, and the word root *fin* relates to ending. (Think about *final*, *finite*, or *finish*, all of which relate to ending.)

10. **(B) duration** *Longevity* means the length of time of a life or project, so choice (B) *duration* is correct. The word part *long* relates to length, so look for an answer choice that also relates to length.

11. **(A) trick** Remember where you have heard "to fall for" used before. One "falls for" a *trick*. Answer choice (A) is correct.

12. **(C) announced** To *proclaim* is to *announce* or *decree*, especially in regards to a prominent figure like a president. Think about word parts: *pro–* means in *front* or *before* and *claim* is an English word meaning *state* or *assert*. To *proclaim* therefore means something like "state in front of." Answer choice (C), *announced*, is a synonym for the given word.

13. **(A) basic** *Rudimentary* means *simple* or *easy*. Of the answer choices given, only (A) *basic* applies to something simple or rudimentary. On this question, try rereading the sentence in your mind and substitute each answer choice. When you do so, choice (A) makes the most sense.

14. **(B) defeat** The word *rout* means to "defeat soundly" in battle or competition. You may have heard it in the context of stories about sports on the news. Answer choice (B) is correct.

15. **(C) rebellious** A *rogue* is someone who does not play by the rules. Use context: the sentence draws a contrast between the idea of a *team* and the idea of *rogue*. Answer choice (C) most directly contradicts the idea of teamwork or team spirit.

16. **(C) evoke agreement** *Resonate* has the word root *son*, which relates to sound, and the prefix *re–* which means *again* or *back*. So *resonate* means something like "sound again" or "sound back." From there, you could think about which answer choice both fits that meaning and fits the context. *Irritation*, (A), and *discussion*, (D), do not seem to fit with the idea of sounding back. Choice (B) doesn't make any sense in the context. (In fact, to *resonate* in music means "to have a full sound that seems to reverberate." In other contexts, *resonate* has come to mean "harmoniously correspond" or "generate agreement.")

17. **(A) repetitive** *Redundant* starts with the prefix *re–*, or *again*. The first choice you have available is *repetitive*, which means again and again. Context is also useful here: the sentence suggests that *redundant* has something to do with *duplicating*.

18. **(B) dictate** *Prescribe* means to "state authoritatively or recommend." Remembered real-life context may be useful here. Think about when your doctor prescribes a medicine: she is urging you, with some authority, to take the medicine.

19. **(D) sensible** *Logical* means "based on logic" or "sensible." Answer choice (D) is the only choice that has anything to do with logic. If you weren't sure what *sensible* meant, you could have used remembered real-life context. Think about the phrase "sensible shoes," which are shoes that display good sense or judgement on the part of the wearer.

20. **(A) requirement** Note that choice (A) *requirement* has the same root as the given word. A *prerequisite* is something that is required before a project or course of study begins. Answer choice (A) is correct.

21. **(A) be drawn** *Gravitate* is based on a word it looks like: *gravity*. Even if you didn't know that *gravitate* means "moving towards," you might be able to make a guess based on the knowledge that the force of gravity pulls things closer together. Choice (A) is correct.

22. **(D) careful** *Cautious* means *careful*. Mentally rereading the sentence with each answer choice substituted for the underlined word would likely have been helpful here.

23. **(C) boaster** The meaning of *braggart* is "someone who brags." While many (B) *leaders* might brag, the correct answer choice here is (C) *boaster* (one who boasts).

24. **(A) rumor** *Hearsay* is a term often used in law. It means "secondhand stories": literally, what someone has heard someone else say. A *rumor* is a story that has been heard from others without proof. Choice (A) is correct.

25. **(C) ridicule** To *taunt* is to *ridicule* or *tease*. Even if you didn't remember a context in which you've heard the word *taunt*, you might have a sense that it has a negative charge. Only choice (C) also has a negative charge.

26. **(B) productive** The term *constructive criticism* refers to criticism that constructs or *builds up* the listener rather than destroying her. Think of the word *construction*, which means *building*. While (D) *flattering* words might build some people up, *flattery* contradicts the idea of *criticism*. The correct answer here is choice (B) *productive*.

27. **(C) countable** To *quantify* something is to "be able to count" it. Think about the root *quant*, which relates to numbers and which gives us the word *quantity*, or *amount*.

28. **(C) useful** Remembered real-life context might be helpful here. If you think about where you have heard the word *effective*, advertisements for cosmetics, drugs, and supplements might come to mind. Now ask yourself: what do the companies who sell cosmetics or drugs want to tell us about those products? They're most likely to want to tell us that their products are useful, productive, or successful.

29. **(C) disinterest** *Apathy* means the state of not caring or *disinterest*. Think about word parts: the prefix *a–* means *not*, and the word root *path* relates to feeling. Think about *sympathetic*, which means sharing the feelings of others. Thus, *apathy* means something like the state of not entering into others' feelings, *or disinterest*.

30. **(B) defect** A *blemish* is a glaring problem or defect. Try mentally rereading the sentence with each choice substituted. You wouldn't describe an (A) *impact* as *glaring*, so that one is out. Moreover, *glaring* has a negative context that does not fit with (C) *commendation* or (D) *compliment*.

31. **(A) agreement** *Consensus* has a root of *con–* meaning "with" (don't confuse this with *con–* meaning not). Only answer choice (A) suggests togetherness. Another way of tackling this question is to read the sentence carefully and glean that to get a measure passed, one would need *agreement*, not (C) *infighting* (that is, fighting within a group). (A) is the correct answer.

32. **(D) insubordination** Think about word parts: *dis–* means "not." The word root *sent* relates to perceiving or feeing and also appears in the word *consent*, which means agreement. Thus, you can guess that *dissent* is the opposite of agreement. Even if you don't know what *insubordination* means, you can eliminate the other choices because none of them suggest the opposite of agreement.

33. **(D) speed** *Tempo* means "the speed of something." Remembered real-life context might be helpful here: if you have heard the word *tempo*, it was likely in a discussion about music. And it makes far more sense to say that music has a speed than to say that it has a *heartbeat*, *safety*, or *modernity*.

34. **(B) comfort** Try to fit some of the words into the sentence given to test their context. *Glee* and *protection* may be tempting at first glance, but neither fit the construction of the sentence if you replace them for the word *solace*. But *comfort*, (B), would make her feel better.

35. **(B) traps** *Pitfalls* are "traps and problems that await." Try mentally rereading the sentence and substituting each choice. While climbers might experience (A) *joys* and probably do work with (C) *ropes*, neither of those ideas relate specifically to inexperienced climbers.

Part 4: Paragraph Comprehension Answers and Explanations

1. **(A) can be very difficult to track down** This is an Inference question, so the correct answer may be implied but not spelled out directly. The first paragraph talks about the participants of urban legends. It turns out that those participants can never be found.

2. **(D) a basis in reality** This is an EXCEPT question, so the correct answer is the one that does not characterize successful urban legends. If you were able to determine that urban legends are fictional, then answer choice (D), *a basis in reality*, is clearly the correct answer. Otherwise, you could use the process of elimination on the wrong answer choices. The passage states that urban legends "persist both for their entertainment value and for the transmission of popular values and beliefs," so choices (A) and (B), which paraphrase this, are wrong. Finally, the very first sentence describes urban legends as a new form of folktale, so (C) is out.

3. **(B) Darwin's theory accounted for many examples of evolution** The author, while maintaining that there were other scientists with similar evolutionary ideas, notes toward the end that Darwin's eloquence and his ability "to account for many diverse examples" helped him establish his theories as the benchmark by which others stacked up. Thus, choice (B) is correct. None of the other choices are supported by the passage.

4. **(B) informative** In questions that ask about the author's tone, beware of answer choices that exaggerate the author's point of view. Choice (B) is correct because the passage merely conveys relevant facts about an issue. While the author does describe two opposing reactions that parents have had to the changes that have taken place in elementary schools, no preference is implied. Therefore choices (A) and (C) are incorrect. Be cautious with choices such as these that use extreme language.

5. **(D) light from the Sun** Although the wording of this question indicates that this is an Inference question, the correct answer choice, (D), can be taken directly from the first sentence. If most life is "dependent on organisms that store radiant energy from the Sun," it is dependent on light from the Sun. Remember that on Inference questions you are looking for the one answer that must be true based on what is stated in the passage.

6. **(A) Both perform similar functions in different food chains.** The first paragraph describes ecosystems that are dependent on photosynthetic organisms, while the second paragraph describes ecosystems that are dependent on chemosynthetic organisms. In both cases, the organisms serve similar functions as primary producers within their different food chains, so choice (A) is correct. Choice (B) is wrong because only chemosynthetic organisms are described in the passage as supporting higher organisms "at great ocean depths." (C) is wrong because chemosynthetic organisms do not rely on sunlight for their basic source of energy. Choice (D) is never discussed in the passage.

7. **(D) confirmed Halley's calculations** The last sentence of the passage implies that Halley accurately predicted the 1758 appearance of the comet. Thus, you can infer that the appearance of the comet confirmed Halley's theory and accorded with his calculations.

8. **(D) by the narrower snout found on a crocodile** The passage discusses a few ways to distinguish a crocodile from an alligator. The first and most easily observed of these is the fact that the crocodile's head and jaws are longer and narrower. In other words, it has a narrower snout, choice (D). (A) is out, because the animals' mouths do not necessarily have to be closed to distinguish one from the other. Choice (B) is out because both animals can be found in the Everglades National Park, and (C) is never discussed as one of the ways to distinguish one giant reptile from the other.

9. **(A) certain actors are inappropriate for certain roles** On this question you are asked to infer what the agent believes. She must assume that certain actors are inappropriate for certain parts, because if she didn't believe this her conclusion would make no sense; it wouldn't be possible for an actor to audition for an inappropriate part. The actors' talent, (B), isn't questioned by the agent; she focuses on the types of roles. The possibility that an actor's appropriateness for the part may be difficult to predict, (C), is never assumed by the agent; in fact, if anything, such a belief would weaken her conclusion. Major roles, (D), are never discussed in the passage.

10. **(A) a grouping of syllables** To answer this Detail question, scan the passage for a reference to *foot* or *feet*. The first such mention is in the second sentence, which states: "Pentameter refers to the number of feet, or groupings of syllables." So a foot is just that, a grouping of syllables, choice (A).

11. **(D) the sinking of the *Lusitania* prompted increased American involvement in World War I** This is a Global question, so reading the passage for details is not necessary, although paying attention to the overall effect of those details is. The author notes that public opinion on the war was swayed when the *Lusitania* was attacked, prompting increased military involvement from the United States. Choice (D) is a paraphrase of this main idea. Of the wrong answer choices, (A) distorts a detail found in the passage, implying a causal link between America's support for Britain and France and Germany's sinking of the *Lusitania* that is not supported by the passage. Choice (B) refers to a minor detail in the passage, and choice (C) is simply not true according to the passage, particularly after the sinking of the *Lusitania*.

12. **(D) When critics and general audiences disagree about a movie's quality, the critics' opinion is not necessarily more valid.** In this Inference question, you are asked to supply a conclusion that is best supported by the passage. The author seems to believe that critics' opinions aren't necessarily more significant than those of the average person, so (D), which restates this idea, is a valid inference. Though the author says the critics' opinions aren't more valid than those of average moviegoers, he doesn't imply (A), that the critics' opinions are *usually* incorrect. Choice (B) is out because the author does not say that the critics' judgments are meaningless—merely that critics don't always make the right judgments. And the passage doesn't state how frequently critics dislike popular movies, (C).

13. **(C) they believed it to be poisonous** This Detail question asks why Americans did not start using the tomato until the mid-nineteenth century. According to the passage, the English considered the tomato poisonous, and this myth continued to hold sway in America until the mid-nineteenth century. In other words, they, too, believed it to be poisonous, choice (C). None of the other answer choices are supported by the statements in the passage.

14. **(B) tired** The word "drowsy" is used to describe Daria. In the following sentence, you learn that Daria looks forward to going to sleep each night. From this information, you can determine that "drowsy" means *tired* or sleepy. Nothing in the passage suggests that Daria was *drunk*, (A). Throughout the passage, Daria is presented as the opposite of (C) *happy* and (D) *satisfied*.

15. **(A) They provide adequate shade.** The author discusses how she enjoys escaping the sun by sitting under the variety of trees in the forest. This implies that the trees provide adequate shade for her enjoyment as stated in choice (A). Choice (B) is the opposite of the meaning the author conveys. The author is thankful for the shade from the trees, so it's not accurate to say they are impairing her enjoyment, (C). While it may be true that trees provide a home for many animals, this concept is never discussed or implied in the passage, so choice (D) is incorrect.

Part 5: Mathematics Knowledge Answers and Explanations

1. **(D) $12x^2 + 7x - 12$** Use FOIL to solve:

$(3x + 4)(4x - 3)$

$= (3x)(4x) + (3x)(-3) + (4)(4x) + (4)(-3)$

$= 12x^2 - 9x + 16x - 12$

$= 12x^2 + 7x - 12$

2. **(A) $\frac{60}{n}$** Pick numbers to solve this one. If $n = 2$, the tree grew 5 feet, or 60 inches, in 2 years, which means it grew at a rate of 30 inches per year. Plug in 2 for n into the answer choices, and only (A) gives the target number, 30.

3. **(D) 192** If the perimeter of $ABCD$ is 56, then half that, or 28, is the sum of the length and the width, since the perimeter of a rectangle = 2(length + width). If $AD = 16$, then $AB = 28 - 16 = 12$, and the area of $ABCD$ must be $16 \times 12 = 192$.

4. **(A) equal** A square is a rectangle with four equal sides. Choice (B) may be tempting, but in fact, any two opposing sides of a square are parallel. It would be inaccurate to say that a square has four sides, all of which are parallel. Choice (D), *curvilinear*, means "curved" and is not correct.

5. **(D) $a + b < 2c$** The question gives two inequalities here: $a < b$ and $b < c$, which can combine into one, $a < b < c$. Go through the answer choices to see which *must* be true. If you add the corresponding sides of these inequalities: $a + b < c + c$, or $a + b < 2c$. This statement is *always* true, so it must be the correct answer.

This question could also be approached by picking numbers. That might look like this:

Choose $a = 2$, $b = 3$, and $c = 4$, to conform to the information in the question stem. Now try each choice to see whether it is a true statement about those numbers:

(A) $3 + 4 < 4$ No way! Incorrect.

(B) $2 + 3 < 4$ Also not true. Incorrect.

(C) $2 - 3 < 3 - 4$ Nope. Incorrect.

(D) $2 + 3 < 8$ Definitely true. This choice is correct.

If more than one answer choice had produced a true statement, you would have tried again using different numerical values until you had only one true statement.

6. **(C) 36** Since the angle marked $x°$ and the angle marked $y°$ together form a straight angle, their measures must sum to 180°. Substitute in $2z$ for x and $3z$ for y, and solve for z:

$x + y = 180$

$2z + 3z = 180$

$5z = 180$

$z = 36$

7. **(B) 7.5** There are 360 degrees in a circle and 60 minutes in an hour, so you could solve this question by setting up a proportion in which n is the number of minutes $\frac{45}{360} = \frac{n}{60}$. Then cross-multiply and reduce the fraction:

$n = \frac{45 \times 60}{360}$

$n = \frac{45 \times 60}{360} \div \frac{60}{60} = \frac{45}{6}$

$n = \frac{45}{6} \div \frac{3}{3} = \frac{15}{2}$

$n = \frac{15}{2} = 7.5$

8. **(C) 12** The area of a triangle is $\frac{1}{2} \times$ base $\times$ height:

$\frac{1}{2} \times 6 \times 4 = 3 \times 4 = 12.$

9. **(B) 115** Since lines p and q are parallel, use the rule about alternate interior angles. The angle marked $y°$ is supplementary to a 65° angle, so $y = 180 - 65 = 115$.

10. **(A) 4.5** You could either turn the percentages into decimals and multiply: $0.15 \times 0.15 \times 200 = 0.15 \times 30 = 4.5$ and get the answer (A), or you could take it one step at a time. 15% of 200 is $\frac{15}{100} \times 200 = 15 \times 2 = 30$. 15% of 30 is

$\frac{15}{100} \times 30 = \frac{3}{20} \times 30 = \frac{3}{2} \times 3 = \frac{9}{2} = 4.5.$

11. **(B) $\frac{1}{24}$** Don't fall for the answer choice trap and assume that the average of $\frac{1}{20}$ and $\frac{1}{30}$ is $\frac{1}{25}$. Instead, use the average formula: Average $= \frac{\text{Sum of the terms}}{\text{Number of terms}}$. In this case:

$$\text{Average} = \frac{\frac{1}{20} + \frac{1}{30}}{2} = \frac{\frac{3}{60} + \frac{2}{60}}{2} = \frac{\frac{5}{60}}{2} = \frac{5}{120} = \frac{1}{24}.$$

12. **(C) $(180 - y)°$** This question is testing your knowledge of definitions. The *supplement* of an angle is the angle that when added to the original angle equals 180°. So if an angle measures $y°$, its supplement is $180 - y$.

13. **(B) 5** If Diane has $\frac{2}{3}$ of her room still to paint, she'll need $2 \times 2\frac{1}{2}$ or 5 more cans to do the job.

14. **(D) −12** You can't find the value of either variable alone, but you don't need to. Rearranging the equation, you get:

$13 + a = 25 + b$

$13 = 25 + b - a$

$13 - 25 = b - a$

$-12 = b - a$

15. **(C) $15x^5$** When you multiply terms that have exponents over the same base, you add the exponents and multiply the coefficients:

$3x^2 \times 5x^3 = (3 \times 5)x^{2+3} = 15x^5$

16. **(D) 21%** You can pick numbers to make sense of this geometry problem. You are asked to increase the sides of a square by 10%, so you want to pick a number for the original sides of the square of which it's easy to take 10%. For instance, you could say that the original square is 10 by 10. 10% of 10 is 1, so the dimensions of the increased square are 11 by 11. In this case, the area of the original square is 100, and the area of the new square is 121, which represents a 21% increase.

17. **(B) 10** If $\frac{m}{2} = 15$, then $m = 15 \times 2 = 30$

Plug 30 in for m to solve the expression asked for in the question:

$$\frac{m}{3} = \frac{30}{3} = 10$$

18. **(C) $3x - 2$** To factor $6x^2 - 13x + 6$, you need a pair of binomials whose "first" terms will give you a product of $6x^2$ and whose "last" terms will give you a product of 6. Since the middle term of the result is negative and the last term is positive, the two last terms of the factors must both be negative. One of the factors is among the answer choices, so you can use the answer choices in your trial-and-error effort to factor. You're looking for a factor with a minus sign in it, so the answer's either (C) or (D). Try (C) first: Its first term is $3x$, so the other factor's first term would have to be $2x$ (to get that $6x^2$ in the product). (C)'s last term is -2, so the other factor's last term would have to be -3.

Check to see if $(3x - 2)(2x - 3)$ works:

$(3x - 2)(2x - 3)$

$= (3x)(2x) + 3x(-3) + (-2)(2x) + (-2)(-3)$

$= 6x^2 - 9x - 4x + 6$

$= 6x^2 - 13x + 6$

It works. There's no need to check (D).

19. **(D) $3x - y + 2 = 0$** Just be careful and translate the English into math: "the product of 3 and x is equal to 2 less than y" becomes $3x = y - 2$. But all of the equations in the answer choices set the right side of the equation to zero, so do that to your equation:

$3x = y - 2$

$3x - y = -2$

$3x - y + 2 = 0$

20. **(A) $\frac{x}{x-1}$** Since $\frac{b}{a}$ is the reciprocal of $\frac{a}{b}$, $\frac{b}{a}$ must be the reciprocal of $1-\frac{1}{x}$ as well. Combine the terms in $1-\frac{1}{x}$ and then find the reciprocal:

$$\frac{a}{b} = 1 - \frac{1}{x} = \frac{x}{x} - \frac{1}{x} = \frac{x-1}{x}$$

Therefore, $\frac{b}{a} = \frac{x}{x-1}$.

21. **(C) 32** This figure is an unfamiliar solid, so don't try to calculate the volume directly. The question says that the solid in question is half of a cube. Imagine the other half lying on top of the solid, forming a complete cube.

Notice that the diagonal with length $4\sqrt{2}$ and two of the cube's edges form an isosceles right triangle. In an isosceles right triangle, the hypotenuse is $\sqrt{2}$ times the length of a leg. Here the hypotenuse has length $4\sqrt{2}$, so the legs have length 4. Thus, the edges of the cube have a length of 4. The volume of a cube is the length of an edge cubed. So the volume of the whole cube is $4 \times 4 \times 4$, or 64. The volume of the solid in question is $\frac{1}{2}$ of this, or 32.

22. **(B) 400** Begin by writing the division problem as a fraction: $\frac{7.6}{0.019} = x$. Now move the decimal points on the top and the bottom of the fraction the same number of places to the right until you are dealing with whole numbers: $\frac{7.6}{0.019} = \frac{7{,}600}{19}$. Now divide: $\frac{7{,}600}{19} = 400$.

23. **(B) $\frac{16}{\pi}$** The circumference of a circle $= 2\pi$(radius), so radius $= \frac{\text{circumference}}{2\pi}$. The circumference is 8: $\frac{8}{2\pi} = \frac{4}{\pi}$. The area of a circle $= \pi(\text{radius})^2$, so the area of a circle with a radius of $\frac{4}{\pi}$ is: $\pi\left(\frac{4}{\pi}\right)^2 = \pi\left(\frac{16}{\pi^2}\right) = \frac{16}{\pi}$

24. **(B) 36** You could translate the English into math to get:

$0.36 \times 18 = 0.18 \times n$

$$n = \frac{0.36 \times 18}{0.18} = 2 \times 18 = 36$$

However, you don't have to go through all that work if you realize that x percent of $y = y$ percent of x.

25. **(C) 300 and 400** Here you're looking for an extremely rough approximation (the answer choices all have a range of 100), so square the upper bounds of the ranges in the answer choices, until you find the range that encompasses 104,906. Start with the upper bound of (A): $(200)^2 = 40{,}000$, which is less than 104,906. Now try (B): $(300)^2 = 90{,}000$, which is still too low. Now check (C): $(400)^2 = 160{,}000$. 104,906 is between 90,000 and 160,000, so (C) is the correct answer.

Part 6: Electronics Information Answers and Explanations

1. **(B) ohms** Resistance is measured in ohms, and the symbol for an ohm is Ω (omega).

2. **(B) positive, neutrally** Two types of particles are found within the nucleus: protons and neutrons. Protons have a positive charge, and neutrons are neutrally charged.

3. **(A) alternating current** Alternating current (AC) is when the electric current reverses direction many times in a second.

4. **(D) short circuit** This symbol represents a short circuit, which is an accidental path of low resistance that passes an irregularly high amount of current.

5. **(B) current** *I* represents current. Current is the rate of flow of electrons, or the **I**ntensity of the flow. (Specifically, *I* is the rate of charge flow.)

6. **(D) anode** The P-type material in a diode is the anode, and the N-type material is known as the cathode.

7. **(D) doping** Pure silicon must be doped in order to generate holes or free electrons. This process creates the P and N materials needed to make diodes and transistors.

8. **(A) capacitor** Capacitors stop direct current flow but not alternating current. The higher the frequency of an alternating current, the less resistant effect a capacitor has. Standard resistors, both fixed and variable, definitely resist AC flow, as do inductors.

9. **(D) 20 Ω** When wired in series, the resistance of several loads can be added together to determine the effective resistance. The sum of the four resistances is 20 Ω.

10. **(C) the arrow in the transistor symbol is always placed on the collector** The arrow in a transistor symbol is always placed on the emitter, and the arrow always points toward the N-type material at that junction.

11. **(A) voltmeter** This is the circuit symbol for a voltmeter, which is used to measure voltage.

12. **(B) watt** There are three essential components of an electrical circuit: a voltage source, a load, and conductors to connect the load to the voltage source. When these three components are connected so that current can flow, it is a closed circuit.

13. **(A)** 

A doorbell is typically operated by a push switch, which allows current to flow only when a button is pressed. (A) is the symbol for a push switch.

14. **(B)** 

This is the schematic symbol for an alternating current (AC) voltage source.

15. **(C) condensers** Capacitors that are electrical storage units are also known as condensers.

16. **(C) an insulator** An insulator is a material that does not conduct electricity.

17. **(A) rubber** Rubber is an example of an insulator. All of the other examples in this question are conductors.

18. **(C) 500 mA** Four 6-ohm resistors wired in series will have a total resistance of 24 ohms. Using $I = V \div R$, a total of 0.5 A will flow in this circuit. 0.5 A is the same as 500 mA.

19. **(A) silicon** The term *semiconductor* refers to an element that has four electrons in its valence shell. Since the bonds between these four electrons and the nucleus are somewhat strong, these elements are neither good conductors nor good insulators. One element that is widely recognized as a semiconductor is silicon.

20. **(D) galvanometer** This is the circuit symbol for a galvanometer, which is used to measure extremely small currents, usually of 1 mA voltage or less.

Part 7: Auto and Shop Information Answers and Explanations

1. **(D) miter box** While a *miter box* might be used with a backsaw to make even cuts in wood at specific angles, by itself, it is not capable of shaping wood. All of the other choices are tools capable of making cuts or smoothing edges on wood.

2. **(B) suspension** Of all the aspects of vehicle operation, two that are a high priority for most drivers are ride comfort and handling. The suspension system, choice (B), is responsible for the ride quality of the vehicle.

3. **(D) kerf** The cut or slot left by a saw in a piece of wood is also known as a *kerf*.

4. **(D) transfer case** The *transfer case* is located between the transmission and the drive axles on a four-wheel drive vehicle and is the piece that splits the engine's power between the front- and rear-drive axles.

5. **(B) a cut across the grain of the wood** This is a crosscut saw. It is designed to, as its name suggests, cut across the grain of the wood. Crosscut saw teeth are unique in that they cut like a knife, in contrast to a ripsaw, which is made to cut with the grain of wood.

6. **(C) compression stroke** This image depicts the compression stroke, which is the second stroke in the four-stroke cycle.

7. **(A) counting the number of threads per inch** Fractional measurement fasteners (measured in fractions of inches) use threads that are identified by the number of threads per inch.

8. **(D) the tools shown in both (B) and (C)** To loosen tight fasteners, it is good to use a box-end wrench. Choice (B) depicts a combination wrench, which has an open end wrench on one side and a box end on the other. Choice (C) shows a wrench with box ends on both sides. Therefore, both (B) and (C) are suitable for loosening tight fasteners; choice (D) is the correct answer. The box end wraps completely around the head of a bolt, and therefore makes greater surface contact.

9. **(A) coil spring** Of the answer choices given, only (A), *coil spring,* is not a part associated with the steering system. Coil springs are what aid the wheels in moving up and down while the vehicle chassis stays steady. The other pieces listed—steering knuckle, tie rod end, and idler arm—are all parts of the steering system.

10. **(C) 3** The camshaft, indicated in the diagram by the number 3, is responsible for the opening and closing of the engine's intake and exhaust valves. The camshaft turns at one-half the speed of the engine's crankshaft.

11. **(D) in the fuel tank** In fuel-injected engines, the electric fuel pump is located in the vehicle's fuel tank and is what supplies fuel under pressure to the fuel injectors.

12. **(B) fewer** The two basic thread classifications—*Unified National Coarse (UNC)*, and *Unified National Fine (UNF)*—have different numbers of threads per inch. A UNC or coarse thread would have relatively few threads per inch, where a UNF or fine thread would have a larger number of threads per inch.

13. **(C) both the engine intake manifold vacuum and fluid pressure from the power steering system** Power brake systems use both the engine intake manifold vacuum and fluid pressure from the power steering system to increase pressure. Engine intake manifold vacuum is utilized to generate greater force on the master cylinder, so higher hydraulic pressures can be generated in the brake system. Some brake booster systems use fluid pressure from the power steering system to increase braking power.

14. **(D) rotor** The primary is the low-voltage part of the system and contains a battery, an ignition switch, a primary coil winding, an ignition module, a reluctor and pickup coil, and a distributor. A rotor is not a part of the ignition primary system.

15. **(A) the ideal air-fuel mixture in a gasoline engine** *Stoichiometric* is the ideal ratio of air to fuel in the engine and it is the responsibility of the engine's fuel system to maintain this balance.

16. **(D) viscosity** *Viscosity* is resistance to flow, and is expressed as a number that is directly proportional to the thickness of the oil.

17. **(C) both direct and alternating current** Because they serve different purposes within an automobile, both negative ground current and direct current electrical systems are incorporated into cars.

18. **(D) work** Stick welding involves the use of an electric arc welding machine and two cables: one that attaches to the work being welded through a ground clamp, and the other going to an electrode that is sometimes referred to as a *stinger*. For safety reasons, always clamp to the work.

19. **(D) both flat and horizontally opposed** The last major cylinder arrangement that has been utilized by engine designers is the horizontally opposed or flat design. This has all of the cylinders lying on the horizontal plane, with half of the cylinders facing away from the other half and the crankshaft located between them. Some refer to this design as a "boxer" engine because the pistons move back and forth like a boxer throwing punches.

20. **(B) striking an object where it is important that the hammer itself not come in direct contact with the work** The tool pictured is a drift. It is used for striking an object where it is important that the hammer itself does not come in direct contact with the work. A chisel normally has a long edge and is used for cutting, (A), where a punch is narrow and is used for driving small fasteners and making layout marks, (C). To guide a saw when cutting across the grain, (D), you might use a miter box, but you would certainly not use a drift.

21. **(A) single overhead cam or SOHC** This design is known as single overhead cam or SOHC. In a V-type engine with two cylinder heads, there would be two camshafts, with one installed above each cylinder head.

22. **(D) the catalytic converter** Any air passing through to the car's combustion chamber must naturally pass through the intake port in the cylinder head, the intake manifold, and the air filter. Since the intake valve is open and is allowing atmospheric air to enter the combustion chamber, higher atmospheric air pressure pushes air through the engine's intake system and toward the low-pressure area above the piston. As the air is traveling through the intake system, fuel is injected into the air stream before it enters the combustion chamber. Gases do not pass through the catalytic converter until they are almost ready to be expelled from the car.

23. **(D) 6-point socket and breaker bar** Since most nuts are smaller sized, and because the 6-point socket is a stronger design, it is usually the mechanic's first choice in the smaller socket drive sizes. And a breaker bar attached to a socket can give enough twisting force to loosen very tight fasteners.

24. **(D) external hole-type** This is an external hole-type retaining ring. Retaining rings (or snap rings) are used to prevent end-movement of cylindrical parts in bores or parts mounted on shafts. External snap rings are installed in grooves on shafts, whereas internal snap rings install in grooves inside a bore. Snap rings are installed and removed using snap-ring pliers.

25. **(B) They are converted to lead sulfate.** A car's battery contains lead plates immersed in an electrolyte made up of sulfuric acid and water. When the battery discharges energy, the sulfur binds with the lead plates to make lead sulfate, and the electrolyte is reduced to water. Even though water is a result of this process, it is not true to say that (A) the lead plates release water, because the water is already part of the electrolyte mixture.

Part 8: Mechanical Comprehension Answers and Explanations

1. **(A) its potential energy is being converted into kinetic energy** The principle of conservation of mechanical energy says that the total mechanical energy of an object ($PE + KE$) remains constant as an object moves. As the car rolls down the hill, its speed increases because its potential energy (*PE*) is being converted into kinetic energy (*KE*).

2. **(C) its velocity is zero** During the ball's journey upward after being thrown, the ball has its maximum kinetic energy when it leaves the thrower's hand. This kinetic energy is converted into potential energy as it rises. Maximum potential energy (*PE*) occurs when the ball stops rising, and this is where its velocity is zero.

3. **(B) inertia** *Inertia* is a function of an object's mass. A large mass will require a large amount of force to cause it to accelerate. The larger the mass, the greater the inertia of the object. *Momentum*, (A), is a quality of an object in motion and describes how difficult it is to change the velocity of a moving object. *Torque*, (C), is a rotational force. *Pressure*, (D), is force per unit area.

4. **(C) law of universal gravitation** Isaac Newton discovered that every particle in the universe is attracted to every other particle. This concept became a physical law known as Newton's law of universal gravitation. The gravitational force that each body exerts on the other grows stronger as the bodies get closer. The force also increases as the mass of the bodies increase.

5. **(D) 6 m/s²** Based on Newton's second law, $F = ma$. Rearrange that equation to solve for acceleration: $a = \frac{F}{m}$, where units of force are in Newtons and mass is expressed in kg. Here: $a = \frac{30\text{ N}}{5\text{ kg}} = \frac{6\text{ m}}{s^2}$.

6. **(C) the same amount of work** Work is defined as force times distance. Lifting a box straight up to the top of a ramp will require the same amount of work as pushing the box up the ramp (neglecting friction). The ramp allows a smaller force to be applied over a larger distance. However, the end result or net work done is the same in both cases.

7. **(B) it is possible to get more work out of a machine than what is put in** It is not possible to get more work out of a machine than what is put in. Machines do not reduce work, they simply manage it and make it easier by diminishing the amount of force that is required.

8. **(A) wedge** A chisel is an example of a wedge.

9. **(B) a wedge is made to move, whereas an inclined plane stays stationary** The wedge is a variation of the inclined plane, but is made to move whereas the inclined plane is made to stay in one place.

10. **(A) post A** The distance from post A to the load is less than the distance from post B to the load; therefore, it holds the greater part of the load.

11. **(C) joules** Work and kinetic energy are both measured in joules. This is not a coincidence, as any work that is done to accelerate an object at rest will be converted into the kinetic energy of that object. This principle is known as the work-energy theorem.

12. **(B) wheel and axle** A screwdriver gains mechanical advantage through the use of a large wheel (screwdriver handle) rotating a small wheel (screwdriver tip), much like a wheel and axle. Large movements converted into small movements will amplify the force applied.

13. **(B) a decrease** Any time that force is increased through mechanical advantage, there will always be a proportional decrease in the distance moved.

14. **(A) first-class lever** A prybar can increase mechanical advantage, and can change the direction of motion, so it is a first-class lever.

15. **(B) 746** One horsepower is equal to 550 foot-pounds of work per second or 746 watts.

16. **(C) moved toward where the object is moved** Moving the fulcrum toward the object will increase the force applied to the object, but will decrease the distance that the object can be moved.

17. **(C) The work put out by a simple machine can never exceed the work put in.** If a machine were able to produce more work than was put into the machine, there would be a net increase in energy. This would violate the laws of thermodynamics. The work put out by a simple machine can never exceed the work put into it.

18. **(D) reduction in rotational speed** If a small gear drives a large gear, a speed reduction takes place. The large gear will turn more slowly than the small gear and the speed of the output will be slower.

19. **(B) multiple pulleys** A block-and-tackle requires a minimum of two pulleys, with a rope, belt, or chain used to operate them. There is at least one fixed and one moveable pulley in the block and tackle.

20. **(A) the weight of the object** The force exerted to lift an object using a single pulley is the same as the weight of the object itself.

21. **(B) wheel and axle** A doorknob is similar to a wheel and axle in that a large wheel (the doorknob or handle) is used to operate a small wheel (the latch mechanism).

22. **(A) the same** Mass is a measure of the amount of matter in a substance. It does not change from place to place. Weight, however, depends on the gravitational constant, and does change depending on the location.

23. **(C) inversely proportional** Consider "Torque" and "Speed" to be two people on a teeter-totter. Whenever "Torque" goes up, "Speed" goes down, and vice versa. These two quantities can be described as inversely proportional.

24. **(C)** $\boldsymbol{PE = m \times g \times h}$ An object's potential energy (*PE*) can be calculated using the formula

$PE = mgh$

where m is mass in kilograms (kg), g is acceleration due to gravity (9.8 m/s^2), h is the height of the object in meters (m), and PE is gravitational potential energy measured in joules (J).

25. **(A) 0 J** No work is done because the box does not move. $W = Fd$, where F is force (newtons), and d is distance (m). If an object experiences no change in position, then no net work is done.

Part 9: Assembling Objects Answers and Explanations

1. **C**

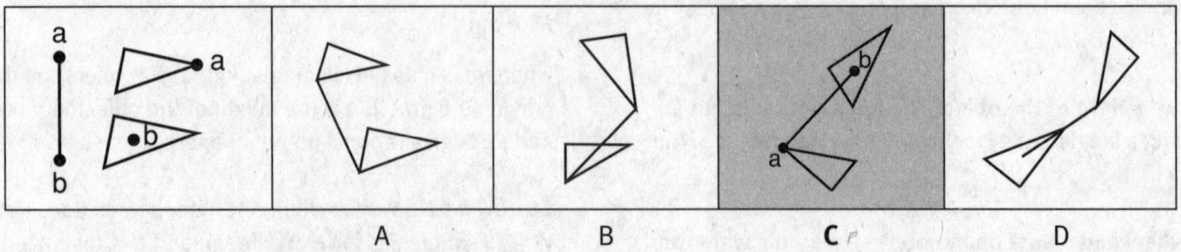

2. **A**

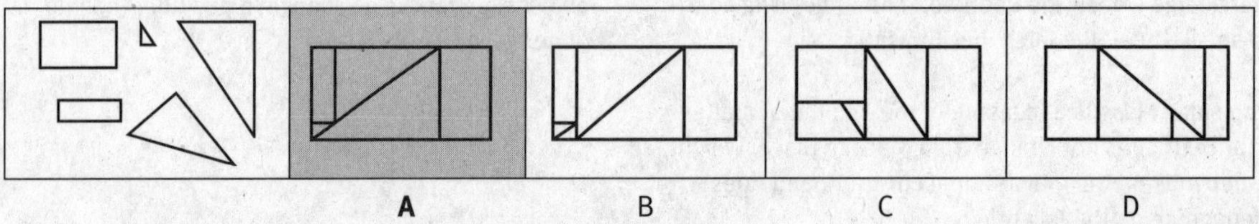

3. **D**

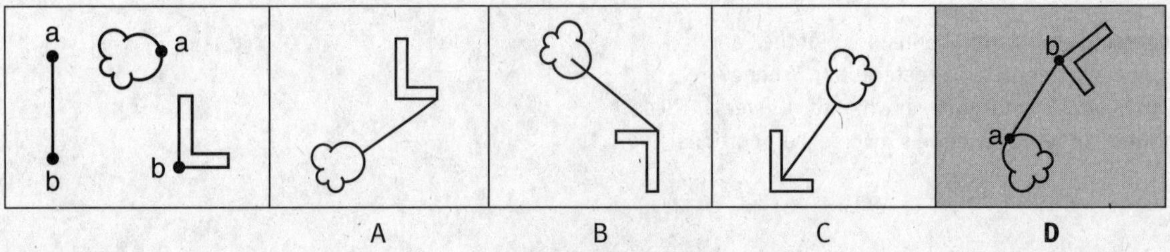

4. **B**

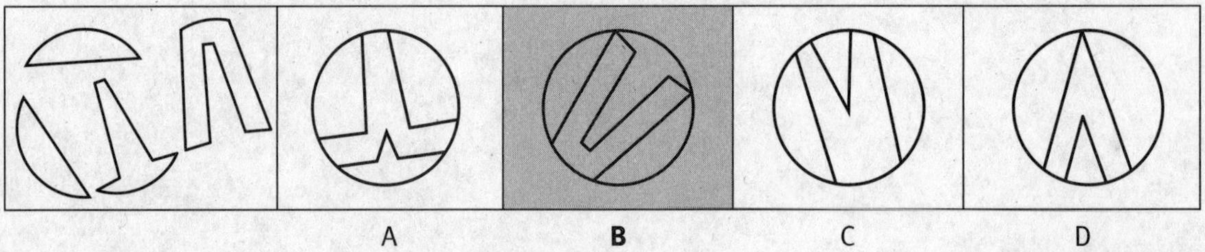

5. **D**

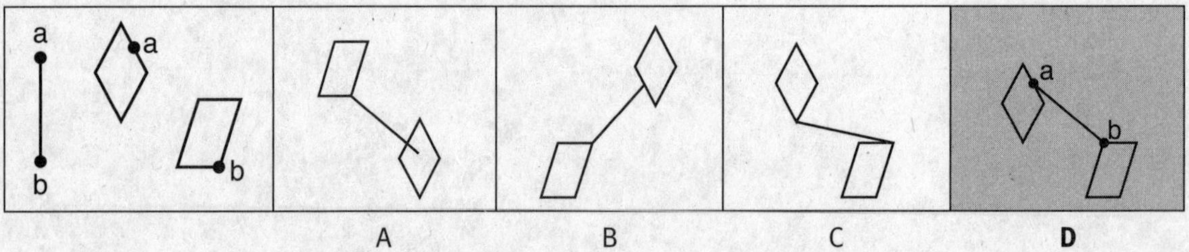

6. **C**

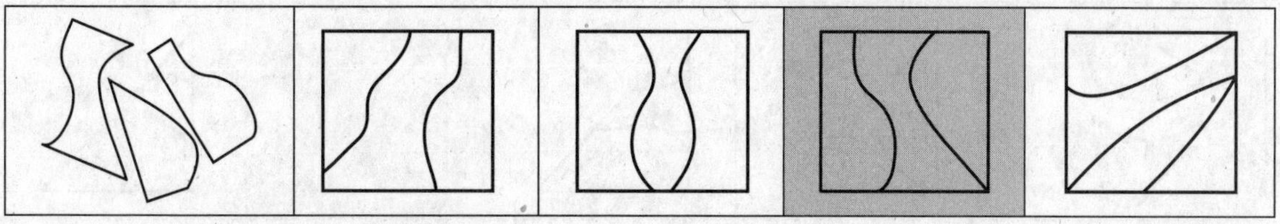

A B **C** D

7. **A**

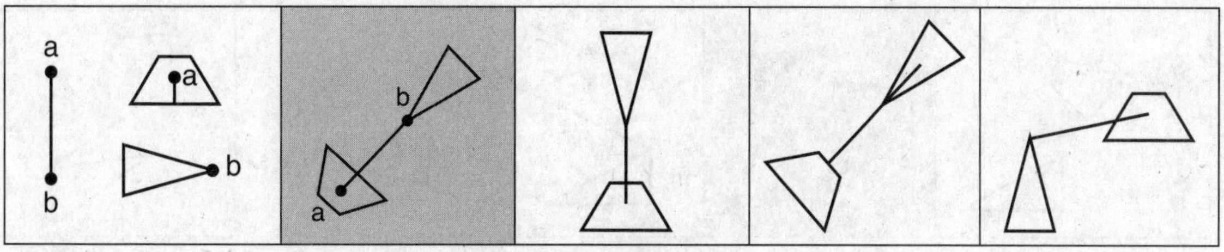

A B C D

8. **B**

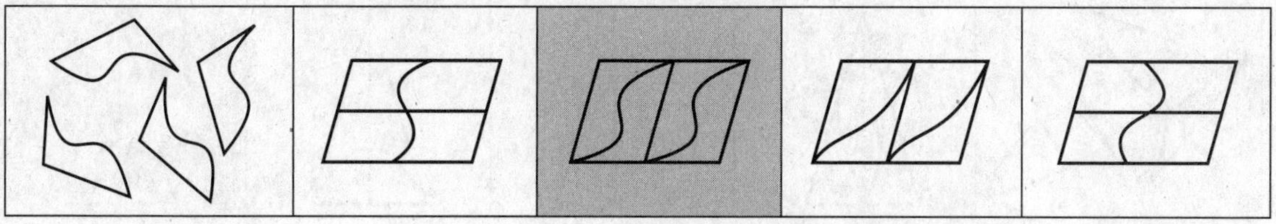

A **B** C D

9. **C**

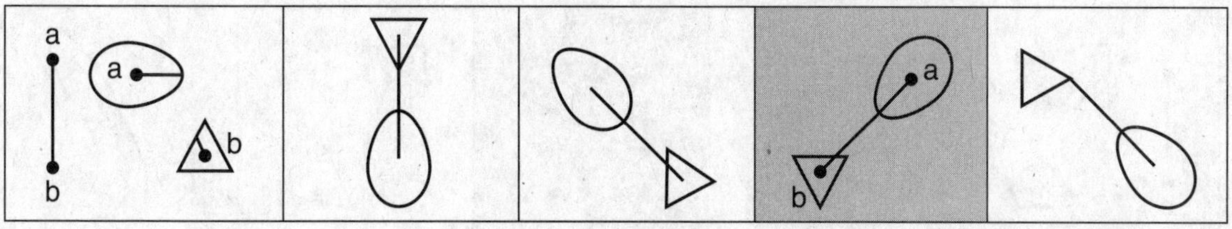

A B **C** D

10. **D**

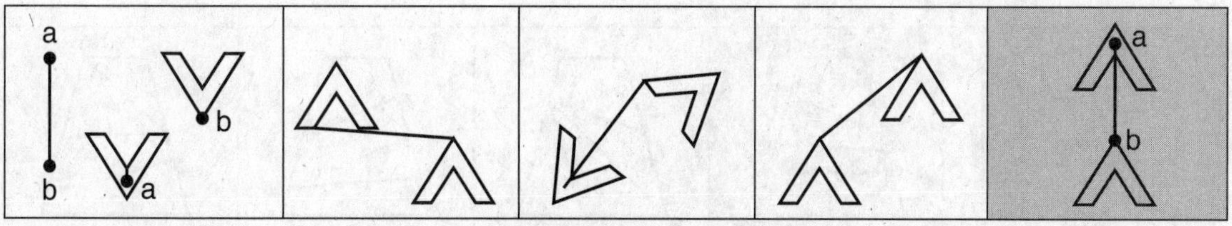

A B C **D**

11. **B**

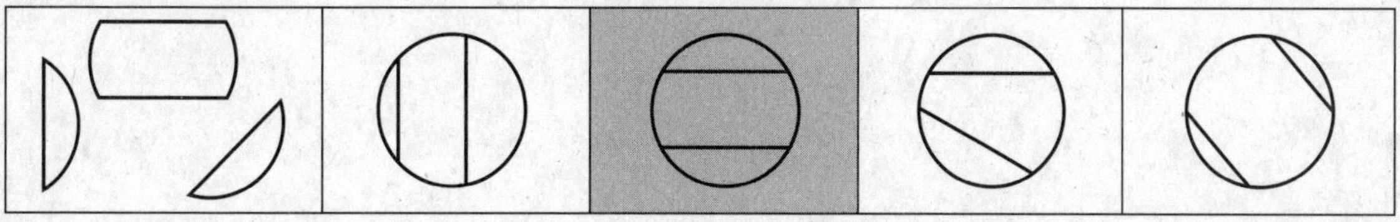

A **B** C D

12. **C**

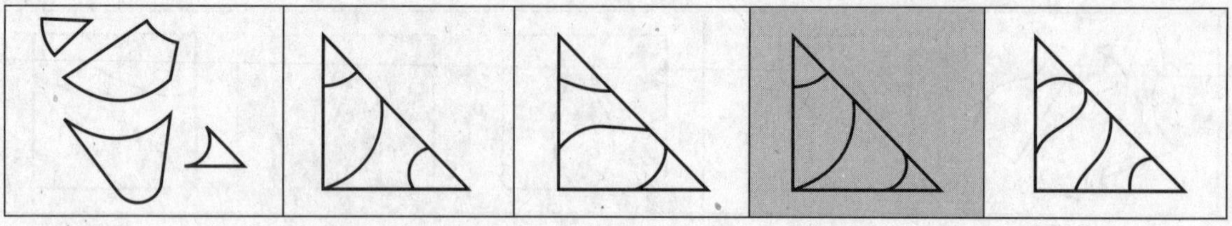

A B **C** D

13. **A**

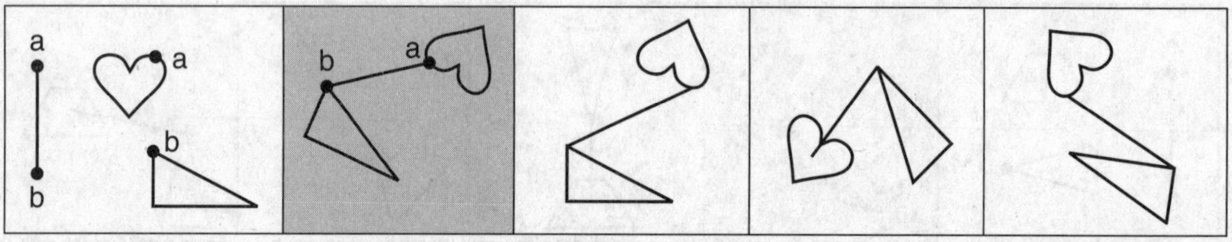

A B C D

14. **D**

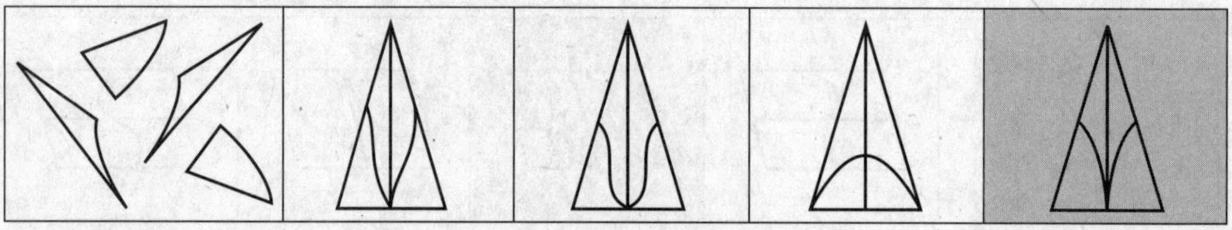

A B C **D**

15. **B**

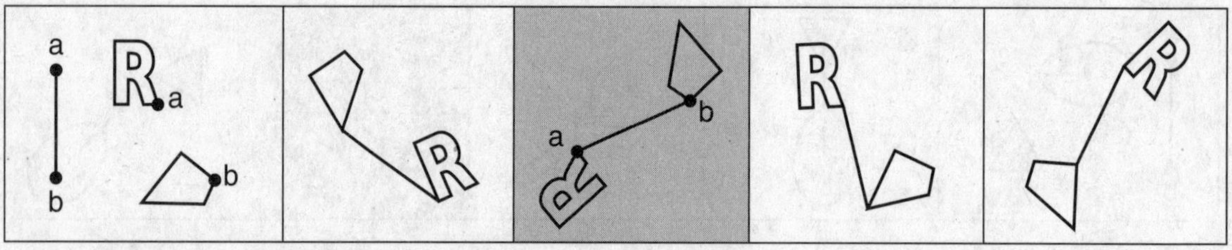

A **B** C D

16. **A**

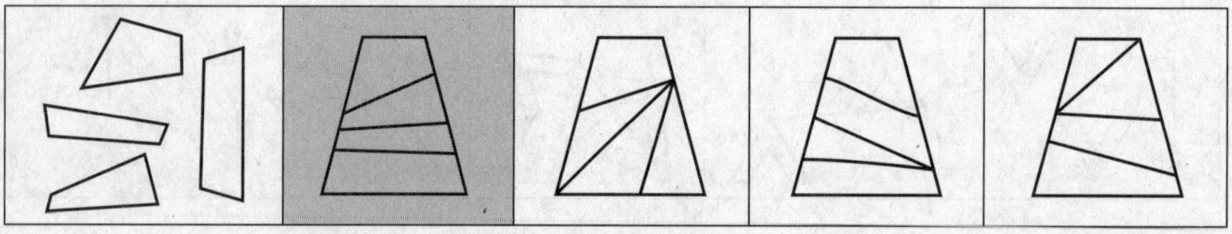

A B C D

17. **C**

A B **C** D

18. **D**

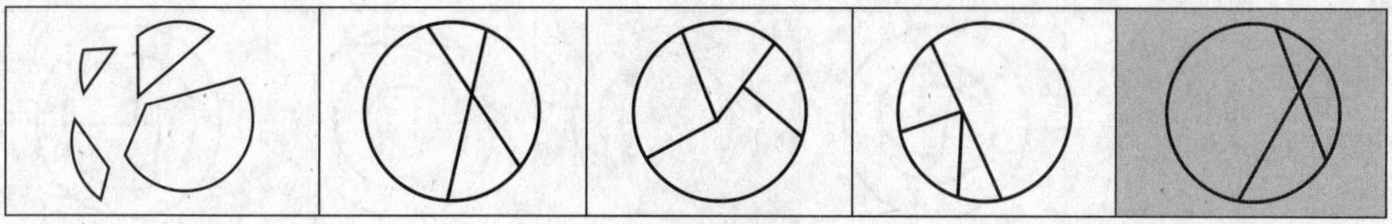

A B C **D**

19. **B**

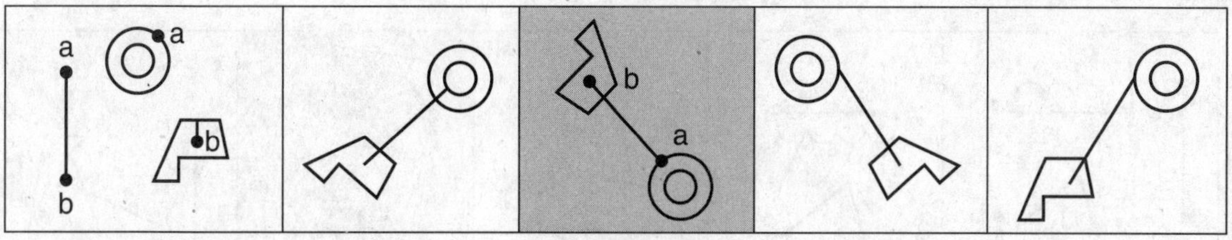

A **B** C D

20. **B**

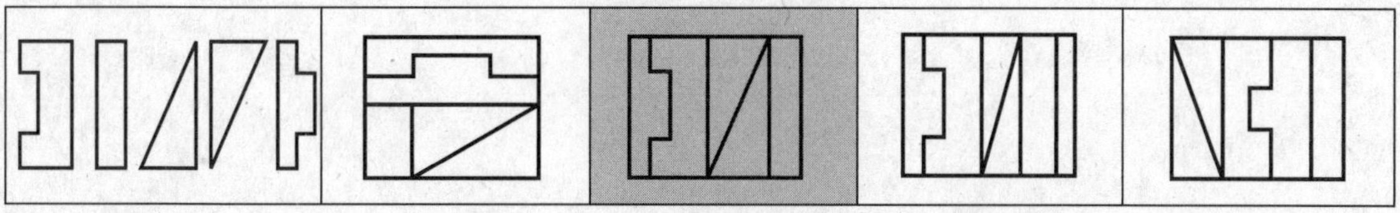

A **B** C D

21. **C**

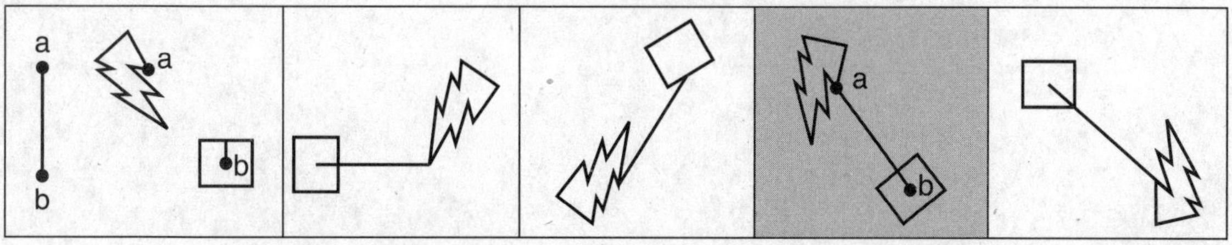

A B **C** D

22. **A**

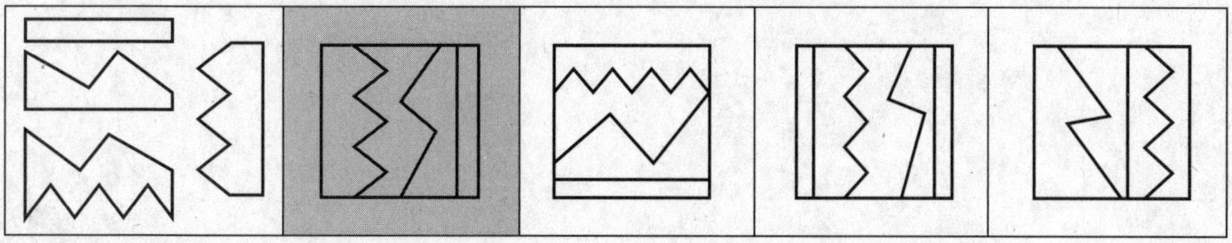

A B C D

23. **D**

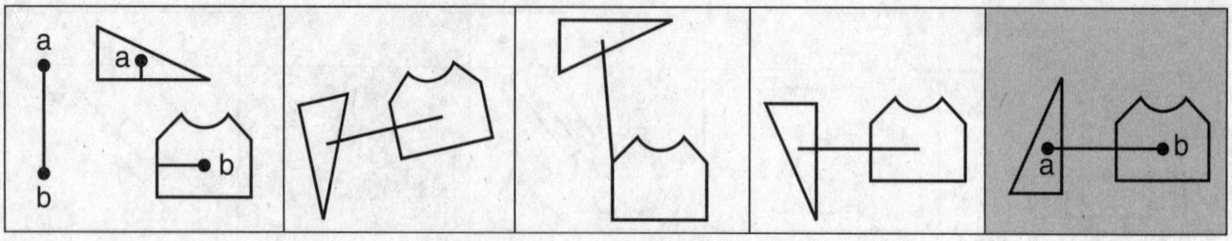

A B C **D**

24. **B**

A **B** C D

25. **C**

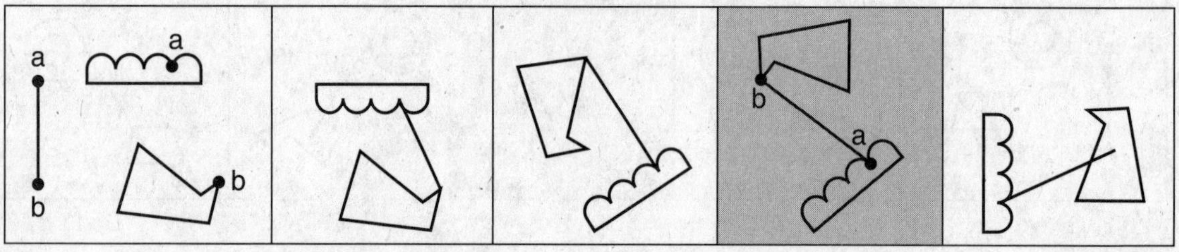

A B **C** D

ANSWER SHEET

Part 1: General Science (GS)

1 Ⓐ Ⓑ Ⓒ Ⓓ	6 Ⓐ Ⓑ Ⓒ Ⓓ	11 Ⓐ Ⓑ Ⓒ Ⓓ	16 Ⓐ Ⓑ Ⓒ Ⓓ	21 Ⓐ Ⓑ Ⓒ Ⓓ
2 Ⓐ Ⓑ Ⓒ Ⓓ	7 Ⓐ Ⓑ Ⓒ Ⓓ	12 Ⓐ Ⓑ Ⓒ Ⓓ	17 Ⓐ Ⓑ Ⓒ Ⓓ	22 Ⓐ Ⓑ Ⓒ Ⓓ
3 Ⓐ Ⓑ Ⓒ Ⓓ	8 Ⓐ Ⓑ Ⓒ Ⓓ	13 Ⓐ Ⓑ Ⓒ Ⓓ	18 Ⓐ Ⓑ Ⓒ Ⓓ	23 Ⓐ Ⓑ Ⓒ Ⓓ
4 Ⓐ Ⓑ Ⓒ Ⓓ	9 Ⓐ Ⓑ Ⓒ Ⓓ	14 Ⓐ Ⓑ Ⓒ Ⓓ	19 Ⓐ Ⓑ Ⓒ Ⓓ	24 Ⓐ Ⓑ Ⓒ Ⓓ
5 Ⓐ Ⓑ Ⓒ Ⓓ	10 Ⓐ Ⓑ Ⓒ Ⓓ	15 Ⓐ Ⓑ Ⓒ Ⓓ	20 Ⓐ Ⓑ Ⓒ Ⓓ	25 Ⓐ Ⓑ Ⓒ Ⓓ

Part 2: Arithmetic Reasoning (AR)

1 Ⓐ Ⓑ Ⓒ Ⓓ	7 Ⓐ Ⓑ Ⓒ Ⓓ	13 Ⓐ Ⓑ Ⓒ Ⓓ	19 Ⓐ Ⓑ Ⓒ Ⓓ	25 Ⓐ Ⓑ Ⓒ Ⓓ
2 Ⓐ Ⓑ Ⓒ Ⓓ	8 Ⓐ Ⓑ Ⓒ Ⓓ	14 Ⓐ Ⓑ Ⓒ Ⓓ	20 Ⓐ Ⓑ Ⓒ Ⓓ	26 Ⓐ Ⓑ Ⓒ Ⓓ
3 Ⓐ Ⓑ Ⓒ Ⓓ	9 Ⓐ Ⓑ Ⓒ Ⓓ	15 Ⓐ Ⓑ Ⓒ Ⓓ	21 Ⓐ Ⓑ Ⓒ Ⓓ	27 Ⓐ Ⓑ Ⓒ Ⓓ
4 Ⓐ Ⓑ Ⓒ Ⓓ	10 Ⓐ Ⓑ Ⓒ Ⓓ	16 Ⓐ Ⓑ Ⓒ Ⓓ	22 Ⓐ Ⓑ Ⓒ Ⓓ	28 Ⓐ Ⓑ Ⓒ Ⓓ
5 Ⓐ Ⓑ Ⓒ Ⓓ	11 Ⓐ Ⓑ Ⓒ Ⓓ	17 Ⓐ Ⓑ Ⓒ Ⓓ	23 Ⓐ Ⓑ Ⓒ Ⓓ	29 Ⓐ Ⓑ Ⓒ Ⓓ
6 Ⓐ Ⓑ Ⓒ Ⓓ	12 Ⓐ Ⓑ Ⓒ Ⓓ	18 Ⓐ Ⓑ Ⓒ Ⓓ	24 Ⓐ Ⓑ Ⓒ Ⓓ	30 Ⓐ Ⓑ Ⓒ Ⓓ

Part 3: Word Knowledge (WK)

1 Ⓐ Ⓑ Ⓒ Ⓓ	8 Ⓐ Ⓑ Ⓒ Ⓓ	15 Ⓐ Ⓑ Ⓒ Ⓓ	22 Ⓐ Ⓑ Ⓒ Ⓓ	29 Ⓐ Ⓑ Ⓒ Ⓓ
2 Ⓐ Ⓑ Ⓒ Ⓓ	9 Ⓐ Ⓑ Ⓒ Ⓓ	16 Ⓐ Ⓑ Ⓒ Ⓓ	23 Ⓐ Ⓑ Ⓒ Ⓓ	30 Ⓐ Ⓑ Ⓒ Ⓓ
3 Ⓐ Ⓑ Ⓒ Ⓓ	10 Ⓐ Ⓑ Ⓒ Ⓓ	17 Ⓐ Ⓑ Ⓒ Ⓓ	24 Ⓐ Ⓑ Ⓒ Ⓓ	31 Ⓐ Ⓑ Ⓒ Ⓓ
4 Ⓐ Ⓑ Ⓒ Ⓓ	11 Ⓐ Ⓑ Ⓒ Ⓓ	18 Ⓐ Ⓑ Ⓒ Ⓓ	25 Ⓐ Ⓑ Ⓒ Ⓓ	32 Ⓐ Ⓑ Ⓒ Ⓓ
5 Ⓐ Ⓑ Ⓒ Ⓓ	12 Ⓐ Ⓑ Ⓒ Ⓓ	19 Ⓐ Ⓑ Ⓒ Ⓓ	26 Ⓐ Ⓑ Ⓒ Ⓓ	33 Ⓐ Ⓑ Ⓒ Ⓓ
6 Ⓐ Ⓑ Ⓒ Ⓓ	13 Ⓐ Ⓑ Ⓒ Ⓓ	20 Ⓐ Ⓑ Ⓒ Ⓓ	27 Ⓐ Ⓑ Ⓒ Ⓓ	34 Ⓐ Ⓑ Ⓒ Ⓓ
7 Ⓐ Ⓑ Ⓒ Ⓓ	14 Ⓐ Ⓑ Ⓒ Ⓓ	21 Ⓐ Ⓑ Ⓒ Ⓓ	28 Ⓐ Ⓑ Ⓒ Ⓓ	35 Ⓐ Ⓑ Ⓒ Ⓓ

Part 4: Paragraph Comprehension (PC)

1 Ⓐ Ⓑ Ⓒ Ⓓ	4 Ⓐ Ⓑ Ⓒ Ⓓ	7 Ⓐ Ⓑ Ⓒ Ⓓ	10 Ⓐ Ⓑ Ⓒ Ⓓ	13 Ⓐ Ⓑ Ⓒ Ⓓ
2 Ⓐ Ⓑ Ⓒ Ⓓ	5 Ⓐ Ⓑ Ⓒ Ⓓ	8 Ⓐ Ⓑ Ⓒ Ⓓ	11 Ⓐ Ⓑ Ⓒ Ⓓ	14 Ⓐ Ⓑ Ⓒ Ⓓ
3 Ⓐ Ⓑ Ⓒ Ⓓ	6 Ⓐ Ⓑ Ⓒ Ⓓ	9 Ⓐ Ⓑ Ⓒ Ⓓ	12 Ⓐ Ⓑ Ⓒ Ⓓ	15 Ⓐ Ⓑ Ⓒ Ⓓ

Part 5: Mathematics Knowledge (MK)

1 Ⓐ Ⓑ Ⓒ Ⓓ	6 Ⓐ Ⓑ Ⓒ Ⓓ	11 Ⓐ Ⓑ Ⓒ Ⓓ	16 Ⓐ Ⓑ Ⓒ Ⓓ	21 Ⓐ Ⓑ Ⓒ Ⓓ
2 Ⓐ Ⓑ Ⓒ Ⓓ	7 Ⓐ Ⓑ Ⓒ Ⓓ	12 Ⓐ Ⓑ Ⓒ Ⓓ	17 Ⓐ Ⓑ Ⓒ Ⓓ	22 Ⓐ Ⓑ Ⓒ Ⓓ
3 Ⓐ Ⓑ Ⓒ Ⓓ	8 Ⓐ Ⓑ Ⓒ Ⓓ	13 Ⓐ Ⓑ Ⓒ Ⓓ	18 Ⓐ Ⓑ Ⓒ Ⓓ	23 Ⓐ Ⓑ Ⓒ Ⓓ
4 Ⓐ Ⓑ Ⓒ Ⓓ	9 Ⓐ Ⓑ Ⓒ Ⓓ	14 Ⓐ Ⓑ Ⓒ Ⓓ	19 Ⓐ Ⓑ Ⓒ Ⓓ	24 Ⓐ Ⓑ Ⓒ Ⓓ
5 Ⓐ Ⓑ Ⓒ Ⓓ	10 Ⓐ Ⓑ Ⓒ Ⓓ	15 Ⓐ Ⓑ Ⓒ Ⓓ	20 Ⓐ Ⓑ Ⓒ Ⓓ	25 Ⓐ Ⓑ Ⓒ Ⓓ

ANSWER SHEET

Part 6: Electronics Information (EI)

1 A B C D	5 A B C D	9 A B C D	13 A B C D	17 A B C D
2 A B C D	6 A B C D	10 A B C D	14 A B C D	18 A B C D
3 A B C D	7 A B C D	11 A B C D	15 A B C D	19 A B C D
4 A B C D	8 A B C D	12 A B C D	16 A B C D	20 A B C D

Part 7: Auto and Shop Information (AS)

1 A B C D	6 A B C D	11 A B C D	16 A B C D	21 A B C D
2 A B C D	7 A B C D	12 A B C D	17 A B C D	22 A B C D
3 A B C D	8 A B C D	13 A B C D	18 A B C D	23 A B C D
4 A B C D	9 A B C D	14 A B C D	19 A B C D	24 A B C D
5 A B C D	10 A B C D	15 A B C D	20 A B C D	25 A B C D

Part 8: Mechanical Comprehension (MC)

1 A B C D	6 A B C D	11 A B C D	16 A B C D	21 A B C D
2 A B C D	7 A B C D	12 A B C D	17 A B C D	22 A B C D
3 A B C D	8 A B C D	13 A B C D	18 A B C D	23 A B C D
4 A B C D	9 A B C D	14 A B C D	19 A B C D	24 A B C D
5 A B C D	10 A B C D	15 A B C D	20 A B C D	25 A B C D

Part 9: Assembling Objects (AO)

1 A B C D	6 A B C D	11 A B C D	16 A B C D	21 A B C D
2 A B C D	7 A B C D	12 A B C D	17 A B C D	22 A B C D
3 A B C D	8 A B C D	13 A B C D	18 A B C D	23 A B C D
4 A B C D	9 A B C D	14 A B C D	19 A B C D	24 A B C D
5 A B C D	10 A B C D	15 A B C D	20 A B C D	25 A B C D

ASVAB PRACTICE TEST B

Part 1: General Science (GS)

Time: 11 minutes; 25 questions

Directions: In this section, you will be tested on your knowledge of concepts in science generally reviewed in high school. For each question, select the best answer and mark the corresponding oval on your answer sheet.

1. Those wishing to reduce the "bad" cholesterol levels in their blood should avoid

(A) polyunsaturated fats
(B) carbohydrates
(C) monounsaturated fats
(D) saturated fats

2. Water at sea level freezes at what temperature?

(A) 373 K
(B) 0°F
(C) 100°C
(D) 32°F

3. Which of the following subatomic particles has the largest mass?

(A) proton
(B) electron
(C) positron
(D) neutrino

4. Which of the following is an example of an autotroph?

(A) a vulture
(B) an apple tree
(C) a toadstool
(D) a sea anemone

5. The period from 4.6 billion years to 570 million years ago is called the

(A) Phanerozoic eon
(B) Mesozoic era
(C) Paleozoic era
(D) Precambrian eon

6. Which of the following planets is larger than Earth?

(A) Mars
(B) Mercury
(C) Uranus
(D) Venus

7. When these fall into the Earth's gravitational field, they are seen as "falling stars."

(A) asteroids
(B) meteoroids
(C) comets
(D) craters

8. The study of interactions between organisms and their interrelationships with the physical environment is known as

(A) cytology
(B) ecology
(C) physiology
(D) embryology

9. Most absorption of nutrients from food happens in the human body's

(A) stomach
(B) pylorus
(C) small intestine
(D) large intestine

10. High tide occurs

(A) four times a year
(B) once a day
(C) once a year
(D) twice a day

11. Which is NOT a fact according to Newton's Law of Gravitation?

(A) If the mass of the Earth were doubled, the gravitational force on the Earth would double.
(B) The force exerted on the Earth by the Sun is equal and opposite to the force exerted on the Sun by the Earth.
(C) If the mass of the Earth were doubled, the gravitational force on the Earth would quadruple.
(D) If the mass of the Sun were doubled, the force it exerts on the Earth would double.

12. Which of the following types of electromagnetic radiation has the longest wavelength and lowest frequency?

(A) radio waves
(B) microwaves
(C) gamma rays
(D) visible light

13. The ratio by which light is slowed down by a medium is known as the medium's

(A) electromagnetic spectrum
(B) angle of incidence
(C) reflection
(D) refractive index

14. Which of the following has the highest pH?

(A) vinegar
(B) water
(C) baking soda
(D) cranberry juice

15. Foods rich in carbohydrates include all of the following except

(A) potatoes
(B) pasta
(C) fruits
(D) almonds

16. Members of an order are more alike than members of a

(A) class
(B) family
(C) genus
(D) species

17. The number of protons in the nucleus of an atom is its

(A) atomic symbol
(B) atomic weight
(C) atomic number
(D) atomic radius

18. The system of muscles that allows the lungs to expand and contract, drawing air in and out is the

(A) diaphragm
(B) pharynx
(C) trachea
(D) bronchioles

19. The atom of an element with an atomic number of 17 must have

(A) 17 electrons
(B) 17 protons
(C) 17 neutrons
(D) an atomic mass of 17

20. Because mushrooms absorb nutrients from decaying leaves, they are classified as

(A) autotrophs
(B) anaerobes
(C) saprophytes
(D) protozoans

21. The blood vessel with the LEAST oxygenated blood is the

(A) pulmonary artery
(B) aorta
(C) pulmonary vein
(D) arterioles

22. Saliva in the mouth begins the process of breaking down

(A) starch
(B) fat
(C) protein
(D) vitamins

23. In which of the following kingdoms do cells lack nuclei?

(A) Fungi
(B) Monera
(C) Plantae
(D) Protista

24. An atom that is not electrically neutral is called a(n)

(A) isotope
(B) positron
(C) ion
(D) allotrope

25. Which of the following is an example of a metamorphic rock?

(A) granite
(B) marble
(C) limestone
(D) shale

STOP. IF YOU FINISH BEFORE THE TIME IS UP, YOU MAY CHECK OVER YOUR WORK ON THIS PART ONLY.

Part 2: Arithmetic Reasoning (AR)

Time: 36 minutes; 30 questions

Directions: In this section, you are tested on your ability to use arithmetic. For each question, select the best answer and mark the corresponding oval on your answer sheet.

1. A certain machine caps 5 bottles every 2 seconds. At this rate, how many bottles will be capped in 1 minute?

(A) 75
(B) 150
(C) 225
(D) 300

2. Jonah traveled 650 miles on his most recent trip and averaged 25 miles to the gallon. If gasoline cost $1.30 per gallon, how much did he spend on his trip?

(A) $26.00
(B) $27.30
(C) $32.50
(D) $33.80

3. If there are approximately 3.86 liters in a gallon, and gasoline costs $1.54 per gallon, to the nearest penny what is the cost of a liter of gasoline?

(A) $0.35
(B) $0.40
(C) $0.44
(D) $0.47

4. How many minutes are there in one week?

(A) 3,600
(B) 7,200
(C) 10,080
(D) 86,400

5. Pat deposited 15% of last week's take-home pay into a savings account. If she deposited $37.50, what was last week's take-home pay?

(A) $56.25
(B) $112.50
(C) $225.00
(D) $250.00

6. If the ratio of males to females in a group of students is 3:5, which of the following could be the total number of students in the group?

(A) 148
(B) 150
(C) 152
(D) 154

7. A car travels 288 miles in 6 hours. At that rate, how many miles will it travel in 8 hours?

(A) 360
(B) 368
(C) 376
(D) 384

8. Martin's average score after 4 tests is 89. What score on the fifth test would bring Martin's average up to exactly 90?

(A) 91
(B) 92
(C) 93
(D) 94

9. In 2010, the population of town A was 9,400 and the population of town B was 7,600. Since then, each year the population of town A has decreased by 100 and the population of town B has increased by 100. Assuming that in each case the rate continues, in what year will the two populations be equal?

(A) 2019
(B) 2020
(C) 2027
(D) 2028

10. One number is 5 times another number and their sum is −60. What is the lesser of the two numbers?

(A) −10
(B) −12
(C) −48
(D) −50

11. Jane gets paid $6.00 for each of the first 40 toy cars she makes in a week. For any additional toy cars beyond 40, her pay increases by 50%. How much does Jane get paid in a week in which she makes 48 toy cars?

(A) $288
(B) $300
(C) $312
(D) $321

12. If a bora = 2 fedis, and a fedi = 3 glecks, how many boras are equal to 48 glecks?

(A) 8
(B) 16
(C) 32
(D) 96

13. Nine temperature readings were taken, one reading every four hours, with the first reading taken at 12 p.m. (that is, at noon). What will be the time when the final reading is taken?

(A) 4 p.m.
(B) 8 p.m.
(C) 12 a.m.
(D) 8 a.m.

14. A school raised monthly tuition payments from $225.00 per month to $300.00 per month. The percent of the tuition increase is which of the following?

(A) 25%
(B) $33\frac{1}{3}\%$
(C) $66\frac{1}{2}\%$
(D) 75%

15. A truck going at a rate of 20 miles per hour takes 6 hours to complete a trip. How many fewer hours would the trip have taken if the truck were traveling at a rate of 30 miles per hour?

(A) 4
(B) 3
(C) 2
(D) 1

16. If a barrel has the capacity to hold 75 gallons, how many gallons does it contain when it is $\frac{3}{5}$ full?

(A) 45
(B) 48
(C) 54
(D) 60

17. If a salary of $45,000 is subject to a 40% deduction, the new salary is:

(A) $18,000
(B) $27,000
(C) $30,000
(D) $36,000

18. 8! =

(A) 8
(B) 64
(C) 40,000
(D) 40,320

19. Fifteen people chip in $40 each to throw a birthday bash for a mutual friend. If 35% of the money is spent on a group gift and the rest is spent on the party, how much was spent on the party?

(A) $210
(B) $300
(C) $350
(D) $390

20. If an employee worked a total of $33\frac{1}{2}$ hours over five days, what was the average amount of time that the employee worked each day?

(A) 6 hours, 35 minutes
(B) 6 hours, 40 minutes
(C) 6 hours, 42 minutes
(D) 6 hours, 45 minutes

21. If a train starting out at point A travels 180 miles at a rate of 60 miles per hour and then 150 miles at a rate of 75 miles per hour before arriving at point B, what was the average rate, in miles per hour, for the entire trip?

(A) 66
(B) 67.5
(C) 68.5
(D) 70

22. After being discounted by 20%, a bicycle sells for \$140.00. What was the original price of the bicycle?

(A) \$155.00
(B) \$162.00
(C) \$175.00
(D) \$180.00

23. A checking account has a balance of \$1,162.76. After a check for \$352.68 is deposited into the account and three checks are drawn from the account for the amounts of \$152.45, \$82.85, and \$255.50, what is the new balance for the checking account?

(A) \$1,024.54
(B) \$1,024.64
(C) \$1,025.54
(D) \$1,025.64

24. Harold works 4.5 hours a day, 3 days each week after school. He is paid \$4.25 per hour. How much is his weekly pay rounded to the next highest cent?

(A) \$13.50
(B) \$19.13
(C) \$54.00
(D) \$57.38

25. Convert $\frac{6}{24}$ into a decimal.

(A) 0.2
(B) 0.25
(C) 0.4
(D) 4.0

26. What is the total cost of 4 sheets of 23-cent stamps, 2 sheets of 37-cent stamps, and 3 sheets of 60-cent stamps if each sheet has 100 stamps?

(A) \$336
(B) \$346
(C) \$356
(D) \$366

27. The Pacific standard time (PST) in Los Angeles is 3 hours earlier than the Eastern standard time (EST) in New York. A plane leaves Los Angeles at 5:30 p.m. PST and arrives in New York City at 12:45 a.m. EST. How long did the flight take?

(A) 4 hours, 15 minutes
(B) 4 hours, 45 minutes
(C) 6 hours, 45 minutes
(D) 7 hours, 15 minutes

28. After the price of a digital camera is discounted 25%, the camera sells for \$120. What was the original price of the camera?

(A) \$145
(B) \$150
(C) \$160
(D) \$180

29. The square root of 225 is

(A) 5
(B) 15
(C) 25
(D) 112

30. If it takes 20 minutes to type 5 pages, how long will it take to type a 162-page document?

(A) 10 hours, 12 minutes
(B) 10 hours, 36 minutes
(C) 10 hours, 45 minutes
(D) 10 hours, 48 minutes

STOP. IF YOU FINISH BEFORE THE TIME IS UP, YOU MAY CHECK OVER YOUR WORK ON THIS PART ONLY.

Part 3: Word Knowledge (WK)

Time: 11 minutes; 35 questions

<u>Directions</u>: In this section, you are tested on the meaning of words. Each of the following questions has an underlined word. Select the answer that most nearly means the same as the underlined word and mark the corresponding oval on your answer sheet.

1. <u>Listless</u> most nearly means
   - (A) capable
   - (B) lethargic
   - (C) talkative
   - (D) bungled

2. His ability to <u>comprehend</u> complex ideas made him a great teacher.
   - (A) create
   - (B) loosen
   - (C) understand
   - (D) compute

3. Whichever one scores the most points will be the <u>victor</u>.
   - (A) singer
   - (B) manufacturer
   - (C) best
   - (D) winner

4. <u>Lackadaisical</u> most nearly means
   - (A) talented
   - (B) relaxed
   - (C) actual
   - (D) personable

5. It was his <u>contention</u> that more money should be spent on education.
   - (A) celebration
   - (B) position
   - (C) ideal
   - (D) sacrifice

6. The arrest wasn't even <u>newsworthy</u>.
   - (A) notable
   - (B) defaming
   - (C) credible
   - (D) safe

7. His attendance in class was, if anything, <u>sporadic</u>.
   - (A) latent
   - (B) constant
   - (C) irregular
   - (D) common

8. <u>Potential</u> most nearly means
   - (A) scary
   - (B) trustable
   - (C) telling
   - (D) possible

9. <u>Blacklisted</u> most nearly means
   - (A) sacred
   - (B) barred
   - (C) old-fashioned
   - (D) barren

10. It is <u>presumptuous</u> to expect a hundred volunteers for such a tough task.
    - (A) exciting
    - (B) predictable
    - (C) satisfying
    - (D) overconfident

11. A <u>memento</u> can provide a way to remember your vacation once you are home.
    - (A) sacrifice
    - (B) emblem
    - (C) souvenir
    - (D) symbol

12. The school's president is far too concerned with the <u>legacy</u> he will leave.
    - (A) memory
    - (B) salary
    - (C) tension
    - (D) grace

13. The economy has some parents worried that their children's future is in <u>jeopardy</u>.

(A) limbo
(B) danger
(C) custody
(D) paradise

14. <u>Integrity</u> most nearly means

(A) ability
(B) coercion
(C) togetherness
(D) honesty

15. He was punished too harshly for such a minor <u>infraction</u>.

(A) aptitude
(B) belief
(C) gift
(D) violation

16. <u>Collective</u> most nearly means

(A) group
(B) rare
(C) soft
(D) free

17. <u>Hoax</u> most nearly means

(A) trick
(B) story
(C) treat
(D) complication

18. The <u>genesis</u> of the idea was scribbled on a napkin.

(A) cancellation
(B) detail
(C) origin
(D) craft

19. The brothers' <u>feud</u> goes back to their childhood days.

(A) love
(B) dispute
(C) sensitivity
(D) belief

20. <u>Fraudulent</u> most nearly means

(A) pretty
(B) fake
(C) disgusting
(D) sanitary

21. <u>Adept</u> most nearly means

(A) older
(B) talkative
(C) cautious
(D) skilled

22. She is an <u>ardent</u> supporter of families' rights.

(A) passionate
(B) testy
(C) unaffiliated
(D) tired

23. The town suffered from <u>rampant</u> overcrowding and crime.

(A) overstated
(B) widespread
(C) old
(D) predictable

24. Such a <u>puny</u> fence stands no chance with that big dog.

(A) satisfying
(B) colorful
(C) ghastly
(D) weak

25. <u>Sinister</u> most nearly means

(A) brotherhood
(B) believable
(C) gaudy
(D) evil

26. <u>Witty</u> most nearly means

(A) sulking
(B) bloody
(C) clever
(D) careless

27. The witness's <u>vivid</u> account of the crime certainly opened the jury's eyes.

(A) clear
(B) unobstructed
(C) false
(D) condescending

28. <u>Relinquish</u> most nearly means

(A) concede
(B) color
(C) trick
(D) berate

29. I can only hope that these good times <u>recur</u>.

(A) disappear
(B) satiate
(C) return
(D) let go

30. <u>Incite</u> most nearly means

(A) induce
(B) intrigue
(C) blow
(D) react

31. It was wise of George to <u>implement</u> such a bold strategy while his critics were distracted.

(A) enact
(B) suggest
(C) trivialize
(D) investigate

32. <u>Query</u> most nearly means

(A) telling
(B) question
(C) bogus
(D) captive

33. No one should be in awe of such a <u>timid</u> animal.

(A) crazy
(B) cowardly
(C) dangerous
(D) salty

34. <u>Colleague</u> most nearly means

(A) enemy
(B) savior
(C) peer
(D) pedant

35. His opponent proved to have a <u>potent</u> mixture of toughness and intelligence.

(A) costly
(B) derelict
(C) sad
(D) powerful

STOP. IF YOU FINISH BEFORE THE TIME IS UP, YOU MAY CHECK OVER YOUR WORK ON THIS PART ONLY.

Part 4: Paragraph Comprehension (PC)

Time: 13 minutes; 15 questions

Directions: This section contains paragraphs followed by incomplete statements or questions. For each question, read the paragraph and select the answer that best completes the statements or answers the question that follows. Mark the corresponding oval on your answer sheet.

The character of Sherlock Holmes illustrates author Arthur Conan Doyle's admiration for the logical mind. In each case that Holmes investigates, he is able to use the most seemingly insignificant evidence to track down his opponent. In fact, Holmes's painstaking attention to detail often reminds the reader of Charles Darwin's *On the Origin of the Species*, published some 20 years earlier.

1. The author compares Sherlock Holmes to Charles Darwin in order to

(A) show Holmes's educational background
(B) explain evolution
(C) show that both were logical and meticulous
(D) praise Darwin for his research skills

The climate of a major city is often very different from the climate of the surrounding areas. However, the geographic differences between the city and the country do not have to be dramatic to show major climatic differences. Even between the center of a city and its suburbs, there are often differences in air temperature, humidity, wind speed, and direction. Tall buildings, paved streets, and parking lots affect such patterns as wind flow and water runoff.

2. According to the passage, the relationship between cities and climate is

(A) based on pollution controls
(B) based on urban structures and the environment
(C) based on city politics
(D) based on rural areas

Marianne had seen it all on television—the Cuban Missile Crisis, man's first steps on the Moon, the fall of the Berlin Wall, even the economic boom of the 1990s. So when she won the lottery and the news vans started gathering outside her house, it didn't yet dawn on her just how much her life would be changed. A little attention doesn't hurt you, she thought. But when she stepped outside to an army of eager young reporters, she began to wonder if there weren't bigger events somewhere in need of coverage.

3. The character in the passage would probably do which of the following?

(A) wear revealing clothing
(B) watch news on TV
(C) dye her hair bright colors
(D) refuse the money publicly

In the "Gunpowder Plot" of 1605, a group of conspirators planned to blow up the English king and Parliament. The conspirators were a group of English Catholics who objected to the government's religious policies and who decided to carry out a daring assassination. They rented a cellar under the Palace of Westminster and hid 20 barrels of gunpowder there, intending to explode the gunpowder when the king and Parliament met. However, the conspirators were unable to keep the plot secret and their infamous plan was foiled.

4. The word *infamous* most nearly means

(A) heroic
(B) funny
(C) secret
(D) notorious

Just as Johnny hit yet another red light, his cell phone began to ring. He looked at his watch and saw that he was already 20 minutes late to what might be the most important meeting of his career. Reaching for his phone, he knocked over his drink, spilling piping hot coffee all over his lap. He screamed loudly in his car, and though it looked like no one could hear him, a woman nearby ran away from the edge of the car quickly.

5. Johnny's attitude is

(A) perturbed
(B) tender
(C) excited
(D) snooty

The presence or absence of water can have a dramatic effect on the plants in a given environment. The desert, for example, is a biologically complex area where rain is often violent and unpredictable. Since desert plants are dependent on the occasional rain for survival, they must act quickly to make the most of it when it appears. Many desert plants, as a result, can complete an entire life cycle in a matter of months or even weeks. The more barren a desert is, and thus the less rain, the rarer and more astounding these life cycles and the blooms that accompany them will be.

6. According to the passage, the life cycle of desert plants

(A) is unrelated to rain levels
(B) is directly related to their color
(C) is akin to that of local wildlife
(D) may be short and dynamic

It is a common belief among writers that great art is born from experience. However, some of the greatest writers in literary history have been people with a very limited knowledge of the world. Novelist Jane Austen, for example, didn't venture far beyond her circle of family and friends. And yet, just by observing the people around her, she was able to write acclaimed comedies about love and marriage. Similarly, Robert Louis Stevenson, the author of classic adventures such as *Treasure Island* and *Kidnapped*, wrote many of his adventures without having similar experiences. Their achievements illustrate that if you have a good imagination, you can write a novel, no matter how unadventurous your life may seem.

7. Based on the information in the passage, some of the greatest writers
   - (A) always write about people and their social scenes
   - (B) write only adventure stories
   - (C) rely on imagination and not just experience
   - (D) get advice from close family and friends

Tania is an extremely versatile actress. In a recent series of plays, she played a schoolteacher, two divorcees, three best friends, and a stepmother. She played a divorcee throughout the first play, but in subsequent plays she often played more than one part. Who would have suspected her acting prowess a few years ago when she ran crying off the stage with a terrible case of stage fright?

8. The word *prowess* most nearly means
   - (A) knowledge
   - (B) disaster
   - (C) talent
   - (D) bias

Many people dream about living on a coral island, but probably few of us would be able to describe one with any accuracy. Popular books and films create a romantic image of these islands, but it is not always as nice if seen from the land. Beneath the waves, however, the coral island is a fantastic and very beautiful world, depending entirely upon a complex web of interrelationships between plants and animals. The environment of the coral reef is formed over thousands of years by the life cycle of vast numbers of animals.

9. One logical conclusion to draw from this passage is
   - (A) coral animals are more colorful than other animals
   - (B) the true beauty of coral islands is underwater
   - (C) all coral islands are tropical paradises
   - (D) coral islands are formed mostly by volcanos

Questions 10 and 11 both refer to the following passage.

Few of the immigrants of the period just before the turn of the century found life in America easy. Many of those who lacked professional skills and did not speak English found themselves living in slum conditions in the busy cities of the Northeast, exploited by their employers and trapped at poverty level. Around the turn of the century, a number of different organizations made efforts to help these newly arrived immigrants adapt to American life. While many groups emerged as leaders in these efforts to aid immigration, one organization preached a much colder answer to adapting to America.

A conservative group called the Daughters of the American Revolution approached immigrants with the expectation that newcomers should completely adopt American customs and culture. Consequently, they supported laws that required immigrants to take oaths of loyalty and to pass English language tests. They also tried to discourage the use of languages other than English in the schools.

10. Which of the following best tells what this passage is about?
    - (A) how some sought to assimilate immigrants into America
    - (B) the cultural contributions offered by immigrants
    - (C) the history of immigration in the United States
    - (D) how immigrant life affected U.S. cities

11. The author implies that the Daughters of the American Revolution
    - (A) understood the difficulties facing new immigrants
    - (B) were ruthless in their intentions toward immigrants
    - (C) believed that immigrants would not survive in America
    - (D) preached fitting in, not standing out

Some animals use coloring to safeguard themselves from predators. In certain cases, an animal adapts in color, shape, and behavior in order to blend into its environment. The camouflage of the pale green tree frog is a good example of this. The tree frog blends so perfectly into its surroundings that when it sits motionless it is all but invisible against a background of leaves. Another type of camouflage, shown by zebras and leopards, is a pattern that diverts the eye from the outline of the animal. The chameleon, even more versatile than these, changes color in just a few minutes to match whatever surface it happens to be lying on or clinging to.

12. According to the passage, a reason that animals use camouflage is

    (A) to divert the eye from the animal's outline
    (B) to make nesting more safe
    (C) so that frogs' and chameleons' habitats remain protected
    (D) to ensure predatory dominance

Perhaps the biggest surprise of the Voyager mission was the discovery on the moons of Jupiter of intense volcanic activity. First seen on the moon Io, the eruptions were recognized as plumes of dust and gas and were immediately noticeable on the Voyager photographs. Further inspection revealed at least seven such events occurring all at once on Io's otherwise frigid surface. At other points, scientists detected three hot spots believed to be ponds of molten lava, sulphur, or sodium. The largest of these hot lakes was estimated to have a greater surface area than the state of Hawaii.

13. What was the most unexpected fact to emerge from the Voyager photographs?

    (A) the size of Io's molten lakes
    (B) the discovery of volcanic activity on Io
    (C) the disappearance of impact sites on Europa
    (D) the evidence of asteroid bombardment on all four moons

Questions 14 and 15 both refer to the following passage.

Most people think that the Hula Hoop was a fad born in the 1950s, but in fact people were doing much the same thing with circular hoops made from grapevines and stiff grasses all over the ancient world. More than 3,000 years ago, children in Egypt played with large hoops of dried grapevines. The toy was propelled along the ground with a stick or swung around at the waist. During the fourteenth century, a "hooping" craze swept England, and was equally popular among adults and children.

The word *hula* became associated with the toy in the early 1800s when British sailors visited the Hawaiian Islands and noted the similarity between hooping and hula dancing. In 1957, an Australian company began making wood rings for sale in retail stores. The item attracted the attention of Wham-O, a fledgling California toy manufacturer. The plastic Hula Hoop was introduced in 1958 and was an instant hit.

14. According to the passage, all of the following statements are true EXCEPT

    (A) most people do not appreciate the origins of the Hula Hoop
    (B) early precursors of the Hula Hoop were made of grape leaves and stiff grasses
    (C) early precursors to the Hula Hoop were primarily children's toys
    (D) the Hula Hoop was an early success for the toy maker Wham-O

15. The author's primary purpose in this passage is to

    (A) describe the way that fads like the Hula Hoop come and go
    (B) discuss the origins of the Hula Hoop
    (C) explain how the Hula Hoop got its name
    (D) question the reasons for the Hula Hoop's popularity

STOP. IF YOU FINISH BEFORE THE TIME IS UP, YOU MAY CHECK OVER YOUR WORK ON THIS PART ONLY.

Part 5: Mathematics Knowledge (MK)

Time: 24 minutes; 25 questions

Directions: In this section, you will be tested on your knowledge of basic mathematics. For each question, select the best answer and mark the corresponding oval on your answer sheet.

1. If $\frac{3}{4} + 1\frac{1}{3} = \frac{5}{6} - x$, then $x =$

(A) $-2\frac{11}{12}$
(B) $-1\frac{1}{4}$
(C) $\frac{1}{6}$
(D) $1\frac{1}{4}$

2. For all a, $(3a + 4)(4a - 3) =$

(A) $3a^2 - 4$
(B) $9a^2 - 4$
(C) $12a^2 + 7a - 12$
(D) $9a^2 - 16$

3. If $2m > 24$ and $3m < 48$, which of the following could NOT be a possible value for m?

(A) 13
(B) 14
(C) 15
(D) 16

4. $\sqrt{75} + \sqrt{108} =$

(A) $11\sqrt{3}$
(B) $6\sqrt{5}$
(C) $5\sqrt{6}$
(D) $\sqrt{183}$

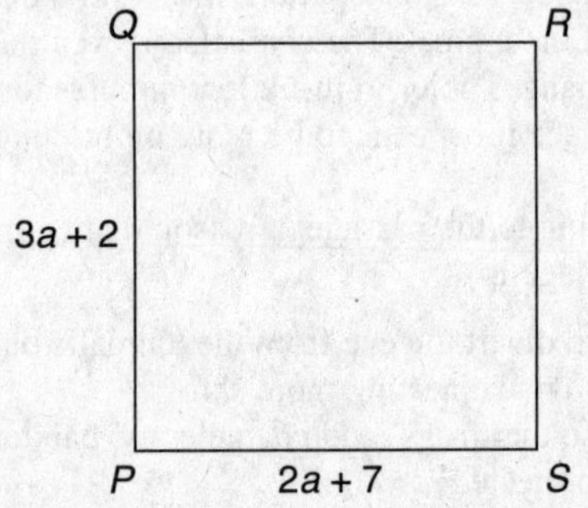

5. In the figure above, if $PQRS$ is a square, what is the value of a?

(A) $\frac{9}{2}$
(B) 5
(C) 7
(D) 9

6. In triangle ABC, the degree measures of the three interior angles are in the ratio of 1:2:3. What is the difference in the degree measures between the largest and the smallest angles?

(A) 30
(B) 60
(C) 90
(D) 120

7. What is the perimeter of a right triangle with perpendicular sides of lengths 3 and 6?

(A) $9 + 3\sqrt{5}$
(B) $9 + 5\sqrt{3}$
(C) $12 + 3\sqrt{3}$
(D) 18

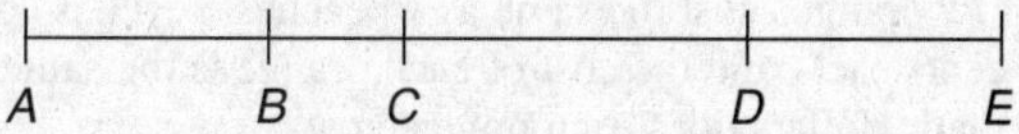

8. In the diagram above, if $AD = BE = 6$, $AE = 8$ and $CD = 3BC$, then $BC =$

(A) 4
(B) 3
(C) 2
(D) 1

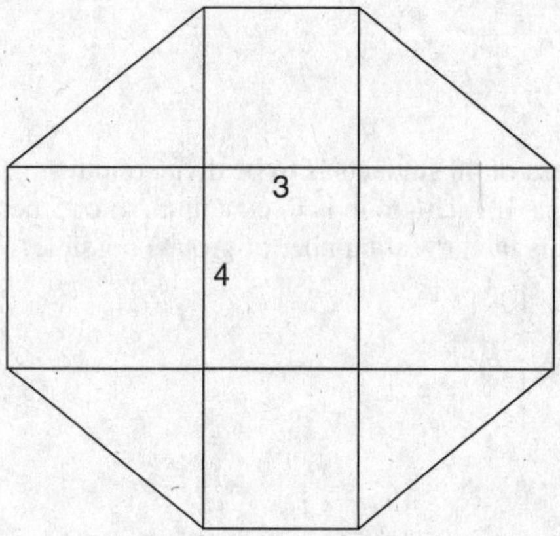

9. The figure above is composed of nine regions: four squares, four triangles, and one rectangle. If the rectangle has sides of length 4 and width 3, what is the perimeter of the entire figure?

(A) 30
(B) 34
(C) 40
(D) 48

10. If $\frac{\sqrt{n}}{3}$ is an even integer, which of the following could be the value of n?

(A) 27
(B) 48
(C) 81
(D) 144

11. Gheri is n years old. Carl is 6 years younger than Gheri and 2 years older than Jean. What is the sum of the ages of all three?

(A) $3n + 4$
(B) $3n - 4$
(C) $3n - 8$
(D) $3n - 14$

12. Which of the following is NOT a prime number?

(A) 2
(B) 37
(C) 51
(D) 67

13. What is the area of a circle whose circumference is 2π?

(A) $\frac{\pi}{2}$
(B) π
(C) 2π
(D) 4π

14. A high school band is composed of 13 freshmen, 20 sophomores, 16 juniors, and 15 seniors. What is the probability that a band member chosen at random will be a sophomore?

(A) $\frac{2}{5}$
(B) $\frac{1}{3}$
(C) $\frac{4}{13}$
(D) $\frac{5}{16}$

15. If $k - 3 = -\frac{5}{3}$, then $3 - k =$

(A) $-\frac{5}{3}$
(B) $-\frac{3}{5}$
(C) $\frac{3}{5}$
(D) $\frac{5}{3}$

16. If $x \neq 0$ and $\frac{3x}{5} = 3x^2$, then $x =$

(A) $\frac{1}{5}$
(B) $\frac{3}{5}$
(C) 3
(D) 5

17. If a, b, c, and d are all positive numbers, $a = 2b$, $\frac{1}{2}b = c$, and $4c = 3d$, what is the value of $\frac{d}{a}$?

(A) $\frac{1}{3}$
(B) $\frac{3}{4}$
(C) $\frac{4}{3}$
(D) 3

18. How many rectangular tiles with dimensions of 3 inches by 4 inches are needed to cover a rectangular area with dimensions of 2 feet and 3 feet?

(A) 60
(B) 72
(C) 120
(D) 144

19. For how many integer values of x will $\frac{7}{x}$ be greater than $\frac{1}{4}$ and less than $\frac{1}{3}$?

(A) 6
(B) 7
(C) 12
(D) 28

20. A circular hole is covered by a circular cover that has a diameter of 14 inches. If the hole has a diameter of 12 inches, how much greater than the area of the hole is the area of the cover, in square inches?

(A) π
(B) 13π
(C) 36π
(D) 52π

21. Which of the following is closest in value to the decimal 0.40?

(A) $\frac{1}{3}$
(B) $\frac{3}{8}$
(C) $\frac{1}{2}$
(D) $\frac{4}{7}$

22. If x oranges cost the same as y peaches, peaches cost 39 cents each, and the cost of each orange is the same, how many dollars does each orange cost?

(A) $\frac{39y}{100x}$
(B) $\frac{39x}{100y}$
(B) $\frac{3{,}900}{xy}$
(D) $\frac{39x}{y}$

23. A class of 40 students is to be divided into smaller groups. If each group is to contain 3, 4, or 5 people, what is the largest number of groups possible?

(A) 10
(B) 12
(C) 13
(D) 14

24. If the area of a triangle is 36 and its base is 9, what is the length of the altitude to that base?

(A) 4
(B) 6
(C) 8
(D) 12

25. If $x = -5$, then $2x^2 - 6x + 5 =$

(A) 15
(B) 25
(C) 85
(D) 135

STOP. IF YOU FINISH BEFORE THE TIME IS UP, YOU MAY CHECK OVER YOUR WORK ON THIS PART ONLY.

Part 6: Electronics Information (EI)

Time: 9 minutes; 20 questions

Directions: In this section, you will be tested on your knowledge of electronics basics. For each question, select the best answer and mark the corresponding oval on your answer sheet.

1. Voltage and current are

(A) inversely proportional
(B) directly proportional
(C) unrelated
(D) sometimes proportional

2. For a current-carrying wire with a DC source providing a constant 100 amps, which of the following set-ups will produce the largest magnetic field strength at any given point?

(A) a simple single-loop circuit made of two meters of wire
(B) a simple single-loop circuit made of five meters of wire
(C) wrapping the two meters of current-carrying wire in five coiled loops around an iron nail
(D) wrapping the five meters of current-carrying wire in five coiled loops

3. Which of the following causes the least resistance to a consistent direct current?

(A) a capacitor
(B) a fixed resistor
(C) a variable resistor
(D) an inductor

4. Which is the base in the transistor symbol below?

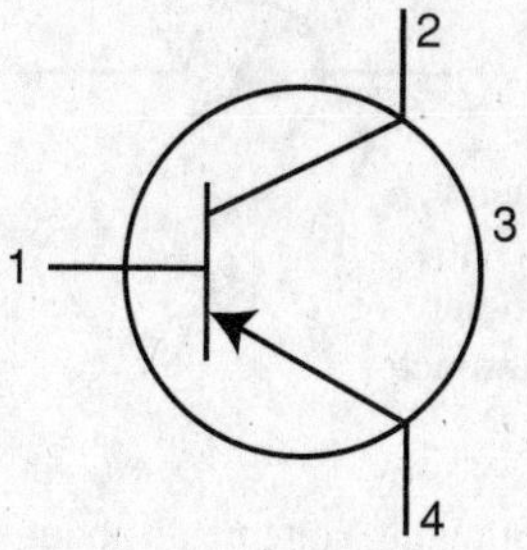

(A) 1
(B) 2
(C) 3
(D) 4

5. If each resistor in this circuit equals 750 ohms, what is the total resistance in this circuit?

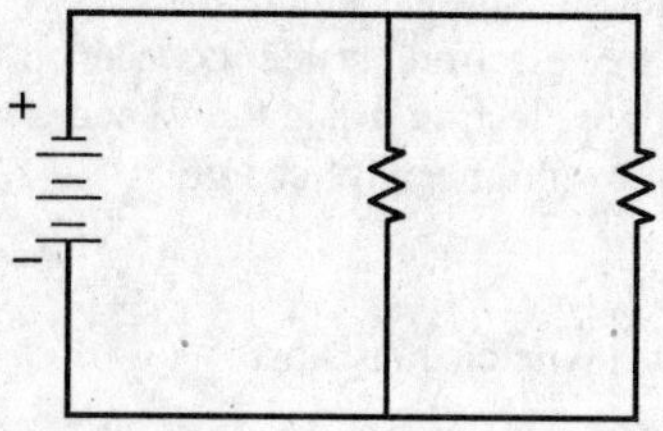

(A) 250 ohms
(B) 375 ohms
(C) 750 ohms
(D) 1,500 ohms

6. Counter-emf is produced by a process known as

(A) inductive resistance
(B) capacitive resistance
(C) self-induction
(D) doping

7. In a simple series circuit, what resistance will allow 5 mA to flow when 50 volts is applied to it?

(A) 0.1 Ω
(B) 250 Ω
(C) 10 K Ω
(D) 250 K Ω

8. What is the proper term for the circuit pictured?

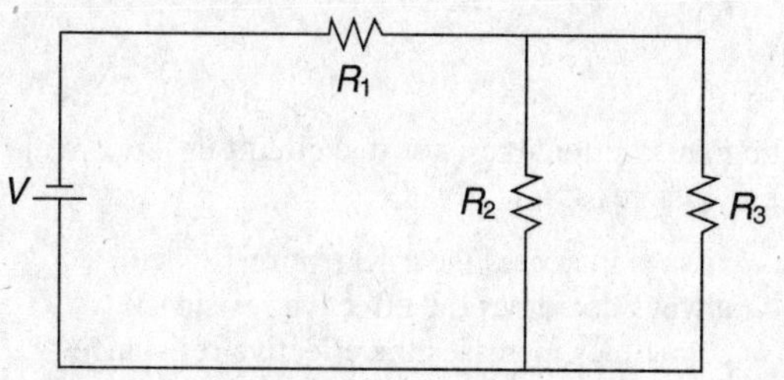

(A) series circuit
(B) parallel circuit
(C) series-parallel circuit
(D) integrated circuit

9. Adding a resistor in parallel with an existing load
   (A) always increases the effective resistance
   (B) always decreases the effective resistance
   (C) sometimes increases the effective resistance
   (D) sometimes decreases the effective resistance

10. N-type material can conduct electricity because of
   (A) silicon's unique structure
   (B) a free electron outside the valence shell
   (C) a free electron inside the valence shell
   (D) an overall negative charge

11. Which is a type of transistor?
   (A) NPN
   (B) NNP
   (C) PPN
   (D) NNN

12. What is the effective resistance of a 2 Ω resistor and 4 Ω resistor if they are wired in parallel?
   (A) $\frac{3}{4}$ Ω
   (B) $1\frac{1}{3}$ Ω
   (C) 3 Ω
   (D) 6 Ω

13. A circuit contains two 3 Ω resistors in series. What is their effective resistance?
   (A) $\frac{2}{3}$ Ω
   (B) $\frac{3}{2}$ Ω
   (C) 3 Ω
   (D) 6 Ω

14. Adding an additional resistor to a circuit in series with the previous resistor
   (A) always increases the effective resistance
   (B) always decreases the effective resistance
   (C) sometimes increases the effective resistance
   (D) sometimes decreases the effective resistance

15. An element with five electrons in its valence shell is a(n)
   (A) conductor
   (B) semiconductor
   (C) insulator
   (D) metal

16. An ampere is defined as one ______ of electrical charge flowing past a point in one second.
   (A) hertz
   (B) coulomb
   (C) ohm
   (D) voltage

17. Unlike capacitors, inductors
   (A) allow DC current to pass easily
   (B) store charge.
   (C) turn DC power around to increase power
   (D) have two parallel plates

18. A circuit initially contains only one load which has a resistance of 5 Ω. Assuming this is the only component of the circuit which provides any resistance initially, which of the following additions would decrease the effective resistance of the circuit by half?
   (A) adding a resistor of 2.5 Ω in series with the original load
   (B) adding a resistor of 2.5 Ω in parallel with the original load
   (C) adding a resistor of 5 Ω in series with the original load
   (D) adding a resistor of 5 Ω in parallel with the original load

19. This is the symbol for which type of meter?

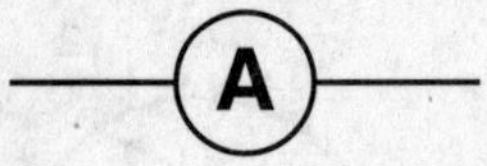

   (A) voltmeter
   (B) ammeter
   (C) ohmmeter
   (D) galvanometer

20. All of the following statements about series circuits are true, EXCEPT
   (A) current flow is the same throughout
   (B) there is only one path for current to follow
   (C) current flow varies throughout the circuit
   (D) voltage drops across the loads depend on their resistance

STOP. IF YOU FINISH BEFORE THE TIME IS UP, YOU MAY CHECK OVER YOUR WORK ON THIS PART ONLY.

Part 7: Auto and Shop Information (AS)

Time: 11 minutes; 25 questions

Directions: In this section, you will be tested on your knowledge of automotive and shop basics. For each question, select the best answer and mark the corresponding oval on your answer sheet.

1. What is detonation?
   (A) when an air-fuel mixture explodes, rather than burns
   (B) if combustion is started by something other than electricity at the spark plug
   (C) rapid, thorough burning of a compressed air-fuel mixture, initiated by a spark from the engine's ignition system
   (D) slow burning of an air-fuel mixture

2. The ignition ________ coil generates the high voltage for the spark plugs.
   (A) primary
   (B) preliminary
   (C) secondary
   (D) tertiary

3. Engine coolant is normally made up of a ______ mix of antifreeze and water.
   (A) 60/40
   (B) 75/25
   (C) 50/50
   (D) 25/75

4. A socket can be used with which tool?
   (A) a ratchet only
   (B) an impact wrench only
   (C) either a ratchet or an impact wrench
   (D) none of the above

5. An automotive technician who wants to measure the width of an object might use any of the following tools EXCEPT
   (A) an outside micrometer
   (B) a tape measure
   (C) a steel rule
   (D) a Crescent wrench

6. This image depicts what stroke in the four-stroke cycle?

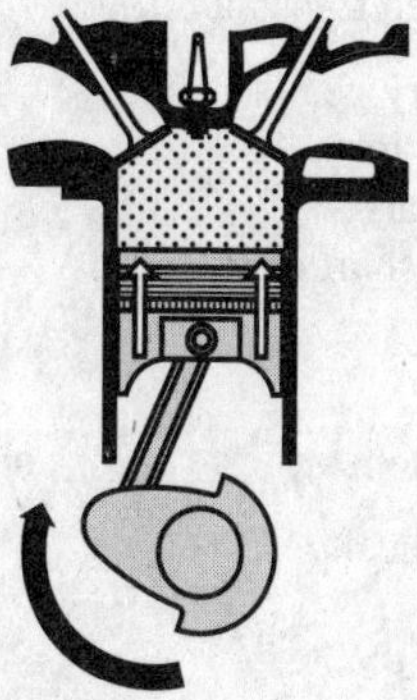

   (A) intake stroke
   (B) exhaust stroke
   (C) power stroke
   (D) compression stroke

7. A fuel injector sprays fuel into the intake air stream as it receives electrical signals from the
   (A) fuel rail
   (B) intake manifold
   (C) fuel pressure regulator
   (D) powertrain control module

8. A general purpose claw hammer would have a ______ oz head.
   (A) 7
   (B) 13
   (C) 40
   (D) 212

9. What would this tool be used to do?

   (A) stop a rotating motor
   (B) start a hole for drilling
   (C) fasten two pieces of metal
   (D) stabilize a car jack

10. Spark plugs are threaded into the
   (A) battery
   (B) distributor
   (C) cylinder head
   (D) coil wire

11. Automatic transmissions use __________ to create speed ratios and transmit torque.
   (A) planetary gear sets
   (B) idler gears
   (C) spur gears
   (D) countershaft gears

12. The brake system is __________ operated.
   (A) hydraulically
   (B) cable
   (C) electrically
   (D) wire

13. Which of the following saws is designed to cut metal?

(A) 

(B) 

(C) 

(D) 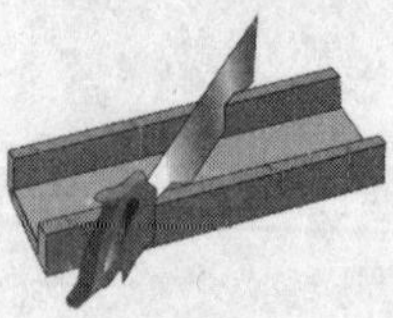

14. MIG welding is also known as
   (A) Russian welding
   (B) coaxle welding
   (C) electric-bit welding
   (D) wire-feed welding

15. What is the name for the tool below?

   (A) carpenter's level
   (B) machinist's level
   (C) line level
   (D) striding level

16. Contact points would be found in __________ ignition systems.
   (A) coil-on-plug
   (B) distributor-less
   (C) electronic
   (D) none of the above

17. To remove a small sliver of wood in order to create a better fit, use
   (A) a plane
   (B) a flat file
   (C) needle-nosed pliers
   (D) Johnson bars

18. The ______________ connects the steering box to the steering linkage.
   (A) center link
   (B) idler arm
   (C) tie rod end
   (D) pitman arm

19. The catalytic converter is part of the vehicle's __________ system.
   (A) fuel intake
   (B) braking
   (C) emission control
   (D) steering

20. Which of the following additions often increases the versatility of slip-joint pliers?
   (A) ratchet head
   (B) extension capability
   (C) wire cutter
   (D) variable speeds

21. Advancing the ignition timing means that the spark takes place ___________ in the combustion cycle.

(A) earlier
(B) later
(C) at the same time
(D) none of the above

22. Which of the following hammers is used for forming soft metal?

(A) 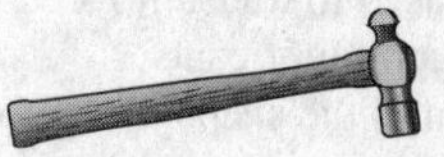

(B) 

(C) 

(D) 

23. Long-short arm suspensions involve all of the following EXCEPT

(A) ball joints
(B) upper control arms
(C) lower control arms
(D) torque converters

24. To cut off a bolt head, you would most likely use which of the following?

(A) miter box
(B) pin punch
(C) coping saw
(D) cold chisel

25. A loss of compression in an engine cylinder can be caused by all of the following, EXCEPT

(A) worn engine bearings
(B) worn piston rings/cylinder wall
(C) burned valves
(D) blown head gasket

STOP. IF YOU FINISH BEFORE THE TIME IS UP, YOU MAY CHECK OVER YOUR WORK ON THIS PART ONLY.

Part 8: Mechanical Comprehension (MC)

Time: 19 minutes; 25 questions

Directions: In this section, you will be tested on your knowledge of mechanics and basic physics. Select the best answer for each question and mark the corresponding oval on your answer sheet.

1. The force of friction between two surfaces moving past each other DOES NOT depend on
   - (A) the normal force
   - (B) the nature of the surfaces in contact with each other
   - (C) the area of the surfaces in contact with each other
   - (D) the speed with which the surfaces are moving past each other

2. The formula for work is
   - (A) $w = m \times v$
   - (B) $w = m \times g \times h$
   - (C) $w = m \times a$
   - (D) $w = F \times d$

3. A small pulley drives a large pulley with a belt. Which of the following statements about this arrangement are true?
   - (A) The pulleys turn in opposite directions.
   - (B) The smaller pulley turns faster.
   - (C) Speed output is greater than the input.
   - (D) The smaller pulley is above the larger pulley.

4. Any work that is done to accelerate an object at rest to speed v will be converted into the kinetic energy of that object. This principle is known as
   - (A) the work-energy theorem
   - (B) gravitational potential energy
   - (C) the principle of conservation of mechanical energy
   - (D) Newton's second law

5. Performing work at the rate of 1 joule per second is the same as expending
   - (A) 100 foot-pounds
   - (B) 10 psi
   - (C) 1 meter per second
   - (D) 1 watt of power

6. All of the following statements about liquids are true EXCEPT
   - (A) liquids are practically incompressible
   - (B) liquids can be used to transmit force
   - (C) all liquids are good lubricants
   - (D) liquids conform to the shape of their container

7. A crate with a weight of 250 N is resting on a rough, horizontal surface with a coefficient of static friction of 0.8. What will happen if an applied horizontal force of 200 N acts on the crate?
   - (A) The crate will move and continually accelerate while the force acts.
   - (B) The crate will move at a constant speed.
   - (C) The crate will not move.
   - (D) The crate will move but then decelerate back to rest as friction increases.

8. If 100 pounds of force are applied over an area of 2 square inches, the pressure would be
   - (A) 5 psi
   - (B) 50 psi
   - (C) 100 psi
   - (D) 200 psi

9. The advantage gained by the use of a mechanism in transmitting force is known as
   - (A) pressure
   - (B) hydraulics
   - (C) torque multiplication
   - (D) mechanical advantage

10. The point that the lever pivots on is the
    - (A) gear
    - (B) torque
    - (C) fulcrum
    - (D) axle

11. A 30 N force is applied to 5 kg box. Assuming no friction, what is the acceleration gained by the box?
    (A) $0.17 \frac{m}{s^2}$
    (B) $6 \frac{m}{s^2}$
    (C) $15 \frac{m}{s^2}$
    (D) $150 \frac{m}{s^2}$

12. When a net force acts on an object,
    (A) the object will slow to a stop
    (B) the force will give the object a cumulative force
    (C) the object will accelerate in the direction of the net force
    (D) constant speed is impossible

13. If pulley X turns in a clockwise direction, in which direction will pulley Y turn?

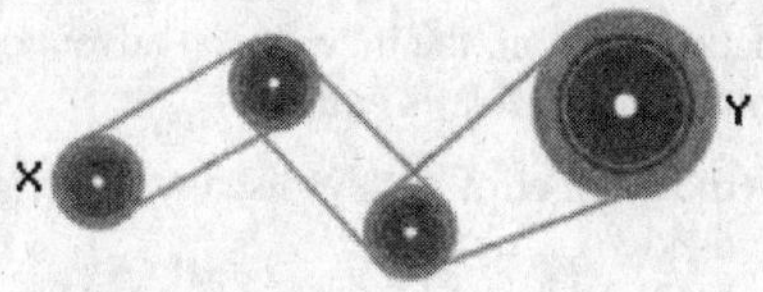

    (A) clockwise only
    (B) counterclockwise only
    (C) either clockwise or counterclockwise
    (D) the direction cannot be determined based on the information provided

14. The illustration below represents

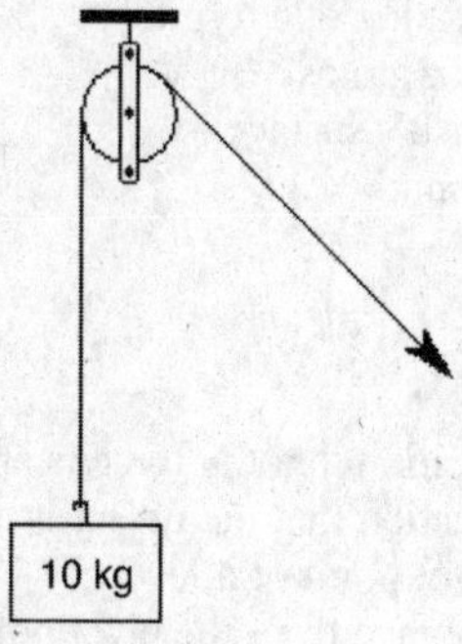

    (A) a fixed pulley
    (B) a moving pulley
    (C) a rear axle
    (D) a pendulum

15. In order for a piston with a total surface area of 10 square inches to support 100 pounds, how much pressure will have to be applied?
    (A) 1,000 psi
    (B) 50 psi
    (C) 10 psi
    (D) 5 psi

16. Given the following forces acting on a block, what are the net or resultant forces? (Assume that "left" means your left.)

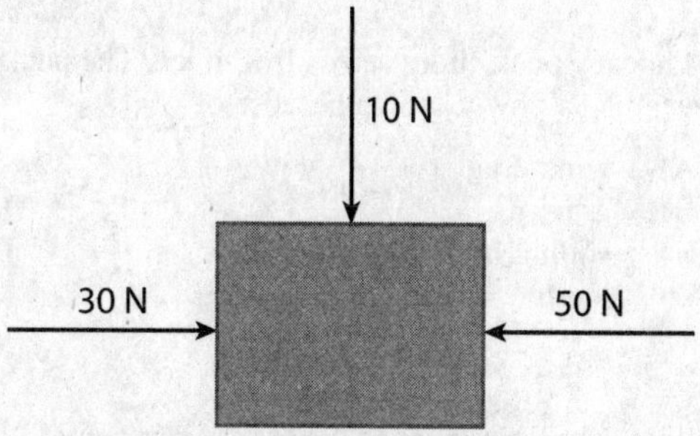

    (A) 80 N to the right and 10 N down
    (B) 80 N to the left and 10 N down
    (C) 20 N to the left and 10 N down
    (D) 20 N to the right and 10 N down

17. Given a block and tackle with four ropes to lift a load one foot, barring friction, what is the mechanical advantage gained?

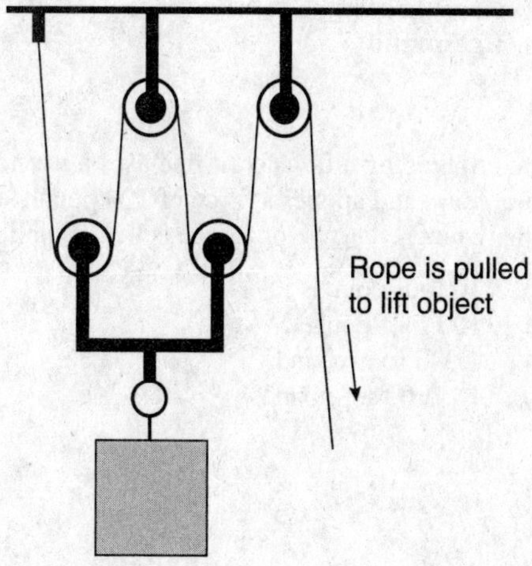

    (A) 24:1
    (B) 10:1
    (C) 8:1
    (D) 4:1

18. A child's teeter-totter, of the sort shown below, is an example of which kind of lever?

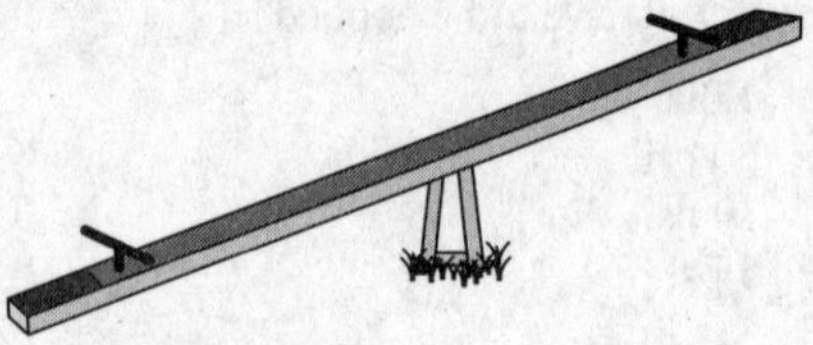

(A) first-class
(B) second-class
(C) third-class
(D) fourth-class

19. A hockey puck slides across frozen ice. The puck will eventually slow to a stop because of

(A) wind drag
(B) gravity
(C) sliding frictional force
(D) inertia

20. What is being applied to the bolt as it is being tightened?

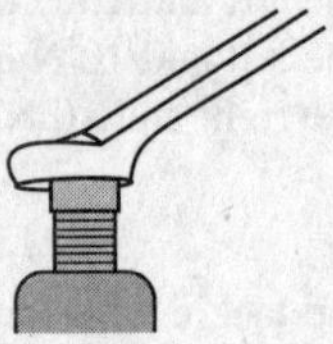

(A) torque
(B) friction
(C) kinetic energy
(D) gear ratio

21. When tightening a fastener, if one uses a wrench that is 1 foot long, and applies a force of 100 pounds to the end of the wrench, ______ of torque is being applied.

(A) 10 foot-pounds
(B) 100 foot-pounds
(C) 1,000 foot-pounds
(D) 10,000 foot-pounds

22. Given the following pulley system, suppose a person is holding the other end of the rope, so that the box is stationary and the system is in equilibrium. What is the Tension force T that develops in the rope?

(A) 10 N
(B) 25 N
(C) 50 N
(D) 100 N

23. Neglecting friction, the mechanical advantage provided by a block and tackle is equal to

(A) the number of pulleys used in the block and tackle
(B) half of the number of pulleys used in the block and tackle
(C) twice the number of pulleys used in the block and tackle
(D) the diameter of the pulleys divided by the number of pulleys used in the block and tackle

24. The symbol *g*, according to Newtonian laws, refers to

(A) acceleration due to gravity
(B) mass, in grams
(C) the Earth's surface
(D) the ground

25. The fulcrum is

(A) the location that the force is applied to a lever
(B) the location that the object is applied to a lever
(C) the pivot point of a lever
(D) the distance from the pivot point of a lever to the location the force is applied to the lever

STOP. IF YOU FINISH BEFORE THE TIME IS UP, YOU MAY CHECK OVER YOUR WORK ON THIS PART ONLY.

Part 9: Assembling Objects (AO)

Time: 15 minutes; 25 questions

Directions: In this section, you will be tested on your ability to determine how an object will look when its parts are put together. For each question, select the best answer and mark the corresponding oval on your answer sheet.

1.

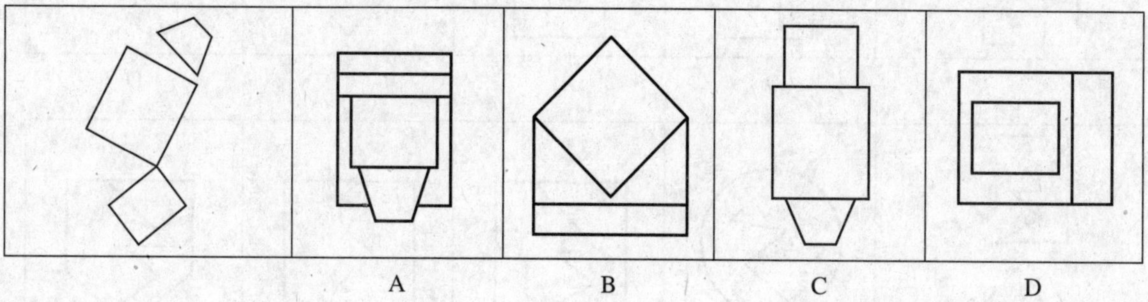

2.

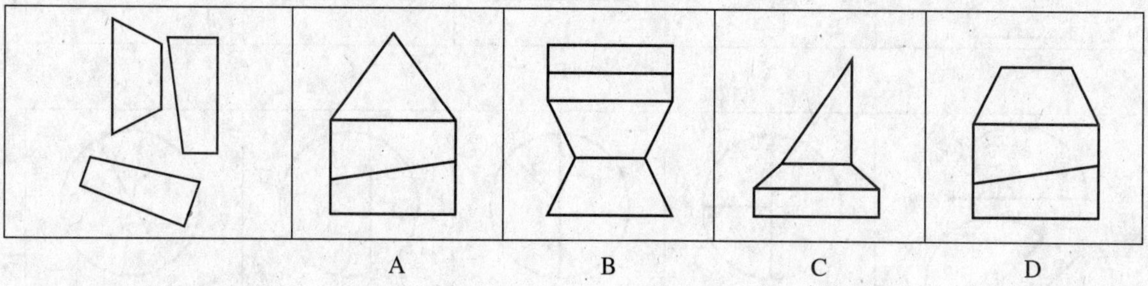

3.

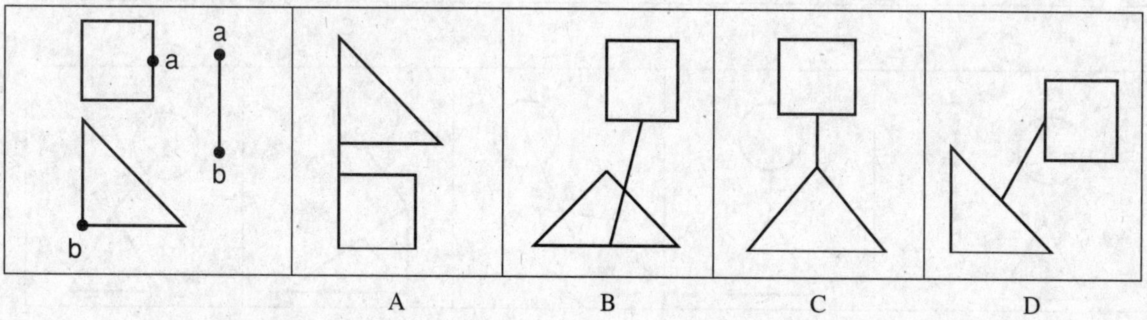

4.

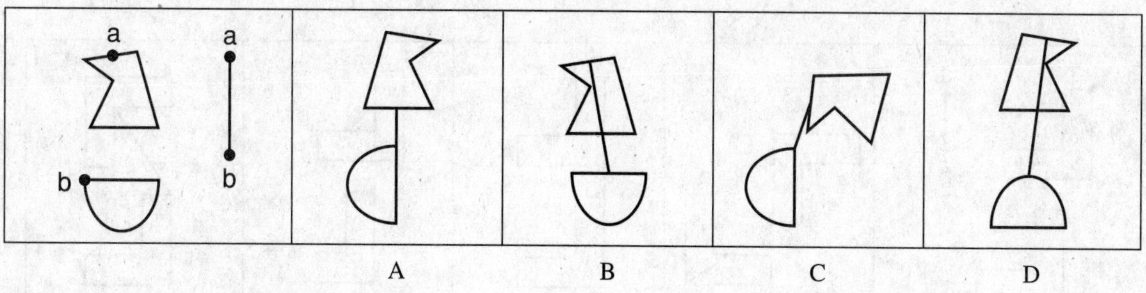

5.

A B C D

6.

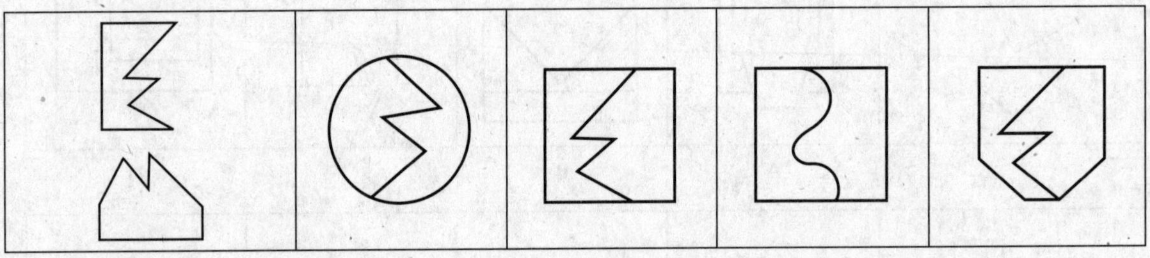

A B C D

7.

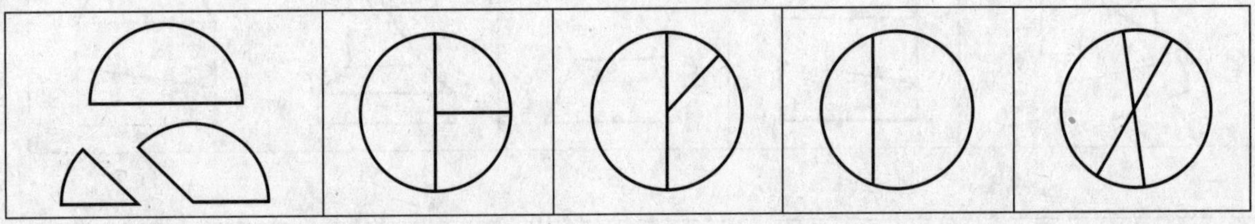

A B C D

8.

A B C D

9.

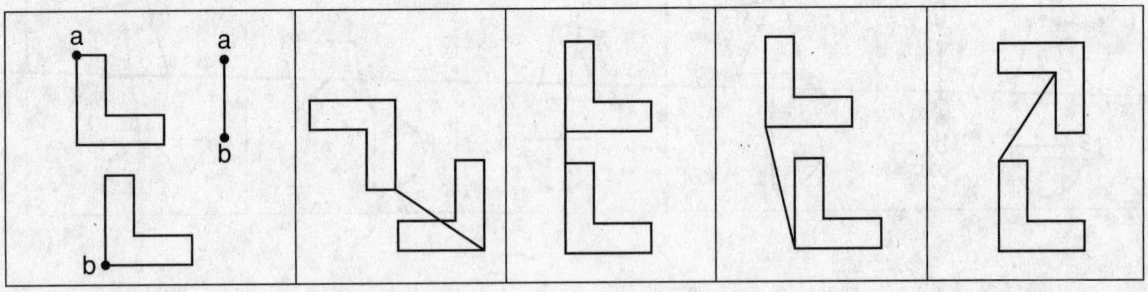

A B C D

10.

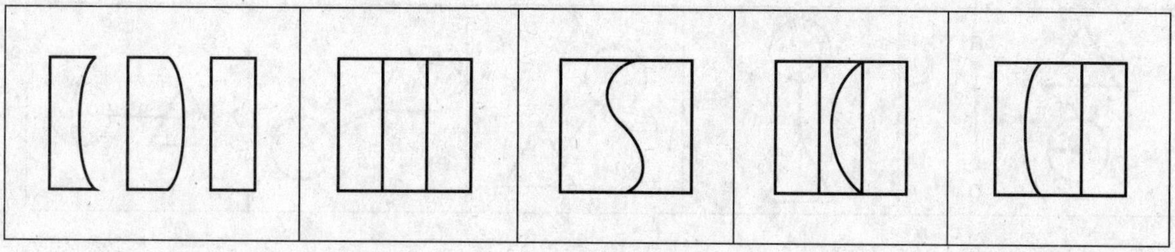

A B C D

11.

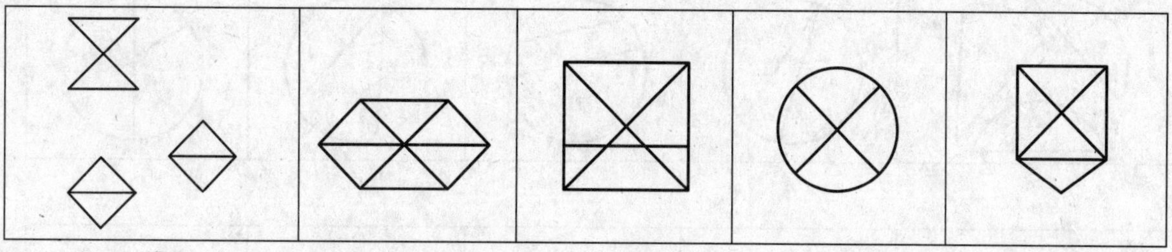

A B C D

12.

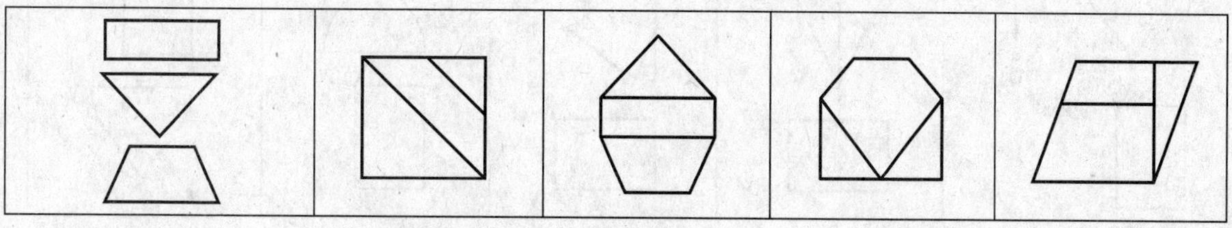

A B C D

13.

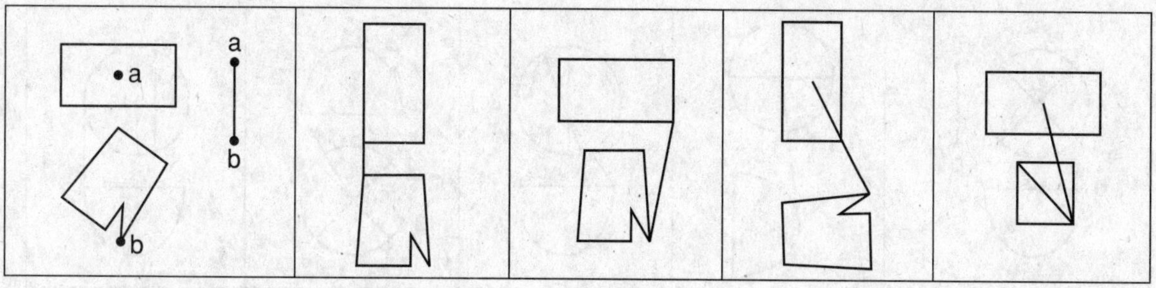

A B C D

14.

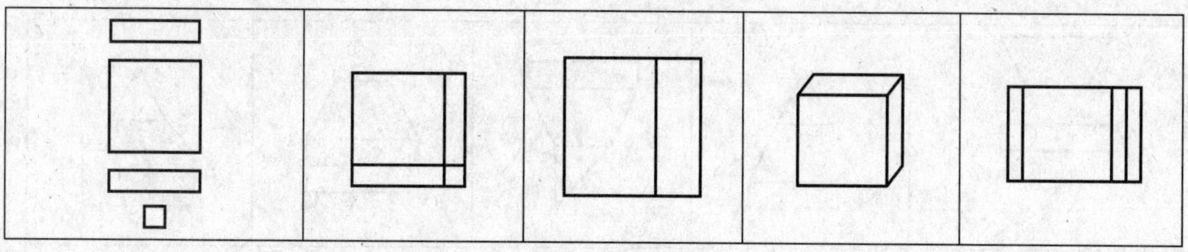

A B C D

15.

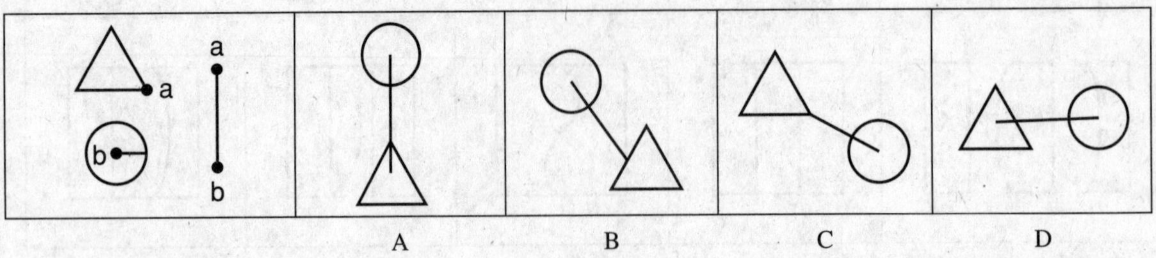

16.

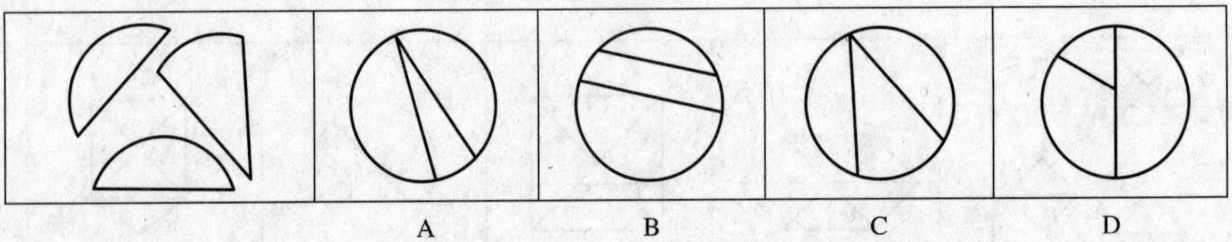

17.

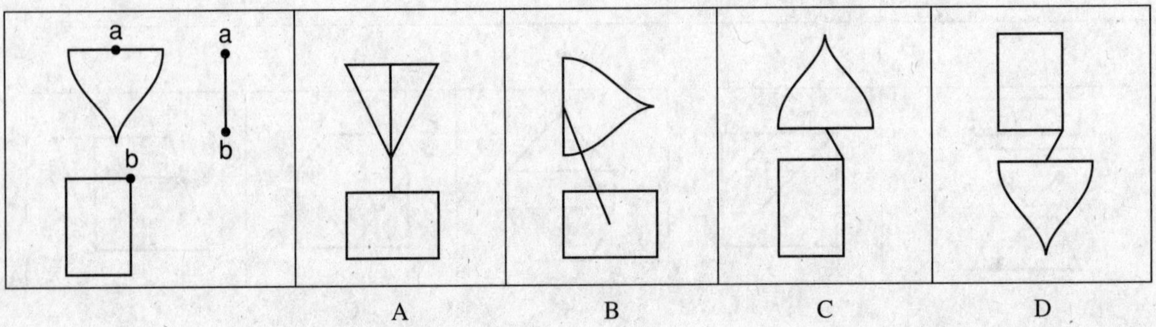

18.

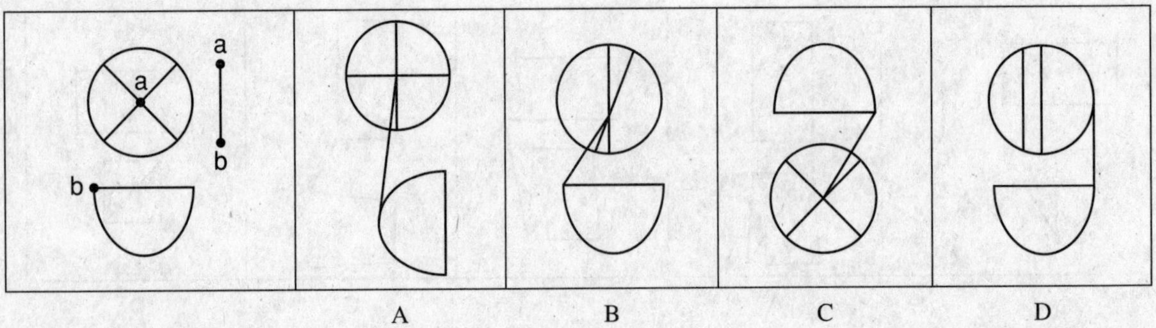

19.

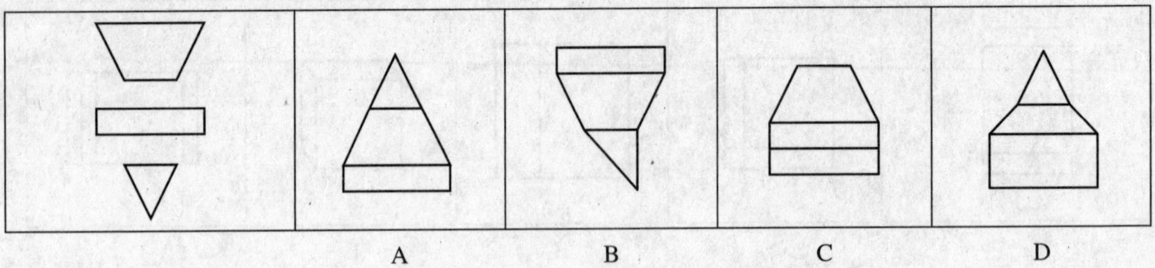

20.

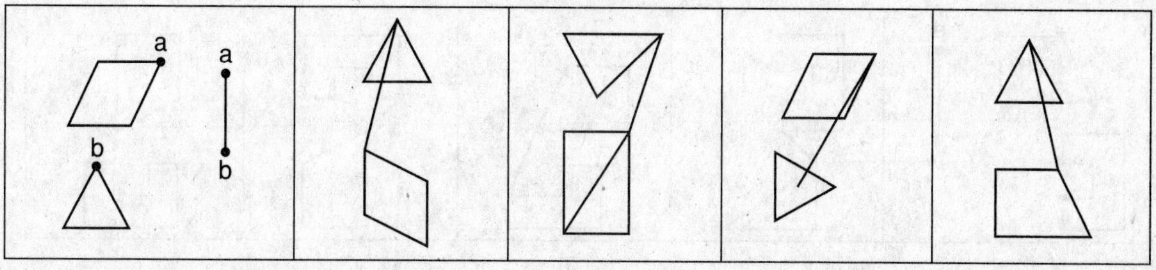

A B C D

21.

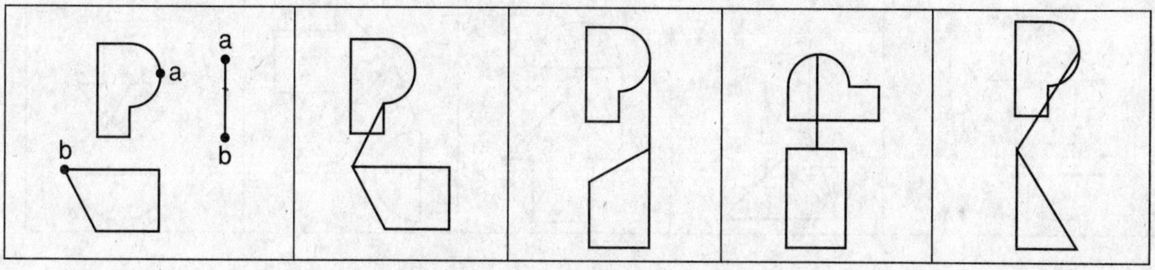

A B C D

22.

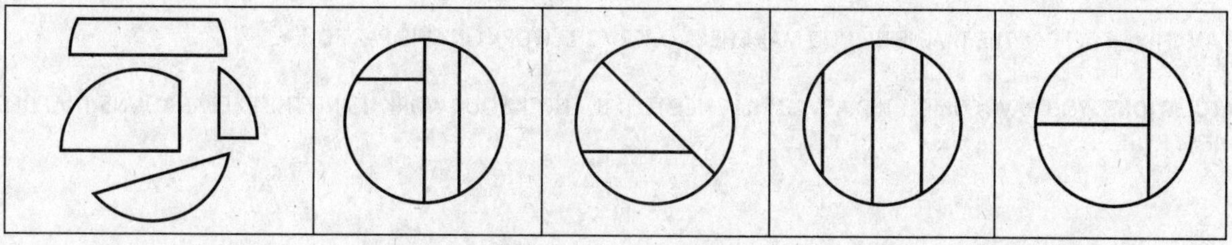

A B C D

23.

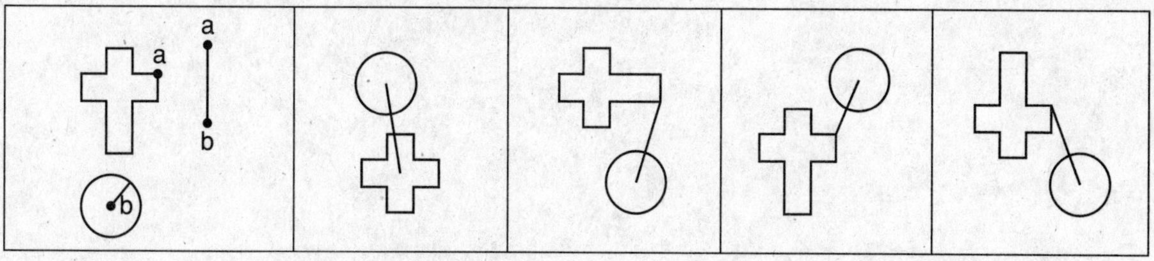

A B C D

24.

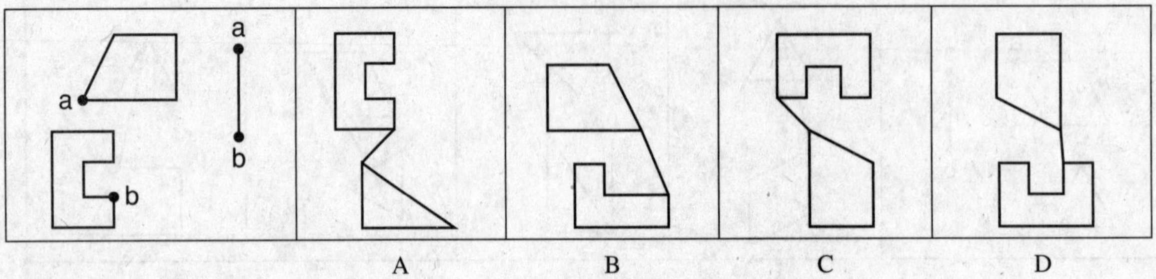

25.

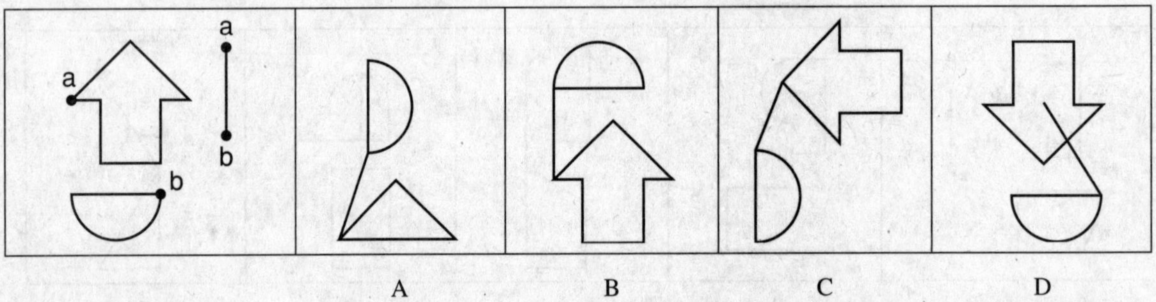

STOP. IF YOU FINISH BEFORE THE TIME IS UP, YOU MAY CHECK OVER YOUR WORK ON THIS PART ONLY.

CONGRATULATIONS! YOU HAVE COMPLETED ASVAB PRACTICE TEST B. CHECK YOUR WORK USING THE EXPLANATIONS THAT BEGIN ON THE NEXT PAGE.

ASVAB PRACTICE TEST B ANSWERS AND EXPLANATIONS

Part 1: General Science Answers and Explanations

1. **(D) saturated fats** Those wishing to reduce the bad cholesterol levels in their blood should avoid saturated fats. Sources of saturated fats include meats, shellfish, eggs, milk, and milk products.

2. **(D) 32°F** Water at sea level freezes at 32°F. This can also be expressed as 0°C and 273 K.

3. **(A) proton** Of the subatomic particles listed, the proton has the largest mass. Among subatomic particles, protons and neutrons have the most mass. Electrons, positrons, and neutrinos all have negligible masses relative to protons and neutrons.

4. **(B) an apple tree** An "autotroph" is an organism capable of synthesizing its own organic substances from inorganic compounds, as through photosynthesis. Another term for an autotroph is a producer. An apple tree is an example of an autotroph. A vulture is a type of consumer known as a scavenger, which means it feeds on dead or decaying flesh. A toadstool is a type of fungi. It feeds on and breaks down dead organic matter. A sea anemone is a consumer that feeds on small marine organisms and organic matter.

5. **(D) Precambrian eon** The Earth is approximately 4.6 billion years old, and for most of that time, very few fossil traces were left. This is why the period from 4.6 billion years to 570 million years ago is called the Precambrian eon, meaning the period before the fossil record began.

6. **(C) Uranus** Mercury, Venus, and Mars are all smaller than Earth. Of the planets listed, only Uranus is larger than Earth.

7. **(B) meteoroids** Falling stars are meteoroids that fall into the Earth's gravitational field. They leave streaks of light, called meteors, as they are burned up in Earth's atmosphere.

8. **(B) ecology** The study of interactions between organisms and their interrelationships with the physical environment is known as *ecology*. *Cytology*, (A), is the study of cells, *physiology*, (C), is the study of processes in the organism, and *embryology*, (D), is the study of embryos and their development.

9. **(C) small intestine** The small intestine absorbs most of the nutrients in food.

10. **(D) twice a day** Tides are caused by the gravitational pull that exists between the Moon and Earth. High tides occur twice a day, when the Moon is at the points closest to and farthest from the coastline experiencing a high tide.

11. **(C) If the mass of the Earth were doubled, the gravitational force on the Earth would quadruple.** According to Newton's Law of Gravitation, if the mass of the Earth were doubled, the gravitational force on the Earth would double, not quadruple.

12. **(A) radio waves** Of the states of electromagnetic radiation listed, radio waves have the longest wavelength and lowest frequency.

13. **(D) refractive index** The speed of light can slow down, depending on the material through which the waves move. For example, light travels more slowly through water or glass than through air. The ratio by which light is slowed down is called the refractive index of that medium.

14. **(C) baking soda** A high pH indicates a substance that is basic rather than acidic. Of the food products listed, baking soda is the most basic and has the highest pH.

15. **(D) almonds** Carbohydrates include starches (potatoes and pasta) and sugars (fruits). Almonds are a type of nut and have a high fat content.

16. **(A) class** The levels of classification, from largest to smallest, are: domain, kingdom, phylum, class, order, family, genus, species. Members of a family, genus and species are all more similar than members of an order. Of the answer choices given, members of an order are only more alike than members of a class.

17. **(C) atomic number** The number of protons in the nucleus of an atom gives it its atomic number. The number of electrons can also be used to determine atomic number, but only if the atom is neutral (in a neutral atom, proton number = electron number). The atomic symbol is the letter designation of the element and is based on the element's Latin name. Atomic weight is the mass of one atom expressed in atomic mass units. Atomic radius gives the size of the atom.

18. **(A) diaphragm** The diaphragm is a system of muscles that allow the lungs to expand and contract, drawing air in and out.

19. **(B) 17 protons** The atom of an element with an atomic number of 17 must have 17 protons. The number of *electrons*, (A), however, is only the same as the number of protons if the atom is neutral. The number of *neutrons*, (C), is often similar to but not necessarily equal to the number of protons. For example, hydrogen often has one proton and no neutrons. And since the *atomic mass*, (D), is contributed to by both protons and neutrons, it is almost always larger than the atomic number.

20. **(C) saprophytes** Because mushrooms absorb nutrients from decaying leaves, they are classified as saprophytes. Another term for a plant living on dead or decaying organic matter is decomposer.

21. **(A) pulmonary artery** The vessel with the least oxygenated blood is the pulmonary artery, which sends oxygen-depleted blood from the heart into the lungs to be oxygenated.

22. **(A) starch** Saliva in the mouth begins the process of breaking down starch. Fats, proteins and vitamins begin breaking down later in the digestive process.

23. **(B) Monera** The Moneran kingdom is composed of prokaryotic organisms; that is, organisms with cells that lack nuclei.

24. **(C) ion** An atom that is not electrically neutral is called an *ion*. It possesses a positive or negative charge. *Isotopes*, (A), are two or more atoms of the same element that have the same atomic number but different mass numbers; i.e., they have the same number of protons but a different number of neutrons. A *positron*, (B), is a positively charged particle having the same mass and magnitude of charge as the electron. An *allotrope*, (D), is an element with two or more different forms; e.g., carbon is an allotrope and can exist as either graphite or diamond.

25. **(B) marble** A metamorphic rock is one that has undergone significant changes due to pressure, heat, and/or water. This results in a more compact and more highly crystalline state. Marble is an example of a metamorphic rock.

Part 2: Arithmetic Reasoning Answers and Explanations

1. **(B) 150** If a machine caps 5 bottles every 2 seconds, then it would cap 30 times as many bottles in one minute (since a minute is 60 seconds). $5 \times 30 = 150$ bottles per minute.

2. **(D) $33.80** Jonah traveled 650 miles at 25 miles per gallon. To determine how many gallons he used, divide 650 by 25. $650 \div 25 = 26$ gallons. Each gallon costs $1.30, so his cost for the trip would be $26 \times \$1.30 = \$33.80$.

3. **(B) $0.40** On problems that involve long calculations, use Backsolving. Start with either (B) or (C). Since (B), $0.40, is the easier number, start there: If there are 3.86 liters in a gallon, and a liter of gasoline costs $0.40, a gallon of gasoline costs $3.86 \times \$0.40 = \$1.544$, or roughly $1.54 a gallon, which is the correct price of a gallon.

4. **(C) 10,080** To determine the number of minutes in a week, begin with the number of minutes in an hour and work your way up from there. There are 60 minutes per hour and 24 hours per day, so there are $60 \times 24 = 1{,}440$ minutes per day. There are 7 days per week, so there are $1{,}440 \times 7 = 10{,}080$ minutes per week.

5. **(D) $250.00** 15% of take home pay = $37.50. So,

$0.15(x) = \$37.50.\ x = \frac{37.5}{0.15} = \frac{3{,}750}{15} = 250.$

You could also use Backsolving, starting with either (B) or (C). Trying (C) would yield:

$\$225 \times .15 = \$33.75$, which is less than the $37.50 we are told Pat deposited. Thus, (C) is too small, and the correct answer must be (D).

6. **(C) 152** If the ratio of males to females is 3:5, then there are 8 total parts in the ratio. The total number of students must be a multiple of 8. Only choice (C), 152, is a multiple of 8. To test if a number is a multiple of 8, divide the number by 2, then by 2 again. If the result is even, the number is divisible by 8.

7. **(D) 384** A car travels 288 miles in 6 hours. Divide 288 by 6 to find the rate in miles per hour. $288 \div 6 = 48$ miles per hour. Then, 48 miles per hour multiplied by 8 hours equals 384 miles.

8. **(D) 94** If Martin's average score after 4 tests is 89, then the sum for his 4 tests would be:

$4 \times 89 = 356.$

To average 90 on 5 tests, he would have to reach a sum of:

$5 \times 90 = 450.$

His score on the fifth test has to be the difference between the two sums:

$450 - 356 = 94.$

9. **(A) 2019** If the first year begins with town A at 9,400 and town B at 7,600, the populations are: $9{,}400 - 7{,}600 = 1{,}800$ apart. Each year after 2010 beginning in 2011, the gap will close by 200. It would take 9 years for the gap to close entirely, so by 2019 the populations will be equal.

10. **(D) −50** One number is 5 times another number, and their sum is −60. You could rewrite that sentence as two equations:

$x = 5y$

$x + y = -60$

Substitute $5y$ in for x in the second equation and solve:

$5y + y = -60$

$6y = -60$

$y = -10$

If $y = -10$ and $x = 5y$:

$x = 5(-10) = -50.$

-50 is less than -10, so -50 is correct.

11. **(C) $312** There are two rates by which Jane gets paid: one for the first 40 cars ($6 per car) and another for all other cars after those first 40 ($6 per car plus 50% of $6, which is $\$6 + \$3 = \$9$ per car). Since she makes 48 cars, she gets paid $\$6 \times 40$ cars ($240) plus 8 extra cars at $9 per car ($72), which totals $312.

12. **(A) 8** 48 glecks at 3 glecks per fedi would equal $\frac{48}{3} = 16$ fedis. 16 fedis at 2 fedis per bora would equal $\frac{16}{2} = 8$ bora.

13. **(B) 8 p.m.** Work through the clock systematically: First reading at 12 p.m., second at 4 p.m., third at 8 p.m., fourth at midnight, fifth at 4 a.m., sixth at 8 a.m., seventh at 12 p.m., eighth at 4 p.m., ninth and final reading at 8 p.m.

14. **(B) $33\frac{1}{3}$%** If monthly tuition payments were raised from \$225 to \$300, use the percent increase formula to determine how much the tuition went up:

$$\text{Percent increase} = \frac{\text{New} - \text{Old}}{\text{Old}} = 100\%$$

$$= \frac{300 - 225}{225}$$

$$= \frac{75}{225} = \frac{75}{225} \div \frac{5}{5} = \frac{15}{45}$$

$$= \frac{15}{45} \div \frac{5}{5} = \frac{3}{9} = \frac{1}{3}$$

$$\frac{1}{3} \times 100\% = 33\frac{1}{3}\%$$

15. **(C) 2** If a truck takes 6 hours to complete a trip at 20 miles per hour, the total distance for the trip must be $6 \times 20 = 120$ miles. When traveling 120 miles at a rate of 30 miles per hour, the truck would take $\frac{120}{30} = 4$ hours. So instead of taking 6 hours, the truck would take 4 hours. That saves 2 hours.

16. **(A) 45** If a full barrel can hold 75 gallons, when it is $\frac{3}{5}$ full it will hold $\frac{3}{5}$ of 75 gallons. $\frac{3}{5} \times 75 = 3 \times 15 = 45$ gallons.

17. **(B) \$27,000** A salary of \$45,000 is reduced by 40%; that is the same as reducing it by $\frac{2}{5}$. You may find it easier to find $\frac{1}{5}$ first: $\frac{1}{5}$ of \$45,000 is \$9,000, so $\frac{2}{5}$ would be \$18,000. The new salary is \$45,000 – \$18,000 = \$27,000.

18. **(D) 40,320** $8! = 8 \times 7 \times 6 \times 5 \times 4 \times 3 \times 2 \times 1 = 40{,}320$ To solve long multiplication problems without a calculator, group the numbers strategically and multiply in pieces. The calculation can become:

$$(8 \times 7) \times (6 \times 5) \times (4 \times 3) \times (2 \times 1)$$

$$= 56 \times 30 \times 12 \times 2$$

$$= 56 \times 60 \times 12$$

$$= 56 \times 720$$

$$= 40{,}320$$

If you are running short on time or don't feel confident with long multiplication, guess strategically using logic. 8! will be a huge number and it probably won't be a nice round number. So (D) is the best choice.

19. **(D) \$390** If 15 people contribute \$40 each, that would amount to $15 \times \$40 = \$600$. Thirty-five percent of the money is spent on a group gift, so the remaining 65% of the money is spent on the party. $0.65 \times \$600 = \$390$.

20. **(C) 6 hours, 42 minutes**

$\text{Average} = \frac{\text{Sum of the terms}}{\text{Number of terms}}$. In this question, an employee works for a total of 33.5 hours over 5 days. The average hours worked per day would be determined as follows:

$$\text{Average} = \frac{\text{Sum of the terms}}{\text{Number of terms}} = \frac{33.5}{5} = 6 + \frac{3.5}{5}\text{ hours}$$

All the answer choices are expressed in minutes, not hours. So you need to convert 0.7 hours into minutes. There are 60 minutes in an hour, so 0.7 hours = $60 \times 0.7 = 42$ minutes. The average amount of time worked on a given day is 6 hours and 42 minutes.

21. **(A) 66** To find the average speed for an entire trip, divide the total distance by the total time. If a train travels 180 miles at an average speed of 60 miles per hour, that leg of the journey must have taken $\frac{180}{60} = 3$ hours. The second leg of 150 miles at an average speed of 75 miles per hour would take $\frac{150}{75} = 2$ hours.

$$\text{Average speed} = \frac{\text{Total distance}}{\text{Total time}}$$

$$= \frac{180 + 150}{3 + 2}$$

$$= \frac{330}{5} = 66\text{ mph}$$

22. (C) $175.00 Try Backsolving on a question like this one. Start with choice (B). If a bicycle starts at $162 and is discounted 20%, its price would be reduced by (0.2)($162) = $32.40. The selling price would then be $129.60. This is too low. Eliminate answer choices (A) and (B). Then try answer choice (C). If a bicycle starts at $175.00 and is discounted 20%, its price would be reduced by (0.20)($175) = $35. So the new price would be $175 − $35 = $140. That's the discount price you're looking for, so it's the correct answer.

23. (B) $1,024.64 This is an addition and subtraction question. Begin by adding the deposit to the balance:

$1{,}162.76 + 352.68 = 1{,}515.44$

Then sum up the withdrawals:

$152.45 + 82.85 + 255.50 = 490.80$

Find the answer by subtracting the sum of the withdrawals from the balance:

$1{,}515.44 - 490.80 = 1{,}024.64$

24. (D) $57.38 If Harold works 4.5 hours per day for 3 days, then he works $4.5 \times 3 = 13.5$ hours at $4.25 per hour. He earns $13.5 \times \$4.25 = \$57.375$. That rounds up to $57.38.

25. (B) 0.25 To convert a fraction to a decimal, divide the denominator into the numerator. $6 \div 24 = 0.25$.

You can also reduce the fraction until it represents one you know the conversion for:

$$\frac{6}{24} \div \frac{6}{6} = \frac{1}{4} = 0.25$$

26. (B) $346 There are 100 stamps per sheet, which means that each sheet of 100 23-cent stamps would cost $23, each sheet of 37-cent stamps would cost $37, and each sheet of 60-cent stamps would cost $60. Thus, the calculations look like this:

$(4 \times \$23) + (2 \times \$37) + (3 \times \$60)$

$= \$92 + \$74 + \$180 = \$346$

27. (A) 4 hours, 15 minutes Pacific time is three hours behind Eastern time. So if a plane leaves Los Angeles at 5:30 p.m. PST, it leaves at 8:30 p.m. EST. If it arrives in New York at 12:45 a.m. EST, the total time of the flight would be 4 hours, 15 minutes.

28. (C) $160 Try Backsolving here. Start with choice (C) because you are looking for a 25% discount, which is $\frac{1}{4}$ and 160 is divisible by 4. If the original cost of the camera was $160, and it was reduced by 25% or $\frac{1}{4}$, the new cost would be:

$\$160 - 0.25(\$160)$

$= \$160 - \$40 = \$120$

That's the selling price given in the problem, so $160 is the correct original price.

29. (B) 15 A square root is a number that, when multiplied by itself, produces the given quantity. $15 \times 15 = 225$.

30. (D) 10 hours, 48 minutes If it takes 20 minutes to type 5 pages, then it takes 4 minutes to type one page. Then the total document takes $4 \times 162 = 648$ minutes to type. 600 minutes is 10 hours, so 648 minutes is 10 hours and 48 minutes.

Part 3: Word Knowledge Answers and Explanations

1. **(B) lethargic** *Listless* means "tired, sluggish, or lacking energy"; thus *lethargic*, (B), is the correct answer. Word parts are not terribly helpful on this question, but you might have a sense that *listless* has a negative charge. If so, that narrows your choices down to either (B) or (D).

2. **(C) understand** Use the context of the sentence to guess the underlined word's meaning. What are great teachers able to do with complex ideas? They explain them, understand them, or simplify them. Only choice (C) reflects one of those concepts.

3. **(D) winner** Think of words that sound like *victor*. The first word that comes to mind may be *victory*, which means "the act of winning or defeating others," and this fits with the sentence. A *victor* is defined as "one who defeats or vanquishes a foe," so (D) *winner* is the best answer.

4. **(B) relaxed** Think about what the word sounds like. Something that is *lacking* is *missing*. The answer that has something missing is (B) *relaxed*, which means a lack of stress or action. The definition of *lackadaisical* is "lacking life, spirit, or zest."

5. **(B) position** Mentally reread the sentence, substituting each answer choice for the underlined word. Choices (A) and (D) make no sense when you do so. Don't fall for choice (C): one might have an *idea* that more money should be spent on education, but it isn't correct usage to say that one "has an *ideal* that." In fact, *contention* is defined as a "point advanced in a debate or argument," so the answer (B) *position* is closest in meaning to the given word.

6. **(A) notable** Break the word into parts: *news* and *worthy* are both English words, so you can guess that *newsworthy* means "interesting enough to warrant news coverage." Answer choice (A), *notable*, is the closest in meaning.

7. **(C) irregular** Try mentally rereading the sentence and substituting each answer choice for the underlined word. It doesn't make sense to say that someone's class attendance was (A) *latent* (hidden), or (D) *common*. Class attendance could possibly be (B) *constant*, but that doesn't jibe with the phrase "if anything," which suggests that the person's class attendance is minimal or scanty. In fact, something *sporadic* is something that is "occurring occasionally," and (C) *irregular* is the best fit.

8. **(D) possible** Remembered real-life context may be helpful here. Have you ever heard a young person described as "having a lot of potential"? That phrase is used to describe a young person who has many promising possibilities before her. Here, potential is used in its adjective form (you can tell because the answer choices are all adjectives), and as an adjective *potential* means *possible*.

9. **(B) barred** This word is made up of other English words: *black* and *list*. You can guess that being put on a blacklist is probably a metaphor for something fairly negative. In fact, the word *blacklisted* refers to a way of keeping out undesired people by creating a list of banned individuals. Thus, answer choice (B) is the only correct answer, as *banned* and *barred* have similar meanings.

10. **(D) overconfident** Use the context of the sentence to make a prediction. You're told that the task is a tough one, so perhaps not many people would want to volunteer for it. That suggests that *presumptuous* means something like: "unrealistic." Choice (D) *overconfident*, or too confident, is the best match.

11. **(C) souvenir** Use the context of the sentence to make a prediction. The sentence tells you that a *memento* helps you to remember something. Only choice (C) *souvenir* is something that helps you remember.

12. **(A) memory** Use the sentence's context to make a prediction. A *legacy* is something that the school president will leave behind him when he's gone. Of the choices, only (A) *memory* is something that remains behind after someone is gone.

13. **(B) danger** Use the context to make a prediction. The parents are worried, so you can infer that something bad may be coming in their children's future. Only choice (B) *danger* is sufficiently negative to fit the context.

14. **(D) honesty** There's no context, and word parts aren't going to be helpful here. Perhaps remembered real-life context could help. If you have ever heard someone referred to as a "man of integrity" or a "woman of integrity," you may remember that's a compliment. That knocks out choice (B), since *coercion* has a negative charge, and choice (C), since *togetherness* is not a trait that a single individual can have. (A family or other close-knit group might have togetherness.) That leaves only choices (A) *ability* and (D) *honesty*, and only (D) conveys the same sense of virtue as the phrase "man of integrity." In fact, *integrity* means the state of being whole or the state of being very moral or upright.

15. **(D) violation** Use the context to make a prediction. An *infraction* is something for which a person is punished, so it must be a crime or act of breaking the rules. Choice (D) *violation* is the best fit.

16. **(A) group** Think about words that sound like the underlined word: *collective* sounds like a noun or adjective form of the word *collect*. When you *collect*, you gather together things or people to form a group. That should point you to answer choice (A).

17. **(A) trick** There's no context in the question, and it would be difficult to see word parts in the word *hoax*. Perhaps remembered real-life context will help. You may have heard the word *hoax* in connection with the Loch Ness Monster, Bigfoot, UFO photos, or other fake things that people have claimed are real. In fact, a *hoax* is a trick meant to convince someone that what is false is true.

18. **(C) origin** Word parts may be helpful here: the root *gen*, which relates to beginnings, also gives us *generate*. You might also have heard the word *genesis* used in real life to refer to beginnings. In fact, *genesis* means the "origin or coming into being" of something.

19. **(B) dispute** Remembered real-life context may be helpful here. The famous feud between the Hatfields and McCoys was a prolonged dispute between two families. In fact, a *feud* is defined as a "quarrel that is often prolonged," and choice (B) fits that definition.

20. **(B) fake** There is no context in the question stem, and word parts may not be helpful if you aren't sure what *fraud* means. Remembered real-life context may be helpful. If you have heard news stories about people who have committed *tax fraud*, you know that those people lied to the government about their taxes. Thus, something *fraudulent* must have to do with lying; (B) *fake* is the best choice.

21. **(D) skilled** There's no context, and word parts may not be helpful. However, think of words that sound like *adept*. If you have ever heard someone described as *inept*, you may remember that person is ineffective or clumsy. Thus, you might guess that *adept* means the opposite. An *adept* person is someone highly skilled or well-trained.

22. **(A) passionate** The context here may be helpful: think about what words might be used to describe how supportive of a cause someone is. Only choice (A) is a word that might be used in that way. Word parts may also be helpful here. The root *ard* or *ars* relates to fire and gives us words like *arson* and *ardor*. *Ardent* is the state of being "fiery or intense in feeling," and (A) *passionate* is the best match.

23. **(B) widespread** Use context to make a prediction: if the town was suffering, the overcrowding must have been pretty severe. Only choice (B) *widespread* suggests the same idea.

24. **(D) weak** Make a prediction based on the context. The speaker predicts that in a contest between the big dog and the fence, the big dog is going to prevail. Thus, the fence must not be very strong. Choice (D) *weak* is the best fit.

25. **(D) evil** There's no context, and word parts are unlikely to be helpful. Remembered real-life context may be helpful. If you think about where you've heard the word *sinister*, the villains in movies or comic books may come to mind. This suggests that *sinister* means "wicked or evil."

26. **(C) clever** You may have heard the word *witty* used to describe people who tell good jokes. Then, ask yourself which of the answer choices would also describe good jokes. Typically, good jokes are *clever*. None of the other choices make sense.

27. **(A) clear** Make the most of context here. An account that is said to "open the jury's eyes" would must have clarified or illuminated a situation. Thus, the last two choices are not appropriate, as they are negative responses. Of the other two, only choice (A) fits the context of the sentence.

28. **(A) concede** There's no context, but the prefix *re*–, which means *back* or *again*, might be helpful. It's difficult to see how (B) *color* or (C) *trick* would involve doing something *back* or *again*. In fact, to *relinquish* something is to "give it up" or "give it back," so of the answer choices, only (A) *concede* fits.

29. **(C) return** Make the most of context here: a person is most likely to hope that good times either continue or come back again. Only choice (C) *return* makes sense.

30. **(A) induce** Remembered real-life context may be helpful. You may have heard news stories in which someone "incited" others to violence. That suggests that *incite* means something like "cause to do." Of the choices given, only (A) *induce* gets that meaning across.

31. **(A) enact** Use the context in the sentence to make a prediction. George's critics were paying attention to something else, so that might have been a good time for George to do whatever he thought was best. Thus, in *implementing* the strategy, George went ahead with it or set it in motion. Only answer choice (A) *enact* has that meaning.

32. **(B) question** *Query* has the same root as *question* and actually means the same thing.

33. **(B) cowardly** Make the most of context: you are told that a *timid* animal is not one you should fear. Thus, that animal is reluctant to attack you; perhaps it's likely to run away instead. In fact, *timid* means "shy or lacking in courage." The correct answer choice is (B), *cowardly*.

34. **(C) peer** Word parts can be helpful here. *Col–* is a form of *con–*, together. A *colleague* is someone you are together in a league or group with. For example, at work, a *colleague* is a coworker. Of the choices given, only *peer* is an equivalent answer.

35. **(D) powerful** Make the most of context: an opponent who is both tough and smart would be a powerful enemy. In fact, *potent* means "having or wielding authority or influence." Choice (D) is the best fit.

Part 4: Paragraph Comprehension Answers and Explanations

1. **(C) show that both were logical and meticulous** The question asks you to read the passage with an eye toward its use of examples. Here, the author very clearly sets up the theme that Conan Doyle's character Holmes is a logical character. To illustrate that, he brings up a recognized scientific contemporary. Choices (A) and (B) have no bearing on the paragraph. Choice (D) brings up the idea of Darwin's meticulous research abilities, but does not reference the main focus of the paragraph (Holmes). Only choice (C) includes both men.

2. **(B) based on urban structures and the environment** This Detail question asks about the specific relationship between cities and climate as illustrated in the text. The paragraph introduces its subject first, and then gives some description of the changes that might take place. The last sentence in the paragraph finally gives the information needed. Structures and man-made items affect the weather. Of the answer choices, choice (B) is the clearest paraphrase of that sentence.

3. **(B) watch news on TV** This question asks you to make a prediction based on information given in the paragraph. From the tone of the passage, you can tell that Marianne is a generally quiet woman thrust into a high-profile situation. You can make an assumption based on the tone of the piece that she is overwhelmed by the attention given to her newfound wealth. Choices (A) and (C) highlight the idea that she is a gaudy or flamboyant person, an assumption that seems all wrong given the final sentence. Choice (D) is not reflected in the passage. Only answer choice (B) is a fair assessment of Marianne's potential behavior.

4. **(D) notorious** For Vocabulary-in-Context questions, head back to the original text and find the word in question. Examine the context within which the word *infamous* is found. The group was plotting to blow up the King and the Parliament but was foiled. The gist of the paragraph is that the conspirators were up to no good, quite the opposite of *heroic*, choice (A). You can also use word parts: *in* means "not," and *famous* means "well thought of." *Infamous*, therefore, means something like "not well thought of," or *notorious*.

5. **(A) perturbed** Take a look at the language choices used: *screamed*, *knocked*, *ran*. Now take a look at the answer choices. Choice (A) makes sense, since the action is so vivid and aggressive. Save it and check the other choices. Answer choice (B) is wrong because nothing in the passage is *tender*. Choice (C) has a positive connotation and Johnny is not positive in this situation. Nothing in the passage is conceited or *snooty*, so choice (D) is incorrect. Choice (A) is correct.

6. **(D) may be short and dynamic** Eliminate answer choices that disagree with the passage, or that may or may not be true based on the passage. Choice (A) cannot be correct because it contradicts the passage: the life cycle is a direct result of the rain levels. (B) may or may not be true: although the passage mentions "astounding" blooms, it doesn't say anything about their color. *Wildlife*, (C), isn't mentioned in the passage at all. (D) is a near-paraphrase of what the passage says about desert plant life cycles, so it is the correct answer.

7. **(C) rely on imagination and not just experience** The passage argues that many famous writers had narrow life experiences, but vivid imaginations, so look for an answer choice that makes a specific contrast between the need for imagination and experience in a writer's life. (C) makes that contrast. (A) contradicts the example of Robert Louis Stevenson, who wrote about things outside of his social scene. (B) similarly contradicts the example of Jane Austen, who did not write adventure stories. *Family and friends*, (D), are mentioned in the passage, but not advice from friends or family.

8. **(C) talent** The passage says that Tania was a good actress and gives examples of her variety of roles, so acting *prowess* describes a quality of good actors that she possesses. (C) is a positive quality. Eliminate (B) and (D) because they are negative words. *Knowledge*, (A), is positive, but is not necessarily related to acting versatility.

9. **(B) the true beauty of coral islands is underwater** Eliminate any answer choices that do not have to be true based on the passage. The color of coral animals, (A), is not mentioned in the passage at all. (B) is practically a paraphrase of sentence two, "Beneath the waves . . . is a fantastic and very beautiful world," so (B) is the correct answer. (C) is contradicted by the passage when the author writes "not always as nice if seen from the land." (D) is contradicted by the last sentence, which says that vast numbers of animals—not volcanoes—form coral islands.

10. **(A) how some sought to assimilate immigrants into America** In a question about the passage as a whole, the wrong answer choices will often be either too broad or too narrow, or will mention subjects that aren't in the passage. Eliminate answer choices that don't fit the scope of the passage. (B) is incorrect because cultural contributions aren't mentioned in the passage. (C) is too broad: the passage is about groups of immigrants in a particular area and time, not the whole history of immigration. (D) is too narrow: the effect on cities is only mentioned in one sentence. (A) is the correct answer: the idea of assimilation is referred to throughout the passage.

11. **(D) preached fitting in, not standing out** The passage says that the Daughters of the American Revolution (DAR) thought that immigrants should "completely adopt American customs and culture." The rest of the paragraph about the DAR just gives examples of what they did to support that belief, so look for an answer choice that describes how they thought immigrants should conform. (D) matches that idea. Notice that, while other answer choices may or may not be true, they don't fit specifically with what the author is saying about the DAR in the passage.

12. **(A) to divert the eye from the animal's outline** In the second sentence, the passage says that an animal adapts "color, shape and behavior in order to blend into its environment." The rest of the paragraph is a list of examples about how different animals become "all but invisible" with camouflage, so look for an answer that says camouflage makes animals hard to see. (A) matches that idea. The passage says nothing about *nesting*, (B), or *habitats*, (C), and most of the animals mentioned are prey animals, not *predators*, (D).

13. **(B) the discovery of volcanic activity on Io** The key to this question is in the first sentence: "... the biggest surprise... was the discovery... of intense volcanic activity." Choice (B), the correct answer, is a paraphrase of the sentence about surprise. Don't be distracted by the rest of the details listed in the paragraph or by the other answer choices. The key idea here is the observers' surprise.

14. **(C) early precursors to the Hula Hoop were primarily children's toys** In an "all are true EXCEPT" question, concentrate on eliminating any answer choices that are true according to the passage. Eliminate choice (A) because it is true according to the first sentence. The second sentence says that choice (B) is true, so eliminate it too. Sentence four says that adults enjoyed the early Hula Hoops, so (C) must be false and is the correct answer. The last two sentences of the passage show how (D) must be true since they describe how the Hula Hoop was a hit for the "fledgling" toy company.

15. **(B) discuss the origins of the Hula Hoop** Eliminate answer choices that are too broad, too narrow, or not mentioned in the passage. (A) is too broad: the passage doesn't include any information on fads besides the Hula Hoop. (B) is the correct answer: the passage starts with the earliest origins of the hoop toy and ends up with the popularization of the Hula Hoop. (C) is too narrow: only one sentence in the passage explains how the name came about. The passage says that the Hula Hoop is popular, (D), but doesn't question why.

Part 5: Mathematics Knowledge Answers and Explanations

1. **(B) $-1\frac{1}{4}$** In order to make this equation manageable, first turn the mixed number into an improper fraction, and then multiply both sides of the equation by 12 to clear the denominators:

$$\frac{3}{4}+1\frac{1}{3}=\frac{5}{6}-x$$
$$\frac{3}{4}+\frac{4}{3}=\frac{5}{6}-x$$
$$12\left(\frac{3}{4}+\frac{4}{3}\right)=12\left(\frac{5}{6}-x\right)$$
$$9+16=10-12x$$
$$12x=10-9-16$$
$$12x=-15$$
$$x=-\frac{15}{12}=-\frac{5}{4}=-1\frac{1}{4}.$$

2. (C) $12a^2+7a-12$ Use FOIL:

$(3a+4)(4a-3)$

$(3a \times 4a)+[3a \times (-3)]+(4 \times 4a)+[4 \times (-3)]$

$=12a^2-9a+16a-12$

$=12a^2+7a-12$

3. (D) 16 If $2m > 24$, then $m > 12$. But so are all the answer choices, so check out the other inequality. If $3m < 48$, then $m < 16$. Thus, (D), 16, is not a possible value for m.

4. (A) $11\sqrt{3}$ (D) is out because when you add square roots, you can't simply combine the sums under the radical sign, as (D) does. But what you can do is factor out perfect squares from the number under the radical signs:

$\sqrt{75}+\sqrt{108}$

$=\sqrt{25\times 3}+\sqrt{36\times 3}$

You can then separate the terms under the radicals because they are connected by multiplication:

$=\sqrt{25}\times\sqrt{3}+\sqrt{36}\times\sqrt{3}$

$=5\sqrt{3}+6\sqrt{3}$

Because the number under the radical is the same in both terms, you can add the terms together by adding the coefficients:

$5\sqrt{3}+6\sqrt{3}=11\sqrt{3}$

5. (B) 5 Since the lengths of each side of a square are the same, $PQ=PS$. Therefore:

$3a+2=2a+7$

$a=5$

6. (B) 60 The three interior angles of any triangle add up to 180°, and the parts of the ratio here add up to $1+2+3=6$. To get the amount of one part, divide 180 by 6, which gives you 30. Now you know that the three angles have degree measures of $1\times 30=30$, $2\times 30=60$, and $3\times 30=90$. The difference in the degree measures between the largest and the smallest angles is:

$90-30=60$

7. (A) $9+3\sqrt{5}$ The Pythagorean theorem states that in a right triangle, $a^2+b^2=c^2$, where a and b are the two perpendicular side lengths, called the legs, and c is the length of the hypotenuse. To figure out the perimeter of this triangle, first find the length of the hypotenuse using the Pythagorean theorem:

$$(3)^2+(6)^2=c^2$$
$$9+36=c^2$$
$$45=c^2$$
$$c=\sqrt{45}=\sqrt{9}\times\sqrt{5}=3\sqrt{5}$$

Now add together the sides to get the perimeter:

$3+6+3\sqrt{5}=9+3\sqrt{5}$

8. (D) 1 Since AE is a line segment, all the lengths are additive, so $AE=AD+DE$. The question states that $AD=6$ and $AE=8$. So $DE=AE-AD=8-6=2$. The question also states that $BE=6$. So $BD=BE-DE=6-2=4$. The question asks for the length of BC. Since $CD=3(BC)$, the relationship can be written like this:

$BC+3BC=4$

$4BC=4$

$BC=1$

9. (B) 34 The central rectangle shares a side with each of the four squares, and the four squares form the legs of the four right triangles. Two of the rectangle's sides have a length of 4, so the two squares that share these sides must also have sides of length 4. The other two sides of the rectangle have a length of 3, so the other two squares, which share these sides, must also have sides of length 3. Each triangle shares a side with a small square and a side with a large square, so the legs of each triangle have lengths of 3 and 4, respectively. Since the legs are of length 3 and 4, the hypotenuse of each triangle must have a length of 5.

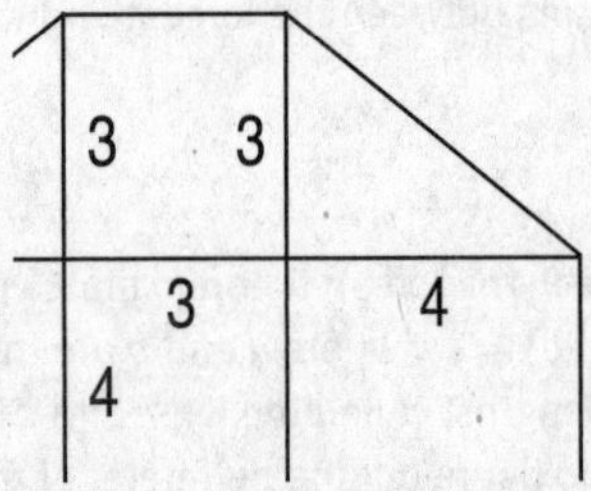

The perimeter is the sum of the hypotenuses of the triangles and a side from each square:

$$\begin{aligned}\text{Perimeter} &= 4(5) + 2(4) + 2(3)\\ &= 20 + 8 + 6\\ &= 34\end{aligned}$$

10. (D) 144 If $\frac{\sqrt{n}}{3}$ is an even integer, then $\frac{\sqrt{n}}{3} = 2k$, where k is an integer because any even integer is equal to 2 multiplied by an integer. Since $\frac{\sqrt{n}}{3} = 2k$, $\sqrt{n} = 6k$. Squaring both sides of $\sqrt{n} = 6k$ gives: $(\sqrt{n})^2 = (6k)^2$, and $n = 36k^2$. So n must be a multiple of 36. Only choice (D), 144, is a multiple of 36. $144 = 4 \times 36$. Checking choice (D), $\frac{\sqrt{144}}{3} = \frac{12}{3} = 4$ and 4 is an even integer.

If all that math seems too time consuming, you could use a combination of logic and Backsolving to get the correct answer. Since the problem deals with integers, then n must be a perfect square; only 81 and 144 among the answer choices are perfect squares. You could then backsolve by plugging one of those choices into the equation to determine that choice (D) 144 is correct.

11. (D) $3n - 14$ This is a good opportunity to pick numbers. For instance, you could pick 10 for n, so that Gheri is 10 years old. In that case, Carl is 4 years old and Jean is 2 years old. Then the sum of their ages is: $10 + 4 + 2 = 16$. Now, plug 10 in for n into the answer choices, to see which gives you the target number of 16:

$3n + 4 = 3 \times 10 + 4 = 34$	Too big.
$3n - 4 = 3 \times 10 - 4 = 26$	No good.
$3n - 8 = 3 \times 10 - 8 = 22$	Still too big.
$3n - 14 = 3 \times 10 - 14 = 16$	That's correct!

12. (C) 51 2, 37, and 67 are all prime. $51 = 3 \times 17$, so it is not prime. Knowing the *rules of divisibility* can help you when dealing with problems involving prime numbers, multiples, or factors. For instance, 51 is divisible by 3, because its digits—5 and 1—add up to 6, which is a multiple of 3.

13. (B) π The circumference of a circle $= 2\pi$ (radius), so a circle with a circumference of 2π has a radius of 1. The area of a circle $= \pi(\text{radius})^2$ so the area of a circle with a radius of 1 is $\pi(1)^2 = \pi$.

14. (D) $\frac{5}{16}$ To figure out the probability that a band member chosen at random will be a sophomore, apply the probability formula:

$$\text{Probability} = \frac{\text{Number of desired outcomes}}{\text{Number of possible outcomes}}$$

In this case:

$$\text{Probability} = \frac{\text{Number of sophomores}}{\text{Number of band members}} = \frac{20}{13 + 20 + 16 + 15} = \frac{20}{64} = \frac{5}{16}$$

15. (D) $\frac{5}{3}$ You could use an insight about math to solve this problem:

$a - b = -(b - a)$. So if $k - 3 = -\frac{5}{3}$, then $3 - k = -\left(-\frac{5}{3}\right) = \frac{5}{3}$.

You could also have simply done the calculations:

$$\begin{aligned}k - 3 &= -\frac{5}{3}\\ 3k - 9 &= 5\\ 3k &= 4\\ k &= \frac{4}{3}\end{aligned}$$

Now substitute that value into the second equation:

$$3 - k = 3 - \frac{4}{3} = \frac{9}{3} - \frac{4}{3} = \frac{5}{3}$$

16. (A) $\frac{1}{5}$ Solve the equation $\frac{3x}{5} = 3x^2$ for x. Since $x \neq 0$, you can divide both sides by x. Then $\frac{3}{5} = 3x$. Dividing both sides by 3: $\frac{1}{5} = x$.

17. (A) $\frac{1}{3}$ You can pick numbers or use substitution for this problem. The best place to start picking numbers is with c and d. The question states that $4c = 3d$, so make $c = 3$ and $d = 4$. If $c = 3$ and $\frac{1}{2}b = c$, then $b = 6$. If $b = 6$ and $a = 2b$, then $a = 12$. Therefore, using the numbers picked: $\frac{d}{a} = \frac{4}{12} = \frac{1}{3}$.

18. (B) 72 Begin by converting the dimensions of the rectangular area to be covered to inches: 2 feet by 3 feet = 24 inches by 36 inches. 3 inches goes into 24 inches 8 times, and 4 inches go into 36 inches 9 times, so to cover the area, $8 \times 9 = 72$ tiles are needed.

Alternatively, you could determine that each tile has an area of $\frac{1}{4} \times \frac{1}{3} = \frac{1}{12}$ square foot. So you need 12 tiles to cover one square foot. The area to be covered is 6 square feet, so $6 \times 12 = 72$ tiles needed.

19. (A) 6 If $\frac{7}{x} > \frac{1}{4}$, then multiply both terms by $4x$ and the result is $x < 28$. (When multiplying inequalities the sign must be flipped when multiplying by a negative number. You will remember that that means, in most cases, that you can't multiply both sides of an inequality by a variable, since you don't know whether a variable is negative or positive. However, in this case, you know from the construction of the problem that x must be positive.) Similarly, if $\frac{7}{x} < \frac{1}{3}$, then $x > 21$. Therefore, 22, 23, 24, 25, 26, and 27 (a total of 6 numbers) are the allowable integer values of x.

20. (B) 13π The hole cover has a diameter of 14 inches, so its radius is 7 inches and its area is $\pi(7)^2 = 49\pi$ square inches. The diameter of the hole is 12 square inches, so its radius is 6 inches and its area is $\pi(6)^2 = 36\pi$ square inches. Thus, the cover is $49\pi - 36\pi = 13\pi$ square inches greater than the area of the hole.

21. (B) $\frac{3}{8}$ Some answer choices you can eliminate right away. $\frac{4}{7}$ is greater than $\frac{1}{2}$, so that's out, as is $\frac{1}{2}$. You're left with $\frac{1}{3}$, which is about 0.333, and $\frac{3}{8}$, which is 0.375 and is the closest in value to 0.40.

22. (A) $\frac{39y}{100x}$ Since the question asks for the answer in dollars, start by converting cents to dollars. There are 100 cents in a dollar, so 39 cents $= \frac{39}{100}$ dollars. Since each peach costs $\frac{39}{100}$ dollars, y peaches cost $\frac{39}{100}y$ dollars. If x oranges cost as much as y peaches, x oranges also cost $\frac{39}{100}y$ dollars or $\frac{39y}{100}$ dollars. Then one orange costs $\frac{1}{x}$ as much: $\frac{39y}{100x}$ dollars.

Alternately, this is an ideal problem to solve by picking numbers. Say that 5 oranges and 10 peaches cost the same; that is $x = 5$ and $y = 10$. If peaches are 39 cents each, 10 of them will cost \$3.90 and that would equal the cost of 5 oranges. That means each orange costs $\frac{\$3.90}{5}$ or \$0.78. Try the numbers in each answer choice, looking for the target number:

Choice (A): $\frac{(39)(10)}{(100)(5)} = \frac{390}{500} = \frac{39}{50} = 0.78$	This may be the answer; hold onto it for now.
Choice (B): $\frac{(39)(5)}{(100)(10)} = \frac{195}{1{,}000}$	Discard.
Choice (C): $\frac{3{,}900}{(5)(10)} = \frac{3{,}900}{50} = 78$	This is 78 dollars, not 78 cents, so discard.
Choice (D): $\frac{(39)(5)}{10} = \frac{195}{10} = 19.5$	Discard.

23. (C) 13 To come up with the largest number of groups, you should minimize the number of students in each group. Each group must contain at least three students, so that means you could have a total of 12 groups with 3 students in them, and one final group with 4 students in it, for a total of 13 groups.

24. (C) 8 The altitude at the base of the triangle is the same thing as the height of the triangle, and the area of a triangle $= \frac{1}{2}$(base)(height). If the area of the triangle is 36 and its base is 9, the altitude is found as follows:

$$36 = \frac{1}{2}(9)(\text{altitude})$$

$$\text{altitude} = \frac{36}{9} \times 2 = 4 \times 2 = 8$$

25. (C) 85 Plug in -5 for x into the equation and solve:

$2(-5)^2 - 6(-5) + 5$

$= 2(25) + 30 + 5$

$= 85$

Part 6: Electronics Information Answers and Explanations

1. **(B) directly proportional** Voltage and current are directly proportional. Anytime voltage increases, current will also increase as long as the resistance remains constant, in accordance with Ohm's Law.

2. **(C) wrapping the wire around an iron nail** As electric current passes through a wire, a magnetic field is generated around the wire. The strength of the field at any given point at a certain distance around a straight wire will be constant for a given current value (and increasing the circumference of a non-coiled circuit will likewise have little effect). However, winding the wire into loops will cause the field to overlap and increase within the loops. Wrapping the wire in a coil around a ferrous (that is, made of iron) object like a nail, will lead to an even greater increase in the magnetic field. That's because the material's many microscopic magnets will align with and strengthen it. What about the greater wire length in answer choice (D)? Increasing the length of the wire while holding the current constant will not result in a greater field if the number of loops isn't increased. In fact, having the same number of loops spread out over a greater wire length will actually result in a smaller increase in magnetic field strength inside the loops.

3. **(D) an inductor** Inductors resist change in current, and thus act as resistors for alternating current, increasing their apparent resistance with the AC frequency. However, for a continuous direct current, an inductor is the one option listed that will offer little resistance to current flow.

4. **(A) 1** In the symbol, the number 1 indicates the base, which is the middle piece of semiconductor material inside a transistor.

5. **(B) 375 ohms** For a simple, parallel circuit with two resistors of equal value, divide the resistance by two to find the equivalent resistance. In this question, two 750 ohm resistors in a parallel circuit gives an equivalent resistance of $\frac{750 \text{ ohms}}{2} = 375$ ohms

6. **(C) self-induction** When current first flows through a coil, the magnetic field builds relatively slowly. This is because the expanding magnetic field generates a voltage in the coil that opposes the original current flow. This is known as counter-emf, and it is produced by a process known as self-induction.

7 **(C) 10 K Ω** Via Ohm's Law, $R = V/I$. But before plugging in the current, it must be converted into amperes. 5 mA = 0.005 A. Thus $R = \frac{50 \text{ V}}{0.005 \text{ A}} = 10{,}000$ ohms. 10,000 ohms is the same as 10 k Ω.

8. **(C) series-parallel circuit** Resistors R2 and R3 are in parallel with each other, while R1 is in series with them both. Since this simple circuit combines series and parallel loads, it is a series-parallel circuit.

9. **(B) always decreases the effective resistance** Adding resistors in parallel always lowers the resistance, while adding resistors in series always increases it. The rule can be tested or rediscovered by making up a simple question of two resistors in parallel and calculating the effective resistance, but it's easy to remember if you just imagine more and more barriers in a row (series resistors) adding to the difficulty in moving forward, and the addition of more pathway choices (parallel resistors) making it easier by reducing crowding of electrons.

10. **(B) a free electron outside the valence shell** The crystalline structure of pure silicon is very stable. The four valence electrons in each silicon atom bond with the valence electrons in the atoms around it, so no free electrons exist to allow current flow. This can be changed by "doping" the silicon's crystal structure with phosphorus, arsenic, or antimony. Since these elements all have five electrons in their valence shell, they will bond themselves to the other silicon atoms, but leave one free electron that is able to migrate throughout the crystal. This changes the silicon crystal into an N-type material. This new material is still electrically neutral, but is able to conduct electricity due to the presence of free electrons.

11. **(A) NPN** There are two types of transistors: an NPN transistor, and a PNP transistor. An NPN transistor is made up of a thin piece of P-type material sandwiched between two pieces of N-type material. A PNP transistor is the opposite: two pieces of P-type material that have a piece of N-type material between them.

12. **(B) $1\frac{1}{3}$ Ω** The reciprocals of the resistance values are $\frac{1}{2}$ and $\frac{1}{4}$. Since $\frac{1}{2}$ is equivalent to $\frac{2}{4}$, their sum can be simplified as $\frac{3}{4}$. The reciprocal is $\frac{4}{3}\Omega$, which can also be written as $1\frac{1}{3}\Omega$.

13. **(D) 6 Ω** The effective resistance is the sum of the resistances of each resistor in series, so the answer is $3 + 3 = 6\ \Omega$.

14. **(A) always increases the effective resistance** Adding a resistor when a resistor is already present increases the effective resistance.

15. **(C) insulator** More than four electrons in the valence shell means that the element is an insulator. Insulators do not conduct electricity well, and therefore are useful for creating electrical barriers. Thus, an element with five electrons in its valence shell is an insulator.

16. **(B) coulomb** An ampere is defined as one coulomb of electrical charge flowing past a point in one second. One coulomb is the amount of charge in 6.25310^{18} electrons. This is the same as 6,253,100,000,000,000,000 electrons. If this many electrons flow past a point in a conductor in one second, one ampere of current is flowing.

17. **(A) allow DC current to pass easily** Inductors work exactly opposite to capacitors, in the sense that they allow DC to pass easily, but resist the flow of AC. This is known as *inductive reactance*, and it will rise in direct proportion to the frequency of the current flowing through the inductor.

18. **(D) adding a resistor of 5 Ω in parallel with the original load** Only adding a resistor in parallel will decrease the overall resistance, since it allows more paths for the current to travel through. This means the answer is either (B) or (D). Answer choice (D) works, because $\frac{1}{5}+\frac{1}{5}=\frac{2}{5}$, and the reciprocal is $\frac{5}{2}=2.5\ \Omega$. In contrast, answer choice (B) would result in an effective resistance of 1.6 Ω, so it does not work. You could also have solved this problem using logic. When two loads of equal resistance value are wired in parallel, their effective resistance is half of their individual resistances (likewise, with three identical resistors in parallel, their effective resistance is $\frac{1}{3}$ of their individual resistances, and with four resistors in parallel, effective resistance is decreased to $\frac{1}{4}$, etc. If you know this rule, (D) is the clear choice without doing any math.

19. **(B) ammeter** This is the circuit symbol for an ammeter, which is used to measure current.

20. **(C) current flow varies throughout the circuit** In a series circuit, *the current flow will be the same in all parts of the circuit.* The current that leaves the voltage source must return to the voltage source, and since there is only one path for current to follow in a series circuit, current will be the same throughout the circuit. Voltage measured across each of the components, on the other hand, may be different depending on their resistance.

Part 7: Auto and Shop Information Answers and Explanations

1. **(A) when an air-fuel mixture explodes, rather than burns** Detonation is when an air-fuel mixture explodes, rather than burns. It can often take place when an engine's air-fuel mixture is lean.

2. **(C) secondary** The current that flows through the ignition primary winding builds a strong magnetic field that surrounds both it and the *secondary coil winding*. This is the device that creates high voltage for the spark plugs.

3. **(C) 50/50** Engine coolant is normally made up of a 50/50 mix of antifreeze and water.

4. **(C) either a ratchet or an impact wrench** The most common drive tool for sockets is the ratchet, which turns the fastener in only one direction as the handle is moved back and forth through a narrow arc. Ratchets are reversible, so they can be set to tighten or loosen a fastener. Sockets can also be used with pneumatic (compressed air) power tools, such as an air impact wrench. The air impact wrench can remove fasteners quickly by applying tremendous amounts of torque (twisting force) using a "hammering" action that vibrates fasteners loose.

5. **(D) a Crescent wrench** Automotive technicians will sometimes use a steel rule or tape measure to determine distances or the width of objects. When accuracy down to one-thousandths of an inch is required, an outside micrometer is used. The outside micrometer is made to measure the outside diameter of cylindrically shaped objects, as well as the thickness of flat objects. While a Crescent wrench can be adjusted to accommodate different widths, it is not used for measurement.

6. **(C) power stroke** The diagram indicates that the intake and exhaust valves are closed, and the piston is headed for BDC. That means that the engine is in the power stroke phase of its cycle.

7. **(D) powertrain control module** A fuel injector sprays fuel into the intake stream as it receives electrical signals from the powertrain control module (or PCM).

8. **(B) 13** Carpenters often will use a claw hammer, which serves a dual purpose. The hammer head has two ends: one to drive nails and the other to remove nails. Claw hammers come in a variety of sizes, and these are determined by the weight of the hammer head. A general purpose claw hammer would have a 13 oz head.

9. **(B) start a hole for drilling** This is a center punch. Center punches are used to make small indentations that serve as starting marks for drilling operations. Making a small indentation with a center punch can help the drill bit stay on target long enough to get a hole started.

10. **(C) cylinder head** Spark plugs are threaded into the cylinder head where the spark plug protrudes into the combustion chamber and generates the spark to initiate combustion.

11. **(A) planetary gear sets** Automatic transmissions do all of the gear selection for the driver. This is done in the transmission using hydraulics and *planetary gear sets*. The newest automatic transmissions are controlled electronically by the vehicle's powertrain control module.

12. **(A) hydraulically** Brake systems are *hydraulically operated*. A pumping piston, located in the master cylinder, is operated by the brake pedal and puts pressure on the system's brake fluid.

13. **(B)**

A hacksaw is used for cutting metals such as steel, aluminium, or copper. The blades in a hacksaw are replaceable, and it is important to choose the right blade for the material that is going to be cut.

14. **(D) wire-feed welding** MIG welding is also known as *wire-feed welding* because the electrode used for the weld process is a wire that is automatically fed from a spool.

15. **(A) carpenter's level** The carpenter's level has three level vials, which are mounted horizontally, vertically, and at a 45-degree angle. It is used in construction for checking for true vertical, true horizontal, and 45-degree angles.

16. **(D) none of the above** None of the ignition systems listed would utilize contact points. Only older cars with breaker points would have contact points.

17. **(A) a plane** When working with wood, there are many occasions when it is necessary to remove a small amount of material to make a piece fit properly or to make a surface smooth. One tool that can be used for this purpose is a *plane*.

18. **(D) pitman arm** The *pitman arm* is a short lever arm connected to the steering gear that transmits the steering force from the cross shaft to the steering linkage system. Long-short arm suspensions most commonly use linkage steering to connect the steering column to the pitman arm, center link, and idler arm.

19. **(C) emission control** The catalytic converter is responsible for converting the toxic components of engine exhaust into relatively harmless compounds such as carbon dioxide and water and is, thus, part of the emission control system.

20. **(C) wire cutter** The most common type of pliers is the combination slip-joint. These are adjustable at the joint of the two handles of the pliers. With two different positions to choose from, these pliers can grip objects in a wide range of sizes. Sometimes, this design also incorporates a *wire cutter* for increased versatility.

21. **(A) earlier** For higher engine speeds, the flame must be started *earlier* in order to generate the most effective downward push on the piston. This is known as advancing the timing.

22. **(A)** 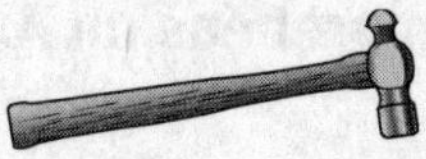

A ball-peen hammer is used for forming soft metal, peening rivet heads, and striking metal in out-of-the-way places.

23. **(D) torque converters** A common type of suspension system, long-short arm suspension systems include (A), *ball joints*, (B), *upper control arms,* and (C), *lower control arms*, among other components. A (D) *torque converter* is part of an automatic steering system, not a suspension system.

24. **(D) cold chisel** The most common chisel is the *cold chisel*, which has a straight, sharp edge for cutting off bolt heads or separating two pieces of an assembly.

25. **(A) worn engine bearings** Of the given choices, only *worn engine bearings* would not be the cause of a loss of compression, as the engine bearings do not weigh in on compression issues.

Part 8: Mechanical Comprehension Answers and Explanations

1. **(D) the speed with which the surfaces are moving past each other** Kinetic friction is exhibited when surfaces/objects move past one another. The normal force, which is a factor of the weight of the object, will affect friction. (For example, move your hand across a table lightly and then again with more force; increased force leads to increased friction.) The nature and area of the surfaces in contact with each other will also affect the friction that develops between two surfaces. (For example, moving your hand across a desk with oil on it is a lot easier than with honey; also, moving one finger across a desk develops less friction than moving your entire hand.) The speed at which the surfaces move past each other however, does not impact the frictional forces (though it may generate more heat).

2. **(D) $w = F \times d$** Work is accomplished when force is applied against an object. This is summarized by the formula $W = Fd$ where F is force in newtons (N), d is distance in meters (m), and W is work in joules (J).

3. **(B) The smaller pulley turns faster.** The smaller pulleys *always* turn faster than larger pulleys in a system.

4. **(A) the work-energy theorem** Any work that is done to accelerate an object at rest to speed v will be converted into the kinetic energy of that object. This principle is known as the work-energy theorem.

5. **(D) 1 watt of power** Performing work at the rate of 1 joule per second is the same as expending 1 watt of power.

6. **(C) all liquids are good lubricants** There are several properties of liquids that are unchanging. First, they are effectively incompressible. Even when extremely high pressure is applied to a liquid, the volume of the liquid will decrease only a very small amount. This property makes liquids very effective for transmitting force. And the second overarching principle of hydraulics is that liquids conform themselves to the shape of their container. Whether in a pipe or a pump, liquids will always change their shape to fill the space completely. But not all liquids are by nature good lubricants.

7. **(C) The crate will not move.** The force of static friction can increase in response to an applied force, resisting movement, but only up to a maximum value. If the applied force is greater than the maximum value of static friction, the crate will move. Since the applied force is given, the maximum force of static friction is what needs to be calculated. The force of static friction has a maximum value equal to the product of the normal force and the coefficient. The coefficient is given, but the normal force is not. However, since the object is at rest on a flat, horizontal surface, the normal force will be equal and opposite to the object's weight. Therefore $F_N = 250$ N. The maximum force of static friction then is: $F_{fs} = (0.8)(250) = 200$ N. The applied force is 200 N, but the maximum force of static friction is also 200 N. Since friction always opposes movement, it is opposite the applied force, so the net force can be calculated like this: $F_{net} = (200\text{ N}) + (-200\text{ N}) = 0$ N. Since the net force is zero, the object will not move. Another way to put it is that the applied force is enough to take the resisting static friction to its limit, but not enough to overcome it.

8. **(B) 50 psi** Using the formula for pressure $\left(P = \frac{F}{A}\right)$, if 100 pounds of force are applied over an area of 2 square inches, then the resulting pressure is 50 pounds per square inch (psi).

9. **(D) mechanical advantage** Mechanical advantage is defined as "the advantage gained by the use of a mechanism in transmitting force." Using the proper equipment, it is possible to increase or even multiply force many times over what is initially applied.

10. **(C) fulcrum** The fulcrum is the point that the lever pivots on. The position of the fulcrum will define whether any mechanical advantage is gained. If the fulcrum is closer to the object being lifted, less effort is required to do the work.

11. **(B) $6\frac{\text{m}}{\text{s}^2}$** Based on Newton's second law, $F = ma$, rearrange the equation for acceleration: $a = \frac{F}{m}$. Units of force must be in Newtons and mass in kg. Taking 30 N and dividing by 5 kg equals $6\frac{\text{m}}{\text{s}^2}$.

12. **(C) the object will accelerate in the direction of the net force** According to Newton's Second Law of Motion, when a net force acts on an object, the object will accelerate in the direction of the net force. The acceleration will be less if the mass of the object is greater. (A) is not necessarily true. An object may speed up or slow down when a net force acts on it depending on the direction of the force and the object's initial motion. The cumulative force is the same as the net force, so (B) does not make much sense. As for (D), in uniform circular motion, speed (magnitude) is constant, but the direction of motion changes constantly. Thus the net force applied leads to acceleration but no change in speed.

13. **(A) clockwise only** Belted pulleys, unlike meshed gears, turn in the same direction.

14. **(A) a fixed pulley** The illustration shows a fixed pulley.

15. **(C) 10 psi** The load piston has a total surface area of 10 square inches. Using the formula $P = \frac{F}{A}$, 100 pounds ÷ 10 square inches = 10 pounds per square inch (psi) of pressure that should be applied to the load piston.

16. **(C) 20 N to the left and 10 N down** Forces are vector quantities, which means they have both magnitude and direction. The 30 N and 50 N forces are acting in opposite directions on the horizontal dimension. Since 50 N > 30 N, it will dominate and the resultant is: 50 N – 30 N = 20 N in the direction of the greater force (i.e., to the left). There is nothing opposing the 10 N force, so the result is 10 N acting vertically downward.

17. **(D) 4:1** Note that in order to lift the load 1 foot, it is necessary to pull the rope a total of 4 feet. This is because each of the four rope links must shorten by 1 foot to get the lower block to move 1 foot. Neglecting friction, this gives a total mechanical advantage of 4:1, so if 4 pounds of force are required to lift a load, only 1 pound needs to be applied to the rope.

18. **(A) first-class** A child's teeter-totter is an example of a first-class lever because the fulcrum is located between the load and the applied force.

19. **(C) sliding frictional force** In the example of the hockey puck sliding along an ice surface, while ice has much less friction than most surfaces, it will still exert a sliding frictional force on the puck that eventually causes it to stop moving.

20. **(A) torque** Torque is *twisting force*. When a bolt is being tightened, torque is being applied to the bolt. For example, when tightening a fastener, if a wrench is used that is 1 foot long, and a force of 100 pounds applied to the end of the wrench in a direction perpendicular to the wrench, 100 foot-pounds (ft-lb) of torque is being applied.

21. **(B) 100 foot-pounds** When tightening a fastener, if a person uses a wrench that is 1 foot long, and applies a force of 100 pounds to the end of the wrench, 100 foot-pounds (ft-lb) of torque is being applied. Simply multiply the applied force by the length of the wrench to determine the torque.

22. **(C) 50 N** A 50 N weight will produce a 50 N force acting downward. Since the system is in equilibrium, the upward tension force must exactly balance the downward weight, which gives a tension of 50 N acting upwards.

23. **(A) the number of pulleys used in the block and tackle** The pulleys of a block-and-tackle system multiply the force exerted in pulling the rope of the system by the number of pulleys in the system. Therefore, the lifting force that the pulleys exert on the object is the force exerted in pulling the rope multiplied by the number of pulleys used.

24. **(A) acceleration due to gravity** According to Newton's Gravitational Law, since gravity exerts a force on bodies that pulls them together, this force causes an acceleration due to gravity, which is represented by the symbol g.

25. **(C) the pivot point of a lever** The fulcrum is the point on which a lever pivots.

Part 9: Assembling Objects Answers and Explanations

1. **C**

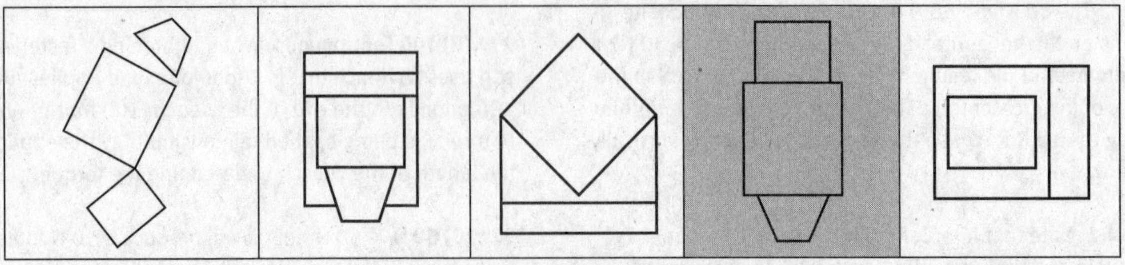

A B **C** D

2. **D**

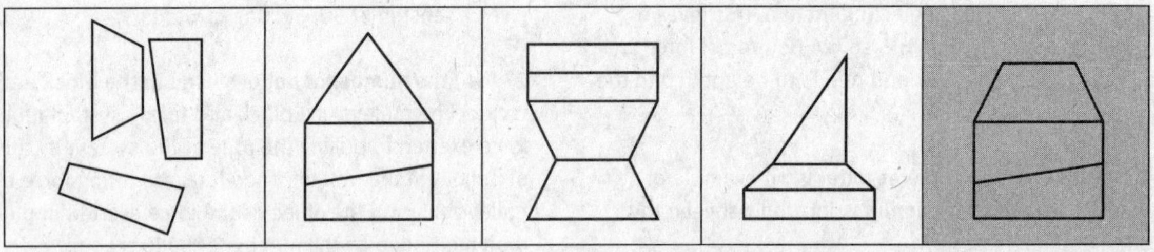

A B C **D**

3. **C**

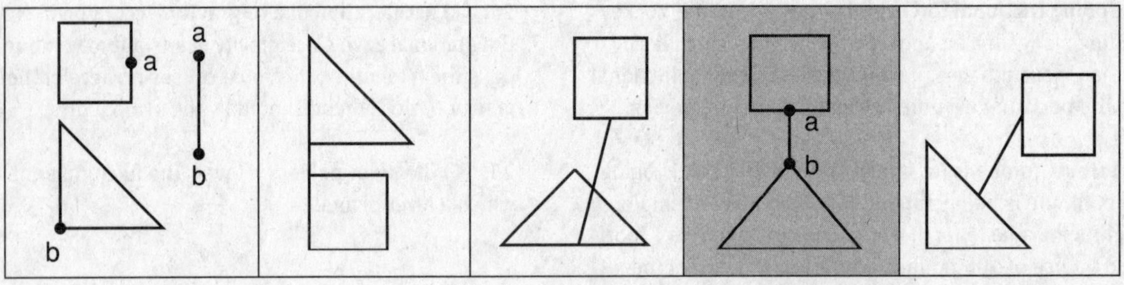

A B **C** D

4. **C**

A B **C** D

5. **C**

A B **C** D

6. **B**

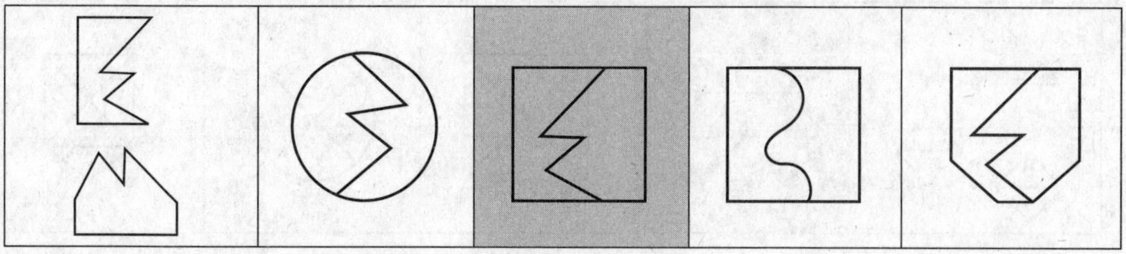

A **B** C D

7. **B**

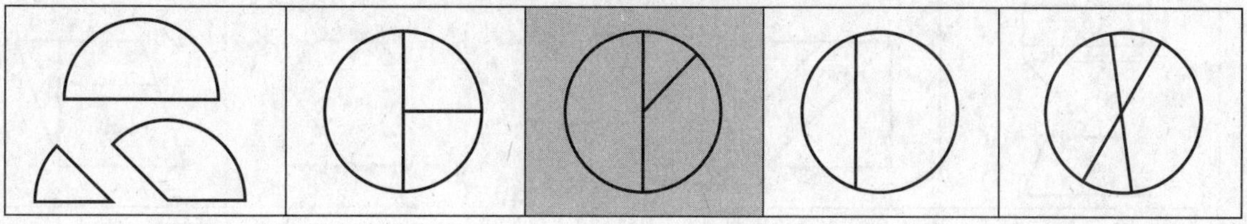

A **B** C D

8. **D**

A B C **D**

9. **A**

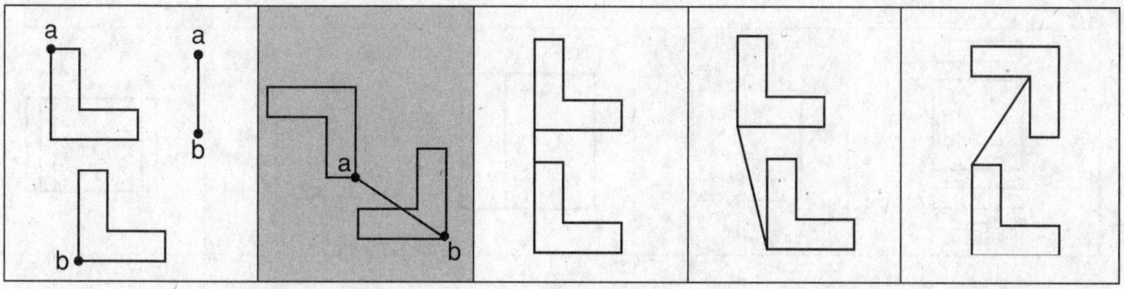

A B C D

10. **D**

A B C **D**

11. A

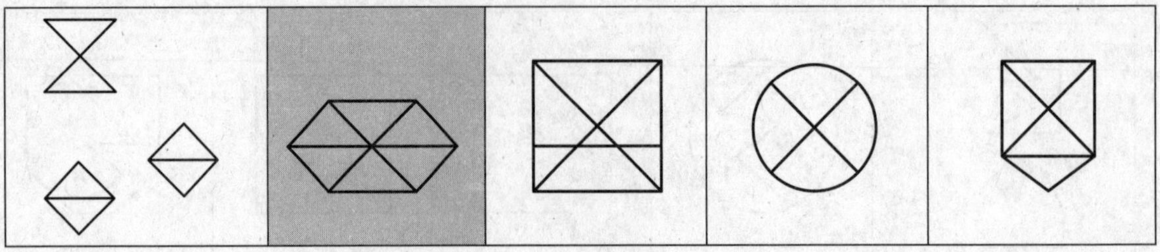

A B C D

12. B

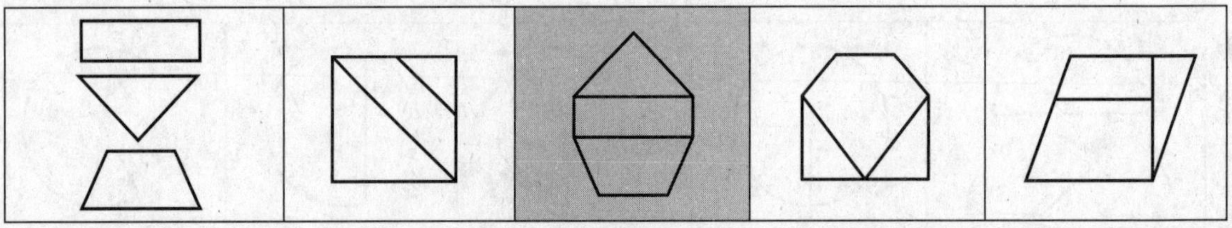

A **B** C D

13. C

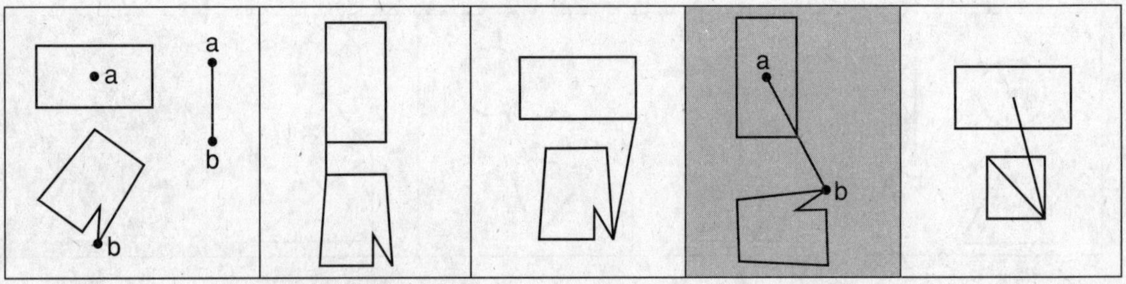

A B **C** D

14. A

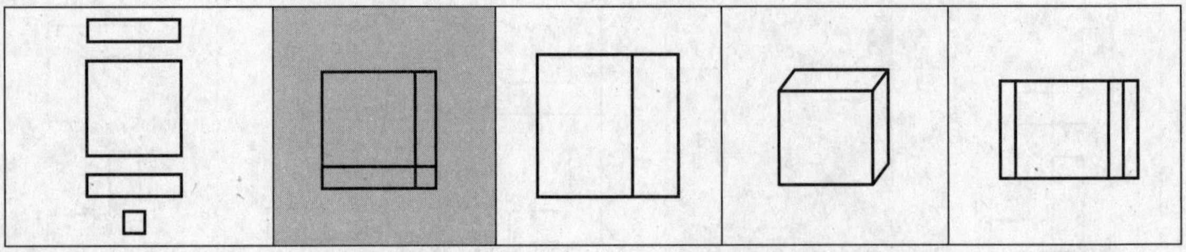

A B C D

15. C

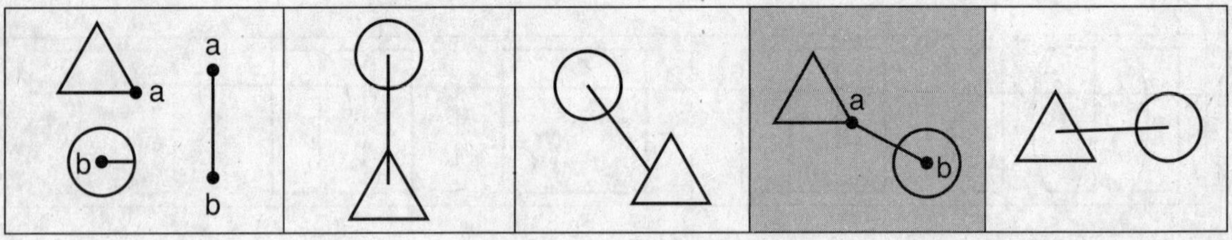

A B **C** D

16. **C**

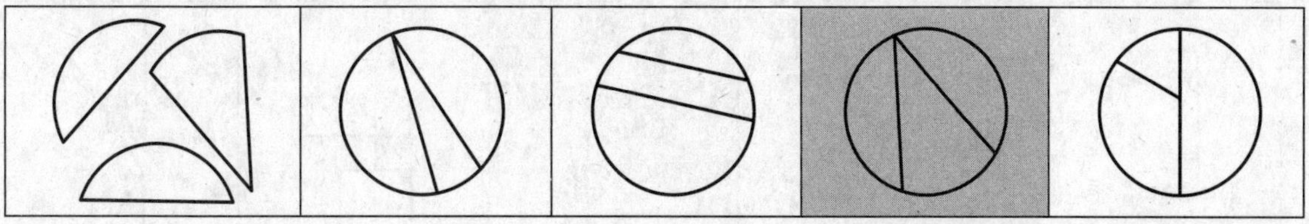

A B **C** D

17. **C**

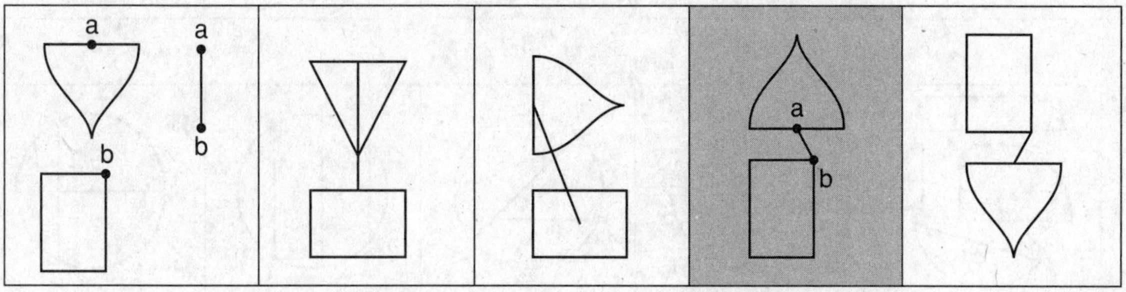

A B **C** D

18. **C**

A B **C** D

19. **A**

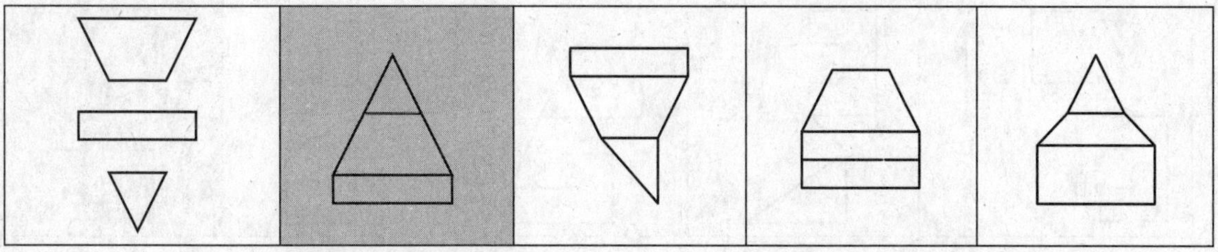

A B C D

20. **A**

A B C D

21. **B**

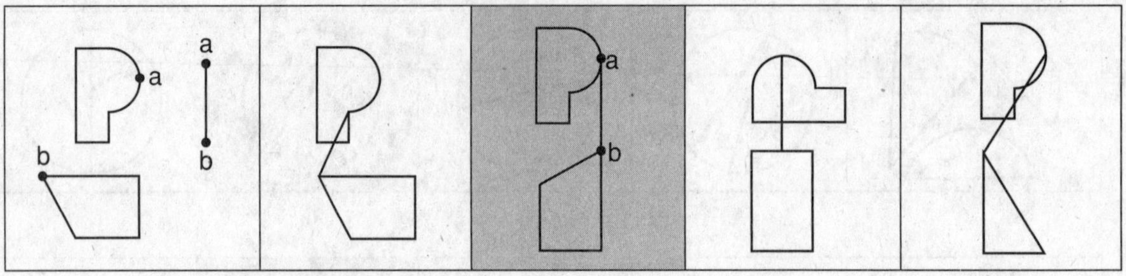

A **B** C D

22. **A**

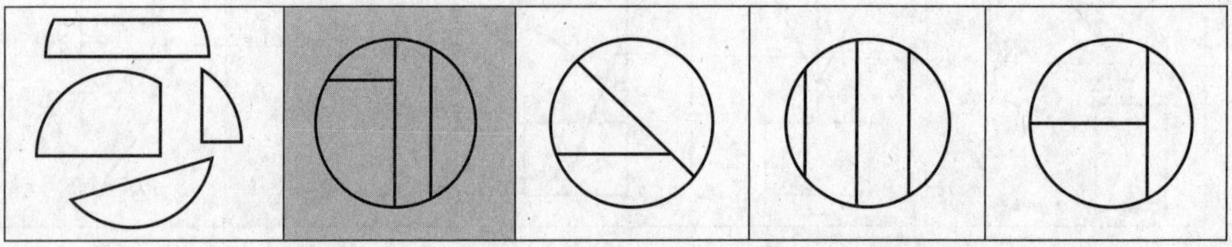

A B C D

23. **C**

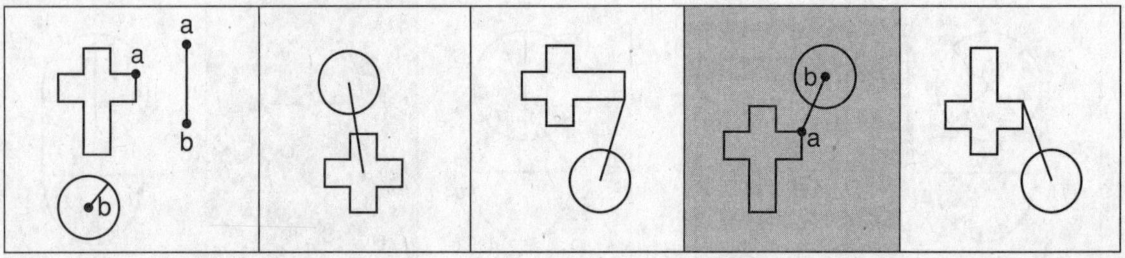

A B **C** D

24. **D**

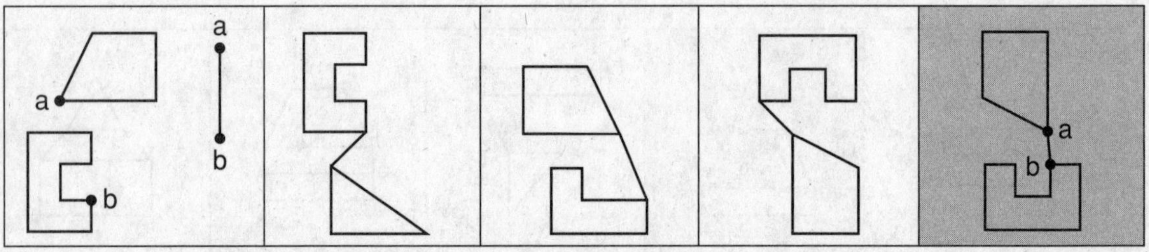

A B C **D**

25. **B**

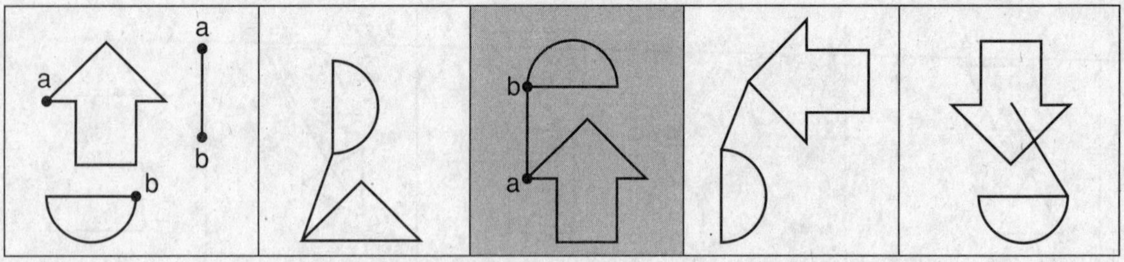

A **B** C D

APPENDIX: WORD PARTS

This appendix contains comprehensive lists and definitions of prefixes, suffixes, and word roots, along with examples of words that contain each word part.

- Part I: prefixes
- Part II: suffixes
- Part III: word roots

If you are struggling to raise your performance on the Word Knowledge subtest, you may wish to consider making flashcards out of the word parts in this chapter, and then use them to quiz yourself as part of your ASVAB studies. In studying this word list, you will want to be familiar with the following abbreviations:

- (n) noun: a word that names a person, place, thing, or idea
- (v) verb: an action word
- (adj) adjective: a word that modifies a noun
- (adv) adverb: a word that modifies a verb, adjective, or other adverb

PART I: PREFIXES

The table below lists the most common prefixes in English, along with the meaning of each and examples.

Note: Sometimes the word parts listed below appear in the middle of words, but they appear most commonly as prefixes. Also, remember that not all words have prefixes.

Prefix	Meaning	Examples
a, an	not, without	agnostic: (n) one who believes that it cannot be known whether or not God exists; (adj) claiming no knowledge amoral: (adj) not related to morality; lacking regard for morality anomaly: (n) an irregularity anonymous: (adj) lacking a named author; having an unknown author apathy: (n) lack of feeling, interest, or emotional investment atheist: (n) one who does not believe in God atypical: (adj) not typical
ab	away from, apart from, down	abduct: (v) to take away by force abhor: (v) to hate; to detest abject: (adj) cast down; degraded abnormal: (adj) not normal; not conforming to a standard abolish: (v) to do away with, ban, or make void abstract: (v) to draw or pull away, remove; (adj) theoretical; related to ideas rather than to specific instances or objects
ad	toward, near	(Sometimes the *d* is dropped and the first letter to which *a* is prefixed is doubled.) adapt: (v) to adjust in response to new circumstances addict: (v) to cause (someone) to become dependent on a substance or activity; (n) a person who is dependent on a substance or activity address: (n) a speech; (v) to give a speech or direct a statement to adhere: (v) to stick fast; to cleave; to cling adjacent: (adj) next to, close to, or connected to adjoin: (v) to be next to, close to, or connected to admire: (v) to look up to; to look at with approval or pleasure advocate: (v) to plead for; to argue in favor of attract: (v) to draw by physical force or by an appeal to emotions or senses
ambi, amphi	both, on both sides	ambidextrous: (adj) able to use both hands equally well ambiguous: (adj) open to various interpretations amphibian: (n) an animal that lives part of its life in water and part of its life on land; a person with a twofold nature; (adj) having the traits of an amphibian
ant, ante	before	antebellum: (adj) before the war (especially the American Civil War) antecedent: (adj) existing, being, or going before antedate: (v) to precede in time anterior: (adj) placed before
anti	against	antidote: (n) a remedy intended to counteract a poison antifreeze: (n) a chemical that lowers the freezing point of a liquid such as water antiseptic: (adj) free from germs; particularly clean or neat antithetical: (adj) opposed to, contrary to

Prefix	Meaning	Examples
bi, bin	two	biennial: (adj) happening every two years bilateral: (adj) related to both sides bilingual: (adj) able to speak one's native language and another with equal facility binocular: (adj) involving two eyes bipartisan: (adj) representing two parties combination: (n) the joining of two or more things into a whole
cent	hundred	centimeter: (n) one hundredth of a meter centipede: (n) a creature with many legs century: (n) one hundred years percent: (adj) out of every hundred
circu, circum	around	circuit: (n) a path or journey around an area; a path traveled by electrical current circuitous: (adj) roundabout, indirect circumference: (n) the outer boundary of a circular area circumstances: (n) the state of affairs that exist around a particular time or person; the factors that influence a person or situation
co, col, com, con	together, completely	coerce: (v) to force (another to do an action) by using fear, authority, or violence collaborate: (v) to work with another; to cooperate collide: (v) (used of two or more objects or persons) to crash together; to make contact forcefully or violently commensurate: (adj) suitable in measure, proportionate compatible: (adj) able to exist together with someone or something else; capable of harmonious coexistence conciliate: (v) to placate; to win over connect: (v) to bind or fasten together
contra, contro, counter	against	contradict: (v) to oppose; to speak against contrary: (adj) opposed to; opposite controversy: (n) a prolonged debate or disagreement about a topic counterfeit: (adj) fake; (n) a false imitation encounter: (v) a meeting, often with an opponent
de	away, off, down, reversal	decipher: (v) to interpret; to decode; to discern a hidden meaning defame: (v) to slander; to publicly speak ill of delineate: (v) to draw the outlines of; to sketch; to describe descend: (v) to move from a higher to a lower place
deca	ten	decade: (n) ten years decathlon: (n) a sports competition composed of ten events
di, dia	in two, across, through	diagnose: (v) to determine the nature of (a sickness or problem) by examining symptoms dialogue: (n) conversation between two or more persons diameter: (n) a line going through a circle, dividing it in two dichotomy: (n) division into two parts, kinds, etc.

Prefix	Meaning	Examples
di, dis	away from, reversal, not	diffuse: (v) to pour out and spread, as in a fluid; (adj) spread out dilate: (v) to widen; to expand disperse: (v) to spread over a wide area; to drive away in various directions disseminate: (v) to scatter or spread widely; to promulgate dissuade: (v) to persuade (someone) against (a course of action); to deter; to advise against
dys	faulty, abnormal	dysfunctional: (adj) poorly functioning dyslexia: (n) an impairment of reading ability due to a neurological problem
e, ex	out of, from, former	evade: (v) to escape from; to avoid exclude: (v) to shut out; to leave out exonerate: (v) to free or declare free from blame expire: (v) to breathe one's last; to die; to reach the end of viability
em, en	inside, into	embrace: (v) to clasp in the arms; to include or contain enclose: (v) to close in on all sides
extra	outside, beyond	extract: (v) to take out; to obtain against a person's will extradite: (v) to send (a person accused of a crime) to another state or nation for trial or punishment extraordinary: (adj) beyond the ordinary; unusual extrasensory: (adj) outside the senses; coming from or pertaining to knowledge that cannot be gained through normal sense perception
fore	before	foreshadow: (v) to warn of or indicate (a future event) foresight: (n) the act of foreseeing; care for the future; prudence forestall: (v) to prevent by advance action
hemi	half	hemiplegia: (n) paralysis of the arm, leg, and trunk on one side of the body hemisphere: (n) half a sphere; half of the Earth
hetero	different, other	heterogeneous: (adj) made up of different kinds heterosexual: (adj) pertaining to different sexes; having a sexual orientation toward members of the opposite sex
homeo, homo	same, similar	homogeneous: (adj) having a uniform nature or substance homonym: (n) one of two or more words spelled and pronounced alike but different in meaning homosexual: (adj) pertaining to the same sex; having a sexual orientation toward members of the same sex
hyper	over, above, more than, excessive	hyperactive: (adj) excessively active hyperbole: (n) extreme exaggeration used to create an effect
hypo	under, beneath, less than	hypochondriac: (n) one who imagines physical ailments; one who is overly preoccupied with physical health hypocritical: (adj) pretending to have virtues or qualities one does not have hypothesis: (n) assumption subject to proof
in, im	not, without	immoral: (adj) not moral impartial: (adj) unbiased; fair inactive: (adj) not active indolence: (adj) a tendency to avoid work or exertion; laziness innocuous: (adj) harmless, inoffensive

Prefix	Meaning	Examples
in, im	inside, into	implicit: (adj) not stated; inherent incarnate: (adj) having a body or physical form indigenous: (adj) native to a place influx: (n) the pouring or flowing of one thing into another intrinsic: (adj) natural; innate
inter	between, among	interim: (n) time period between one event and another; (adj) not permanent interloper: (n) one who intrudes in the domain of others intermittent: (adj) happening on and off; not constant intersperse: (v) to scatter; to put things among other things interstate: (adj) involving two or more states
intra	inside, within	intramural: (adj) within a school; inside a city intrastate: (adj) within a state intravenous: (adj) inside the veins
macro	great, large	macroeconomics: (n) study of the economy on a large scale macroscopic: (adj) large enough to be seen without magnification
mal, male	bad, evil, wrong	maladroit: (adj) clumsy, tactless malady: (n) an illness malediction: (n) a curse malfunction: (n) the act of not working correctly; (v) to work incorrectly; to break malicious: (adj) intended to hurt someone malign: (v) to tell lies with the intent of hurting someone's reputation; to slander
med, medi	middle	immediate: (adj) nearest; having nothing in between intermediate: (adj) between the beginning and end mediate: (v) to serve as a go-between; to try to settle an argument medieval: (adj) related to the Middle Ages mediocre: (adj) of only so-so quality medium: (n) size between small and large; a substance or agency that things travel through (as, for example, light travels through air, and news is conveyed by television and newspapers)
mega, megalo	very large	megalith: (n) a very big stone megalomania: (n) a mental condition involving delusions of greatness; an obsession with achieving great things megalopolis: (n) a very large city megaphone: (n) a device for magnifying the sound of one's voice megaton: (n) explosive power equal to 1,000,000 tons of TNT
micro	very small	microbe: (n) a very small organism microcosm: (n) a small system that reflects a larger whole microorganism: (n) a very small organism microscope: (n) a device that magnifies very small things for viewing
min, mini	small	diminish: (v) to lessen diminution: (n) reduction; the act of reducing miniature: (n) a copy or model that represents something in greatly reduced size minute: (n) one-sixtieth of an hour; (adj) very small minutiae: (n) small or trivial details

Prefix	Meaning	Examples
mis	bad, wrong, hateful	misadventure: (n) bad luck; an unlucky accident misanthrope: (n) one who hates people or humanity misapply: (v) to use something incorrectly mischance: (n) bad luck; an unlucky accident mischief: (n) naughty or annoying behavior misconstrue: (v) to understand something incorrectly misfit: (n) somebody or something that doesn't fit in
mon, mono	one, single	monarchy: (n) rule by a single person monk: (n) a man in a religious order living apart from society monogram: (n) a design made up of letters combined into one shape monograph: (n) a scholarly paper on one topic monologue: (n) a speech or other dramatic composition recited by one person monomania: (n) an obsession with a single subject monotonous: (adj) boring; spoken using only one tone
multi	many	multiple: (adj) many; having many parts; a number containing some quantity of a smaller number without remainder multiplex: (adj) having more than one part; (n) a building with many separate units multiply: (v) to increase; to become many multitudinous: (adj) very many; containing very many; having very many forms
non	not	nonconformist: (n) one who does not conform to a church or other societal institution nonentity: (n) something that doesn't exist; something that is unimportant nonpartisan: (adj) not affiliated with a political party
nov, neo, nou	new	innovate: (v) to develop a new way of doing something neologism: (n) a newly invented word or phrase neophyte: (n) a beginner; a new convert; a new worker novice: (n) a person new to any field or activity renovate: (v) to repair something so it is like new or does not show as much wear and tear
oct	eight	octagon: (n) a shape with eight sides octogenarian: (n) a person whose age is 80–89
omni	all	omnibus: (n) an anthology of the works of one author or of writings on related subjects omnipotent: (adj) all-powerful omnipresent: (adj) being everywhere at one time omniscient: (adj) knowing everything
pan, pant	all, everyone	pandemic: (adj) widespread panoply: (n) an impressive or vast group or display panorama: (n) a view or scene that extends a long way pantheon: (n) the group made up of all the gods of a particular culture; a building that honors dead heroes

Prefix	Meaning	Examples
para	next to, beside	parable: (n) a story that teaches a lesson through allegory paragon: (n) an example of excellence to be emulated parallel: (adj) being side by side and the same distance apart at all points; having similar paths or structures paranoid: (adj) suffering from a baseless distrust of others parasite: (n) a living thing that draws its nutrients from another on which it lives; a person who lives off another without providing anything in return parody: (v) to satirize through imitation; (n) an imitative satire
pent	five	pentagon: (n) a five-sided shape pentathlon: (n) a sports competition with five events
peri	around	perimeter: (n) the distance around a shape; a border around a shape peripatetic: (adj) not stationary, moving about periscope: (n) an optical instrument used to view objects that otherwise couldn't be seen
poly	many	polyandry: (n) the practice of having multiple husbands polygamy: (n) the practice of having multiple spouses, often wives polyglot: (n) someone who speaks many languages polygon: (n) a figure with many sides polytheism: (n) belief in many gods
post	behind, after	post facto: (adv) after the fact posterior: (adj) situated at the rear posthumous: (adj) after death
pre	before, in front	precedent: (n) an act that serves as an example for subsequent situations precept: (n) a rule to govern behavior precocious: (adj) unusually advanced at a young age premonition: (n) a feeling or intuition that something is going to happen before it does
pro	in front, before, much, for	proceed: (v) to go forward profuse: (adj) occurring in large amounts; abounding; overly giving prolific: (adj) highly fruitful proselytize: (v) to convert, recruit, or attempt to convert provident: (adj) possessing foresight
prot, proto	first	protagonist: (n) the main character in a play or story protocol: (n) diplomatic etiquette; a system of proper conduct; the original record of a treaty or other negotiation prototype: (n) the first version of an invention, on which later models are based
pseud, pseudo	false	pseudonym: (n) a false name; a pen name pseudopod: (n) part of a single-celled organism that can be stuck out (like a foot) and used to move around pseudoscience: (n) false science; something believed to be based on the scientific method but actually is not

Prefix	Meaning	Examples
quad, quar, quat	four	quadrant: (n) a quarter of a circle; a 90-degree arc quadruple: (adj) four times as many quadruplets: (n) four children born in one birth quart: (n) one-fourth of a gallon
quin, quint	five	quintile: (n) one-fifth quintuple: (adj) five times as many
re	back, again	recline: (v) to lean back; to lie down regain: (v) to gain again; to take back remain: (v) to stay behind; to be left; to continue to be reorganize: (v) to organize again request: (v) to ask (originally: to seek again)
retro	backward	retroactive: (adj) extending to things that happened in the past retrofit: (v) to install newer parts into an older device or structure retrograde: (adj) moving backward; appearing to move backward retrospective: (adj) looking back at the past; (n) a review of past events
se	apart, away	secede: (v) to withdraw formally from an association sedition: (n) provocation of rebellion against a government seduce: (v) to lead astray segregate: (v) to separate a larger group into two or more groups; to set an individual or subgroup apart from the larger group select: (v) to choose one thing over another separate: (v) to keep apart; to divide sequester: (v) to isolate or set apart from a larger group
semi	half	semicircle: (n) half a circle semiconscious: (adj) only partly conscious; half awake
sept	seven	septennial: (adj) occurring every seven years septuplet: (n) one of seven children born together
sex, hex	six	hexagon: (n) a shape with six sides sextet: (n) a band with six musicians
sub, sup	below, under	subliminal: (adj) existing beneath consciousness submissive: (adj) obedient; not dominant subsidiary: (adj) supplemental; secondary subterfuge: (n) a trick or stratagem used to deceive, hide something, or avoid punishment subtle: (adj) not direct; difficult to understand suppose: (v) to put down as a hypothesis; to use as the underlying basis of an argument; to assume
super, sur	over, above	superfluous: (adj) extra; more than necessary superlative: (n) the highest or best of its kind supersede: (v) to replace in power or preference by another surmount: (v) to overcome an obstacle or prevail over a problem surpass: (v) to exceed in amount or degree surveillance: (n) a watch kept over someone or something

Prefix	Meaning	Examples
sym, syn	the same, together	symbiosis: (n) the act of living together in a mutually beneficial relationship symmetry: (n) balanced proportions; having opposite parts that mirror one another sympathy: (n) the attempt to understand the feelings of others symposium: (n) a meeting at which ideas are discussed (originally: a party at which people drink together) synonym: (n) a word that means the same thing as another synthesis: (n) the act of combining things to create a new whole
trans	across, beyond	transaction: (n) an exchange, especially a business deal involving buying and selling transcendent: (adj) going beyond ordinary limits transgress: (v) to disobey or violate a law; to sin transition: (n) a change from one way of being to another transparent: (adj) easy to see through; easy to perceive
un	not	unseen: (adj) not seen unusual: (adj) not usual; exceptional; strange
uni, un	one	reunion: (n) a meeting that brings people back together unanimous: (adj) in complete agreement unicorn: (n) a mythical animal with a single horn uniform: (adj) of one kind; consistent universe: (n) all things considered as one whole

PART II: SUFFIXES

As we discussed in the Word Knowledge chapter, suffixes do two things: They affect the meaning of a word, and often they also indicate the word's part of speech. The table here lists the most common suffixes in English. Each is listed with the part of speech the suffix usually indicates, as well as examples demonstrating its usage.

Note: Sometimes the word parts listed below appear in the middle of words, but they appear most commonly as suffixes. Also, remember that not all words have suffixes.

Suffix	**Usually indicates a(n) . . .**	**Meaning**	**Examples**
a, ia	noun	state of, condition of, thing that has the quality of	insomnia: (n) condition of being unable to sleep
able, ible	adjective	capable of, worthy of	changeable: (adj) able to be changed combustible: (adj) capable of being burned; easily inflamed
age	noun	action or process, amount or rate of, place of	brokerage: (n) business of negotiating mileage: (n) number of miles orphanage: (n) place where orphans are housed
al, ial	adjective	relating to	hormonal: (adj) relating to hormones literal: (adj) relating to the usual or exact meaning of words residential: (adj) relating to a residence
an, ian	adjective or noun	adjective: relating to noun: one who relates to or is skilled at	comedian: (n) one who is funny musician: (n) one who is skilled in music reptilian: (adj) like a reptile
ance, ence, ancy, ency	noun	state of, action of	consistency: (n) state of being consistent furtherance: (n) act of furthering something
ard	noun	person who has a particular quality	coward: (n) one who is afraid, easily cowed dullard: (n) one who is dull
ary	adjective	relating to	budgetary: (adj) relating to a budget literary: (adj) relating to literature
ate	adjective, noun, or verb	adjective: having the quality of noun: one who has a quality verb: to impart or display a quality	inviolate: (adj) having not been violated renovate: (v) to make something like new again vertebrate: (n) an animal with vertebrae
cide	noun	related to killing	homicide: (n) act of murder suicide: (n) act of killing oneself insecticide: (n) product that kills insects

Suffix	Usually indicates a(n) . . .	Meaning	Examples
cracy	noun	ruling order, system of government	plutocracy: (n) rule by the wealthy class theocracy: (n) rule of God or gods
crat	noun	person who rules or has authority	plutocrat: (n) wealthy person of great influence technocrat: (n) skilled person of great influence
cy	noun	state of or quality of	bankruptcy: (n) state of being without money dormancy: (n) state of being asleep
dom	noun	state of, realm of	kingdom: (n) realm of a king wisdom: (n) state of being wise
ee	noun	person or thing that receives an action	payee: (n) one who receives payment tutee: (n) one who receives tutelage (that is, teaching)
er, or	noun	one who does	exterminator: (n) one who exterminates robber: (n) one who steals or robs
escence	noun	process of becoming or process of taking on a quality	obsolescence: (n) process of becoming obsolete or out-of-date luminescence: (n) act of emitting light
escent	adjective	having a quality, or being in the process of becoming or taking on a quality	effervescent: (adj) bubbling; lively; high-spirited luminescent: (adj) possessing the quality of emitting light
ese	adjective or noun	adjective: relating to a place or language noun: the language of a nation	Japanese: (adj) related to Japan or its language; (n) the language of Japan
esque	adjective	having or being in the style of, resembling, imitating	picturesque: (adj) resembling a painting; visually pleasing or striking statuesque: (adj) tall, shapely, and attractive, like a statue
etic	adjective	relating to, tending to, or having the quality of	apologetic: (adj) relating to apology or tending to apologize apoplectic: (adj) displaying apoplexy (extreme anger)
ette, et, let	noun	diminutive: makes small	kitchenette: (n) a small kitchen booklet: (n) a small book
ful	adjective	having the quality of	hopeful: (adj) having hope peaceful: (adj) having the quality of peace
hood	noun	state of or quality of, place with or group of people with a specific quality	brotherhood: (n) state of being brothers, or group of people who call themselves brothers neighborhood: (n) place where neighbors live

Suffix	Usually indicates a(n) . . .	Meaning	Examples
ial, ian, arian	adjective or noun	relating to or characterized by	antiquarian: (n) one who studies valuable old things centenarian: (n) one who is a hundred or more years old ceremonial: (adj) relating to ceremony disciplinarian: (n) one who is strict librarian: (n) one who works in a library vegetarian: (n) one who does not eat meat
ic, ical	adjective	relating to, having the quality of	comic: (adj) relating to comedy hypocritical: (adj) saying one thing and doing another
ify	transitive verb	to make something into something else	deify: (v) to make into a god magnify: (v) to make larger
ish	adjective	resembling, being related to, having the quality of	youngish: (adj) somewhat young fiendish: (adj) resembling a fiend or devil outlandish: (adj) resembling something strange or unfamiliar
ism, asm	noun	action of; state of being; state of having a quality, doctrine, or religion; bias or hatred	capitalism: (n) a belief in private ownership criticism: (n) the act of criticizing gigantism: (n) state of being abnormally large phantasm: (n) product of fantasy sexism: (n) bias against women or men
ist	noun	person who does	harpist: (n) one who plays the harp linguist: (n) one who studies language
ite	noun	person who has a trait, person who adheres to a school of thought	socialite: (n) person who is known for socializing
itis	noun	inflammation of, condition resembling a disease	arthritis: (n) disease of the joints laryngitis: (n) inflammation of the larynx
ity	noun	state of, quality of	density: (n) state of being dense; mass based on unit per volume ferocity: (n) state of being ferocious
ive	adjective or noun	adjective: having a quality of or tending to noun: a person or thing with the quality of	creative: (adj) tending to create things digestive: (n) a substance that aids digestion
ize	verb (transitive or intransitive)	cause to become, treat a certain way	eulogize: (v) to speak well of terrorize: (v) to cause to become terrified

Suffix	Usually indicates a(n) . . .	Meaning	Examples
less	adjective	without, lacking	fearless: (adj) without fear graceless: (adj) lacking grace
logy	noun	the study of	biology: (n) study of life geology: (n) study of Earth's structure
ly	adverb	(makes a word into an adverb)	frighteningly: (adv) in a frightening manner rapidly: (adv) in a rapid manner
ment	noun	action of, state of being	contentment: (n) state of being content development: (n) action of developing something
ness	noun	state of being, condition, or quality	goodness: (n) state of being good sweetness: (n) quality of being sweet
nomy	noun	rules, system of law or order, field of study	astronomy: (n) scientific study of the universe beyond the Earth autonomy: (n) independence, self-governance
oid	adjective or noun	adjective: resembling noun: something that resembles something else	humanoid: (n) a being that resembles a human planetoid: (n) a small celestial body resembling a planet
osis	noun	act of, or disease of	hypnosis: (n) the act of hypnotizing someone psychosis: (n) disease of the mind
ous, ious	adjective	having the quality of	anxious: (adj) having anxiety poisonous: (adj) having the quality of poison
ship	noun	state of being	hardship: (n) difficult circumstances ownership: (n) the state of being the owner of something
some	adjective	having the quality of; causing a feeling or condition	fearsome: (adj) causing fear quarrelsome: (adj) tending to quarrel
tion, sion, ion	noun	act or process; result of a process; state or condition	promotion: (n) act of promoting something or someone tension: (n) state of being tense
tive	adjective	having a quality or tendency	emotive: (adj) having the quality of causing strong emotions talkative: (adj) tending to talk
tude	noun	state of being	certitude: (n) state of being certain gratitude: (n) state of being grateful
ular	adjective	of, relating to, or resembling	muscular: (adj) having well-developed muscles popular: (adj) of or relating to the people

Suffix	Usually indicates a(n) . . .	Meaning	Examples
ure, eur	noun	action of, condition of	exposure: (n) act of exposing something grandeur: (n) condition of being grand
ward	adverb	in the direction of	eastward: (adv) toward the east heavenward: (adv) toward heaven
y	adjective	having, having a quality of, or resembling	dirty: (adj) not clean; relating to dirt grouchy: (adj) resembling a grouch

PART III: WORD ROOTS

The roots below, most of which come from Greek or Latin, form the basis for many common words in English. Not all words you will see on the ASVAB are based on the word parts below. In fact, many short English words do not have Greek or Latin roots. However, knowing the word roots in this list will usually help you determine the definitions of longer words.

AC/ACR: sharp, bitter, sour
acid: (n) something that is sharp, sour, or ill-natured
acrimonious: (adj) bitter; hostile
acumen: (n) mental sharpness; quickness of wit
acute: (adj) sharp at the end; ending in a point
exacerbate: (v) to make (a problem or conflict) worse

ACOU: hearing
acoustic: (adj) related to sound or hearing

ACT/AG: to do, to drive, to force, to lead
activate: (v) to cause something to act
agile: (adj) having good coordination and quick movements
agitate: (v) to stir up or roil

AL/ALI/ALTER: other, another
alias: (n) an assumed name; another name
alibi: (n) an excuse, often that a person was at another place at the time a crime was committed
alien: (n) one born in another country; a foreigner
alter ego: (n) the second self; a substitute or deputy
alternative: (n) a possible choice
altruist: (n) someone who cares about others or gives (money or resources) in support of others

AM: love
amateur: (n) someone who does an activity for the love of it, rather than as part of her job; a hobbyist; an inexperienced person; (adj) lacking or displaying a lack of experience; non-professional
amiable: (adj) friendly, likable
amicable: (adj) characterized by exhibiting good will
amity: (n) friendship; peaceful harmony
amorous: (adj) inclined to love, esp. sexual love
enamored: (adj) inflamed with love; charmed; captivated

AMBL/AMBUL: to go, to walk
ambulance: (n) a vehicle equipped for carrying sick people (from a phrase meaning "walking hospital")
ambulatory: (adj) able to walk; having to do with walking
preamble: (n) an introductory statement (originally: to walk in front)

ANIM: of the life, mind, soul, breath
animal: (n) a living being
animosity: (n) hostility
equanimity: (n) ability to remain calm under pressure
magnanimous: (adj) generous; forgiving
unanimous: (adj) of one mind; in complete accord

ANNUI/ENNI: year
annals: (n) a log or record, often with yearly entries
anniversary: (n) yearly observance
annual: (adj) yearly
annuity: (n) periodic cash payments or receipts
perennial: (adj) lasting for an indefinite amount of time

ANTHRO/ANDR: man, human
androgynous: (adj) having both male and female traits or anatomy
android: (n) robot; mechanical man
anthropology: (n) a branch of science dealing with human origins, traits, and behavior

AQUA/AQUE: water
aquamarine: (n) a bluish-green color
aquarium: (n) a tank for keeping fish and other underwater creatures
aquatic: (adj) having to do with water
aqueduct: (n) a channel for transporting water

ARCH/ARCHI/ARCHY: chief, principal, ruler
anarchy: (n) the absence of a formal government; a state of lawlessness
archenemy: (n) chief enemy
architect: (n) the devisor, maker, or planner of anything
monarchy: (n) rule by a single person
oligarchy: (n) a state or society ruled by a select group

AUTO: self
autocrat: (n) an absolute ruler
automatic: (adj) self-moving or self-acting
autonomy: (n) independence or freedom

BEN/BENE: good
benediction: (n) act of uttering a blessing
benefit: (n) anything advantageous to a person or thing
benevolent: (adj) desiring to do good to others
benign: (adj) gracious, kindly; not harmful

BON/BOUN: good, generous
bona fide: (adj) true, authentic (literally, in good faith)
bonus: (n) something of value given beyond what is expected
bountiful: (adj) generous

BREV/BRID: short, small
abbreviate: (v) to shorten
abridge: (v) to shorten
brevity: (n) shortness
brief: (adj) short

BURS: purse, money
bursar: (n) treasurer
disburse: (v) to pay
reimburse: (v) to pay back

CANT/CENT/CHANT: to sing
accent: (n) emphasis or stress
chant: (n) a song; singing
enchant: (v) to fascinate or captivate
incantation: (n) a magical chant or spell
recant: (v) to take back something spoken or written

CAP/CIP/CEPT: to take, to get
anticipate: (v) to look forward to something; to foresee
capture: (v) to take by force
precept: (n) a rule to govern behavior
susceptible: (adj) capable of receiving, admitting, undergoing, or being affected by something

CARN: flesh
carnage: (n) slaughter, widespread killing
carnivorous: (adj) eating flesh
incarnation: (n) a being invested with a bodily form
reincarnation: (n) rebirth in a new body

CAUS/CAUT: to burn
caustic: (adj) burning or corrosive
cauterize: (v) to burn or deaden

CED/CEED/CESS: to go, to yield, to stop
accede: (v) to yield to a demand; to give in
antecedent: (adj) existing or going before
cessation: (n) the act of stopping or discontinuing (either temporarily or permanently)
concede: (v) to admit the truth of a statement (especially in a debate); to yield; to admit defeat
incessant: (adj) without stop
predecessor: (n) a person who holds an office or position prior to another person; one who comes before

CELER: speed
accelerant: (n) something used to speed up a process
accelerate: (v) to increase in speed
decelerate: (v) to decrease in speed

CENTR: center
centrist: (adj) pertaining to moderate political or social ideas; (n) a person who holds moderate views
concentrate: (v) to converge; to bring together in a central location; to focus one's attention; to intensify
concentric: (adj) (of circles or spheres) having the same center point
eccentric: (adj) off-center

CERN/CERT/CRET/CRIM/CRIT: to separate, to judge, to distinguish, to decide
ascertain: (v) to make sure of; to determine
certitude: (n) freedom from doubt
criterion: (n) a standard used in judging something
discreet: (adj) careful to keep secrets; prudent in speech; careful to avoid causing embarrassment
discrete: (adj) detached from others, separate
discriminate: (v) to distinguish between; to show preference
hypocrite: (n) someone who pretends to have beliefs or virtues he does not actually have

CHROM: color
chromatic: (adj) having to do with color
chrome: (n) a metallic element (chromium) used to make vivid colors; something plated with chromium
monochromatic: (adj) having only one color

CHRON: time
anachronism: (n) something or someone that seems to belong to a different historical time than the one it is in
chronic: (adj) constant, habitual
chronology: (n) the sequence in which events in the past occurred; a history recounted sequentially
chronometer: (n) a highly accurate clock or watch
synchronize: (v) to occur at the same time or agree in time

CIS: to cut
incision: (n) a cut, gash, or notch
incisive: (adj) penetrating, cutting
precise: (adj) definitely stated or defined
scissors: (n) cutting instrument for paper, cloth, or other material

CLA/CLO/CLU: to shut, to close
claustrophobia: (n) an abnormal fear of enclosed places
cloister: (n) enclosed courtyard; secluded place; monastery or convent
conclusive: (adj) final; closing an argument
disclose: (v) to reveal or make known
exclude: (v) to shut out from inclusion or consideration; to omit; to be incompatible with
preclude: (v) to prevent; to make impossible; to exclude

CLAIM/CLAM: to shout, to cry out
clamor: (n) a loud uproar
disclaim: (v) to claim to have no interest in; to deny; to renounce
exclaim: (v) to cry out; to say loudly or emphatically
proclaim: (v) to announce or declare in an official way
reclaim: (v) to claim or demand the return of a right or possession

CLI: to lean toward
climax: (n) the highest point; the most intense point in a story or development
decline: (v) to refuse (an offer); to withhold (from another)
disinclination: (n) aversion, distaste
recline: (v) to lean back

COGN/CONN: to know
cognition: (n) the process of knowing
incognito: (adj) with one's name or identity concealed
recognize: (v) to identify as already known

CORP/CORS: body
corporation: (n) a company legally treated as an individual
corps: (n) an organized body of troops
corpse: (n) a dead body
incorporation: (n) the act of combining into a single body; the act of forming a corporation

COSM: order, universe, world
cosmetic: (adj) improving the appearance; (n) substance used to beautify
cosmic: (adj) relating to the universe
cosmology: (n) a theory of the universe as a whole
cosmopolitan: (adj) worldly
cosmos: (n) the universe; an orderly system; order
microcosm: (n) a small system that reflects a larger whole

COUR/CUR: running, a course
concur: (v) to accord in opinion; to agree
courier: (n) a messenger
curriculum: (n) the regular course of study
cursive: (n) a fluid style of handwriting in which letters are joined
cursory: (adj) superficial (of a review of material); hasty
excursion: (n) a short journey or trip
incursion: (n) a hostile, sudden entrance into a place or group
recur: (v) to happen again

CRE/CRESC/CRET: to grow
accretion: (n) an increase by natural growth
accrue: (v) to gain by natural growth; to accumulate
creation: (n) the act of creating, inventing, or producing
increase: (v) to make greater or more in number; to grow in number or size
increment: (n) an amount added or increased; an addition; profit

CRED: to believe, to trust
credentials: (n) written proof of authority or status
credit: (n) trustworthiness
credo: (n) any formula of belief
credulous: (adj) too willing to trust in or believe
incredible: (adj) unbelievable

CRYPT: hidden
apocryphal: (adj) having questionable authenticity; possibly not genuine
crypt: (n) an underground vault or tomb
cryptology: (n) the science of interpreting secret writings, codes, ciphers, and the like

CUB/CUMB: to lie down
cubicle: (n) a small, partitioned space
incubate: (v) to cause to develop or hatch (as when hens sit on eggs); to develop or take form
incumbent: (n) a person who holds an office; (adj) holding an office or position
succumb: (v) to yield; to give way

CULP: fault, blame
culpable: (adj) blameworthy; at fault; guilty
culprit: (n) a person guilty of an offense
inculpate: (v) to accuse; to incriminate

DAC/DOC: to teach
didactic: (adj) related to instruction; prone to instruct too often
docile: (adj) tame, teachable
doctor: (n) a medical practitioner; someone with an advanced degree
doctrine: (n) principles or teachings relating to a specific principle or group
indoctrinate: (v) to teach a doctrine to a person

DELE: to erase
delete: (v) to erase; to remove
indelible: (adj) impossible to erase; lasting

DEM: people
democracy: (n) government by the people
demographics: (n) vital and social statistics of populations
epidemic: (adj) affecting a large number of people simultaneously; (n) a widespread disease or problem

DEXT: right hand, right side, deft
ambidextrous: (adj) able to use both hands (e.g., for writing)
dexterity: (n) quality of being skilled at working with one's hands

DI: day
dial: (n) a surface that displays information in a circular form (such as a clock face); a rotatable knob
diary: (n) a record of one's days
dismal: (adj) gloomy (from "bad days")

DIC/DICT/DIT: to say, to tell, to use words
dictate: (v) to give a directive; to talk while being recorded
dictionary: (n) a compilation of the meanings of words
interdict: (v) to forbid; to prohibit
predict: (v) to foresee; to give a forecast of
verdict: (n) a judgment or decision

DIGN: worth
deign: (v) to think fit or in accordance with one's dignity
dignitary: (n) a person who holds a high rank or office
dignity: (n) the state of being titled or privileged; worthiness

DOG/DOX: opinion, belief
dogma: (n) a strongly held specific belief
orthodox: (adj) holding customary or traditional beliefs
paradox: (n) a situation that seems to contradict itself

DOL: to suffer, to pain, to grieve
condolence: (n) expression of sympathy
doleful: (adj) sorrowful; mournful
dolorous: (adj) sorrowful; causing sorrow
indolence: (n) tendency to be inactive; laziness

DON/DOT/DOW: to give
anecdote: (n) a short narrative about an interesting event
antidote: (n) a remedy intended to counteract a poison
donate: (v) to give; to contribute
endow: (v) to give in a way that provides future income or future benefits
pardon: (n) forgiveness; (v) to forgive

DORM: sleep
dormant: (adj) sleeping; inactive
dormitory: (n) a place for sleeping; a residence hall

DORS: back
dorsal: (adj) having to do with the back
endorse: (v) to sign on the back; to vouch for

DUB: doubt
dubious: (adj) doubtful
indubitable: (adj) unquestionable

DUC/DUCT: to lead
abduct: (v) to lead away by force or coercion
conducive: (adj) contributive; helpful
conduct: (n) behavior; (v) to lead
induce: (v) to cause an action by influence
induct: (v) to admit to a group or position, usually with formal ceremonies
produce: (v) to make; to cause to be; to bring into existence; (n) (usually used collectively) things brought into existence; agricultural products

DUR: hard, lasting
dour: (adj) sullen, gloomy, stern
durable: (adj) able to resist deterioration
duration: (n) the length of time something exists
duress: (n) compulsion by threat or force
endure: (v) to last, to sustain under pressure

EGO: self
ego: (n) oneself; the part of oneself that is self-aware
egocentric: (adj) focused on oneself
egoism/egotism: (n) selfishness; self-absorption

EQU: equal, even
adequate: (adj) equal to the requirement or occasion
equation: (n) the act of comparing two equal things; the act of equalizing
equivalent: (n) the same
iniquity: (n) injustice; act of evil

ERR: to wander
err: (v) to be mistaken; to make a mistake
errant: (adj) wandering, unsteady, deviating
erratic: (adj) deviating from the proper or usual course of conduct
error: (n) something that is incorrect or untrue; a mistake

ESCE: becoming
adolescent: (adj) between childhood and adulthood
convalescent: (adj) recovering from illness
incandescent: (adj) glowing with heat; shining
reminiscent: (adj) reminding of; suggestive of

FAB/FAM: to speak
affable: (adj) friendly; courteous
defame: (v) to slander; to publicly speak ill of
fable: (n) fictional tale, esp. legendary
famous: (adj) well known; celebrated
ineffable: (adj) inexpressible

FAC/FIC/FIG/FAIT/FEIT/FY: to do, to make
configuration: (n) manner of arrangement; shape
counterfeit: (n) imitation; forgery
deficient: (adj) incomplete or insufficient
faction: (n) a subgroup within a larger group or party; strife; dissention
factory: (n) building used for manufacturing
prolific: (adj) producing many offspring or much output

FAL: to err, to deceive
default: (v) to fail
fail: (v) to be insufficient; to be unsuccessful; to die out
fallacy: (n) a flawed argument
false: (adj) not true; erroneous; lying
infallible: (adj) incapable of being wrong or being deceived

FATU: foolish
fatuity: (n) foolishness; stupidity
fatuous: (adj) foolish; stupid
infatuated: (adj) swept up in a fit of passion, impairing one's reason

FER: to bring, to carry, to bear
confer: (v) to grant; to bestow
offer: (v) to present for acceptance, refusal, or consideration
proffer: (v) to offer
proliferate: (v) to reproduce; to produce rapidly
referendum: (n) a vote on a political question by the entire electorate; the process of referring political questions to such a vote

FERV: to boil, to bubble
effervescent: (adj) bubbling; lively; high-spirited
fervid: (adj) ardent; intense
fervor: (n) passion; zeal

FI/FID: faith, trust
affidavit: (n) written statement made under oath
confide: (v) to entrust with a secret
fidelity: (n) faithfulness; loyalty
infidel: (n) one who does not accept a particular religion

FIN: end
confine: (v) to keep or restrict within certain limits; to imprison
definitive: (adj) decisive; unconditional; final
final: (adj) at the end; coming last
infinite: (adj) boundless; endless
infinitesimal: (adj) infinitely or very small

FLAGR/FLAM: to burn
conflagration: (n) a large destructive fire
flagrant: (adj) blatant; scandalous
inflame: (v) to set on fire

FLECT/FLEX: to bend, to turn
deflect: (v) to bend; to cause to turn aside; to dissuade from a purpose
flexible: (adj) able to bend without breaking
inflect: (v) to bend; to change pitch
reflect: (v) to throw back

FLU/FLUX: to flow
confluence: (adj) merging into one
fluctuation: (n) continual change between one state and another; wavelike motion
fluid: (n) a substance that is capable of flowing freely; (adj) able to flow freely

FORT: chance
fortuitous: (adj) happening by luck
fortunate: (adj) lucky; auspicious
fortune: (n) chance or luck in human affairs

FORT: strength
forte: (n) a person's best skill or talent; strong suit
fortify: (v) to strengthen; to strengthen defenses against attack

FRA/FRAC/FRAG/FRING: to break
fractious: (adj) irritable; peevish
fracture: (n) breakage, esp. of a bone
fragment: (n) an incomplete or broken-off part
infringe: (v) to break or violate (a law, etc.)

FUG: to flee, to fly
fugitive: (adj) on the run; (n) someone who flees
refuge: (n) a haven for those fleeing
refugee: (n) a person who flees in search of safety or freedom
subterfuge: (n) a trick or stratagem used to deceive, hide something, or avoid punishment

FUM: smoke
fume: (n) smoke; (v) to emit smoke or vapors
fumigate: (v) to treat with smoke or vapors
perfume: (n) scents, from burning incense or other sources of fragrance

FUS: to pour
diffuse: (v) to spread over a wide area; to scatter; to pour out
fusillade: (n) sustained delivery of gunshots or criticism
infusion: (n) the act of introducing or pouring; the act of steeping or soaking in liquid; the resulting liquid
profuse: (adj) occurring in large amounts; abounding; overly giving
suffuse: (v) to spread over or throughout

GEN: birth, creation, race, kind
carcinogenic: (adj) producing cancer
congenital: (adj) (of a disease or defect) existing from birth
generous: (adj) giving or given freely
genetics: (n) study of heredity and variation among animals and plants
progeny: (n) offspring; descendants

GN/GNO: to know
agnostic: (n) one who believes that it cannot be known whether or not God exists
diagnose: (v) to determine the nature of (a sickness or problem) by examining symptoms
ignorant: (adj) possessing inadequate knowledge
ignore: (v) to overlook; to refuse to consider
prognosis: (v) to forecast, especially of disease

GRAD/GRESS: to step
aggressive: (adj) given to hostile acts or feelings
degrade: (v) to humiliate; to dishonor; to reduce to lower rank
digress: (v) to depart from the main subject
progress: (n) forward movement
regress: (v) to move backward; to revert to an earlier state

GRAM/GRAPH: to write, to draw
diagram: (n) a figure made by drawing lines; an illustration
epigram: (n) a short poem; a pointed statement
grammar: (n) a system of language and its rules
graph: (n) a diagram used to convey mathematical information
graphite: (n) mineral used for writing, as the "lead" in pencils
photograph: (n) a picture, originally made by exposing chemically treated film to light

GRAT: pleasing
gracious: (adj) kindly, esp. to inferiors; merciful; courteous
grateful: (adj) thankful
gratuity: (n) money given above what is due for a service or product
ingratiate: (v) to bring oneself into favor

GREG: flock
aggregate: (n) a number of things considered as a collective whole
congregate: (v) to come together in a group
gregarious: (adj) sociable; enjoying spending time with others
segregate: (v) to separate a larger group into two or more groups; to set an individual or subgroup apart from the larger group

HAP: by chance
haphazard: (adj) at random
hapless: (adj) without luck
happen: (v) to occur (originally: to occur by chance)
happy: (adj) pleased, as by good fortune
mishap: (n) an unlucky accident
perhaps: (adv) maybe; possibly

HER/HES: to stick
adherent: (adj) able to adhere; (n) believer or advocate of a particular school of thought
adhesive: (adj) sticky; coated with glue; (n) an adhesive substance
coherent: (adj) logically consistent; hanging together (as an argument)
inherent: (adj) existing as a permanent, inborn, or essential trait

HOL: whole
catholic: (adj) universal
holistic: (adj) considering something as a unified whole
holocaust: (n) complete destruction by fire or other means
hologram: (n) a two-dimensional picture that appears to be three-dimensional

ICON: image, idol
icon: (n) a symbolic, religious picture; something seen as representative of a culture or movement
iconic: (adj) representative of a culture or movement
iconology: (n) symbolism

IT/ITER: way, journey
ambition: (n) desire to achieve
itinerant: (adj) traveling
itinerary: (n) travel plans
reiterate: (v) to repeat
transit: (n) means of transportation

JECT: to throw, to throw down
abject: (adj) miserable; wretched; contemptible
conjecture: (n) the act of forming an opinion based on inadequate information; an opinion so formed
dejected: (adj) sad; depressed
eject: (v) to throw out; to expel
inject: (v) to insert into a space or body; to introduce

JOIN/JUG/JUNCT: to meet, to join
adjoin: (v) to be next to and joined with
conjugal: (adj) related to marriage
conjunction: (n) the act or instance of joining; an instance of events occurring together; a connecting word
injunction: (n) a command; a legal order
junction: (n) the act of joining; a combination; a place where multiple paths join
rejoinder: (n) a reply or retort
subjugate: (v) to make subservient; to place under a yoke

JUD: to judge
adjudicate: (v) to act as a judge
judiciary: (n) a system of courts; members of a court system
judicious: (adj) having good judgment
prejudice: (n) a previous or premature judgment; bias

JUR: law, to swear
abjure: (v) to renounce on oath
adjure: (v) to beg or command
jurist: (n) a legal expert; a judge
perjury: (n) willful lying while under oath

JUV: young
juvenile: (adj) young; immature
rejuvenate: (v) to refresh; to make young again

LANG/LING: tongue
bilingual: (adj) speaking two languages
language: (n) a system of spoken or written communication
linguistics: (n) the study of language

LAUD: praise, honor
applaud: (v) to give praise
laudable: (adj) praiseworthy
laudatory: (adj) expressing praise

LAV/LAU/LU: to wash
deluge: (n) a great flood of water
dilute: (v) to thin (a liquid) with water
laundry: (n) items of clothing or linens collected for washing; a building where clothing is washed
lavatory: (n) a room for washing (esp. hand-washing)

LEC/LEG/LEX: to read, to speak
dialect: (n) a regional variety of a language
lectern: (n) a reading desk
lecture: (n) an instructional speech
legend: (n) a story; a written explanation of a map or illustration
legible: (adj) readable
lexicon: (n) dictionary

LECT/LEG: to select, to choose
collect: (v) to gather; to assemble; to accumulate
eclectic: (adj) selecting ideas, etc. from various sources
elect: (v) to choose; to decide
predilection: (n) preference; liking
select: (v) to choose one thing over another

LEV: to lift, to rise, light (weight)
alleviate: (v) to ease or lessen (pain, discomfort, or problems)
levee: (n) embankment against river flooding
levitate: (v) to rise in the air or cause to rise
levity: (n) humor; frivolity; gaiety
relevant: (adj) bearing on the current situation or topic; pertinent
relieve: (v) to ease a person of, or separate a person from, a burden or problem; to take away pain or responsibility

LI/LIG: to tie, to bind
ally: (v) to unite; (n) one in an alliance
league: (n) an association; a group of nations, teams, etc. that have agreed to work for a common cause
liable: (adj) legally responsible; bound by law
liaison: (n) a connection; one who serves to connect
lien: (n) the right to hold a property due to an outstanding debt
ligament: (n) a band holding bones together; a bond
oblige: (v) to make a person obligated; to make indebted or form personal bonds by doing a favor
rely: (v) to depend upon (originally: to come together; to rally)

LIBER: free
deliver: (v) to set free; to save; to hand over
liberal: (adj) generous; giving away freely
liberality: (n) generosity
liberate: (v) to set free
liberty: (n) freedom

LOC/LOG/LOQU: word, speech, thought
colloquial: (adj) of ordinary or familiar conversation
dialogue: (n) conversation, esp. in a literary work
eulogy: (n) a live speech or a written piece in praise of a specific person, often one who has recently passed away
loquacious: (adj) talkative
prologue: (n) an introduction to a written work such as a novel or play

LUC/LUM/LUS: light (brightness)
illuminate: (v) to make bright or fill with light; to clarify an idea
illustrate: (v) to clarify an idea by providing alternate explanations, examples, or ways of expressing it
illustrious: (adj) highly distinguished
lackluster: (adj) lacking brilliance or radiance
lucid: (adj) easily understood; intelligible
luminous: (adj) bright; brilliant; glowing
translucent: (adj) having the quality of allowing light to pass through but without perfect clarity

MAG/MAJ/MAX: big, great
magnanimous: (adj) lacking in pettiness or spite; forgiving and understanding
magnate: (n) a powerful or influential person
magnify: (v) to make an image larger than the actual object
magnitude: (n) measurement, value, or strength of a quantity
maxim: (n) an accepted principle or code within a system of thought
maximum: (n) the greatest value possible for a quantity

MAN/MANU: hand
emancipate: (v) to free from bondage
manifest: (adj) apparent; evident
manual: (adj) operated by hand
manufacture: (v) to make objects by hand or using tools or machines

MAND/MEND: to command, to order, to entrust
command: (v) to order; (n) an order; control
commend: (v) to give something over to the care of another; to praise
countermand: (v) to retract an order
demand: (v) to strongly ask for; to claim; to require
mandatory: (adj) commanded; required
recommend: (v) to praise and suggest the use of; to advise
remand: (v) to send back

MIS/MIT: to send
emissary: (n) a messenger or agent sent to represent the interests of another
intermittent: (adj) operating at intervals rather than continuously
remission: (n) a lessening; a cessation; in medicine, a period when the symptoms of a disease disppear
remit: (v) to send money
transmit: (v) to cause an item to travel from its sender to a recipient

MOB/MOM/MOT/MOV: to move
automobile: (n) a motorized vehicle
demote: (v) to move downward in an organization
immovable: (adj) impossible to move
locomotion: (n) moving from place to place, or the ability to do so
mobile: (adj) movable
mobilize: (v) to make ready for movement; to assemble
moment: (n) an instant; importance
momentous: (adj) of great importance (originally: having the power to move)
momentum: (n) the force driving a moving object to keep moving; a growing force
motion: (n) movement
motive: (n) a reason for action; what moves a person to do something
motor: (n) a device that makes something move
promote: (v) to move to a higher rank in an organization
remove: (v) to take away; to move away

MON/MONIT: to remind, to warn
admonish: (v) to mildly criticize or warn against an action
monitor: (n) a person who supervises or reminds; (v) to watch over or supervise
monument: (n) a structure, such as a building, tower, or sculpture, erected as a memorial
premonition: (n) a feeling or intuition that something is going to happen before it does
remonstrate: (v) to plead, argue, or object
summon: (v) to call together

MOR/MORT: death
immortal: (adj) undying; not subject to dying
morbid: (adj) susceptible to preoccupation with unwholesome matters

MORPH: shape
amorphous: (adj) having an indefinite shape; shapeless
metamorphosis: (n) a transformation in which a person or thing completely changes form

MUT: to change
commute: (v) to substitute; to exchange; to interchange
immutable: (adj) unchangeable, invariable
mutation: (n) the process of being changed
permutation: (n) a complete change; transformation
transmute: (v) to change from one form into another

NAT/NAS/NAI/GNA: birth
cognate: (adj) related by blood; deriving from a common origin
naive: (adj) ignorant or credulous
native: (adj) indigenous or natural to a place
natural: (adj) present due to nature, not to artificial or human-made means

NAU/NAV: ship, sailor
astronaut: (n) one who travels in outer space
nauseous: (adj) causing a squeamish feeling (originally: sea-sickness)
nautical: (adj) related to sailing or sailors
naval: (adj) related to the navy
nave: (n) the main and central section of a church (resembling the shape of a ship)
navy: (n) a military force consisting of ships and sailors

NIHIL: nothing, none
annihilate: (v) wipe out; reduce to nothing
nihilism: (n) denial of all moral beliefs; denial that existence has any meaning

NOC/NOX: harm
innocent: (adj) not guilty; uncorrupted
innocuous: (adj) harmless, inoffensive
noxious: (adj) injurious or harmful to health or morals
obnoxious: (adj) highly disagreeable or offensive

NOM/NYM/NOUN/NOWN: name
acronym: (n) a word formed by the initials of other words
anonymous: (adj) lacking a named author; having an unknown author
nominal: (adj) existing in name only; negligible
nominate: (v) to suggest that someone be considered as a candidate or member
noun: (n) a word that names a person, place, or thing
renown: (n) fame; reputation
synonym: (n) a word having a meaning similar to that of another word of the same language

NULL: nothing
annul: (v) to cancel; to make into nothing
null: (adj) worth nothing; lacking; nonexistent
nullify: (v) to cancel; to make into nothing

ONER: burden
exonerate: (v) to free from blame; originally: to relieve of a burden
onerous: (adj) burdensome; difficult
onus: (n) a burden; a responsibility

PAC/PEAC: peace
appease: (v) to satisfy or pacify; to soothe
pacifier: (n) something or someone that eases anger or agitation
pacify: (v) to ease anger or agitation
pact: (n) a formal agreement, as between nations

PAR: equal
disparage: (v) to belittle; to speak disrespectfully about someone or something
disparate: (adj) dissimilar; different
par: (n) a state of equality; an average amount or trait
parity: (n) equal representation; equivalent

PAS/PAT/PATH: feeling, suffering, disease
compassion: (n) a deep feeling of sympathy for another
dispassionate: (adj) lacking personal feeling or passion
empathy: (n) the identification with the feelings or thoughts of others
impassive: (adj) showing or feeling no emotion
sociopath: (n) a person whose behavior is antisocial and who lacks a sense of moral responsibility
sympathy: (n) agreement in feeling; the attempt to share in the feelings of others

PAU/PO/POV/PU: few, little, poor
impoverish: (v) to deplete
paucity: (n) small quantity; scarcity
pauper: (n) a poor person
poverty: (n) the condition of being poor

PED: child, education
encyclopedia: (n) a collection of texts that contains a broad range of, or extensively detailed, information
pedant: (n) one who displays learning ostentatiously
pediatrician: (n) a doctor who primarily has children as patients

PED/POD: foot
expedite: (v) to do something more quickly; to cause something to be done more quickly
impede: (v) to hinder or obstruct
pedal: (n) a platform upon which one places a foot to work an instrument
pedestrian: (n) a person who walks
podium: (n) a raised platform used by a speaker or conductor

PEL: to drive, to push
compel: (v) to force; to command
dispel: (v) to drive away; to disperse
expel: (v) to drive out; to banish; to eject
impel: (v) to set in motion, to push forward
propel: (v) to drive forward

PEN/PUN: to pay, to compensate
penal: (adj) pertaining to criminal punishment
penalty: (n) a punishment for a crime or transgression
penitent: (adj) wanting to repent or confess
punitive: (adj) having to do with punishment

PEND/PENS: to hang, to weigh, to pay
appendage: (n) a limb or other subsidiary part that diverges from the central structure
appendix: (n) additional material found at the end of a text; a small attachment to the colon in some animals
compensate: (v) to offset with something equivalent
depend: (v) to rely; to place trust in
indispensable: (adj) necessary
stipend: (n) a payment that occurs at regular intervals

PET/PIT: to go, to seek, to strive
appetite: (n) a desire for food, drink, or something else
compete: (v) to strive to outdo another
impetuous: (adj) impulsive; acting quickly without thinking
petition: (n) a formal request addressed to a position of power

PHIL: love
bibliophile: (n) one who loves or collects books
philharmonic: (n) a lover of music; (adj) devoted to music
philology: (n) the study of literary texts to establish their authenticity and determine their meaning
philosophy: (n) the study of knowledge and being; a type of knowledge

PHOB: fear
claustrophobia: (n) fear of enclosed places
hydrophobia: (n) fear of water, which is a symptom of rabies; rabies
phobia: (n) fear; an irrational fear

PHON: sound
megaphone: (n) a device for magnifying the sound of one's voice
phonetics: (n) the study of the sounds used in speech
telephone: (n) a device for transmitting sound at a distance

PHOTO: light
photograph: (n) a picture, originally made by exposing chemically treated film to light
photon: (n) a packet of electromagnetic radiation
photosynthesis: (n) the process through which plants turn light into energy

PLAC: to please
complacent: (adj) self-satisfied; unconcerned
implacable: (adj) unable to be pleased
placebo: (n) an inert substance which a patient believes to be a medicine
placid: (adj) pleasantly calm or peaceful

PLE/PLEN: to fill, full
complete: (adj) whole, entire
deplete: (v) to reduce or exhaust a supply
implement: (n) an instrument or tool; (v) to use as a tool; to put to use
plethora: (n) excess; overabundance
replete: (adj) abundant
supplement: (n) something added to remedy a shortage; (v) to add to

PON/POS/POUND: to put, to place
component: (n) a part or ingredient
expose: (v) to uncover or reveal; to put in the way of danger
expound: (v) to explain in detail
juxtapose: (v) to place next to; to contrast with
repository: (n) a place where things are kept or stored

PORT: to carry
deportment: (n) conduct, behavior
disport: (v) to amuse oneself
export: (v) to transmit out of the country
import: (v) to bring in from another country
portable: (adj) easily carried

POT: to drink
potable: (adj) drinkable; safe to drink; a drink
potion: (n) a drinkable substance, often medical, poisonous, or magical

PREHEND/PRISE: to take, to get, to seize
apprehend: (v) to take into custody
comprise: (v) to include or contain
enterprise: (n) a project or business
reprehensible: (adj) worthy of blame or disgust
reprisal: (n) act of retaliation
surprise: (v) to act in a way that produces astonishment or shock; (n) something unexpected

PRI/PRIM: first
primal: (adj) original; most important
primary: (adj) first; most important
prime: (n) first in quality; best
primeval: (adj) ancient; going back to the first age of the world
pristine: (adj) original; like new; unspoiled; pure

PROP/PROX: near
approximate: (adj) near; close to being accurate
proximate: (adj) nearby; coming just before or just after
proximity: (n) nearness; distance

PUG: to fight
pugilist: (n) a fighter or boxer
pugnacious: (v) to quarrel or fight readily
repugnant: (adj) objectionable or offensive

PUNC/PUNG/POIGN: to point, to prick, to pierce
compunction: (n) a feeling of uneasiness for doing wrong
expunge: (v) to erase, eliminate completely
poignant: (adj) heartfelt, moving, distressing, tending to stir emotions
puncture: (n) the act of piercing; (v) to pierce
pungent: (adj) caustic, sharp or biting (as a smell or idea)

PYR: fire
pyre: (n) a pile of combustible material to be set afire, often for burning a dead body
pyromania: (n) an urge to start fires
pyrotechnics: (n) fireworks

QUE/QUIS: to seek
acquire: (v) to obtain
conquest: (n) the act of gaining control by force
exquisite: (adj) of near-perfect workmanship
inquisitive: (adj) curious; hungry for knowledge
query: (n) a question; an inquiry

QUIE/QUIT: quiet, rest
acquiesce: (v) to comply; to give in
disquiet: (n) lack of calm or peace; (v) to upset or irritate
tranquil: (adj) calm, peaceful

RECT: straight, right
correct: (v) to set right
direct: (v) to guide; to put straight
erect: (adj) upright; (v) to construct
rectangle: (n) a four-sided figure in which every angle is a right angle

REG: king, rule
realm: (n) a kingdom; a domain
regal: (adj) kingly; royal
regent: (n) one who serves on behalf of a king; one who rules
regiment: (n) a military unit; (v) to subject to a rule
regular: (adj) having a structure following some rule; orderly; normally used; average

ROG: to ask
arrogant: (adj) conceited
derogatory: (adj) belittling; disparaging
interrogate: (v) to ask a series of questions of, often formally or aggressively
surrogate: (n) a person acting in place of another

SACR/SANCT: holy
sacrament: (n) a holy symbol or practice
sacred: (adj) holy
sacrifice: (n) an offering to a deity; the act of giving up something of value; (v) to give up (something of value)
sacrilege: (n) actions or words contrary to sacred practice
sanctify: (v) to make holy
sanction: (n) approval or permission from one in authority

SALV: to save
salvage: (v) to save; (n) something saved or recovered
salvation: (n) being saved
salve: (n) a substance that aids healing and is applied to the skin
savior: (n) one who saves

SAN: healthy
sane: (adj) mentally healthy
sanitarium: (n) a place of healing
sanitary: (n) promoting health; clean; sterile

SCI: to know
conscience: (n) an instinctive sense of moral right and wrong
conscious: (adj) aware; awake
omniscient: (adj) knowing everything
prescient: (adj) able to predict future events
unconscionable: (adj) very evil

SCRIBE/SCRIPT: to write
ascribe: (v) to assign, attribute, or credit to
circumscribe: (v) to encircle, either by drawing a circle around or by traveling in a circle around
conscription: (n) military draft
describe: (v) to explain or create an image with words
postscript: (n) a written addition or addendum, usually to a letter
scribble: (v) to write hastily or sloppily; (n) a hastily or sloppily done piece of writing or drawing
script: (n) handwriting
transcript: (n) a written record of oral speech

SEC/SEQU/SUE/SUI: to follow
obsequious: (adj) fawning
prosecute: (v) to seek to prove guilt by legal action
pursue: (v) to chase after
second: (adj) next after the first
sequence: (n) the order of events
suite: (n) a series; a set; originally, a train of followers

SED/SESS/SID: to sit, to settle
dissident: (adj) disagreeing with prevailing opinion or authority (literally, "sitting apart"); (n) one who disagrees with prevailing opinion or authority
preside: (v) to exercise control; to lead (as a meeting)
resident: (n) one who resides; a dweller in a place; (adj) living in a place
residual: (adj) remaining; leftover
sediment: (n) matter (usually small particles) that has settled to the bottom of a liquid or that has been deposited by a body of water
session: (n) a meeting at which people sit together in discussion

SEM: seed, to sow
disseminate: (v) to spread; to scatter around
semen: (n) seed (of male animals)
seminary: (n) a school, esp. for religious training (originally: a place for raising plants)

SEN: old
senate: (n) the highest legislative body (from "council of elders")
senile: (adj) relating to old age; experiencing memory loss or other age-related mental impairments
sire: (n) a title for a king; a father (originally: an important person; an old man)

SENS/SENT: to feel, to be aware
dissent: (v) to disagree with the prevailing opinion or majority
presentiment: (n) a foreboding; a feeling that something will occur
resent: (v) to feel displeasure
sense: (v) to become aware of through a faculty such as sight, hearing, taste, smell, touch, or intuition; (n) one of the avenues through which one perceives stimuli
sensory: (adj) related to the senses
sentiment: (n) an attitude or feeling about something
sentinel: (n) a person or thing that watches over, especially over a border

SOL: alone
desolate: (adj) deserted; laid waste; left alone; (v) to lay waste
isolate: (v) to separate from the group; to make alone or apart
soliloquy: (n) a long speech delivered by one actor
solitude: (n) the state of being alone

SOL: sun
parasol: (n) an umbrella that protects from the Sun
solar: (adj) related to the Sun
solstice: (n) one of two days when the Sun reaches its highest point at noon and seems to stand still

SOMN: sleep
insomnia: (n) inability to sleep
somnolent: (adj) sleep-inducing; sleepy; drowsy

SOURC/SURG/SURRECT: to rise
insurgent: (adj) rising up in revolution; rushing in; (n) a person rising in revolution
insurrection: (n) armed rebellion
resurrection: (n) the act of coming back to life; the act of rising again
source: (n) where something comes from (such as spring water rising out of the ground)
surge: (v) to rise up forcefully, as ocean waves

SPEC/SPIC: to look, to see
conspicuous: (adj) easy to see or notice
perspective: (n) one's mental view of facts, ideas, and their interrelationships
retrospective: (n) a review of the past; (adj) reflecting on or reviewing the past
specious: (adj) deceptively attractive
spectrum: (n) a broad range of related things that form a continuous series
speculation: (n) the act of making guesses based on known ideas; the act of hypothesizing

SPIR: breath
aspire: (v) to desire to achieve something
expire: (v) to breathe one's last; to die; to reach the end of viability
spirit: (n) the breath of life; the soul; an incorporeal supernatural being; a lively quality; an essence

STA/STI: to stand, to be in place
constitute: (v) to make up
destitute: (adj) extremely poor
obstinate: (adj) stubborn; unwilling to change
stasis: (n) a state of inaction; a state of equilibrium
static: (adj) unmoving or in equilibrium

STRICT/STRING/STRAN: to tighten, to bind
constrain: (v) to confine; to bind within certain limits
restriction: (n) a limitation
strangle: (v) to kill by suffocation, usually by tightening a cord or one's hand around the throat

SUA: sweet, pleasing, to urge
assuage: (v) to ease, lessen, or relieve
dissuade: (v) to deter; to advise against
persuade: (v) to encourage; to convince
suave: (adj) smooth, agreeable, pleasing

SUMM: highest, total
consummate: (adj) highly qualified; complete; perfect
sum: (n) total; amount of money
summary: (n) concisely stating the total findings on a subject; comprehensive
summit: (n) highest point

TAC/TIC: to be silent
reticent: (adj) reluctant to speak
tacit: (adj) unspoken
taciturn: (adj) uncommunicative

TACT/TAG/TAM/TANG: to touch
contact: (v) to touch; to get in touch
contagious: (adj) able to spread by contact, as disease
contaminate: (v) to corrupt, taint, or otherwise damage the integrity of something by contact or mixture
intact: (adj) untouched; whole
intangible: (adj) unable to be touched; not physical or material
tactile: (adj) touchable; having to do with the sense of touch

TAIN/TEN/TENT/TIN: to hold
abstention: (n) the act of refraining voluntarily
detain: (v) to delay or restrain
pertain: (v) to be relevant to
sustenance: (n) nourishment, means of livelihood
tenable: (adj) workable, maintainable
tenacious: (adj) holding fast
tenure: (n) length of time holding a position

TEND/TENS/TENT/TENU: to stretch, to thin
contentious: (adj) quarrelsome, disagreeable, belligerent
distend: (v) to lengthen or distort by stretching
extenuating: (adj) making less serious by offering excuses
tension: (n) tautness, stress, strain
tentative: (adj) not certain, hesitant, not finalized

TEST: to bear witness
attest: (v) to bear witness
contest: (v) to dispute (from bringing a lawsuit by calling witnesses); (n) event in which people compete
detest: (v) to hate (originally: to curse something by calling upon God to witness it)
protest: (n) a dissent; a declaration of disagreement; (v) to express dissent
testament: (n) a witnessed statement of wishes to be carried out after one's death; a will
testify: (v) to bear witness

THEO: god
atheist: (adj) one who does not believe in a deity or divine system
theocracy: (n) a form of government in which a deity is recognized as the supreme ruler
theology: (n) the study of divine things and the divine faith

THERM: heat
thermal: (adj) relating to heat; retaining heat
thermometer: (n) a device for measuring temperature
thermostat: (n) a device for regulating temperature

TOR/TORQ/TORT: to twist
contort: (v) to twist; to distort
distort: (v) to pull out of shape, often by twisting; to twist or misrepresent facts
extort: (v) to get something of value by threats or force
torque: (n) twisting force; a force that creates rotation
torture: (v) to inflict pain (including by twisting instruments like the rack or wheel)

TOX: poison
intoxication: (n) the state of being poisoned; drunkenness
toxic: (adj) poisonous
toxin: (n) poisonous or harmful substance

TRACT: to drag, to pull, to draw
abstract: (v) to draw or pull away; to remove; (adj) theoretical; related to ideas rather than to specific instances or objects
attract: (v) to draw toward by force or enticement
contract: (n) a legally binding document; (v) to enter into a contract
detract: (v) to take away from, esp. a positive thing
protract: (v) to prolong; to draw out; to extend
tractor: (n) a powerful vehicle used to pull farm machinery

ULT: last, beyond
ulterior: (adj) beyond what is immediately present; future; beyond what is stated; hidden
ultimate: (adj) last; final
ultimatum: (n) final offer; final terms
ultraviolet: (adj) beyond the spectrum of visible light (on the violet end of the spectrum)

URB: city
suburb: (n) a residential area just outside a city; an outlying area of a city
urban: (adj) relating to a city
urbane: (adj) polite; refined; polished (considered characteristic of those in cities)
urbanization: (n) the process through which a place becomes more like a city

VAIL/VAL: strength, use, worth
ambivalent: (adj) caught between contradictory feelings of equal power or worth
avail: (v) to have force; to be useful; to be of value
convalescent: (adj) recovering strength; healing
equivalent: (adj) of equal worth, strength, or use
evaluate: (v) to determine the worth of
invalid: (adj) having no force or strength; void; sickly; (n) a chronically sickly person
valid: (adj) having force; legally binding; effective; useful
value: (n) worth

VEN/VENT: to come or to move toward
contravene: (v) to conflict with; to go against
convene: (v) to assemble for some public purpose
intervene: (v) to step in between parties in a dispute; to mediate

VER: truth
aver: (v) to affirm; to declare to be true
veracious: (adj) habitually truthful
verdict: (n) a judgment or decision
verity: (n) truthfulness

VERB: word
proverb: (n) an adage; a byword; a short, commonly known saying
verbatim: (adj) exactly as stated; word-for-word
verbiage: (n) excessive use of words; diction
verbose: (adj) wordy

VERS/VERT: to turn
aversion: (n) dislike
avert: (v) to turn away from
controversy: (n) a prolonged debate or disagreement over a topic
diverse: (adj) of different kinds
extrovert: (n) an outgoing person
introvert: (n) a person concerned primarily with inner thoughts and feelings
revert: (v) to return to a previous state, habit, or position

VI: life
viable: (adj) capable of living
vivacity: (adj) the quality of being lively, animated, spirited
vivid: (adj) bright or intense

VID/VIS: to see
adviser: (n) one who gives counsel
evident: (adj) plain or clear
survey: (v) to view in a general or comprehensive way; (n) a comprehensive, high-level account
video: (n) the elements of television or other media pertaining to the image
vista: (n) a view, as of a landscape

VIL: base, mean
revile: (v) to criticize with harsh language
vile: (adj) loathsome; unpleasant
vilify: (v) to slander, to defame
villain: (n) the bad person in a story

VOC/VOK: call, word
advocate: (v) to support or argue in favor of; (n) one who supports or argues in favor of
avocation: (n) something one pursues, outside one's main vocation or career
convoke: (v) to call together
equivocate: (v) to speak ambiguously; to avoid taking a position
invoke: (v) to call on for assistance (as from a deity)
vocabulary: (n) a collection of words used by a group or person
vocation: (n) occupation; calling

VOL: wish
benevolent: (adj) characterized by or expressing good wishes
malevolent: (adj) characterized by or expressing bad wishes
volition: (n) free choice; free will; act of choosing
voluntary: (adj) freely undertaken; not forced

VOLU/VOLV: to roll, to turn
convolution: (n) a twisting or folding
evolve: (v) to develop naturally; literally, to unfold or unroll
revolt: (v) to rebel; to turn against those in authority; (n) a rebellion
revolve: (v) to rotate; to turn around
voluble: (adj) easily turning; fluent; changeable
volume: (n) a book; size or dimensions (originally: of a book)

VOR: to eat
carnivorous: (adj) meat-eating
omnivorous: (adj) eating or absorbing everything
voracious: (adj) having a great appetite

INDEX

D

ASVAB
Verbal Strategy Sheet
KAPLAN

The Kaplan Method for Word Knowledge

STEP 1 Identify the word's meaning, or apply decoding strategies to guess its meaning.

STEP 2 Make a prediction.

STEP 3 Look for your prediction among the answer choices, or strategically eliminate wrong answer choices.

Two Types of Word Knowledge Questions:

1) No-Context Questions:
Use decoding strategies (looking at the word's prefix, suffix, and root), remembered context, or positive and negative charge to determine the meaning of the word.

2) In-Context Questions:
Refer to the clues and keywords in the sentence to determine the meaning of the word.

Using Word Parts: Many (but not all) long words are composed of the following parts, though not every long word has all of these parts.

Prefixes are found at the beginnings of words. Examples:

contra– against (contradictory, contraindication)

micro– very small (microscope, microchip)

pre– before, in front (predetermined, preface)

Suffixes are found at the ends of words. Examples:

–ist person who does (scientist, artist)

–less without, lacking (humorless, friendless)

–ment action of, state of being (establishment, merriment)

Roots can form the bases of words. Examples:

dict to say, to tell, to use words (dictation, dictionary)

test to bear witness (attest, testify)

vi life (vivisection, vivacious)

Consult the Appendix for a comprehensive list of prefixes, suffixes, and word roots.

The key to success in Word Knowledge on the ASVAB is a strong vocabulary.

Improve your vocabulary by:

- learning common prefixes, suffixes, and word roots
- reading as many books, newspapers, magazines, and web articles as possible
- making vocabulary flashcards when you encounter words you don't know
- using the words you are learning in daily conversations and daily correspondence

The Kaplan Method for Paragraph Comprehension

STEP 1 Read the question stem to identify your task.

STEP 2 Read the passage strategically.

STEP 3 Make a prediction.

STEP 4 Find the correct answer.

Strategy Tip

The ASVAB Paragraph Comprehension section is an open-book test. You don't need any outside knowledge to answer these questions correctly. Always refer back to the passage and find support for the correct answer.

Global Questions: In a question that asks you to identify the main idea of a passage, separate supporting details from opinions and recommendations to determine the author's overall point.

Which of the following is the main idea of the passage?

The purpose of the passage above is to

The passage above is primarily concerned with

Common Wrong Answer Types

- deal with just one small detail of the passage
- go beyond the scope of the passage
- contradict the information presented in the passage
- distort the author's opinion by overstating it

Detail Questions: In a question that asks you to find specific information stated in the passage, do not overanalyze or read too much into the question. The correct answer to a Detail Question will almost always be a paraphrase of something found directly in the passage.

According to the author, which of the following is a type of rug found in French castles?

The second step in constructing a picket fence is to

Which of the following is cited in the passage as an advantage of a retirement account?

Common Wrong Answer Types

- paraphrase more of the passage than is necessary
- focus on the wrong detail in the passage
- distort what the passage states

Inference Questions: In a question that asks for what is implied or suggested by the statements in a passage, piece together multiple pieces of information to make an inference or deduction.

Which of the following is implied by the passage?

The author apparently feels that

It can be inferred from the passage that

Based on the passage, which of the following is most likely true of 19th century paintings?

Common Wrong Answer Types

- contradict information in the passage
- bring in outside information that is not discussed in the passage
- distort the information presented in the passage
- exaggerate the author's view

Vocabulary-in-Context Questions: In a question that asks you for the meaning of a word used in context, find a word that can replace that word without altering the meaning of the sentence.

As it is used in the passage, "breakthrough" most nearly means

Which of the following is the closest in meaning to "illuminating" as it is used in the passage above?

Common Wrong Answer Types

- provide one correct definition of the word, but not the way the word is used in the passage
- sound correct in the sentence, but not the passage as a whole

Conclusion Keywords

These words often signal an author's main point.

thus
in sum
in conclusion
therefore
hence
so
clearly
obviously
as a result
I suggest (or propose, or believe)
it follows that

Contrast Keywords

These words can signal the author's main point, if the author is disagreeing with someone else.

but
yet
however
on the other hand
instead
on the contrary
rather
nevertheless
in contrast

Evidence Keywords

These words often signal the facts, details, or other information an author uses to support the main point.

after all
because
since
for
due to
for example
an illustration of this
this can be seen from

ASVAB
Math Strategy Sheet

KAPLAN

The Kaplan Method for ASVAB Math

STEP 1 Analyze the information given.
STEP 2 Identify what you are being asked for.
STEP 3 Solve strategically.
STEP 4 Confirm your answer.

Strategies for Solving Multiple-Choice Math Problems

▶ Backsolving

Use when there is an unknown in the question stem and numbers in the answer choices. Plug one of the choices into the question stem and see if the result is too large, too small, or just right. If the answer choices are arranged in numerical order, start with (B) or (C).

▶ Picking Numbers

Use when (1) there are variables in the question stem and variables in the choices or (2) there's a number (other than the answer) that the test maker doesn't provide and that would be helpful in solving the problem. Pick numbers that are permissible (allowed, given the problem) and manageable (easy to work with).

▶ Strategic Guessing Using Logic

Use when the answer choice has to have a certain property (such as being a perfect square, or being below a certain value). Eliminate answer choices that do not display that property.

Math How-Tos

Use correct order of operations:

Remember PEMDAS: Parentheses, Exponents (and radicals), Multiplication and Division, Addition and Subtraction. Work from left to right.

$9 - 2 \times (5 - 3)^2 + 6 \div 3 =$
$9 - 2 \times (2)^2 + 6 \div 3 =$
$9 - 2 \times 4 + 6 \div 3 =$
$9 - 8 + 2 =$
$1 + 2 = 3$

Add or subtract fractions:

Find a common denominator. $\frac{1}{13} + \frac{1}{2} = \frac{2}{26} + \frac{13}{26} = \frac{15}{26}$

Multiply or divide fractions:

Multiply the numerators, then the denominators. To divide, invert the second fraction before multiplying. $\frac{3}{20} \times \frac{2}{5} = \frac{3 \times 2}{20 \times 5} = \frac{6}{100} = \frac{3}{50}$

$\frac{3}{20} \div \frac{2}{5} = \frac{3}{20} \times \frac{5}{2} = \frac{3 \times 5}{20 \times 2} = \frac{15}{40} = \frac{3}{8}$

Determine whether a number is divisible by another number:

Divisor	Divisibility Rule
2	All even numbers are divisible by 2.
3	Add up the individual digits of the number. If the total is divisible by 3, then the number itself is divisible by 3; for example, 243 is divisible by 3 because the sum of its digits is 2 + 4 + 3 = 9. However, 367 is not divisible by 3 because the sum of its digits is 3 + 6 + 7 = 16 and 16 is not a multiple of 3.
4	Take the last two digits and divide them by 2. If the result is even, the number is divisible by 4. If the result is odd, then the number is not divisible by 4.
5	All numbers ending in 5 or 0 are divisible by 5.
6	All even numbers that meet the test for divisibility by 3 are divisible by 6.
8	Divide the number by 2 twice; if the result is even, then the number is divisible by 8.
9	Add up the digits of the number; if the total is divisible by 9, then the number is divisible by 9.

Multiply or divide decimals:

To multiply decimals, multiply both numbers. Count how many digits are to the right of the decimal in both numbers and move the decimal over to the left that number of spaces in the result.
To divide decimals, move the decimal to the right in both numbers until each is a whole number.
To solve 3.11 × 2.2, multiply 311 × 22 = 6,842
Then move the decimal to the left three places: 6.842
To divide 5.12 by 6.4, move the decimal in both numbers over to make 5.12 into a whole number:
5.12 ÷ 6.4 = 512 ÷ 640 = 0.8

Convert fractions to decimals or percents:

To express a fraction as a decimal, divide the numerator by the denominator. Multiply that by 100 to express as a percent. Remember these common fraction-decimal equivalencies:

$\frac{1}{2} = 0.5 = 50\%$ $\quad \frac{1}{3} = 0.33\overline{3} = 33.\overline{3}\%$
$\frac{1}{4} = 0.25 = 25\%$ $\quad \frac{1}{5} = 0.20 = 20\%$
$\frac{1}{8} = 0.125 = 12.5\%$ $\quad \frac{1}{9} = 0.11\overline{1} = 11.\overline{1}\%$
$\frac{2}{3} = 0.66\overline{6} = 66.\overline{6}\%$

Perform operations with exponents:

$a^x \times a^y = a^{x+y}$
$a^x \div a^y = a^{x-y}$
$(a^x)^y = a^{xy}$
$(a^x)(b^x) = ab^x$
$a^{-x} = \frac{1}{a^x}$

When a negative number is raised to an even exponent, the result is positive. When a negative number is raised to an odd exponent, the result is negative.
When multiplying a number by a power of 10, as in scientific notation, the exponent simply represents the number of places to move the decimal. For example:
$1.75 \times 10^5 = 175{,}000$
$2.4 \times 10^{-4} = 0.00024$

Perform operations with radicals:

$\sqrt{a^2} = |a|$
$\sqrt{ab} = \sqrt{a} \times \sqrt{b}$
$\sqrt{\frac{a}{b}} = \frac{\sqrt{a}}{\sqrt{b}}$
$\sqrt{a+b}$ does NOT equal $\sqrt{a} + \sqrt{b}$
(and the same applies to subtraction)
$a^{\frac{1}{2}} = \sqrt{a}$

Simplify a radical:

Look for factors of the number under the radical sign that are perfect squares; then find the square root of those perfect squares. Pull those roots outside of the radical sign.

$\sqrt{200} = \sqrt{25 \times 8} = \sqrt{25 \times 4 \times 2}$
$= \sqrt{25} \times \sqrt{4} \times \sqrt{2} = 5 \times 2 \times \sqrt{2} = 10\sqrt{2}$

Calculate a factorial

Multiply the number by each positive integer smaller than that number.

8! = 8 × 7 × 6 × 5 × 4 × 3 × 2 × 1 = 40,320

Find the prime factorization of a number:

Begin by identifying one prime factor, and then break the number down into smaller and smaller factors until only primes remain. For example, to find the prime factorization of 60:

60 = 30 × 2 = 15 × 2 × 2 = 5 × 3 × 2 × 2

Predict whether the result of a calculation will be positive or negative, even or odd:

pos + pos = pos
pos + neg = (outcome will take the sign of whichever addend has the larger absolute value)
pos × or ÷ pos = pos
pos × or ÷ neg = neg
neg × or ÷ neg = pos

odd ± odd = even
even ± even = even
even ± odd = odd
odd × odd = odd
even × even = even
even × odd = even

(There are no rules for predicting whether the result of division will be even or odd, since division does not always result in a whole number, and only whole numbers are even or odd.)

Find the mean (arithmetic average) of a set of numbers:

$\frac{\textit{sum of terms}}{\textit{number of terms}}$

The mean (arithmetic average) of 3, 4, 4, 6, and 8:

$\frac{3+4+4+6+8}{5} = \frac{25}{5} = 5$

Find the range of a set of numbers:

largest number − smallest number
The range of 3, 4, and 8: 8 − 3 = 5

Find the median of a set of numbers:

Find the middle number or (if the set has an even number of items) the mean of the two middle numbers.
The median of 3, 4, 4, 6, and 8: 4
The median of 3, 5, 8, and 9: $\frac{5+8}{2} = 6.5$

Find the mode of a set of numbers:

Find the number that repeats most often.
The mode of 3, 4, 4, 6, and 8: 4

Calculate probability:

$\dfrac{\textit{number of desired possible outcomes}}{\textit{number of total possible outcomes}}$

To calculate the probability that two events will occur, multiply the probabilities of each.
To calculate the probability that at least one of two independent events will occur, add the probabilities of each.

Calculate percent, whole, or part:

whole × percent (expressed as a decimal or fraction) = part
Given two parts of that equation, you can solve for the third. E.g., 18 is 60% of what number?
whole × 0.6 = 18
whole = 18 ÷ 0.6 = 30

Perform calculations involving rates:

distance = rate (that is, speed) × time
work performed = rate of work × time
Given two parts of that equation, you can solve for the third.
Given incomplete information and two journeys or work projects, set up a proportion to solve for the unknowns. For example, Joe can travel 62 miles on his bike in 5 hours. At that rate, how far could he travel in 7 hours?

$\frac{62}{5} = \frac{x}{7}$

Cross-multiply to solve for x.

Solve multi-part-journey problems:

total distance = average rate × total time

Multiply binomials:

Remember FOIL: first, outer, inner, last.
$(x + 3)(x - 2) = x^2 - 2x + 3x - 6 = x^2 + x - 6$

Solve a quadratic equation:

Manipulate the quadratic so that it is in the form $ax^2 + bx + c = 0$ (where a, b, and c can be positive or negative numbers). Then perform reverse-FOIL: that is, find two binomials that multiply to make the quadratic. This will usually yield two possible values for x. For example:

$$x^2 - 3x = 54$$
$$x^2 - 3x - 54 = 0$$

Find two numbers that add to –3 and multiply to –54.

$$(x - 9)(x + 6) = 0$$
$$x = 9 \text{ or } -6$$

These shortcuts may help on some quadratics:
$a^2 - b^2 = (a + b)(a - b)$
$a^2 + 2ab + b^2 = (a + b)(a + b) = (a + b)^2$
$a^2 - 2ab + b^2 = (a - b)(a - b) = (a - b)^2$

Solve a linear equation:

Isolate the variable you're solving for on one side of the equation. You may add, subtract, multiply, or divide both sides of the equation by variables or numbers, as long as you do the same thing to both sides.

Solve an inequality:

Follow the same steps as solving a linear equation, but remember these two rules:
(1) If you multiply or divide both sides of an inequality by a negative number, you must reverse the inequality sign.
(2) You cannot multiply or divide both sides of an inequality by a variable (since you do not know the sign of the variable).

Describe an angle as right, acute, obtuse, straight, or reflex:

angle measure = 90: right
angle measure < 90: acute
90 < angle measure < 180: obtuse
angle measure = 180: straight
angle measure > 180: reflex

Find angle measures, given two parallel lines and a transversal:

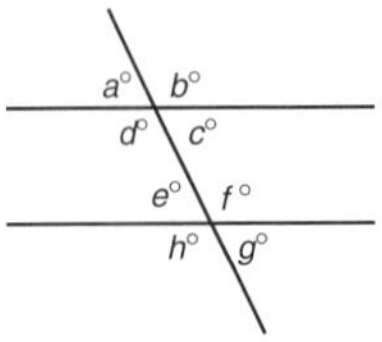

Corresponding angles are equal. For example, $a = e$.
Same-side interior angles are supplementary. For example, $c + f = 180$.
Opposite angles are equal, and alternate interior angles are equal. For example, $b = d = f$.
All four acute angles are equal. All four obtuse angles are equal.

Find the perimeter or area of a quadrilateral:

perimeter = 2(length) + 2(width)
area = length × width

Find the measure of an interior or exterior angle of a triangle:

The interior angles of a triangle add to 180.
An exterior angle of a triangle is equal to the sum of the two opposite interior angles.

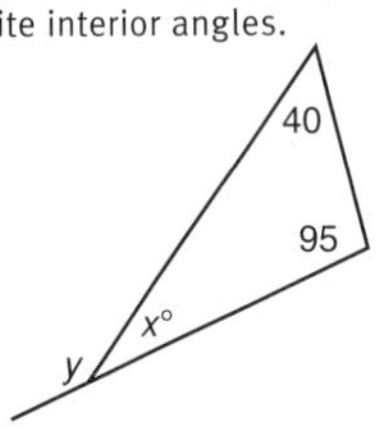

$x + 40 + 95 = 180$
$y = 40 + 95$

Find the perimeter or area of a triangle:

perimeter: side + side + side
area: $A = \frac{1}{2}bh$, where b is the base of the triangle and h is the height (which is the perpendicular distance between the base and the vertex opposite the base)

Find the circumference (that is, perimeter) or area of a circle:

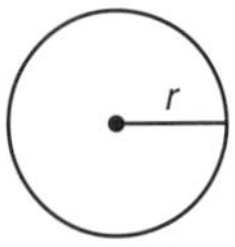

radius: half the distance across the middle of the circle
diameter: the distance across the middle of the circle, or $2r$
circumference: $C = 2\pi r$
area: $A = \pi r^2$

Use the Pythagorean theorem:

if you know the lengths of two sides of a right triangle (that is, a triangle with a 90 degree angle), you can calculate the third side.

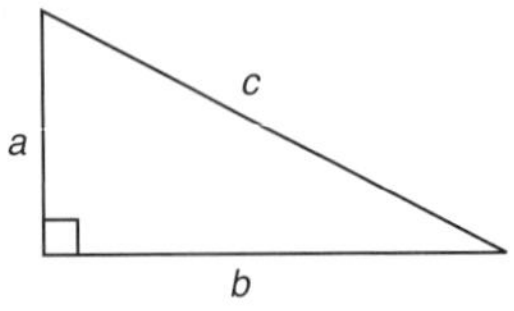

$a^2 + b^2 = c^2$
Memorize these ratios, often (but not always) displayed by the sides of right triangles:
3:4:5
5:12:13

Also, remember these proportions displayed by some right triangles:

The legs of a right triangle with angle measures 45:45:90 display this proportion: $x:x:x\sqrt{2}$

The legs of a right triangle with angle measures 30:60:90 display this proportion: $x: x\sqrt{3}:2x$

Use similar triangles to solve problems:

Similar triangles have the same angle measures as each other. Corresponding sides of these triangles are proportional.

$\frac{3}{4} = \frac{x}{y} = \frac{6}{s}$

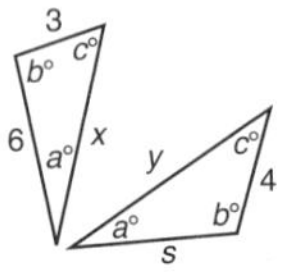

Find the volume or surface area of a cube or rectangular prism:

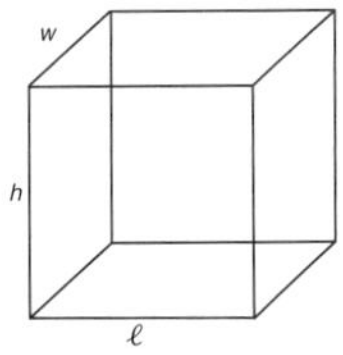

$V = \ell wh$
$SA = 2\ell w + 2wh + 2\ell h$

Find the volume or surface area of a cylinder:

$V = \pi r^2 h$
$SA = 2\pi r^2 + 2\pi rh$

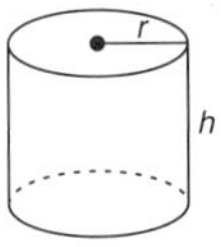

Express the equation of a line in slope-intercept form:

$y = mx + b$, where m is the slope and b is the y-intercept

Calculate the slope of a line:

slope $= \dfrac{\textit{rise}}{\textit{run}}$ or $\dfrac{\textit{change in } y}{\textit{change in } x}$

A horizontal line has a slope of zero.
The slope of a vertical line is undefined.

Find the distance between two points on the coordinate plane:

Use the Pythagorean theorem. For example, to find the distance between points A and B below, first draw a right triangle with A and B as the two ends of the hypotenuse.

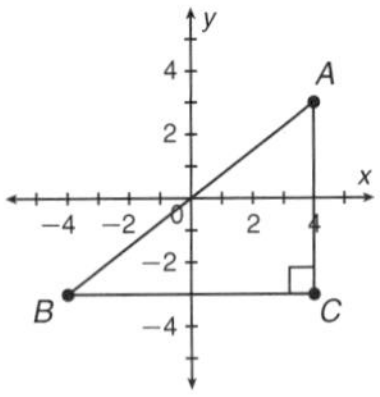

Find the distance from A to C and from B to C, and plug those distances into the Pythagorean theorem:

$$\left(\overline{AC}\right)^2 + \left(\overline{BC}\right)^2 = \left(\overline{AB}\right)^2$$
$$6^2 + 8^2 = \left(\overline{AB}\right)^2$$
$$100 = \left(\overline{AB}\right)^2$$
$$\overline{AB} = 10$$

General Science

The diaphragm is a system of muscles that allows _____.

General Science

A convex mirror _____ light waves; a convex lens _____ light waves.

General Science

The state of matter occurring at lower temperatures is the _____ state.

General Science

What is the difference between physical change and chemical change?

General Science

Put these in order to describe how blood flows from the heart to the body: body, left atrium, right ventricle, left ventricle, lungs, right atrium.

General Science

An acid gives up _____ charged hydrogen ions when dissolved in water.

General Science

When genetic signals in a cell fail to tell it to stop reproducing, the result may be _____.

General Science

17
Cl
35.5

This is the element box for chlorine within the periodic table of elements. What do the numbers 17 and 35.5 indicate?

General Science

A type of matter is classified as an element when _____.

General Science

The speed of light in a vacuum is approximately _____ meters per second.

diverges; converges

breathing

A physical change may alter the form, size, and/or shape of matter; a chemical change alters its molecules.

solid

positively

right atrium, right ventricle, lungs, left atrium, left ventricle, body

The atomic number is 17: there are 17 protons in a chlorine atom's nucleus and 17 electrons in the electron cloud. The atomic mass of chlorine is 35.5.

cancer

300 million

it cannot be separated into different types of matter by ordinary chemical methods

General Science

What is inertia?

General Science

What is a vector quantity?

General Science

What formula is used to calculate velocity?

General Science

Which two bodily functions are enhanced by calcium?

General Science

The Moon affects which activity on Earth?

General Science

The asteroid belt closest to the Earth lies between which two planets?

General Science

Define biosphere.

General Science

Why are the four planets closest to the Sun called the terrestrial planets?

General Science

A stationary front produces _____ weather.

General Science

Earth is approximately _____ years old.

a quantity defined by both a magnitude and a direction

the tendency of all matter to resist changes in motion

bone growth, muscle function

$\vec{v} = \frac{\vec{d}}{t}$, where $\vec{v}$ = velocity (a vector quantity), $\vec{d}$ = displacement (a vector quantity), and t = time

Mars and Jupiter

tides

The composition of these planets is similar to that of Earth.

The biosphere consists of the layers of Earth (crust, surface, atmosphere) where life naturally occurs.

4.6 billion

cloudy and wet

General Science

What are the five kingdoms of living things?

K

General Science

Put these classifications in order from broadest to narrowest: class, domain, family, genus, kingdom, species, order, phylum.

K

General Science

Define biome.

K

General Science

The genetic material of a eukaryotic cell is contained in its _____.

K

General Science

Diploid cells have _____ times the number of chromosomes as do haploid cells.

K

General Science

How does vaccination (immunization) prevent disease?

K

General Science

What is a zygote?

K

General Science

1 kilometer is how many times larger than 1 millimeter?

K

General Science

Ova are carried from the ovaries to the uterus through the _____.

K

General Science

The _____ is the major connecting center between the brain and the nerve network.

K

domain, kingdom, phylum, class, order, family, genus, species

Monera (sometimes split into Bacteria and Archaea), Protista, Fungi, Plantae, Animalia

nucleus

A biome is a major life zone of interrelated species bound together by a similar climate and similar plant and animal life.

It causes the immune system to react to a deactivated, weakened, or partial pathogen so that the immune system will react to future exposure to a similar pathogen.

two

1,000,000 (1 million) times larger

$1 \text{ mm} \times 1{,}000 = 1 \text{ meter};$
$1 \text{ meter} \times 1{,}000 = 1 \text{ km}$

a fertilized egg

spinal cord

fallopian tubes

General Science

What is another name for the large intestine?

K

General Science

A magnetic compass's _____ pole points to the Earth's north pole.

K

General Science

What does a positive or negative sign in a blood type indicate?

K

General Science

The aorta is the largest _____ in the body.

K

General Science

The respiratory system enables blood cells to absorb _____ and eliminate _____ and water vapor.

K

General Science

Bananas are a source of which important mineral?

K

General Science

What is a unique feature of vitamin D?

K

General Science

In chemical reactions, the molecules and atoms entering the reaction are called _____, and those resulting from the reaction are called _____.

K

General Science

Which particle of an atom carries a positive charge? A negative charge? No charge?

K

General Science

The _____ nervous system controls voluntary actions, and the _____ nervous system controls involuntary activity.

K

south

colon

artery

Rh factor

potassium

oxygen; carbon dioxide

reactants; products

One of its best sources comes not from diet but from the Sun. The skin generates vitamin D in response to ultraviolet rays.

somatic; autonomic

proton; electron; neutron

General Science

Earth's core is nearly twice as dense as its mantle because the core is _____.

General Science

Define molecule.

General Science

What role do proteins play in nutrition?

General Science

What is the common name for the larynx?

General Science

Carnivores are also known as _____ consumers.

General Science

What is the lowest level of Earth's atmosphere?

General Science

A solar eclipse occurs when the _____ passes between the _____ and the Sun.

General Science

Name the three major types of clouds.

General Science

Sound waves travel _____ through metal than through water, and _____ through water than through air.

General Science

What is the name of the process by which gametes are produced from diploid cells?

A molecule is the smallest particle of a substance that has the properties of that substance. A molecule is composed of two or more atoms.

metallic

voice box

Proteins enable the body's growth and repair of tissues.

troposphere

secondary

stratus, cirrus, cumulus

Moon; Earth

meiosis

faster; faster

General Science

Water vapor that has entered the atmosphere _____ to form clouds.

K

General Science

Where in the body is the radius bone located?

K

General Science

Define evaporation.

K

General Science

In the water cycle, water soaks into the Earth in a process known as _____.

K

General Science

If someone has type A+ blood, the plus sign indicates the _____.

K

General Science

When a warm front meets a cold front, a likely result is _____.

K

General Science

Identify the value and abbreviation of this metric system prefix: mega

K

General Science

Elements whose outer shells are completely filled with electrons are _____.

K

General Science

The theory of _____ explains how chunks of the Earth's crust and upper mantle slowly move.

K

General Science

What is the common name for the coccyx?

K

forearm

condenses

infiltration

Evaporation is the process by which liquid water becomes water vapor.

precipitation

Rh factor

noble gases

base times 1,000,000; M

tailbone

plate tectonics

General Science

The pitch of the sound produced by an ambulance's siren seems to change as it passes you. This is an example of the _____.

General Science

What are some compounds produced by the pancreas?

General Science

Decomposers, which break down organic matter and release minerals into the soil, are also known as _____.

General Science

_____ rock is formed when magma cools.

General Science

In cellular respiration, cell organelles called _____ process sugar and oxygen in order to produce energy, water, and carbon dioxide.

General Science

Where in the body is the humerus bone located?

General Science

Where in the body is the femur located?

General Science

In females, the sex chromosomes are _____ and are designated by the letters _____. In males, the sex chromosomes are _____ and are designated by the letters _____.

General Science

What is the name for nerve cells?

General Science

Producers, which make their own food through photosynthesis, are also known as _____.

the enzymes pancreatic amylase, lipase, and trypsin and the hormone insulin (as well as other less well-known hormones)	Doppler effect
Igneous	saprotrophs
upper arm	mitochondria
(females) homozygous; XX (males) heterozygous; XY	thigh
autotrophs	neurons

General Science

Define sublimation (as it is used in chemistry).

General Science

What are some minerals that are required for proper nutrition?

General Science

The process by which a cell divides, producing two cells both of which are identical to the parent cell, is called _____.

General Science

Define genotype.

General Science

Gametes are called _____ because they have half as many chromosomes as normal cells.

General Science

Identify the value and abbreviation of this metric system prefix: kilo

General Science

Each row of the periodic table is called a(n) _____.

General Science

When an acid is combined with a base, they produce _____.

General Science

_____ are animals that eat other organisms.

General Science

_____ rock is formed when existing rock material is altered by changes in temperature, pressure, or other processes.

There are many possible answers. Examples are: iron, calcium, magnesium, potassium, phosphorous, and zinc

Sublimation is the process by which a solid turns into a gas (without passing through a liquid phase).

A person's genotype is the entire genetic makeup of that person, including recessive genes.

mitosis

base times 1,000; k

haploid (think *HAploid* = *HAlf*)

water and a salt

period

Metamorphic

Consumers

General Science

The _____ scale is based on a measurement known as absolute zero.

K

General Science

The gaseous state of a substance occurs at a(n) _____ temperature than the liquid state, which occurs at a(n) _____ temperature than the solid state.

K

General Science

Is strep throat caused by a bacterium or a virus?

K

General Science

Small objects that enter the Earth's surface and burn up, creating a visible bright streak in the night sky, are called _____. If one makes it to Earth's surface, it is called a(n) _____.

K

General Science

What is the difference between an exoskeleton and an endoskeleton?

K

General Science

Can light travel through a vacuum?

K

General Science

A recessive trait is displayed in an organism only if _____.

K

General Science

When light reflects off a surface, the angle of _____ is equal to the angle of _____.

K

General Science

_____ rock is formed when mineral deposits, such as those carried by a river, harden over time.

K

General Science

Where in the body is the fibula located?

K

higher; higher

Kelvin

meteors; meteorite

a bacterium (specifically, *Streptococcus pyogenes*; hence the name “strep”)

yes

An exoskeleton is a hard exterior structure; an endoskeleton is a frame located inside the body.

incidence; reflection (or reflection; incidence)

both alleles that the organism has for that trait are recessive

lower leg

Sedimentary

General Science

Viruses can replicate only _____.

General Science

Identify the value and abbreviation of this metric system prefix: micro

General Science

Define ovulation.

General Science

What is an element's atomic number?

General Science

Rows of the periodic table refer to _____. Columns of the periodic table refer to _____.

General Science

When the northern hemisphere is tilted toward the Sun, the season is _____ in the northern hemisphere and _____ in the southern hemisphere.

General Science

Name the levels of taxonomic classification, from the broadest to the most specific category.

General Science

What is the common name for hypertension?

General Science

When the Earth is closest to the Sun, it is _____ in the northern hemisphere and _____ in the southern hemisphere.

General Science

The ozone layer is part of the _____.

base times 0.000001; μ (the Greek letter mu)

inside a host's cells (that is, viruses cannot replicate outside a host)

the number of protons in one of the element's atoms

Ovulation is the release of an egg from an ovary through the fallopian tube and into the uterus.

summer; winter

the number of electron shells in atoms; the number of electrons in the outer shell of atoms

high blood pressure

domain, kingdom, phylum, class, order, family, genus, species (Remember: *Dear King Philip Came Over For Good Soup*.)

stratosphere

winter; summer

General Science

What is the difference between bone and cartilage?

K

General Science

Where in the body are blood cells created?

K

General Science

Carnivores eat _____, while omnivores eat both _____.

K

General Science

If a person has two alleles that are different for a particular gene, that person is _____ for that particular gene.

K

General Science

When heat is transferred through electromagnetic waves, this is known as _____.

K

General Science

Is the flu caused by a bacterium or a virus?

K

General Science

Define endometrium.

K

General Science

When one atom bonds to another atom by "borrowing" an electron from it, a(n) _____ bond is formed.

K

General Science

Two atoms with the same number of protons and electrons but different numbers of neutrons are called _____.

K

General Science

In the heart, _____ are chambers that collect blood, and _____ are chambers that pump blood.

K

bone marrow	Bones are more rigid and provide the body's framework. Cartilage is more flexible and is found at the ends of bones, in joints, and in the nose and ears.
heterozygous	consumers; producers and consumers. Another acceptable answer is: animals; plants and animals
a virus	radiation
ionic	The endometrium is the lining of the uterus.
atria; ventricles	isotopes

General Science

Name the three ways heat can be transferred from one object or medium to another.

K

General Science

Water freezes at _____ Kelvin.

K

General Science

What role does pepsin play in the body?

K

General Science

A(n) _____ lens curves outward; a(n) _____ lens curves inward.

K

General Science

Where in the body are the tarsal bones located?

K

General Science

When a gas changes to a liquid, this process is known as _____.

K

General Science

What is produced in the testes?

K

General Science

Name the four major components of blood.

K

General Science

_____ carry urine from the kidneys to the bladder. The _____ carries urine from the bladder outside the body.

K

General Science

High-pitched sounds have a(n) _____ frequency than low-pitched sounds.

K

273

conduction, convection, radiation

convex; concave

Pepsin breaks down proteins in the stomach.

condensation

the ankles

plasma, red blood cells, white blood cells, platelets

sperm

higher

Ureters; urethra

General Science

A person's physical characteristics, taken all together, are known as that person's _____.

K

General Science

Each person has _____ alleles for every trait.

K

General Science

Which blood type is the universal recipient?

K

General Science

Acidic substances have a pH _____ than 7; basic substances have a pH _____ than 7.

K

General Science

What is the formula for converting a temperature expressed in degrees Fahrenheit into a temperature expressed in degrees Celsius?

K

General Science

Define vertebrate.

K

General Science

A convection oven works by _____.

K

General Science

The central nervous system is composed of the _____ and the _____.

K

General Science

The _____ is the portion of the brain that controls involuntary actions such as breathing.

K

General Science

Antibiotic drugs are not effective against illnesses caused by _____.

K

two (one from each parent)

phenotype

lower; higher

type AB positive

A vertebrate is an animal with a spine.

$C^\circ = \frac{5}{9}(F^\circ - 32)$

brain; spinal cord

circulating hot air

There are several possible answers; examples include viruses, diet and lifestyle choices, and genetic disorders.

medulla

General Science

What are some ways that carbon is removed from the atmosphere?

K

General Science

What are the four blood types?

K

General Science

Where in the body are the carpal bones located?

K

General Science

The _____ carries sperm from the testes to the penis by way of the _____.

K

General Science

Prokaryotic cells are defined as cells that do not have a(n) _____.

K

General Science

Name the planets, in order from the Sun outward.

K

General Science

Define gamete.

K

General Science

What do white blood cells do?

K

General Science

When water evaporates from the leaves of plants, this is known as _____.

K

General Science

What type of clouds produce precipitation?

K

A, B, AB, O

There are several possible answers, but the major ones are photosynthesis and absorption by rain and by bodies of water.

vas deferens; prostate

the wrists

Mercury, Venus, Earth, Mars, Jupiter, Saturn, Uranus, Neptune (Remember: *My Very Educated Mother Just Served Us Nachos*.)

nucleus

fight infection

A gamete is a reproductive cell.

stratus and cumulus

transpiration

General Science

Which organ in the body employs gastric acid?

K

General Science

If a person has two alleles that are alike for a particular gene, that person is _____ for that particular gene.

K

General Science

Define photosynthesis.

K

General Science

What does DNA do?

K

General Science

In terms of the food web, vultures and wolverines are examples of _____.

K

General Science

Primary consumers are also known as _____.

K

General Science

In the metric system, the basic unit of mass or weight is the _____.

K

General Science

Identify the value and abbreviation of this metric system prefix: deci

K

General Science

Define pathogen.

K

General Science

When two atoms bond by sharing a pair of electrons, a(n) _____ bond is formed.

K

homozygous

stomach

DNA contains instructions for the construction of proteins. It determines an organism's genetic makeup and traits.

Photosynthesis is the process by which plants take in light energy, water, and carbon dioxide and use these substances to produce sugar and oxygen.

herbivores

scavengers

base times 0.1; d

gram

covalent

A pathogen is a disease-causing agent, such as a bacterium or virus.

General Science

What is the formula for converting a temperature expressed in degrees Celsius into a temperature expressed in degrees Fahrenheit?

K

General Science

What is the difference between tendons and ligaments?

K

General Science

Each column of the periodic table is called a(n) _____.

K

General Science

In the metric system, the basic unit of volume is the _____.

K

General Science

On the Richter scale, an earthquake of magnitude 2 is _____ times greater than an earthquake of magnitude 1.

K

General Science

What gases are exchanged in human respiration?

K

General Science

The ratio by which light is slowed when passing through a medium is referred to as the medium's _____.

K

General Science

In general, _____ carry oxygenated blood, while _____ carry deoxygenated blood.

K

General Science

When heat is transferred through direct contact between two objects, this is known as _____.

K

General Science

The _____ is the portion of the brain that controls balance and muscle coordination.

K

Both tendons and ligaments are connective tissues. Tendons connect muscles to bones, while ligaments connect bones to other bones.

$$F^{\circ} = \frac{9}{5}C^{\circ} + 32$$

liter

group

In respiration, humans take in oxygen and breathe out carbon dioxide.

ten

arteries; veins

refractive index

cerebellum

conduction

General Science

What is the Kuiper Belt?

General Science

In the metric system, the basic unit of length is the _____.

General Science

What are some ways that carbon is released into the atmosphere?

General Science

What type of star is the Sun?

General Science

What are some factors that can contribute to heart disease?

General Science

When light bends due to passing through a medium, this bending is referred to as _____.

General Science

Which organ produces bile?

General Science

Define capillaries.

General Science

What is the name of the two largest veins in the body?

General Science

Identify the value and abbreviation of this metric system prefix: deka

meter	a loose group of asteroids and other material that lies beyond Neptune
G2V or yellow dwarf	There are several possible answers, but the major sources are manufacturing emissions, vehicle emissions, respiration, and decomposing organic matter.
refraction	There are several possible answers; examples include high cholesterol, high blood pressure, lack of exercise, smoking, and excess weight.
Capillaries are the smallest blood vessels, which deliver oxygenated blood to the body's tissues and carry away deoxygenated blood.	liver
base times 10; da	venae cavae (plural of vena cava)

General Science

A(n) _____ is a ball of ice, frozen gases, and dust that orbits the Sun.

K

General Science

Name some types of electromagnetic radiation.

K

General Science

Define esophagus.

K

General Science

What is the name of the portion of the brain generally considered to be the center of intelligence?

K

General Science

Can sound travel through a vacuum?

K

General Science

Identify the value and abbreviation of this metric system prefix: centi

K

General Science

Red light has a(n) _____ frequency than violet light.

K

General Science

What is the common name for the cranium?

K

General Science

What is the pH of pure water?

K

General Science

Scurvy is caused by a lack of which nutrient?

K

Possible answers include: visible light, radio waves, microwaves, ultraviolet waves, infrared waves, X-rays, and gamma rays.

comet

cerebrum

The esophagus is the tube that carries food from the mouth to the stomach.

base times 0.01; c

no

skull

lower

vitamin C

7

General Science

Plant cells have a rigid _____, while animal cells do not.

General Science

What does DNA stand for?

General Science

Identify the value and abbreviation of this metric system prefix: milli

General Science

Which have a larger diameter: bronchi or bronchioles?

General Science

Absolute zero expressed in degrees Celsius is _____.

General Science

The thick layer that lies between the Earth's crust and the Earth's core is called the _____.

General Science

Define atomic mass.

General Science

Identify the value and abbreviation of this metric system prefix: hecto

General Science

How many chromosomes do human gametes have?

General Science

Where in the body is the tibia located?

deoxyribonucleic acid

cell wall

bronchi

base times 0.001; m

mantle

−273°C

base times 100; h

An element's atomic mass is the mass of a single atom. It can be approximated by adding together the atom's protons and neutrons. One proton or one neutron has a mass roughly equal to one atomic mass unit (amu).

lower leg

23

General Science

The hottest and uppermost layer of Earth's atmosphere is the _____.

K

General Science

Define trachea.

K

General Science

Which blood type is the universal donor?

K

General Science

Above the stratosphere is the _____.

K

General Science

What role do ribosomes play in a eukaryotic cell?

K

General Science

Variants of genes are known as _____.

K

General Science

Our solar system's asteroid belt lies between the orbits of _____ and _____.

K

General Science

A lunar eclipse occurs when the _____ passes between the _____ and the Sun.

K

General Science

Where in the body is the ulna located?

K

General Science

An acid with a pH of 5 is _____ times more acidic than an acid with a pH of 6.

K

The trachea is the windpipe—the passageway from the throat to the lungs.

thermosphere

mesosphere

type O

alleles

Ribosomes produce proteins.

Earth; Moon

Mars and Jupiter

10

forearm

General Science

How many chromosomes do nonreproductive cells have?

K

General Science/Mechanical Comprehension

Variables are inversely proportional when an increase in the independent variable results in a(n) _____ in the dependent variable.

K

General Science/Mechanical Comprehension

As an object falls, its _____ energy is converted to _____ energy.

K

General Science/Mechanical Comprehension

What is the definition of mass?

K

General Science/Mechanical Comprehension

What is the difference between mass and weight?

K

General Science/Mechanical Comprehension

The basic unit of force is the _____, which represents _____.

K

General Science/Mechanical Comprehension

Identify the common name for this rule: An object at rest tends to stay at rest, and an object in motion tends to stay in motion at constant velocity unless acted on by an unbalanced force.

K

General Science/Mechanical Comprehension

What is the formula for acceleration?

K

General Science/Mechanical Comprehension

The basic unit of work is the _____, and it represents _____.

K

General Science/Mechanical Comprehension

What is the formula used to express Newton's law of universal gravitation?

K

decrease

46

a measure of the total quantity of matter in an object

potential; kinetic

newton (N); the force required to accelerate 1 kg by 1 meter per second per second: kg • m/s^2

Mass is the amount of matter an object has, while weight is the amount of force exerted on the object's mass by gravity.

acceleration =
change in velocity ÷ change in time

$$a = \frac{\Delta v}{t}$$

Newton's first law of motion

$F_g = \frac{Gm_1m_2}{r^2}$, where F_g is the force of gravity, G is a constant equal to 6.67×10^{-11}, m_1 and m_2 are the masses of the two objects, and r is the distance between the two objects

joule (J); newtons times meters (or kg • m^2/s^2)

General Science/Mechanical Comprehension

_____ can be defined as the capacity to do work.

General Science/Mechanical Comprehension

Power is defined as _____ and measured in _____.

General Science/Mechanical Comprehension

The term _____ refers to the total of various pushes and pulls acting on an object.

General Science/Mechanical Comprehension

What is inertia?

General Science/Mechanical Comprehension

Acceleration due to gravity is approximately _____ meters per second per second, not counting air resistance.

General Science/Mechanical Comprehension

What is the formula for work?

General Science/Mechanical Comprehension

What is the formula for momentum?

General Science/Mechanical Comprehension

Rewrite the formula for momentum in order to solve for mass.

General Science/Mechanical Comprehension

What is the formula for power?

General Science/Mechanical Comprehension

_____ energy is the energy possessed by a moving object.

the rate at which work is performed; watts (or sometimes horsepower)

Energy

Inertia is the tendency of objects to resist changes in their motion.

net force

$w = fd$

work = force × displacement

9.8

$m = \frac{p}{v}$

$p = mv$, where p = momentum, m = mass, and v = velocity

Kinetic

$p = \frac{w}{t}$

power = work divided by time

General Science/Mechanical Comprehension

Newton's third law of motion states that _____.

K

General Science/Mechanical Comprehension

In science, acceleration is often measured in _____.

K

General Science/Mechanical Comprehension

$F = ma$ is the mathematical formula representing Newton's _____.

K

General Science/Mechanical Comprehension

_____ energy derives from an object's position, shape, or state, rather than from its motion.

K

General Science/Mechanical Comprehension

In science, the delta symbol (Δ) represents _____.

K

General Science/Mechanical Comprehension/Electronics Information

A watt is defined as _____.

K

General Science/Electronics Information

A neutral atom has the same number of _____ and electrons.

K

General Science/Electronics Information

The outer shell of an atom is known as its _____.

K

General Science/Electronics Information

Frequency is usually measured in _____.

K

General Science/Electronics Information

In an atom, protons have a(n) _____ charge, while electrons have a(n) _____ charge.

K

meters per second per second, or $\frac{m}{s^2}$

for every action, there is an equal and opposite reaction

Potential

second law of motion

one joule per second, or one ampere times one volt

change in a quantity

valence shell

protons

positive; negative

hertz (Hz)

General Science/Electronics Information

Electrons orbit an atom in a set of layers or levels called _____.

General Science/Electronics Information

A(n) _____ is a subatomic particle that bears no charge.

Mechanical Comprehension

A lever pivots upon a(n) _____.

Mechanical Comprehension

No machine is 100 percent efficient because all machines are affected by _____.

Mechanical Comprehension

The most likely place for a cable to snap is nearest *the pulling device/the object being pulled.*

Mechanical Comprehension

Aircraft are designed to be sleek to _____ the effect of drag; parachutes are designed with large areas to _____ the effect of drag.

Mechanical Comprehension

Static friction only arises in response to an attempt to _____.

Mechanical Comprehension

Why is an inclined plane a good solution for a lifting task?

Mechanical Comprehension

On horizontal surfaces, the normal force is usually equal to _____.

Mechanical Comprehension

Slippery surfaces have _____ coefficients of both kinetic and static friction.

neutron

shells

friction

fulcrum

minimize; maximize

the pulling device

The force required will be less than the force required to lift the object directly.

move an object along a surface

low

the weight of the object

Mechanical Comprehension

Liquids are effective at transmitting force because they are difficult to _____.

Mechanical Comprehension

In a hydraulic system, _____ force is applied over a greater distance to produce a greater force over _____ distance.

Mechanical Comprehension

What does torque measure?

Mechanical Comprehension

Torque is directly proportional to _____ and _____.

Mechanical Comprehension

What is the advantage of a pulley?

Mechanical Comprehension

How is a gear ratio determined?

Mechanical Comprehension

What unit is used to express torque?

Mechanical Comprehension

In a third-class lever, where is the fulcrum, and where is the force applied?

Mechanical Comprehension

What unit is used to express kinetic or potential energy, and what does that unit represent?

Mechanical Comprehension

Torque is *directly/inversely* proportional to the distance from the center of rotation to the point where the force acts.

a smaller; a smaller

compress

force; distance

how much a force is able to cause an object to rotate

By comparing the number of teeth on the larger gear to the number of teeth on the smaller gear

A pulley enables someone to lift an object while exerting downward, rather than upward, force.

In a third-class lever, the fulcrum is positioned near one end of the lever, and an upward force is applied to the middle of the lever in order to lift an object positioned on the other end of the lever.

either the newton-meter or the foot-pound

directly

the joule (J), which represents

$$\frac{kg \times m^2}{s^2}$$

Fluid pressure is also known as _____ pressure.

What is the formula for calculating gravitational potential energy?

What is the formula for calculating friction?

In hydraulics, Pascal's law states that _____.

What is the formula for calculating pressure applied evenly to a surface?

What primarily determines the mechanical advantage of a block and tackle?

_____ is the common term for the force that opposes movement through air or liquid.

What does the work-energy theorem say?

How would you increase the mechanical advantage of a wheel and axle arrangement?

When two surfaces rub together, _____ friction results.

$PE = mgh$, where m represents the mass of the object in kg, g is acceleration due to gravity (roughly 9.8 m/s^2), and h is the height of the object in meters

hydraulic

the pressure in any part of an enclosed fluid is the same and points in all directions

$F_f = \mu \bullet F_n$, where F_f represents the force of friction, μ represents the coefficient of friction (static or kinetic), and F_n represents the normal force (the force a surface exerts when an object presses against it)

the number of pulleys

$P = \frac{F}{A}$, where P represents pressure expressed in pounds of force per square inch, F represents force in pounds, and A represents area in square inches

The work-energy theorem states that any work done to move an object from a resting position to a certain velocity will be converted into the kinetic energy of that object.

Drag

kinetic

by increasing the ratio of the diameter of the large wheel to that of the small wheel

Mechanical Comprehension

What is the formula for calculating torque?

Mechanical Comprehension

Define mechanical advantage.

Mechanical Comprehension

In a wheel and axle arrangement, force is usually applied to the *larger/smaller* wheel in order to move the *larger/smaller* wheel.

Mechanical Comprehension

When two or more pulleys are arranged to work together, the resulting apparatus is called a _____.

Mechanical Comprehension

What are the two types of mechanical energy?

Mechanical Comprehension

Increasing the length of a wedge relative to its height _____ the mechanical advantage provided by the wedge.

Mechanical Comprehension

What is the formula for calculating kinetic energy?

Mechanical Comprehension

_____ friction prevents an object from moving unless overcome by a sufficiently large force.

Mechanical Comprehension

In a first-class lever, where is the fulcrum, and where is the force applied?

Mechanical Comprehension

What is one horsepower equal to?

Mechanical advantage is the advantage gained by using a mechanism to help transmit force.

$\tau = rF$, where τ (the Greek letter tau) represents torque, r represents the length of the lever arm of the object being turned, and F represents a force being applied perpendicularly to that lever arm

block and tackle

larger; smaller

increases

kinetic energy and potential energy

Static

$KE = \frac{1}{2}mv^2$, where m is mass in kg and v is speed in meters per second

550 foot-pounds per second, or 746 watts

In a first-class lever, the fulcrum is positioned in the middle of the lever. Force is applied to one end of the lever to lift an object on the other end.

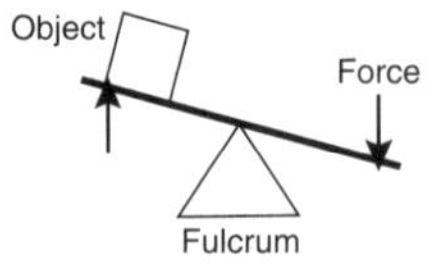

Mechanical Comprehension

A pulley changes the _____ but not the magnitude of a force.

K

Mechanical Comprehension

For any given surface, the coefficient of _____ friction is higher than the coefficient of _____ friction.

K

Mechanical Comprehension

How is the efficiency of a machine defined?

K

Mechanical Comprehension

A _____ is a type of inclined plane that can be moved in order to gain mechanical advantage in lifting, splitting, or tightening objects.

K

Mechanical Comprehension

Potential energy derives from what and has the potential to do what?

K

Mechanical Comprehension

_____ is the term for the ability of a force to cause an object to rotate.

K

Mechanical Comprehension

_____ is the term for the responsive force exerted within an object as the object is stretched.

K

Mechanical Comprehension

What is the principle of the conservation of mechanical energy?

K

Mechanical Comprehension

In a second-class lever, where is the fulcrum, and where is the force applied?

K

Mechanical Comprehension

A wheel and axle arrangement links two wheels of different _____.

K

static; kinetic

direction

wedge

The efficiency of a machine is an expression of how much of the force inputted into the machine is turned into movement or force.

Torque

Potential energy derives from an object's position or shape, and it has the potential to be converted into kinetic energy (that is, energy of movement).

The principle of the conservation of mechanical energy states that the total mechanical energy (PE + KE) of an object remains constant as long as no other force is applied.

Tension

sizes

In a second-class lever, the fulcrum is positioned near one end of the lever underneath it, and the force is applied upward to the other end in order to move an object resting on the lever between the force and the fulcrum.

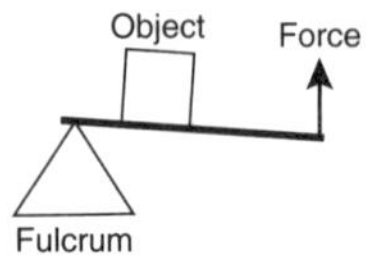

Mechanical Comprehension

How would you increase the mechanical advantage of a second-class lever?

K

Mechanical Comprehension

How would you increase the mechanical advantage of a first-class lever?

K

Mechanical Comprehension

If a small gear drives a large gear, the large gear will turn with a _____ speed than the small gear, but _____ will be increased.

K

Mechanical Comprehension

A ramp is a type of _____.

K

Electronics Information

What is the common term for electrical potential difference or electromotive force?

K

Electronics Information

Why is copper used as a conductor more frequently than is silver or aluminum?

K

Electronics Information

Current will flow only if it can follow a(n) _____ circuit.

K

Electronics Information

A circuit with only one path is a(n) _____ circuit.

K

Electronics Information

Why are parallel circuits usually superior to series circuits?

K

Electronics Information

What is the equation for frequency?

K

by increasing the ratio of the distance between the force and the fulcrum to the distance between the fulcrum and the object to be lifted

by moving the object to be lifted closer to the fulcrum, perhaps by making the lever longer

inclined plane

slower; torque (or rotational force)

Copper is much less expensive than silver and has a lower resistance than does aluminum.

voltage

series

closed

$f = \frac{1}{T}$, where f = frequency in Hz and T = the amount of time (in seconds) for one cycle

If one path has a break or gap, the circuit will remain closed and current can flow.

Electronics Information

Capacitive reactance is measured in _____.

Electronics Information

When silicon's crystalline structure is doped with a five-electron element, the silicon crystal becomes a(n) _____ material.

Electronics Information

The _____ is a solid-state device that has revolutionized the construction of electronic devices.

Electronics Information

What is the difference between these connection symbols?

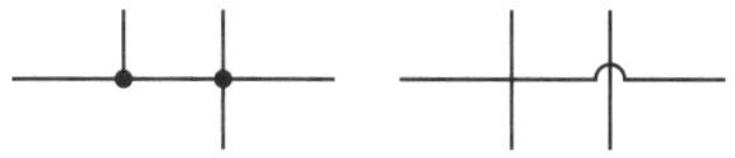

Electronics Information

Identify this symbol:

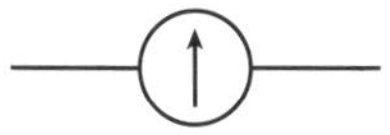

Electronics Information

Why are transformers especially useful in the transmission of electricity from power plants to residences?

Electronics Information

How can the magnetic field formed by a current-carrying wire be strengthened?

Electronics Information

What are the two types of transistors?

Electronics Information

If multiple loads are connected in a series circuit, how is the overall voltage of the circuit calculated?

Electronics Information

What is the formula for electrical power?

N-type (also known as the cathode)

ohms

Left: indicates that the vertical wires are connected to the horizontal wire

Right: indicates that the vertical wires are not connected to the horizontal wire

transistor

It is more efficient to transmit high-voltage electricity, but most household appliances use low-voltage electricity.

galvanometer

NPN transistor and PNP transistor

by winding the wire into a coil around a ferrous (iron) core

$P = IV$

power = current × voltage

by adding the voltage drops across the various loads: $V_{tot} = V_1 + V_2 + V_3 \ldots$

Electronics Information

Identify this symbol:

Electronics Information

What is the formula for Ohm's law?

Electronics Information

If an element's atoms have valence shells that are more than half full of electrons, that element is a(n) _____.

Electronics Information

Identify this symbol:

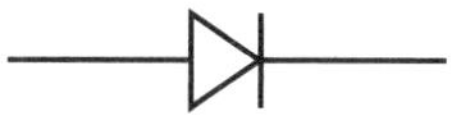

Electronics Information

When a current is passed through a wire coiled around a ferrous material, the result is a(n) _____.

Electronics Information

If an element's atoms have valence shells that are exactly half full of electrons, that element is a(n) _____.

Electronics Information

Do household electrical outlets in North America use AC or DC current?

Electronics Information

What does an electrical ground do?

Electronics Information

If an element's atoms have valence shells that are less than half full of electrons, that element is a(n) _____.

Electronics Information

The conversion of AC to DC is called _____.

$V = IR$
voltage = current × resistance

on/off switch

diode

insulator

semiconductor

electromagnet

A ground provides a low-resistance path to guide excess electrical energy away from panels and loads, to protect them in the case of lightning or circuit failure.

AC

rectification

conductor

Electronics Information

What do DC and AC stand for?

Electronics Information

Electrical current flowing through a conductor produces a(n) _____ field.

Electronics Information

What is the difference between direct current and alternating current?

Electronics Information

The basic unit of electrical charge is the _____.

Electronics Information

If multiple loads are connected in a parallel circuit, how is the overall resistance of the circuit calculated?

Electronics Information

In a series circuit, does the current vary, alternate, or remain the same throughout the circuit?

Electronics Information

Identify this symbol:

Electronics Information

Identify this symbol:

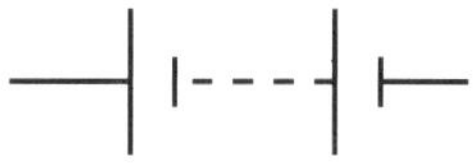

Electronics Information

Identify this symbol:

Electronics Information

Current and resistance are *directly/ inversely* proportional.

magnetic

direct current; alternating current

coulomb (C)

Direct current flows in one direction only; alternating current changes its direction many times a second.

current remains the same

by adding the inverse of the resistance of the various loads, and then taking the inverse of the result:

$$\frac{1}{R_{tot}} = \frac{1}{R_1} + \frac{1}{R_2} + \frac{1}{R_3} + \dots$$

battery

AC supply

inversely

resistor

Electronics Information

A(n) _____ is used to increase or decrease voltage in a circuit.

K

Electronics Information

Induction is measured in _____.

K

Electronics Information

If multiple loads are connected in a series circuit, how is the overall resistance of the circuit calculated?

K

Electronics Information

Define *load*, as it relates to electrical systems.

K

Electronics Information

Define electrical power.

K

Electronics Information

A(n) _____ is used to raise or lower current in a circuit.

K

Electronics Information

A(n) _____ is a thin wire that melts when current exceeds a certain amount.

K

Electronics Information

If one battery's terminal supplies greater voltage than that of another battery, the first battery terminal has more _____.

K

Electronics Information

Resistance is measured in _____.

K

Electronics Information

Voltage and current are *directly/inversely* proportional.

K

henries

transformer

In electricity, a load is a source of resistance that turns electrical energy into some other type of energy.

by adding the resistances of the various loads: $R_{tot} = R_1 + R_2 + R_3 \ldots$

resistor

Electrical power is the amount of energy delivered to and used by an electric circuit.

electrons (or negative charge)

fuse

directly

ohms (Ω)

Electronics Information

The process of adding impurities to silicon or germanium in order to improve its usefulness as a semiconductor is called _____.

Electronics Information

Electrical current is measured in _____.

Electronics Information

In most home electrical systems, loads are wired in parallel with a fuse wired in *series/parallel*.

Electronics Information

Identify this symbol:

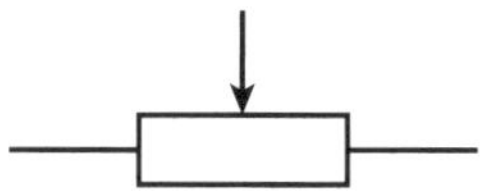

Electronics Information

What metal is most often used as a conductor in household electrical systems?

Electronics Information

Identify this symbol:

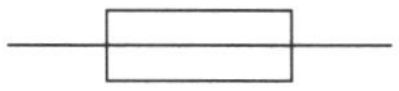

Electronics Information

Identify this symbol:

 or 

Electronics Information

In North America, AC is delivered at _____ Hz.

Electronics Information

Identify this symbol:

Electronics Information

Identify this symbol:

amperes (A)	doping
potentiometer	series
fuse	copper
60	ground
ammeter	circuit breaker

Electronics Information

Poor conductors have _____ resistance than good conductors.

Electronics Information

The name for the imaginary flow of positive charge, opposite to the actual flow of electrons, is _____.

Electronics Information

A(n) _____ is constructed of two metal plates and used to store electricity.

Electronics Information

If the resistance provided by a load were to decrease, the current flowing through that circuit would _____.

Electronics Information

What is voltage?

Electronics Information

Identify this symbol:

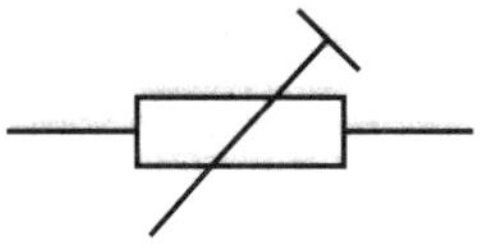

Electronics Information

A(n) _____ resists changes in current flow by applying a voltage that runs against increased current.

Electronics Information

Identify this symbol:

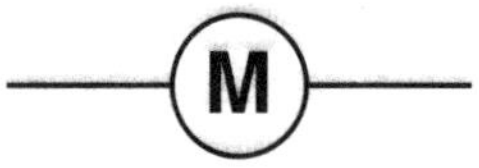

Electronics Information

If an AC current is delivered at 20 Hz, what is it that happens 20 times per second in that circuit?

Electronics Information

Identify this symbol:

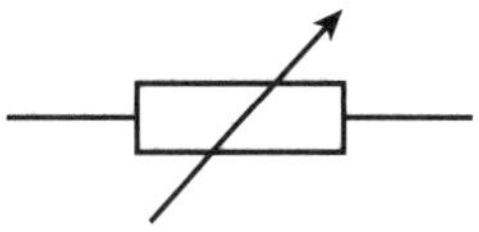

conventional current	higher
increase	capacitor (or condenser)
variable resistor	Voltage is the electric "pressure" produced by a large negative charge.
motor	inductor
rheostat	The voltage begins at zero, increases to a maximum level, decreases to zero again, switches direction and increases to a maximum level in that direction, and returns to zero again.

Electronics Information

When a P-type semiconductor is joined to an N-type semiconductor, the result is a(n) _____.

Electronics Information

The three major parts of a transistor are _____.

Electronics Information

Identify this symbol:

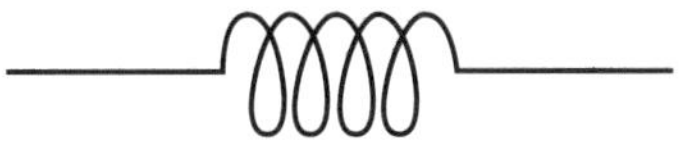

Electronics Information

Identify this symbol:

Electronics Information

Identify this symbol:

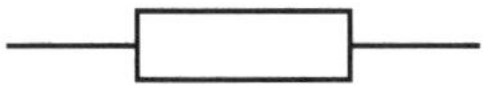

Electronics Information

A(n) _____ serves the same function as a fuse but can be used multiple times.

Electronics Information

Does a battery supply AC or DC current?

Electronics Information

Identify this symbol:

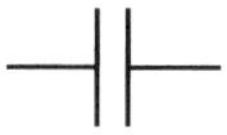

Electronics Information

What does an ammeter do?

Electronics Information

Resistance can be measured using a device called a(n) _____.

the collector, the emitter, and the base

diode

DC supply

inductor

circuit breaker

fixed resistor

capacitor (or condenser)

DC

ohmmeter

An ammeter measures electrical current.

Electronics Information

What is a short circuit?

Auto Information

What three things must be present before combustion can occur in an internal combustion engine?

Auto Information

What function does a cylinder perform?

Auto Information

What are the four strokes of a four-stroke cycle from first to last?

Auto Information

What are TDC and BDC?

Auto Information

What are the three most common cylinder arrangements in internal combustion engines?

Auto Information

In a double overhead cam arrangement, what does each cam do?

Auto Information

What are the two advantages of a four-valve cylinder head over a two-valve cylinder head?

Auto Information

25 degrees BTDC (before top dead center) is an example of _____ ignition timing.

Auto Information

What are two likely causes of pre-ignition?

air; fuel; a heat source

A short circuit occurs when a load is bypassed by a conductor, allowing more current to flow through the circuit.

intake; compression; power; exhaust

A cylinder forms a guide for the piston, allowing the piston to move up and down as the engine completes its cycle.

inline; horizontally opposed (flat); V-type

top dead center; bottom dead center. Travel from TDC to BDC and back constitutes one stroke.

higher engine operating speeds; more complete combustion of the air-fuel mixture

One cam operates the exhaust valves; the other operates the intake valves.

a glowing spark plug; a hot piece of carbon

advanced

Auto Information

What is the most likely cause of engine "knock"?

K

Auto Information

What is the function of the radiator?

K

Auto Information

What is the optimal range of pressure in a cooling system?

K

Auto Information

When the thermostat is _____, coolant does not pass through the radiator.

K

Auto Information

What are the two main components of engine oil?

K

Auto Information

Which area of the engine is most reliant on lubrication?

K

Auto Information

The most common types of fuel injection systems are _____ and _____.

K

Auto Information

What function does the distributor perform?

K

Auto Information

What parts have been eliminated in coil-on-plug ignition systems?

K

Auto Information

What is the basic function of the drivetrain system?

K

to transmit heat from the coolant to the outside air

The air-fuel mixture explodes (detonates) rather than burns, often due to a lean air-fuel mixture.

closed

9 to 16 psi (pounds per square inch)

the reciprocating assembly (pistons, connecting rods, crankshaft)

base oil; additive package

The distributor is responsible for timing the ignition spark and distributing it to the correct cylinder.

multiport fuel injection; direct injection

to transmit power from the engine to the wheels

spark plug wires

Auto Information

The transaxle in a front-wheel drive vehicle combines which two components that are normally distinct in a rear-wheel drive vehicle?

Auto Information

The _____ allows the left and right wheels to turn at different speeds while cornering.

Auto Information

What does an automatic transmission use instead of a clutch?

Auto Information

The _____ generates the fluid pressure to operate the brake assemblies at the wheels.

Auto Information

What five functions does an engine's lubrication system perform?

Auto Information

What is the function of the intake valve in an internal combustion engine?

Auto Information

What are a car's water jackets?

Auto Information

What does an intake manifold do?

Auto Information

Define stoichiometric ratio.

Auto Information

Between the drive axle and the CV joints is a short drive shaft called the _____.

differential

transmission; drive axle

master cylinder

a torque converter

The intake valve allows the air-fuel mixture to be drawn into the combustion chamber.

The lubrication system lubricates, cools, seals, cleans, and quiets.

An intake manifold is the structure that delivers air to the intake ports of the engine's cylinders.

Water jackets are hollow sections of the engine block through which coolant flows.

half shaft

The stoichiometric ratio is the ideal air-fuel mixture for insertion into the engine cylinder. For gasoline engines, this is 14.7:1 by weight.

Auto Information

Spent fuel gases exit the cylinder and pass into the _____.

Auto Information

Diesel engines lack a _____ (a common element in gasoline-burning engines).

Auto Information

Does an alternator produce DC or AC power?

Auto Information

What does a camshaft do?

Auto Information

The low-voltage winding in a car's ignition system is called the _____.

Auto Information

What is chiefly responsible for generating fluid pressure to operate a car's brakes?

Auto Information

What does a fuel rail do?

Auto Information

Brakes work through _____ pressure.

Auto Information

The camshaft is driven by the crankshaft by means of a _____.

Auto Information

Where is the oil filter positioned in an engine?

spark-ignition system

exhaust manifold

A camshaft opens and closes the cylinders' intake and exhaust valves.

AC

the master cylinder

primary coil winding

hydraulic (or fluid)

A fuel rail delivers fuel under pressure to the engine's fuel injectors.

The oil filter filters oil after it leaves the oil pump and before it circulates through the engine.

timing belt or timing chain

Auto Information

_____ work by adjusting the brake pressure in each wheel depending on the wheel's speed.

K

Auto Information

Diesel engines employ _____ to ignite the air-fuel mixture.

K

Auto Information

What is a PCM?

K

Auto Information

Engine coolant is usually composed of a mixture of _____% antifreeze and _____% water.

K

Auto Information

The _____ matches engine speed to the desired speed of the vehicle.

K

Auto Information

Internal combustion engines take _____ energy, convert it into _____ energy and then into _____ energy.

K

Auto Information

Most internal combustion engines employ a _____-stroke cycle.

K

Auto Information

Define viscosity.

K

Auto Information

What is the difference between disc brakes and drum brakes?

K

Auto Information

A radiator cap has two valves: a(n) _____ valve and a(n) _____ valve.

K

a high compression ratio (or high pressure)

Anti-lock brakes (ABS)

50; 50

powertrain control module—that is, a car's computer

chemical; heat; mechanical

transmission

The viscosity of a liquid is its resistance to flow; liquids with higher viscosity are thicker and flow less easily.

four

pressure; vacuum (or vacuum; pressure)

In disc brakes, brake pads on both sides of a rotating disc are pinched together to slow the disc's rotation. In drum brakes, brake shoes inside a rotating drum expand to make contact with the interior of the drum, slowing its motion.

Auto Information

Define engine block.

Auto Information

In most cars, exhaust gases pass from the catalytic converter and through the _____ before leaving the car.

Auto Information

_____ can transmit power through very steep angles in order to transfer mechanical energy to the car's wheels.

Auto Information

What are oil galleries?

Auto Information

What flows through radiator hoses?

Auto Information

What does a connecting rod do?

Auto Information

What are ball joints?

Auto Information

OHV stands for _____.

Auto Information

What do piston rings do?

Auto Information

Does a car's battery provide DC or AC electricity?

muffler

The engine block is a framework in which the cylinders are arranged.

Oil galleries are passages throughout the engine that deliver oil to various engine components.

Constant velocity (CV) joints

Connecting rods connect each piston to the engine's crankshaft.

coolant flowing from the radiator to the engine and returning from the engine to the radiator

overhead valve

Ball joints are ball-and-socket arrangements that allow the steering knuckle to turn and move up and down.

DC

Piston rings seal the cylinder so that exhaust gases do not leak out and oil does not leak in.

Auto Information

_____ protect other car parts from the energy produced by the up-and-down movement of the wheels.

K

Auto Information

The high-voltage winding in a car's ignition system is called the _____.

K

Auto Information

What does a crankshaft do?

K

Auto Information

Why is a coolant recovery reservoir necessary?

K

Auto Information

What are the two major types of engine cooling systems?

K

Auto Information

The _____ is the main component of the charging system and converts mechanical energy into electrical energy.

K

Auto Information

_____ send data to a car's computer; _____ receive commands from a car's computer.

K

Auto Information

The AC electricity produced by the alternator is converted to DC by the _____.

K

Auto Information

What happens in an engine's oil pan?

K

Auto Information

Define piston.

K

secondary coil winding

Shock absorbers

A coolant recovery reservoir is needed to capture overflow when the radiator pressure cap valve opens, because the flow of coolant between the engine and radiator varies based on the heat of the engine.

A crankshaft converts the linear motion of the piston into rotary motion.

alternator

air-cooling and water-cooling

rectifier (or rectifier bridge)

Sensors; actuators

The piston is a cylindrical object that moves up and down inside an engine's cylinder.

Oil drips from various engine parts into the pan, cools, and is picked up by the oil tube and sent back to the oil pump.

Auto Information

Instead of carburetors, most cars now have _____.

Auto Information

What does a catalytic converter do?

Auto Information

A car's battery has plates made from _____ immersed in a mixture of _____ and water.

Auto Information

When a driver turns on the car's lights or radio, this causes a decrease in system voltage. The _____ responds to this by increasing the output of the alternator.

Auto Information

How is a car's oil pump powered?

Auto Information

The camshaft turns at _____ the speed of the engine's crankshaft.

Auto Information

What happens in an engine's cylinder?

Shop Information

What fastener is used to create a semipermanent assembly?

Shop Information

What tool is used to drive pins out of holes and follow the pins through the holes?

Shop Information

What tool is used to make a small indentation as a starting mark for a drilling operation?

A catalytic converter processes the toxic gases in engine exhaust, turning them into carbon dioxide, water, and other less harmful substances.

electronic fuel injection

voltage regulator

lead; sulfuric acid

one-half

The oil pump is usually powered by the crankshaft.

rivet

Fuel and air are ignited, pushing the piston downward, and then exhaust gases are expelled when the piston moves upward.

center punch

pin punch

Shop Information

What is a common distinction between a screw and a bolt?

K

Shop Information

_____ thread onto bolts to clamp assemblies together.

K

Shop Information

The UNC (Unified National Coarse) and UNF (Unified National Fine) classifications define fasteners based on which measure?

K

Shop Information

A hacksaw is made to cut on the _____ stroke.

K

Shop Information

Why is the diameter of a fastener critical to a task?

K

Shop Information

Before soldering an electrical connection, it is wise to make a(n) _____ connection.

K

Shop Information

A welder will require more electric current if either the _____ or the _____ is larger.

K

Shop Information

Why must welders wear helmets and protective clothing?

K

Shop Information

MIG (metal inert gas) welding is also known as _____ welding.

K

Shop Information

When would a technician use adjustable joint pliers?

K

Nuts	A screw typically has a round head with an indentation matching the appropriate screwdriver; a bolt typically has a hexagonal or square head.
forward	number of threads per inch
mechanical	The diameter of a fastener determines the size of the hole the fastener is installed in.
for protection from ultraviolet light	pieces being welded; electrode in use
when large diameter objects must be gripped or twisted	wire-feed

Shop Information

Which tool is frequently used to cut wire and small cables?

Shop Information

Which tool is often used with a back saw to make even cuts at specific angles?

Shop Information

A right-hand drill bit is made to cut while rotating in a(n) _____ direction.

Shop Information

A file is made to cut on the _____ stroke.

Shop Information

What is the main advantage of a soldering gun over a soldering iron?

Shop Information

Why does a combination wrench have both a box end and an open end?

Shop Information

What type of screwdriver would be used with a screw with this type of head?

Shop Information

What type of saw is embedded within a table and has a rotating blade?

Shop Information

What type of screwdriver would be used with a screw with this type of head?

Shop Information

_____ are smoothing tools that come in flat, half-round, and triangular shapes, among others.

miter box

diagonal cutters

forward

clockwise

The box end is used to loosen a tight bolt, and the open end is used to quickly remove a bolt that has been loosened.

The soldering gun has a very rapid warm-up cycle.

a table saw, which is a type of circular saw

Phillips head screwdriver

Files

flat tip screwdriver

Shop Information

Aside from oxyacetylene welding, the other major type of welding is _____.

Shop Information

_____ pliers are used for cutting and bending heavy-gauge wire.

Shop Information

Locking pliers are also known as _____.

Shop Information

Binding surfaces together with a molten metal alloy is called _____.

Shop Information

A _____ saw is used for making fine, straight cuts.

Shop Information

Which are used for finer finishing—rasps or files?

Shop Information

What type of screwdriver would be used with a screw with this type of head?

Shop Information

What type of saw is used for cutting metal?

Shop Information

_____ calipers measure the exterior dimensions of an object; _____ calipers measure interior dimensions.

Shop Information

Oxyacetylene welding uses a torch fueled by a mixture of _____ and _____.

Lineman

electric arc welding

soldering

Vise-Grip pliers

files

back

hacksaw

Robertson screwdriver

oxygen; acetylene

Outside; inside

A _____ can be struck with a hammer in order to drive home a fastener without damaging the work surface.

A *nut/bolt* has external threads, while a *nut/bolt* has internal threads.

_____ saws are used to cut across the grain of a piece of wood, while _____ saws are used to cut with the grain.

A(n) _____ can be used to smooth a large wooden surface.

A(n) _____ can be struck with a hammer in order to make holes or to drive home fasteners.

A(n) _____ saw can be used to bore circular cuts in wood.

Crosscut saws have teeth that are _____, while rip saws have teeth that are _____.

An adjustable wrench is also called a(n) _____.

Why would someone use a rubber or wooden mallet?

Adjustable joint pliers are used to grip larger-diameter objects and are also known as _____.

bolt; nut

drift

plane (or jack plane)

Crosscut; rip

hole

punch

Crescent wrench

vertically straight; angled in alternating directions

Channellock pliers or water pump pliers

Rubber or wooden mallets are most frequently used to protect the work surface from dents or scratches. Rubber mallets are often used on delicate metal surfaces, while wooden mallets are usually used in carpentry.

Shop Information

Combination slip joint pliers are adjustable at the _____.

Shop Information

What is the difference between a ball-peen hammer and a claw hammer?

Shop Information

A castellated nut uses a(n) _____ to hold it in place.

Shop Information

An air wrench combines torque with _____ to loosen fasteners.

Shop Information

_____ pliers are used for gripping very small objects.

Shop Information

_____ nuts can be loosened by hand.

Shop Information

A worker would use a _____ to join two pieces of wood while glue used to bind them dries.

Shop Information

What is the difference between a tubular spirit level and a bullseye spirit level?

Shop Information

What type of screwdriver would be used with a screw with this type of head?

Shop Information

A drill's chuck is identified by the _____ diameter drill bit it will hold.

Both a ball-peen hammer and a claw hammer have striking surfaces. A ball-peen hammer has a rounded steel end opposite the striking surface, while a claw hammer has a bifurcated (that is, split) lever for removing nails.

joint of the two handles (or joint of the two "jaws")

vibration or vibrating action

cotter pin

Wing

Needle nose

A tubular spirit level is useful for measuring levelness in one dimension, whereas a bullseye spirit level can measure the levelness of a plane in all directions.

clamp

largest

Torx screwdriver

Shop Information

A(n) _____ can be struck with a hammer in order to make long cuts.

K

Shop Information

A(n) _____ is used to clean a surface that will be soldered.

K

Shop Information

Snap rings are usually installed and removed using _____.

K

Shop Information

A(n) _____ holds a piece of work to a surface such as a worktable.

K

Shop Information

A(n) _____ saw is used to make fine, curving cuts.

K

Shop Information

A(n) _____ saw is a powered saw that sits on a table and can be used to make angled cuts.

K

Shop Information

A(n) _____ is the best tool for measuring small objects to the thousandths of an inch.

K

Shop Information

Instead of using a wrench to loosen a fastener in a tight space, a worker could use a _____ with a _____.

K

Shop Information

A measuring device with two arms that meet at a 90-degree angle is called a(n) _____.

K

Shop Information

A hammer with a 16- to 20-ounce head is typically used for what purpose?

K

flux

chisel

vise

snap-ring pliers

miter

coping

socket; ratchet handle

micrometer

This weight of hammer is a rough-framing hammer and is often used for framing wooden houses.

steel square, carpenter’s square, or framing square